State Variables for Engineers

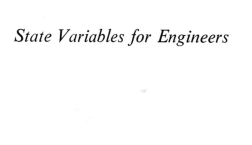

State Variables for Engineers

PAUL M. DERUSSO ROB J. ROY CHARLES M. CLOSE

Department of Electrical Engineering
Rensselaer Polytechnic Institute

John Wiley & Sons, Inc. *New York London Sydney*

 Library of Congress Catalog Card Number: 65-21443. Printed in the United States of America.

Preface

The feedback control field has been given a strong impetus in the theoretical direction. Stability theory, as formulated by Lyapunov, and modern optimization theory, as developed by Bellman and Pontryagin, and the works of Kalman, LaSalle, Merriam, and others are responsible for the changes. These works rely heavily upon "state variable" formulations, and it is apparent that advanced presentations of control theory must be given from this viewpoint. Furthermore, practicing control engineers must learn this viewpoint, because most of the current technical papers in this field utilize a state variable formulation. Otherwise, the current divergence of theory and practice will increase.

Acquiring this body of knowledge is difficult, since most of the topics that are published in book form are contained in advanced books on mathematics. Many engineering students and practicing engineers do not have the level of mathematical sophistication necessary to comprehend these advanced treatments. The purpose of this book is to provide these people with the necessary self-contained transitional presentation. It is intended to follow a conventional first course in feedback control systems, and to provide the necessary background for advanced presentations of nonlinear theory, adaptive system theory, sampled-data theory, and optimization theory. Furthermore, it attempts to unite the state variable approach and the more usual transfer function concepts, so that the reader can relate the new material to what he already knows. For these reasons, extensive, complicated proofs are omitted in favor of numerous examples. We have not tried to include all possible topics, but rather have attempted to cover basic principles so that the reader can subsequently investigate the literature for himself. Thus some treatments are not as advanced as can be found in the literature. After completing this book, the reader requiring more depth in state variables, Lyapunov theory, or optimization theory is encouraged to read Reference 1 of Chapter 5, References 26 and 28 of Chapter 7, and Reference 25 of Chapter 8, respectively, and the recent technical papers on these topics.

This book is an outgrowth of notes prepared to meet the needs of a first-year graduate course in modern automatic control theory initiated in the Electrical Engineering Department of Rensselaer Polytechnic Institute in 1962. Since its original presentation, the course has also been given to advanced undergraduates. The book contains more material than can be presented in one semester. At Rensselaer, we omit Sections 1.1 to 1.6, 2.1 to 2.6, and 3.1 to 3.8 because they have been covered in previous courses, and we briefly cover Chapters 7 and 8 in our one-semester course. The material of Chapter 7 is included in a subsequent one-semester graduate course entitled "Nonlinear and Adaptive Systems," and extensive coverage of optimization theory is given in a subsequent two-semester graduate course.

We are indebted to many people. First of all, Dr. Bernard A. Fleishman of Rensselaer's Department of Mathematics reviewed the entire manuscript and made many suggestions. Drs. Chi S. Chang and Imsong Lee reviewed Chapter 7, and Drs. Charles W. Merriam III and Frederick J. Ellert reviewed and contributed to Chapter 8. In addition, Drs. Dean N. Arden, Robert W. Miller, Dean K. Frederick, Howard Kaufman, and William G. Tuel, Jr., and Mr. Hugh J. Dougherty made numerous suggestions leading to improvement of the manuscript. Furthermore, Drs. Ellert and Tuel supplied computer results used in several examples. The entire final manuscript was patiently typed by Miss Rosana Laviolette, and portions of the preliminary effort were typed by Miss Sandra Elliott and Mrs. Joan Hayner.

Paul M. DeRusso
Rob J. Roy
Charles M. Close

Troy, New York
July, 1965

Contents

1 Time-Domain Techniques

2 Classical Techniques

3 Transform Techniques

4 Matrices and Linear Spaces

5 State Variables and Linear Continuous Systems

8 *Introduction to Optimization Theory*

State Variables for Engineers

1

Time-Domain Techniques

1.1 INTRODUCTION

Physical systems are customarily represented by models consisting of idealized components which can be precisely defined mathematically. The choice of a suitable model, embodying all the features of a physical system that are critical to its performance, may be difficult. If an overly simplified model is used, the results obtained from it will not closely approximate the behavior of the physical system. If an unnecessarily complicated model is used, it may be difficult or even impossible to analyze. Once a model is chosen, its characteristics are determined purely mathematically. In some cases, a system may be directly characterized in some way, without reference to any particular model. A number of different ways of utilizing these mathematical characteristics or models are discussed in this book. Like most textbooks, this one says very little about the choice of a model, or about how a specific system should be characterized. Instead, its purpose is to demonstrate the application of mathematical tools after the system has been properly represented by a model, or in some equivalent, purely mathematical manner.

The principal concern in the analysis of a system is the relationship between certain inputs (or sources) and outputs (or responses). These may be electrical, hydraulic, mechanical, thermal, or other types of quantities, and they are normally real functions of time. In Fig. 1.1-1, the inputs are denoted by $v_1(t)$, $v_2(t)$, . . . , $v_m(t)$, and the outputs by $y_1(t)$, $y_2(t)$, . . . , $y_n(t)$. The desired relationship is one of cause and effect, so that, when the inputs are known, the resulting outputs can be deter-

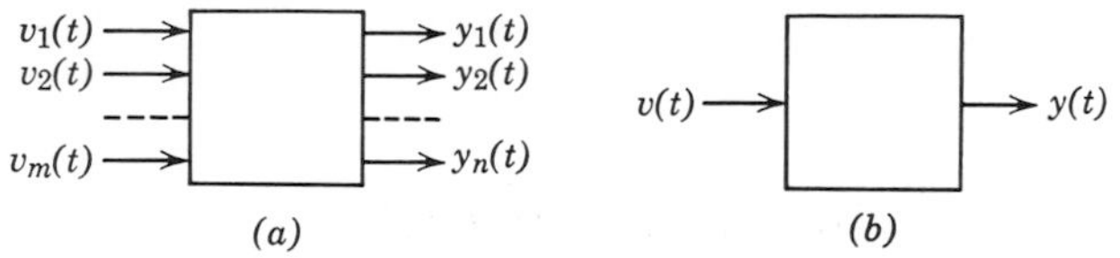

Fig. 1.1-1

mined. The system is often, however, characterized in some implicit way, instead of by a direct cause and effect relationship.

Systems with several inputs and outputs are conveniently handled by matrices, which are discussed in Chapter 4. The first three chapters, therefore, are in general restricted to systems with a single input and a single output. All the techniques of the first three chapters can, however, be readily extended to multiple input-output systems, as is done in Chapters 5 and 6.

The first three chapters are also in general restricted to linear deterministic systems having no internal energy prior to the application of the input. As shown in Section 1.4, any initial energy stored within the system can be represented by added inputs. Nonlinear systems are considered in the last two chapters.

Each of the first three chapters is concerned with one of the three common methods of characterizing linear deterministic systems. Chapter 1 is based upon the system's response to singularity or delta functions; Chapter 2, upon the system's differential or difference equation; and Chapter 3, upon the system function related to the Fourier, Laplace, or Z transforms. Both continuous and discrete and both fixed and varying systems are treated. A major attempt is made to correlate the three methods of characterizing systems.

Matrices are introduced in Chapter 4. Thus the basic mathematics to be presented for the state variable approach are completed at this point and are utilized throughout the remaining chapters. Chapter 5 introduces state variable techniques from an engineering viewpoint, as applicable to linear continuous systems. Linear discrete systems are similarly considered in Chapter 6. Lyapunov stability theory, as applied to nonlinear systems, is introduced in Chapter 7. Chapter 8 is an introductory consideration of system design and optimization.

In essence, this book may be divided into three parts. The first consists of Chapters 1 through 4, which present the basic mathematics utilized in later chapters. Chapters 5 and 6 comprise the second part of the book and present the basic concepts of state variables and some applications. The third part of the book consists of Chapters 7 and 8, which are further applications of the state variable viewpoint.

1.2 CLASSIFICATION OF SYSTEMS

Systems may be classified in a number of different ways. Consider the system of Fig. 1.1-1*b*, with the single input $v(t)$ and the single output $y(t)$. Assume that the system does not contain any independent sources and is at rest, with no internal energy, before the signal is applied. The input-output relationship is often indicated symbolically by

$$y(t) = Lv(t) \tag{1.2-1}$$

where L is an operator that characterizes the system. It may be a function of v, y, and t, may include operations such as differentiation and integration, and may be given in probabilistic language. Equation 1.2-1 is really nothing more than a shorthand way of saying that there is some cause and effect relationship between $v(t)$ and $y(t)$.

A system is *deterministic* if, for each input $v(t)$, there is a unique output $y(t)$.† In a nondeterministic or probabilistic system, there may be several possible outputs, each with a certain probability of occurrence, for a given input. Inputs to a system may also be given either as known functions or as random functions. Random functions, such as noise, can be de-described only in a statistical or probabilistic sense. If such a signal is an input to a deterministic system, the output is not deterministic.

A system is *nonanticipative* if the present output does not depend on future values of the input. In such a case, $y(t_0)$ is completely determined by the characteristics of the system and by values of $v(t)$ for $t \leq t_0$. In particular, if $v(t) \equiv 0$ for all $t < t_0$, then $y(t) \equiv 0$ for $t < t_0$. An anticipative system would, on the other hand, violate the normal cause and effect relationship.

A system is *realizable* if it is nonanticipative, and if $y(t)$ is a real function of time for all real $v(t)$. This definition does not imply that there is necessarily a known procedure for combining components to yield a given realizable system.

Assume that the responses to two different inputs $v_1(t)$ and $v_2(t)$ are $y_1(t)$ and $y_2(t)$, respectively. Let c_1 and c_2 denote two constants. A system is *linear* if the response to $v(t) = c_1v_1(t) + c_2v_2(t)$ is $y(t) = c_1y_1(t) + c_2y_2(t)$ for all values of v_1, v_2, c_1, and c_2. Expressing this definition symbolically,

$$L[c_1v_1(t) + c_2v_2(t)] = c_1L[v_1(t)] + c_2L[v_2(t)] \tag{1.2-2}$$

Equation 1.2-2 is also known as the *superposition principle*. If it holds only for inputs within a certain range, the system is linear only within that range. Unless otherwise stated, a linear system is understood to be linear for all inputs.

† See Section 5.4 for further discussion.

A system is *time-invariant* or *fixed* if the relationship between the input and output is independent of time. If the response to $v(t)$ is $y(t)$, then the response to $v(t - \lambda)$ is $y(t - \lambda)$. In such a system, the size and shape of the output are independent of the time at which the input is applied. Symbolically,

$$L[v(t - \lambda)] = y(t - \lambda) \tag{1.2-3}$$

Several examples of these definitions are given below.

Example 1.2-1. A differentiator is characterized by the relationship

$$y(t) = \frac{d}{dt} v(t)$$

The system is linear since

$$\frac{d}{dt}[c_1v_1(t) + c_2v_2(t)] = c_1 \frac{d}{dt} v_1(t) + c_2 \frac{d}{dt} v_2(t)$$

The system is also realizable and time-invariant.

Example 1.2-2. A squarer is characterized by the relationship

$$y(t) = v^2(t)$$

The system is nonlinear because

$$[c_1v_1(t) + c_2v_2(t)]^2 \neq c_1v_1^2(t) + c_2v_2^2(t)$$

It is realizable and time-invariant.

Example 1.2-3. A system is characterized by the relationship

$$y(t) = t \frac{d}{dt} v(t)$$

The system is linear since

$$t \frac{d}{dt}[c_1v_1(t) + c_2v_2(t)] = c_1t \frac{d}{dt} v_1(t) + c_2t \frac{d}{dt} v_2(t)$$

It is realizable but time-varying because

$$t \frac{d}{dt} v(t - \lambda) \neq (t - \lambda) \frac{dv(t - \lambda)}{d(t - \lambda)}$$

Example 1.2-4. A system is characterized by the relationship

$$y(t) = v(t) \frac{d}{dt} v(t)$$

It is nonlinear since

$$[c_1v_1(t) + c_2v_2(t)] \frac{d}{dt} [c_1v_1(t) + c_2v_2(t)]$$

$$\neq c_1v_1(t) \frac{d}{dt} v_1(t) + c_2v_2(t) \frac{d}{dt} v_2(t)$$

It is realizable and time-invariant.

Example 1.2-5. Prove or disprove the following statement. In a linear system, if the response to $v(t)$ is $y(t)$, then the response to Re $[v(t)]$ is Re $[y(t)]$. The symbol Re is read "the real part of."

$$\text{Re}\,[v] = \frac{v_1 + v_1^*}{2}$$

where the asterisk superscript denotes the complex conjugate.

If $y(t) = L[v(t)]$, the response to Re $[v(t)]$ is

$$L\left[\frac{v_1 + v_1^*}{2}\right] = \tfrac{1}{2}L[v_1] + \tfrac{1}{2}L[v_1^*]$$

This equals Re $[y(t)]$ if and only if

$$L[v_1^*] = \{L[v_1]\}^*$$

Although the last equation is not necessarily true, it is true if L is a real operator, i.e., if the responses to all real inputs are real. The proposed statement is valid, therefore, for all realizable linear systems. By a similar proof, the phases "the real part of" can all be replaced by "the imaginary part of."

Example 1.2-6. Prove or disprove the following statement. In a linear system, if the response to $v(t)$ is $y(t)$, the response to $(d/dt)v(t)$ is $(d/dt)y(t)$.

If $y(t) = L[v(t)]$, the response to dv/dt is

$$L\left[\frac{dv}{dt}\right] = \frac{d}{dt}L[v], \quad L \text{ not a function of } t$$

For a time-invariant system, L is not a function of time (although it may include differentiation or integration with respect to time), and the last step is justified. The proposed statement is therefore valid for time-invariant linear systems. By a similar proof, differentiation may be replaced by integration.

For time-varying systems, the last step is not justified. A specific counter example is the system of Example 1.2-3. The question is whether two linear operators commute. The answer is, in general, no.

Example 1.2-7. Prove or disprove the following statement. The cascade connection of two linear systems, as shown in Fig. 1.2-1, is linear.

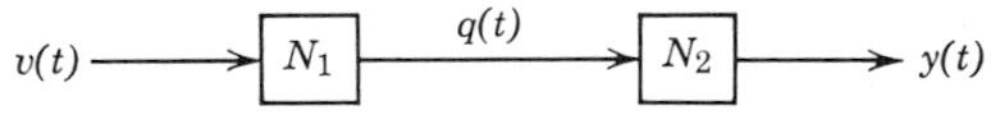

Fig. 1.2-1

$q(t)$ is the response of N_1 to $v(t)$, and $y(t)$ is the response of N_2 to $q(t)$.

$$q(t) = L_1[v(t)]$$
$$y(t) = L_2[q(t)] = L_2\{L_1[v(t)]\}$$

Since the two component systems are linear, the responses to $v(t) = c_1v_1(t) + c_2v_2(t)$ are given by

$$q(t) = L_1[c_1v_1(t) + c_2v_2(t)] = c_1L_1[v_1(t)] + c_2L_1[v_2(t)]$$
$$y(t) = L_2\{c_1L_1[v_1(t)] + c_2L_1[v_2(t)]\}$$
$$= c_1L_2\{L_1[v_1(t)]\} + c_2L_2\{L_1[v_2(t)]\}$$

By the definition of linearity, the last equation proves that the cascaded combination of N_1 and N_2 forms a linear system. It is true, in fact, that any system composed only of linear components is itself linear.

Systems may be further classified according to the types of signals that are present. A *continuous* signal is one which is a function of a continuous independent variable t. The signal must be uniquely defined at all values of t within a given range, except possibly at a denumerable set of points.† The definition of a continuous signal is broader than the mathematical definition of a continuous function. The function shown in Fig. 1.2-2*a*, for example, is not continuous for $a < t < b$ because of the discontinuity at $t = t_0$, but it does represent a continuous signal for $a < t < b$ by the definition above.

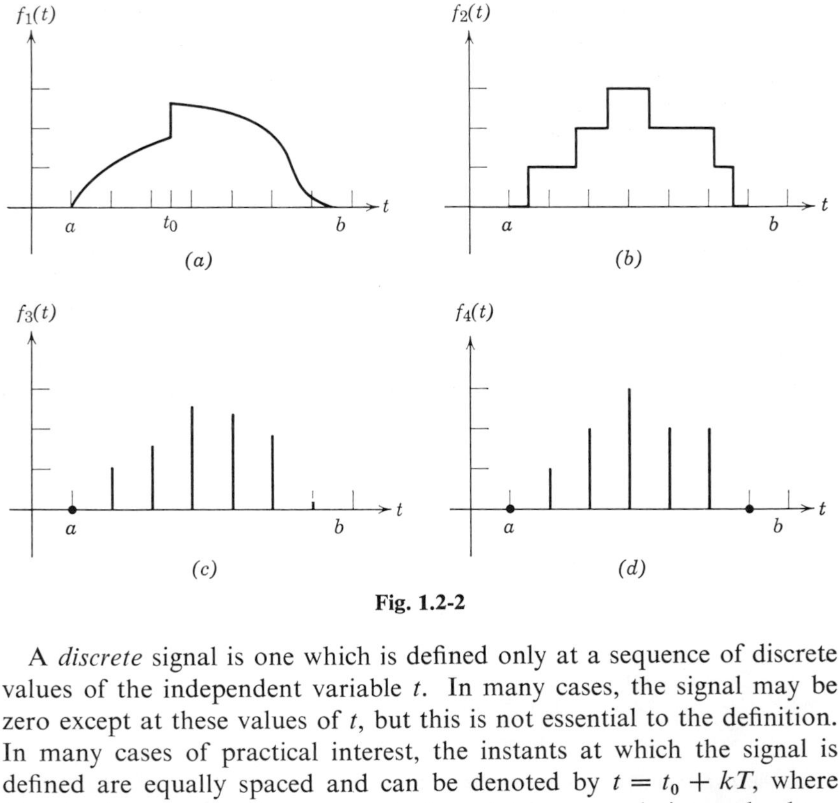

Fig. 1.2-2

A *discrete* signal is one which is defined only at a sequence of discrete values of the independent variable t. In many cases, the signal may be zero except at these values of t, but this is not essential to the definition. In many cases of practical interest, the instants at which the signal is defined are equally spaced and can be denoted by $t = t_0 + kT$, where T is the time between instants, and where k takes on only integral values. The signal is a function of the discrete independent variable k.

† A set of points is denumerable if the points can be listed on a one-to-one basis with the positive integers.

A *quantized* signal is one which can assume only a denumerable number of different values. Quantizing is simply rounding off to the nearest acceptable value, similar to rounding off a number to the nearest integer. Quantized signals may be either continuous or discrete.

Example 1.2-8. The function $f_1(t)$ in Fig. 1.2-2*a* represents a continuous signal. For use in the other parts of the figure, seven discrete instants of time and four discrete signal levels are shown. Rounding off the signal to the nearest level produces the quantized continuous signal $f_2(t)$. Sampling $f_1(t)$ at the seven discrete instants of time produces the discrete signal $f_3(t)$, while sampling $f_2(t)$ yields the quantized discrete signal $f_4(t)$.

Example 1.2-9. Continuous signals that change only at discrete instants may be completely described by equivalent discrete signals. Figure 1.2-3*a* shows a staircase

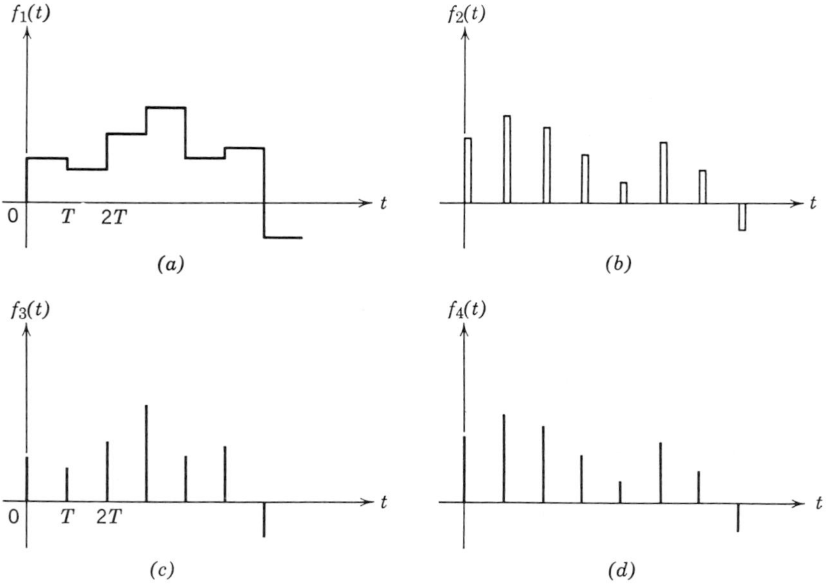

Fig. 1.2-3

function whose value changes only at $t = kT\,(k = 0, 1, 2, \ldots)$, and Fig. 1.2-3*b* shows a series of pulses of known width. These continuous signals are equivalent to the discrete signals in parts (*c*) and (*d*) of the figure, respectively, in the sense that the original signals can ideally be regenerated.

Consider the discrete signal produced by sampling *any* finite bandwidth, continuous signal at regular intervals of T seconds. Shannon's famous sampling theorem says that the discrete signal is equivalent to the continuous signal, provided that all frequency components of the latter are less than $1/2T$ cycles per second.[1] In this event, the original continuous signal can theoretically be recovered from the discrete signal by linear filtering. In practice, this cannot be precisely accomplished.

Example 1.2-10. The output of a digital computer is a quantized discrete signal.

Example 1.2-11. In pulse code modulation, *each* time increment of signal is quantized, often to the nearest of eight levels, including zero. Each quantized increment is then

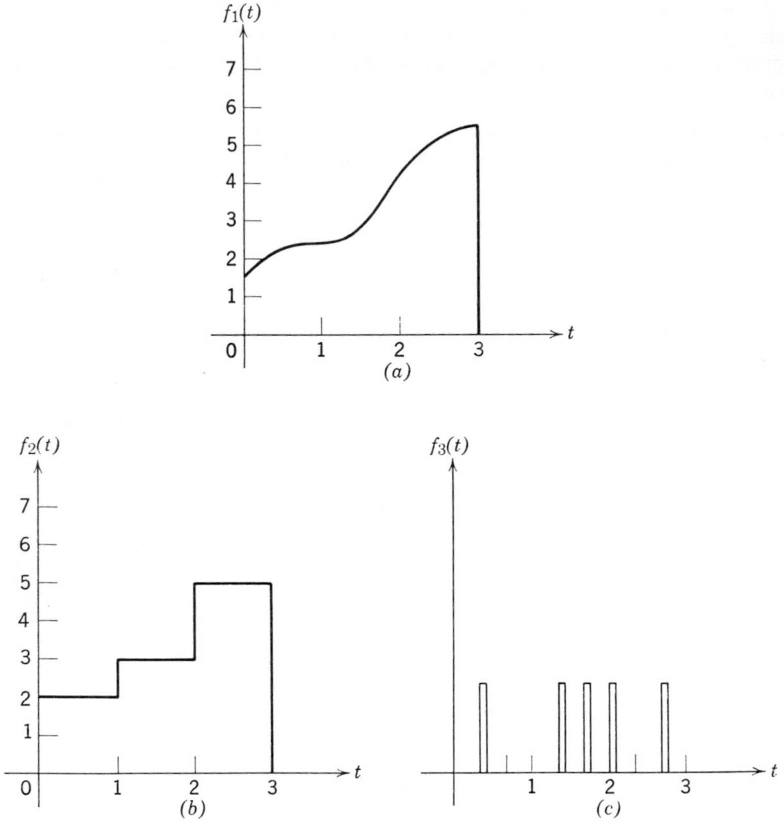

Fig. 1.2-4

represented by three two-level pulses using the binary number system. Thus the signal of Fig. 1.2-4*a* is quantized and coded as shown in parts (*b*) and (*c*), respectively.

1.3 RESOLUTION OF SIGNALS INTO SETS OF ELEMENTARY FUNCTIONS

The next chapter discusses the classical solution for the response $y(t)$ to a given input $v(t)$. Depending upon the nature of both the system and the input, classical techniques may involve considerable effort. One would certainly expect, and it is true, that the responses to certain classes of input functions could be determined more easily than the response to an arbitrary input. Accordingly, a sensible procedure for linear systems is to try to express a given arbitrary input as the sum of elementary functions. If the response to each of the elementary functions is known, or if it can be easily found, then the response to the arbitrary function can be found by the superposition principle of Eq. 1.2-2.

Although this scheme is satisfactory for linear systems, whether they are time-varying or not, it cannot be extended to nonlinear systems. Its validity is based upon Eq. 1.2-2, which does not apply to nonlinear systems. This section describes the scheme in general terms, and the next few sections and parts of Chapters 2 and 3 deal with the important special cases. The reader may wish to return to this section later.

Computational ease is not the only reason behind the proposed scheme. The characteristics of a system may be expressed in several different ways, e.g., by differential equations, or by the response to certain elementary functions. Although it is usually possible to convert from one method of characterization to another, it is not always easy to do so. Furthermore, different methods of characterization give different insights into system analysis, and certain problems are much more conveniently treated by a particular approach, such as the one discussed here.

Suppose that the input functions under consideration can be decomposed into a denumerable number of elementary functions denoted as $k_i (i = 0, 1, 2, \ldots)$. Then

$$v(t) = \sum_{i=1}^{\infty} a_i k_i(t)$$

Each of the set of elementary functions $\{k_i\}$ is a function of time t. To distinguish one of the elementary functions from its neighbors, the parameter λ is introduced. k_i is then regarded as a continuous function of t and a discrete function of λ and is written $k(t, \lambda)$. Each of the set of coefficients $\{a_i\}$ is a constant with respect to t. They represent the relative strength of the elementary functions composing $v(t)$ and are, of course, different for different elementary functions. Since a_i is a discrete function of λ, it is written $a(\lambda)$. The input is then written

$$v(t) = \sum_{\lambda} a(\lambda) k(t, \lambda) \tag{1.3-1}$$

$k(t, \lambda)$ is called the component or elementary function, and $a(\lambda)$ is known as the spectral function of $v(t)$ relative to $k(t, \lambda)$. The response of a system to the elementary function will be another function of t and λ, denoted by

$K(t, \lambda)$. By superposition, the response of a linear system to $v(t)$ is

$$y(t) = \sum_{\lambda} a(\lambda)K(t, \lambda) \tag{1.3-2}$$

As discussed in Chapter 3, the Fourier series for a periodic function has the form of Eq. 1.3-1. For nonperiodic functions, however, the resolution of $v(t)$ into a denumerable set of elementary functions is not possible. A continuous set of elementary functions is then required. $k(t, \lambda)$ and $a(\lambda)$ become continuous functions of the parameter λ, and Eq. 1.3-1 is replaced by

$$v(t) = \int a(\lambda)k(t, \lambda)\, d\lambda \tag{1.3-3}$$

$k(t, \lambda)$ and $a(\lambda)$ are still called the elementary and spectral functions, respectively. If the response of a linear system to $k(t, \lambda)$ is again denoted by $K(t, \lambda)$, the response to $v(t)$ is

$$y(t) = \int a(\lambda)K(t, \lambda)\, d\lambda \tag{1.3-4}$$

The parameter λ may or may not be a real variable. If it is a complex variable, Eqs. 1.3-3 and 1.3-4 require integration in a complex plane, a matter discussed in Chapter 3. The limits of summation and integration in Eqs. 1.3-1 through 1.3-4 depend upon the set of elementary functions used, but in general the integration is over the entire range of the parameter λ. In the case of a real parameter, the limits would in general be from $-\infty$ to $+\infty$. Finally, by use of the impulse function discussed in the next section, Eqs. 1.3-1 and 1.3-2 can be regarded as a special case of Eqs. 1.3-3 and 1.3-4.

The approach summarized by Eqs. 1.3-3 and 1.3-4 is practical only if the following two requirements are met.

(i) There must be an easy way of finding the coefficients $a(\lambda)$ for any arbitrary function of time.

(ii) There must be an easy way of finding $K(t, \lambda)$, the response to the elementary function, for an arbitrary linear system.

An appropriate class of elementary functions $k(t, \lambda)$ must be chosen with these two requirements in mind.

The unit step and unit impulse functions, discussed in the next section, are two different sets of $k(t, \lambda)$ that meet both requirements. In each case, all the functions in the set have exactly the same size and shape when they are plotted versus t, regardless of the value of λ. The only difference between different functions of the same set is that they occur at different times. Because of this, the $K(t, \lambda)$ functions are different functions of t

for different values of the parameter λ. Since time t is explicitly involved in all steps of the solution, the use of the step and impulse functions leads to "time-domain analysis."

In Chapter 3, the set of functions $k(t, \lambda)$ has the exponential form $\epsilon^{\lambda t}$, where λ is a complex parameter. Each $\epsilon^{\lambda t}$ is nonzero over the entire time range of interest, which is either $-\infty < t < \infty$, or $0 < t < \infty$. Two functions corresponding to two different values of λ do, however, have different sizes and shapes. The $K(t, \lambda)$ functions are, for time-invariant systems, independent of t, so that t is not explicitly involved in all steps of the solution. The use of the exponential functions leads to "complex-frequency-domain" analysis, which includes the Fourier and Laplace transforms. The inverse Fourier and Laplace transforms are special cases of Eq. 1.3-4.

Further general discussion of the resolution of functions into sets of elementary functions is best deferred until after a careful consideration of time-domain and complex-frequency-domain analysis. Additional remarks are found in Section 3.10. The remainder of this chapter is concerned with the resolution of signals into sets of singularity functions.

1.4 THE SINGULARITY FUNCTIONS

The singularity functions are a class of functions that forms the basis for the time-domain analysis of systems. Every singularity function and all its derivatives are continuous functions of time for all real values of t except one. Furthermore, the singularity functions can be obtained from one another by successive differentiation or integration. The singularity function that is perhaps the most familiar is the unit step function, which is denoted by U_{-1} and is shown in Fig. 1.4-1. As can be seen from the figure,

$$U_{-1}(t) = \begin{cases} 0 & \text{for} \quad t < 0 \\ 1 & \text{for} \quad t > 0 \end{cases} \tag{1.4-1}$$

Although the function is sometimes defined to have the values $\frac{1}{2}$ or 1 at the point of discontinuity, it is more commonly left undefined at this point.

For any function of time, $f(t - a)$ represents the function $f(t)$ displaced a units to the right. Hence it is logical to denote a unit step function with its discontinuity at $t = a$ by

$$U_{-1}(t - a) = \begin{cases} 0 & \text{for} \quad t < a \\ 1 & \text{for} \quad t > a \end{cases} \tag{1.4-2}$$

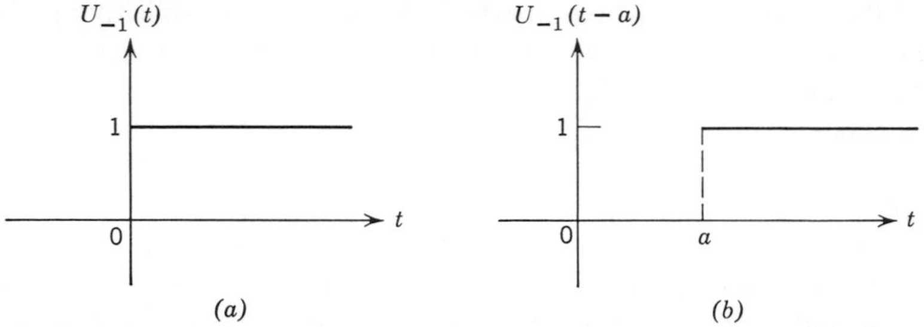

Fig. 1.4-1

Note specifically that the discontinuity occurs when the quantity in parentheses is zero. The unit step function is zero or 1, when its argument (the quantity in parentheses) is negative or positive, respectively.

If the unit step function is successively integrated, other singularity functions are obtained. The unit ramp function U_{-2} is

$$U_{-2}(t) = \int_{-\infty}^{t} U_{-1}(t)\, dt = \begin{cases} 0 & \text{for} \quad t < 0 \\ t & \text{for} \quad t > 0 \end{cases} \tag{1.4-3}$$

and is shown in Fig. 1.4-2*a*. Another singularity function is

$$U_{-3}(t) = \int_{-\infty}^{t} U_{-2}(t)\, dt = \begin{cases} 0 & \text{for} \quad t < 0 \\ t^2/2 & \text{for} \quad t > 0 \end{cases} \tag{1.4-4}$$

and is shown in Fig. 1.4-2*b*. The subscript notation is suggested by the following symbolism, sometimes used in mathematics.

$$f^{-1}(t) = \int f(t)\, dt$$

$$f^{-2}(t) = \int f^{-1}(t)\, dt$$

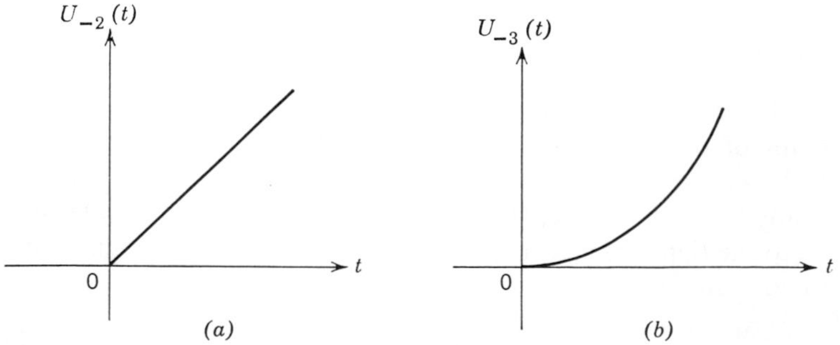

Fig. 1.4-2

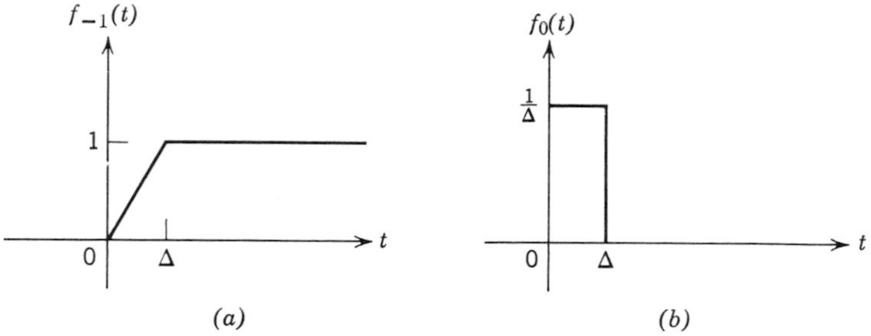

Fig. 1.4-3

As is evident in Chapter 3, the subscript also indicates the power of s in the Laplace transform of the singularity function.

The Impulse Function

Successive integration, as in Eqs. 1.4-3 and 1.4-4, yields an infinite number of singularity functions. The next logical step is to attempt to form other singularity functions by successive differentiation of the unit step function. The derivative of $U_{-1}(t)$, however, is zero for $t \neq 0$ and does not exist for $t = 0$. To gain some insight into such a result, consider a function that only approximates a unit step function, such as $f_{-1}(t)$ of Fig. 1.4-3*a*. The derivative of $f_{-1}(t)$ is the function $f_0(t)$ in part (*b*) of the figure.

$$f_0(t) = \frac{d}{dt} f_{-1}(t)$$
$$f_{-1}(t) = \int_{-\infty}^{t} f_0(t)\, dt \qquad (1.4\text{-}5)$$

As the dimension Δ becomes smaller, $f_{-1}(t)$ more closely approximates the unit step. $f_0(t)$ becomes a narrower and higher pulse, with the total area underneath the curve remaining equal to unity. As least as long as Δ remains nonzero, $f_0(t)$ is the derivative of $f_{-1}(t)$. Certainly

$$\lim_{\Delta \to 0} f_{-1}(t) = U_{-1}(t)$$

except possibly at $t = 0$. The limit of $f_0(t)$ is denoted by

$$U_0(t) = \lim_{\Delta \to 0} f_0(t) = \begin{cases} 0 & \text{for} \quad t \neq 0 \\ \text{infinity} & \text{for} \quad t = 0 \end{cases}$$

$U_0(t)$ is called the unit impulse function, and its properties are shown by the representation of Fig. 1.4-4. The number 1 alongside the arrow is

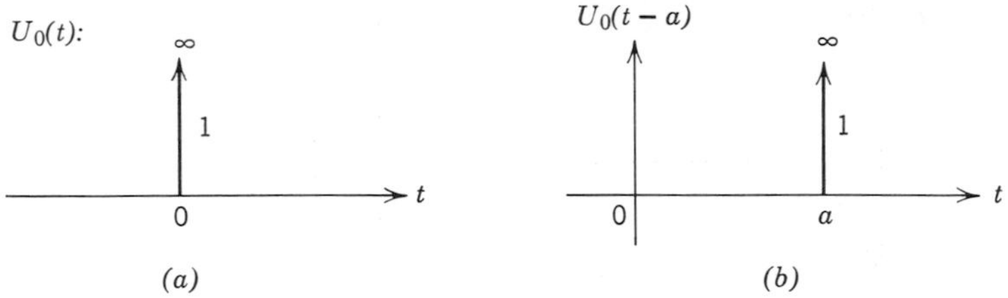

Fig. 1.4-4

intended to infer that the total *area* underneath the "curve" is unity. In the limit, Eq. 1.4-5 becomes

$$U_0(t) = \lim_{\Delta \to 0} \frac{d}{dt} f_{-1}(t)$$

$$U_{-1}(t) = \lim_{\Delta \to 0} \int_{-\infty}^{t} f_0(t)\, dt$$

It would seem that these could in turn be rewritten as

$$\begin{aligned} U_0(t) &= \frac{d}{dt} U_{-1}(t) \\ U_{-1}(t) &= \int_{-\infty}^{t} U_0(t)\, dt \end{aligned} \qquad (1.4\text{-}6)$$

This last step does not, however, necessarily follow by rigorous mathematics. Differentiation and integration are both limiting processes themselves, and Eq. 1.4-6 is valid only if the differentiation and integration are commutative with taking the limit as Δ approaches zero. That interchanging the order of two limiting processes is not always valid can be seen by noting that

$$\lim_{x \to 0} \lim_{y \to 0} \frac{x^2 - y^2}{x^2 + y^2} \neq \lim_{y \to 0} \lim_{x \to 0} \frac{x^2 - y^2}{x^2 + y^2}$$

The mathematical difficulties with the impulse function can be traced to the fact that it is not a function at all, in the normal mathematical use of the word. If f has a unique value corresponding to each value of t lying in some domain D, f is said to be a function of t for that domain of t. The function f is defined by listing in some way, as by equations, tables, or graphs, the values of f corresponding to all values of t within the domain D. A functional relationship is a point by point relationship and can be visualized as a black box with an input and output slot, as in Fig. 1.4-5. If any t_k within the domain D is thrown into the input slot, a unique f_k falls out of the output slot. The domain D is often, but not always, the set of all real numbers. The definition above makes the terms function and single-valued function identical. A "multivalued function" may be represented by two or more single-valued functions.

The unit step function is a perfectly good mathematical function. If one starts to define the unit impulse function by the relationship

$$U_0(t) = 0 \quad \text{for} \quad t \neq 0 \quad \text{and does not exist at } t = 0$$

this is satisfactory also. But the definition of the unit impulse by taking the limit of $f_0(t)$, shown in Fig. 1.4-3*b*, must include the fact that the area under the curve is unity. This can be indicated by

$$\int_{-\infty}^{t} U_0(t)\,dt = U_{-1}(t)$$

or, equivalently,

$$\int_{-\epsilon}^{\epsilon} U_0(t)\,dt = 1 \quad \text{for all } \epsilon > 0$$

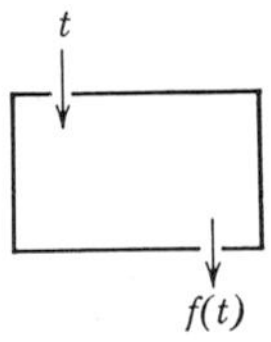

Fig. 1.4-5

But neither of the last two equations is a permissible way of defining a function as a point by point relationship.

The use of the impulse in system analysis produces results that can invariably be justified by more conventional and advanced mathematics. This involves, however, detailed consideration of changing the order of limiting processes, and the use of Stieltjes and Lebesgue integrals in place of the usual Riemann integral. It is now customary to regard the unit impulse as a "distribution" or "generalized function."† The theory includes the ordinary mathematical functions as special cases of distributions. A distribution having the properties desired of the unit impulse is possible. These properties are summarized as

$$\begin{aligned} U_0(t-\tau) &= 0 \quad \text{for } t \neq \tau \\ \int_{-\infty}^{t} U_0(t-\tau)\,dt &= U_{-1}(t-\tau) \qquad (1.4\text{-}7) \\ \int_{-\infty}^{\infty} f(t)\,U_0(t-\tau)\,dt &= f(\tau) \quad \text{if } f(t) \text{ is a continuous function} \end{aligned}$$

The first two properties have already been discussed. The third can be proved by noting that, since the integrand is zero except at $t = \tau$,

$$\int_{-\infty}^{\infty} f(t)\,U_0(t-\tau)\,dt = \int_{\tau-\epsilon}^{\tau+\epsilon} f(t)\,U_0(t-\tau)\,dt$$

† The standard work on this subject is Reference 2. A simpler and more readable treatment is given in Reference 3.

Using the mean-value theorem, this becomes

$$f(\tau)\int_{\tau-\epsilon}^{\tau+\epsilon} U_0(t-\tau)\,dt = f(\tau)$$

This relationship is sometimes called the sampling property. As one application, consider Eqs. 1.3-3 and 1.3-4. If $k(t, \tau)$ is a train of impulses, the integrand is zero except at discrete times, and the integration can be replaced by a summation. This explains the statement that Eqs. 1.3-1 and 1.3-2 can be considered as a special case of Eqs. 1.3-3 and 1.3-4.

Representing Initial Stored Energy by Impulses

When a unit impulse is the input to a system with energy storing elements, it serves to change the system's stored energy instantaneously. If a unit impulse of force is applied to a fixed mass M that has previously been at rest, the velocity of the mass is

$$v(t) = \frac{1}{M}\int_{-\infty}^{t} U_0(t)\,dt = \frac{1}{M}\,U_{-1}(t)$$

Thus $1/2M$ units of kinetic energy are instantaneously imparted to the mass at $t = 0$. If a unit impulse of current flows into a fixed, initially uncharged capacitance, the voltage is

$$e(t) = \frac{1}{C}\int_{-\infty}^{t} U_0(t)\,dt = \frac{1}{C}\,U_{-1}(t) \text{ volts}$$

Thus the current impulse instantaneously places a charge of 1 coulomb on the capacitance, and $1/2C$ joules of energy in its electrostatic field. Similarly, a source of $U_0(t)$ volts placed across an inductance causes a current

$$i(t) = \frac{1}{L}\int_{-\infty}^{t} U_0(t)\,dt = \frac{1}{L}\,U_{-1}(t) \text{ amperes}$$

corresponding to $1/2L$ joules of energy instantaneously placed in the magnetic field.

These examples suggest that the effects of any initial energy stored within a system may be represented by impulsive inputs. Recall that the basic definitions of Section 1.2, such as those for linear and time-invariant systems, assumed no internal energy before the external inputs are applied.

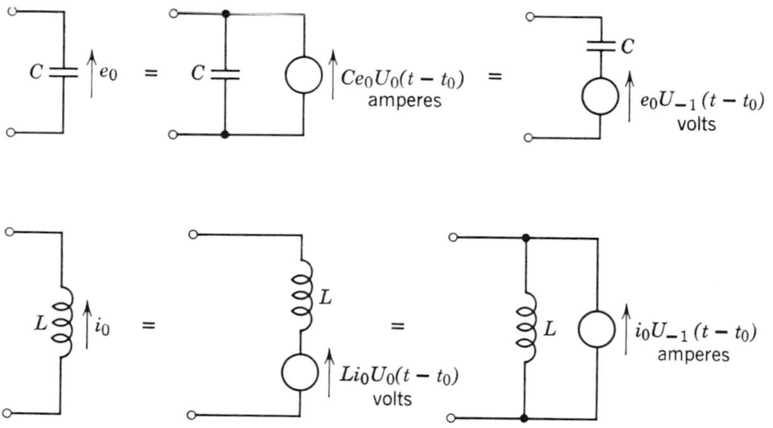

Fig. 1.4-6

If these definitions are to be extended to systems with initial internal energy, such energy must be represented by added inputs. A capacitance or inductance with initial energy at time t_0 may be replaced by the equivalent circuits of Fig. 1.4-6, as far as the response external to its terminals for $t > t_0$ is concerned. Figure 1.4-6 also gives equivalent circuits using step functions. Their equivalence follows directly from Thévenin's and Norton's theorems.

Another Use of the Impulse Function

The presentation of the unit impulse as the limit of a rectangular pulse of unit area, as the width approaches zero, suggests that a system's response to a pulse can be approximated by the response to an impulse with the same area. The idea is worthwhile, because the impulse response is more easily calculated than is the pulse response. The approximation is a good one if the width of the input pulse is small compared to the time constants of the system. If, for example, a rectangular pulse of force is applied to a mass, and if the mass does not move appreciably while the pulse is applied, then the response of the system which includes the mass will be the same as to an impulse of force of equal area. Any input pulse, whether rectangular or not, serves to introduce energy into a system in some way. If all the energy is introduced before the output has a chance to change appreciably, the output can be closely approximated by the impulse response. Clearly, for a given system, the approximation becomes progressively better as the input pulse width becomes narrower. This is the basis for the subject matter in the next section.

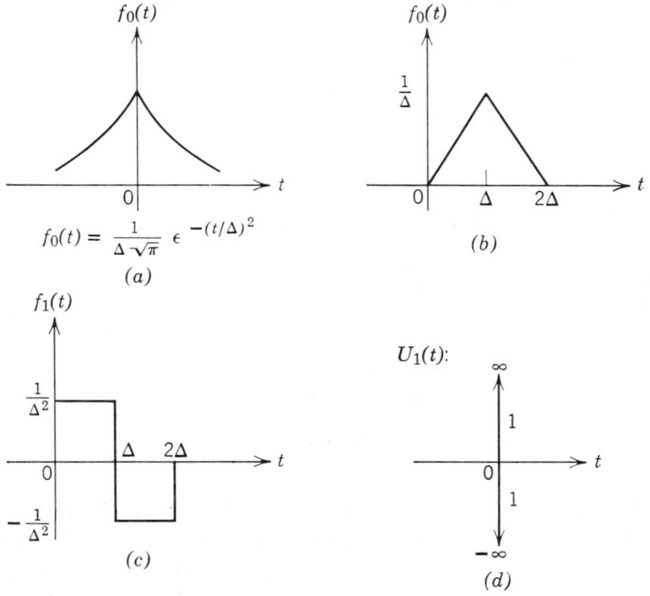

Fig. 1.4-7

Other Singularity Functions

The limit of the rectangular pulse of Fig. 1.4-3*b* is not the only way of arriving at the unit impulse. Any function which possesses the desired properties in the limit may be used in place of a rectangular pulse. Two such functions are shown in Figs. 1.4-7*a* and 1.4-7*b*. In each case, as Δ approaches zero, the unit impulse results. The derivative of the function in part (*b*) of the figure is shown in part (*c*). In the limit as Δ approaches zero, $f_0(t)$ becomes the unit impulse, and $f_1(t)$ becomes the unit doublet of part (*d*). The unit doublet has the characteristics

$$U_1(t) = \begin{cases} 0 & \text{for } t \neq 0 \\ -\infty \text{ and } +\infty & \text{for } t = 0 \end{cases}$$
$$U_0(t) = \int_{-\infty}^{t} U_1(t)\, dt \qquad (1.4\text{-}8)$$

The same complications regarding the rigorous treatment of the unit impulse carry over to the unit doublet.

Successive differentiation of the unit doublet leads to an infinite number of other singularity functions. In summary, all the singularity functions are completely characterized by describing any one of them and using the relationships

$$U_{n+1}(t-a) = \frac{d}{dt} U_n(t-a)$$
$$U_{n-1}(t-a) = \int_{-\infty}^{t} U_n(t-a)\,dt \qquad (1.4\text{-}9)$$

The singularity functions that are the most useful as elementary functions are the unit step and the unit impulse.

Determination of the Impulse and Step Response

Later sections of this chapter show how the response of a linear system to any arbitrary input can be determined, once the response to a unit impulse or unit step function is known. The terms impulse and step response are always understood to mean the response to *unit* singularity functions when the system contains *no* stored energy prior to the application of the input.

Example 1.4-1. If the input to Fig. 1.4-8 is $i_1(t) = U_0(t)$, it acts like an open circuit except at $t = 0$, when it instantaneously inserts some energy into the circuit. Because of the presence of the inductance, the impulse of current will all flow through the right-hand resistance, causing an impulse in the response $e_0(t)$. Since the current through the *RL* branch will remain finite, the impulse of voltage will appear directly across the inductance, creating a current of $\frac{1}{2}$ ampere. For $t > 0$, the circuit may be redrawn as in Fig. 1.4-9, where $i(0+) = \frac{1}{2}$. The circuit has a time constant of 1 second, so

$$e_0(t) = -i(t) = -[\tfrac{1}{2}\epsilon^{-t}] \quad \text{for } t > 0$$

The complete impulse response is

$$e_0(t) = U_0(t) - [\tfrac{1}{2}\epsilon^{-t}]\,U_{-1}(t)$$

Since the circuit is time-invariant, the step response is the integral of the impulse response, namely,

$$1 - \int_0^t [\tfrac{1}{2}\epsilon^{-t}]\,dt = \tfrac{1}{2}(1 + \epsilon^{-t}) \quad \text{for } t > 0$$

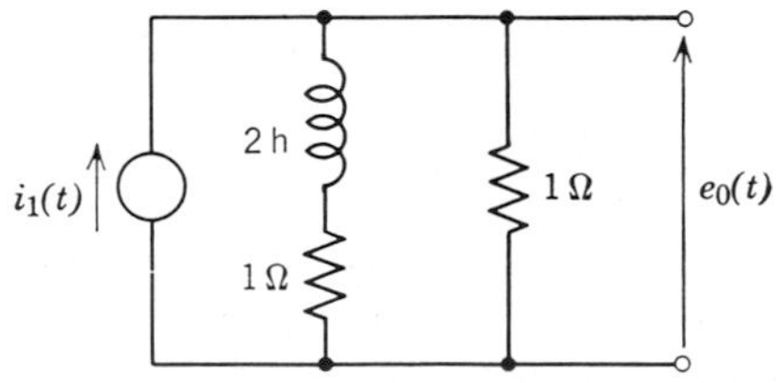

Fig. 1.4-8

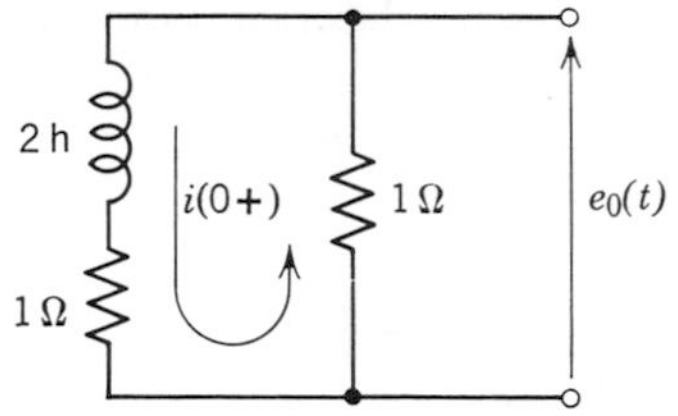

Fig. 1.4-9

For most systems, the impulse and step responses are determined analytically from the differential equations describing the system. The relationship of the impulse response to the system's differential equation is discussed in detail in Chapter 2. Its relationship to the system function of Laplace transform theory is covered in Chapter 3. In some cases, the impulse or step response may be given or may be approximately determined experimentally.

1.5 RESOLUTION OF A CONTINUOUS SIGNAL INTO SINGULARITY FUNCTIONS

Consider the function $v(t)$ shown by the solid curve in Fig. 1.5-1*a*. It can be crudely approximated by the staircase function, shown by the broken line. The staircase function is the superposition of the five step functions in part (*b*) of the figure.

$$v(t) \doteq U_{-1}(t) + 2U_{-1}(t-1) + U_{-1}(t-2) - 3U_{-1}(t-3) - U_{-1}(t-4)$$

where $\doteq$ stands for "approximately equal to." The approximation becomes progressively better as smaller time intervals and more steps are chosen in the staircase function.

The area underneath $v(t)$ has been divided into four sections in Fig. 1.5-1*c*. In part (*d*) of the figure, each section is approximated by an impulse function having an area equal to the area of the corresponding section in part (*c*). Then

$$v(t) \doteq 2U_0(t) + 3.5U_0(t-1) + 3U_0(t-2) + 0.5U_0(t-3)$$

The last approximation is certainly not a good representation of $v(t)$, since parts (*c*) and (*d*) of the figure bear little resemblance to each other. An approximation having values of only zero and infinity does not closely represent the finite function $v(t)$. If, however, the aim is to approximate the response of a linear system to the input $v(t)$, the procedure may be satisfactory. As discussed in the last section, the intervals of time chosen

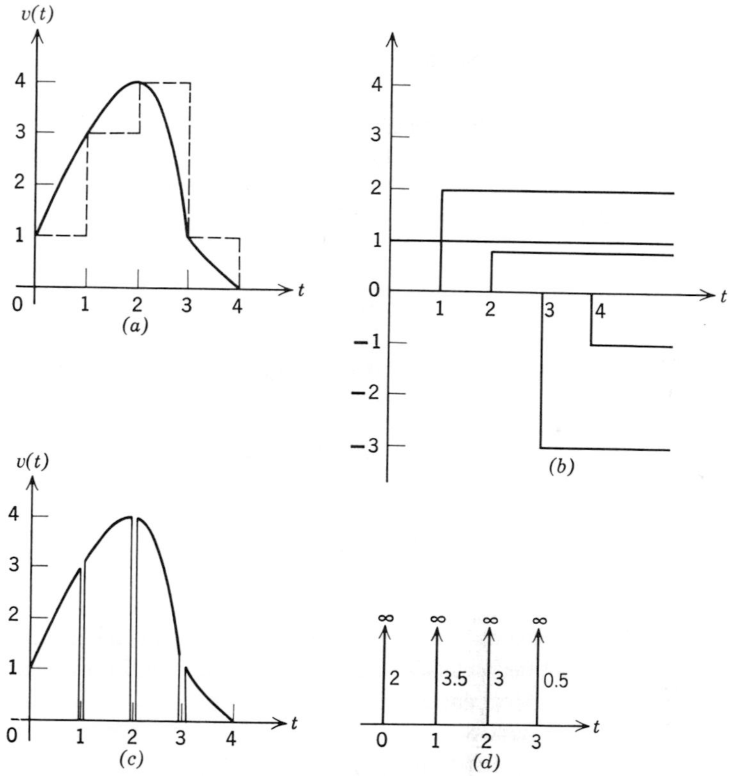

Fig. 1.5-1

in the approximation must be small compared to the time constants of the system.

The approximations illustrated in Fig. 1.5-1 should become exact as the time intervals become infinitesimal in size. Consider the arbitrary function $v(\lambda)$ shown in Fig. 1.5-2*a*. The value of the function at some arbitrary point t can be approximated by a series of step or impulse functions. The dummy variable λ has been introduced so that it is possible to distinguish between the particular point t and the variable representing the general distance from the origin along the abscissa. In terms of step functions,

$$v(t) \doteq \sum_{k=-\infty}^{\infty} \Delta v(k\,\Delta\lambda)U_{-1}(t - k\,\Delta\lambda)$$

where $\Delta v(k\,\Delta\lambda)$ denotes the jump at the point $k\,\Delta\lambda$ in the staircase approximation of Fig. 1.5-2*b*. Since the factor $U_{-1}(t - k\,\Delta\lambda) = 0$ for $t < k\,\Delta\lambda$, it ensures that the jumps in the staircase function to the right

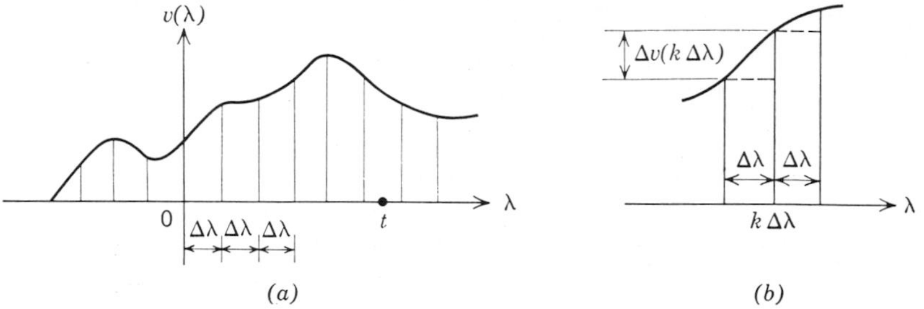

Fig. 1.5-2

of point t have no bearing on the value of $v(\lambda)$ at the point t. Note also that

$$\Delta v(k\,\Delta\lambda) \doteq \left[\frac{dv(\lambda)}{d\lambda}\right]_{k\,\Delta\lambda} \Delta\lambda$$

Thus

$$v(t) \doteq \sum_{k=-\infty}^{\infty} \left[\frac{dv(\lambda)}{d\lambda}\right]_{k\,\Delta\lambda} U_{-1}(t - k\,\Delta\lambda)\,\Delta\lambda \tag{1.5-1}$$

All three of the approximations above become exact as $\Delta\lambda$ approaches zero. In the limit, $\Delta\lambda$ becomes $d\lambda$, $k\,\Delta\lambda$ becomes the continuous variable λ, and the summation is replaced by an integration.

$$v(t) = \int_{-\infty}^{\infty} \frac{dv(\lambda)}{d\lambda} U_{-1}(t - \lambda)\, d\lambda$$

Since $U_{-1}(t - \lambda) = 0$ for $\lambda > t$,

$$v(t) = \int_{-\infty}^{t} \frac{dv(\lambda)}{d\lambda}\, d\lambda \tag{1.5-2}$$

a result that could have been written down directly.

In terms of impulse functions,

$$v(t) \doteq \sum_{k=-\infty}^{\infty} [v(k\,\Delta\lambda)]\, U_0(t - k\,\Delta\lambda)\,\Delta\lambda \tag{1.5-3}$$

where the quantity $[v(k\,\Delta\lambda)]\,\Delta\lambda$ is the area of the rectangular pulse beginning at the point $k\,\Delta\lambda$ in Fig. 1.5-2*b*. The approximation becomes exact as $\Delta\lambda$ approaches zero, yielding

$$v(t) = \int_{-\infty}^{\infty} v(\lambda)\, U_0(t - \lambda)\, d\lambda \tag{1.5-4}$$

This result is simply a restatement of the sampling property of the unit impulse, previously given in Eq. 1.4-7. As discussed in the last section, Eq. 1.5-3 does not really represent a good approximation to $v(t)$ for any nonzero value of $\Delta\lambda$. As $\Delta\lambda$ approaches zero, however, both the area of the impulses and the spacing between them become zero, yielding exactly the function $v(t)$.

While Eqs. 1.5-2 and 1.5-4 are rather trivial results, the concepts lead to very important expressions for the response of a linear system. Let the input of a linear system be approximated by Eq. 1.5-1. Denote the response of the system to a unit step occurring at $t = \beta$ by the symbol $r(t, \beta)$. Thus, when $v(t) = U_{-1}(t - \beta)$, $y(t) = r(t, \beta)$. Then, by the superposition principle, the approximate response to an arbitrary input is

$$y(t) \doteq \sum_{k=-\infty}^{\infty} \left[\frac{dv(\lambda)}{d\lambda}\right]_{k\,\Delta\lambda} r(t, k\,\Delta\lambda)\,\Delta\lambda$$

The exact output is given by the limit as $\Delta\lambda$ approaches zero.

$$y(t) = \int_{-\infty}^{\infty} \frac{dv(\lambda)}{d\lambda} r(t, \lambda)\,d\lambda \tag{1.5-5}$$

For nonanticipatory systems, $r(t, \lambda) = 0$ for $t < \lambda$, so

$$y(t) = \int_{-\infty}^{t} \frac{dv(\lambda)}{d\lambda} r(t, \lambda)\,d\lambda \tag{1.5-6}$$

Next let the input be approximated by Eq. 1.5-3. Denote the response of the system to a unit impulse occurring at $t = \beta$ by the symbol $h(t, \beta)$. By superposition, the approximate response to an arbitrary input is

$$y(t) \doteq \sum_{k=-\infty}^{\infty} v(k\,\Delta\lambda)h(t, k\,\Delta\lambda)\,\Delta\lambda$$

The exact output is given by the limit as $\Delta\lambda$ approaches zero.

$$y(t) = \int_{-\infty}^{\infty} v(\lambda)h(t, \lambda)\,d\lambda \tag{1.5-7}$$

Since, for nonanticipatory systems, $h(t, \lambda) = 0$ for $t < \lambda$,

$$y(t) = \int_{-\infty}^{t} v(\lambda)h(t, \lambda)\,d\lambda \tag{1.5-8}$$

In many problems, the input is zero for negative values of time, in which case the lower limit for the integrals in Eqs. 1.5-5 through 1.5-8 becomes zero.

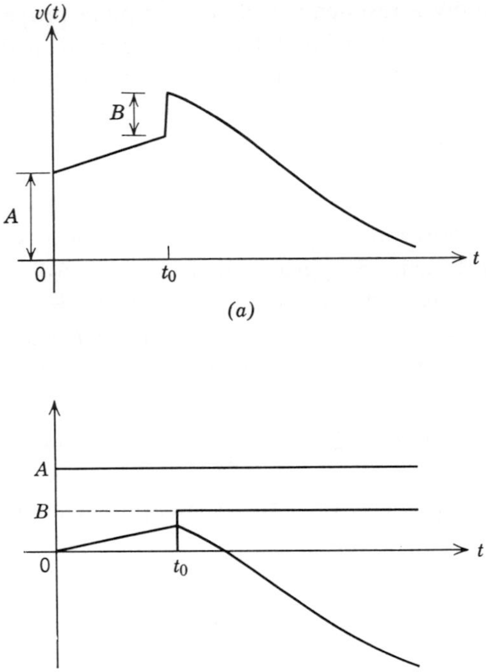

(a)

(b)

Fig. 1.5-3

There are two situations which sometimes cause difficulty in the use of these equations. If the input $v(t)$ has discontinuities, its derivative, which appears in the integrand of Eq. 1.5-6, will contain impulses. In the evaluation of the integral, the sampling property of Eqs. 1.4-7 and 1.5-4 is useful. The derivative of the input shown in Fig. 1.5-3a has impulses of value A and B, respectively, at $t = 0$ and t_0. Thus Eq. 1.5-6 reduces to

$$y(t) = Ar(t, 0) + Br(t, t_0) + \int_0^t \frac{dv_1(\lambda)}{d\lambda} r(t, \lambda)\, d\lambda \qquad (1.5\text{-}9)$$

where $dv_1/d\lambda$ stands for $dv/d\lambda$ with the impulses removed. Their effect is accounted for by separate terms. The last equation follows directly from the sampling property of impulses, or alternatively from breaking the function of Fig. 1.5-3a up into the three components shown in part (b).

The impulse response of a system may occasionally contain an impulse itself. Then the integrand of Eq. 1.5-8 contains an impulse. In this case, the impulse response can be written as the sum of two terms,

$$h(t, \lambda) = kU_0(t - \lambda) + h_1(t, \lambda)$$

where h_1 does not contain an impulse. Then the response to an arbitrary input $v(t)$ is given by

$$y(t) = kv(t) + \int_{-\infty}^{t} v(\lambda)h_1(t, \lambda)\, d\lambda \qquad (1.5\text{-}10)$$

1.6 THE CONVOLUTION INTEGRAL FOR TIME-INVARIANT SYSTEMS

The preceding derivations are valid for all linear systems. For time-invariant systems, a further simplification is possible. The symbols $h(t)$ and $r(t)$ are customarily used to stand for the responses to a unit impulse and unit step, respectively, applied at $t = 0$.

$$h(t) = h(t, 0) \quad \text{and} \quad r(t) = r(t, 0)$$

By the definition of a time-invariant system given in Eq. 1.2-3,

$$\begin{aligned} h(t, \lambda) &= h(t - \lambda, 0) = h(t - \lambda) \\ r(t, \lambda) &= r(t - \lambda, 0) = r(t - \lambda) \end{aligned} \qquad (1.6\text{-}1)$$

Equations 1.5-5 through 1.5-8 then lead to Eqs. 1.6-2 and 1.6-3 in Table 1.6-1. Equations 1.6-4 and 1.6-5 in the table are easily obtained from the first two by the substitution of variable $\tau = t - \lambda$. For nonanticipatory

Table 1.6-1

General Equation	Simplification when System is Nonanticipatory	Simplification when Input Is Zero for $t < 0$	Equation Number
$y(t) = \int_{-\infty}^{\infty} v(\lambda)h(t - \lambda)\, d\lambda$	Upper limit is t	Lower limit is 0	(1.6-2)
$y(t) = \int_{-\infty}^{\infty} \frac{dv(\lambda)}{d\lambda} r(t - \lambda)\, d\lambda$			(1.6-3)
$y(t) = \int_{-\infty}^{\infty} v(t - \tau)h(\tau)\, d\tau$	Lower limit is 0	Upper limit is t	(1.6-4)
$y(t) = \int_{-\infty}^{\infty} \frac{dv(t - \tau)}{d(t - \tau)} r(\tau)\, d\tau$			(1.6-5)

systems with $v = 0$ for $t < 0$, the limits on all the integrals become 0 and t. Even in this case, however, it is not incorrect to use the wider limits, and this is occasionally done to facilitate the proofs of some theorems.

Equations 1.6-4 and 1.6-5 suggest another interpretation for time-domain analysis. Consider Eq. 1.6-4 for a nonanticipatory system with an input that is zero for $t < 0$.

$$y(t) = \int_0^t v(t - \tau)h(\tau)\, d\tau$$

$y(t)$ and $v(t)$ represent the output and input, respectively, at the instant of time t. Since $v(t - \tau)$ represents the input τ seconds before this instant, τ is sometimes called the age variable. As τ increases, $v(t - \tau)$ represents the input further and further back into the past. For $\tau = 0$ and t, respectively, $v(t - \tau)$ is the input at the instants t and 0, respectively. The equation above indicates that the entire past history of the input contributes to the output at the instant t. The past history is weighted, however, by the factor $h(\tau)$. In fact, $h(\tau)$ is sometimes called the system-weighting function instead of the impulse response. This interpretation in terms of a weighting function will, perhaps, become clearer later when the graphical interpretation of the equation is considered. For stable systems, $h(\tau)$ approaches zero as τ approaches infinity. For such systems, the more recent past is weighted more heavily than the far distant past.

Equations 1.6-2 and 1.6-4 may be rewritten in still another way if desired. Simply use the fact that, for time-invariant linear systems,

$$h(\tau) = \frac{dr(\tau)}{d\tau} \quad \text{and} \quad h(t - \lambda) = \frac{dr(t - \lambda)}{d(t - \lambda)}$$

For the case of time-invariant systems with zero inputs for $t < 0$, all the formulas can be summarized as follows:

$$y(t) = \frac{dv(t)}{dt} * r(t) = v(t) * \frac{dr(t)}{dt} = v(t) * h(t) \tag{1.6-6}$$

The asterisk symbolizes the convolution of the functions on either side of it. Convolution is mathematically defined as

$$f_1(t) * f_2(t) = \int_0^t f_1(t - \lambda) f_2(\lambda)\, d\lambda$$
$$= \int_0^t f_1(\lambda) f_2(t - \lambda)\, d\lambda \tag{1.6-7}$$

When the more general limits of integration of $-\infty$ and $+\infty$ are used, the term convolution and the symbol $f_1(t) * f_2(t)$ are still used by some workers.

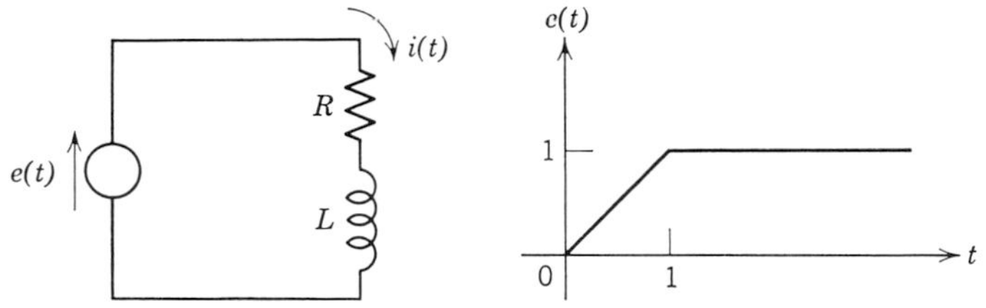

Fig. 1.6-1

Example 1.6-1. For the circuit and voltage source shown in Fig. 1.6-1, find the current $i(t)$. Assume that the circuit has no stored energy for $t < 0$.

In the preceding terminology, $v(t) = e(t)$ and $y(t) = i(t)$. The current response to a step of voltage is

$$r(t) = \frac{1}{R}[1 - \epsilon^{-(R/L)t}] \quad \text{for } t > 0$$

The equation

$$i(t) = \int_0^t \frac{de(\lambda)}{d\lambda} r(t - \lambda)\, d\lambda$$

will be used. Since the character of the input is different for $0 < t < 1$ and for $1 < t$, the solution must be carried out in two parts. For $0 < t < 1$,

$$i(t) = \int_0^t (1)\frac{1}{R}[1 - \epsilon^{-(R/L)(t-\lambda)}]\, d\lambda$$

$$= \frac{1}{R}\left[t - \frac{L}{R}(1 - \epsilon^{-(R/L)t})\right]$$

For $1 < t$,

$$i(t) = \int_0^1 (1)\frac{1}{R}[1 - \epsilon^{-(R/L)(t-\lambda)}]\, d\lambda + \int_1^t 0\, d\lambda$$

$$= \frac{1}{R}\left[1 - \frac{L}{R}(\epsilon^{R/L} - 1)\epsilon^{-(R/L)t}\right]$$

Note that the two solutions are identical for $t = 1$, as expected. Also note that the second solution is *not* the first solution with t replaced by 1.

Example 1.6-2. Figure 1.6-2 shows a cascade connection of two identical circuits. Assuming that the impulse response of each one individually is $h(t) = t\epsilon^{-t}$ for $t \geq 0$, and that the second circuit does not load down the first, find the impulse response of the entire combination.

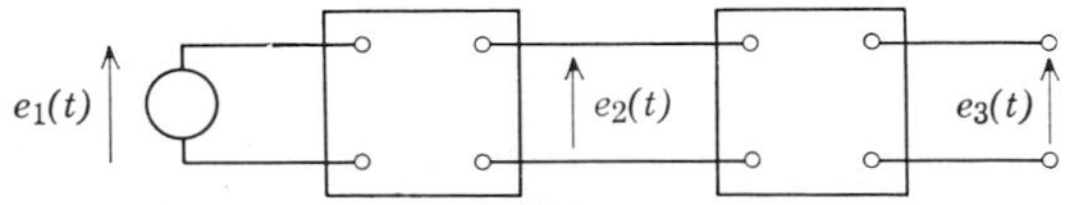

Fig. 1.6-2

When

$$e_1(t) = U_0(t)$$

then

$$e_2(t) = t\epsilon^{-t} \quad \text{for } t \geqslant 0$$

and

$$e_3(t) = \int_0^t e_2(\lambda)h(t-\lambda)\,d\lambda$$

$$= \int_0^t (\lambda\epsilon^{-\lambda})[(t-\lambda)\epsilon^{-(t-\lambda)}]\,d\lambda = \frac{t^3\epsilon^{-t}}{6} \quad \text{for } t \geqslant 0$$

The last expression represents the impulse response of the entire combination.

Example 1.6-3. Assume that the individual circuits of Fig. 1.6-2 are amplifiers, each having the step response $r(t) = -A\epsilon^{-t/RC}$ for $t \geq 0$. Find the step response for the two stages in cascade, assuming that the second does not load down the first. Extend the result to n stages.

When

$$e_1 = U_{-1}(t)$$

then

$$e_2(t) = -A\epsilon^{-t/RC} \quad \text{for } t \geqslant 0$$

and

$$e_3(t) = \int_0^t \frac{de_2(\lambda)}{d\lambda} r(t-\lambda)\,d\lambda$$

$$= e_2(0)r(t) + \int_{0+}^t \frac{de_2(\lambda)}{d\lambda} r(t-\lambda)\,d\lambda$$

The last expression has a separate term as a result of the discontinuity in $e_2(t)$ at $t = 0$. This is similar to Eq. 1.5-9.

$$e_3(t) = (-A)(-A\epsilon^{-t/RC}) + \int_0^t \left[\frac{A}{RC}\epsilon^{-\lambda/RC}\right][-A\epsilon^{[-(t-\lambda)/RC]}]\,d\lambda$$

$$= A^2\epsilon^{-t/RC}\left[1 - \frac{t}{RC}\right] \quad \text{for } t \geqslant 0$$

The last expression is the step response for two stages in cascade. For three stages in cascade, the step response is

$$A^2(-A\epsilon^{-t/RC}) + \int_0^t A^2 \frac{d}{d\lambda}\left[\epsilon^{-\lambda/RC}\left(1 - \frac{\lambda}{RC}\right)\right][-A\epsilon^{-(t-\lambda)/RC}]d\lambda$$

$$= -A^3\epsilon^{-t/RC}\left[1 - 2\left(\frac{t}{RC}\right) + \frac{1}{2}\left(\frac{t}{RC}\right)^2\right] \quad \text{for } t \geqslant 0$$

For n stages in cascade, the step response is

$$(-A)^n\epsilon^{-t/RC}\left[1 + \sum_{k=1}^{n-1} \frac{(n-1)!\,(-t/RC)^k}{(n-k-1)!(k!)^2}\right] \quad \text{for } t \geqslant 0$$

The last expression can be more easily obtained by means of the Laplace transform, and it is derived in Chapter 3. The expression has been plotted for various values of n.[4]

Graphical Interpretation of the Convolution Integrals

The graphical evaluation of the quantity

$$f_1(t) * f_2(t) = \int_0^t f_1(\lambda) f_2(t - \lambda)\, d\lambda \tag{1.6-8}$$

is based on the fact that the integral of a function between two limits represents the area under a curve between those limits. For concreteness, let

$$f_1(t) = \begin{cases} 1 & \text{for } 0 < t < 1 \\ 0 & \text{elsewhere} \end{cases}$$

$$f_2(t) = \frac{1}{R}[1 - e^{-(R/L)t}] \quad \text{for } t \geqslant 0$$

The convolution of these two functions was found analytically in Example 1.6-1. As shown in Fig. 1.6-3, $f_2(t - \lambda)$ plotted versus $(\lambda - t)$ is the

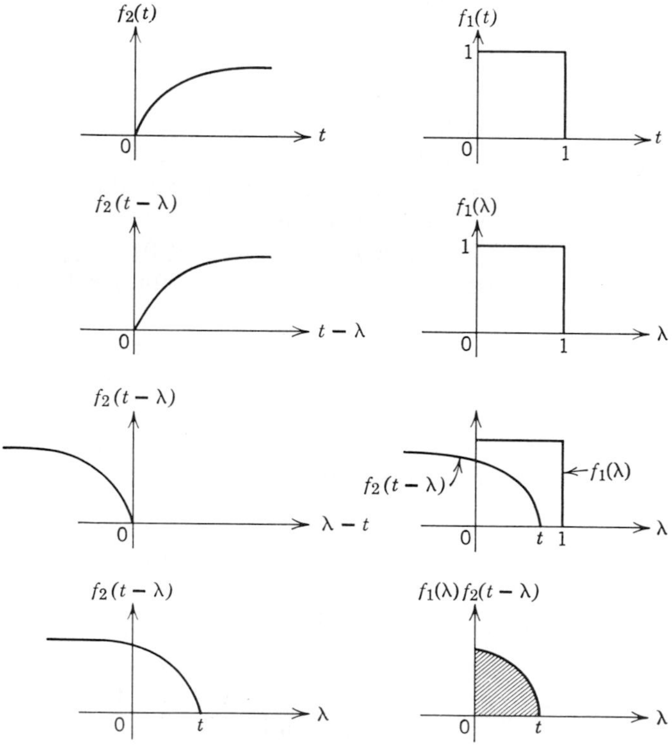

Fig. 1.6-3

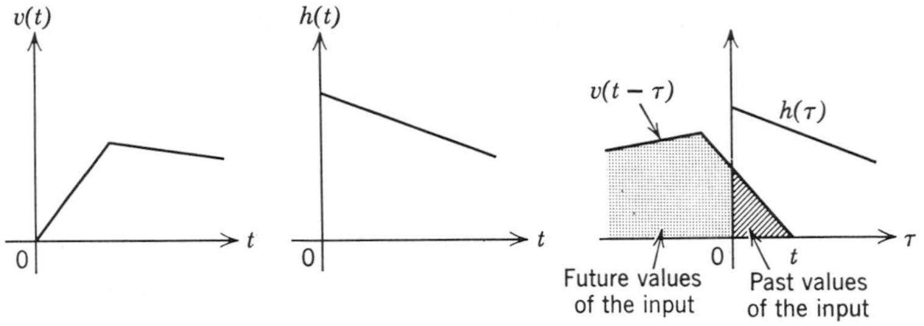

Fig. 1.6-4

mirror image of the original function about the vertical axis. In plotting this quantity versus λ, it is shifted t units to the right. The multiplication of $f_1(\lambda)$ and $f_2(t - \lambda)$ produces the last curve in Fig. 1.6-3. The shaded area represents the integral in Eq. 1.6-8 for a typical value of t. In summary, to find $f_1(t) * f_2(t)$, fold one of the functions about the vertical axis, slide it a distance t into the other function, and take the area underneath the product curve.

Graphically determining this area for different values of t will lead to a plot of $f_1(t) * f_2(t)$ versus t. In the example above, both the analytical and the graphical approaches indicate that the response monotonically increases from 0 at $t = 0$ to $1/R$ as $t \to \infty$, but that the rate of increase suddenly decreases at $t = 1$. Note that the result would not be changed if the limits of integration were changed to $-\infty$ and $+\infty$. The integrand is zero except for $0 < \lambda < t$, since $f_1(\lambda) = 0$ for $\lambda < 0$, and $f_2(t - \lambda) = 0$ for $\lambda > t$. For anticipatory systems with inputs for $t < 0$, the limits would have to be $-\infty$ and $+\infty$.

The graphical evaluation of the convolution integrals is particularly useful when the analytic evaluation proves to be difficult, or when the input or impulse response is given graphically instead of analytically. It also reinforces the previous interpretation of the impulse response as a weighting function. Consider the formula

$$y(t) = \int_0^t v(\lambda)h(t - \lambda)\, d\lambda = \int_0^t v(t - \tau)h(\tau)\, d\tau$$

Typical v and h functions are shown in Fig. 1.6-4. $v(t - \tau)$ represents the input τ seconds before the "present" time t. Thus $\tau = 0$ corresponds to the present; $\tau > 0$, to past time; and $\tau < 0$, to future time, as indicated in the figure. Since $v(t - \tau)$ is multiplied by $h(\tau)$, the past values of the input are weighted by the impulse response.

A Useful Property of the Convolution Integrals

The convolution of two integrable functions $f_1(t)$ and $f_2(t)$ was given by

$$f_1(t) * f_2(t) = \int_{-\infty}^{\infty} f_1(t-\lambda) f_2(\lambda)\, d\lambda$$
$$= \int_{-\infty}^{\infty} f_1(\lambda) f_2(t-\lambda)\, d\lambda \qquad (1.6\text{-}9)$$

where the more general limits of $-\infty$ and $+\infty$ have been used. Differentiating both formulas with respect to t, it is found that $f_1(t) * f_2'(t) = f_1'(t) * f_2(t)$, where the prime denotes differentiation. In general,

$$\begin{aligned} f_1(t) * f_2^{(k)}(t) &= f_1^{(k)}(t) * f_2(t) \\ f_1(t) * f_2^{(-k)}(t) &= f_1^{(-k)}(t) * f_2(t) \end{aligned} \qquad (1.6\text{-}10)$$

where the superscripts (k) and $(-k)$ denote the kth derivative and integral, respectively.[5] The first of these two expressions is written out explicitly below.

$$\int_{-\infty}^{\infty} f_1(\lambda) f_2^{(k)}(t-\lambda)\, d\lambda = \int_{-\infty}^{\infty} f_1^{(k)}(\lambda) f_2(t-\lambda)\, d\lambda \qquad (1.6\text{-}11)$$

where

$$f_1^{(k)}(\lambda) = \frac{d^k}{d\lambda^k} f_1(\lambda)$$
$$f_2^{(k)}(t-\lambda) = \frac{d^k}{d(t-\lambda)^k} f_2(t-\lambda)$$

The use of Eq. 1.6-11 for $k = 1$, together with the fact that $h(t) = (dr/dt)$, enables Eqs. 1.6-3 and 1.6-5 to be derived from Eqs. 1.6-2 and 1.6-4, respectively. Equation 1.6-11 can also be helpful in the evaluation of the convolution integrals, whether using analytical, graphical, or computer techniques.[5] Another application is the generalization of the impulse sampling property of Eq. 1.5-4 to higher-order singularity functions. Letting $f_2(t) = U_0(t)$, Eqs. 1.6-11 and 1.5-4 yield

$$\int_{-\infty}^{\infty} f_1(\lambda) U_k(t-\lambda)\, d\lambda = \frac{d^k f_1(t)}{dt^k} \quad \text{for } k = 0, 1, 2, \ldots \qquad (1.6\text{-}12)$$

A modification of Eq. 1.6-11 is used in Section 2.8 to relate a system's impulse response to its differential equation. Note that

$$\frac{d}{d\lambda}[f_2(t-\lambda)] = \frac{df_2(t-\lambda)}{d(t-\lambda)} \frac{d(t-\lambda)}{d\lambda} = (-1)\frac{df_2(t-\lambda)}{d(t-\lambda)}$$

In general,

$$\frac{d^k}{d\lambda^k}[f_2(t-\lambda)] = (-1)^k \frac{d^k f_2(t-\lambda)}{d(t-\lambda)^k}$$

Equation 1.6-11 can then be written

$$\int_{-\infty}^{\infty} f_1(\lambda)\left[\frac{d^k}{d\lambda^k} f_2(t-\lambda)\right] d\lambda = (-1)^k \int_{-\infty}^{\infty}\left[\frac{d^k}{d\lambda^k} f_1(\lambda)\right][f_2(t-\lambda)]\, d\lambda \tag{1.6-13}$$

For the special case where $f_2(t) = U_0(t)$, this becomes

$$\int_{-\infty}^{\infty} f_1(\lambda)\left[\frac{d^k}{d\lambda^k} U_0(t-\lambda)\right] d\lambda = (-1)^k \frac{d^k f_1(t)}{dt^k} \tag{1.6-14}$$

1.7 SUPERPOSITION INTEGRALS FOR TIME-VARYING SYSTEMS

The formulas of Section 1.5 are valid for all linear systems, whether they be fixed or varying. The formulas of Section 1.6 are based upon Eq. 1.6-1, which is valid only for fixed systems. The analogous results are now established for varying systems.

Equation 1.6-1 indicated that for fixed systems the impulse and step responses are a function only of τ, the time that has elapsed since the application of the impulse or step function. For varying systems, these responses are functions of two variables. These may be taken as the present time t and the time λ at which the singularity function was applied or, alternatively, the present time t and the elapsed time τ. It must be clearly understood which pair of variables is being used. Thus $h(t, \beta)$ could be interpreted as the response at time t resulting from an impulse applied at either time β or time $t - \beta$. Since both interpretations are used in the literature, this causes some confusion. In this book, the symbol $h(t, \beta)$ is always given the first of the two interpretations. When the second interpretation is intended, the symbol $h_*(t, \beta)$ is used.† Specifically,

$$\begin{aligned} h(t, \lambda) &= \text{response at time } t \text{ to a unit impulse at time } \lambda \\ h_*(t, \tau) &= h(t, t-\tau) = \text{response at time } t \text{ to a unit impulse } \\ &\qquad \tau \text{ seconds earlier (time } t-\tau) \\ r(t, \lambda) &= \text{response at time } t \text{ to a unit step at time } \lambda \\ r_*(t, \tau) &= r(t, t-\tau) = \text{response at time } t \text{ to a unit step } \tau \\ &\qquad \text{seconds earlier (time } t-\tau) \end{aligned} \tag{1.7-1}$$

† The asterisk subscript is not a common notation. In most references, the correct interpretation must be deduced from the context in which the symbol appears.

Equations 1.5-5 through 1.5-8 lead to the results in the first half of Table 1.7-1. The last half of the table is obtained by the substitution of variable $\tau = t - \lambda$. In Eqs. 1.7-2 and 1.7-3 the $h(t, \lambda)$ and $r(t, \lambda)$ functions are more commonly used, while in Eqs. 1.7-4 and 1.7-5 the $h_*(t, \tau)$ and $r_*(t, \tau)$ functions lead to more compact results. Since the term convolution is restricted to formulas like those in Table 1.6-1, the equations in Table 1.7-1 are

Table 1.7-1

General Equation	Simplification When System Is Nonanticipatory	Simplification when $v = 0$ for $t < 0$	Equation Number
$y(t) = \int_{-\infty}^{\infty} v(\lambda)h(t, \lambda)\, d\lambda$			
$y(t) = \int_{-\infty}^{\infty} v(\lambda)h_*(t, t - \lambda)\, d\lambda$			(1.7-2)
	Upper limit is t	Lower limit is 0	
$y(t) = \int_{-\infty}^{\infty} \frac{dv(\lambda)}{d\lambda} r(t, \lambda)\, d\lambda$			
$y(t) = \int_{-\infty}^{\infty} \frac{dv(\lambda)}{d\lambda} r_*(t, t - \lambda)\, d\lambda$			(1.7-3)
$y(t) = \int_{-\infty}^{\infty} v(t - \tau)h(t, t - \tau)\, d\tau$			
$y(t) = \int_{-\infty}^{\infty} v(t - \tau)h_*(t, \tau)\, d\tau$			(1.7-4)
	Lower limit is 0	Upper limit is t	
$y(t) = \int_{-\infty}^{\infty} \frac{dv(t - \tau)}{d(t - \tau)} r(t, t - \tau)\, d\tau$			
$y(t) = \int_{-\infty}^{\infty} \frac{dv(t - \tau)}{d(t - \tau)} r_*(t, \tau)\, d\tau$			(1.7-5)

usually called superposition integrals. Table 1.6-1 for fixed systems can, of course, be regarded as a special case of Table 1.7-1. For fixed systems, $h(t, \lambda)$ reduces to $h(t - \lambda)$, and $h_*(t, \tau)$ to $h(\tau)$.

$h_*(t, \tau)$ can be interpreted as a weighting function for varying systems, exactly corresponding to the weighting function $h(\tau)$ for fixed systems. The graphical interpretation of the convolution integrals can also be extended to the general superposition integrals. The discussion associated with Fig. 1.6-4 applies to varying systems if $h(\tau)$ is replaced by $h_*(t, \tau)$. To find

$$y(t) = \int_0^t v(t - \tau)h_*(t, \tau)\, d\tau$$

fold $v(t)$ about the vertical axis, and slide it forward a distance t to form $v(t-\tau)$ versus τ. Multiply the resulting curve by $h_*(t, \tau)$, and take the area underneath the product curve. The only difference from the fixed system procedure is that, since $h_*(t, \tau)$ is a function of time, a different $h_*(t, \tau)$ curve must be used for each value of t considered. The adjoint technique of Chapter 5 proves helpful when using $h_*(t, \tau)$ for a fixed t and varying τ.

Equations 1.6-9 and 1.6-11 are mathematical identities that may be used whenever the integrands have the proper form. One of the functions must be an explicit function of λ, and the other an explicit function of $t - \lambda$, where t is treated as a parameter. Starting with Eq. 1.7-5

$$y(t) = \int_{-\infty}^{\infty} \frac{dv(t-\tau)}{d(t-\tau)}\, r_*(t, \tau)\, d\tau$$

Eq. 1.6-11 gives

$$y(t) = \int_{-\infty}^{\infty} v(t-\tau) \frac{d}{d\tau}[r_*(t, \tau)]\, d\tau$$

A comparison of this result with Eq. 1.7-4 infers, as is true, that

$$\begin{aligned} h_*(t, \tau) &= \frac{d}{d\tau}[r_*(t, \tau)] \\ r_*(t, \tau) &= \int_{-\infty}^{\tau} h_*(t, \sigma)\, d\sigma \end{aligned} \tag{1.7-6}$$

where the lower limit becomes zero for nonanticipatory systems. Also

$$\begin{aligned} h(t, \lambda) &= -\frac{d}{d\lambda}[r(t, \lambda)] \\ r(t, \lambda) &= \int_{\lambda}^{\infty} h(t, \sigma)\, d\sigma \end{aligned} \tag{1.7-7}$$

where the upper limit becomes t for nonanticipatory systems. Equation 1.7-7 serves to relate Eqs. 1.7-2 and 1.7-3. The reader should be reminded, however, that replacing the input $v(t)$ of a varying system by $dv(t)/dt$ does not necessarily produce a new output of $dy(t)/dt$. (See Examples 1.2-6 and 1.2-3.)

Example 1.7-1. If the impulse response of a linear system is $h(t, \lambda) = \dfrac{t-\lambda}{t^2}$ for $\lambda \le t$, find the response to $v(t) = (t\epsilon^{-t})U_{-1}(t)$.

By Eq. 1.7-2,

$$y(t) = \int_0^t (\lambda\epsilon^{-\lambda}) \left(\frac{t-\lambda}{t^2}\right) d\lambda = \frac{1}{t}\left[\left(1 - \frac{2}{t}\right) + \left(1 + \frac{2}{t}\right)e^{-t}\right]$$

Finding the impulse response of a varying system can be difficult. It is shown in Section 2.8 that the impulse response in this example describes a system characterized by the differential equation

$$t^2 \frac{d^2y}{dt^2} + 4t \frac{dy}{dt} + 2y = v(t)$$

The same example is solved in Example 3.9-3 using the Laplace transform.

1.8 RESOLUTION OF DISCRETE SIGNALS INTO SETS OF ELEMENTARY FUNCTIONS

The only discrete signals considered in this book are those which are defined at equally spaced instants, denoted by $t = t_0 + kT$. In this chapter, k is a time index which takes on only integral values, and T is the time between instants. While t_0 may have any value, the time origin may be chosen to make $t_0 = 0$, and this is consistently done. It is also possible to choose the time scale so that $T = 1$, as in Sections 2.9 through 2.12, but this is not done in this chapter.

A discrete signal consisting of a series of finite numbers may arise naturally, as in the case of a digital computer, or it may result from sampling a continuous function, as in Figs. 1.2-2*c* and 1.8-1*c*. If the ideal periodic switch in Fig. 1.8-1*b* is assumed to close every T seconds for an infinitesimal time, $f_2(t)$ is zero except at the sampling instants. In terms of the unit delta function

$$\delta(t - kT) = \begin{cases} 1 & \text{for} \quad t = kT \\ 0 & \text{for} \quad t \neq kT \end{cases} \tag{1.8-1}$$

the signal $f_2(t)$, shown in part (*c*) of the figure, can be expressed as

$$f_2(t) = f_1(t) \sum_{k=-\infty}^{\infty} \delta(t - kT) = \sum_{k=-\infty}^{\infty} f_1(kT)\delta(t - kT) \tag{1.8-2}$$

Consider the more practical situation where the ideal switch in Fig. 1.8-1*b* remains closed for ΔT seconds. The signal $f_2(t)$, shown in part (*d*) of the figure, is a continuous signal that may be approximated by a discrete signal. As discussed in Sections 1.4 and 1.5, a narrow pulse may be approximated by an impulse of equal area, provided that the time constants of the system which follows are large compared to the pulse width. The signal of Fig. 1.8-1*d* can be approximated by

$$f_2(t) = \sum_{k=-\infty}^{\infty} [f_1(kT)\, \Delta T] U_0(t - kT) \tag{1.8-3}$$

Note that, if $\Delta T = T$, Eq. 1.8-3 describes the staircase function shown in

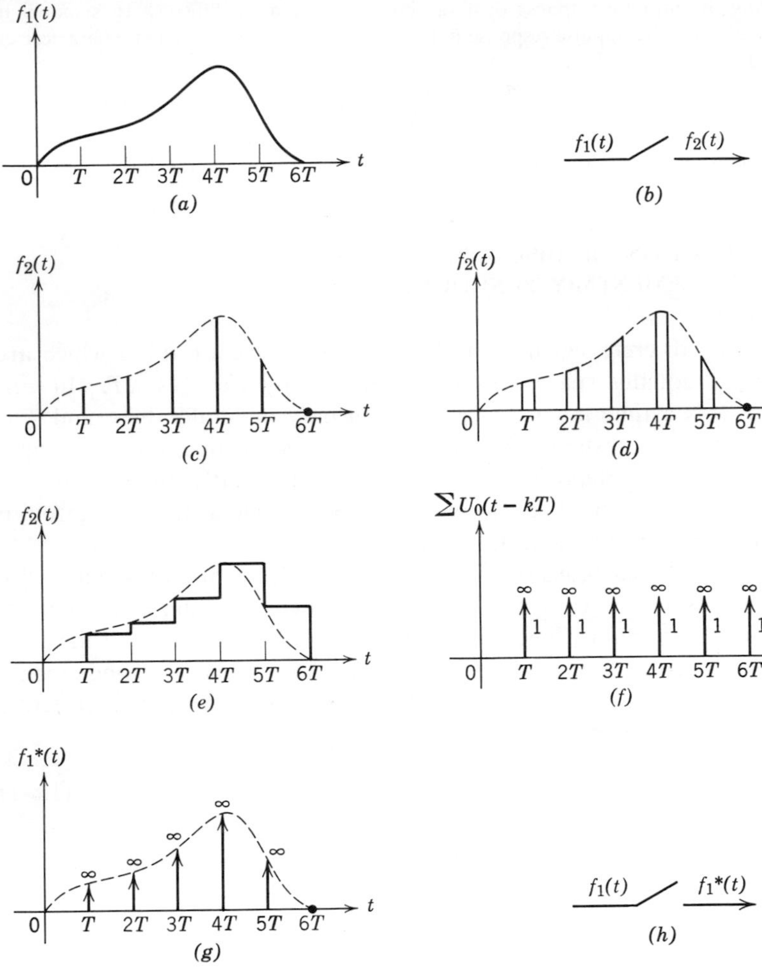

Fig. 1.8-1

part (*e*). Such a result can be produced in practice by following the sampling switch by a "zero order hold" or "boxcar generator" circuit. This might be done to approximately recover the original signal if only the sampled signal were available. Equation 1.8-3 is then essentially identical with Eq. 1.5-3.

In Eq. 1.8-3, the time scale may be adjusted so that $\Delta T = 1$, or the constant multiplying factor of ΔT may be incorporated into the gain of the system which follows. In either case, the factor ΔT disappears from Eq. 1.8-3, and the signal of Fig. 1.8-1*d* can be approximated by the signal of part (g), which in turn is given by

$$f_2(t) = f_1{}^*(t) = f_1(t) \sum_{k=-\infty}^{\infty} U_0(t - kT) = \sum_{k=-\infty}^{\infty} f_1(kT)\, U_0(t - kT) \qquad (1.8\text{-}4)$$

where $\sum_{k=-\infty}^{\infty} U_0(t - kT)$ represents the impulse train shown in part (f). The idealized sampling switch of part (h), together with the symbol $f_1{}^*(t)$, is defined by Eq. 1.8-4. Such a switch modulates the input by a train of unit impulses to produce a discrete output signal composed of impulses of varying area. This is the sampling device assumed in books on sampled-data control systems, and it is useful in later parts of this book. The degree of approximation involved in going from part (d) to part (g) of Fig. 1.8-1 is best discussed in terms of transform techniques.[6] It should be noted that all of the ideal switches discussed in this section are linear, time-varying components.

1.9 SUPERPOSITION SUMMATIONS FOR DISCRETE SYSTEMS

If the input to a linear system initially at rest is the discrete signal given in either Eq. 1.8-2 or 1.8-4, the output can be found by superposition. In this section, any sampling switch needed to produce the discrete input is not considered to be a part of the system under consideration. The only information needed about the system is its response to a unit delta or impulse function, whichever is appropriate.

Time-Invariant Systems

The response of a fixed system depends only upon the time that has elapsed since the application of the input. Let $h(t)$ and $d(t)$ denote the response to a unit impulse and delta function, respectively, applied at $t = 0$. Then the response to $U_0(t - kT)$ or to $\delta(t - kT)$ is $h(t - kT)$ or $d(t - kT)$, respectively.

Except in trivial cases, the delta response $d(t)$ is identically zero for *continuous* systems. Any input for which $\int_{-\infty}^{\infty} |v(t)|\, dt = 0$ does not insert any energy into the system and hence cannot cause an output to occur. In *discrete* systems, such as a digital computer, a finite discrete input is converted to a finite discrete output. In such a case, $d(t)$ is defined only at the instants $t = t_0 + nT$, where n is an integer. The term t_0 represents a possible constant time delay between the input and output sequences. Since the chief concern is with the form of the output, t_0 will be ignored

Table 1.9-1

Input	Output	Simplification when System is Nonanticipatory	Simplification when Input is Zero for $t < 0$	Equation Number
$v(t) = \sum_{k=-\infty}^{\infty} v(kT)\delta(t - kT)$	$y(nT) = \sum_{k=-\infty}^{\infty} v(kT)\,d(nT - kT)$	Upper limit is $k = n$	Lower limit is zero	(1.9-1)
	$y(nT) = \sum_{k=-\infty}^{\infty} v(nT - kT)\,d(kT)$	Lower limit is zero	Upper limit is $k = n$	(1.9-2)
$v^*(t) = \sum_{k=-\infty}^{\infty} v(kT)U_0(t - kT)$	$y(t) = \sum_{k=-\infty}^{\infty} v(kT)h(t - kT)$	Upper limit is $kT = t$	Lower limit is zero	(1.9-3)
	$y(t) = \sum_{k=-\infty}^{\infty} v(t - kT)h(kT)$	Lower limit is zero	Upper limit is $kT = t$	(1.9-4)
$v^*(t) = \sum_{k=-\infty}^{\infty} v(kT)U_0(t - kT)$	$y(nT) = \sum_{k=-\infty}^{\infty} v(kT)h(nT - kT)$	Upper limit is $k = n$	Lower limit is zero	(1.9-5)
	$y(nT) = \sum_{k=-\infty}^{\infty} v(nT - kT)h(kT)$	Lower limit is zero	Upper limit is $k = n$	(1.9-6)

in this chapter, and the delta response will be written $d(nT)$. For nonanticipatory systems, $d(nT) = 0$ for $n < 0$. By superposition, the response of a discrete system to the input of Eq. 1.8-2 is given by Eq. 1.9-1 in Table 1.9-1. Equation 1.9-2 is obtained by replacing the dummy variable k by $n - k$. Note that these equations define the output only at discrete instants of time. For nonanticipatory systems with $v(t) = 0$ for $t < 0$, the limits of summation become 0 and n for both equations, although it is never incorrect to use the wider limits of $-\infty$ and $+\infty$.

Consider a continuous system whose input is given by Eq. 1.8-4. The output is the continuous function of time given by Eq. 1.9-3 or 1.9-4. In the simplifications listed for nonanticipatory systems and for inputs that are zero for $t < 0$, the statement "Upper limit is $kT = t$" is literally correct only when t is an integral multiple of T. The summation is only over discrete values of k. At time $t = (n + \gamma)T$, where n is an integer and $0 < \gamma < 1$, the statement above should be interpreted as $k = n$. Equations 1.9-3 and 1.9-4 give the exact output of a continuous system to a train of weighted impulses. When such an input is used to approximate the output of a physical sampler, however, these equations likewise involve an approximation. The degree of approximation depends upon the width of the pulses from the sampler compared to the time constants of the system.

If the output of the continuous system is desired only when t is an integral multiple of T, Eqs. 1.9-3 and 1.9-4 reduce to Eqs. 1.9-5 and 1.9-6, respectively. The latter equations would be used if, for example, the output of the system were to be followed by another sampling switch. For nonanticipatory systems with $v(t) = 0$ for $t < 0$, the limits of summation again become 0 and n. The reader has no doubt noticed the similarities between Tables 1.6-1 and 1.9-1. In fact, Eqs. 1.9-3 through 1.9-6 can be treated as special cases of Eqs. 1.6-2 and 1.6-4.

Equations 1.9-1 and 1.9-2 are identical with Eqs. 1.9-5 and 1.9-6, respectively, except for the use of different symbols. It is therefore instructive to compare the discrete and continuous systems shown in parts (*a*) and (*b*) of Fig. 1.9-1. The relationship of $y(t)$ to $v(t)$ in part (*b*)

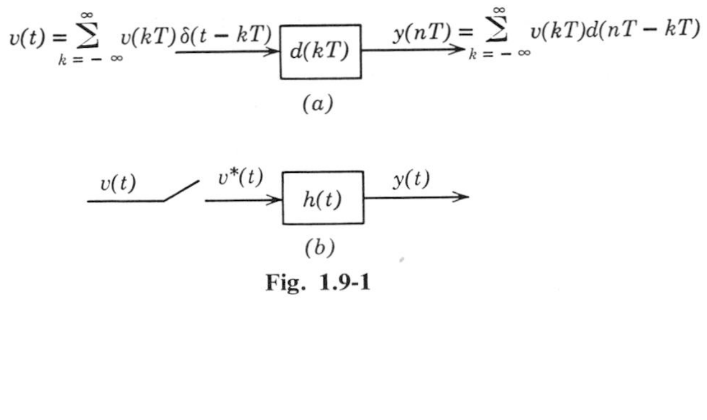

Fig. 1.9-1

at instants which are integral multiples of T is exactly the same as the relationship of $y(nT)$ to $v(kT)$ in part (a), provided that $h(kT) = d(kT)$. It is shown in Chapter 2 that a discrete system can be described by a difference equation, and thus so can the configuration of Fig. 1.9-1b, provided that the output is desired only at $t = nT$. Difference equations may be solved classically, as in Chapter 2, or by the Z transform, as in Chapter 3.

Example 1.9-1. In Fig. 1.9-1a, $T = 1$,

$$d(k) = \begin{cases} 2^k & \text{for } k \geqslant 1 \\ 0 & \text{for } k \leqslant 0 \end{cases}$$

and the input is given by

$$v(k) = \begin{cases} k & \text{for } k \geqslant 0 \\ 0 & \text{for } k < 0 \end{cases}$$

Find an expression for the discrete output signal $y(n)$.

Using Eq. 1.9-2,

$$y(n) = \sum_{k=1}^{n} (n - k)2^k \quad \text{for } n \geqslant 1$$

Thus

$$y(n) = 0 \quad \text{for } n \leqslant 1$$
$$y(2) = 2$$
$$y(3) = 8$$
$$y(4) = 22$$
$$\cdots\cdots\cdots$$

The calculation of $y(n)$ for large values of n can be greatly simplified if the expression can be put in closed form. It can be shown, although not without some thought, that

$$\sum_{k=1}^{n} (n - k)2^k = 2[2^n - n - 1]$$

It is shown in Example 2.12-6 that the delta response $d(k)$ in the last example describes a system characterized by the difference equation

$$y(k + 2) - 3y(k + 1) + 2y(k) = 2v(k + 1) - 2v(k)$$

One disadvantage of the superposition summations in Table 1.9-1 is that it is difficult and often impossible to express the answer in closed form. When the preceding problem is solved in Example 3.13-2 by the use of the Z transform, the solution is easily obtained in closed form.

Example 1.9-2. If, in Fig. 1.9-1b, $T = 1$,

$$h(t) = \begin{cases} 2^t & \text{for } t \geqslant 1 \\ 0 & \text{for } t \leqslant 0 \end{cases}$$

and

$$v(t) = tU_{-1}(t)$$

then the output at the instants $t = n$ is

$$y(n) = 2[2^n - n - 1] \quad \text{for } n = 0, 1, 2, \ldots$$

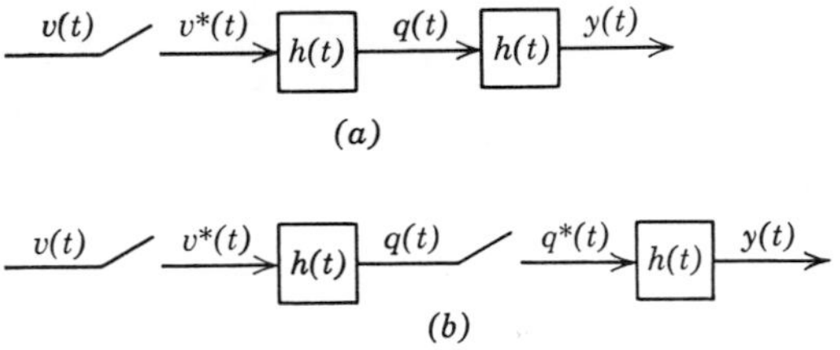

Fig. 1.9-2

Example 1.9-3. Figure 1.9-2*a* shows a sampling switch followed by two identical systems in cascade, each having the impulse response

$$h(t) = t\epsilon^{-t} \quad \text{for } t > 0$$

Assuming that the second system does not load down the first, and that the sampling interval is $T = 1$, find the output $y(t)$ when

$$v(t) = \epsilon^{-t}U_{-1}(t)$$

From Example 1.6-2, the impulse response of the two systems in cascade is $t^3\epsilon^{-t}/6$. Using Eq. 1.9-3,

$$y(t) = \sum_{k=0}^{t} \epsilon^{-k}\frac{(t-k)^3\epsilon^{-(t-k)}}{6} = \frac{\epsilon^{-t}}{6}\sum_{k=0}^{t}(t-k)^3 \quad \text{for } t > 0$$

Thus

$$y(t) = t^3\epsilon^{-t}/6 \qquad \text{for } 0 \leqslant t \leqslant 1$$
$$y(t) = [t^3 + (t-1)^3]\epsilon^{-t}/6 \quad \text{for } 1 \leqslant t \leqslant 2$$

. . . .

Example 1.9-4. Repeat the previous example, with the added sampling switch shown in Fig. 1.9-2*b*.

$$q(t) = \sum_{k=0}^{t} \epsilon^{-k}(t-k)\epsilon^{-(t-k)} = \epsilon^{-t}\sum_{k=0}^{t}(t-k)$$

$$y(t) = \sum_{m=0}^{t}\left[\epsilon^{-m}\sum_{k=0}^{m}(m-k)\right](t-m)\epsilon^{-(t-m)}$$

Using this approach, $y(t)$ is not easily evaluated in closed form. For present considerations, it is sufficient to note that

$$y(t) = 0 \qquad \text{for } 0 \leqslant t \leqslant 1$$

$$y(t) = \epsilon^{-t}(t-1) \quad \text{for } 1 \leqslant t \leqslant 2$$

. . . .

It should be observed that the results of the last two examples are not identical, not even at the sampling instants of $t = n$. Expressions for $y(t)$ in closed form are easily found by means of the modified Z transform in Example 3.14-1.

Table 1.9-2

Input	Output	Simplification when System is Nonanticipatory	Simplification when Input is Zero for $t < 0$	Equation Number
$v(t) = \sum_{k=-\infty}^{\infty} v(kT)\delta(t - kT)$	$y(nT) = \sum_{k=-\infty}^{\infty} v(kT)d(nT, kT)$	Upper limit is $k = n$	Lower limit is zero	(1.9-7)
	$y(nT) = \sum_{k=-\infty}^{\infty} v(nT - kT) \times d_*(nT, kT)$	Lower limit is zero	Upper limit is $k = n$	(1.9-8)
$v^*(t) = \sum_{k=-\infty}^{\infty} v(kT)U_0(t - kT)$	$y(t) = \sum_{k=-\infty}^{\infty} v(kT)h(t, kT)$	Upper limit is $kT = t$	Lower limit is zero	(1.9-9)
	$y(t) = \sum_{k=-\infty}^{\infty} v(t - kT)h_*(t, kT)$	Lower limit is zero	Upper limit is $kT = t$	(1.9-10)
$v^*(t) = \sum_{k=-\infty}^{\infty} v(kT)U_0(t - kT)$	$y(nT) = \sum_{k=-\infty}^{\infty} v(kT)h(nT, kT)$	Upper limit is $k = n$	Lower limit is zero	(1.9-11)
	$y(nT) = \sum_{k=-\infty}^{\infty} v(nT - kT) \times h_*(nT, kT)$	Lower limit is zero	Upper limit is $k = n$	(1.9-12)

Graphical Interpretation of the Convolution Summations

The equations in Table 1.9-1 are known as convolution or superposition summations. Their graphical interpretation is similar to the graphical interpretation of the convolution integrals in Section 1.6. The graphical evaluation of Eq. 1.9-3 is carried out by folding $h(t)$ about the vertical axis, sliding it a distance t into the discrete signal $v(kT)$, and then summing up all values of the discrete product curve (instead of taking the area underneath the product curve). In the case of the convolution summations, however, the graphical evaluation is really identical with the analytic evaluation and is therefore seldom used. The interpretation of $h(t - kT)$ as a weighting function may, however, be useful.

Time-Varying Systems

The equations in Table 1.9-1 are restricted to fixed linear systems, except for the presence of ideal sampling switches. The extension of these results to time-varying systems is parallel to the discussion of Section 1.7. Define

$d(nT, kT) =$ the response at time nT to $\delta(t - kT)$

$d_*(nT, kT) =$ the response at time nT to a unit delta function occurring kT seconds earlier [time $(n - k)T$]

$h(t, kT) =$ the response at time t to $U_0(t - kT)$

$h_*(t, kT) =$ the response at time t to a unit impulse occurring kT seconds earlier [time $(t - kT)$]

With these definitions, the equations of Table 1.9-1 are replaced by those of Table 1.9-2 for time-varying systems. The interpretation of $h_*(t, kT)$ as a weighting function is still valid.

REFERENCES

1. C. E. Shannon, "Communication in the Presence of Noise," *Proc. IRE*, Vol. 37, January 1949, p. 11.
2. L. Schwartz, *Théorie des Distributions*, Hermann et Cie., Paris, 1950–51.
3. P. W. Ketchum and R. Aboudi, "Schwartz Distributions," Second Midwest Symposium on Circuit Theory, 1956.
4. L. B. Arguimbau, *Vacuum-Tube Circuits*, John Wiley and Sons, New York, 1948, p. 161.
5. R. E. Scott, *Linear Circuits*, Addison-Wesley Publishing Company, Reading, Mass., 1960, Section 13–5.
6. E. I. Jury, *Sampled-Data Control Systems*, John Wiley and Sons, New York, 1958, Chapter 9.

Problems

1.1 Under what conditions does the differential equation

$$a_n \frac{d^n y}{dt^n} + \cdots + a_1 \frac{dy}{dt} + a_0 y = b_m \frac{d^m v}{dt^m} + \cdots + b_1 \frac{dv}{dt} + b_0 v$$

describe a linear system? a time-invariant system? a nonanticipatory system?

1.2 Assume that two different linear systems are connected in cascade, as in Fig. 1.2-1. If the systems are time-invariant, and if the second system does not load down the preceding one, will interchanging their order affect the relationship between $y(t)$ and $v(t)$? Repeat for time-varying systems. Prove your answers.

1.3 Show that $\lim_{\Delta \to 0} f_0(t) = U_0(t)$, where $f_0(t)$ is the function in Fig. 1.4-7*a*.

1.4 Find the step response of the circuit in Fig. P1.4.

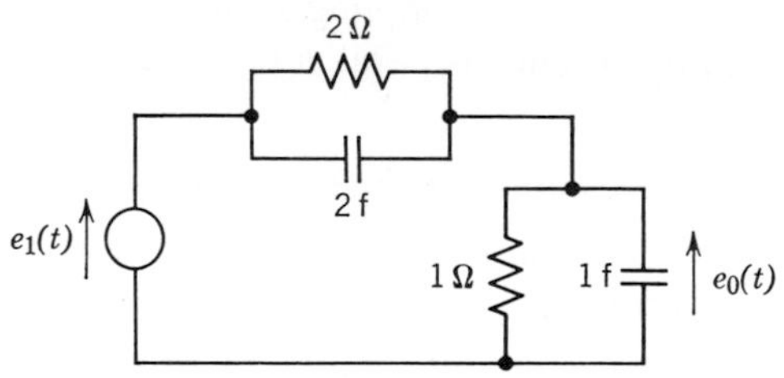

Fig. P1.4

1.5 Find the response of Fig. 1.4-8 when the current source has the waveform shown in Fig. P1.5, and also when $i_1(t) = 3\epsilon^{-t}$ for $t \geqslant 0$.

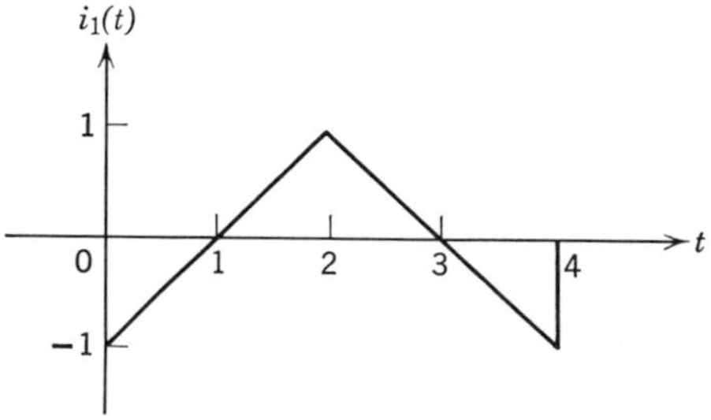

Fig. P1.5

1.6 Check the results of Example 1.6-1 by considering the voltage source to be the sum of two singularity functions, and by finding the response to each one.

1.7 A fixed linear system containing no initial stored energy has the input and impulse response shown in Fig. P1.7. Find the response at $t = 4$

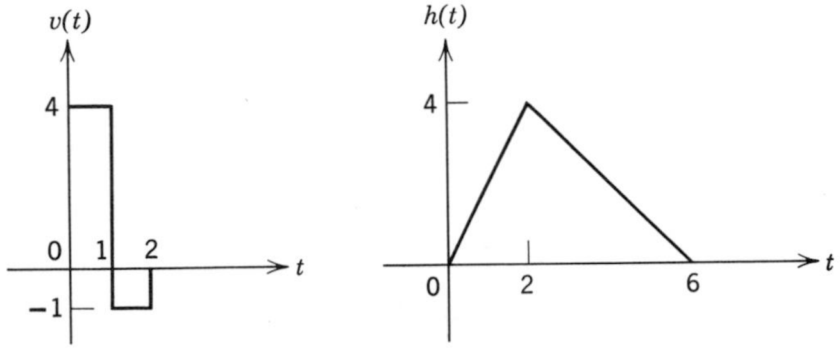

Fig. P1.7

seconds both analytically and graphically. Roughly sketch $y(t)$ versus t for $t \geqslant 0$.

1.8 The cross-correlation functions

$$\phi_{12}(t) = \int_{-\infty}^{\infty} f_1(\tau) f_2(t + \tau)\, d\tau$$

$$\phi_{21}(t) = \int_{-\infty}^{\infty} f_2(\tau) f_1(t + \tau)\, d\tau$$

are important in the analysis of systems subjected to aperiodic inputs. Prove that $\phi_{21}(t) = \phi_{12}(-t)$. If

$$f_1(t) = (t - 1)U_{-1}(t - 1) - (t - 1)U_{-1}(t - 2)$$

and $f_2(t) = 2\epsilon^{-t}U_{-1}(t)$, find and sketch $\phi_{12}(t)$.

1.9 Prove Eq. 1.6-10.

1.10 Derive Eq. 1.7-7 by letting $v(t) = U_{-1}(t - \tau)$ and $y(t) = r(t, \tau)$ in Eq. 1.7-2. Then derive Eq. 1.7-6 by using the substitution $\lambda = t - \tau$.

1.11 A system is described by the differential equation

$$(t + 1)\frac{dy}{dt} + y = (t + 1)v(t)$$

Noting that the left-hand side of this equation is the derivative of $(t + 1)y$, find $h(t, \lambda)$, $h_*(t, \tau)$, $r(t, \lambda)$, and $r_*(t, \tau)$. Find the response to $v(t) = \epsilon^{-t}U_{-1}(t)$.

1.12 Repeat the previous problem for the differential equation

$$\frac{dy}{dt} + \frac{y}{t + 1} = \frac{dv}{dt} + v$$

1.13 In the cascade connection of Fig. 1.2-1, assume that the impulse responses of N_1 and N_2 are denoted by $h_1(t, \lambda)$ and $h_2(t, \lambda)$, respectively. Derive an expression for the impulse response of the cascaded combination.

1.14 Assume that three linear components are placed in cascade. If the first is an ideal integrator, the third an ideal differentiator, and the middle one characterized by the impulse response $h_*(t, \tau) = t\epsilon^{-(\tau+t)}$ for $\tau > 0$, find the impulse and step responses of the combination.

1.15 In Fig. P1.15, $h_1(t) = (t - 1)\epsilon^{-t}U_{-1}(t)$ and $h_2(t) = \epsilon^{-t}[\cos(\pi/2)t - \sin(\pi/2)t]U_{-1}(t) = \sqrt{2}\,\epsilon^{-t}\cos[(\pi/2)t + (\pi/4)]U_{-1}(t)$. The sampling switch has a period $T = 1$ second. Find $y(nT)$ when $v(t) = U_{-1}(t)$.

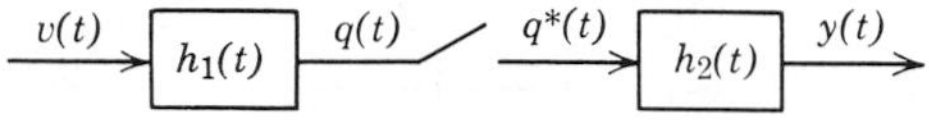

Fig. P1.15

1.16 A sampling switch is added just before $h_1(t)$ in Fig. P1.15. The switches operate synchronously with a period $T = 1$ second, and $h_1(t) = (1/2)^t$ and $h_2(t) = 2\cos(\pi/3)t$ for $t \geqslant 0$. Find $y(nT)$ when $v(t) = U_{-2}(t)$, and also when $v(t) = \sin(\pi/3)t$.

1.17 In Problem 1.15, find $y(t)$ for $0 \leqslant t \leqslant 3$.

1.18 A certain feedback control system samples the input $v(t)$ at $t = nT$, where $\epsilon^{-T} = \frac{1}{2}$, and $n = 0, 1, 2, \ldots.$ The response to $v(t) = U_{-1}(t)$ is $y(t) = (\epsilon^{-\gamma T}/6)[1 - \epsilon^{-nT}]$, where $t = (n + \gamma)T$, and $0 \leqslant \gamma \leqslant 1$. Find and sketch $y(t)$ for $0 \leqslant t \leqslant 2T$ when $v(t) = \epsilon^{-2t}U_{-1}(t)$.

2

Classical Techniques

2.1 INTRODUCTION

The classical method of describing a linear system is by a differential or difference equation relating the output to the input. A differential equation is used for continuous systems, and a difference equation for discrete systems. Since a continuous system may be initially described by a *set* of simultaneous equations, Section 2.3 considers solving such a set for the single desired differential equation.

An explicit method exists for solving both time-invariant and time-varying differential equations of the first order, and it is presented in Section 2.5. For higher order equations, a general and explicit method of solution exists only for the time-invariant case. This method is given in Section 2.6, and higher order time-varying equations are deferred to the following section. Section 2.8 relates the material of Chapters 1 and 2. It presents methods of finding the impulse response from the differential equation for both fixed and varying systems.

The treatment of difference equations is roughly parallel to that for differential equations, except that a preliminary section on difference and antidifference operations is included. The entire chapter is restricted to linear systems. Only analytical methods of solution are considered, although in many applications computer techniques are the most practical approach.[1]

2.2 REPRESENTING CONTINUOUS SYSTEMS BY DIFFERENTIAL EQUATIONS

A continuous system is often described by a differential equation relating the output $y(t)$ to the input $v(t)$. The general form of such an equation† is

$$a_n \frac{d^n y}{dt^n} + a_{n-1} \frac{d^{n-1} y}{dt^{n-1}} + \cdots + a_1 \frac{dy}{dt} + a_0 y = b_m \frac{d^m v}{dt^m} + \cdots + b_1 \frac{dv}{dt} + b_0 v \tag{2.2-1}$$

Since the input $v(t)$ is presumably known, the right side of this equation can be represented by $F(t)$, which is often called the forcing function.

$$a_n \frac{d^n y}{dt^n} + a_{n-1} \frac{d^{n-1} y}{dt^{n-1}} + \cdots + a_1 \frac{dy}{dt} + a_0 y = F(t) \tag{2.2-2}$$

For linear systems, the a_i's and b_j's cannot be functions of v or y but may be functions of t. For *fixed* linear systems, these coefficients must be constants. The proof of these two statements is similar to Examples 1.2-3 and 1.2-4.

A system's differential equation may be given, or it may have to be found from a model of the system. In the latter case, a set of simultaneous differential equations are written directly from the model. The next section gives an example and discusses the solution of the simultaneous equations to yield a single equation relating $y(t)$ and $v(t)$.

The operator p is often used to indicate differentiation and is defined by‡

$$p[v(t)] = \frac{d}{dt}[v(t)] \tag{2.2-3}$$

If c_1 and c_2 are constants,

$$\begin{aligned} p^m(p^n v) &= p^{m+n} v = \frac{d^{m+n} v}{dt^{m+n}} \\ p^m(c_1 v_1 + c_2 v_2) &= c_1 p^m v_1 + c_2 p^m v_2 \\ (p + c_1)(p + c_2)v &= [p^2 + (c_1 + c_2)p + c_1 c_2]v \end{aligned} \tag{2.2-4}$$

where m and n are non-negative integers. In fact, in most respects the operator p may be treated as an algebraic quantity. The most notable exception is that it does not, in general, commute with functions.

$$p(tv) \neq t(pv)$$
$$p(v_1 v_2) \neq v_1(pv_2)$$

Using the operator p, Eqs. 2.2-1 and 2.2-2 become

$$\begin{aligned} &(a_n p^n + a_{n-1} p^{n-1} + \cdots + a_1 p + a_0) y(t) \\ &\qquad = (b_m p^m + \cdots + b_1 p + b_0) v(t) = F(t) \end{aligned} \tag{2.2-5}$$

† For a differential equation to uniquely describe a system, it must be understood that the system is nonanticipative.

‡ D is another commonly used symbol for the differentiating operator. It is not used in this book, however.

The quantities in parentheses preceding y and v are themselves operators. For fixed linear systems, where the coefficients are constants, the last equation is written symbolically as

$$A(p)y(t) = B(p)v(t) = F(t) \tag{2.2-6}$$

For varying linear systems, where the coefficients are functions of time, A and B are time-varying operators. This is shown by writing

$$A(p, t)y(t) = B(p, t)v(t) = F(t) \tag{2.2-7}$$

The formal definition of the operators A and B follows from a comparison of the last three equations.

2.3 REDUCTION OF SIMULTANEOUS DIFFERENTIAL EQUATIONS

A continuous model can be described mathematically by a set of simultaneous differential equations. Basically, one set of equations is needed to characterize each of the components, and another set to describe their interconnection. These two types of equations are usually, however, combined by inspection when mathematically describing the model. The resulting set of differential equations can then be reduced, although not always without difficulty, to yield a single equation relating the output and input. The solution of simultaneous time-invariant equations is far easier than that of time-varying equations and is considered first.

Example 2.3-1. An equation relating the output voltage e_2 to the source voltage e_1 is desired for the circuit of Fig. 2.3-1.

Summing up the currents leaving nodes 3 and 2, respectively,

$$\left[C_1\frac{de_3}{dt} + \left(\frac{1}{R_1} + \frac{1}{R_3}\right)e_3\right] - \left[C_1\frac{de_2}{dt} + \frac{1}{R_1}e_2\right] = \frac{1}{R_3}e_1$$

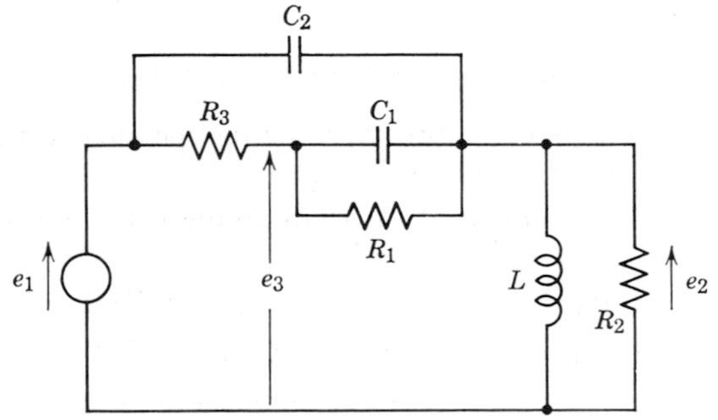

Fig. 2.3-1

and

$$-\left[C_1\frac{de_3}{dt}+\frac{1}{R_1}e_3\right]+\left[(C_1+C_2)\frac{de_2}{dt}+\left(\frac{1}{R_1}+\frac{1}{R_2}\right)e_2+\frac{1}{L}\int e_2\,dt\right]$$
$$=C_2\frac{de_1}{dt}$$

Differentiating the second equation term by term to remove the integral sign, and using the operator $p = d/dt$, these equations become, after assuming for simplicity that all resistances, capacitances, and the inductance have values of 1 ohm, 1 farad, and 1 henry, respectively,

$$(p+2)e_3-(p+1)e_2=e_1$$
$$-(p^2+p)e_3+(2p^2+2p+1)e_2=(p^2)e_1$$

Premultiply each term in the first equation by the operator $p^2 + p$, and each term in the second equation by $p + 2$, and then add the two equations. Since

$$(p^2+p)(p+2)e_3=(p+2)(p^2+p)e_3$$

the terms involving e_3 are eliminated. Then

$$(p+2)(2p^2+2p+1)e_2-(p^2+p)(p+1)e_2=[(p^2+p)+p^2(p+2)]e_1$$

or

$$(p^3+4p^2+4p+2)e_2=(p^3+3p^2+p)e_1$$

which is the desired result.

The procedure used in the preceding example is valid for any two time-invariant equations. If L represents an operator which is a function only of p, the equations can be written symbolically as

$$\begin{aligned}L_{11}(p)y_1(t)+L_{12}(p)y_2(t)&=F_1(t)\\L_{21}(p)y_1(t)+L_{22}(p)y_2(t)&=F_2(t)\end{aligned}\qquad(2.3\text{-}1)$$

Premultiply the first equation by L_{21}, and the second by L_{11}, and subtract. Since $L_{21}L_{11}y_1 = L_{11}L_{21}y_1$,

$$(L_{11}L_{22}-L_{21}L_{12})y_2=-L_{21}F_1+L_{11}F_2\qquad(2.3\text{-}2)$$

Similarly,

$$(L_{11}L_{22}-L_{21}L_{12})y_1=L_{22}F_1-L_{12}F_2\qquad(2.3\text{-}3)$$

Each of the last two equations has only one independent variable and can be solved by the methods of Section 2.6.

Consider next the general case of n simultaneous equations. Symbolically,

$$\begin{aligned}L_{11}y_1+L_{12}y_2+\cdots+L_{1n}y_n&=F_1(t)\\L_{21}y_1+L_{22}y_2+\cdots+L_{2n}y_n&=F_2(t)\\\cdots\cdots\cdots&\cdots\cdots\\L_{n1}y_1+L_{n2}y_2+\cdots+L_{nn}y_n&=F_n(t)\end{aligned}\qquad(2.3\text{-}4)$$

As long as the L_{ij} operators depend only upon p, the solution obtained by Cramer's rule can be shown to be valid.†

$$\Delta(p)y_i = \sum_{k=1}^{n} \Delta_{ki}(p)F_k(t) \tag{2.3-5}$$

where $\Delta(p)$ is the differential operator found by evaluating the determinant

$$\Delta(p) = \begin{vmatrix} L_{11} & L_{12} & \cdots & L_{1n} \\ L_{21} & L_{22} & \cdots & L_{2n} \\ \cdots & \cdots & \cdots & \cdots \\ L_{n1} & L_{n2} & \cdots & L_{nn} \end{vmatrix} \tag{2.3-6}$$

$\Delta_{ki}(p)$ is the operator found by evaluating the kith cofactor of this determinant, i.e., the determinant $\Delta(p)$ with the kth row and ith column removed, multiplied by $(-1)^{k+i}$.

This section has thus far been restricted to time-invariant differential equations. Obtaining a single differential equation relating the output and input of a varying system is more difficult. Suppose that a system is represented by the simultaneous equations

$$t(py_1) + p(t^2y_2) = F_1(t)$$
$$p(ty_1) + t(py_2) = F_2(t)$$

To eliminate y_1 from these equations, one might try premultiplying the first equation by pt and the second by tp, and then subtracting the two. If this is done,

$$pt^2(py_1) + pt(pt^2y_2) = ptF_1(t)$$
$$tp(pty_1) + tp(tpy_2) = tpF_2(t)$$

However, $pt^2(py_1) = (t^2p^2 + 2tp)y_1$, while $tp(pty_1) = t(tp^2 + 2p)y_1$; hence subtracting the last two equations will not eliminate y_1.

Equations 2.3-2 and 2.3-3 are not valid for varying systems, because

$$L_{21}(p, t)L_{11}(p, t)y_1(t) \neq L_{11}(p, t)L_{21}(p, t)y_1(t)$$

Similarly, Eq. 2.3-5 does not hold. While a single differential equation relating the output and input can often be obtained, no simple formula of the type of Eq. 2.3-5 exists which is generally applicable. Also, the difficulty in determining a single differential equation often makes other methods advisable. One of these methods is discussed in Chapter 5.

† Those in need of a review of determinants and Cramer's rule should read Sections 4.3 and 4.6, respectively.

2.4 GENERAL PROPERTIES OF LINEAR DIFFERENTIAL EQUATIONS

Any nth order linear differential equation can be written in the form of Eq. 2.2-5, repeated below.

$$(a_n p^n + a_{n-1} p^{n-1} + \cdots + a_1 p + a_0) y(t) = F(t) \qquad (2.4\text{-}1)$$

Because all the remarks of this section are valid for both fixed and varying systems, the a_i's can, in general, be functions of t. If the right side of the last equation is identically zero,

$$(a_n p^n + a_{n-1} p^{n-1} + \cdots + a_1 p + a_0) y(t) = 0 \qquad (2.4\text{-}2)$$

which is called a homogeneous differential equation. Equation 2.4-1 is called a nonhomogeneous or inhomogeneous equation.

Equation 2.4-2 can have, at most, n linearly independent solutions. The concept of linear independence is precisely defined in Chapter 4. In brief, n objects are said to be linearly dependent (or just dependent) if at least one of them can be expressed as a linear combination of the remaining ones. If the objects are not dependent, they are said to be independent. A necessary and sufficient condition for n solutions of Eq. 2.4-2 to be independent is that their Wronskian does not vanish. If $y_1, y_2, \ldots, y_n$ represent n solutions, the Wronskian is given by the determinant

$$W(t) = \begin{vmatrix} y_1 & y_2 & \cdots & y_n \\ py_1 & py_2 & \cdots & py_n \\ \cdots & \cdots & \cdots & \cdots \\ p^{n-1}y_1 & p^{n-1}y_2 & \cdots & p^{n-1}y_n \end{vmatrix} \qquad (2.4\text{-}3)$$

The most general solution of Eq. 2.4-2 is

$$y_H = K_1 y_1 + K_2 y_2 + \cdots + K_n y_n \qquad (2.4\text{-}4)$$

where the K_i's are arbitrary constants. The subscript H refers to the solution of the homogeneous equation.

The last equation says that, once n independent solutions are known, any other solution can be expressed as a linear combination of these n solutions. For fixed systems, there is a general method of finding n independent solutions to the homogeneous differential equation. For varying systems there is, unfortunately, no such general method.

The most general (or "complete") solution of the nonhomogeneous Eq. 2.4-1 is

$$y = y_H + y_P \qquad (2.4\text{-}5)$$

where y_H is given in Eq. 2.4-4 and is the solution of the related homogeneous equation. y_P is any one solution, no matter how arrived at, which satisfies Eq. 2.4-1, and is known as the particular or particular integral solution. y_H is called the complementary solution. Any method of finding y_P, including guesswork, is allowable. In fixed circuits with a sinusoidal source, for example, a-c steady-state circuit theory might be employed. The next section shows an explicit general method of finding y_P once y_H is known.

As y_P contains no arbitrary constants, y and y_H both contain n such constants. Their evaluation requires a knowledge of initial or boundary conditions, and it is discussed in Section 2.6.

2.5 SOLUTION OF FIRST ORDER DIFFERENTIAL EQUATIONS

Any linear first order differential equation can be written in the form of Eq. 2.5-1.

$$\frac{dy}{dt} + a(t)y = F(t) \tag{2.5-1}$$

For convenience, it is assumed that the coefficient of dy/dt has been made equal to unity. The coefficient a may in general be a function of t. Such an equation may always be solved by the introduction of the integrating factor $\epsilon^{\int a(t)\,dt}$. Both sides of Eq. 2.5-1 are multiplied by this factor.

$$\frac{dy}{dt}\epsilon^{\int a(t)\,dt} + a(t)y\epsilon^{\int a(t)\,dt} = F(t)\epsilon^{\int a(t)\,dt} \tag{2.5-2}$$

The left side of the last equation is the time derivative of $y\epsilon^{\int a(t)\,dt}$, so

$$y\epsilon^{\int a(t)\,dt} = \int F(t)\epsilon^{\int a(t)\,dt}\,dt + c, \qquad c = \text{constant} \tag{2.5-3}$$

Although the integration on the right side of Eq. 2.5-3 may sometimes be difficult, the method constitutes an explicit procedure for solving the differential equation, for both the time-invariant and the time-varying cases. It will give both the complementary and the particular solutions in the answer.

In evaluating $\int a(t)\,dt$, it is not necessary to include an arbitrary constant of integration. The reader should convince himself that including such a constant will not increase the generality of the final solution.

Example 2.5-1. Solve $dy/dt - ty = t$ for $y(t)$.

The integrating factor is $\epsilon^{-\int t\,dt} = \epsilon^{-t^2/2}$. Multiplying the original equation by the integrating factor,

$$\frac{d}{dt}(y\epsilon^{-t^2/2}) = t\epsilon^{-t^2/2}$$

$$y\epsilon^{-t^2/2} = -\int \epsilon^{-t^2/2}(-t\,dt) + c = -\epsilon^{-t^2/2} + c$$

$$y = -1 + c\epsilon^{t^2/2}$$

Although the well-known separation of variables method is sometimes simpler than the use of the integrating factor, it is not applicable to all first order equations. It could not be used in this example.

2.6 SOLUTION OF DIFFERENTIAL EQUATIONS WITH CONSTANT COEFFICIENTS

Linear time-invariant systems are represented by linear differential equations with constant coefficients. The nth order nonhomogeneous and homogeneous equations are given by Eqs. 2.4-1 and 2.4-2, respectively, where the a_i's are constants. First order equations were discussed in Section 2.5, while the methods of this section are applicable to all orders.

Homogeneous Differential Equations

Consider the nth order equation

$$(a_np^n + a_{n-1}p^{n-1} + \cdots + a_1p + a_0)y = 0 \tag{2.6-1}$$

Assume that the solutions are of the form $y = \epsilon^{rt}$, where r is a constant to be determined. Substituting the assumed solution into Eq. 2.6-1, there results

$$(a_nr^n + a_{n-1}r^{n-1} + \cdots + a_1r + a_0)\epsilon^{rt} = 0$$

For the last equation to be satisfied for all values of t,

$$a_nr^n + a_{n-1}r^{n-1} + \cdots + a_1r + a_0 = 0 \tag{2.6-2}$$

Equation 2.6-2 is called the characteristic or auxiliary equation and can be written down directly from Eq. 2.6-1. The left side is an nth order algebraic polynomial, so Eq. 2.6-2 has n roots. Denoting these roots by r_1, $r_2, \ldots, r_n$, the corresponding solutions to Eq. 2.6-1 are

$$y_1 = \epsilon^{r_1t}, \quad y_2 = \epsilon^{r_2t}, \quad \ldots, \quad y_n = \epsilon^{r_nt}$$

If these n solutions are linearly independent, the most general solution to the homogeneous differential equation is

$$y_H = K_1\epsilon^{r_1t} + K_2\epsilon^{r_2t} + \cdots + K_n\epsilon^{r_nt} \tag{2.6-3}$$

If the r_i's are all different, it turns out that $W(t)$ in Eq. 2.4-3 does not vanish, so that the n individual solutions are independent. If $r_1 = r_2$, then $y_1 = \epsilon^{r_1t}$ and $y_2 = t\epsilon^{r_1t}$ are independent solutions. If a root, say r_1, is repeated k times, so that $r_1 = r_2 = \cdots = r_k$, then the most general solution is

$$y_H = K_1\epsilon^{r_1t} + K_2t\epsilon^{r_1t} + \cdots + K_kt^{k-1}\epsilon^{r_1t} + K_{k+1}\epsilon^{r_{k+1}t} + \cdots + K_n\epsilon^{r_nt} \tag{2.6-4}$$

Thus finding y_H involves only solving for the roots of an nth order equation. In the event that some of the roots are complex, however, the solution should be written in another form.

Since the coefficients in Eq. 2.6-2 are real, any complex roots must occur in complex conjugate pairs. If one root is $r_1 = \alpha + j\beta$, where α and β are real, another root must be $r_2 = \alpha - j\beta$. Then

$$\begin{aligned} K_1\epsilon^{r_1t} + K_2\epsilon^{r_2t} &= \epsilon^{\alpha t}(K_1\epsilon^{j\beta t} + K_2\epsilon^{-j\beta t}) \\ &= \epsilon^{\alpha t}[(K_1 + K_2)\cos\beta t + j(K_1 - K_2)\sin\beta t] \\ &= \epsilon^{\alpha t}[A\cos\beta t + B\sin\beta t] \end{aligned}$$

For a real system, the a_i's are real numbers, and y_H is a real function of time. This means that, when the arbitrary constants are evaluated, A and B must be real numbers, which in turn means that K_1 and K_2 must be complex conjugates. Since two trigonometric terms of the same frequency may be written as a single term with a phase angle, it is also possible to write

$$K_1\epsilon^{r_1t} + K_2\epsilon^{r_2t} = K\epsilon^{\alpha t}\cos(\beta t + \phi)$$

Nonhomogeneous Differential Equations—Method of Undetermined Coefficients

Consider the equation

$$(a_np^n + a_{n-1}p^{n-1} + \cdots + a_1p + a_0)y = F(t) \tag{2.6-5}$$

whose solution is

$$y = y_H + y_P \tag{2.6-6}$$

The complementary solution y_H is found by replacing $F(t)$ by zero and solving the resulting homogeneous equation as before. There are two standard methods of solving for the particular integral solution y_P, the method of undetermined coefficients and the variation of parameters method.

The method of undetermined coefficients can be used only when the forcing function $F(t)$ possesses a finite number of linearly independent derivatives. $F(t)$ may be a polynomial in positive integral powers of t, or a combination of simple exponential, sinusoidal, or hyperbolic functions. If $F(t)$ is, for example, $\ln t$ or $\sqrt{t}$, the method is not applicable (unless a solution in the form of an infinite series is assumed). The basic procedure is to assume that y is a linear combination of the terms in $F(t)$ and its derivatives, each term being multiplied by an undetermined constant. The assumed solution is then substituted into Eq. 2.6-5, and the undetermined constants are so chosen that the equation is satisfied for all values of t.

A modification of this procedure is necessary if a term in $F(t)$ has exactly the same form as a term in the complementary solution. Physically, this is the familiar resonance phenomenon, where the system is being excited at one of its natural modes. For example, the equation

$$\frac{d^2y}{dt^2} + 3\frac{dy}{dt} + 2y = 1 + \epsilon^{-t}$$

for which $y_H = K_1\epsilon^{-t} + K_2\epsilon^{-2t}$, is not identically satisfied by $y_P = A + B\epsilon^{-t}$, regardless of the choice of A and B. It is reasonable to expect, however, that the term resulting from exciting the system at a natural mode would decay more slowly with time than would otherwise be the case. It is thus logical to try as a solution

$$y_P = A + Bt\epsilon^{-t}$$

When this solution is substituted into the differential equation, it is found that the equation is satisfied identically for $A = \frac{1}{2}$ and $B = -1$.

The general procedure when a term in $F(t)$ is identical in form with a term in y_H is to multiply by t the corresponding terms in the assumed y_P. The same procedure is followed when a term in $F(t)$ is t^n times a term in y_H. If, however, the term in $F(t)$ corresponds to a *repeated* root of the characteristic equation (an mth order root, for example), then the corresponding terms in y_P are multiplied by t^m.

Example 2.6-1. Find the general solution of

$$\frac{d^2y}{dt^2} + 2\frac{dy}{dt} + y = t\epsilon^{-t}$$

The characteristic equation is $r^2 + 2r + 1 = 0$, so $r_1 = r_2 = -1$, and $y_H = K_1t\epsilon^{-t} + K_2\epsilon^{-t}$. Although normally $y_P = At\epsilon^{-t} + B\epsilon^{-t}$, in this example the characteristic equation has a double root at -1; hence

$$y_P = At^3\epsilon^{-t} + Bt^2\epsilon^{-t}$$

Note that, although repeated differentiation of $At^3\epsilon^{-t}$ also leads to the terms $Ct\epsilon^{-t}$ and $D\epsilon^{-t}$, these terms are not included in the assumed y_P. The reason is that these terms are solutions of the related homogeneous differential equation and would therefore vanish when they are substituted in the left side of the original differential equation. Substituting y_P as determined above,

$$\frac{d^2y_P}{dt^2} + 2\frac{dy_P}{dt} + y_P = (0)t^3\epsilon^{-t} + (0)t^2\epsilon^{-t} + 6At\epsilon^{-t} + 2B\epsilon^{-t} = t\epsilon^{-t}$$

Thus $A = \frac{1}{6}$, $B = 0$, and the complete solution is

$$y = K_1t\epsilon^{-t} + K_2\epsilon^{-t} + \tfrac{1}{6}t^3\epsilon^{-t}$$

Nonhomogeneous Differential Equations—Variation of Parameters

Unlike the method of undetermined coefficients, the variation of parameters method of finding y_P will work whether or not the forcing function $F(t)$ has a finite number of independent derivatives. Also unlike the first method, it will work even if the a_i's in Eq. 2.6-5 are functions of time, although this is of no concern in this section.

The variation of parameters method attempts to find the particular integral solution in terms of the complementary solution. Since this method may be less familiar to the reader, consider first the first order differential equation

$$(a_1p + a_0)y = F(t) \tag{2.6-7}$$

The complementary solution satisfying the homogeneous differential equation

$$(a_1p + a_0)y = 0 \tag{2.6-8}$$

has only the single term $y_H = Ky_1$. The particular integral solution is assumed to have the form

$$y_P = uy_1 \tag{2.6-9}$$

where all three symbols in Eq. 2.6-9 are functions of t. To find u, Eq. 2.6-9 is substituted into Eq. 2.6-7, giving

$$a_1(u\dot{y}_1 + \dot{u}y_1) + a_0uy_1 = F(t)$$

where the dots denote derivatives with respect to t. Rearranging yields

$$a_1\dot{u}y_1 + u(a_1\dot{y}_1 + a_0y_1) = F(t)$$

In the last equation, the factor in parentheses is zero, because y_1 satisfies Eq. 2.6-8. Therefore

$$\frac{du}{dt} = \frac{1}{a_1y_1}F(t) \tag{2.6-10}$$

Example 2.6-2. Find the complete solution to

$$\frac{dy}{dt} + 3y = \frac{\epsilon^{-3t}}{t}$$

The homogeneous solution is $y_H = K\epsilon^{-3t}$, i.e., $y_1 = \epsilon^{-3t}$. Assuming $y_P = u\epsilon^{-3t}$,

$$\frac{du}{dt} = \epsilon^{3t}\frac{\epsilon^{-3t}}{t} = \frac{1}{t}$$

$$u = \ln t \quad \text{and} \quad y_P = \epsilon^{-3t} \ln t$$

The complete solution is

$$y = K\epsilon^{-3t} + \epsilon^{-3t} \ln t$$

Next consider the second order differential equation

$$(a_2p^2 + a_1p + a_0)y = F(t) \tag{2.6-11}$$

The complementary solution satisfying the homogeneous differential equation

$$(a_2p^2 + a_1p + a_0)y = 0 \tag{2.6-12}$$

consists of two parts.

$$y_H = K_1y_1 + K_2y_2$$

The particular integral solution is assumed to be

$$y_P = u_1y_1 + u_2y_2 \tag{2.6-13}$$

where u_1 and u_2 are undetermined functions of t. To evaluate u_1 and u_2, two conditions are needed. One of these is that Eq. 2.6-13 must satisfy Eq. 2.6-11, but the other condition may be chosen in any way that appears convenient.

$$\dot{y}_P = u_1\dot{y}_1 + \dot{u}_1y_1 + u_2\dot{y}_2 + \dot{u}_2y_2$$

The expressions for $\dot{y}_P$ and $\ddot{y}_P$ are less cumbersome if

$$\dot{u}_1y_1 + \dot{u}_2y_2 = 0 \tag{2.6-14}$$

Equation 2.6-14 is therefore taken as the second of the two needed conditions.

$$\dot{y}_P = u_1\dot{y}_1 + u_2\dot{y}_2$$

$$\ddot{y}_P = u_1\ddot{y}_1 + \dot{u}_1\dot{y}_1 + u_2\ddot{y}_2 + \dot{u}_2\dot{y}_2$$

Substituting into Eq. 2.6-11,

$$a_2(u_1\ddot{y}_1 + \dot{u}_1\dot{y}_1 + u_2\ddot{y}_2 + \dot{u}_2\dot{y}_2) + a_1(u_1\dot{y}_1 + u_2\dot{y}_2) + a_0(u_1y_1 + u_2y_2) = F(t)$$

Rearranging,

$$a_2(\dot{u}_1\dot{y}_1 + \dot{u}_2\dot{y}_2) + u_1(a_2\ddot{y}_1 + a_1\dot{y}_1 + a_0y_1) + u_2(a_2\ddot{y}_2 + a_1\dot{y}_2 + a_0y_2) = F(t)$$

Since y_1 and y_2 both satisfy Eq. 2.6-12,

$$\dot{u}_1\dot{y}_1 + \dot{u}_2\dot{y}_2 = \frac{F(t)}{a_2} \tag{2.6-15}$$

To obtain explicit formulas for u_1 and u_2, Eqs. 2.6-14 and 2.6-15 are solved simultaneously. There results

$$\dot{u}_1 = \frac{-y_2F(t)}{a_2(y_1\dot{y}_2 - \dot{y}_1y_2)}, \quad \dot{u}_2 = \frac{y_1F(t)}{a_2(y_1\dot{y}_2 - \dot{y}_1y_2)} \tag{2.6-16}$$

It is worth noting that, since y_1 and y_2 are two independent solutions of Eq. 2.6-12, Eq. 2.4-3 implies that $y_1\dot{y}_2 - \dot{y}_1y_2 \neq 0$. Since the denominator of Eq. 2.6-16 is not zero, $\dot{u}_1$ and $\dot{u}_2$ can always be explicitly found.

Example 2.6-3. Find the complete solution to

$$\frac{d^2y}{dt^2} + y = \frac{1}{t}$$

The homogeneous solution can be written as

$$y_H = K_1 \cos t + K_2 \sin t$$

Thus

$$y_1 = \cos t \quad \text{and} \quad y_2 = \sin t$$

In Eq. 2.6-16,

$$y_1\dot{y}_2 - \dot{y}_1y_2 = \cos^2 t + \sin^2 t = 1$$

$$\dot{u}_1 = -\frac{\sin t}{t}, \quad \dot{u}_2 = \frac{\cos t}{t}$$

The complete solution is

$$y = K_1 \cos t + K_2 \sin t - (\cos t)\int \frac{\sin t}{t}\,dt + (\sin t)\int \frac{\cos t}{t}\,dt$$

The integrals appearing in the answer cannot be expressed in terms of a finite number of elementary functions. The integrals can be expressed by infinite series, however, and happen to be well-tabulated functions—the "sine integral" and "cosine integral" functions that have many applications.

Consider finally the nth order differential equation shown in Eq. 2.6-5. The complementary solution has the form

$$y_H = K_1y_1 + K_2y_2 + \cdots + K_ny_n$$

The assumed particular integral solution is

$$y_P = u_1y_1 + u_2y_2 + \cdots + u_ny_n \tag{2.6-17}$$

where the u_i's are functions of t. The derivatives of the u_i's are found by solving simultaneously the following n equations.

$$\dot{u}_1 y_1 + \dot{u}_2 y_2 + \cdots + \dot{u}_n y_n = 0$$
$$\dot{u}_1 \dot{y}_1 + \dot{u}_2 \dot{y}_2 + \cdots + \dot{u}_n \dot{y}_n = 0$$
$$\cdots\cdots\cdots\cdots\cdots\cdots\cdots\cdots$$
$$\dot{u}_1 y_1^{n-1} + \dot{u}_2 y_2^{n-1} + \cdots + \dot{u}_n y_n^{n-1} = \frac{F(t)}{a_n}$$

where dots and superscripts denote derivatives with respect to t. The first $n - 1$ of these conditions are arbitrarily selected to put the result in a tractable form. The last of the equations is found by substituting the assumed y_P into Eq. 2.6-5 while making use of the first $n - 1$ equations.

The set of equations above can be solved in terms of determinants by Cramer's rule.

$$\dot{u}_i = \frac{W_{ni}(t)F(t)}{a_n W(t)} \quad \text{for } i = 1, 2, \ldots, n \tag{2.6-18}$$

where W is the determinant

$$W(t) = \begin{vmatrix} y_1 & y_2 & \cdots & y_n \\ \dot{y}_1 & \dot{y}_2 & \cdots & \dot{y}_n \\ \cdots & \cdots & \cdots & \cdots \\ y_1^{n-1} & y_2^{n-1} & \cdots & y_n^{n-1} \end{vmatrix} \tag{2.6-19}$$

and where $W_{ni}(t)$ is the nith cofactor. By Eq. 2.4-3, $W(t)$ is the Wronskian, which does not vanish if y_1 through y_n are independent solutions of the homogeneous equation.

Example 2.6-4. Find the complete solution to

$$\frac{d^3y}{dt^3} + 3\frac{d^2y}{dt^2} + 3\frac{dy}{dt} + y = \epsilon^{-t}$$

The homogeneous solution is

$$y_H = K_1\epsilon^{-t} + K_2 t\epsilon^{-t} + K_3 t^2\epsilon^{-t}$$

$$W(t) = \epsilon^{-3t}\begin{vmatrix} 1 & t & t^2 \\ -1 & (-t+1) & (-t^2+2t) \\ 1 & (t-2) & (t^2-4t+2) \end{vmatrix} = 2\epsilon^{-3t}$$

$$\dot{u}_1 = (\tfrac{1}{2}\epsilon^{3t})(t^2\epsilon^{-2t})(\epsilon^{-t}) = \tfrac{1}{2}t^2$$
$$\dot{u}_2 = (\tfrac{1}{2}\epsilon^{3t})(-2t\epsilon^{-2t})(\epsilon^{-t}) = -t$$
$$\dot{u}_3 = (\tfrac{1}{2}\epsilon^{3t})(\epsilon^{-2t})(\epsilon^{-t}) = \tfrac{1}{2}$$
$$u_1 = \tfrac{1}{6}t^3$$
$$u_2 = -\tfrac{1}{2}t^2$$
$$u_3 = \tfrac{1}{2}t$$
$$y_P = (\tfrac{1}{6}t^3)(\epsilon^{-t}) + (-\tfrac{1}{2}t^2)(t\epsilon^{-t}) + (\tfrac{1}{2}t)(t^2\epsilon^{-t}) = \tfrac{1}{6}t^3\epsilon^{-t}$$

Thus the complete solution is

$$y = K_1\epsilon^{-t} + K_2t\epsilon^{-t} + K_3t^2\epsilon^{-t} + \tfrac{1}{6}t^3\epsilon^{-t}$$

A comparison of the two standard methods of finding y_P reveals that the method of undetermined coefficients is often the simpler. It has the disadvantages, however, of being valid for only a limited class of forcing functions and of not applying for time-varying systems.

The variation of parameters method *always* gives an explicit expression for y_P once y_H is known. This expression takes the form of a single time integration of a known function, which is one reason why y_P is commonly called the particular *integral* solution. In some cases, the integral may be difficult to evaluate, and numerical methods may have to be used. The important point, however, is that the method is valid for all forcing functions and can be extended to the solution of time-varying systems.

The Particular Solution for the Input ϵ^{st}

The general time-invariant system is described by Eq. 2.2-1 or 2.2-6. Using the A and B operators,

$$A(p)y(t) = B(p)v(t) \tag{2.6-20}$$

If $v(t) = \epsilon^{st}$, the particular solution has the form $y_P(t) = H(s)\epsilon^{st}$, where H is not a function of t. Substituting these expressions into the differential equation gives

$$A(p)[H(s)\epsilon^{st}] = B(p)[\epsilon^{st}]$$

Since $p^k\epsilon^{st} = s^k\epsilon^{st}$, $B(p)\epsilon^{st} = \epsilon^{st}B(s)$, where $B(s)$ is a function formed by replacing the differentiation operator p by s.

$$H(s)\epsilon^{st}A(s) = \epsilon^{st}B(s) \quad \text{or} \quad H(s) = \frac{B(s)}{A(s)} \tag{2.6-21}$$

Thus $H(s)$ can be written down by inspection of the differential equation. Also

$$H(s) = \left[\frac{y_P(t)}{v(t)}\right]_{v(t)=\epsilon^{st}} \tag{2.6-22}$$

$H(s)$ is usually called the system function or transfer function. The fact that the response to ϵ^{st} can be found so easily suggests decomposing an arbitrary input into a set of ϵ^{st} functions. In the terminology of Section 1.3, the elementary function would be $k(t, \lambda) = \epsilon^{\lambda t}$, and the response to the elementary function would be $K(t, \lambda) = H(\lambda)\epsilon^{\lambda t}$. Carrying out this suggestion leads to the Laplace transform technique of Chapter 3.

For a time-invariant circuit, the system function $H(s)$ can be determined directly from the model, instead of first writing the differential equation.

In the particular solution for $v(t) = \epsilon^{st}$, all currents and voltages have the form $e(t) = E(s)\epsilon^{st}$ and $i(t) = I(s)\epsilon^{st}$. The three passive circuit elements are shown in Table 2.6-1. When the preceding expressions for e and i are

Table 2.6-1

Element	Defining Equation	$Z(s)$
R	$e = Ri$	R
L	$e = L \dfrac{di}{dt}$	sL
C	$i = C \dfrac{de}{dt}$	$\dfrac{1}{sC}$

inserted into the defining equation for the circuit element, the ratio of the voltage to the current can be found. This ratio, called the impedance and denoted by $Z(s)$, is independent of t. Because of this fact, the relationship between $y_P(t)$ and $v(t)$ can be found by the same basic rules as apply to d-c circuits.

Example 2.6-5. For the circuit of Fig. 2.6-1, find the particular solution $e_P(t)$ when the input current is $i(t) = \epsilon^{st}$.

$$E(s) = \frac{(R + sL)(1/sC)}{R + sL + 1/sC} I(s) = \frac{sL + R}{s^2LC + sRC + 1} I(s)$$

Thus

$$H(s) = \frac{sL + R}{s^2LC + sRC + 1}$$

and the particular solution when $i(t) = \epsilon^{st}$ is

$$e_P(t) = \frac{sL + R}{s^2LC + sRC + 1} \epsilon^{st}$$

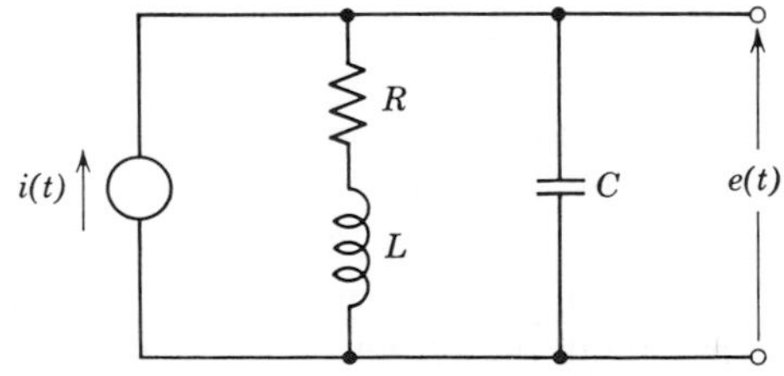

Fig. 2.6-1

Equation 2.6-21 can be used to construct the differential equation relating the input and output from a knowledge of the system function. For the example above, the differential equation is

$$(LCp^2 + RCp + 1)e(t) = (Lp + R)i(t)$$

The variable s is usually called the complex frequency. When s is replaced by $j\omega$, the expressions for $Z(j\omega)$ in Table 2.6-1 are those used in a-c steady-state analysis. The particular solution when $v = \epsilon^{j\omega t}$ can be found from

$$H(j\omega) = \left[\frac{y_P(t)}{v(t)}\right]_{v(t)=\epsilon^{j\omega t}} = \frac{B(j\omega)}{A(j\omega)} \qquad (2.6\text{-}23)$$

The particular solution to a sinusoidal input can be found directly from $H(j\omega)$ by using Example 1.2-5. Since $\cos \omega t = \text{Re}\,[\epsilon^{j\omega t}]$, and since the response to $v(t) = \epsilon^{j\omega t}$ is $y_P(t) = H(j\omega)\epsilon^{j\omega t}$, the response to $v(t) = \cos \omega t$ is

$$y_P = \text{Re}\,[H(j\omega)\epsilon^{j\omega t}]$$

If the complex quantity $H(j\omega)$ is expressed in polar form by $|H(j\omega)|\,\epsilon^{j\theta}$, the response can be written

$$y_P = |H(j\omega)| \cos (\omega t + \theta)$$

$H(j\omega)$ is often called the frequency spectrum of the system and is basic to a-c steady-state circuit theory.

The Evaluation of the Arbitrary Constants

The arbitrary constants in the complementary solution are evaluated from a knowledge of boundary conditions or initial conditions. In most system problems, the values at $t = 0+$ of $y(t)$ and the first $n - 1$ derivatives are used to evaluate the arbitrary constants in the solution to an nth order differential equation. The $0+$ indicates that the value of the dependent variable and its derivatives are known an infinitesimal interval after the reference time $t = 0$. The quantities are normally found by examining the stored energy in the system at $t = 0$. It is important to emphasize that the arbitrary constants depend upon the forcing function used and *cannot* be evaluated until after y_P has been found.

Example 2.6-6. A simplified model of a vacuum-tube amplifier with high-frequency compensation is shown in Fig. 2.6-2. The voltage applied to the grid is

$$e_1(t) = U_{-1}(t) \text{ volts}$$

For negative values of t there is no stored energy associated with L and C. Find the output voltage $e_2(t)$.

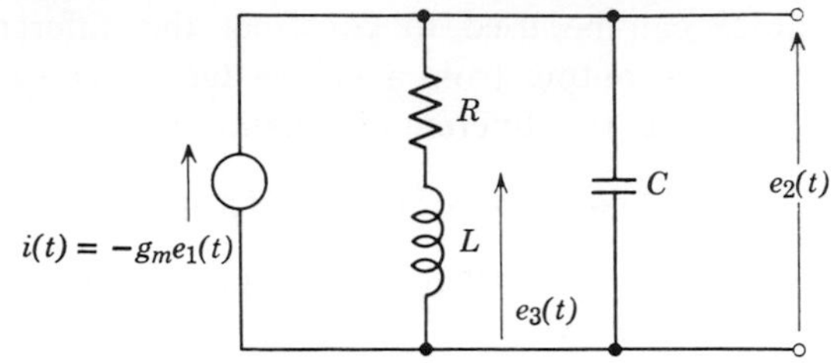

Fig. 2.6-2 ($g_m = 10^{-2}$ mho; $C = 25 \times 10^{-12}$ farad; $L = 50 \times 10^{-6}$ henry; $R =$ 1800 ohms.)

The circuit can be described mathematically by writing two node equations. Applying Kirchhoff's current law at nodes 2 and 3,

$$\left(C\frac{de_2}{dt} + \frac{1}{R}e_2\right) - \frac{1}{R}e_3 = -g_m e_1$$

$$-\frac{1}{R}e_2 + \left(\frac{1}{R}e_3 + \frac{1}{L}\int e_3\,dt\right) = 0$$

The second equation can be differentiated term by term to remove the integral sign. Solving the first equation algebraically for e_3 and substituting the result into the second equation gives

$$\frac{d^2e_2}{dt^2} + \frac{R}{L}\frac{de_2}{dt} + \frac{1}{LC}e_2 = -\frac{g_m}{C}\left(\frac{de_1}{dt} + \frac{R}{L}e_1\right)$$

As expected, this is the same differential equation as the one found for Fig. 2.6-1 by a different method. Inserting the given numerical values and expressing t in microseconds rather than seconds, there results

$$\frac{d^2e_2}{dt^2} + 36\frac{de_2}{dt} + 800e_2 = -400\frac{de_1}{dt} - 14{,}400e_1$$

For $t > 0$,

$$\frac{d^2e_2}{dt^2} + 36\frac{de_2}{dt} + 800e_2 = -14{,}400$$

The characteristic equation is

$$r^2 + 36r + 800 = 0$$

which has roots at

$$r = -18 \pm j21.8$$

Hence

$$(e_2)_H = \epsilon^{-18t}[K_1 \cos 21.8t + K_2 \sin 21.8t]$$

The particular integral solution can be found by examining the differential equation, or by using d-c steady-state theory.

$$(e_2)_P = -\frac{14{,}400}{800} = -18$$

The complete solution is

$$e_2 = -18 + \epsilon^{-18t}[K_1 \cos 21.8t + K_2 \sin 21.8t] \quad \text{for } t > 0$$

Also

$$\frac{de_2}{dt} = \epsilon^{-18t}[(21.8K_2 - 18K_1)\cos 21.8t - (21.8K_1 + 18K_2)\sin 21.8t] \quad \text{for } t > 0$$

K_1 and K_2 can be evaluated from the last two equations if e_2 and de_2/dt are known at $t = 0+$. From the problem statement, the voltage e_2 and the current through L were zero for negative values of t. These values cannot change instantaneously unless there is an impulse of current into C or an impulse of voltage across L, so they must remain zero at $t = 0+$. To find the value of de_2/dt at $t = 0+$, note that, since the inductor current remains zero, the current through C must be $i_c = i = -g_m = -0.01$. Since $i_c = C\,(de_2/dt)$,

$$\frac{de_2}{dt}(0+) = \frac{-0.01}{25 \times 10^{-12}} = -400 \times 10^6 \frac{\text{volts}}{\text{second}} = -400 \frac{\text{volts}}{\text{microsecond}}$$

Thus

$$0 = -18 + K_1$$

$$-400 = 21.8K_2 - 18K_1$$

giving

$$K_1 = 18 \quad \text{and} \quad K_2 = -3.5$$

The solution is

$$e_2 = -18 + \epsilon^{-18t}[18\cos 21.8t - 3.5\sin 21.8t]$$

Example 2.6-7. Repeat the previous example with $e_1(t) = (\cos 4 \times 10^7 t)U_{-1}(t)$.

The form of the complementary solution is not changed, and the particular solution can be found from the system function

$$H(j\omega) = \frac{j\omega L + R}{1 - \omega^2 RC + j\omega RC}$$

For the given values of L, R, and C, and for $\omega = 4 \times 10^7$,

$$H(j\omega) = (-13.1)\epsilon^{-j71^\circ}$$

The complete solution, when t is in microseconds, is

$$e_2(t) = -13.1\cos(40t - 71^\circ) + \epsilon^{-18t}[K_1\cos 21.8t + K_2\sin 21.8t]$$

The arbitrary constants can again be evaluated from $e_2(0+)$ and $(de_2/dt)(0+)$.

$$0 = -13.1\cos 71^\circ + K_1$$

$$-400 = -13.1\,(40)\sin 71^\circ + 21.8K_2 - 18K_1$$

giving $K_1 = 4.3$ and $K_2 = 7.9$.

Homogeneous Initial Conditions

In some important problems, the initial conditions are homogeneous. For the nth order differential Eq. 2.6-5, homogeneous initial conditions are

$$y(0) = \frac{dy}{dt}(0) = \cdots = \frac{d^{n-1}y}{dt^{n-1}}(0) = 0 \qquad (2.6\text{-}24)$$

The variation of parameters method, previously used to find a particular solution, can be easily extended to give the complete solution which satisfies these homogeneous initial conditions. For $n = 1$, Eqs. 2.6-9 and 2.6-10 are combined as

$$y(t) = y_1(t)[u(t) - u(0)] = y_1(t)\int_0^t \frac{F(z)}{a_1(z)y_1(z)}\,dz \tag{2.6-25}$$

where $y_1(t)$ is the complementary solution, and z is a dummy variable of integration. The upper limit yields the particular solution previously discussed, while the lower limit yields a specific arbitrary constant for the complementary solution. Note that $y(0) = 0$.

For $n = 2$, Eqs. 2.6-13 and 2.6-16 are combined as

$$\begin{aligned} y(t) &= y_1(t)[u_1(t) - u_1(0)] + y_2(t)[u_2(t) - u_2(0)] \\ &= -y_1(t)\int_0^t \frac{y_2(z)F(z)}{a_2(z)W(z)}\,dz + y_2(t)\int_0^t \frac{y_1(z)F(z)}{a_2(z)W(z)}\,dz \end{aligned} \tag{2.6-26}$$

where $y_1(t)$ and $y_2(t)$ are two solutions of the homogeneous equation, and where $W(t)$ is the Wronskian

$$W(t) = y_1(t)\dot{y}_2(t) - \dot{y}_1(t)y_2(t)$$

The previous comments about the upper and lower limits of integration still apply. Note that $y(0) = 0$, and

$$\dot{y}(t) = \dot{y}_1(t)[u_1(t) - u_1(0)] + \dot{y}_2(t)[u_2(t) - u_2(0)] + [y_1(t)\dot{u}_1(t) + y_2(t)\dot{u}_2(t)]$$

The last bracket is always zero by Eq. 2.6-14, so $\dot{y}(0) = 0$ as required.

In the general case of an nth order differential equation with homogeneous initial conditions, Eqs. 2.6-17 and 2.6-18 yield

$$\begin{aligned} y(t) &= y_1(t)[u_1(t) - u_1(0)] + \cdots + y_n(t)[u_n(t) - u_n(0)] \\ &= \sum_{i=1}^{n}\left[y_i(t)\int_0^t \frac{W_{ni}(z)}{a_n(z)W(z)}F(z)\,dz\right] \end{aligned} \tag{2.6-27}$$

where $W(t)$ and $W_{ni}(t)$ are defined in Eq. 2.6-19. Since they are based on the variation of parameters method, Eqs. 2.6-24 through 2.6-27 are valid for varying as well as for fixed systems. There is an important application in Section 2.8, where these equations are used to find the impulse response of a time-varying system.

The Physical Significance of the Complementary and Particular Solutions

The form of the complementary solution depends only upon the system and not upon the input. The characteristic equation depends only upon the parameters of the system, and the roots of the characteristic equation determine the kind of terms appearing in the complementary solution. In the event that there is no external source (e.g., the system may be excited only by some initial stored energy within it), the complementary solution becomes the complete solution. Thus the complementary solution represents the natural behavior of the system, when it is left unexcited. For this reason, the complementary solution is also called the free or unforced response.

If the free response of a system increases without limit as t approaches infinity, the system is said to be unstable. This is the case if the characteristic equation has a root with a positive real part, since the complementary solution then contains a term which increases exponentially with t. Roots with negative real parts, on the other hand, lead to terms that become zero as t approaches infinity. Purely imaginary roots, if they are simple, lead to sinusoidal terms of constant amplitude in the complementary solution. This case of constant amplitude oscillation in the complementary solution, which is characteristic, for example, of LC circuits, is usually considered a stable and not an unstable response. Repeated imaginary roots lead to terms of the form $t^n \cos(\omega t + \phi)$, which is an unstable response. If the roots of the characteristic equation are plotted in a complex plane, the following statement can be made. For a stable system, none of the roots can lie in the right half-plane, and any roots on the imaginary axis must be simple. If all the roots of the characteristic equation lie in the left half-plane, the complementary solution approaches zero as t approaches infinity and is identical with the transient response of the system.

The magnitudes of the terms in the complementary solution, i.e., the arbitrary constants, depend upon two things, one of which is the input. The other is the past history of the system before the input was applied. This history can be completely summarized by a knowledge of the energy stored within the system at the time the input is applied.

The form of the particular solution is dictated by the forcing function, as can be clearly seen from the method of undetermined coefficients. The only time the system has any influence upon this form is when a term in the forcing function duplicates a term in y_H. In this event, the system is being excited in one of its natural modes, a phenomenon called resonance.

Since the form of the particular solution depends upon the input, it is also called the forced response. If all the roots of the characteristic equation lie in the left half-plane, the forced solution is identical with the steady-state solution. The magnitudes of the terms in the forced solution depend upon both the input and the system parameters.

It is customary to think of the forced component of the solution as being immediately established by the application of the input. The free component, which is the complementary solution, adjusts itself through the proper evaluation of the arbitrary constants to provide for the proper transition from the unexcited system to a system under the dominance of the input. Some like to think of the system as initially resisting the wishes of the input, by means of the complementary solution. The magnitudes of the arbitrary constants depend upon how greatly the character of the input differs from the natural behavior of the system.

2.7 SOLUTION OF DIFFERENTIAL EQUATIONS WITH TIME-VARYING COEFFICIENTS

The differential equation describing a time-varying system is given by Eqs. 2.2-1, 2.2-2, or 2.2-7, where the a_i and b_i coefficients are now functions of time.

$$a_n(t)\frac{d^n y}{dt^n} + \cdots + a_1(t)\frac{dy}{dt} + a_0(t)y$$

$$= b_m(t)\frac{d^m v}{dt^m} + \cdots + b_1(t)\frac{dv}{dt} + b_0(t)v = F(t) \quad (2.7\text{-}1)$$

$$A(p, t)y(t) = B(p, t)v(t) = F(t)$$

For first order equations, the method of Section 2.5 always yields a solution. For higher order equations, however, an explicit solution cannot be found, in general.

The Complementary Solution

Consider Eq. 2.7-1 with $F(t)$ replaced by zero, and examine the method of Section 2.6 for finding y_H. Solutions of the form $y = \epsilon^{rt}$, where r is a constant, were assumed, resulting in the equation

$$a_n r^n + \cdots + a_1 r + a_0 = 0$$

But, since the a_i's are now functions of time, the roots of this equation are functions of time, violating the assumption just made. In fact, there is no general method of finding y_H in terms of elementary functions.

It is sometimes helpful to know that, if all but one of the n independent solutions to an nth order homogeneous equation are known, the remaining one can be found by a modification of the variation of parameters method.[2]

One of the special cases which can be solved is Euler's equation

$$[a_n(b + ct)^n p^n + a_{n-1}(b + ct)^{n-1} p^{n-1} + \cdots + a_1(b + ct)p + a_0]y(t) = 0 \quad (2.7\text{-}2)$$

where the a_i's, b, and c are constants. By using the substitution of variable

$$(b + ct) = \epsilon^z, \quad \text{i.e., } z = \ln(b + ct)$$

the equation can be reduced to one with constant coefficients.

Example 2.7-1. Find the general solution to

$$t^2 \frac{d^2y}{dt^2} + t\frac{dy}{dt} + y = 0$$

Let $\epsilon^z = t$ and $\epsilon^z\, dz = dt$. Then

$$\frac{dy}{dt} = \frac{dy}{dz}\epsilon^{-z}$$

$$\frac{d^2y}{dt^2} = -\frac{dy}{dz}\epsilon^{-2z} + \frac{d^2y}{dz^2}\epsilon^{-2z}$$

The new equation becomes

$$\frac{d^2y}{dz^2} + y = 0$$

the solution of which is

$$y = K_1 \cos z + K_2 \sin z = K_1 \cos(\ln t) + K_2 \sin(\ln t)$$

Occasionally, an nth order homogeneous differential equation can be reduced to differential equations of lower order. An equation in which y and its first $(k - 1)$ derivatives are absent can be reduced to an equation of order $n - k$ by the substitution of variable $z = d^ky/dt^k$. Another case is that in which the operator can be expressed in factored form, as in the following example.

Example 2.7-2. The differential equation

$$[tp^2 + (t^2 + 1)p + 2t]y = 0$$

may be written as

$$(tp + 1)(p + t)y = 0$$

This is equivalent to the two first order equations

$$(tp + 1)v = 0, \quad (p + t)y = v$$

The first of these two equations may be solved for v as a function of t, and the second then solved for y, using the method of Section 2.5.

Considerable work has been done on differential equations with coefficients that are periodic in time. A second order equation of this nature can be transformed into Mathieu's or Hill's equation.[3] Higher order equations may be handled by the Floquet Theory.[4]

When the a_i coefficients are polynomials in t, an infinite series solution may be obtained. Bessel's and Legendre's equations are the best-known examples. When the coefficients are rational fractions (i.e., the quotient of two polynomials), an infinite series solution can still be found. The infinite series has, in such a case, only a limited region of convergence; hence, if a solution is required for all time, a number of different series must be used.

Finally, a number of methods for approximating y_H are available.[5]† These methods usually assume that the a_i coefficients either vary slowly with t or have a small variation compared to their mean value.

The Particular Solution

Although the method of undetermined coefficients is not suitable for time-varying differential equations, the variation of parameters method is. All steps in the derivations of Eqs. 2.6-10, 2.6-16, and 2.6-18 are valid whether or not the a_i's are functions of time. Thus y_P can always be found once y_H is known. Unfortunately, as has just been pointed out, y_H cannot, in general, be found.

Example 2.7-3. Find the complete solution to

$$t^2\frac{d^2y}{dt^2} + t\frac{dy}{dt} - y = \frac{1}{t}$$

The homogeneous equation

$$t^2\frac{d^2y}{dt^2} + t\frac{dy}{dt} - y = 0$$

has the form of Euler's equation. Using $\epsilon^z = t$,

$$y_H = K_1\epsilon^z + K_2\epsilon^{-z} = K_1t + \frac{K_2}{t}$$

Using Eq. 2.6-16, with $y_1 = t$ and $y_2 = 1/t$,

$$y_1\dot{y}_2 - \dot{y}_1y_2 = -\frac{2}{t}$$

$$\dot{u}_1 = \frac{(-1/t)(1/t)}{t^2(-2/t)} = \frac{1}{2t^3} \quad \text{and} \quad u_1 = -\frac{1}{4t^2}$$

$$\dot{u}_2 = \frac{(t)(1/t)}{t^2(-2/t)} = -\frac{1}{2t} \quad \text{and} \quad u_2 = -\tfrac{1}{2}\ln t$$

$$y_P = u_1y_1 + u_2y_2 = -\frac{1}{4t} - \frac{1}{2t}\ln t$$

$$y = K_1t + \frac{1}{t}(K_2 - \tfrac{1}{4} - \tfrac{1}{2}\ln t)$$

† See Section 5.9.

The identification of the complementary and particular solutions with the free and forced response is still valid for time-varying differential equations.

2.8 OBTAINING THE IMPULSE RESPONSE FROM THE DIFFERENTIAL EQUATION

Chapter 1 presents methods of finding the response of a system to an arbitrary input, once the impulse response is known. Although the impulse response is sometimes given directly, it is important to be able to find it from the system's differential equation. Consider, therefore, the equation

$$a_n \frac{d^n y}{dt^n} + a_{n-1} \frac{d^{n-1} y}{dt^{n-1}} + \cdots + a_1 \frac{dy}{dt} + a_0 y = U_0(t) \qquad (2.8\text{-}1)$$

Since the system is at rest for $t < 0$, $y = 0$ for $t < 0$. Because the forcing function $U_0(t)$ is zero except at $t = 0$, the solution for all $t > 0$ consists only of the complementary solution

$$y = K_1 y_1 + K_2 y_2 + \cdots + K_n y_n \qquad (2.8\text{-}2)$$

To evaluate the n arbitrary constants, n initial conditions are needed. For a fixed system, where the a_i's are constants, these can be found directly from the coefficients. The nth derivative of the solution, but none of the lower order derivatives, contains an impulse at $t = 0$. This is the only way Eq. 2.8-1 can be satisfied at $t = 0$. If one of the lower order derivatives were to contain an impulse, then $d^n y/dt^n$ would contain a singularity of higher order. Since $a_n(d^n y/dt^n)$ does contain a unit impulse at $t = 0$, $d^{n-1}y/dt^{n-1}$ must jump from 0 to $1/a_n$ at $t = 0$, and all lower order derivatives must be continuous at the origin. The n initial conditions must be

$$y(0+) = \dot{y}(0+) = \cdots = y^{n-2}(0+) = 0, \quad y^{n-1}(0+) = \frac{1}{a_n} \qquad (2.8\text{-}3)$$

A fixed linear system is characterized by Eq. 2.2-1 or 2.2-6 as

$$a_n \frac{d^n y}{dt^n} + \cdots + a_1 \frac{dy}{dt} + a_0 y = b_m \frac{d^m v}{dt^m} + \cdots + b_1 \frac{dv}{dt} + b_0 v$$

or

$$A(p)y(t) = B(p)v(t) = F(t) \qquad (2.8\text{-}4)$$

Unless $b_m = b_{m-1} = \cdots = b_1 = 0$, the procedure above must be modified. When $v(t) = U_0(t)$, the right side of Eq. 2.8-4 contains singularity functions of several orders. One convenient approach is to assume that, for small non-negative values of time, $y(t)$ can be represented by a Taylor series, as in the following example.

Example 2.8-1. Find the impulse response for

$$\frac{d^2y}{dt^2} + 3\frac{dy}{dt} + 2y = \frac{1}{2}\frac{dv}{dt} + v$$

Let $v = U_0(t)$.

$$y = [c_0 + c_1 t + c_2 t^2 + \cdots]U_{-1}(t)$$

where the c_i's are constants to be determined.

$$\frac{dy}{dt} = c_0 U_0(t) + [c_1 + 2c_2 t + \cdots]U_{-1}(t)$$

$$\frac{d^2y}{dt^2} = c_0 U_1(t) + c_1 U_0(t) + [2c_2 + \cdots]U_{-1}(t)$$

Substituting into the differential equation and collecting terms,

$$U_1(t)[c_0] + U_0(t)[c_1 + 3c_0] + U_{-1}(t)[2c_2 + 3c_1 + 2c_0] + \cdots = \tfrac{1}{2}U_1(t) + U_0(t)$$

Equating corresponding coefficients,

$$c_0 = \tfrac{1}{2}$$

$$c_1 + 3c_0 = 1, \quad \text{so } c_1 = -\tfrac{1}{2}$$

Immediately after the impulse,

$$y(0+) = c_0 = \frac{1}{2}, \quad \frac{dy}{dt}(0+) = c_1 = -\frac{1}{2}$$

From the original differential equation, the complementary solution for $t > 0$ is

$$y = K_1\epsilon^{-t} + K_2\epsilon^{-2t}$$

Evaluating K_1 and K_2 by the two initial conditions above, $K_1 = \frac{1}{2}$ and $K_2 = 0$. The impulse response is

$$h(t) = \tfrac{1}{2}\epsilon^{-t}$$

The reader familiar with the Laplace transform may wish to compare this example with Example 3.5-2.

The Impulse Response for Time-Varying Systems

The previous approach cannot be readily extended to time-varying systems, where the operators A and B are functions of time.

$$a_n(t)\frac{d^ny}{dt^n} + \cdots + a_0(t)y = b_m(t)\frac{d^mv}{dt^m} + \cdots + b_0(t)v \qquad (2.8\text{-}5)$$

$$A(p, t)y(t) = B(p, t)v(t) = F(t)$$

One method which is valid for varying as well as for fixed systems is closely related to Eqs. 2.6-24 through 2.6-27. For nonanticipatory systems having zero input for $t < 0$, the response to any arbitrary input is given by Eq. 1.7-2.

$$y(t) = \int_0^t v(\lambda)h(t, \lambda)\, d\lambda \qquad (2.8\text{-}6)$$

Note the similarity of this equation and Eqs. 2.6-25 through 2.6-27. Since $y_i(t)$ is not a function of z, and $F(z)$ and $W(z)$ do not depend upon i, Eq. 2.6-27 may be rewritten as

$$y(t) = \int_0^t F(z)\left[\frac{1}{a_n(z)W(z)} \sum_{i=1}^{n} y_i(t)W_{ni}(z)\right] dz \tag{2.8-7}$$

where the y_i's are n independent solutions to the homogeneous differential equation. The quantity in brackets is known as the one-sided Green's function and is denoted by[6]

$$g(t, z) = \frac{1}{a_n(z)W(z)} \sum_{i=1}^{n} y_i(t)W_{ni}(z) \tag{2.8-8}$$

Then

$$y(t) = \int_0^t F(z)g(t, z)\, dz \tag{2.8-9}$$

First, compare Eqs. 2.8-6 and 2.8-9 for the special case $F(t) = v(t)$, which occurs when the operator $B(p, t) = 1$. Then

$$h(t, \lambda) = g(t, \lambda) \quad \text{for} \quad 0 < \lambda < t$$

Of course,

$$h(t, \lambda) = 0 \quad \text{for} \quad t < \lambda$$

$W_{ni}(z)/a_n(z)W(z)$ then represents the proper values of the arbitrary constants in the impulse response. Green's function is frequently used in mathematical literature, and it must satisfy a number of useful properties.[6] The factors in Eq. 2.8-8 may be written explicitly as determinants by Eq. 2.6-19.

$$W(z) = \begin{vmatrix} y_1(z) & y_2(z) & \cdots & y_n(z) \\ \dot{y}_1(z) & \dot{y}_2(z) & \cdots & \dot{y}_n(z) \\ \cdots & \cdots & \cdots & \cdots \\ y_1^{n-1}(z) & y_2^{n-1}(z) & \cdots & y_n^{n-1}(z) \end{vmatrix} \tag{2.8-10}$$

$$\sum_{i=1}^{n} y_i(t)W_{ni}(z) = \begin{vmatrix} y_1(z) & y_2(z) & \cdots & y_n(z) \\ \dot{y}_1(z) & \dot{y}_2(z) & \cdots & \dot{y}_n(z) \\ \cdots & \cdots & \cdots & \cdots \\ y_1^{n-2}(z) & y_2^{n-2}(z) & \cdots & y_n^{n-2}(z) \\ y_1(t) & y_2(t) & \cdots & y_n(t) \end{vmatrix}$$

$$= (-1)^{n-1} \begin{vmatrix} y_1(t) & y_2(t) & \cdots & y_n(t) \\ y_1(z) & y_2(z) & \cdots & y_n(z) \\ \dot{y}_1(z) & \dot{y}_2(z) & \cdots & \dot{y}_n(z) \\ \cdots & \cdots & \cdots & \cdots \\ y_1^{n-2}(z) & y_2^{n-2}(z) & \cdots & y_n^{n-2}(z) \end{vmatrix} \tag{2.8-11}$$

In general, $F(t) \neq v(t)$, and Green's function and the impulse response are not identical. Let the only restriction on Eq. 2.8-5 now be $m < n$.

$$F(t) = B(p, t)v(t)$$
$$= [b_m(t)p^m + \cdots + b_1(t)p + b_0(t)]v(t) \qquad (2.8\text{-}12)$$

where $p = d/dt$. Equation 2.8-9 can be used to find the impulse response, once Green's function is known. If $v(t) = U_0(t - \lambda)$, then $y(t) = h(t, \lambda)$.

$$h(t, \lambda) = \int_0^t [B(p, z)U_0(z - \lambda)]g(t, z)\, dz \qquad (2.8\text{-}13)$$

where now $p = d/dz$. Equation 2.8-13 is used only to find the impulse response for $0 < \lambda < t$, and the integrand is zero except at $z = \lambda$. The limits on the integral may therefore be changed to $-\infty$ and $+\infty$.

This equation is not nearly so formidable as it looks. A typical term in $B(p, z)U_0(z - \lambda)$ is

$$b_k(z)\frac{d^k}{dz^k}U_0(z - \lambda) = b_k(z)\frac{d^k}{dz^k}U_0(\lambda - z)$$

The integral involving this typical term can be evaluated by Eq. 1.6-14 as follows.

$$\int_{-\infty}^{\infty} g(t, z)b_k(z)\frac{d^k}{dz^k}[U_0(\lambda - z)]\, dz = (-1)^k\frac{d^k}{d\lambda^k}[g(t, \lambda)b_k(\lambda)] \qquad (2.8\text{-}14)$$

Equation 2.8-13 can therefore be rewritten as

$$h(t, \lambda) = \sum_{k=0}^{m}(-1)^k\frac{d^k}{d\lambda^k}[g(t, \lambda)b_k(\lambda)] \qquad (2.8\text{-}15)$$

The same result can be obtained in a more elegant way by the concept of the adjoint operator.[6]

In summary, Eq. 2.8-8 gives $h(t, \lambda) = g(t, \lambda)$ for $0 < \lambda < t$ when $F(t) = v(t)$. In general, $h(t, \lambda)$ is found from Eq. 2.8-13 or 2.8-15. These equations constitute a general method of finding the impulse response from a linear differential equation. The principal limitation is that there is no perfectly general method of finding the solutions $y_1, \ldots, y_n$ to a varying, homogeneous differential equation.

Example 2.8-2. Find the impulse response of a system described by the following differential equation.[7]

$$t^2\frac{d^2y}{dt^2} + 4t\frac{dy}{dt} + 2y = v$$

Since this is a form of Euler's equation, the complementary solution can be found by the substitution of variable $t = \epsilon^z$, as discussed in Section 2.7.

$$y_H = \frac{K_1}{t} + \frac{K_2}{t^2}$$

Thus $y_1(t) = 1/t$ and $y_2(t) = 1/t^2$. To determine Green's function, note that

$$W(z) = \begin{vmatrix} \frac{1}{z} & \frac{1}{z^2} \\ \frac{-1}{z^2} & \frac{-2}{z^3} \end{vmatrix} = -\frac{1}{z^4}$$

$$\sum_{i=1}^{n} y_i(t) W_{ni} = (-1) \begin{vmatrix} \frac{1}{t} & \frac{1}{t^2} \\ \frac{1}{z} & \frac{1}{z^2} \end{vmatrix} = \frac{z - t}{t^2 z^2}$$

$$g(t, z) = \left(\frac{1}{z^2}\right)(-z^4)\left(\frac{z - t}{t^2 z^2}\right) = \frac{t - z}{t^2}$$

Since the operator $B(p, t) = 1$ in this example,

$$h(t, \lambda) = \frac{t - \lambda}{t^2} \quad \text{for} \quad t \geqslant \lambda$$

This result is used in Examples 1.7-1 and 3.9-4.

2.9 DIFFERENCE AND ANTIDIFFERENCE OPERATORS

Difference equations arise when functions of a discrete variable are considered. Let $y(k)$ denote a function that is defined only for integral values of k.† As discussed in Chapter 1, any function that is defined only at equal intervals can be expressed in this way by a suitable change of scale.

The shifting operator E is defined by

$$E[y(k)] = y(k + 1) \tag{2.9-1}$$

The repeated application of this operator gives

$$E^2[y(k)] = E[Ey(k)] = y(k + 2)$$

or, in general,

$$E^n[y(k)] = y(k + n) \tag{2.9-2}$$

for any positive integer n. The difference operator‡ Δ is defined by

$$\Delta y(k) = y(k + 1) - y(k) \tag{2.9-3}$$

Since the last equation may be rewritten as

$$\Delta y(k) = (E - 1)y(k)$$

† In many books, the function is written y_k.

‡ This is the forward difference operator, as distinguished from the backward difference operator ∇, which is occasionally used, and which is defined by $\nabla y(k) = y(k) - y(k - 1)$.

the operators Δ and E are related by

$$\Delta = E - 1 \tag{2.9-4}$$

$\Delta y(k)$ is called the first difference of the function $y(k)$. Higher order differences are defined as

$$\Delta^2 y(k) = \Delta[\Delta y(k)] = y(k+2) - 2y(k+1) + y(k)$$
$$\Delta^3 y(k) = \Delta[\Delta^2 y(k)] = y(k+3) - 3y(k+2) + 3y(k+1) - y(k)$$

or, in general,

$$\Delta^n y(k) = \sum_{r=0}^{n} (-1)^r \binom{n}{r} y(k+n-r) \tag{2.9-5}$$

where $\binom{n}{r}$ represents the binomial coefficients. The last equation is consistent with Eqs. 2.9-2 and 2.9-4, for

$$\Delta^n y(k) = (E-1)^n y(k) = \sum_{r=0}^{n} (-1)^r \binom{n}{r} E^{n-r} y(k)$$

The operators E and Δ obey the usual algebraic laws, such as

$$\begin{aligned} \Delta[cy(k)] &= c\,\Delta y(k) \\ \Delta^m[y(k) + z(k)] &= \Delta^m y(k) + \Delta^m z(k) \\ \Delta^m\,\Delta^n y(k) &= \Delta^n\,\Delta^m y(k) = \Delta^{m+n} y(k) \end{aligned} \tag{2.9-6}$$

where c is a constant, and where m and n are positive integers. Equations 2.9-6 remain valid if every Δ is replaced by E. The operators commute with each other, but not, in general, with functions.

$$\Delta[y(k)z(k)] \neq y(k)\,\Delta z(k)$$

The operator Δ for discrete functions is somewhat analogous to the differentiation operator $p = d/dt$ for continuous functions. To make this clear, consider the continuous function $f(t)$, whose derivative is

$$\frac{df(t)}{dt} = \lim_{T\to 0} \frac{f(t+T) - f(t)}{T} \tag{2.9-7}$$

If the function is considered only at the discrete instants $t = nT (n = 0, 1, 2, \ldots)$, the shifting and differencing operators are defined by

$$\begin{aligned} Ef(t) &= f(t+T) \\ \Delta f(t) &= f(t+T) - f(t) \end{aligned} \tag{2.9-8}$$

Then

$$\frac{df(t)}{dt} = \lim_{T\to 0} \frac{\Delta f(t)}{T} \tag{2.9-9}$$

and, in general,

$$\frac{d^m f(t)}{dt^m} = \lim_{T\to 0} \frac{\Delta^m f(t)}{T^m} \tag{2.9-10}$$

It is not surprising, therefore, that there are differencing formulas similar to but not identical with the common differentiation formulas. For example,

$$\Delta[y(k)z(k)] = y(k+1)\,\Delta z(k) + z(k)\,\Delta y(k)$$
$$\Delta\left[\frac{y(k)}{z(k)}\right] = \frac{z(k)\,\Delta y(k) - y(k)\,\Delta z(k)}{z(k)z(k+1)} \tag{2.9-11}$$

The differentiation of polynomials is analogous to the differencing of "factorial polynomials."† The factorial polynomial of order m is defined as

$$(k)^{(m)} = k(k-1)(k-2)\cdots(k-m+1) \tag{2.9-12}$$

where m is any positive integer. Note that $(k)^{(m)}$ contains exactly m factors. Using the definition of the difference operator, it is found that

$$\Delta(k)^{(m)} = m(k)^{(m-1)} = mk(k-1)(k-2)\cdots(k-m+2) \tag{2.9-13}$$

Because of the analogy between Δ and p, it is logical to ask if there is an operator for discrete functions that is analogous to the integration operator p^{-1}, where

$$p^{-1}[f(t)] = \int f(t)\,dt + c$$
$$= \int_{t_0}^{t} f(\lambda)\,d\lambda + K \tag{2.9-14}$$

where c and K are constants of integration. The lower limit t_0 is arbitrary and forms part of the constant of integration. Specifically,

$$c = K - \left[\int f(t)\,dt\right]_{t=t_0}$$

The quantity

$$y(t) = p^{-1}[f(t)] \tag{2.9-15}$$

is the general solution to the equation

$$py(t) = f(t) \tag{2.9-16}$$

Equivalently,

$$pp^{-1}f(t) = f(t) \tag{2.9-17}$$

† Any ordinary polynomial can be expressed as the sum of factorial polynomials, if desired. See, for example, Reference 8.

By analogy, an antidifference operator Δ^{-1} should exist such that

$$y(k) = \Delta^{-1}f(k) \tag{2.9-18}$$

is the general solution to

$$\Delta y(k) = f(k) \tag{2.9-19}$$

or, equivalently, such that

$$\Delta\, \Delta^{-1}f(k) = f(k) \tag{2.9-20}$$

Since

$$\Delta\left[\sum_{n=0}^{k-1} f(n) + K\right] = [f(k) + f(k-1) + \cdots + f(0) + K] - [f(k-1) + \cdots + f(0) + K] = f(k)$$

the antidifference operator satisfying Eqs. 2.9-19 and 2.9-20 must be

$$\Delta^{-1}f(k) = \sum_{n=0}^{k-1} f(n) + K \tag{2.9-21}$$

The antidifference operator may be rewritten as

$$\Delta^{-1}f(k) = \sum^{n=k-1} f(n) + c = \sum^{n=k} f(n-1) + c \tag{2.9-22}$$

where the summation is still with respect to the dummy variable n. The lower limit in Eq. 2.9-22 is unspecified because any *fixed* number of the terms $f(0), f(1), f(2), \ldots$ in Eq. 2.9-21 can be combined with the constant of summation, K, to form the new constant c. The lower limit may be selected in whatever manner appears to be the most convenient. The arbitrary lower limit in Eq. 2.9-22 is the analog of the arbitrary value of t_0 in Eq. 2.9-14.

The reader may recall that the exact evaluation of integrals may be tricky or difficult, and in some cases downright impossible. Since there is no conceptual difficulty in summing a finite number of terms, he may be tempted to conclude that a similar problem does not exist for the operator Δ^{-1}. A brute force calculation of $\Delta^{-1}f(k)$ from Eq. 2.9-22 over a wide range of values of k is often, however, unnecessarily tedious. As k increases without limit, so does the number of terms in the summation. Whenever possible, $\Delta^{-1}f(k)$ should be expressed in closed form.

The summation of a finite series is one of the topics usually included in books on finite differences. Among the available techniques are short tables of summation formulas, summation by parts (analogous to but not identical with integration by parts), the use of Bernoulli polynomials, and the use of partial fractions. For example, the telescopic series

$$\sum_{n=2}^{k} \frac{1}{n(n-1)}$$

can be easily summed by noting that

$$\frac{1}{n(n-1)} = \frac{1}{n-1} - \frac{1}{n}$$

The result is

$$\sum_{n=2}^{k} \frac{1}{n(n-1)} = 1 - \frac{1}{k} \qquad (2.9\text{-}23)$$

The summation of some of the very simple function (e.g., $\sum^{n=k} 1/n$) cannot be expressed in closed form in terms of elementary functions. In some of these cases, an approximate summation can be obtained.

It is useful to note that, when factorial polynomials are used, Eq. 2.9-13 gives

$$\Delta^{-1}(k)^{(m)} = \frac{1}{m+1}(k)^{(m+1)} + K$$

or

$$\begin{aligned}\Delta^{-1}[k(k-1)\cdots(k-m+1)]& \\ &= \sum^{n=k-1} n(n-1)\cdots(n-m+1) \\ &= \frac{1}{m+1}k(k-1)\cdots(k-m) + K \qquad (2.9\text{-}24)\end{aligned}$$

In the following sections, the reader will notice a striking resemblance between the solution of differential and difference equations.† This has already been suggested by the analogy of the operators p and Δ, and p^{-1} and Δ^{-1}. Although there are some dissimilarities, the major concepts in the solution of differential equations have their parallel in the solution of difference equations. For this reason, the treatment of difference equations is somewhat shorter and more concise than the treatment of differential equations.

2.10 REPRESENTING DISCRETE SYSTEMS BY DIFFERENCE EQUATIONS

The general form of a difference equation relating the output $y(k)$ to the input $v(k)$ of a discrete system is

$$\begin{aligned}a_n y(k+n) + a_{n-1}y(k+n-1) + \cdots + a_1 y(k+1) + a_0 y(k)& \\ = b_m v(k+m) + \cdots + b_1 v(k+1) + b_0 v(k)& \qquad (2.10\text{-}1)\end{aligned}$$

† In fact, difference equations may be used to obtain approximate solutions of ordinary and partial differential equations. From another point of view, they form a bridge between ordinary and partial differential equations.

or, equivalently,

$$[c_n \Delta^n + c_{n-1} \Delta^{n-1} + \cdots + c_1 \Delta + c_0]y(k) = [d_m \Delta^m + \cdots + d_1 \Delta + d_0]v(k) \quad (2.10\text{-}2)$$

Either of the two equations may be easily obtained from the remaining one by Eqs. 2.9-1 through 2.9-5. The second is a closer analog to the differential equation 2.2-1, but the first is easier to use and is the one more commonly given. For a linear system, the a_i's, b_j's, c_i's, and d_j's cannot be functions of y or v, but they may be functions of k. For *fixed* linear systems, these coefficients must be constants.

Using the shifting operator E, Eq. 2.10-1 may be rewritten as

$$(a_n E^n + a_{n-1} E^{n-1} + \cdots + a_1 E + a_0)y(k) = (b_m E^m + \cdots + b_1 E + b_0)v(k) = F(k) \quad (2.10\text{-}3)$$

Since the input $v(k)$ is presumably known, the right side of Eq. 2.10-1 is represented by the known forcing function $F(k)$. For fixed linear systems, where the coefficients are constants, the last equation is written symbolically as

$$A(E)y(k) = B(E)v(k) = F(k) \quad (2.10\text{-}4)$$

Difference equations arise when functions of a discrete variable are considered. Systems which can be described by difference equations include computers, sequential circuits, and systems with time-delay components. Any system whose input $v(t)$ and output $y(t)$ are defined only at the equally spaced intervals $t = kT$ is described by the difference equation

$$a_n y(kT + nT) + a_{n-1} y(kT + nT - T) + \cdots + a_1 y(kT + T) + a_0 y(kT) = b_m v(kT + mT) + \cdots + b_1 v(kT + T) + b_0 v(kT)$$

where k, m, and n are integers. By this equation, the value of the output at the instant $t = (k + n)T$ is expressed in terms of the past outputs from $t = kT$ to $(k + n - 1)T$, and in terms of the inputs from $t = kT$ to $(k + m)T$. For nonanticipatory systems, $m \leqslant n$. If the time scale is adjusted so that $T = 1$, this equation is identical with Eq. 2.10-1.

Difference equations also arise in ways other than those mentioned above. The right side of Eqs. 2.10-3 and 2.10-4 need not always be a direct function of the input, and the discrete variable k may be an index of position instead of an index of time. The following two examples† show how difference equations result whenever there is a repetition at equal intervals of position or at equal intervals of time.

† These examples are somewhat similar to those of Reference 9.

Example 2.10-1. An example of the repetition of structure at equal intervals of position is the cascade connection of m identical T-sections shown in Fig. 2.10-1. The currents are a function of the position index k, and $i(k)$ is desired for $k = 0, 1, 2, \ldots, m$.

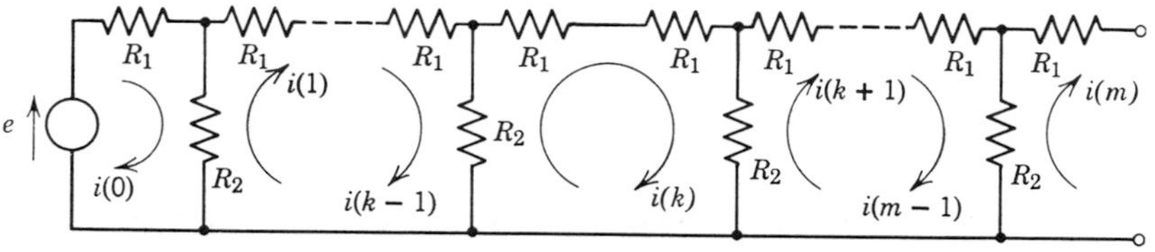

Fig. 2.10-1

The equation for the kth mesh is

$$-R_2 i(k-1) + 2(R_1 + R_2)i(k) - R_2 i(k+1) = 0$$

This is a second order, homogeneous difference equation which is solved for i as a function of k in Example 2.12-1. If the resistance R_2 is replaced by a capacitance C, the equation for the rth mesh becomes a differential-difference equation. The solution for i as a function of the continuous variable t and the discrete variable k is carried out in Example 3.7-3.

Example 2.10-2. Systems subjected to periodic inputs or periodic switching are examples of a repetition at equal intervals of time. Figure 2.10-2 shows a circuit subjected to a square wave of voltage. Assuming that there is no initial stored energy, find an expression for the current.

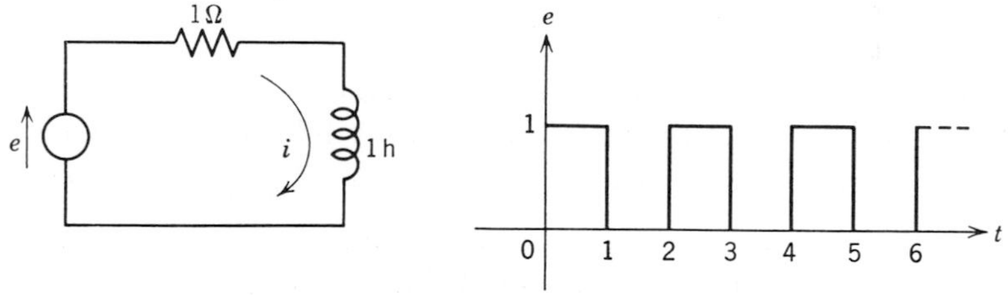

Fig. 2.10-2

While the solution could be found by calculating the response for each cycle in succession, this would be a very tedious process. The steady-state solution could be obtained by use of the Fourier series, but the answer would be an infinite series instead of being in closed form. Examine, therefore, the kth cycle, and for convenience move the time origin to the start of the kth cycle. The differential equations describing the circuit are

$$\frac{di_1}{dt} + i_1 = 1 \quad \text{for } 0 < t < 1$$

$$\frac{di_2}{dt} + i_2 = 0 \quad \text{for } 1 < t < 2$$

where the subscripts 1 and 2 denote the first and second halves of the cycle, respectively. Since the current is a function of the cycle under consideration, it is again a function of the discrete variable k as well as the continuous variable t. The solution is carried out in Example 2.12-3. A first order difference equation must be solved, along with the above differential equations.

2.11 GENERAL PROPERTIES OF LINEAR DIFFERENCE EQUATIONS

An nth order linear difference equation can be written in the form of Eq. 2.10-3, repeated below.

$$(a_nE^n + a_{n-1}E^{n-1} + \cdots + a_1E + a_0)y(k) = F(k) \qquad (2.11\text{-}1)$$

where $a_n \neq 0$, $a_0 \neq 0$, and where all the a_i's are defined for all integral values of k of interest. The remarks of this section are valid for both fixed and varying equations; hence the a_i's can in general be functions of k. If $a_0 = 0$, $a_1 \neq 0$, and $a_n \neq 0$, the equation is of order $n - 1$. In contrast to differential equations, the order of a difference equation is defined as the difference between the lowest and highest power of E. If the operator Δ is used, as in

$$(c_n \Delta^n + c_{n-1} \Delta^{n-1} + \cdots + c_1 \Delta + c_0)y(k) = F(k) \qquad (2.11\text{-}2)$$

the order of the equation cannot be determined by inspection. For example, the equation $(\Delta^2 + 3\Delta + 2)y(k) = 0$ is equivalent to

$$(E^2 + E)y(k) = 0$$

which is a first and not a second order equation.

Equation 2.11-1 is called a nonhomogeneous difference equation, while

$$(a_nE^n + a_{n-1}E^{n-1} + \cdots + a_1E + a_0)y(k) = 0 \qquad (2.11\text{-}3)$$

is an nth order homogeneous equation. Difference equations are also called recurrence formulas. Equation 2.11-1 can be rewritten as

$$y(k + n) = -\frac{1}{a_n}[a_{n-1}y(k + n - 1) + \cdots + a_1y(k + 1) + a_0y(k) - F(k)] \qquad (2.11\text{-}4)$$

If $y(0)$ through $y(n - 1)$ are known, $y(k)$ may be explicitly found for all $k \geqslant n$ by the repeated application of Eq. 2.11-4. Thus, in contrast to a differential equation, $y(k)$ may be calculated from the difference equation itself for any value of k (given enough patience or a computer) in terms of the first n values of $y(k)$.† The values of $y(0)$ through $y(n - 1)$, or equivalent information, must be given in order for the solution of Eq. 2.11-1 to be unique.

† This is the basis for one method of obtaining numerical solutions of differential equations.

Use of the iterative process suggested by Eq. 2.11-4 is not usually considered to constitute a "solution" of Eq. 2.11-1. Most of the remainder of this chapter is concerned with finding an expression for $y(k)$ in closed form that satisfies Eq. 2.11-1 or 2.11-3.

A homogeneous difference equation of order n has n linearly independent solutions. If $a_n \neq 0$, and $a_0 \neq 0$, the independent solutions of Eq. 2.11-3 can be denoted $y_1(k)$, $y_2(k)$, . . . , $y_n(k)$. A necessary and sufficient condition for the solutions to be independent is that

$$W(k) = \begin{vmatrix} y_1 & y_2 & \cdots & y_n \\ Ey_1 & Ey_2 & \cdots & Ey_n \\ \cdots & \cdots & \cdots & \cdots \\ E^{n-1}y_1 & E^{n-1}y_2 & \cdots & E^{n-1}y_n \end{vmatrix} \neq 0\dagger \tag{2.11-5}$$

The most general solution of Eq. 2.11-3 is

$$y_H = C_1y_1(k) + C_2y_2(k) + \cdots + C_ny_n(k) \tag{2.11-6}$$

where the C_i's are arbitrary constants that are independent of k.

The most general solution of Eq. 2.11-1 is

$$y = y_H + y_P$$

where y_H is given in Eq. 2.11-6, and where y_P is any one solution satisfying Eq. 2.11-1. The components y_H and y_P are called the complementary and particular solutions, respectively. Since y_P contains no arbitrary constants, y contains n such constants, which must be evaluated by initial or boundary conditions, as in Examples 2.12-1 and 2.12-4.

2.12 SOLUTION OF DIFFERENCE EQUATIONS WITH CONSTANT COEFFICIENTS

Explicit procedures exist for finding the closed-form solution of difference equations with constant coefficients. The nth order nonhomogeneous and homogeneous difference equations are given by Eqs. 2.11-1 and 2.11-3, respectively, where the a_i's are constants.

Homogeneous Difference Equations

Consider the nth order equation

$$(a_nE^n + a_{n-1}E^{n-1} + \cdots + a_1E + a_0)y(k) = 0 \tag{2.12-1}$$

† The function $W(k)$ is analogous to the Wronskian, but it is often called Casorati's determinant and given the symbol $C(k)$.

By analogy with Section 2.6, it is reasonable to assume solutions of the form

$$y(k) = \epsilon^{rk} \tag{2.12-2}$$

where r is a constant to be determined.

Example 2.12-1. Solve Example 2.10-1 for i as a function of k.

The equation for the rth mesh in Fig. 2.10-1 is

$$-R_2 i(k-1) + 2(R_1 + R_2)i(k) - R_2 i(k+1) = 0 \tag{2.12-3}$$

There must be two independent solutions to this second order difference equation. For the assumed solution $i(k) = \epsilon^{rk}$,

$$[-R_2\epsilon^{-r} + 2(R_1 + R_2) - R_2\epsilon^{r}]\epsilon^{rk} = 0$$

The expression in the brackets, which is a polynomial in ϵ^r, is zero if

$$\cosh r = \frac{R_1 + R_2}{R_2}$$

Since $\cosh(-\theta) = \cosh(\theta)$, the two allowable values of r are given by $r = \pm\theta$ where

$$\cosh \theta = \frac{R_1 + R_2}{R_2} \tag{2.12-4}$$

The general solution is

$$i(k) = C_1\epsilon^{k\theta} + C_2\epsilon^{-k\theta}$$

The two boundary conditions needed to evaluate the arbitrary constants C_1 and C_2 are found by examining the first and last meshes in Fig. 2.10-1.

$$\begin{aligned} (R_1 + R_2)i(0) - R_2 i(1) &= e \\ i(m) &= 0 \end{aligned} \tag{2.12-5}$$

Using the second boundary condition, $C_1\epsilon^{m\theta} + C_2\epsilon^{-m\theta} = 0$; hence

$$C_2 = -C_1\epsilon^{2m\theta}$$

and

$$i(k) = C_1[\epsilon^{k\theta} - \epsilon^{2m\theta}\epsilon^{-k\theta}] = -2C_1\epsilon^{m\theta}\sinh(m-k)\theta$$

Using the first boundary condition,

$$(R_1 + R_2)[-2C_1\epsilon^{m\theta}\sinh m\theta] - R_2[-2C_1\epsilon^{m\theta}\sinh(m-1)\theta] = e$$

Since

$$R_1 + R_2 = R_2\cosh\theta$$

and

$$\sinh(m-1)\theta = \sinh m\theta\cosh\theta - \sinh\theta\cosh m\theta$$

the last equation becomes

$$2C_1R_2\epsilon^{m\theta}[-\sinh m\theta\cosh\theta + \sinh m\theta\cosh\theta - \sinh\theta\cosh m\theta] = e$$

Thus

$$2C_1\epsilon^{m\theta} = \frac{-e}{R_2\sinh\theta\cosh m\theta}$$

The final answer is then

$$i(k) = \frac{e \sinh (m - k)\theta}{R_2 \sinh \theta \cosh m\theta} \tag{2.12-6}$$

for $k = 0, 1, 2, \ldots, m$, where θ is given by Eq. 2.12-4. Example 3.7-3 gives the solution for i as a function of t and k when the resistance R_2 in Fig. 2.10-1 is replaced by a capacitance C.

While the form of the assumed solution given in Eq. 2.12-2 happens to be the most convenient one for Example 2.12-1, it is not the form that is usually assumed. Recall from the example that the allowable values of r were the roots of a polynomial in ϵ^r. If $\beta = \epsilon^r$, the assumed solution becomes

$$y(k) = \beta^k \tag{2.12-7}$$

where β is a constant to be determined. Substituting Eq. 2.12-7 into Eq. 2.12-1 and using the fact that $E^m\beta^k = \beta^m\beta^k$, the following characteristic or auxiliary equation results.

$$a_n\beta^n + a_{n-1}\beta^{n-1} + \cdots + a_1\beta + a_0 = 0 \tag{2.12-8}$$

In practice, this equation is written down directly from an inspection of the difference equation, Eq. 2.12-1.

If the n roots of the characteristic equation are distinct and are denoted by $\beta_1, \beta_2, \ldots, \beta_n$, the most general solution of the homogeneous difference equation is

$$y_H = C_1\beta_1{}^k + C_2\beta_2{}^k + \cdots + C_n\beta_n{}^k \tag{2.12-9}$$

If the β_i's are distinct, it can be shown that the individual solutions $y_1 = \beta_1{}^k$, $y_2 = \beta_2{}^k, \ldots, y_n = \beta_n{}^k$ do satisfy Eq. 2.11-5 and are therefore independent.[10] If a root, β_1 for example, is repeated m times, then the most general solution is

$$y_H = C_1\beta_1^k + C_2k\beta_1^k + \cdots + C_mk^{m-1}\beta_1^k + C_{m+1}\beta_{m+1}^k + \cdots + C_n\beta_n^k \tag{2.12-10}$$

Any complex roots of a characteristic equation with real coefficients must occur in complex conjugate pairs. If there are first order roots at $\beta_1 = \rho\epsilon^{j\theta}$ and $\beta_2 = \rho\epsilon^{-j\theta}$, where ρ and θ are real numbers, the solution contains the terms

$$\begin{aligned} C_1\beta_1{}^k + C_2\beta_2{}^k &= \rho^k[A \cos \theta k + B \sin \theta k] \\ &= C\rho^k \cos (\theta k + \phi) \end{aligned} \tag{2.12-11}$$

where A, B, C, and ϕ are arbitrary real constants.

Roots of the characteristic equation at $\beta = 0$ deserve special mention. If $a_0 = 0$, $a_1 \neq 0$, and $a_n \neq 0$ in Eq. 2.12-1, the characteristic equation has a first order root at the origin. Since the order of the difference equation is only $n - 1$, and the order of the characteristic polynomial is n, the root at the origin is extraneous and should be disregarded. Higher order roots at $\beta = 0$ should likewise be disregarded.

Nonhomogeneous Difference Equations—Method of Undetermined Coefficients

Consider the nth order difference equation

$$(a_n E^n + a_{n-1} E^{n-1} + \cdots + a_1 E + a_0) y(k) = F(k) \qquad (2.12\text{-}12)$$

whose solution has the form

$$y(k) = y_H + y_P \qquad (2.12\text{-}13)$$

The complementary solution y_H is found by solving the related homogeneous equation, Eq. 2.12-1. The particular solution y_P can be found by the same two methods used for differential equations, i.e., undetermined coefficients, and variation of parameters.

The method of undetermined coefficients can be used only when the repeated application of the operator E to the forcing function $F(k)$ produces a finite number of linearly independent terms. $F(k)$ may be a polynomial, exponential, sinusoidal, or hyperbolic function (e.g., $k^m + \cdots + b_1 k + b_0$, c^k, $\sin \theta k$ or $\cos \theta k$, $\sinh \theta k$ or $\cosh \theta k$), or it may be a product or linear combination of such functions. The solution is assumed to be a linear combination of the terms in $F(k)$, $F(k+1)$, $F(k+2)$, . . . , each term being multiplied by an undetermined constant. The undetermined constants are chosen so that the assumed solution satisfies Eq. 2.12-12 for all values of k.

If a term in $F(k)$, $F(k + 1)$, $F(k + 2)$, . . . has exactly the same form as a term in y_H, a modification must be made in the form of the particular solution that would otherwise be assumed. All the terms in y_P corresponding to the duplicated term in y_H must be multiplied by the lowest power of k that will remove the duplication.

Example 2.12-2. Find the general solution to

$$(E^2 + 2E + 1) y(k) = k(-1)^k$$

The characteristic equation is

$$\beta^2 + 2\beta + 1 = 0$$

so $\beta_1 = \beta_2 = -1$, and $y_H = C_1(-1)^k + C_2 k(-1)^k$. Although normally $y_P = [Ak + B](-1)^k$, the double root of the characteristic equation at -1 requires

$$y_P = [Ak^3 + Bk^2](-1)^k$$

Then

$$E y_P = -[A(k + 1)^3 + B(k + 1)^2](-1)^k$$
$$E^2 y_P = [A(k + 2)^3 + B(k + 2)^2](-1)^k$$

Substituting these expressions into the left side of the difference equation yields

$$[(6k + 6)A + 2B](-1)^k = k(-1)^k$$

so that $A = \frac{1}{6}$ and $B = -\frac{1}{2}$. The complete solution is

$$y(k) = [C_1 + C_2k - \tfrac{1}{2}k^2 + \tfrac{1}{6}k^3](-1)^k$$

Example 2.12-3. Solve Example 2.10-2 for i as a function of t and k.

The response of the circuit in Fig. 2.10-2 during the kth cycle is given by

$$\begin{aligned} \frac{di_1}{dt} + i_1 &= 1 \quad \text{for } 0 < t < 1 \\ \frac{di_2}{dt} + i_2 &= 0 \quad \text{for } 1 < t < 2 \end{aligned} \tag{2.12-14}$$

where the subscripts 1 and 2 denote the first and second halves of the cycle, respectively. The solutions of these differential equations are

$$\begin{aligned} i_1(t, k) &= 1 + C_1(k)\epsilon^{-t} \quad &\text{for } 0 < t < 1 \\ i_2(t, k) &= C_2(k)\epsilon^{-t} \quad &\text{for } 1 < t < 2 \end{aligned} \tag{2.12-15}$$

The factors C_1 and C_2 are constants with respect to t but may be functions of k, the cycle under consideration. Since the current cannot change instantaneously,

$$i_1(1, k) = i_2(1, k) \quad \text{and} \quad i_1(0, k + 1) = i_2(2, k)$$

Since the current is zero at the start of the first cycle,

$$i_1(0, 1) = 0$$

Inserting each of these three conditions into Eq. 2.12-15 gives, respectively,

$$C_2(k) = C_1(k) + \epsilon \tag{2.12-16}$$

$$1 + C_1(k + 1) = C_2(k)\epsilon^{-2} \tag{2.12-17}$$

$$C_1(1) = -1 \tag{2.12-18}$$

Equations 2.12-16 and 2.12-17 yield the first order difference equation

$$C_1(k + 1) - \epsilon^{-2}C_1(k) = -1 + \epsilon^{-1}$$

The complementary solution is

$$C_{1_H}(k) = c\epsilon^{-2k}$$

Assuming a particular solution of

$$C_{1_P}(k) = a$$

and substituting it into the difference equation yields

$$a(1 - \epsilon^{-2}) = -1 + \epsilon^{-1}$$

Hence

$$a = \frac{-1}{1 + \epsilon^{-1}}$$

The complete solution is

$$C_1(k) = c\epsilon^{-2k} - \frac{1}{1 + \epsilon^{-1}}$$

By Eq. 2.12-18,

$$c = -\frac{\epsilon^2}{1+\epsilon}$$

Finally,

$$C_1(k) = -\frac{\epsilon + \epsilon^{2(1-k)}}{1+\epsilon}$$

and

$$i_1(t, k) = 1 - \frac{\epsilon + \epsilon^{2(1-k)}}{1+\epsilon}\,\epsilon^{-t} \quad \text{for } 0 < t < 1$$

$$i_2(t, k) = \frac{\epsilon^2 - \epsilon^{2(1-k)}}{1+\epsilon}\,\epsilon^{-t} \quad \text{for } 1 < t < 2$$

If k approaches infinity, the steady-state response is seen to be

$$\begin{aligned} i_1(t) &= 1 - \frac{\epsilon}{1+\epsilon}\,\epsilon^{-t} \quad \text{for } 0 < t < 1 \\ i_2(t) &= \frac{\epsilon^2}{1+\epsilon}\,\epsilon^{-t} \quad \text{for } 1 < t < 2 \end{aligned} \tag{2.12-19}$$

The number of cycles needed to reach the steady state is given by

$$\epsilon^{2(1-k)} \ll \epsilon$$

Nonhomogeneous Difference Equations—Variation of Parameters

If the complementary solution has been found, variation of parameters is an explicit method of finding y_P regardless of the nature of $F(k)$. Whereas the application of the method to differential equations resulted in the integration of a known function of t, its application to difference equations results in the summation of a known function of k.

Consider the first order equation

$$(a_1E + a_0)y(k) = F(k) \tag{2.12-20}$$

The complementary solution has only the single term $y_H = Cy_1(k)$. The particular solution is assumed to have the form

$$y_P = \mu(k)y_1(k) \tag{2.12-21}$$

Substituting Eq. 2.12-21 into Eq. 2.12-20 gives

$$a_1\mu(k+1)y_1(k+1) + a_0\mu(k)y_1(k) = F(k)$$

Rearranging,

$$a_1[\mu(k+1)y_1(k+1) - \mu(k)y_1(k+1)] + \mu(k)[a_1y_1(k+1) + a_0y_1(k)] = F(k)$$

The expression in the first bracket is $y_1(k+1)\,\Delta\mu(k)$, while the second bracket vanishes, since $y_1(k)$ is a solution of the related homogeneous equation. Therefore

$$a_1y_1(k+1)\,\Delta\mu(k) = F(k)$$

whose solution, by Eq. 2.9-22, is

$$\mu(k) = \Delta^{-1}\left[\frac{F(k)}{a_1y_1(k+1)}\right] = \sum^{n=k}\frac{F(n-1)}{a_1y_1(n)} \tag{2.12-22}$$

Example 2.12-4. Find the complete solution to

$$(E+3)y(k) = \frac{(-3)^k}{k(k+1)}$$

The complementary solution is

$$y_H = C(-3)^k, \quad \text{i.e., } y_1(k) = (-3)^k$$

The assumed form of the particular solution is

$$y_P = \mu(k)(-3)^k$$

By Eq. 2.12-22,

$$\mu(k) = \sum_{n=2}^{k}\frac{(-3)^{n-1}/(n-1)n}{(-3)^n} = \sum_{n=2}^{k}\frac{1}{-3(n-1)n}$$

By Eq. 2.9-23,

$$\mu(k) = -\frac{1}{3}\left[1-\frac{1}{k}\right]$$

and the complete solution is

$$y(k) = \left[C - \frac{1}{3} + \frac{1}{3k}\right](-3)^k = \left[C_1 + \frac{1}{3k}\right](-3)^k$$

Next consider the second order equation

$$(a_2E^2 + a_1E + a_0)y(k) = F(k) \tag{2.12-23}$$

whose complementary solution has the form

$$y_H = C_1y_1(k) + C_2y_2(k)$$

The particular solution is assumed to be

$$y_P = \mu_1(k)y_1(k) + \mu_2(k)y_2(k) \tag{2.12-24}$$

One of the two conditions needed to evaluate μ_1 and μ_2 is that Eq. 2.12-24 must satisfy Eq. 2.12-23. By analogy to Eq. 2.6-14, the second condition is arbitrarily chosen to be

$$y_1(k+1)\,\Delta\mu_1(k) + y_2(k+1)\,\Delta\mu_2(k) = 0 \tag{2.12-25}$$

Since

$$\mu_i(k+1) = \mu_i(k) + \Delta\mu_i(k)$$

it follows from Eq. 2.12-25 that

$$\begin{aligned} Ey_P &= y_1(k+1)[\mu_1(k) + \Delta\mu_1(k)] + y_2(k+1)[\mu_2(k) + \Delta\mu_2(k)] \\ &= y_1(k+1)\mu_1(k) + y_2(k+1)\mu_2(k) \end{aligned}$$

and

$$E^2y_P = y_1(k+2)[\mu_1(k) + \Delta\mu_1(k)] + y_2(k+2)[\mu_2(k) + \Delta\mu_2(k)]$$

Substituting these expressions into Eq. 2.12-23 and rearranging gives

$$\begin{aligned} a_2[y_1(k+2)\,\Delta\mu_1(k) + y_2(k+2)\,\Delta\mu_2(k)] \\ + \mu_1(k)[a_2y_1(k+2) + a_1y_1(k+1) + a_0y_1(k)] \\ + \mu_2(k)[a_2y_2(k+2) + a_1y_2(k+1) + a_0y_2(k)] = F(k) \end{aligned}$$

Since $y_1(k)$ and $y_2(k)$ are solutions of the related homogeneous equation,

$$y_1(k+2)\,\Delta\mu_1(k) + y_2(k+2)\,\Delta\mu_2(k) = \frac{F(k)}{a_2} \tag{2.12-26}$$

Solving Eqs. 2.12-25 and 2.12-26 simultaneously,

$$\begin{aligned} \Delta\mu_1(k) &= \frac{-y_2(k+1)F(k)}{a_2[y_1(k+1)y_2(k+2) - y_1(k+2)y_2(k+1)]} \\ \Delta\mu_2(k) &= \frac{y_1(k+1)F(k)}{a_2[y_1(k+1)y_2(k+2) - y_1(k+2)y_2(k+1)]} \end{aligned} \tag{2.12-27}$$

so

$$\begin{aligned} \mu_1(k) &= -\sum^{n=k} \frac{y_2(n)F(n-1)}{a_2[y_1(n)y_2(n+1) - y_1(n+1)y_2(n)]} \\ \mu_2(k) &= \sum^{n=k} \frac{y_1(n)F(n-1)}{a_2[y_1(n)y_2(n+1) - y_1(n+1)y_2(n)]} \end{aligned} \tag{2.12-28}$$

If y_1 and y_2 are two independent solutions to the homogeneous equation, as was assumed, then Eq. 2.11-5 proves that the denominators in Eqs. 2.12-27 and 2.12-28 do not vanish. Thus $\mu_1(k)$ and $\mu_2(k)$ can always be explicitly found.

Example 2.12-5. Solve Example 2.12-2 by variation of parameters.

$$(E^2 + 2E + 1)y(k) = k(-1)^k$$

Let

$$y_P = \mu_1(k)y_1(k) + \mu_2(k)y_2(k)$$

where

$$y_1(k) = (-1)^k, \quad y_2(k) = k(-1)^k$$

Note that

$$y_1(k)y_2(k+1) - y_1(k+1)y_2(k) = -1$$

By Eq. 2.12-28,

$$\begin{aligned} \mu_1(k) &= \sum^{n=k} n(-1)^n(n-1)(-1)^{n-1} \\ &= -\sum^{n=k} n(n-1) = -k(k-1) - \sum^{n=k-1} n(n-1) \end{aligned}$$

Using Eq. 2.9-24, with $m = 2$,

$$\mu_1(k) = -k(k-1) - \tfrac{1}{3}k(k-1)(k-2) = -\tfrac{1}{3}(k^3 - k)$$

Similarly,

$$\mu_2(k) = -\sum^{n=k}(-1)^n(n-1)(-1)^{n-1} = \sum^{n=k}(n-1) = \tfrac{1}{2}k(k-1)$$

Then

$$y_P = -\tfrac{1}{3}(k^3-k)(-1)^k + \tfrac{1}{2}(k^2-k)k(-1)^k$$
$$= [\tfrac{1}{6}k^3 - \tfrac{1}{2}k^2 + \tfrac{1}{3}k](-1)^k$$

The complete solution is

$$y(k) = [C_1 + C_2k - \tfrac{1}{2}k^2 + \tfrac{1}{6}k^3](-1)^k$$

agreeing with Example 2.12-2.

Consider finally the nth order difference equation, Eq. 2.12-12, whose complementary solution has the form

$$y_H = C_1y_1(k) + C_2y_2(k) + \cdots + C_ny_n(k)$$

The assumed particular solution is

$$y_P = \mu_1(k)y_1(k) + \mu_2(k)y_2(k) + \cdots + \mu_n(k)y_n(k) \tag{2.12-29}$$

The following $n-1$ conditions are arbitrarily selected to simplify the solution.

$$\begin{aligned}
&y_1(k+1)\,\Delta\mu_1(k) + y_2(k+1)\,\Delta\mu_2(k) + \cdots + y_n(k+1)\,\Delta\mu_n(k) = 0\\
&y_1(k+2)\,\Delta\mu_1(k) + y_2(k+2)\,\Delta\mu_2(k) + \cdots + y_n(k+2)\,\Delta\mu_n(k) = 0\\
&\cdots\cdots\cdots\cdots\cdots\cdots\cdots\cdots\cdots\cdots\cdots\cdots\\
&y_1(k+n-1)\,\Delta\mu_1(k) + y_2(k+n-1)\,\Delta\mu_2(k) + \cdots\\
&\qquad\qquad + y_n(k+n-1)\,\Delta\mu_n(k) = 0
\end{aligned} \tag{2.12-30}$$

Substituting Eq. 2.12-29 into Eq. 2.12-12, and making use of Eqs. 2.12-30, there results

$$y_1(k+n)\,\Delta\mu_1(k) + y_2(k+n)\,\Delta\mu_2(k) + \cdots + y_n(k+n)\,\Delta\mu_n(k) = \frac{F(k)}{a_n} \tag{2.12-31}$$

Solving Eqs. 2.12-30 and 2.12-31 in terms of determinants by Cramer's rule,

$$\Delta\mu_i = \frac{W_{ni}(k+1)F(k)}{a_nW(k+1)} \tag{2.12-32}$$

or

$$\mu_i = \sum^{r=k}\frac{W_{ni}(r)F(r-1)}{a_nW(r)} \tag{2.12-33}$$

where $W(k)$ is the determinant

$$W(k) = \begin{vmatrix} y_1(k) & y_2(k) & \cdots & y_n(k)\\ Ey_1(k) & Ey_2(k) & \cdots & Ey_n(k)\\ \cdots & \cdots & \cdots & \cdots\\ E^{n-1}y_1(k) & E^{n-1}y_2(k) & \cdots & E^{n-1}y_n(k)\end{vmatrix} \tag{2.12-34}$$

and where $W_{ni}(k)$ is the nith cofactor. By Eq. 2.11-5, $W(k)$ does not vanish if $y_1(k)$ through $y_n(k)$ are independent solutions of the homogeneous equation.

Although the summation involved may be difficult to express in closed form, Eq. 2.12-33 gives an explicit solution for y_P once y_H is known. Unlike the method of undetermined coefficients, the variation of parameters method is applicable for all forcing functions and can be extended to the solution of difference equations whose coefficients are functions of k.

The Physical Significance of the Complementary and Particular Solutions

Assume that the input-output relationship of a system is given by the difference equation

$$A(E)y(k) = B(E)v(k) = F(k) \tag{2.12-35}$$

The form of the complementary solution depends only upon the parameters of the system, specifically upon the roots of the characteristic equation

$$A(\beta) = 0$$

If the roots $\beta_1, \beta_2, \ldots, \beta_n$ are distinct, the complementary solution

$$y_H = C_1\beta_1^{\,k} + C_2\beta_2^{\,k} + \cdots$$

increases without limit as k approaches infinity only if one of the roots has a magnitude greater than unity. If the roots of the characteristic equation are plotted in a complex plane, roots inside and outside of a unit circle drawn about the origin correspond to terms that approach zero and that increase without limit, respectively, as k approaches infinity. Simple roots on the unit circle yield terms of constant magnitude, but repeated roots on the unit circle again yield terms that increase without limit.

Since the complementary solution is the complete solution when there is no external source, it is also called the free or unforced response. The system is unstable if the free response increases without limit, i.e., if the characteristic equation has roots outside the unit circle or repeated roots on the unit circle.

The arbitrary constants in the complementary solution depend upon the initial or boundary conditions, which summarize the past history of the system, and upon the input. They cannot be evaluated until both the complementary and the particular solutions have been found.

Since the form of the particular solution depends upon the input, it is also called the forced response. If all the roots of the characteristic equation are inside the unit circle, then the complementary and particular solutions are identical with the transient and steady-state solutions, respectively.

The Particular Solution for the Input z^k

Some insight into the relationship between the classical and transform solutions of difference equations is gained by examining the particular solution of Eq. 2.12-35, which is written out below, when $v(k) = z^k$.

$$\begin{aligned}(a_n E^n + a_{n-1} E^{n-1} + \cdots + a_1 E + a_0) y(k) \\ = (b_m E^m + \cdots + b_1 E + b_0) v(k)\end{aligned} \qquad (2.12\text{-}36)$$

Assuming that $y_P(k) = H(z)z^k$, where $H(z)$ is not a function of k and noting that $E^m y_P(k) = H(z)z^m z^k$, there results

$$H(z)A(z)z^k = B(z)z^k$$

where $A(z)$ is the function formed from $A(E)$ by replacing the operator E by the parameter z. Thus

$$H(z) = \frac{B(z)}{A(z)} = \frac{b_m z^m + \cdots + b_1 z + b_0}{a_n z^n + \cdots + a_1 z + a_0} \qquad (2.12\text{-}37)$$

$H(z)$ is the system function for a discrete system and can be written down by inspection of the difference equation. It plays an important role in the application of the Z transform. Note also that

$$H(z) = \left[\frac{y_P(k)}{v(k)}\right]_{v(k)=z^k} \qquad (2.12\text{-}38)$$

Obtaining the Delta Response from the Difference Equation

The delta response, defined in Section 1.9 and denoted by $d(k)$, is the response of a discrete system initially at rest to

$$v(k) = \begin{cases} 1 & \text{for } k = 0 \\ 0 & \text{for } k \neq 0 \end{cases} \qquad (2.12\text{-}39)$$

In comparing this equation with those of Chapter 1, remember that the difference equations of this chapter assume that the time scale has been adjusted so that $T = 1$. Equation 2.12-36 must be solved subject to the input of Eq. 2.12-39 and to the condition that

$$y(k) = 0 \quad \text{for } k < 0 \qquad (2.12\text{-}40)$$

The general technique is illustrated by the following example, the answer of which is used in Examples 1.9-1 and 3.13-3. The technique is also applicable for finding the impulse response of a continuous system subjected to an input $v^*(t)$, provided that the response is desired only at equally spaced, discrete instants.

Example 2.12-6. Find the delta response to

$$y(k+2) - 3y(k+1) + 2y(k) = 2v(k+1) - 2v(k)$$

The right side of this equation is zero for $k \geqslant 1$, so the delta response is the complementary solution

$$y(k) = C_1 + C_2(2)^k \quad \text{for } k \geqslant 1$$

To evaluate the two arbitrary constants, the values of $y(1)$ and $y(2)$ are needed. Substituting $k = -2$ in the difference equation gives

$$y(0) - 0 + 0 = 0 - 0$$

Hence $y(0) = 0$. For $k = -1$,

$$y(1) - 0 + 0 = 2 - 0$$

Then $y(1) = 2$. For $k = 0$,

$$y(2) - 3(2) + 0 = 0 - 2$$

Thus $y(2) = 4$. Using the last two results to evaluate C_1 and C_2,

$$2 = C_1 + 2C_2$$
$$4 = C_1 + 4C_2$$

so $C_1 = 0$ and $C_2 = 1$. The delta response is

$$d(k) = \begin{cases} 2k & \text{for } k \geqslant 1 \\ 0 & \text{for } k \leqslant 0 \end{cases}$$

2.13 SOLUTION OF LINEAR DIFFERENCE EQUATIONS WITH TIME-VARYING COEFFICIENTS

Consider the nth order difference equation

$$[a_n(k)E^n + \cdots + a_1(k)E + a_0(k)]y(k) = F(k) \qquad (2.13\text{-}1)$$

As with differential equations, a solution can always be found for first order equations but cannot in general be found for higher order equations.

The Complementary Solution

Consider the first order, homogeneous equation

$$a_1(k)y(k+1) + a_0(k)y(k) = 0 \qquad (2.13\text{-}2)$$

where $a_0(k) \neq 0$, and $a_1(k) \neq 0$. This can be rewritten as

$$y(k+1) = a(k)y(k) \tag{2.13-3}$$

Letting $k = 0, 1, 2, \ldots,$

$$y(1) = a(0)y(0)$$
$$y(2) = a(1)a(0)y(0)$$
$$y(3) = a(2)a(1)a(0)y(0)$$

or in general

$$\begin{aligned} y(k) &= [a(0)a(1)\cdots a(k-1)]y(0) \\ &= y(0)\prod_{n=0}^{k-1} a(n) \end{aligned}$$

so that the general solution to Eq. 2.13-3 is

$$y_H(k) = C \prod^{n=k-1} a(n) = C \prod^{n=k} a(n-1) \tag{2.13-4}$$

The lower limit of the indicated product is arbitrary, because any *fixed* number of the factors $a(0)$, $a(1)$, $a(2)$, ... may be combined with the arbitrary constant C.

The solution to a homogeneous equation of order greater than one cannot in general be found in terms of elementary functions, since the procedure based on Eqs. 2.12-7 and 2.12-8 is invalid when the coefficients are functions of k. It is sometimes helpful to know that, if all but one of the independent solutions are known, the remaining one can then be found.[11]

As with differential equations, there are a number of special cases which do have an explicit solution. If the equation can be put in the form

$$a_n f(k+n)y(k+n) + \cdots + a_1 f(k+1)y(k+1) + a_0 f(k)y(k) = 0$$

where the a_i's are constants, the substitution $z(k) = f(k)y(k)$ reduces it to a difference equation with constant coefficients. The procedure is somewhat analogous to that used for Euler's differential equation, but the change of variable is with respect to the dependent rather than the independent variable. This is usually the case when a varying equation is solved by a change of variable.†

Sometimes, the operator can be expressed in factored form, in which case the difference equation can be reduced to equations of lower order. For example,

$$[kE^2 + (k^2 + k + 1)E + k]y(k) = 0$$

† Certain nonlinear equations, such as $y^2(k+2) + y^2(k+1) + y^2(k) = 0$, can likewise be reduced to a linear difference equation by a change of the dependent variable.

may be written as

$$(kE + 1)(E + k)y(k) = 0$$

which is equivalent to the two first order equations

$$(kE + 1)z(k) = 0$$
$$(E + k)y(k) = z(k)$$

each of which may be explicitly solved.

Many of the special cases that have been systematically investigated can be classified according to the nature of the coefficients, such as the case of periodic coefficients.[12] Considerable work has been done in obtaining solutions in the form of an infinite series.[13]

The Particular Solution

The variation of parameters method, including Eqs. 2.12-22, 2.12-28, and 2.12-33, is valid whether or not the a_i's are functions of k. In the right side of these three equations, the symbols a_1, a_2, and a_n are understood to be $a_1(n - 1)$, $a_2(n - 1)$, and $a_n(r - 1)$, respectively. Although y_P can always be found once y_H is known, y_H cannot in general be determined.

Example 2.13-1. Find the complete solution to

$$[(k + 1)E - k]y(k) = k + 1$$

By Eq. 2.13-4, the complementary solution is

$$y_H = C\prod_{n=1}^{k-1} \frac{n}{n + 1} = C\frac{(k - 1)!}{k!} = \frac{C}{k}$$

The particular solution is assumed to have the form $y_P = \mu(k)/k$. By Eq. 2.12-22,

$$\mu(k) = \sum^{n=k} \frac{F(n - 1)}{a_1(n - 1)y_1(n)} = \sum_{n=1}^{k} n = \frac{k(k + 1)}{2}$$

where the last expression follows from the formula for an arithmetical progression or from Eq. 2.9-24. The complete solution is

$$y(k) = \frac{C}{k} + \frac{k + 1}{2}$$

REFERENCES

1. A. Ralston and H. S. Wilf, *Numerical Methods for Digital Computers*, John Wiley and Sons, New York, 1960, Part III.
2. E. L. Ince, *Ordinary Differential Equations*, Dover Publications, New York, 1956, Section 5.22.
3. *Ibid.*, Section 7.4.
4. *Ibid.*, Section 15.7.

5. B. K. Kinariwala, "Analysis of Time-Varying Networks," *IRE Intern. Conv. Record*, Vol. 9, Pt. 4, March 1961, pp. 268–276.
6. K. S. Miller, "Properties of Impulsive Responses and Green's Functions," *IRE Trans. Circuit Theory*, Vol. CT-2, March 1955, pp. 26–31.
7. D. Graham, E. J. Brunelle, Jr., W. Johnson, and H. Passmore, III, "Engineering Analysis Methods for Linear Time Varying Systems," Report ASD-TDR-62-362, Flight Control Laboratory, Wright-Patterson Air Force Base, Ohio, January 1963, pp. 126–127.
8. C. R. Wylie, Jr., *Advanced Engineering Mathematics*, Second Edition, McGraw-Hill Book Company, New York, 1960, p. 137.
9. M. P. Gardner and J. L. Barnes, *Transients in Linear Systems*, John Wiley and Sons, New York, 1942, Chapter IX.
10. Wylie, *op. cit.*, Section 5.5.
11. H. Levy and F. Lessman, *Finite Difference Equations*, Sir Isaac Pitman and Sons, London, 1959, Chapter 6.
12. T. Fort, *Finite Differences and Difference Equations in the Real Domain*, Oxford University Press, London, 1948, Chapters XIII and XIV.
13. L. M. Milne-Thomson, *The Calculus of Finite Differences*, Macmillan and Company, London, 1933, Chapters 14–16.

Problems

2.1 If y_1 and y_2 denote two solutions of a linear, second order differential equation, prove that these solutions are independent if and only if the Wronskian of Eq. 2.4-3 does not vanish.

2.2 Solve the following differential equations. In part (*e*), the substitution $y = (1/x)(dx/dt)$ is helpful.

(*a*) $(p^2 + 3p + 2)y = 0, y(0) = 1, \dot{y}(0) = 4$

(*b*) $(p^2 + 3p + 2)y = \sin(t + \pi/4), y(0) = 1, \dot{y}(0) = 4$

(*c*) $(p^3 + p^2 + p + 1)y = t\epsilon^{-t}, y(0) = \dot{y}(0) = \ddot{y}(0) = 0$

(*d*) $(tp^2 + 2p)y = \epsilon^{-t}, y(0) = \dot{y}(0) = 1$

(*e*) $py - y^2 = 1, y(0) = 0$

2.3 Find the differential equation describing Fig. P2.3 by first determining the system function $H(s)$.

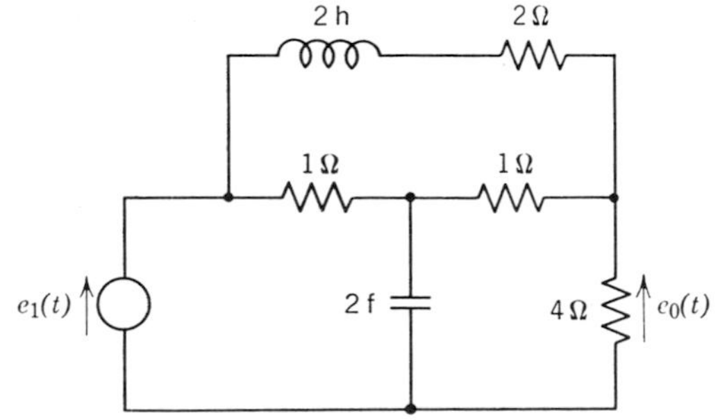

Fig. P2.3

2.4 Find the most general solution of the following homogenous differential equation by noting that one solution is $y_1(t) = t$.

$$(\tfrac{1}{2}t^3p^2 + tp - 1)y = 0$$

2.5 Solve Example 1.6-1, using only the classical solution of differential equations.

2.6 The circuit in Fig. P2.6 is originally operating in the steady state with the switch K open. If the switch closes at $t = 0$, find expressions for the currents for $t \geqslant 0$.

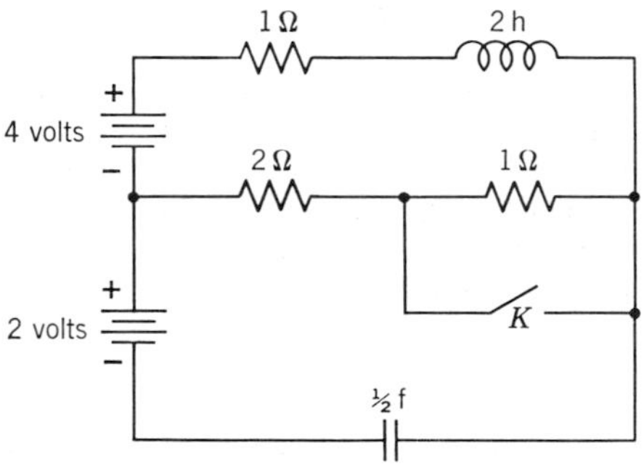

Fig. P2.6

2.7 Find the impulse response of the systems described by the following differential equations.

(*a*) $(p^2 + 2p + 2)y = (p^2 + 3p + 3)v$

(*b*) $(p^3 + 2p^2 + 3p + 2)y = (p + 4)v$

(*c*) $(p + t)y = v$

(*d*) $(tp^2 + 2p)y = v$

2.8 Solve Problem 2.2*d* by using the superposition integrals of Chapter 1 and the result of Problem 2.7*d*.

2.9 Find an expression for $h(t, \lambda)$ if a system is described by $(\alpha p + 1)y = (ap + 1)v$, where α is a constant, but where a may be a function of t.

2.10 Prove Eqs. 2.9-6, 2.9-11, and 2.9-13.

2.11 Sum the series

$$\sum_{n=1}^{k} \frac{1}{n(n+1)(n+2)}$$

2.12 Starting with Eq. 2.9-11*a*, derive the formula for summation by parts:

$$\sum_{n=\alpha}^{k} u(n)\,\Delta v(n) = [u(n)v(n)]_{\alpha}^{k+1} - \sum_{n=\alpha}^{k} v(n+1)\,\Delta u(n)$$

2.13 Find the most general solution of the following difference equations. Which of these equations represent stable systems? Rewrite these equations in terms of the difference operator Δ.

(*a*) $y(k+2) + 5y(k+1) + 6y(k) = 0$

(*b*) $y(k+2) - 2(\sinh \theta)y(k+1) - y(k) = 0$

(*c*) $y(k+2) - Ay(k+1) + y(k) = 0$, where $-1 \leqslant A < 1$

2.14 Solve the following difference equations, and comment on the stability of the systems they represent.

(*a*) $(E^2 + E + \frac{1}{4})y(k) = \sin(k\pi/2)$, $y(0) = 0$, $y(1) = 1$

(*b*) $(\Delta^2 + \Delta + 1)y(k) = 0$, $y(0) = 1$, $\Delta y(0) = 0$

(*c*) $(E^2 + 3E + 2)y(k) = (-1)^k$, $y(0) = y(1) = 1$

(*d*) $(E^2 + 2E + 1)y(k) = (E + 2)v(k)$, $y(k) = v(k) = 0$ for $k < 0$,
$v(k) = 1$ for $k \geqslant 0$

(*e*) $(E^2 + E + 1)y(k) = v(k)$, $y(k) = v(k) = 0$ for $k < 0$,
$v(k) = (-1)^k$ for $k \geqslant 0$.

2.15 Determine the delta response of the system described by

$$(E^3 + 2E^2 + 3E + 2)y(k) = (E + 4)v(k)$$

2.16 Solve Problem 2.14 by the superposition integrals of Chapter 1.

2.17 Find the complete solution of the following difference equations.

(*a*) $(E - k)^2 y(k) = k!$

(*b*) $[E^2 + (2k + 1)E + k^2]y(k) = k!$

3

Transform Techniques

3.1 INTRODUCTION

In Chapter 1, continuous signals are resolved into sets of singularity functions. Once the response of a linear system to a singularity function is known, the response to an arbitrary input can be found by superposition. In Section 3.2, continuous signals are decomposed into sets of sinusoidal functions by means of the Fourier series and integral. A system's response to a sinusoidal function can be easily found by a-c steady-state theory, after which the response to an arbitrary input again follows by superposition. Because of certain limitations in the Fourier methods, the Fourier integral is generalized in Section 3.3 to obtain the Laplace transform. The properties and applications of the Laplace transform are considered in some detail, and the transform methods are related to the convolution integrals of Chapter 1 and the differential equations of Chapter 2.

Section 1.3 suggested that the time domain and transform techniques are only two special cases of the general problem of decomposing signals into sets of elementary functions. This matter is considered further in Section 3.10.

The Z transform of Section 3.11 can be conveniently used for systems with discrete inputs, provided that the output is desired only at certain discrete instants. The Z transform is related to the superposition summations of Chapter 1 and the difference equations of Chapter 2. The modified Z transform of Section 3.14 permits the calculation of the output at any instant of time.

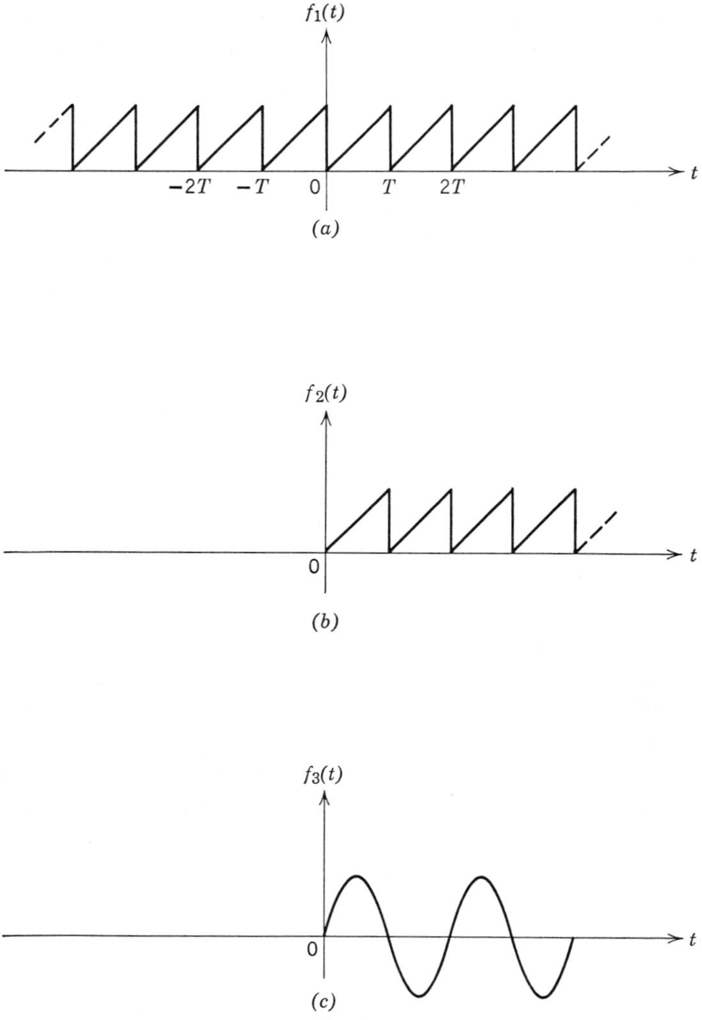

Fig. 3.2-1

3.2 THE FOURIER SERIES AND INTEGRAL

Periodic and nonperiodic functions can be decomposed into sets of sinusoidal functions by the Fourier series and integral, respectively. A function is periodic, with a period T, if it is continually repeated every T seconds over all values of t. Of the functions shown in Fig. 3.2-1, only $f_1(t)$ is periodic.

An infinite series, known as the Fourier series, may be written for any single-valued periodic function which has in any one period only a finite number of maxima, minima, and discontinuities. The area underneath any one cycle of the curve is required to be finite. The Fourier series for a function $f(t)$ with a period T is

$$f(t) = a_0 + \sum_{n=1}^{\infty} (a_n \cos n\omega_0 t + b_n \sin n\omega_0 t) \qquad (3.2\text{-}1)$$

where $\omega_0 = 2\pi/T$ is called the angular frequency of the first harmonic, and where

$$\begin{aligned} a_0 &= \frac{1}{T}\int_{-T/2}^{T/2} f(t)\, dt \\ a_n &= \frac{2}{T}\int_{-T/2}^{T/2} f(t) \cos n\omega_0 t\, dt \qquad (n \neq 0) \\ b_n &= \frac{2}{T}\int_{-T/2}^{T/2} f(t) \sin n\omega_0 t\, dt \end{aligned} \qquad (3.2\text{-}2)$$

The series converges to $f(t)$ for values of t at which the function is continuous. At a finite discontinuity, the series converges to the average of the values of $f(t)$ on either side of the discontinuity. The series converges uniformly for all t, if $f(t)$ satisfies the restrictions of the previous paragraph and in addition remains finite. If only a finite number of terms are taken, the corresponding finite Fourier series approximates $f(t)$ with the least mean square error.

Since two trigonometric terms of the same frequency may be combined into a single term with a phase angle, Eq. 3.2-1 may be rewritten as

$$(t) = a_0 + \sum_{n=1}^{\infty} A_n \cos (n\omega_0 t + \theta_n) \qquad (3.2\text{-}3)$$

where

$$\begin{aligned} A_n &= \sqrt{a_n^{\,2} + b_n^{\,2}} \\ \theta_n &= -\tan^{-1} \frac{b_n}{a_n} \end{aligned} \qquad (3.2\text{-}4)$$

A_n and θ_n represent the magnitude and phase angle (with respect to a pure cosine wave) of the nth harmonic.

The Complex Form of the Fourier Series

The complex form is easily obtained from the trigonometric form by use of the identities

$$\begin{aligned} \cos \phi &= \tfrac{1}{2}(\epsilon^{j\phi} + \epsilon^{-j\phi}) \\ \sin \phi &= \frac{1}{2j}(\epsilon^{j\phi} - \epsilon^{-j\phi}) \end{aligned} \qquad (3.2\text{-}5)$$

Equation 3.2-3 becomes

$$f(t) = \sum_{n=-\infty}^{\infty} c_n \epsilon^{jn\omega_0 t} \tag{3.2-6}$$

where

$$c_n = \frac{1}{T} \int_{-T/2}^{T/2} f(t) \epsilon^{-jn\omega_0 t} \, dt \tag{3.2-7}$$

Equations 3.2-6 and 3.2-3 are related by $c_0 = a_0$ and

$$c_n = \frac{A_n}{2} \epsilon^{j\theta_n} \qquad (n \neq 0)$$

so that the magnitude of c_n is one-half the magnitude of the nth harmonic, and the angle of c_n is the phase angle of the harmonic relative to a cosine wave. Since c_n completely defines $f(t)$ in frequency domain terms, it is called the frequency spectrum of $f(t)$.

Example 3.2-1. In the circuit of Fig. 3.2-2*a* find the steady-state component of the output voltage $e_0(t)$.

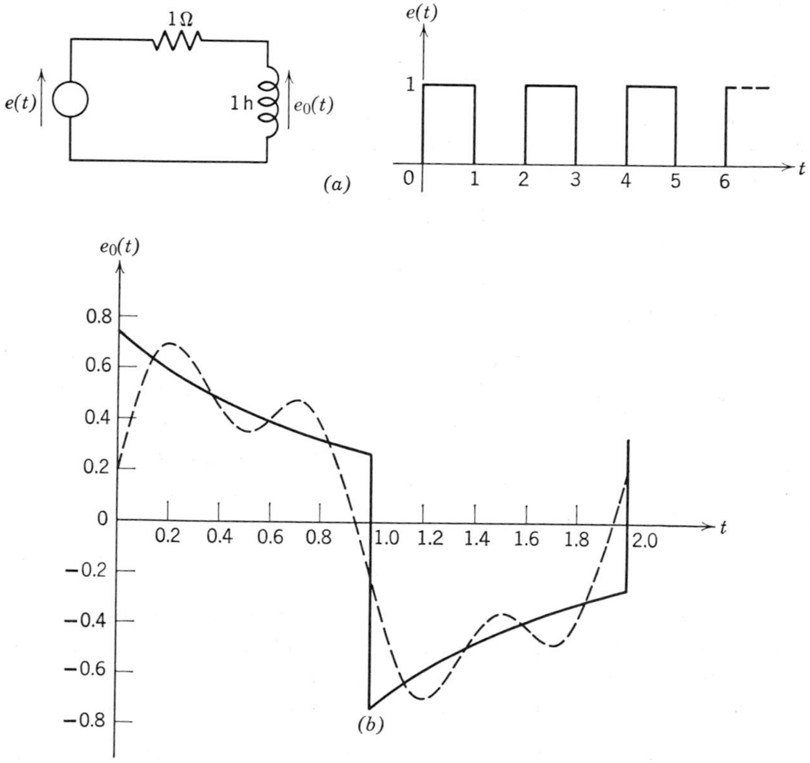

Fig. 3.2-2

For the calculation of the steady-state response, the square wave of input voltage may be assumed to exist for both positive and negative values of time. Then the frequency spectrum of the input, found from Eq. 3.2-7 with $\omega_0 = \pi$, is

$$c_n = \frac{1}{2}\int_0^1 \epsilon^{-jn\pi t}\,dt = \frac{1 - \epsilon^{-jn\pi}}{2jn\pi}$$

$$c_0 = \tfrac{1}{2}$$

$$c_n = 0, \quad n \text{ even}$$

$$c_n = \frac{1}{jn\pi}, \quad n \text{ odd}$$

Thus

$$e(t) = \frac{1}{2} + \sum_{n=\pm1,\pm3,\ldots}^{\pm\infty} \frac{1}{jn\pi}\epsilon^{jn\pi t}$$

By a-c steady-state circuit theory, the response to $\epsilon^{j\omega t}$ is

$$\left[\frac{j\omega}{1 + j\omega}\right]\epsilon^{j\omega t} = \frac{j\omega}{\sqrt{1 + \omega^2}}\epsilon^{j(\omega t - \tan^{-1}\omega)}$$

By superposition, and using Eq. 3.2-5,

$$e_0(t) = \sum_{n=1,3,\ldots}^{\infty} \frac{2}{\sqrt{1 + (n\pi)^2}}\cos(n\pi t - \tan^{-1} n\pi)$$

The broken line in Fig. 3.2-2*b* is the sum of the first two nonzero terms in the last expression. The exact steady-state output, representing the entire infinite series, is

$$e_0(t) = \begin{cases} \dfrac{\epsilon}{1 + \epsilon}\epsilon^{-t} & \text{for } 0 < t < 1 \\[2ex] \dfrac{-\epsilon^2}{1 + \epsilon}\epsilon^{-t} & \text{for } 1 < t < 2 \end{cases}$$

and is shown by the solid line.†

The preceding example contains three essential steps: finding the frequency spectrum of the input, multiplying by the a-c steady-state transfer function to obtain the output frequency spectrum, and constructing the output as a function of time. The frequency spectrum of $f(t)$ is often denoted by $F(\omega)$, rather than by c_n. Then, for a fixed linear system with input $v(t)$ and output $y(t)$,

$$Y(\omega) = V(\omega)H(j\omega) \tag{3.2-8}$$

where $H(j\omega)$ is the a-c steady-state system function defined in Eq. 2.6-23.

Two limitations of the Fourier series method are that $y(t)$ is expressed as an infinite series rather than in closed form, and that only the steady-state component is obtained. Because of this, the method is generally

† This closed-form solution is not obtained directly from the infinite series, but from the classical solution of Example 2.12-3. $e_0 = di/dt$, where i is given in Eq. 2.12-19.

used only when the frequency spectra are desired, as in wave filters, and not to calculate the output waveform as a function of time.

The Fourier Integral

The Fourier series of Eqs. 3.2-6 and 3.2-7 can be rewritten with $\omega = n\omega_0$.

$$\frac{c_n}{\Delta\omega} = \frac{1}{T\,\Delta\omega}\int_{-T/2}^{T/2} f(t)\epsilon^{-j\omega t}\,dt = \frac{1}{2\pi}\int_{-T/2}^{T/2} f(t)\epsilon^{-j\omega t}\,dt$$

$$f(t) = \sum_{\omega=-\infty}^{\infty} \frac{c_n}{\Delta\omega}\,\epsilon^{j\omega t}\,\Delta\omega$$

The symbol $\Delta\omega$ denotes the spacing between lines in the frequency spectrum, and it is equal to $\omega_0 = 2\pi/T$.

A nonperiodic function of time can be obtained from a periodic one by letting T approach infinity. Since $\Delta\omega$ then approaches zero, the formerly discrete frequency spectrum becomes a continuous one, containing all possible frequencies. The quantity $g(\omega) = c_n/\Delta\omega$ is introduced because c_n itself vanishes as T approaches infinity. Taking the limit as T approaches infinity and as $\Delta\omega$ becomes $d\omega$,

$$g(\omega) = \frac{1}{2\pi}\int_{-\infty}^{\infty} f(t)\epsilon^{-j\omega t}\,dt$$

$$f(t) = \int_{-\infty}^{\infty} g(\omega)\epsilon^{j\omega t}\,d\omega$$

To obtain the form most commonly used, let $G(\omega) = 2\pi g(\omega)$.

$$G(\omega) = \int_{-\infty}^{\infty} f(t)\epsilon^{-j\omega t}\,dt = \mathscr{F}[f(t)]$$

$$f(t) = \frac{1}{2\pi}\int_{-\infty}^{\infty} G(\omega)\epsilon^{j\omega t}\,d\omega = \mathscr{F}^{-1}[G(\omega)] \tag{3.2-9}$$

The last two equations are the direct and inverse Fourier transform, respectively. $G(\omega)$ is called the frequency spectrum of $f(t)$, and the expression for $f(t)$ is often called the Fourier integral. The reader may notice that the Fourier integral has the form of Eq. 1.3-3 with $\lambda = \omega$, and $k(t, \lambda) = \epsilon^{j\omega t}$.

Equations 3.2-9 can be used for nonperiodic signals in the same manner that the complex Fourier series is used for periodic signals. The conditions necessary for the existence of a Fourier series carry over to the Fourier integral. The function of time must be single-valued, with a finite number of maxima, minima, and discontinuities, and with $\int_{-\infty}^{\infty} |f(t)|\,dt$ remaining finite.

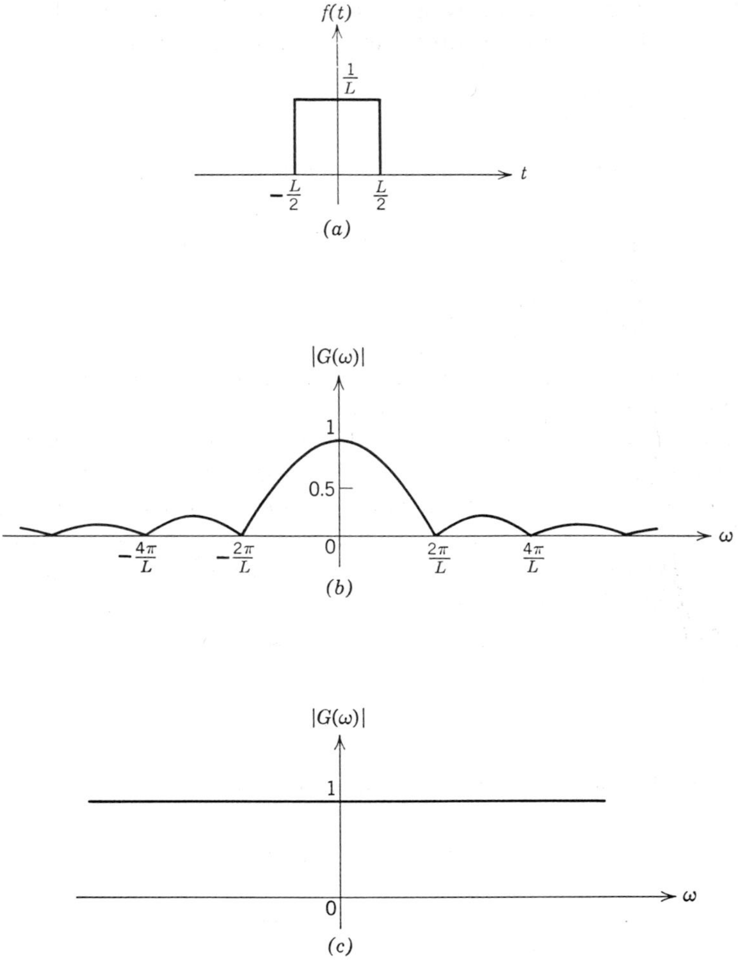

Fig. 3.2-3

Example 3.2-2. Find the frequency spectrum of the rectangular pulse shown in Fig. 3.2-3*a*.

$$G(\omega) = \int_{-L/2}^{L/2} \frac{1}{L} \epsilon^{-j\omega t}\, dt = \frac{\sin(\omega L/2)}{\omega L/2}$$

$G(\omega)$, which is real because $f(t)$ is an even function, is illustrated in Fig. 3.2-3*b*. In the limit as L approaches 0, $f(t)$ becomes the unit impulse, and $G(\omega)$ is shown in part (*c*) of the figure. Thus the unit impulse contains all frequencies in equal strength.

This example demonstrates the principle of reciprocal spreading. The narrower the function of time is made, the more its frequency spectrum spreads out, and the greater is the bandwidth required to reproduce it faithfully. If the sharp corners in the function of time are smoothed, the high-frequency components in its frequency spectrum are reduced. For example if, $f(t) = (1\sqrt{2\pi})\epsilon^{-t^2/2}$, which is the Gaussian-type pulse of unit area shown in Fig. 1.4-7*a*, $G(\omega) = \epsilon^{-\omega^2/2}$.

Example 3.2-3. Find the Fourier transform of the unit step function.

It should be expected that this example cannot be solved since $\int_{-\infty}^{\infty} U_{-1}(t)\,dt$ is not finite. If Eq. 3.2-9 is used,

$$G(\omega) = \int_0^\infty \epsilon^{-j\omega t}\,dt = \left[\frac{\epsilon^{-j\omega t}}{-j\omega}\right]_0^\infty$$

cannot be evaluated, since the upper limit is not uniquely defined, i.e., the defining integral does not converge. If the reader is persistent, he might note that

$$\mathscr{F}[\epsilon^{-at} \quad \text{for } t \geqslant 0] = \int_0^\infty \epsilon^{-at}\epsilon^{-j\omega t}\,dt = \frac{1}{a + j\omega} \qquad (a \neq 0)$$

which, as a approaches 0, reduces to $1/j\omega$. But this limit is not, mathematically, the Fourier transform of the unit step function. To find the function of time corresponding to $G(\omega) = 1/j\omega$, Eq. 3.2-9 gives

$$f(t) = \frac{1}{2\pi}\int_{-\infty}^{\infty} \frac{\epsilon^{j\omega t}}{j\omega}\,d\omega = \frac{1}{2\pi}\int_{-\infty}^{\infty}\left[\frac{\cos \omega t}{j\omega} + \frac{\sin \omega t}{\omega}\right]d\omega$$

$$= \frac{1}{\pi}\int_0^\infty \frac{\sin \omega t}{\omega}\,d\omega = \begin{cases} -\frac{1}{2} & \text{for } t < 0 \\ \frac{1}{2} & \text{for } t > 0 \end{cases}$$

The purpose of this example is to point out that some of the very common functions of time do not have Fourier transforms. The attempted use of the convergence factor ϵ^{-at} suggests the heuristic derivation of the Laplace transform in the next section.

When analyzing a fixed linear system, it is convenient to denote the Fourier transform of the input $v(t)$ and output $y(t)$ by $V(\omega)$ and $Y(\omega)$, respectively. Using the system function defined in Eq. 2.6-23, the response to $v(t) = \epsilon^{j\omega t}$ is $y(t) = H(j\omega)\epsilon^{j\omega t}$. Then, by the superposition principle,

$$Y(\omega) = V(\omega)H(j\omega) \tag{3.2-10}$$

and

$$y(t) = \mathscr{F}^{-1}[V(\omega)H(j\omega)] \tag{3.2-11}$$

In contrast to the Fourier series, the Fourier integral of Eq. 3.2-11 gives the complete response in closed form, provided that the system contains no initial stored energy.

Example 3.2-4. For the circuit of Fig. 3.2-4*a*, plot the frequency spectrum of the input and output, and find $e_2(t)$. Assume that there is no stored energy in the capacitance at $t = 0$.

$$E_1(\omega) = \int_0^\infty \epsilon^{-t}\epsilon^{-j\omega t}\,dt = \frac{1}{1 + j\omega}$$

$$E_2(\omega) = E_1(\omega)\frac{1}{1 + j\omega} = \frac{1}{(1 + j\omega)^2}$$

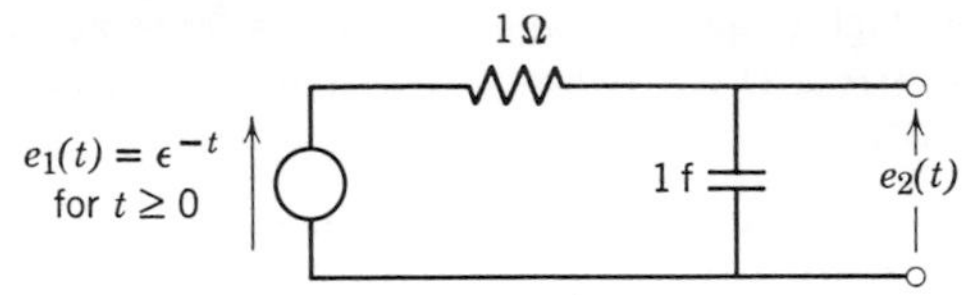

(a)

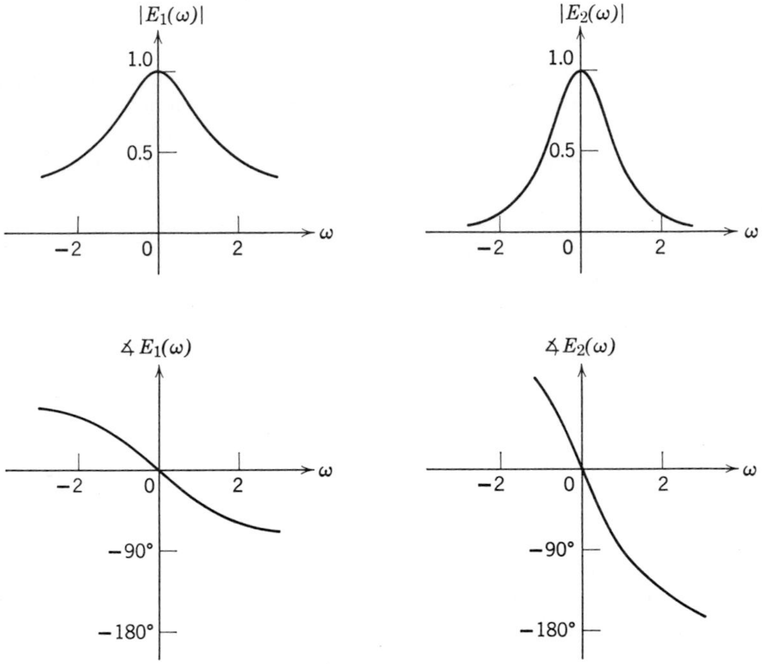

(b)

Fig. 3.2-4

The frequency spectra are plotted in part (b) of the figure. The evaluation of

$$e_2(t) = \frac{1}{2\pi} \int_{-\infty}^{\infty} \frac{\epsilon^{j\omega t}}{(1 + j\omega)^2} \, d\omega$$

involves a difficult integration. Tables of Fourier transforms do exist,[1] and the use of complex variable theory is often helpful. It can be shown by the methods of Section 3.6 that

$$e_2(t) = \frac{1}{2\pi j} \int_{-j\infty}^{j\infty} \frac{\epsilon^{j\omega t}}{(1 + j\omega)^2} \, d(j\omega)$$

$$= \frac{1}{2\pi j} \oint \frac{\epsilon^{st}}{(1 + s)^2} \, ds = t\epsilon^{-t} \quad \text{for } t \geqslant 0$$

where the integration has been carried out with respect to the complex variable s. The details of the integration are not pursued further here, but the reader should realize that the direct evaluation of the inverse Fourier transform is seldom easy.

3.3 THE LAPLACE TRANSFORM

The Laplace transform largely overcomes the chief difficulties encountered in the use of the Fourier transform. The difficulty with the convergence of the direct transform can be remedied by replacing $\epsilon^{-j\omega t}$ by $\epsilon^{-(\sigma+j\omega)t}$ in Eq. 3.2-9. This inserts a convergence factor $\epsilon^{-\sigma t}$ into the integrand of $G(\omega)$ and permits the integral to be evaluated in some cases where it would not otherwise exist. This modification is suggested by the efforts of the previous section to obtain the Fourier transform of the unit step function.

The Two-Sided Laplace Transform

In Eq. 3.2-9, let $G(\omega) = F(j\omega)$, and write

$$F(j\omega) = \int_{-\infty}^{\infty} (t)\epsilon^{-j\omega t}\,dt$$

$$f(t) = \frac{1}{2\pi j}\int_{-j\infty}^{j\infty} F(j\omega)\epsilon^{j\omega t}\,d(j\omega)$$

In the second expression, the variable of integration has been changed to $j\omega$. Replacing $j\omega$ by $s = \sigma + j\omega$,

$$F(s) = \int_{-\infty}^{\infty} f(t)\epsilon^{-st}\,dt = \mathscr{L}_{\mathrm{II}}[f(t)] \tag{3.3-1}$$

and

$$f(t) = \frac{1}{2\pi j}\int_{\sigma-j\infty}^{\sigma+j\infty} F(s)\epsilon^{st}\,ds = \mathscr{L}_{\mathrm{II}}^{-1}[F(s)] \tag{3.3-2}$$

where $\mathscr{L}_{\mathrm{II}}$ and $\mathscr{L}_{\mathrm{II}}^{-1}$ stand for the direct and inverse two-sided Laplace transform.

This presentation is a heuristic one, illustrating the relationship between the Laplace and Fourier transforms, rather than emphasizing mathematical rigor. In Eq. 3.3-1, the value of $\sigma = \mathrm{Re}\ s$ is chosen to make the integral converge, if possible. In Eq. 3.3-2, it is understood that the integration is carried out with σ within the same range that ensures the convergence of Eq. 3.3-1. Since s is a complex variable, Eq. 3.3-2 involves integration in the complex plane. A detailed discussion of this is postponed until after a review of complex variable theory. Two examples of the use of Eq. 3.3-1 are presented below.

Example 3.3-1. Determine the Laplace transform of

$$f(t) = \begin{cases} 0 & \text{for } t < 0 \\ \epsilon^{at} & \text{for } t > 0, \text{ for real } a \end{cases}$$

$$F(s) = \int_0^\infty \epsilon^{at}\epsilon^{-st}\, dt = \left[\frac{\epsilon^{-(s-a)t}}{-(s-a)}\right]_0^\infty = \frac{1}{s-a}$$

with a region of convergence given by $\sigma > a$. The function in this example does not have a Fourier transform if a is positive but does have a Laplace transform.

Example 3.3-2. Determine the Laplace transform of

$$f(t) = \begin{cases} -\epsilon^{at} & \text{for } t < 0, \text{ for real } a \\ 0 & \text{for } t > 0 \end{cases}$$

$$F(s) = -\int_{-\infty}^{0} \epsilon^{at}\epsilon^{-st}\, dt = \left[\frac{\epsilon^{-(s-a)t}}{s-a}\right]_{-\infty}^{0} = \frac{1}{s-a}$$

with a region of convergence given by $\sigma < a$.

It appears from these two examples that two different functions of time have the same Laplace transform. The two regions of convergence, however, are mutually exclusive. The region of convergence must be specified in order to determine which function of t corresponds to the function $(s - a)^{-1}$.

It should be clear that the Fourier and Laplace transforms of a nonperiodic function are essentially identical if the range of convergence for σ includes $\sigma = 0$. In such a case, the Fourier transform is found by replacing s by $j\omega$ in the Laplace transform. The phenomenon of two different functions of time having the same Fourier transform never occurs. If a is positive, $\mathscr{F}^{-1}[1/(j\omega - a)]$ must be the function of Example 3.3-2 and not that of Example 3.3-1. The unbounded function has no Fourier transform and could not therefore be the correct answer. It is the introduction of the convergence factor $\epsilon^{-\sigma t}$ that results in the uniqueness problem.

The One-Sided Laplace Transform

It is usually possible to restrict oneself to functions that are zero for negative values of time. The reason is that the response of a linear system can be determined for all $t > 0$ from a knowledge of the input for $t > 0$ and the energies stored in the system at $t = 0$. The history of the system prior to some reference time is adequately summarized by the stored energies at that time, as far as the system's future behavior is concerned. The two-sided Laplace transform of Eq. 3.3-1 is therefore usually replaced by the normal or one-sided Laplace transform of Eq. 3.3-3. The latter equation simply says that $f(t)$ is zero for negative values of t, or else that $f(t)$ for $t < 0$ is of no concern and can hence be assumed to be zero.

$$F(s) = \int_0^\infty f(t)\epsilon^{-st}\,dt = \mathscr{L}[f(t)] \tag{3.3-3}$$

A subscript I is not used, since the Laplace transform is assumed to be the one-sided transform, unless otherwise indicated. The application of Eq. 3.3-3 to several common functions of time leads to the results in Table 3.3-1.†

Table 3.3-1

$f(t)$ for $t \geqslant 0$	$F(s)$	Region of Convergence
$U_0(t)$	1	$\sigma > -\infty$
$U_{-1}(t)$	$\dfrac{1}{s}$	$\sigma > 0$
t	$\dfrac{1}{s^2}$	$\sigma > 0$
ϵ^{-at}	$\dfrac{1}{s+a}$	$\sigma > -a$
$\sin \beta t$	$\dfrac{\beta}{s^2+\beta^2}$	$\sigma > 0$
$\cos \beta t$	$\dfrac{s}{s^2+\beta^2}$	$\sigma > 0$
$\epsilon^{-at} \sin \beta t$	$\dfrac{\beta}{(s+a)^2+\beta^2}$	$\sigma > -a$
$\epsilon^{-at} \cos \beta t$	$\dfrac{s+a}{(s+a)^2+\beta^2}$	$\sigma > -a$
$\dfrac{\epsilon^{-at}t^{n-1}}{(n-1)!}$, $n = 1, 2, \ldots$	$\dfrac{1}{(s+a)^n}$	$\sigma > -a$

The expression for the inverse transform is the same whether the one-sided or two-sided transform is used.

$$f(t) = \frac{1}{2\pi j}\int_{\sigma-j\infty}^{\sigma+j\infty} F(s)\epsilon^{st}\,ds = \mathscr{L}^{-1}[F(s)] \tag{3.3-4}$$

The integration must still be carried out with σ within the range that ensures the convergence of Eq. 3.3-3.

† A very extensive table appears in Reference 2. Tables of moderate size can be found in several of the other references.

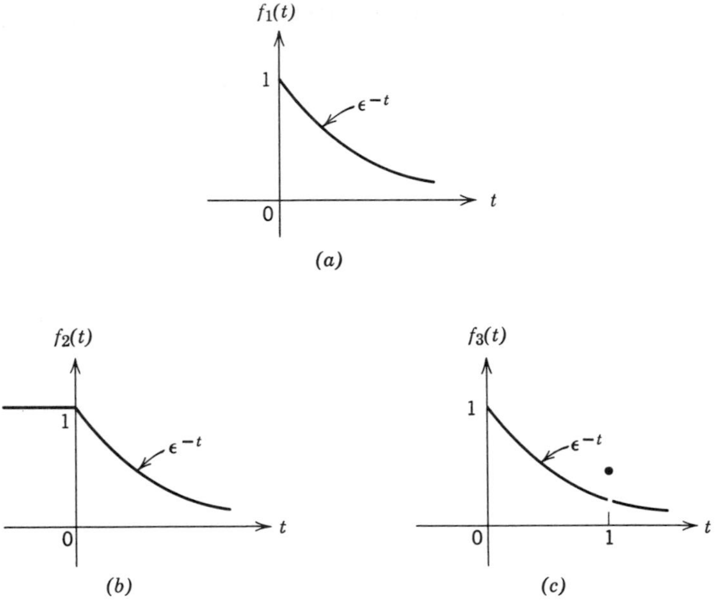

Fig. 3.4-1

3.4 PROPERTIES OF THE LAPLACE TRANSFORM

Several important theorems concerning the existence and uniqueness of the one-sided Laplace transform may be found in standard references and can be partially summarized as follows:[3] A function $f(t)$ must be defined for all $t > 0$, except possibly at a denumerable set of points, in order for its transform $F(s)$ to exist. Every such $f(t)$, which is piecewise continuous and which is of exponential order, has a Laplace transform.† For such functions, the integral in Eq. 3.3-3 converges absolutely for $\sigma > c$, where the constant c is known as the abscissa of absolute convergence.

Two functions of time have the same Laplace transform if and only if they are identical for all $t > 0$, except possibly at a denumerable set of points. In Fig. 3.4-1, all three functions have the same Laplace transform, $(s + 1)^{-1}$. Except for these trivial cases, however, there is a one-to-one

† A function $f(t)$ is of exponential order if some real number k exists such that

$$\lim_{t\to\infty} f(t)\epsilon^{kt} = 0$$

Examples of functions not of exponential order are ϵ^{t^2} and t^t. A function of nonexponential order may or may not have a Laplace transform.

correspondence between $f(t)$ and $F(s)$. This means that Table 3.3-1 may be used to find the inverse as well as the direct transform, which in many, but not all, cases obviates the need for Eq. 3.3-4. Note that the uniqueness problem of Examples 3.3-1 and 3.3-2 no longer exists, because the two functions given there do not have the same *one-sided* transform.

The use of residue theory in connection with Eq. 3.3-4 provides a powerful and relatively simple method of finding $f(t)$. As shown in Section 3.7, the use of Eq. 3.3-4 in the one-sided transform yields a function of time that is zero for $t < 0$, and that is continuous except for possible singularity functions for $t \geqslant 0$. Thus $\mathscr{L}^{-1}[1/(s+1)]$ is identified as the function $f_1(t)$ in Fig. 3.4-1, rather than $f_2(t)$ or $f_3(t)$.

The functions encountered in system theory often have discontinuities at $t = 0$. It is usually convenient to define the value of $f(t)$ at $t = 0$ to be the limit of $f(t)$ as t approaches zero through positive values. This limit is symbolized as $f(0+)$, and for the function $f_1(t)$ in Fig. 3.4-1 has a value of unity. Although it is not necessary to adopt this convention, most engineers do so, for it leads to the fewest difficulties. Consistent with this convention, and with the fact that only the response for $t > 0$ is usually desired, the definition of the direct Laplace transform can be written as

$$F(s) = \lim_{\substack{R\to\infty \\ \epsilon\to 0}} \int_{\epsilon}^{R} f(t)\epsilon^{st}\,dt, \quad R > 0, \quad \epsilon > 0 \tag{3.4-1}$$

Other Properties of the Laplace Transform

Table 3.4-1 lists the most useful properties of the Laplace transform. Proofs in most cases follow directly from Eq. 3.4-1 and can be found in most references.[4] Equations 3.4-2 through 3.4-7 are the basis for the solution of fixed integral-differential equations by means of the Laplace transform, as discussed in the next section. Equations 3.4-8 through 3.4-12 are useful in finding the direct and inverse transforms of given functions.

Example 3.4-1. Find $\mathscr{L}^{-1}[1/(s+2)^2]$.

$$\mathscr{L}^{-1}\left[\frac{1}{s^2}\right] = t \quad \text{for } t \geqslant 0$$

by Table 3.3-1. Using Eq. 3.4-8 with $a = 2$,

$$\mathscr{L}^{-1}\left[\frac{1}{(s+2)^2}\right] = t\epsilon^{-2t}, \quad t \geqslant 0$$

Table 3.4-1

Property	Equation Number
$\mathscr{L}[f_1(t) + f_2(t)] = \mathscr{L}[f_1(t)] + \mathscr{L}[f_2(t)] = F_1(s) + F_2(s)$	(3.4-2)
$\mathscr{L}[af(t)] = a\mathscr{L}[f(t)] = aF(s)$	(3.4-3)
$\mathscr{L}\left[\frac{d}{dt}f(t)\right] = sF(s) - f(0+)$	(3.4-4)
$\mathscr{L}\left[\frac{d^n}{dt^n}f(t)\right] = s^nF(s) - s^{n-1}f(0+) - s^{n-2}\frac{df}{dt}(0+) - \cdots - \frac{d^{n-1}f}{dt^{n-1}}(0+)$	(3.4-5)
$\mathscr{L}\left[\int_0^t f(\tau)\,d\tau\right] = \frac{F(s)}{s}$	(3.4-6)
$\mathscr{L}\left[\int_{-\infty}^t f(\tau)\,d\tau\right] = \frac{F(s)}{s} + \frac{f^{-1}(0+)}{s}$ where $f^{-1}(0+) = \lim_{t\to 0+}\int_{-\infty}^t f(\tau)\,d\tau$	(3.4-7)
$\mathscr{L}[\epsilon^{-at}f(t)] = F(s+a)$	(3.4-8)
$\mathscr{L}[tf(t)] = -\frac{d}{ds}F(s)$	(3.4-9)
$\mathscr{L}\left[\frac{f(t)}{t}\right] = \int_s^\infty F(r)\,dr$	(3.4-10)
$\mathscr{L}[f(t-a)U_{-1}(t-a)] = \epsilon^{-as}F(s)$	(3.4-11)
$\mathscr{L}\left[f\left(\frac{t}{a}\right)\right] = a\,F(as)$	(3.4-12)
$\mathscr{L}\left[\int_0^t f_1(t-\lambda)f_2(\lambda)\,d\lambda\right] = F_1(s)F_2(s)$	(3.4-13)
$\mathscr{L}[f_1(t)f_2(t)] = \frac{1}{2\pi j}\int_{x-j\infty}^{x+j\infty} F_1(w)F_2(s-w)\,dw$ where $w = x + jy$, and where x must be greater than the abscissa of absolute convergence for $f_1(t)$ over the path of integration	(3.4-14)
$\lim_{t\to 0} f(t) = \lim_{s\to\infty} sF(s)$ provided that the limit exists	(3.4-15)
$\lim_{t\to\infty} f(t) = \lim_{s\to 0} sF(s)$ provided that $sF(s)$ is analytic on the $j\omega$ axis and in the right half of the s-plane.	(3.4-16)

Example 3.4-2. Find $\mathscr{L}[t \sin t]$ and $\mathscr{L}[(\sin t)/t]$.

$$\mathscr{L}[\sin t] = \frac{1}{s^2 + 1}$$

by Table 3.3-1. Using Eq. 3.4-9,

$$\mathscr{L}[t \sin t] = -\frac{d}{ds}\left[\frac{1}{s^2 + 1}\right] = \frac{2s}{(s^2 + 1)^2}$$

Using Eq. 3.4-10,

$$\mathscr{L}\left[\frac{\sin t}{t}\right] = \int_s^\infty \frac{1}{r^2 + 1}\, dr = [\tan^{-1} r]_s^\infty$$

$$= \frac{\pi}{2} - \tan^{-1} s = \tan^{-1}\left[\frac{1}{s}\right]$$

Example 3.4-3. Figure 3.4-2 shows a periodic function of time $f(t)$, with period T. Let $f_1(t)$ denote the first cycle only, $f_2(t)$ the second cycle only, etc. If

$$F_1(s) = \mathscr{L}[f_1(t)] = \int_0^T f(t)\epsilon^{-st}\, dt$$

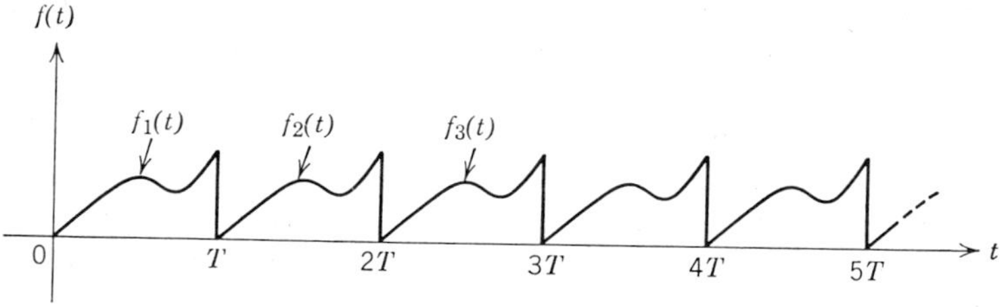

Fig. 3.4-2

find an expression for

$$F(s) = \mathscr{L}[f(t)] = \int_0^\infty f(t)\epsilon^{-st}\, dt$$

Since

$$f(t) = f_1(t) + f_2(t) + f_3(t) + \cdots$$

and since

$$f_2(t) = f_1(t - T)U_{-1}(t - T)$$
$$f_3(t) = f_1(t - 2T)U_{-1}(t - 2T)$$
$$\cdots\cdots\cdots\cdots\cdots\cdots$$

Eq. 3.4-11 yields

$$F(s) = F_1(s) + \epsilon^{-Ts}F_1(s) + \epsilon^{-2Ts}F_1(s) + \cdots$$
$$= F_1(s)[1 + \epsilon^{-Ts} + \epsilon^{-2Ts} + \cdots]$$

Since

$$\frac{1}{1 - x} = 1 + x + x^2 + \cdots \quad \text{for } |x| < 1$$

the expression for $F(s)$ may be written

$$F(s) = \frac{F_1(s)}{1 - \epsilon^{-Ts}} \quad \text{for Re } s > 0$$

This result is very useful when finding the response of a system to a periodic function. Unlike the Fourier series method, both the steady-state and the transient response may be found in closed form.[5]

Equation 3.4-12 is helpful in an understanding of time and frequency scaling. Equations 3.4-13 and 3.4-14 are convolution formulas, and they emphasize the fact that

$$\mathscr{L}[f_1(t)f_2(t)] \neq F_1(s)F_2(s)$$

The last two properties in Table 3.4-1 are known as the initial and final value theorems, respectively. The meaning of the word analytic is discussed in Section 3.6.

Impulses in $f(t)$

The presence of impulses in $f(t)$ introduces some difficulty in the interpretation of Laplace transform theorems. As discussed in Section 1.4, impulses are not functions in the usual mathematical sense. The uniqueness property previously discussed is not valid if there are impulses in one of the two functions at the "denumerable set of points." This does not cause confusion, however, because it is easy to recognize whether or not the inverse transform contains impulses. Perhaps the greatest confusion occurs when dealing with impulses at the time origin. The entry in Table 3.3-1 stating that $\mathscr{L}[U_0(t)] = 1$ assumes that the impulse occurs immediately after $t = 0$. The transform of an impulse occurring immediately before $t = 0$ is zero. These statements are consistent with Eq. 3.4-4. Since the unit impulse is the derivative of the unit step, the transform of an impulse occurring immediately after $t = 0$ is given by $\mathscr{L}[U_0(t)] = s(1/s) - 0 = 1$; for an impulse immediately before $t = 0$, $\mathscr{L}[U_0(t)] = s(1/s) - 1 = 0$.

In the analysis of systems, either viewpoint yields correct results if consistently used. An impulse serves to instantaneously insert stored energy into the system. If $\mathscr{L}[U_0(t)]$ is taken to be unity, which is the usual practice, the initial stored energy in the system should be calculated excluding the effect of the impulse. A similar viewpoint is adopted for higher order singularity functions occurring at the origin. $\mathscr{L}[U_n(t)] = s^n$, and the initial stored energy is calculated excluding the effect of the singularity function.

Partial Fraction Expansions

In many cases, $F(s)$ is the quotient of two polynomials with real coefficients. In such cases, the inverse transform can always be found from

Table 3.3-1, using a partial fraction expansion.[4] Suppose that

$$F(s) = \frac{P(s)}{Q(s)} = \frac{P(s)}{(s - s_1)^r(s - s_{r+1}) \cdots (s - s_n)} \tag{3.4-17}$$

where the denominator polynomial $Q(s)$ has distinct roots $s_{r+1}, \ldots, s_n$, and a root s_1 which is repeated r times. If $Q(s)$ is of higher order than $P(s)$, then

$$F(s) = \frac{K_1}{(s - s_1)} + \cdots + \frac{K_{r-1}}{(s - s_1)^{r-1}} + \frac{K_r}{(s - s_1)^r} + \frac{K_{r+1}}{s - s_{r+1}} + \cdots + \frac{K_n}{s - s_n}$$

where

$$K_i = [(s - s_i)F(s)]_{s=s_i} \qquad (i = r + 1, \ldots, n)$$

$$K_r = [(s - s_1)^r F(s)]_{s=s_1}$$

$$K_{r-1} = \left[\frac{d}{ds}(s - s_1)^r F(s)\right]_{s=s_1}$$

and, in general,

$$K_{r-k} = \frac{1}{k!}\left[\frac{d^k}{ds^k}(s - s_1)^r F(s)\right]_{s=s_1} \qquad (k = 0, 1, 2, \ldots, r - 1)$$

Once the partial fraction expansion has been determined, the inverse transform of each term follows from Table 3.3-1.

Example 3.4-4. Find the inverse transform of

$$F(s) = \frac{s^3 + 1}{s^3 + s}$$

Since the denominator polynomial is not of greater order than the numerator polynomial, a preliminary step of long division is necessary.

$$F(s) = 1 + \frac{-s + 1}{s(s^2 + 1)}$$

The second of the two terms can be expanded in partial fractions as above, yielding

$$F(s) = 1 + \frac{1}{s} - \frac{(\sqrt{2}/2)\epsilon^{-j\pi/4}}{s - j} - \frac{(\sqrt{2}/2)\epsilon^{j\pi/4}}{s + j}$$

Using Table 3.3-1 and Eq. 3.2-5,

$$f(t) = U_0(t) + \left[1 - \sqrt{2}\cos\left(t - \frac{\pi}{4}\right)\right]U_{-1}(t)$$

A partial fraction expansion of $F(s)$ should be a mathematical identity for all values of s. The correctness of a particular expansion can be easily checked by recombining the individual terms over a common denominator.

The expansion in the last example would certainly not be correct without the 1 obtained by long division, since $F(s)$ approaches unity for large values of s. Whenever the order of the numerator equals or exceeds that of the denominator, long division terms are needed to describe properly the function for large values of s.

The Relationship between the Fourier and Laplace Transforms

A further comparison of the Fourier and one-sided Laplace transforms is interesting, partly because tables for one of the transforms can then be applied to the other. Consider a transformable function $f(t)$, with $\mathscr{L}[f(t)] = F(s)$, and $\mathscr{F}[f(t)] = G(\omega)$. How can the direct Fourier transform be found from a table of Laplace transforms? To answer this question, consider the following three cases. If $f(t) = 0$ for $t < 0$, $G(\omega) = F(j\omega)$. If $f(t) = 0$ for $t > 0$,

$$G(\omega) = \int_{-\infty}^{0} f(t)\epsilon^{-j\omega t}\, dt$$

Letting $\gamma = -t$,

$$G(\omega) = \int_{0}^{\infty} f(-\gamma)\epsilon^{j\omega\gamma}\, d\gamma$$

$$= \int_{0}^{\infty} f_R(t)\epsilon^{j\omega t}\, dt = F_R(-j\omega)$$

where $f_R(t) = f(-t)$ and is the original function reflected about the vertical axis. If $f(t)$ is nonzero for both positive and negative values of time, it can be decomposed into two parts, each of which falls into one of the two previous cases.

Example 3.4-5. Find the Fourier transform of the function shown in Fig. 3.4-3a, given by

$$f(t) = \begin{cases} \epsilon^{t} & \text{for } t < 0 \\ \epsilon^{-t} & \text{for } t > 0 \end{cases}$$

Let

$$f(t) = f_1(t) + f_2(t)$$

where the last two functions are shown in parts (b) and (c) of the figure.

$$G_1(\omega) = F_1(j\omega) = \left[\frac{1}{s+1}\right]_{s=j\omega} = \frac{1}{j\omega + 1}$$

$$G_2(\omega) = F_{2R}(-j\omega) = \left[\frac{1}{s+1}\right]_{s=-j\omega} = \frac{1}{-j\omega + 1}$$

Finally,

$$G(\omega) = G_1(\omega) + G_2(\omega) = \frac{2}{\omega^2 + 1}$$

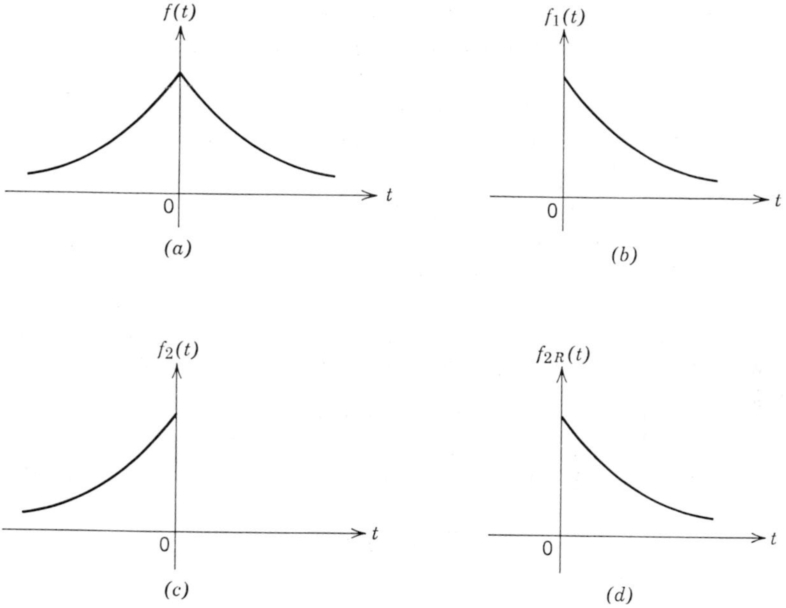

Fig. 3.4-3

3.5 APPLICATION OF THE LAPLACE TRANSFORM TO TIME-INVARIANT SYSTEMS

There are two basic methods of analyzing linear, time-invariant systems using the Laplace transform. Such systems can be described by a set of integral-differential equations with constant coefficients. These equations can be transformed into algebraic equations by Eqs. 3.4-2 through 3.4-7. This approach is illustrated by the following example. A second method, associated with the system function concept, is described later in this section.

Example 3.5-1. Find an expression for the response $e_2(t)$ for $t > 0$ in the circuit of Fig. 3.5-1. The source $e_1(t)$ is a known function of time, and the current i_L and voltage e_c are assumed to be known at the instant $t = 0$.

The circuit is described by the two loop equations

$$\left(2\frac{di_1}{dt} + 3i_1\right) - \left(2\frac{di_2}{dt} + 2i_2\right) = e_1$$

$$-\left(2\frac{di_1}{dt} + 2i_1\right) + \left(2\frac{di_2}{dt} + 3i_2 + 2\int i_2\,dt\right) = 0$$

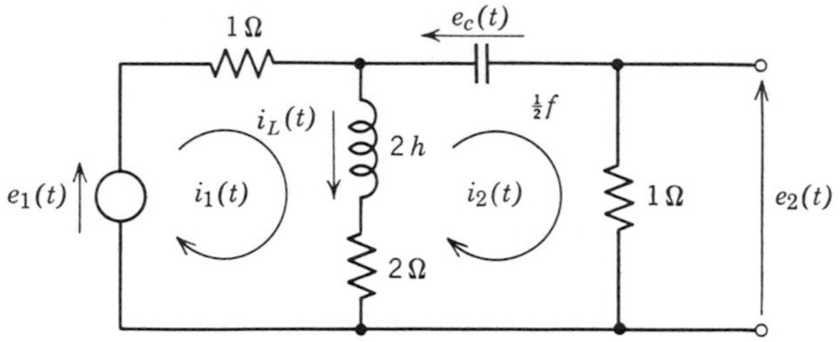

Fig. 3.5-1

Let $I_1(s)$, $I_2(s)$, and $E_1(s)$ stand for the transforms of $i_1(t)$, $i_2(t)$, and $e_1(t)$, respectively. Transforming the differential equations term by term yields the following algebraic equations.

$$(2s + 3)I_1(s) - (2s + 2)I_2(s) = E_1(s) + 2[i_1(0+) - i_2(0+)]$$

$$-(2s + 2)I_1(s) + \left(2s + 3 + \frac{2}{s}\right)I_2(s) = -\frac{2}{s}i_2^{-1}(0+) - 2[i_1(0+) - i_2(0+)]$$

Note that

$$i_1(0+) - i_2(0+) = i_L(0+)$$

and

$$2i_2^{-1}(0+) = 2 \lim_{t\to 0+} \int_{-\infty}^{t} i_2(\tau)\, d\tau = e_c(0+)$$

Since $I_1(s)$ and $I_2(s)$ are the only unknowns, the two algebraic equations may be solved simultaneously for $I_2(s)$. There results

$$E_2(s) = I_2(s) = \frac{(2s^2 + 2s)E_1(s) - 2i_L(0+)s - (2s + 3)e_c(0+)}{4s^2 + 9s + 6}$$

If the circuit contained no stored energy at $t = 0$, $i_L(0+) = e_c(0+) = 0$, and

$$e_2(t) = \mathscr{L}^{-1}\left[\frac{2s^2 + 2s}{4s^2 + 9s + 6}E_1(s)\right]$$

The terms involving initial stored energy summarize the history of the circuit for $t < 0$. Such terms frequently occur when a switching operation is performed at the reference time $t = 0$, adding or disconnecting elements. It is best to transform the differential equations immediately, rather than first differentiating one of them, or solving them simultaneously. If the equations are transformed immediately, the $0+$ terms are always directly related to the initial stored energy, and they vanish if the system contains none.

The Impulse Response

The Laplace transform offers a convenient method of finding the impulse response from the system's differential equation.

Example 3.5-2. A system is described by the differential equation of Example 2.8-1.

$$\frac{d^2y}{dt^2} + 3\frac{dy}{dt} + 2y = \frac{1}{2}\frac{dv}{dt} + v$$

Determine its impulse response.

Assume that the system is initially at rest, and that v is a unit impulse occurring immediately after $t = 0$. Then

$$\frac{dy}{dt}(0+) = y(0+) = 0$$

and the transformed equation is

$$s^2 Y(s) + 3s\,Y(s) + 2\,Y(s) = \frac{s}{2} + 1$$

so

$$Y(s) = \frac{\frac{1}{2}(s+2)}{s^2 + 3s + 2} = \frac{\frac{1}{2}}{s+1}$$

and

$$y(t) = \tfrac{1}{2}\epsilon^{-t}$$

This answer agrees with the impulse response found in Example 2.8-1.

Alternatively, if v is a unit impulse occurring immediately before $t = 0$, the transform of the right side of the differential equation is zero, but $(dy/dt)(0+) = -\frac{1}{2}$ and $y(0+) = \frac{1}{2}$, as in Example 2.8-1. The complete transformed equation is

$$s^2 Y(s) - \frac{s}{2} + \frac{1}{2} + 3s\,Y(s) - \frac{3}{2} + 2\,Y(s) = 0$$

which gives the same result for $y(t)$ as before. This alternative approach is not recommended because of the difficulty of finding the necessary initial conditions.

Systems Containing No Initial Stored Energy

The assumption of no initial stored energy within the system is one that is made throughout much of this book. It is not necessary in such cases to write differential equations describing the system first and then transform them. Consider the typical electrical elements shown in Table 2.6-1. When the element has no stored energy, the ratio $E(s)/I(s) = Z(s)$ is the simple algebraic quantity given in the last column. Each element can therefore be characterized by an impedance $Z(s)$, which is not a function of voltage, current, or time. Thus, when elements are interconnected, the rules for d-c and a-c steady-state analysis carry over to Laplace transform analysis. The relationship between the transformed output and input can be found by the solution of algebraic equations.

Example 3.5-3. Find an expression for $e_2(t)$ for $t > 0$ in the circuit of Fig. 3.5-1, assuming that there is no initial stored energy.

The circuit is redrawn in Fig. 3.5-2, each element being labeled with its impedance, and transformed voltages and currents being used. Any method from d-c circuit theory

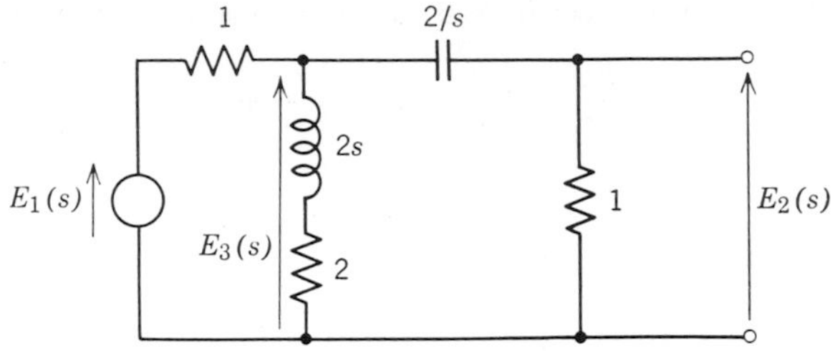

Fig. 3.5-2

may be applied. Perhaps the easiest is to write a single node equation at node 3.

$$E_3(s)\left[1 + \frac{1}{2s + 2} + \frac{1}{1 + 2/s}\right] = E_1(s)$$

or

$$E_3(s) = \frac{(2s + 2)(s + 2)}{4s^2 + 9s + 6} E_1(s)$$

Then

$$E_2(s) = \frac{1}{1 + 2/s} E_3(s)$$

or

$$e_2(t) = \mathscr{L}^{-1}\left[\frac{2s^2 + 2s}{4s^2 + 9s + 6} E_1(s)\right]$$

agreeing with the result of Example 3.5-1.

This algebraic method of solving for the transform of the output may be extended to systems with initial stored energy. Any initial energy may be represented by added sources, as discussed in Section 1.4. Figure 1.4-6 shows how this may be done for energy stored in a capacitance or inductance. When this has been done, each passive element may again be characterized by its impedance.

The System Function $H(s)$

There is a direct relationship between the input $v(t)$ and output $y(t)$ for a linear system with no initial stored energy. The system function (or "transfer function") of a fixed system is defined as the ratio of the transformed output to the transformed input.

$$H(s) = \frac{Y(s)}{V(s)} \tag{3.5-1}$$

or

$$y(t) = \mathscr{L}^{-1}[V(s)H(s)] \tag{3.5-2}$$

assuming that there is no initial stored energy. A review of Section 2.6 indicates that the $Z(s)$ and $H(s)$ of that section are the same impedance and system functions defined in this section. Recall that $H(s)\epsilon^{st}$ is the forced or particular integral response to the input ϵ^{st}. From Eqs. 2.6-21 and 2.6-22,

$$H(s) = \frac{B(s)}{A(s)} = \left[\frac{y_P(t)}{v(t)}\right]_{v(t)=\epsilon^{st}} \tag{3.5-3}$$

where $A(s)$ and $B(s)$ are formed from the operators $A(p)$ and $B(p)$ in the differential equation

$$A(p)y(t) = B(p)v(t) \tag{3.5-4}$$

Thus $H(s)$ can be written down by inspection of the system's differential equation. For the system of Example 3.5-2,

$$H(s) = \frac{\frac{1}{2}s + 1}{s^2 + 3s + 2} = \frac{\frac{1}{2}}{s + 1}$$

which the reader will recognize as the transform of the impulse response.

The use of Eq. 3.5-2 constitutes the same basic three-step procedure used for the complex Fourier series and the Fourier integral. The input function is transformed (this time into the complex frequency domain) and is then multiplied by the system function. The resulting transformed output is then converted back to the time domain. In fact, when s is replaced by $j\omega$ in the system function, $H(j\omega)$ is the same system function that was used with the Fourier series and integral. From Eq. 2.6-23,

$$H(j\omega) = \left[\frac{y_P(t)}{v(t)}\right]_{v(t)=\epsilon^{j\omega t}} = |H(j\omega)|\,\epsilon^{j\theta} \tag{3.5-5}$$

The forced response to $v(t) = \cos \omega t$ is $y_P(t) = |H(j\omega)| \cos (\omega t + \theta)$.

Another useful interpretation of the system function results when Eq. 3.5-2 is used for the special case of $v(t) = U_0(t)$, $y(t) = h(t)$. Since $\mathscr{L}[U_0(t)] = 1$,

$$h(t) = \mathscr{L}^{-1}[H(s)] \tag{3.5-6}$$

Thus the inverse transform of the system function is the impulse response. This is one reason for using the symbol $h(t)$ for the impulse response. Similarly, using Eq. 3.5-2 with $v(t) = U_{-1}(t)$ and $y(t) = r(t)$,

$$r(t) = \mathscr{L}^{-1}\left[\frac{H(s)}{s}\right] \tag{3.5-7}$$

Example 3.5-4. Solve Example 1.6-3 by use of the Laplace transform. The example considers n identical amplifiers in cascade, as in Fig. 3.5-3, each having the step response $r(t) = -A\epsilon^{-t/RC}$ for $t \geqslant 0$.

Since

$$\mathscr{L}[r(t)] = \frac{-A}{s + 1/RC}$$

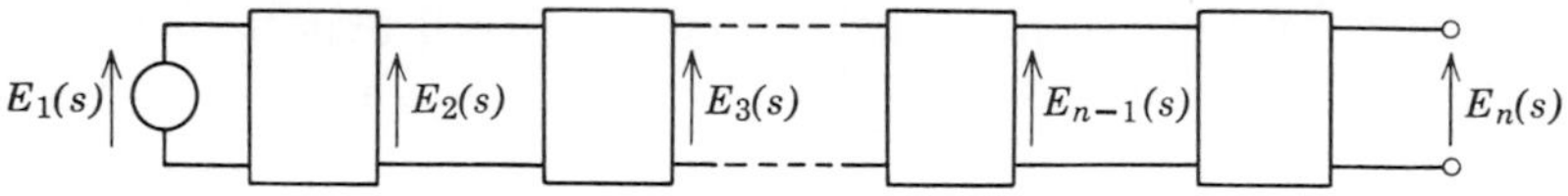

Fig. 3.5-3

the system function for each amplifier is

$$-\frac{As}{s + 1/RC}$$

Assuming that the individual amplifiers are not loaded down by the presence of succeeding stages, the system function for the combination is

$$\frac{E_n(s)}{E_1(s)} = \frac{E_2(s)}{E_1(s)}\frac{E_3(s)}{E_2(s)}\cdots\frac{E_n(s)}{E_{n-1}(s)} = \left[\frac{-As}{s + 1/RC}\right]^n$$

The step response of the combination is

$$\mathscr{L}^{-1}\left[\frac{1}{s}\left(\frac{-As}{s + 1/RC}\right)^n\right]$$

The inverse transform is[6]

$$(-A)^n\epsilon^{-t/RC}\left[1 + \sum_{k=1}^{n-1}\frac{(n-1)!\,(-t/RC)^k}{(n-k-1)!\,(k!)^2}\right]$$

agreeing with the result given in Example 1.6-3.

The Convolution Integrals

By Eqs. 3.5-2, 3.4-13, and 3.5-6,

$$y(t) = \mathscr{L}^{-1}[V(s)H(s)] = \int_0^t v(\lambda)h(t - \lambda)\,d\lambda$$
$$= \int_0^t v(t - \lambda)h(\lambda)\,d\lambda \qquad (3.5\text{-}8)$$

This result is identical with the convolution integrals of Eqs. 1.6-2 and 1.6-4 for nonanticipatory systems with $v(t) = 0$ for $t < 0$. Also,

$$y(t) = \mathscr{L}^{-1}\left[sV(s)\frac{H(s)}{s}\right]$$

The inverse transform of the last factor in the brackets is $r(t)$ by Eq. 3.5-7. By Eq. 3.4-4,

$$\mathscr{L}^{-1}[sV(s)] = \frac{dv}{dt}$$

if $v(0+) = 0$. Using Eq. 3.4-13,

$$y(t) = \int_0^t \frac{dv(\lambda)}{d\lambda}r(t - \lambda)\,d\lambda = \int_0^t \frac{dv(t - \lambda)}{d(t - \lambda)}r(\lambda)\,d\lambda$$

If $v(0+) \neq 0$, the response to the step function $v(0+)U_{-1}(t)$ must be added to the result above, giving

$$y(t) = v(0+)r(t) + \int_0^t \frac{dv(\lambda)}{d\lambda} r(t - \lambda)\, d\lambda$$

$$= v(0+)r(t) + \int_0^t \frac{dv(t - \lambda)}{d(t - \lambda)} r(\lambda)\, d\lambda \tag{3.5-9}$$

Since $v(t)$ is regarded as zero for $t < 0$ in the one-sided transform, $v(0+)$ is the size of a discontinuity at the origin. Then $dv(\lambda)/d\lambda$ will have an impulse of value $v(0+)$ at $\lambda = 0$. To be consistent with the definition of the one-sided transform in Eq. 3.4-1, however, Eq. 3.5-9 should more properly be written

$$y(t) = v(0+)r(t) + \int_{0+}^t \frac{dv(\lambda)}{d\lambda} r(t - \lambda)\, d\lambda$$

so that such an impulse does not contribute to the integral. Equation 3.5-9 is equivalent to the convolution integrals of Eqs. 1.6-3 and 1.6-5 for nonanticipatory systems with $v(t) = 0$ for $t < 0$. It must be remembered that an impulse in $dv(t)/dt$ at the origin is to be included in the integrand of Eqs. 1.6-3 and 1.6-5, but not in the integrand of Eq. 3.5-9. If the reader is troubled by impulses in $dv(t)/dt$, he may wish to review the discussion associated with Eq. 1.5-9.

3.6 REVIEW OF COMPLEX VARIABLE THEORY

There are many readable textbooks devoted to the subject of complex variables.[7] This section presents, without detailed proofs, only the barest skeleton of the part of the theory needed for an understanding of transform techniques. Readers who have not previously been exposed to complex variable theory will probably need to consult one of the references.

Functions of a Complex Variable

Since the theory is applied to the Laplace transform, the independent variable is taken to be

$$s = \sigma + j\omega \tag{3.6-1}$$

where σ and ω are the real and imaginary parts of s, respectively. As for any complex quantity, values of s can be represented graphically in a complex plane, shown in Fig. 3.6-1. It is sometimes convenient to think of an "extended s-plane" as the surface of a sphere of infinite radius, with the

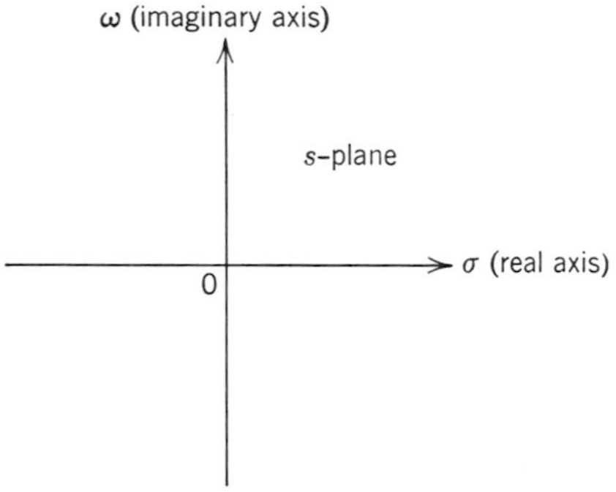

Fig. 3.6-1

point at infinity diametrically opposite the origin, but such an interpretation is not used in this book.

The dependent variable is denoted by

$$F(s) = R + jX \tag{3.6-2}$$

R and X are the real and imaginary parts, respectively, of $F(s)$, and they are real functions of the two real variables σ and ω.

Analytic Functions

The derivative of $F(s)$ is defined as

$$\frac{dF(s)}{ds} = \lim_{\Delta s \to 0} \frac{\Delta F(s)}{\Delta s} = \lim_{\Delta s \to 0} \frac{F(s + \Delta s) - F(s)}{\Delta s} \tag{3.6-3}$$

Δs represents a small distance in the s-plane and may be taken in any arbitrary direction. The usefulness of the derivative concept is greatly increased if the derivative turns out to be independent of the direction of Δs. Equation 3.6-3 is independent of the direction of Δs at a point s_0, if and only if

$$\frac{\partial R}{\partial \sigma} = \frac{\partial X}{\partial \omega} \tag{3.6-4}$$

and

$$\frac{\partial R}{\partial \omega} = -\frac{\partial X}{\partial \sigma} \tag{3.6-5}$$

and if $F(s)$ and these partial derivatives are continuous in some neighborhood of s_0. The last two equations are known as the Cauchy-Riemann conditions for a unique derivative.

A function $F(s)$ is said to be analytic *at a point* in the s-plane if and only if it is single-valued and has a finite and unique derivative at and in the

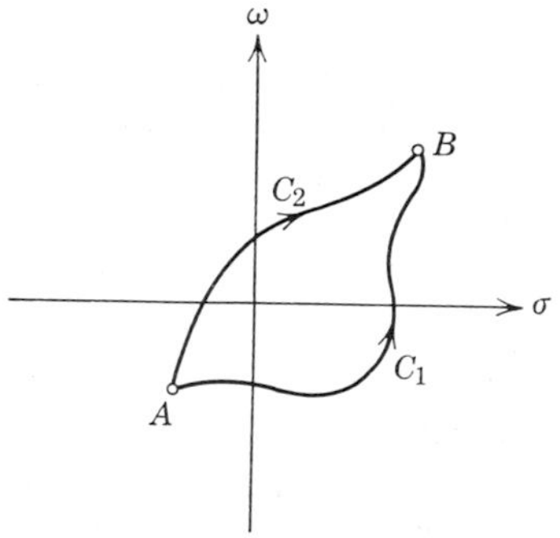

Fig. 3.6-2

neighborhood of that point.† The vast majority of the commonly encountered functions are analytic except at a finite number of points. The quotient of two polynomials in s, for example, is analytic for all finite values of s except where the denominator becomes zero. All the usual rules of differentiation hold for analytic functions.

Integration in the Complex Plane

Integration along a contour C in the s-plane (called a contour integral) can be interpreted in terms of real integrals by writing

$$\int_C F(s)\,ds = \int_C (R + jX)\,(d\sigma + j\,d\omega)$$
$$= \int_C (R\,d\sigma - X\,d\omega) + j\int_C (X\,d\sigma + R\,d\omega)$$

The values of σ and ω are related by the contour along which the integration is carried out.

A contour integral of $F(s)$ between two fixed points in the s-plane is not, in general, independent of the path taken. In Fig. 3.6-2, $\int_{C_1} F(s)\,ds$ and $\int_{C_2} F(s)\,ds$ are not necessarily equal. If the path of integration is a closed contour, this fact is indicated by a circle superimposed upon the integral sign. An arrow is sometimes shown on the circle to indicate that the integration is to be carried out in a specific direction, either clockwise or counterclockwise. The sign of the integral is positive if the direction of the integration is such that the area enclosed lies to the left of the contour.

† The phrase "analytic at and in the neighborhood of a point in the s-plane" is more correct mathematically, but it is also more cumbersome.

In Fig. 3.6-2,

$$\oint F(s)\,ds = \int_{C_1} F(s)\,ds - \int_{C_2} F(s)\,ds$$

The sign of the C_2 integral is negative, since reversing the direction of integration along a path reverses the sign of the contour integral. Since the integrals along contours C_1 and C_2 are not necessarily equal, the integral around a closed path is not necessarily zero.

Two theorems attributed to Cauchy lay the foundation for the practical evaluation of contour integrals. The first theorem states that, if $F(s)$ is analytic on and inside a closed contour C in the s-plane,

$$\oint_C F(s)\,ds = 0 \tag{3.6-6}$$

A corollary to this theorem is that

$$\int_{C_1} F(s)\,ds = \int_{C_2} F(s)\,ds \tag{3.6-7}$$

in Fig. 3.6-2, if $F(s)$ is analytic on and between the two paths. Another corollary deals with multiple connected regions. In Fig. 3.6-3*a*, $F(s)$ is assumed to be analytic on the closed contours C, C_1, and C_2, and in the shaded region R. It is not, however, necessarily analytic inside C_1 and C_2. In Fig. 3.6-3*b*, cuts labeled C_3 through C_6 are constructed so as to give a simply connected region. The distance between corresponding ends of C_3 and C_4, and also C_5 and C_6, is assumed to be infinitesimal. By Cauchy's theorem,

$$\begin{aligned}\int_C F(s)\,ds + \int_{C_3} F(s)\,ds - \int_{C_1} F(s)\,ds \\ + \int_{C_4} F(s)\,ds + \int_{C_5} F(s)\,ds \\ - \int_{C_2} F(s)\,ds + \int_{C_6} F(s)\,ds = 0\end{aligned}$$

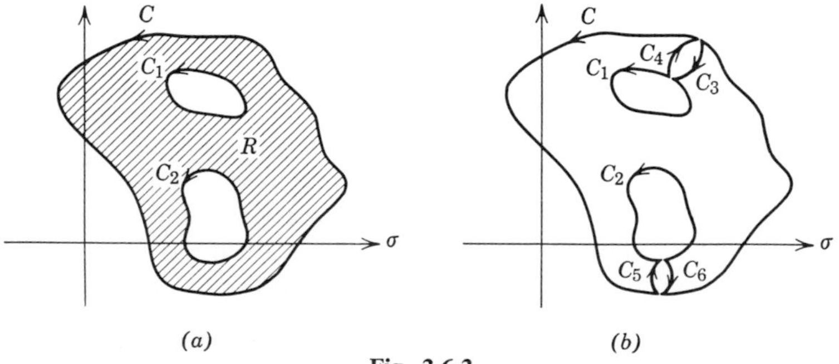

Fig. 3.6-3

since $F(s)$ is analytic on and inside the closed contour formed by the sum of these individual contours. By Eq. 3.6-7, the integration along C_3 and C_4, and also along C_5 and C_6, cancel, so

$$\oint_C F(s)\,ds = \oint_{C_1} F(s)\,ds + \oint_{C_2} F(s)\,ds \tag{3.6-8}$$

A second theorem by Cauchy states that, if $F(s)$ is analytic on and inside a closed contour C, and if s_0 is any point inside C,

$$F(s_0) = \frac{1}{2\pi j}\oint_C \frac{F(s)}{s - s_0}\,ds \tag{3.6-9}$$

and

$$\left.\frac{d^nF}{ds^n}\right|_{s=s_0} = \frac{n!}{2\pi j}\oint_C \frac{F(s)}{(s - s_0)^{n+1}}\,ds \tag{3.6-10}$$

Since the integrands are not analytic at the point s_0, these equations can be used to evaluate integrals that cannot be handled by Eq. 3.6-6. Their primary importance, however, is in the derivation of the Taylor and Laurent series and the residue theorem. Note also that Eq. 3.6-9 leads to the surprising conclusion that, if a function $F(s)$ is defined on a closed contour and is analytic on and inside that contour, then the value of the function is automatically determined at all interior points.

Taylor's Series

If $F(s)$ is analytic at a point s_0, then it can be represented by the following infinite power series in the vicinity of s_0.

$$F(s) = b_0 + b_1(s - s_0) + b_2(s - s_0)^2 + \cdots + b_n(s - s_0)^n + \cdots \tag{3.6-11}$$

where

$$b_0 = F(s_0)$$
$$b_n = \left(\frac{1}{n!}\right)\left.\frac{d^nF(s)}{ds^n}\right|_{s=s_0} \tag{3.6-12}$$

The form of Eq. 3.6-11 is seen to be the same as the Taylor series for a function of a real variable. If s_0 happens to be zero, the series is called Maclaurin's series.

Taylor's series converges to $F(s)$ for all points in the s-plane inside a *circle of convergence.* This is the largest circle that can be drawn about s_0 without enclosing any points at which $F(s)$ is not analytic. For all points outside the circle of convergence, the series diverges.

The coefficients in Taylor's series can always be evaluated by Eqs. 3.6-12, since all the derivatives of an analytic function exist. If $F(s)$

and the first $k - 1$ derivatives are zero at $s = s_0$, the first k terms in Eq. 3.6-11 are missing, and $F(s)$ is said to have a *zero* of order k at s_0.

A uniqueness theorem states that, if two power series represent $F(s)$ in the neighborhood of s_0, then they must be identical. The coefficients may therefore be found in any convenient way, and not necessarily from Eq. 3.6-12. In the examples which follow, long division is used.

The uniqueness theorem of the previous paragraph leads to other useful results. If a series representation of $F(s)$ is valid in any arbitrarily small region, it is the unique representation wherever it converges. Also, if an analytic function is specified throughout any arbitrarily small region, it is then uniquely determined throughout the entire s-plane. The last statement is known as the principle of *analytic continuation.*

Example 3.6-1. Expand $F(s) = 1/(s + 1)$ in a power series about $s = 1$, i.e., in powers of $(s - 1)$.

By long division,

$$\frac{1}{s+1} = \frac{1}{2 + (s-1)} = \frac{1}{2} - \frac{1}{4}(s-1) + \frac{1}{8}(s-1)^2 - \frac{1}{16}(s-1)^3 + \cdots$$

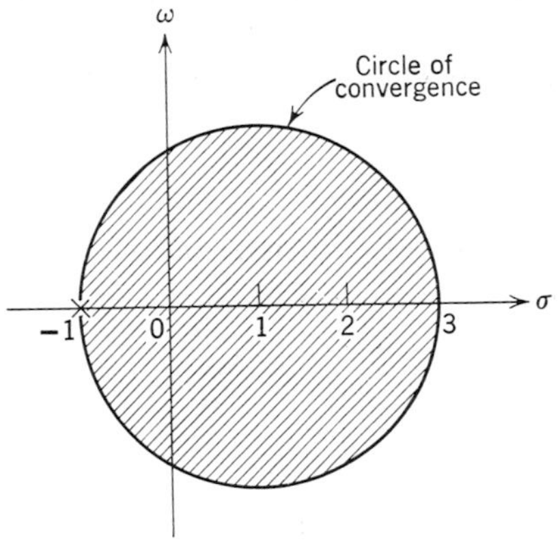

Fig. 3.6-4

The same series results if Eqs. 3.6-11 and 3.6-12 are used with $s_0 = 1$. The function $F(s)$ is analytic except at the point $s = -1$, shown by the cross in Fig. 3.6-4. The circle of convergence indicates that the series converges for $|s - 1| < 2$.

Example 3.6-2. Expand $F(s) = s/[(s - 1)(s - 3)]$ about $s = 1$.

Although $F(s)$ is not analytic at $s = 1$, the function $s/(s - 3)$ is analytic and may be expanded in a Taylor series. By long division,

$$\frac{s}{s-3} = \frac{1 + (s-1)}{-2 + (s-1)} = -\frac{1}{2} - \frac{3}{4}(s-1) - \frac{3}{8}(s-1)^2 - \cdots$$

which converges for $|s - 1| < 2$. Dividing each term of the series by $(s - 1)$,

$$\frac{s}{(s-1)(s-3)} = \frac{-\frac{1}{2}}{s-1} - \frac{3}{4} - \frac{3}{8}(s-1) - \cdots$$

which converges for $0 < |s - 1| < 2$.

Laurent's Series

Even if $F(s)$ is not analytic at the point s_0, it can be represented by an infinite series in powers of $(s - s_0)$. In this case, however, the series contains both positive and negative powers of $(s - s_0)$.

$$F(s) = b_0 + b_1(s - s_0) + b_2(s - s_0)^2 + \cdots + b_{-1}(s - s_0)^{-1} + b_{-2}(s - s_0)^{-2} + \cdots \quad (3.6\text{-}13)$$

where

$$b_k = \frac{1}{2\pi j}\oint_C \frac{F(s)\, ds}{(s - s_0)^{k+1}} \quad \text{for all } k \quad (3.6\text{-}14)$$

The first part of the series is known as the ascending part; the second part, the *principal* or descending part. The series converges between two circles of convergence, both centered at s_0. $F(s)$ must be analytic between these circles, which are labelled C_1 and C_2 in Fig. 3.6-5. If there is an "isolated singularity" of $F(s)$ at s_0, the circle C_2 may shrink to infinitesimal size, and the Laurent series then represents $F(s)$ in the vicinity of s_0. As discussed later in this section, the coefficient b_{-1} is particularly important and is called the *residue* of $F(s)$ at s_0.

In the use of Eq. 3.6-14, the contour C may be any closed contour between C_1 and C_2, shown in Fig. 3.6-5. Since this equation is difficult to evaluate, however, the coefficients in Laurent's series are normally

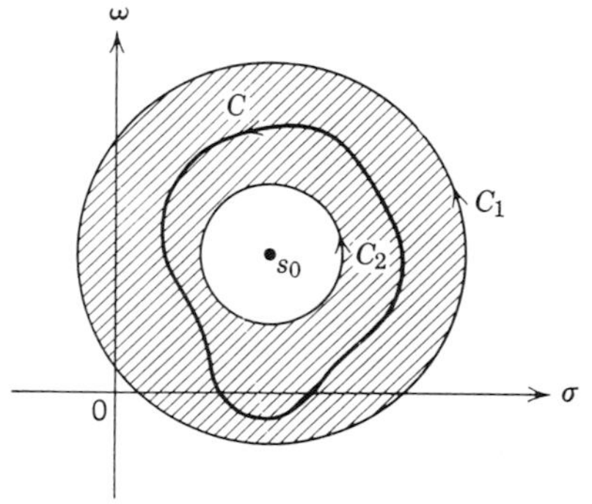

Fig. 3.6-5

found by some other means. Any convenient method may be used, because the representation of $F(s)$ by a series in powers of $(s - s_0)$ in any given region is unique. Example 3.6-2 is an example of a Laurent series, where C_2 is arbitrarily small, and C_1 is a circle of radius just under 2.

Classification of Singularities

Singularities of $F(s)$, also called singular points, are points in the s-plane at which $F(s)$ is not analytic. If nonoverlapping circles, no matter how small, can be drawn around each singular point, the points are called isolated singularities. The function $F(s) = (s + 1)/[s^3(s^2 + 1)]$ has isolated singularities at $s = 0$, $+j$, and $-j$. The function $F(s) = 1/[\sin(\pi/s)]$ has isolated singularities at $s = 1, \frac{1}{2}, \frac{1}{3}, \ldots$, but a nonisolated singularity at the origin. Fortunately, the commonly encountered functions have only isolated singularities. The reader should remember that $F(s)$ can be represented by a Laurent series in the vicinity of every isolated singularity.

An isolated singularity is classified further by examining the Laurent series written about it. If the principal or descending part of the series has an infinite number of terms, the singularity is called an *essential singularity*. Otherwise, the singularity is called a *pole*. The order of the pole is equal to the number of terms in the principal part of the series. If $F(s)$ has a pole of order n at s_0, then there are no terms in the principal part of the power series for $(s - s_0)^n F(s)$, so that $(s - s_0)^n F(s)$ is analytic at s_0.

The Residue Theorem

Letting $k = -1$ in Eq. 3.6-14 gives

$$\oint_C F(s)\, ds = 2\pi j b_{-1} \tag{3.6-15}$$

where b_{-1} is the residue of $F(s)$ at s_0. The contour of integration is shown in Fig. 3.6-5 and is understood to enclose no singularities other than an isolated singularity at s_0. If s_0 is not a singular point, then b_{-1} is zero, and this equation reduces to Eq. 3.6-6. In Example 3.6-2, if C is a circle of radius 2 about the origin,

$$\oint_C F(s)\, ds = 2\pi j(-\tfrac{1}{2}) = -\pi j$$

Suppose the contour of integration encloses several isolated singularities. Since nonoverlapping circles can be drawn about each of them, the integral along contour C can be replaced by the sum of the integrals around the individual circles, as in Eq. 3.6-8 and Fig. 3.6-3. Applying Eq. 3.6-15 to each individual circle,

$$\oint_C F(s)\,ds = 2\pi j \sum \text{residues} \tag{3.6-16}$$

where the summation includes the residues at all the singularities inside C.

The Evaluation of Residues

Because Equation 3.6-16 is very valuable in finding the inverse Laplace transform, it is important to be able to calculate residues quickly and easily. Residues can always be found by writing a Laurent series about each singularity, and selecting the coefficient of the $(s - s_0)^{-1}$ term. Several special formulas are useful, however, and are derived next.

If $F(s)$ has a pole of order n at $s = s_0$, the function $g(s) = (s - s_0)^n F(s)$ is analytic at s_0 and may be expanded in a Taylor series about s_0.

$$g(s) = g(s_0) + g'(s_0)(s - s_0) + \cdots + \left[\frac{g^{(n-1)}(s_0)}{(n-1)!}\right](s - s_0)^{n-1} + \left[\frac{g^{(n)}(s_0)}{n!}\right](s - s_0)^n + \cdots$$

where the primes and superscripts in parentheses denote differentiation with respect to s. Then

$$F(s) = \frac{g(s)}{(s - s_0)^n} = g(s_0)(s - s_0)^{-n} + g'(s_0)(s - s_0)^{-n+1} + \cdots + \left[\frac{g^{(n-1)}(s_0)}{(n-1)!}\right](s - s_0)^{-1} + \left[\frac{g^{(n)}(s_0)}{n!}\right] + \cdots$$

The last expression is the Laurent series for $F(s)$, so the residue at s_0 is

$$b_{-1} = \frac{1}{(n-1)!}\left[\frac{d^{n-1}}{ds^{n-1}}(s - s_0)^n F(s)\right]_{s=s_0} \tag{3.6-17}$$

For a first order pole, where $n = 1$, this reduces to

$$b_{-1} = [(s - s_0)F(s)]_{s=s_0} \tag{3.6-18}$$

The next approach is valid only when $F(s)$ can be expressed as

$$F(s) = \frac{P(s)}{Q(s)} \tag{3.6-19}$$

where both $P(s)$ and $Q(s)$ are analytic at s_0, and where $P(s_0) \neq 0$. If $F(s)$ has a first order pole at s_0, then $Q(s_0) = 0$, but $Q'(s_0) \neq 0$. Writing a

Taylor series for both $P(s)$ and $Q(s)$ and carrying out the indicated long division,

$$F(s) = \frac{P(s_0) + P'(s_0)(s - s_0) + \cdots}{Q'(s_0)(s - s_0) + \dfrac{1}{2!}Q''(s_0)(s - s_0)^2 + \cdots}$$

$$= \frac{P(s_0)}{Q'(s_0)}(s - s_0)^{-1} + \cdots$$

The residue at s_0 is

$$b_{-1} = \frac{P(s_0)}{Q'(s_0)} \tag{3.6-20}$$

This approach becomes cumbersome when applied to higher order poles.

3.7 THE INVERSION INTEGRAL

When the Laplace transform $F(s)$ is known, the function of time can be found by Eqs. 3.3-2 and 3.3-4, rewritten below.

$$f(t) = \frac{1}{2\pi j} \lim_{R\to\infty} \int_{c-jR}^{c+jR} F(s)\epsilon^{st}\, ds \tag{3.7-1}$$

As this equation applies equally well to both the two-sided and the one-sided transform, it is convenient to treat them together.

Recall from Section 3.3 that the defining equation for the direct transform $F(s)$ converges only for certain values of σ, i.e., only within a certain region of the s-plane. Using the principle of analytic continuation, however, this is sufficient to uniquely define $F(s)$ throughout the entire s-plane, except at the singular points. Since the factor ϵ^{st} is analytic throughout the entire finite s-plane, the function $F(s)\epsilon^{st}$ can be integrated without difficulty along any path that does not include any singularities of $F(s)$. If the concept of an extended s-plane mapped on the surface of an infinite sphere is used, the point at infinity must be avoided, since ϵ^{st} has an essential singularity there. In the application of Eq. 3.7-1, a semicircular detour could be made around the point at infinity. An easier interpretation of Eq. 3.7-1, however, is to consider only the finite s-plane, and to apply the limiting process of $R \to \infty$ after the integration has been carried out. This is the procedure that is followed here.

In Section 3.3, it was indicated that the path of integration in Eq. 3.7-1 is restricted to values of σ for which the direct transform formula converges. In the case of the two-sided Laplace transform, the region of convergence must be specified in order to uniquely determine the inverse

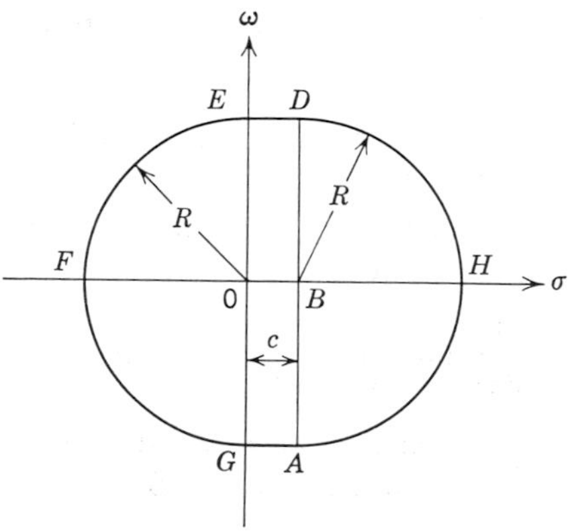

Fig. 3.7-1

transform. In the two-sided transform, the regions of convergence for functions of time that are zero for $t < 0$, or zero for $t > 0$, or that fall into neither category, have the form $\sigma > \sigma_c$, $\sigma < \sigma_c$, or $\sigma_{c_1} < \sigma < \sigma_{c_2}$, respectively. In the one-sided transform, the region of convergence is always given by $\sigma > \sigma_c$, where σ_c is called the abscissa of absolute convergence.

From the point of view of rigorous mathematics, Eqs. 3.3-1 and 3.3-3 should be taken as the formal definition of the Laplace transform. The correctness of Eq. 3.7-1, and the above restriction on the path of integration should then be proved. The proof, in essence, shows that, if Eqs. 3.3-1 and 3.7-1 are applied consecutively to a function $f(t)$, the end result is the same function $f(t)$ at all points of continuity.[8] For a discontinuity at the origin, the result is $\frac{1}{2}[f(0+) + f(0-)]$. In the one-sided transform, the consecutive application of Eqs. 3.3-3 and 3.7-1 yields $f(t)$ for $t > 0$, zero for $t < 0$, and $\frac{1}{2}f(0+)$ for $t = 0$.

The path of integration in Eq. 3.7-1 is usually taken as a straight vertical line, denoted by ABD in Fig. 3.7-1. EFG and DHA are semicircles, whose radius will later become infinite. Since the direct integration of $F(s)\epsilon^{st}$ along ABD is difficult, another procedure is normally used. For most commonly encountered functions,

$$\lim_{R\to\infty} \int_{DEFGA} F(s)\epsilon^{st}\, ds = 0$$

for $t > 0$, and

$$\lim_{R\to\infty} \int_{DHA} F(s)\epsilon^{st}\, ds = 0$$

for $t < 0$. In this event, path ABD may be changed to $ABDEFGA$ for

$t > 0$, and to $ABDHA$ for $t < 0$. Using the residue theorem of Eq. 3.6-16,

$$f(t) = \frac{1}{2\pi j} \lim_{R\to\infty} \oint_{ABDEFGA} F(s)\epsilon^{st}\, ds$$

$$= \sum \begin{matrix}\text{[residues of } F(s)\epsilon^{st} \text{ at the singularities} \\ \text{to the left of } ABD] \quad \text{for } t > 0\end{matrix} \tag{3.7-2}$$

$$f(t) = \frac{1}{2\pi j} \lim_{R\to\infty} \oint_{ABDHA} F(s)\epsilon^{st}\, ds$$

$$= -\sum \begin{matrix}\text{[residues of } F(s)\epsilon^{st} \text{ at the singularities} \\ \text{to the right of } ABD] \quad \text{for } t < 0\end{matrix} \tag{3.7-3}$$

It is necessary to know when Eqs. 3.7-2 and 3.7-3 may be substituted for Eq. 3.7-1. The following theorem helps to answer this question.[9]

Theorem 3.7-1. If s is replaced by the real quantity R, and if $\lim_{R\to\infty} F(R)$ approaches zero at least as fast as N/R, where N is a finite number, then Eqs. 3.7-2 and 3.7-3 are valid.† Note that this theorem includes the special case of the quotient of two polynomials, provided that the order of the denominator exceeds the order of the numerator.

In using Eqs. 3.7-2 and 3.7-3, the residue formulas developed in Section 3.6 prove helpful. For a pole of order n at s_0, the residue of $F(s)\epsilon^{st}$ is

$$\frac{1}{(n-1)!}\left[\frac{d^{n-1}}{ds^{n-1}}(s - s_0)^n F(s)\epsilon^{st}\right]_{s=s_0} \tag{3.7-4}$$

For a first order pole at s_0, the residue of $F(s)\epsilon^{st}$ is given by

$$[(s - s_0)F(s)\epsilon^{st}]_{s=s_0} \tag{3.7-5}$$

or

$$\left[\frac{P(s)\epsilon^{st}}{(d/ds)Q(s)}\right]_{s=s_0} \tag{3.7-6}$$

where $F(s) = P(s)/Q(s)$, and $P(s)$ and $Q(s)$ are analytic at s_0.

Example 3.7-1. $F(s) = -1/[s^2(s - 1)]$. The region of convergence is given by $0 < \sigma < 1$ and is shaded in Fig. 3.7-2. Find $f(t)$.

† The theorem as stated is really unnecessarily restrictive. For example, Eqs. 3.7-2 and 3.7-3 are valid if $F(s)$ is any meromorphic function (the ratio of two functions that are analytic throughout the finite s-plane) that approaches zero uniformly as $|s| \to \infty$. If there is ever any doubt, Eqs. 3.7-2 and 3.7-3 may be used formally to find a function of time. If the transform of this answer is found to be the original $F(s)$, then the validity of the answer is established.

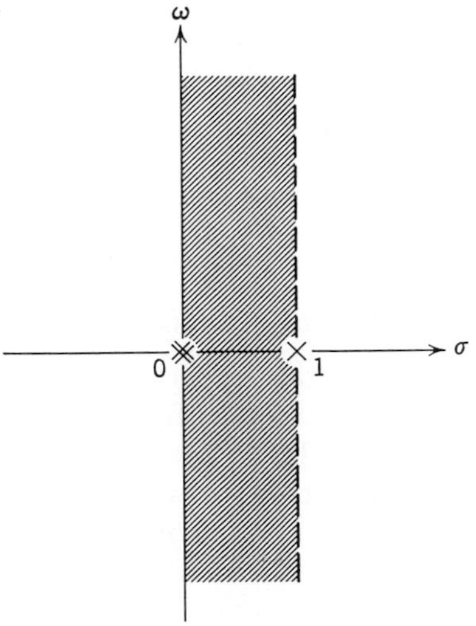

Fig. 3.7-2

Since Theorem 3.7-1 is satisfied, Eqs. 3.7-2 and 3.7-3 may be used. For $t > 0$,

$$f(t) = [\text{residue of } F(s)\epsilon^{st} \quad \text{at } s = 0]$$

$$= \left[\frac{d}{ds}\left(\frac{-\epsilon^{st}}{s-1}\right)\right]_{s=0} = t + 1$$

For $t < 0$,

$$f(t) = -[\text{residue of } F(s)\epsilon^{st} \quad \text{at } s = 1]$$

$$= \left[\frac{\epsilon^{st}}{s^2}\right]_{s=1} = \epsilon^t$$

The complete function is shown in Fig. 3.7-3. The region of convergence in this example indicates that the two-sided Laplace transform must have been used in obtaining $F(s)$. Poles to the left and right of this region yield, respectively, the components of $f(t)$ for positive and negative time.

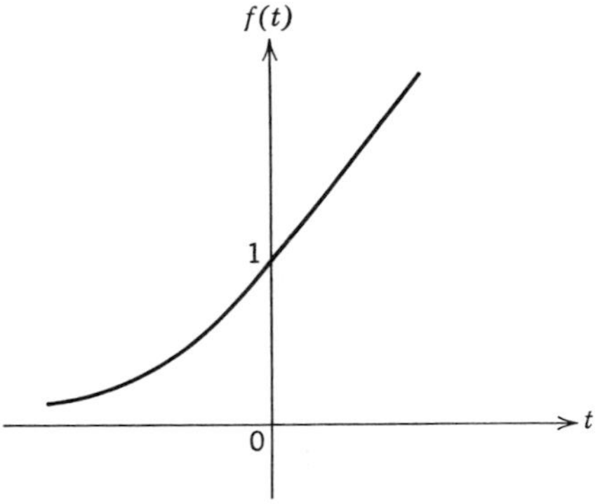

Fig. 3.7-3

Example 3.7-2. $F(s) = (s + a)/[(s + a)^2 + \beta^2] = (s + a)/[(s + a - j\beta)(s + a + j\beta)]$. If the region of convergence is given by $\sigma > -a$, find $f(t)$.

Equations 3.7-2 and 3.7-3 may again be used. For $t > 0$,

$$\begin{aligned} f(t) &= \sum [\text{residues of } F(s)\epsilon^{st} \quad \text{at } s = -a \pm j\beta] \\ &= \left[\frac{(s + a)\epsilon^{st}}{s + a + j\beta}\right]_{s=-a+j\beta} + \left[\frac{(s + a)\epsilon^{st}}{s + a - j\beta}\right]_{s=-a-j\beta} \\ &= \tfrac{1}{2}\epsilon^{(-a+j\beta)t} + \tfrac{1}{2}\epsilon^{(-a-j\beta)t} = \epsilon^{-at} \cos \beta t \end{aligned}$$

For $t < 0, f(t) = 0$, since there are no singularities inside path *ABDHA* in Fig. 3.7-1.

When the one-sided Laplace transform is used, the region of convergence is given by $\sigma > \sigma_c$, as can be expected from Eq. 3.3-1. The abscissa of absolute convergence σ_c must be to the right of all the poles of $F(s)$. Otherwise, the region of convergence would include a singular point, at which $F(s)$ certainly does not exist. For this reason, a statement of the region of convergence is normally omitted in the one-sided transform. Also for this reason, the inversion integral always yields $f(t) = 0$ for $t < 0$. The rest of this section is restricted to the one-sided transform.

For the case where $F(s)$ is the quotient of two polynomials, with the numerator of lower order than the denominator, the residue and partial fraction methods are essentially identical. Although the partial fraction method is restricted to rational functions of s, the more powerful residue method is not. Several typical examples involving nonrational functions are now considered.

Example 3.7-3. Figure 3.7-4 shows an *RC* filter consisting of m identical T-sections in cascade, excited by a unit step of voltage. Find the current $i_r(t)$ at any point on the filter ($r = 0, 1, 2, \ldots, m$). Assume that there is no initial stored energy.

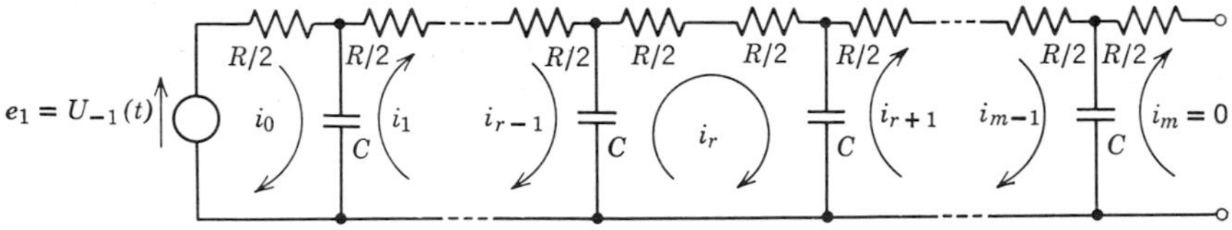

Fig. 3.7-4

Characterizing each element by its impedance, the transformed equation for the rth mesh is

$$-\frac{1}{sC} I_{r-1}(s) + \left(R + \frac{2}{sC}\right) I_r(s) - \frac{1}{sC} I_{r+1}(s) = 0$$

This is a second order, algebraic difference equation. As discussed in Chapter 2, two boundary conditions are needed. These can be obtained by examining the first and last meshes.

$$\left(\frac{R}{2} + \frac{1}{sC}\right) I_0(s) - \frac{1}{sC} I_1(s) = \frac{1}{s}$$

$$I_m(s) = 0$$

By the method of Example 2.12-1, the solution of the difference equation, subject to the boundary conditions above, is

$$I_r(s) = \frac{C \sinh [(m - r)\theta]}{(\sinh \theta)(\cosh m\theta)}$$

where $\cosh \theta = 1 + (RC/2)s$. The hyperbolic functions are analytic for all finite values of the complex variable θ. Whenever $\sinh \theta = 0$, $[\sinh (m - r)\theta]/\sinh \theta$ is finite, so the only poles of $I_r(s)$ occur when $\cosh m\theta = 0$. Letting $\theta = \alpha + j\beta$,

$$\cosh m\theta = \cosh m\alpha \cos m\beta + j \sinh m\alpha \sin m\beta$$

This expression is zero only when $m\beta = \pm(2k - 1)(\pi/2)$ and $m\alpha = 0$, where $k = 1, 2, 3, \ldots$. The poles are at

$$\theta = \pm j \frac{(2k - 1)\pi}{2m}$$

$$\cosh \theta = \cos \frac{(2k - 1)\pi}{2m}$$

$$s = \frac{2}{RC}\left[-1 + \cos \frac{(2k - 1)\pi}{2m}\right]$$

The $\pm$ sign is not needed because $\cos (-\phi) = \cos \phi$. Also, after $k = m$, the values of s in the last equation are repeated. Thus there are m poles, as might be expected from the fact that Fig. 3.7-4 has m energy-storing elements. The poles lie on the negative real axis of the s-plane between the origin and $-4/RC$.

The integral of $F(s)\epsilon^{st}$ along path $DEFGA$ in Fig. 3.7-1 can be shown to vanish for $t > 0$ as $R \to \infty$, so $f(t)$ is the sum of the residues of $F(s)\epsilon^{st}$ at the m poles. Define

$$P(s) = C \sinh (m - r)\theta$$

$$Q(s) = (\sinh \theta)(\cosh m\theta)$$

Note that

$$\frac{dQ(s)}{ds} = \frac{dQ}{d\theta}\frac{d\theta}{ds} = [(\sinh \theta)(m \sinh m\theta) + (\cosh \theta)(\cosh m\theta)]\left[\frac{RC}{2 \sinh \theta}\right]$$

At the poles, where $\cosh m\theta = 0$, and $\sinh m\theta = j \sin [(2k - 1)(\pi/2)] = j(-1)^{k+1}$,

$$\left.\frac{dQ}{ds}\right|_{s=s_0} = j(-1)^{k+1}\left(\frac{RCm}{2}\right)$$

The fact that the last expression is not zero indicates that the poles are of the first order; hence Eq. 3.7-6 may be used.

$$P(s_0) = jC \sin \left[(m - r)\frac{(2k - 1)\pi}{2m}\right] = jC(-1)^{k+1} \cos \frac{(2k - 1)r\pi}{2m}$$

Then

$$f(t) = \frac{2}{Rm}\sum_{k=1}^{m} \cos \frac{(2k - 1)\pi}{2m}\, \epsilon^{\frac{2}{RC}\left[-1+\cos\frac{(2k-1)\pi}{2m}\right]t}$$

Example 3.7-4. Find

$$\mathscr{L}^{-1}\left[\frac{\epsilon^{-s}}{s^2(s^2 + s + 1)}\right]$$

The procedure most commonly used is to find, in the normal manner, that

$$\mathscr{L}^{-1}\left[\frac{1}{s^2(s^2+s+1)}\right] = t - 1 + \frac{2}{\sqrt{3}}\epsilon^{-t/2}\cos\left(\frac{\sqrt{3}}{2}t + \frac{\pi}{6}\right)$$

for $t > 0$. Then, by Eq. 3.4-11,

$$\mathscr{L}^{-1}\left[\frac{\epsilon^{-s}}{s^2(s^2+s+1)}\right] = \left[t - 2 + \frac{2}{\sqrt{3}}\epsilon^{-(t-1)/2}\cos\left(\frac{\sqrt{3}}{2}t - \frac{\sqrt{3}}{2} + \frac{\pi}{6}\right)\right]U_{-1}(t-1)$$

which is the previous function of time shifted along the time axis one unit in the positive t direction.

If, however, the inversion integral were to be used directly,

$$f(t) = \frac{1}{2\pi j}\int_{ABD}\frac{\epsilon^{s(t-1)}}{s^2(s+1-j)(s+1+j)}\,ds$$

where path ABD in Fig. 3.7-1 passes to the right of all the poles.

$$\int\frac{\epsilon^{s(t-1)}}{s^2(s+1-j)(s+1+j)}\,ds$$

would be expected to vanish along path $DEFGA$ only for $t - 1 > 0$, and along path DHA only for $t - 1 < 0$, since the real part of $s(t-1)$ should be negative. The reader should verify that this is indeed the case.

Therefore, for $t < 1$, $f(t) = 0$, while for $t > 1$,

$$f(t) = \sum [\text{residues of } F(s)\epsilon^{st} \text{ at } s = 0,\ -1 \pm j]$$

$$= \left[\frac{d}{ds}\frac{\epsilon^{s(t-1)}}{s^2+s+1}\right]_{s=0} + \left[\frac{\epsilon^{s(t-1)}}{s^2(s+1+j)}\right]_{s=-1+j} + \left[\frac{\epsilon^{s(t-1)}}{s^2(s+1-j)}\right]_{s=-1-j}$$

$$= t - 2 + \frac{2}{\sqrt{3}}\epsilon^{-(t-1)/2}\cos\left(\frac{\sqrt{3}}{2}t - \frac{\sqrt{3}}{2} + \frac{\pi}{6}\right)$$

Example 3.7-5. Find $\mathscr{L}^{-1}[s^{-1/2}]$.

The function $F(s) = 1/\sqrt{s}$ is a double-valued function, as a result of the square root operation. If s is represented in polar form as $r\epsilon^{j\theta}$, then $r\epsilon^{j(\theta+2\pi)}$ is another acceptable representation. However, $\sqrt{r\epsilon^{j(\theta+2\pi)}} = -\sqrt{r\epsilon^{j\theta}}$, giving rise to two different values for $\sqrt{s}$. A double-valued function is not analytic, and the bulk of the theorems from complex variable theory do not apply.

In order to make the function analytic, arbitrarily restrict the angle of s to the range $-\pi < \theta < \pi$, and also exclude the point $s = 0$. This is done formally by the construction of a *branch cut* along the negative real axis, as shown in Fig. 3.7-5. The end of the branch cut, which in this case is at the origin, is called a *branch point*. A branch cut can never be crossed, and so the branch cut ensures that $-\pi < \theta < \pi$, thereby making $F(s)$ single-valued. The basic inversion integral

$$f(t) = \frac{1}{2\pi j}\lim_{R\to\infty}\int_{ABD} F(s)\epsilon^{st}\,ds$$

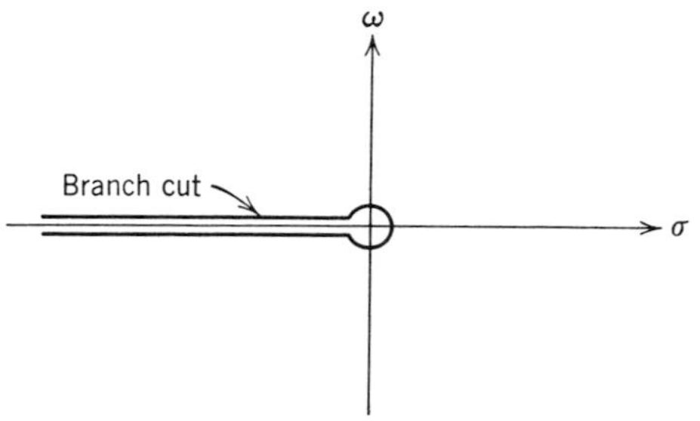

Fig. 3.7-5

still applies, but Fig. 3.7-1 must be modified, as shown in Fig. 3.7-6. By the residue theorem,

$$\int_{ABD} F(s)\epsilon^{st}\,ds + \int_{DEF} F(s)\epsilon^{st}\,ds + \int_{FGHI} F(s)\epsilon^{st}\,ds + \int_{IJA} F(s)\epsilon^{st}\,ds = 0$$

since no singularities are enclosed. For $t > 0$, it can be shown that the second and fourth integrals vanish as R approaches infinity. Then

$$f(t) = -\lim_{R\to\infty} \frac{1}{2\pi j}\int_{FGHI} F(s)\epsilon^{st}\,ds$$

On the infinitesimal circle GH, let $s = r\epsilon^{j\theta} = r\cos\theta + jr\sin\theta$, and

$$\int_{GH} F(s)\epsilon^{st}\,ds = \int_{\pi}^{-\pi} \frac{\epsilon^{r(\cos\theta)t}\epsilon^{jr(\sin\theta)t}}{r^{1/2}\epsilon^{j\theta/2}}\, jr\epsilon^{j\theta}\,d\theta$$

which vanishes as r approaches zero. On the straight line FG, let $s = -u$, $s^{1/2} = ju^{1/2}$, and $ds = -du$, where u and $u^{1/2}$ are real, positive quantities. Note carefully that the branch cut requires $\sqrt{-1} = j$ and not $-j$ for this path. Then

$$\int_{FG} F(s)\epsilon^{st}\,ds = -\int_{\infty}^{0} \frac{\epsilon^{-ut}}{ju^{1/2}}\,du = \frac{1}{j}\int_{0}^{\infty} \frac{\epsilon^{-ut}}{u^{1/2}}\,du$$

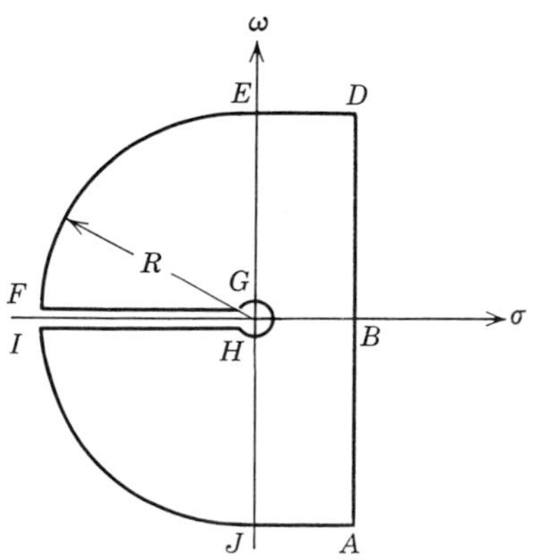

Fig. 3.7-6

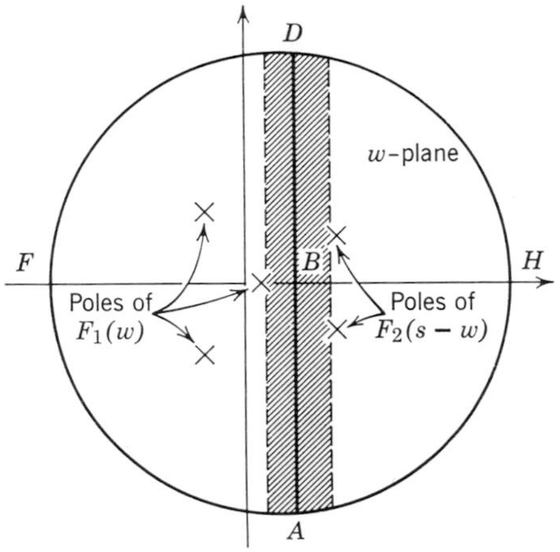

Fig. 3.7-7

For the path HI, let $s = -u$, $s^{1/2} = -ju^{1/2}$ (not $+ju^{1/2}$), and $ds = -du$.

$$\int_{HI} F(s)\epsilon^{st}\, ds = -\int_0^\infty \frac{\epsilon^{-ut}}{-ju^{1/2}}\, du = \frac{1}{j}\int_0^\infty \frac{\epsilon^{-ut}}{u^{1/2}}\, du$$

Combining these results gives

$$f(t) = -\frac{1}{2\pi j}\left[\frac{2}{j}\int_0^\infty u^{-1/2}\,\epsilon^{-ut}\, du\right]$$

$$= \frac{1}{\pi}\int_0^\infty \left(\frac{\gamma}{t}\right)^{-1/2} \epsilon^{-\gamma}\left(\frac{1}{t}\, d\gamma\right) = \frac{1}{\sqrt{\pi t}}$$

The substitution $\gamma = ut$ was made, and the evaluation of the last integral follows either from the definition of the gamma function or from a standard table of definite integrals.

Examples of finding the inverse transform of double-valued functions when the path shown in Fig. 3.7-6 encloses poles of $F(s)$ are discussed in Reference 10.

An important application of the residue theorem is based upon the complex convolution theorem of Eq. 3.4-14. If $F_1(s)$ and $F_2(s)$ represent the one-sided transforms of $f_1(t)$ and $f_2(t)$, respectively, then

$$\mathscr{L}[f_1(t)f_2(t)] = \frac{1}{2\pi j}\int_C F_1(w)F_2(s-w)\, dw \tag{3.7-7}$$

where w is a complex variable, and where s is treated as an independent parameter in carrying out the integration. It is understood that the real part of s must be large enough that all poles of $F_2(s - w)$ in the w-plane are to the right of the poles of $F_1(w)$.† The contour of integration is any path from $c - j\infty$ to $c + j\infty$ in the shaded strip of Fig. 3.7-7.

† This restriction on s ensures that the defining formula for the direct transforms of $f_1(t)$, $f_2(t)$, and $f_1(t)f_2(t)$ converges.

If the integral along the infinite semicircles DFA and DHA vanishes, the contour ABD may be replaced by either of the two closed contours in the figure. Then, by the residue theorem,

$$\begin{aligned}\mathscr{L}[f_1(t)f_2(t)] &= \sum [\text{residues of } F_1(w)F_2(s-w) \text{ at the poles of } F_1(w)] \\ &= -\sum [\text{residues of } F_1(w)F_2(s-w) \text{ at the poles of } F_2(s-w)]\end{aligned} \tag{3.7-8}$$

Example 3.7-6. Find $\mathscr{L}[t\epsilon^{at}]$ by the use of the complex convolution theorem. If $f_1(t) = t$ and $f_2(t) = \epsilon^{at}$, Eq. 3.7-8 gives

$$\begin{aligned}\mathscr{L}[t\epsilon^{at}] &= \left[\text{residue of } \frac{1}{w^2(s-w-a)} \quad \text{at } w = 0\right] \\ &= \left[\frac{d}{dw}\left(\frac{1}{s-w-a}\right)\right]_{w=0} = \frac{1}{(s-a)^2}\end{aligned}$$

Alternatively,

$$\begin{aligned}\mathscr{L}[t\epsilon^{at}] &= -\left[\text{residue of } \frac{1}{w^2(s-w-a)} \quad \text{at } w = s-a\right] \\ &= \left[\frac{1}{w^2}\right]_{w=s-a} = \frac{1}{(s-a)^2}\end{aligned}$$

A very important application of Eq. 3.7-8 is in connection with the Z transform of Section 3.11.

3.8 THE SIGNIFICANCE OF POLES AND ZEROS

Poles and zeros are defined in Section 3.6. In brief, poles and zeros of $F(s)$ are values of s for which $1/F(s)$ and $F(s)$, respectively, become zero. Suppose that $F(s)$ can be written as the quotient of two factored parts.

$$F(s) = \frac{K(s-z_1)(s-z_2)\cdots(s-z_m)}{(s-p_1)(s-p_2)\cdots(s-p_n)} \tag{3.8-1}$$

Then z_1 through z_m are zeros of $F(s)$, and p_1 through p_n are poles. If the factor $(s - z_i)$ or $(s - p_j)$ is repeated r times, $F(s)$ has a zero or pole, respectively, of order r. If $m < n$, $F(s)$ has a zero of order $n - m$ at infinity.

When plotted in the s-plane, poles are denoted by crosses, and zeros by circles. A second order pole or zero is denoted by two superimposed crosses or circles, respectively. The pole-zero plot for

$$F(s) = \frac{10(s^2 - 2s + 2)}{s(s+1)^2}$$

is shown in Fig. 3.8-1. The fact that there are three poles and two zeros in the finite s-plane indicates that $F(s)$ has a first order zero at infinity. The pole-zero plot completely describes $F(s)$, except for the multiplying constant $K = 10$.

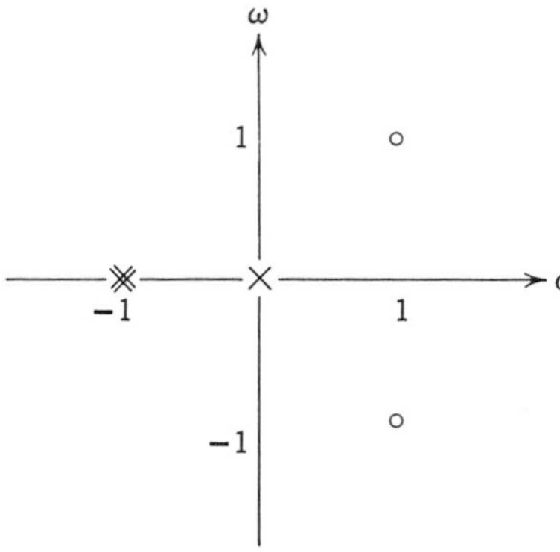

Fig. 3.8-1

The pole-zero plots of greatest interest are those representing inputs, outputs, and system functions. For realizable signals and systems, the inverse transforms of $V(s)$, $Y(s)$, and $H(s)$ must be real functions of time. This in turn requires that any complex zeros or poles must occur in complex conjugate pairs. Furthermore, the residues of complex conjugate poles must themselves be complex conjugates.

Each pole in $F(s)$ gives rise to a term in the inverse transform $f(t) = \mathscr{L}^{-1}[F(s)]$. The form of the term is completely determined by the location and order of the pole, although the size of the term depends upon the location of the other poles and zeros of $F(s)$. Poles in the right half-plane give rise to terms that increase without limit as t approaches infinity, while left half-plane poles yield terms that decay to zero. This is true even for higher order poles. A pole of order r at $s = -a$, for example, will produce a $t^{r-1}\epsilon^{-at}$ term in $f(t)$, but $\lim_{t\to\infty} t^{r-1}\epsilon^{-at} = 0$. The distance of poles from the vertical axis indicates how fast the corresponding terms in $f(t)$ grow or decay as t increases. For first order poles in the left half-plane, this distance is the reciprocal of the time constant. First order poles on the imaginary axis yield sinusoidal terms of constant amplitude, but higher order poles produce terms that increase without limit as t approaches infinity. The distance of first order complex conjugate poles from the horizontal axis equals the angular frequency of oscillation of the corresponding terms in $f(t)$.

The Free and Forced Response

The output of a fixed linear system initially at rest is given by Eq. 3.5-2, repeated below.

$$y(t) = \mathscr{L}^{-1}[V(s)H(s)] \tag{3.8-2}$$

The poles of $V(s)$ depend only upon the input and give rise to the forced response, which is the particular solution in the classical language of Chapter 2. The poles of $H(s)$ depend only upon the system and yield the free or natural response. This is identical with the complementary solution of Chapter 2, with all the arbitrary constants evaluated. Since the size of the terms in the free response depends upon the poles and zeros of both $V(s)$ and $H(s)$, the arbitrary constants depend upon both the input and the system, as stated in Section 2.6.

Example 3.8-1 Find the output voltage $e(t)$ if the circuit of Fig. 3.8-2 has no initial stored energy, and if $i(t) = (1 + \sin t)U_{-1}(t)$.

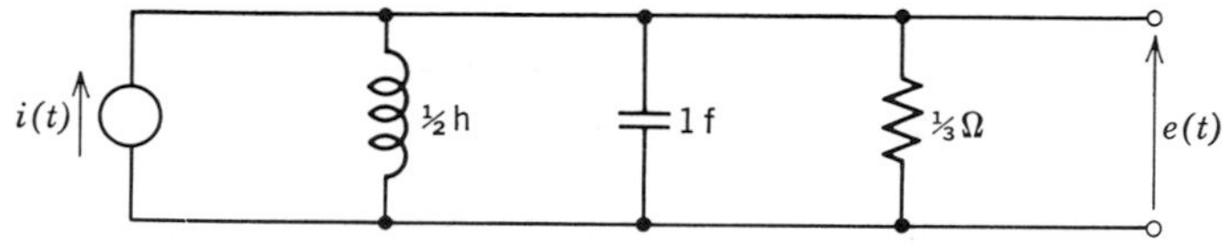

Fig. 3.8-2

The system function is the input impedance

$$Z(s) = \frac{1}{2/s + s + 3} = \frac{s}{(s+1)(s+2)}$$

The Laplace transform of the input is

$$I(s) = \frac{1}{s} + \frac{1}{s^2+1} = \frac{s^2+s+1}{s(s^2+1)}$$

Hence

$$E(s) = \frac{s^2+s+1}{(s+1)(s+2)(s^2+1)}$$

Then

$$e(t) = \mathscr{L}^{-1}[E(s)] = \frac{1}{2}\epsilon^{-t} - \frac{3}{5}\epsilon^{-2t} + \frac{1}{\sqrt{10}}\cos\left(t - \frac{\pi}{2.5}\right)$$

The first two terms in the output constitute the free response and are produced by the poles of $Z(s)$. The last term is the forced response, produced by the poles of $I(s)$ at $s = \pm j$. There is no term in the forced response due to the pole at the origin, because it is cancelled by a zero of $Z(s)$. This can be physically explained from the circuit by noting that the inductance acts like a short circuit to the d-c component of the source. In this example, the free and forced responses are the transient and steady-state components, respectively. This is always the case if $H(s)$ has only left half-plane poles, and if the input does not vanish as t approaches infinity.

Stability

Since $\mathscr{L}^{-1}[H(s)] = h(t)$, the poles of $H(s)$ determine the form of the impulse response. Because an impulsive input simply inserts some energy into the system instantaneously, the nature of $h(t)$ indicates the stability of the system. The system is unstable if $h(t)$ increases without limit as t approaches infinity. Thus a system is unstable if $H(s)$ has poles in the right half-plane or higher order poles on the imaginary axis. If there are first order poles on the imaginary axis, the free response contains oscillations of constant amplitude. Such a system is usually, but not always, considered to be stable. If $H(s)$ has only left half-plane poles, the system is unquestionably stable. These conclusions agree with those reached in Section 2.6. By Eq. 3.5-3, the poles of $H(s)$ are the roots of the characteristic equation $A(s) = 0$.

3.9 APPLICATION OF THE LAPLACE TRANSFORM TO TIME-VARYING SYSTEMS

Previous sections of this chapter have discussed in some detail the use of transform techniques and the system function $H(s)$ in the analysis of fixed systems. This section investigates the application of these techniques to time-varying systems.

The first of the two basic transform methods for fixed systems was to transform the system's differential equations into algebraic equations. Although any differential equation may be transformed term by term, an algebraic equation results only if the coefficients of the differential equation are constants. Otherwise, another differential or integral equation results. Equations 3.4-9 and 3.4-10 are useful in solving homogeneous differential equations whose coefficients are polynomials in t.

Example 3.9-1. Solve the equation

$$t\frac{d^2y}{dt^2} + \frac{dy}{dt} - (t - 1)y = 0$$

when $y(0+) = 1$.

Transforming this equation term by term, using Eq. 3.4-9, yields

$$-\frac{d}{ds}[s^2Y(s) - s - \dot{y}(0+)] + [sY(s) - 1] + \frac{dY(s)}{ds} + Y(s) = 0$$

Simplifying,

$$(s + 1)\frac{dY(s)}{ds} + Y(s) = 0$$

which is a first order differential equation in $Y(s)$. This can be easily solved by the separation of variables, or by the method of Section 2.5, giving

$$Y(s) = \frac{K}{s + 1}$$

Then

$$y(t) = K\epsilon^{-t}$$

Choosing $K = 1$ in order to satisfy the condition $y(0+) = 1$,

$$y(t) = \epsilon^{-t}$$

***Example* 3.9-2.** The differential equation of the previous example is changed slightly to read

$$t\frac{d^2y}{dt^2} + \frac{dy}{dt} + ty = 0$$

with $y(0+) = 1$.

The reader will recognize this equation as Bessel's equation of zero order, whose solution is $y = J_0(t)$. The transformed equation is

$$-\frac{d}{ds}[s^2Y(s) - s - \dot{y}(0+)] + [sY(s) - 1] - \frac{dY(s)}{ds} = 0$$

Simplifying,

$$(s^2 + 1)\frac{dY(s)}{ds} + sY(s) = 0$$

whose solution is

$$Y(s) = \frac{K}{\sqrt{s^2 + 1}}$$

The constant K, needed to satisfy the condition $y(0+) = 1$, can be found by the initial value theorem of Eq. 3.4-15.

$$\lim_{s\to\infty} \frac{sK}{\sqrt{s^2 + 1}} = K$$

so $K = 1$, and

$$y(t) = \mathscr{L}^{-1}\left[\frac{1}{\sqrt{s^2 + 1}}\right]$$

Although the inverse transform appears in standard tables, note instead that, since the solution is known to be $J_0(t)$,

$$\mathscr{L}[J_0(t)] = \frac{1}{\sqrt{s^2 + 1}}$$

The last example gives the easiest method of finding the Laplace transform of the Bessel function. In general, the use of Eq. 3.4-9 yields a differential equation in $Y(s)$ of order equal to the degree of the highest polynomial in t. It should be emphasized that the resulting equation cannot always be easily solved, and that the preceding examples are not typical of the difficulties encountered.

Each term in a differential equation with variable coefficients is the product of one known and one unknown function of time. In transforming such a term, Eq. 3.4-14 can be used. The result is a complicated integral equation in $Y(s)$. Since the resulting equation is very rarely easier to solve than the original one, this method is not pursued further.

At this point, one might seek other integral transforms helpful in solving those differential equations that are not amenable to the Laplace

Table 3.9-1

	Time-Invariant Systems	Equation Number	Time-Varying Systems	Equation Number
Differential equation	$A(p)y(t) = B(p)v(t)$	(3.9-1)	$A(p, t)y(t) = B(p, t)v(t)$	(3.9-9)
Impulse response	$A(p)h(t - \lambda) = B(p)U_0(t - \lambda)$	(3.9-2)	$A(p, t)h(t, \lambda) = B(p, t)U_0(t - \lambda)$	(3.9-10)
			$A(p, t)h_*(t, \tau) = B(p, t)U_0(\tau)$	(3.9-11)
Superposition integrals	$y(t) = \int_{-\infty}^{\infty} v(\lambda)h(t - \lambda)\, d\lambda$	(3.9-3)	$y(t) = \int_{-\infty}^{\infty} v(\lambda)h(t, \lambda)\, d\lambda$	(3.9-12)
	$y(t) = \int_{-\infty}^{\infty} v(t - \tau)h(\tau)\, d\tau$	(3.9-4)	$y(t) = \int_{-\infty}^{\infty} v(t - \tau)h_*(t, \tau)\, d\tau$	(3.9-13)
Relation of system function and impulse response	$H(s) = \mathscr{L}[h(\tau)]$	(3.9-5)	$H(t, s) = \mathscr{L}[h_*(t, \tau)]$	(3.9-14)
	$h(\tau) = \mathscr{L}^{-1}[H(s)]$	(3.9-6)	$h_*(t, \tau) = \mathscr{L}^{-1}[H(t, s)]$	(3.9-15)
Response to an exponential function	$A(p)[H(s)\epsilon^{st}] = B(p)\epsilon^{st}$	(3.9-7)	$A(p, t)[H(t, s)\epsilon^{st}] = B(p, t)\epsilon^{st}$	(3.9-16)
Response by the Laplace transform	$y(t) = \mathscr{L}^{-1}[V(s)H(s)]$	(3.9-8)	$y(t) = \mathscr{L}^{-1}[V(s)H(t, s)]$	(3.9-17)

transform. This has been done, and a number of other transforms are available for certain special cases. The best-known of these are the Mellin and Hankel transforms.[11] General integral transforms are discussed further in Section 3.10.

The System Function

The second basic transform method for fixed systems was based upon the system function $H(s)$. For systems with no initial stored energy, the response to an arbitrary input is given by

$$y(t) = \mathscr{L}^{-1}[V(s)H(s)]$$

Zadeh has shown that the system function concept can be extended to time-varying systems.[12] When the characteristics of a system are changing with time, the system function must be an explicit function of t as well as s. One would expect that the response of a varying system with no initial stored energy would be given by

$$y(t) = \mathscr{L}^{-1}[V(s)H(t, s)]$$

In order to understand how the time-varying system function results from a generalization of $H(s)$, first examine Table 3.9-1. Equations 3.9-1 through 3.9-8 summarize the three principal methods of characterizing fixed systems. They follow directly from Eq. 2.2-6, the definition of the impulse response, Eq. 1.6-2, Eq. 1.6-4, Eq. 3.5-6, Eq. 3.5-6, the discussion following Eq. 2.6-20, and Eq. 3.5-2, respectively. For varying systems, Eqs. 3.9-9 through 3.9-13 follow directly from Eqs. 2.2-7, 1.7-1, 1.7-1, 1.7-2, and 1.7-4, respectively. Equations 3.9-14 through 3.9-17 are derived in this section.

In the differential equations 3.9-1 and 3.9-9, $p = d/dt$. While the operators A and B are functions only of p for fixed systems, they are explicit functions of both p and t for varying systems. For example,

$$A(p, t)y(t) = [a_n(t)p^n + \cdots + a_1(t)p + a_0(t)]y(t)$$

Equations 3.9-2, 3.9-10, and 3.9-11 define the impulse response. A careful distinction must be made between the symbols h and h_* as discussed in Section 1.7. $h(t, \lambda)$ is the response at time t to a unit impulse at time λ, while $h_*(t, \tau)$ is the response at time t to a unit impulse occurring τ seconds earlier. The dummy variables λ and τ in these equations are related by $\tau = t - \lambda$. The limits on the superposition integrals of Eqs. 3.9-3, 3.9-4, 3.9-12, and 3.9-13 may usually be modified, if desired. For nonanticipatory systems with $v(t) = 0$ for $t < 0$, limits of zero to t may be used.

For fixed systems, $H(s) = \mathscr{L}[h(t)]$, where the transformation is with respect to t. Since $h(t)$ is the response at time t to a unit impulse at time zero (i.e., t seconds earlier), the age variable τ has been used in Eq. 3.9-5.

$$H(s) = \mathscr{L}[h(\tau)] \tag{3.9-5}$$

where the transformation is now with respect to τ. By analogy, the system function for the variable case is defined as

$$H(t, s) = \mathscr{L}[h_*(t, \tau)] \tag{3.9-14}$$

where the transformation is still with respect to τ, and where t is regarded as an independent parameter. The system function is *not* the transform of $h(t, \lambda)$ with respect to λ.

$$H(t, s) = \int_0^\infty h_*(t, \tau)\epsilon^{-s\tau}\, d\tau = \int_0^\infty h(t, t - \tau)\epsilon^{-s\tau}\, d\tau$$
$$= \int_{-\infty}^t h(t, \lambda)\epsilon^{-s(t-\lambda)}\, d\lambda \neq \mathscr{L}[h(t, \lambda)]$$

The definition of Eq. 3.9-14 forms the bridge between the time-domain and transform techniques. The response to the input

$$v(t) = \epsilon^{st} \quad \text{for} \quad -\infty < t < \infty$$

can be found by Eq. 3.9-13.

$$y(t) = \int_{-\infty}^\infty \epsilon^{s(t-\tau)} h_*(t, \tau)\, d\tau$$

For nonanticipatory systems, the lower limit may be changed to zero, and

$$y(t) = \epsilon^{st}\int_0^\infty h_*(t, \tau)\epsilon^{-s\tau}\, d\tau = \epsilon^{st}H(t, s)$$

Since the response to ϵ^{st} has been shown to be $\epsilon^{st}H(t, s)$,

$$A(p, t)[H(t, s)\epsilon^{st}] = B(p, t)\epsilon^{st} \tag{3.9-16}$$

Since the exponential input was assumed to exist for all t, $H(t, s)\epsilon^{st}$ really represents only the forced or particular integral response. A similar comment applies to Eq. 3.9-7, as is inferred by Eq. 3.5-3.

From Eq. 3.9-12, the response to an arbitrary input is

$$y(t) = \int_{-\infty}^\infty v(\lambda)h(t, \lambda)\, d\lambda$$
$$= \int_{-\infty}^\infty \left[\frac{1}{2\pi j}\int_C V(s)\epsilon^{s\lambda}\, ds\right] h(t, \lambda)\, d\lambda$$

where C is the contour of integration used for the evaluation of the inversion integral of Section 3.7. Assuming that the order of the two integrations may be reversed,

$$y(t) = \frac{1}{2\pi j}\int_C V(s)\left[\int_{-\infty}^{\infty} \epsilon^{s\lambda}h(t, \lambda)\, d\lambda\right] ds$$

By Eq. 3.9-12, the factor in brackets is the response to ϵ^{st}, namely, $H(t, s)\epsilon^{st}$.

$$y(t) = \frac{1}{2\pi j}\int_C V(s)H(t, s)\epsilon^{st}\, ds = \mathscr{L}^{-1}[V(s)H(t, s)] \qquad (3.9\text{-}17)$$

Once $H(t, s)$ is known, the response to an arbitrary input can be found from the standard tables and techniques of Laplace transform theory. In taking the inverse transform with respect to s, t is treated as an independent parameter. Note that

$$Y(s) \neq V(s)H(t, s)$$

since $Y(s) = \mathscr{L}[y(t)]$ certainly cannot be a function of t.

Example 3.9-3. If $H(t, s) = 1/(t^2s^2)$, find the response to $v(t) = t\epsilon^{-t}U_{-1}(t)$. The next example shows that this system function corresponds to the differential equation

$$t^2\frac{d^2y}{dt^2} + 4t\frac{dy}{dt} + 2y = v$$

Write

$$y(t) = \mathscr{L}^{-1}[V(s)H(t, s)] = \frac{1}{t^2}\mathscr{L}^{-1}\left[\frac{1}{s^2(s+1)^2}\right]$$

The residue of $\epsilon^{st}/[s^2(s+1)^2]$ at $s = 0$ is

$$\left[\frac{d}{ds}\frac{\epsilon^{st}}{(s+1)^2}\right]_{=0} = t - 2$$

and at $s = -1$ is

$$\left[\frac{d}{ds}\left(\frac{\epsilon^{st}}{s^2}\right)\right]_{s=-1} = (t+2)\epsilon^{-t}$$

Thus

$$y(t) = \frac{1}{t}\left[\left(1 - \frac{2}{t}\right) + \left(1 + \frac{2}{t}\right)\epsilon^{-t}\right]$$

which agrees with Example 1.7-1.

Obtaining the System Function from the Differential Equation

Equation 3.9-17 gives a relatively easy method of calculating the response, once the system function is known. Determining the system function is usually the most difficult part of the entire procedure. The three common ways of characterizing a system are by the differential equation, the impulse response, and the system function. There is, of course, no problem if the system function happens to be given. Similarly, if the impulse response is given, $H(t, s)$ can be easily determined by Eq. 3.9-14. If the differential equation is given, one method is to find first the impulse response by the techniques of Section 2.8.

Example 3.9-4. Find $H(t, s)$ for a system characterized by

$$t^2 \frac{d^2y}{dt^2} + 4t \frac{dy}{dt} + 2y = v$$

The impulse response was found in Example 2.8-2.

$$h(t, \lambda) = \frac{t - \lambda}{t^2} \quad \text{for } t \geqslant \lambda$$

Then

$$h_*(t, \tau) = h(t, t - \tau) = \frac{\tau}{t^2} \quad \text{for } \tau \geqslant 0$$

$$H(t, s) = \frac{1}{t^2} \mathscr{L}[\tau] = \frac{1}{t^2 s^2}$$

An alternative method is based upon Eq. 3.9-16. This is a time-varying differential equation which can be solved directly for the system function. The equation is of the same order as the differential equation used to find the impulse response, but in some cases it is easier to solve. Since the repeated differentiation of ϵ^{st} is so simple, the form of Eq. 3.9-16 may be simplified somewhat. A typical term in $B(p, t)\epsilon^{st}$ is

$$b_k(t)p^k\epsilon^{st} = b_k(t)s^k\epsilon^{st}$$

so

$$B(p, t)\epsilon^{st} = \epsilon^{st}B(s, t) \tag{3.9-18}$$

where $B(s, t)$ is a function of s and t, and not an operator. A typical term in $A(p, t)\ [H(t, s)\epsilon^{st}]$ is

$$a_k(t)p^k[H(t, s)\epsilon^{st}]$$

If f and g are two functions of time,

$$p^k(fg) = \sum_{r=0}^{k} \binom{k}{r} (p^{k-r}f)(p^r g)$$

where $\binom{k}{r}$ represents the binomial coefficients, i.e.,

$$(c + d)^k = \sum_{r=0}^{k} \binom{k}{r} c^{k-r}\, d^r$$

Letting $f = H(t, s)$ and $g = \epsilon^{st}$,

$$p^k[H(t, s)\epsilon^{st}] = \sum_{r=0}^{k} \binom{k}{r} [p^{k-r}H(t, s)][s^r\epsilon^{st}]$$

$$= \epsilon^{st} \sum_{r=0}^{k} \binom{k}{r} p^{k-r}s^r H(t, s) = \epsilon^{st}(p + s)^k H(t, s)$$

Thus

$$A(p, t)[H(t, s)\epsilon^{st}] = \epsilon^{st}A(p + s, t)H(t, s) \qquad (3.9\text{-}19)$$

where $A(p + s, t)$ is an operator, as well as a function of s and t. Substituting Eqs. 3.9-18 and 3.9-19 into Eq. 3.9-16 gives

$$A(p + s, t)H(t, s) = B(s, t) \qquad (3.9\text{-}20)$$

The methods of Sections 2.7 and 2.8 are often helpful in solving this equation.

Example 3.9-5. Find $H(t, s)$ for a system characterized by

$$t^2 \frac{d^2y}{dt^2} + 4t \frac{dy}{dt} + 2y = v$$

In operator notation,

$$(t^2p^2 + 4tp + 2)y = v$$

Using Eq. 3.9-20,

$$[t^2(p + s)^2 + 4t(p + s) + 2]H(t, s) = 1$$

or

$$[t^2p^2 + (2st^2 + 4t)p + (t^2s^2 + 4ts + 2)]H(t, s) = 1$$

The reader should verify that the solution

$$H(t, s) = \frac{1}{t^2s^2}$$

does satisfy the last equation and does agree with the result of the previous example. For this particular system, solving the differential equation in $H(t, s)$ is more difficult than solving the original differential equation for the impulse response. For other systems, however, it may be simpler.

The reader should realize that the $H(t, s)$ found in the last example is only the forced or particular integral solution of Eq. 3.9-20. From the derivation of Eq. 3.9-16, however, this is exactly what is desired. Although a different $H(t, s)$ results if a complementary solution is added, it can be shown that the response calculated by Eq. 3.9-17 is not affected.[12]

In many problems, the exact determination of $H(t, s)$ is so difficult that recourse to approximation methods is necessary. Standard approximating techniques may be used when the parameters either vary slowly with time or have a small variation compared to their mean value.[13]

3.10 RESOLUTION OF SIGNALS INTO SETS OF ELEMENTARY FUNCTIONS

Both the superposition integrals of Chapter 1 and the inverse transform methods of this chapter may be regarded as special cases of the resolution of signals into sets of elementary functions.[14] As background for the material of this section, the reader may wish to review Section 1.3. Equations 1.3-3 and 1.3-4 express the input and output of a system initially at rest as

$$v(t) = \int_C a(\lambda)k(t, \lambda)\, d\lambda$$
$$y(t) = \int_C a(\lambda)K(t, \lambda)\, d\lambda \tag{3.10-1}$$

If λ is a real variable, the integration is in general from $\lambda = -\infty$ to $+\infty$. If λ is a complex variable, the integration is over a contour C in the complex plane. $k(t, \lambda)$ represents the family of elementary functions into which $v(t)$ is decomposed. The spectral function $a(\lambda)$ is a measure of the relative strength of the elementary functions comprising $v(t)$. $K(t, \lambda)$ is the system's response to $k(t, \lambda)$.

In order for Eq. 3.10-1 to be useful, there must be a simple method of finding $a(\lambda)$ for an arbitrary input. Because of the linearity of Eq. 3.10-1, one would expect $a(\lambda)$ to be expressible in the following form.

$$a(\lambda) = \int_{-\infty}^{\infty} v(t)k^{-1}(\lambda, t)\, dt \tag{3.10-2}$$

$k^{-1}(\lambda, t)$ is known as the inverse of $k(t, \lambda)$ and is not necessarily easy to find. One helpful relationship between $k(t, \lambda)$ and $k^{-1}(\lambda, t)$ is obtained by letting $v(t) = U_0(t - z)$ in Eq. 3.10-2. The corresponding $a(\lambda)$ is

$$\int_{-\infty}^{\infty} k^{-1}(\lambda, t)U_0(t - z)\, dt = k^{-1}(\lambda, z)$$

Inserting this result into the first of Eq. 3.10-1,

$$U_0(t - z) = \int_C k^{-1}(\lambda, z)k(t, \lambda)\, d\lambda \tag{3.10-3}$$

In order for a given choice of $k(t, \lambda)$ to be fruitful, it must be possible to find a $k^{-1}(\lambda, z)$ satisfying this equation.

Much of Sections 1.5 through 1.7 is based upon the choice of $k(t, \lambda) = U_0(t - \lambda)$, where λ is a real variable. Then $k^{-1}(\lambda, t) = U_0(\lambda - t)$, and Eqs. 3.10-2 and 3.10-1 become

$$a(\lambda) = \int_{-\infty}^{\infty} v(t)U_0(\lambda - t)\, dt = v(\lambda)$$
$$v(t) = \int_{-\infty}^{\infty} a(\lambda)U_0(t - \lambda)\, d\lambda \tag{3.10-4}$$
$$= \int_{-\infty}^{\infty} v(\lambda)U_0(t - \lambda)\, d\lambda$$

The latter rather trivial result is identical with Eq. 1.5-4.

The inverse two-sided Laplace transform† is based upon the choice of $k(t, \lambda) = \epsilon^{\lambda t}/2\pi j$ for $-\infty < t < \infty$, and $k^{-1}(\lambda, t) = \epsilon^{-\lambda t}$, where λ is the complex frequency variable previously denoted by s. Equations 3.10-2 and 3.10-1 become

$$\begin{aligned} a(\lambda) &= \int_{-\infty}^{\infty} v(t)\epsilon^{-\lambda t}\,dt = \mathscr{L}_{\text{II}}[v(t)] \\ v(t) &= \frac{1}{2\pi j}\int_C a(\lambda)\epsilon^{\lambda t}\,d\lambda = \mathscr{L}_{\text{II}}^{-1}[a(\lambda)] \end{aligned} \tag{3.10-5}$$

which agree with the usual equations for the direct and inverse transform. The contour C in the complex λ-plane is described in Section 3.7. When the above choices of $k(t, \lambda)$ and $k^{-1}(\lambda, t)$ are substituted into Eq. 3.10-3, the right side becomes

$$\frac{1}{2\pi j}\int_C \epsilon^{-\lambda z}\epsilon^{\lambda t}\,d\lambda = \mathscr{L}_{\text{II}}^{-1}[\epsilon^{-\lambda z}]$$

This does equal $U_0(t - z)$ as required.

$K(t, \lambda)$, the response to $k(t, \lambda)$ when λ is treated as a parameter, is usually called the characteristic function. When $k(t, \lambda) = U_0(t - \lambda)$, $K(t,\lambda)$ is the impulse response $h(t, \lambda)$. From Eqs. 3.10-1 and 3.10-4, the expression for $y(t)$ then becomes

$$y(t) = \int_{-\infty}^{\infty} v(\lambda)h(t, \lambda)\,d\lambda \tag{3.10-6}$$

agreeing with Eq. 1.5-7. When $k(t, \lambda) = \epsilon^{\lambda t}/2\pi j$, $K(t, \lambda) = (1/2\pi j)H(t, \lambda)\epsilon^{\lambda t}$ by Eq. 3.9-16. Thus, for the Laplace transform approach,

$$\begin{aligned} y(t) &= \frac{1}{2\pi j}\int_C a(\lambda)H(t, \lambda)\epsilon^{\lambda t}\,d\lambda \\ &= \mathscr{L}_{\text{II}}^{-1}[a(\lambda)H(t, \lambda)] \end{aligned} \tag{3.10-7}$$

where $a(\lambda) = \mathscr{L}_{\text{II}}[v(t)]$, agreeing with Eq. 3.9-17.

For any choice of elementary functions $k(t, \lambda)$, the characteristic function $K(t, \lambda)$ can be found once the impulse response $h(t, \lambda)$ is known. Using Eq. 3.10-6 with $v(t) = k(t, z)$ and $y(t) = K(t, z)$,

$$K(t, z) = \int_{-\infty}^{\infty} k(\lambda, z)h(t, \lambda)\,d\lambda \tag{3.10-8}$$

Conversely,

$$h(t, z) = \int_C k^{-1}(\lambda, z)K(t, \lambda)\,d\lambda \tag{3.10-9}$$

† For the Fourier transform, $k(t, \lambda) = \epsilon^{j\lambda t}/2\pi$ for $-\infty < t < \infty$, where λ is the real variable previously denoted by ω. For the one-sided Laplace transform, $k(t, \lambda) = \epsilon^{\lambda t}/2\pi j$ for $0 < t < \infty$. For generality, the discussion is carried out in terms of the two-sided Laplace transform.

For the special case of $k(t, \lambda) = \epsilon^{\lambda t}/2\pi j$ and $k^{-1}(\lambda, t) = \epsilon^{-\lambda t}$, these two equations should relate the system function $H(t, s)$ of Section 3.9 to the impulse response. For this special case, $K(t, z) = (1/2\pi j)H(t, z)\epsilon^{zt}$ by Eq. 3.9-16. Substituting these values into Eq. 3.10-8 gives

$$\frac{1}{2\pi j} H(t, z)\epsilon^{zt} = \frac{1}{2\pi j}\int_{-\infty}^{\infty} \epsilon^{z\lambda} h(t, \lambda)\, d\lambda$$

Thus,

$$H(t, z) = \int_{-\infty}^{\infty} h(t, \lambda)\epsilon^{-z(t-\lambda)}\, d\lambda$$

$$= \int_{-\infty}^{\infty} h_*(t, t - \lambda)\epsilon^{-z(t-\lambda)}\, d\lambda$$

Letting $\xi = t - \lambda$,

$$H(t, z) = \int_{-\infty}^{\infty} h_*(t, \xi)\epsilon^{-z\xi}\, d\xi = \mathscr{L}_{\text{II}}[h_*(t, \xi)] \qquad (3.10\text{-}10)$$

agreeing with Eq. 3.9-14. Equation 3.10-9 becomes

$$h(t, z) = \frac{1}{2\pi j}\int_C \epsilon^{-z\lambda} H(t, \lambda)\epsilon^{\lambda t}\, d\lambda$$

$$= \frac{1}{2\pi j}\int_C H(t, \lambda)\epsilon^{\lambda(t-z)}\, d\lambda$$

Then

$$h_*(t, z) = h(t, t - z) = \frac{1}{2\pi j}\int_C H(t, \lambda)\epsilon^{\lambda z}\, d\lambda$$

$$= \mathscr{L}_{\text{II}}^{-1}[H(t, \lambda)] \qquad (3.10\text{-}11)$$

agreeing with Eq. 3.9-15.

Equations 3.10-1, 3.10-2, 3.10-3, 3.10-8, and 3.10-9 apply to any choice of elementary functions $k(t, \lambda)$. Although the only special cases considered in this book are the superposition integrals of Chapter 1 and the better-known integral transforms, some of the less well-known integral transforms (e.g., the Mellin, Hankel, Hilbert, Euler, and Whittaker transforms) occasionally prove useful. In fact, it might seem worthwhile to make a systematic study of many types of elementary functions $k(t, \lambda)$, with the aim of inventing new integral transforms. Such a study would, however, be severely limited by the general difficulty of finding inverse functions $k^{-1}(\lambda, t)$ that satisfy Eq. 3.10-3. Even then, the characteristic function $K(t, \lambda)$ must be relatively easy to find, if the new transforms are to be useful.

The spectral function $a(\lambda)$ for the input $v(t)$ is often denoted by $V(\lambda)$. Using this notation, Eqs. 3.10-2 and 3.10-1 become

$$
\begin{aligned}
V(\lambda) &= \int_{-\infty}^{\infty} v(t)k^{-1}(\lambda, t)\, dt \\
v(t) &= \int_C V(\lambda)k(t, \lambda)\, d\lambda \\
y(t) &= \int_C V(\lambda)K(t, \lambda)\, d\lambda \\
&= \int_C Y(\lambda)k(t, \lambda)\, d\lambda
\end{aligned}
\tag{3.10-12}
$$

where $Y(\lambda)$ is the spectral function or transform of the output $y(t)$ with respect to the set of elementary functions $k(t, \lambda)$. The relationship between $Y(\lambda)$ and $V(\lambda)$ becomes particularly simple if the characteristic function $K(t, \lambda)$ happens to have the form $K(t, \lambda) = H(\lambda)k(t, \lambda)$. When this occurs, the elementary functions $k(t, \lambda)$ are said to be "eigenfunctions" of the system under consideration.

$$H(\lambda) = \frac{\text{response to } k(t, \lambda)}{k(t, \lambda)}$$

and is called the system function with respect to $k(t, \lambda)$. In such a case,

$$y(t) = \int_C V(\lambda)H(\lambda)k(t, \lambda)\, d\lambda$$

Thus $Y(\lambda) = V(\lambda)H(\lambda)$. Equation 3.10-12 can then be rewritten

$$
\begin{aligned}
V(\lambda) &= \int_{-\infty}^{\infty} v(t)k^{-1}(\lambda, t)\, dt \\
Y(\lambda) &= V(\lambda)H(\lambda) \\
y(t) &= \int_C Y(\lambda)k(t, \lambda)\, d\lambda
\end{aligned}
\tag{3.10-13}
$$

This specifically describes the three steps associated with the Fourier and Laplace transforms: the direct transform, multiplication by the system function, and the inverse transform.

The forced response of a fixed linear system to $\epsilon^{\lambda t}$ always has the form $H(\lambda)\epsilon^{\lambda t}$, as shown in Eq. 2.6-22.† For such systems, the previous equations

† Zadeh points out that the statement is also true for a certain class of nonlinear systems.[15]

become

$$V(\lambda) = \int_{-\infty}^{\infty} v(t)\epsilon^{-\lambda t}\,dt = \mathscr{L}_{\text{II}}[v(t)]$$
$$Y(\lambda) = V(\lambda)H(\lambda) \tag{3.10-14}$$
$$y(t) = \frac{1}{2\pi j}\int_C Y(\lambda)\epsilon^{\lambda t}\,d\lambda = \mathscr{L}_{\text{II}}^{-1}[Y(\lambda)]$$

For varying linear systems, the system function H is a function of t as well as λ, as shown in the previous section.

3.11 THE Z TRANSFORM

The Laplace transform, which in earlier sections was used to solve differential equations and to determine a system's response to continuous signals, is also applicable to the solution of difference equations and the determination of the response to discrete signals.[16] The Z transform is, however, better suited for the latter purposes. Although some early workers, including Laplace himself, had used a similar concept, the Z transform has only recently attracted wide attention from engineers, principally in its application to sampled-data control systems.

To provide motivation, the Z transform is first related to the more familiar Laplace transform and to the discrete signals of Section 1.8. When doing this, it is better to consider a discrete signal consisting of a train of weighted impulses, rather than a series of finite numbers, since the Laplace transform of the latter is identically zero. Any problem involving sequences of finite numbers can be solved by inspection, however, once the corresponding problem with impulses has been solved.

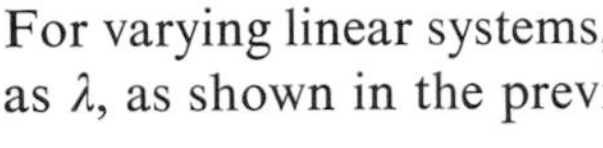

Fig. 3.11-1

Figure 3.11-1 shows a linear system preceded by an ideal sampling switch with a period T. From Eq. 1.8-4,

$$v^*(t) = v(t)\sum_{n=-\infty}^{\infty} U_0(t - nT) = \sum_{n=-\infty}^{\infty} v(nT)U_0(t - nT) \tag{3.11-1}$$

where the lower limit becomes zero if $v(t) = 0$ for $t < 0$. The two-sided and one-sided Laplace transforms of Eq. 3.11-1 are

$$V_{\text{II}}^*(s) = \sum_{n=-\infty}^{\infty} v(nT)\epsilon^{-nTs}$$
$$V^*(s) = \sum_{n=0}^{\infty} v(nT)\epsilon^{-nTs} \tag{3.11-2}$$

Since the variable in both cases is really ϵ^{-nTs}, it is convenient to define

$$z = \epsilon^{sT} \tag{3.11-3}$$

Although it is contrary to normal mathematical nomenclature, it has become customary to define

$$V(z) = [V^*(s)]_{\epsilon^{sT}=z} \tag{3.11-4}$$

$V(z)$ is not $V(s)$ with s replaced by z, but is $V^*(s)$ with s replaced by $(1/T)\ln z$. Equation 3.11-2 then becomes

$$\begin{aligned} V_{II}(z) &= \sum_{n=-\infty}^{\infty} v(nT)z^{-n} \\ V(z) &= \sum_{n=0}^{\infty} v(nT)z^{-n} \end{aligned} \tag{3.11-5}$$

The response of the configuration of Fig. 3.11-1, and those of more complicated configurations, are deferred until the properties of the Z transform have been investigated. As the reader proceeds, he should note the similarities between the properties of the Z and Laplace transforms.

The Two-Sided Z Transform

The formal definition of the two-sided transform is

$$F(z) = \mathscr{Z}_{II}[f(t)] = \mathscr{Z}_{II}[f^*(t)] = \sum_{n=-\infty}^{\infty} f(nT)z^{-n} \tag{3.11-6}$$

or, equivalently,

$$F(z) = \{\mathscr{L}_{II}[f^*(t)]\}_{\epsilon^{sT}=z} \tag{3.11-7}$$

where T is a known constant. Since $F(z)$ depends only upon the value of $f(t)$ at the instants $t = nT$, it can be equally well regarded as the Z transform of either $f(t)$ or $f^*(t)$.

An alternative interpretation is based upon the fact that the exponent of z in Eq. 3.11-6 denotes a particular element in the infinite sequence $\ldots, f(-T), f(0), f(T), \ldots, f(nT), \ldots$. The variable z is often regarded as an "ordering variable" for the infinite sequence.

Equation 3.11-6 defines $F(z)$ only for those values of z for which the infinite series converges. For other values of the complex variable z, $F(z)$ is defined by the principle of analytic continuation. Whenever $F(z)$ is written in the infinite series form of Eq. 3.11-6, the values of $f(nT)$ can be seen by inspection. If $F(z)$ is given in closed form, $f(nT)$ can be found by the techniques of the next section.

Example 3.11-1. Find the Z transform of

$$f(t) = \begin{cases} 0 & \text{for } t < 0 \\ \epsilon^{at} & \text{for } t \geqslant 0 \end{cases}$$

By Eq. 3.11-6,

$$F(z) = \sum_{n=0}^{\infty} \epsilon^{anT} z^{-n} = \sum_{n=0}^{\infty} \left(\frac{\epsilon^{aT}}{z}\right)^n$$

This infinite series converges for

$$\left|\frac{\epsilon^{aT}}{z}\right| < 1, \quad \text{i.e., } |z| > \epsilon^{aT}$$

and can be written in closed form as

$$F(z) = \frac{1}{1 - (\epsilon^{aT}/z)} = \frac{z}{z - \epsilon^{aT}}$$

Example 3.11-2. Find the Z transform of

$$f(t) = \begin{cases} -\epsilon^{at} & \text{for } t < 0 \\ 0 & \text{for } t \geqslant 0 \end{cases}$$

Again using the basic definition,

$$\begin{aligned} F(z) &= -\sum_{n=-\infty}^{-1} \epsilon^{anT} z^{-n} = -\sum_{m=1}^{\infty} (\epsilon^{-aT} z)^m \\ &= -\sum_{n=0}^{\infty} (\epsilon^{-aT} z)^{n+1} = -\epsilon^{-aT} z \sum_{n=0}^{\infty} (\epsilon^{-aT} z)^n \end{aligned}$$

The infinite series converges for $|z| < \epsilon^{aT}$ and can be written in closed form as

$$F(z) = -\epsilon^{-aT} z \frac{1}{1 - \epsilon^{-aT} z} = \frac{z}{z - \epsilon^{aT}}$$

If $F(z)$ is given in closed form, as in the preceding examples, the region of convergence must be specified in order to determine which function of time corresponds to the function of z. Note that the two regions of convergence are mutually exclusive, and that only one of the two functions of time is bounded.

The One-Sided Z Transform

If the history of the system prior to $t = 0$ is summarized by appropriate boundary conditions, $f(t)$ can be considered to be zero for $t < 0$. The one-sided transform, which is the one used henceforth unless otherwise stated, is defined as

$$F(z) = \mathscr{Z}[f(t)] = \sum_{n=0}^{\infty} f(nT) z^{-n} \tag{3.11-8}$$

or, equivalently,

$$F(z) = \{\mathscr{L}[f^*(t)]\}_{\epsilon^{sT}=z} \tag{3.11-9}$$

Based upon this equation, a table of Z transforms such as Table 3.11-1 can be determined. A limitation of Eq. 3.11-8 is that there may be some difficulty in expressing the infinite series in closed form. One method that overcomes this difficulty is discussed in the next section.

Table 3.11-1

$f(t)$	$F(z)$	Region of Convergence
$U_{-1}(t)$	$\dfrac{z}{z-1}$	$\lvert z\rvert > 1$
t	$\dfrac{Tz}{(z-1)^2}$	$\lvert z\rvert > 1$
t^2	$\dfrac{T^2z(z+1)}{(z-1)^3}$	$\lvert z\rvert > 1$
ϵ^{at}	$\dfrac{z}{z-\epsilon^{aT}}$	$\lvert z\rvert > \epsilon^{aT}$
$\sin \beta t$	$\dfrac{z \sin \beta T}{z^2 - 2z \cos \beta T + 1}$	$\lvert z\rvert > 1$
$\cos \beta t$	$\dfrac{z(z - \cos \beta T)}{z^2 - 2z \cos \beta T + 1}$	$\lvert z\rvert > 1$

It is sometimes necessary to find the two-sided transform from a table of one-sided transforms.

$$\mathscr{Z}_{\text{II}}[f(t)] = \sum_{n=-\infty}^{0} f(nT)z^{-n} + \sum_{n=0}^{\infty} f(nT)z^{-n} - f(0)$$

The presence of the last term compensates for the fact that $n = 0$ appears in both of the summations. Letting $m = -n$,

$$\begin{aligned}\mathscr{Z}_{\text{II}}[f(t)] &= \sum_{m=0}^{\infty} f(-mT)z^{m} + \sum_{n=0}^{\infty} f(nT)z^{-n} - f(0)\\ &= \{\mathscr{Z}[f(-t)]\}_{z\to z^{-1}} + \mathscr{Z}[f(t)] - f(0) \qquad (3.11\text{-}10)\end{aligned}$$

For the special case of an even function of time, $f(-t) = f(t)$, and

$$\mathscr{Z}_{\text{II}}[f(t)] = F(z^{-1}) + F(z) - f(0) \qquad (3.11\text{-}11)$$

where $F(z) = \mathscr{Z}[f(t)]$.

3.12 PROPERTIES OF THE Z TRANSFORM

The properties of the one-sided Laplace transform have their parallel in the one-sided Z transform. Although a few properties are proved, most of the proofs are left as exercises for the reader.

Theorem 3.12-1. $f(t)$ must be defined for all $t = nT$ $(n = 0, 1, 2, \ldots)$ in order for $F(z)$ to exist.

Theorem 3.12-2. The definition of Eq. 3.11-8 converges absolutely for $|z| > c$, where c is known as the radius of absolute convergence.

Theorem 3.12-3. Two functions of time have the same Z transform if and only if they are identical for all $t = nT$ $(n = 0, 1, 2, \ldots)$. Thus $F(z)$ contains no information about $f(t)$ except at the sampling instants. The inverse transform $\mathscr{Z}^{-1}[F(z)]$ can be uniquely defined as either $f^*(t)$ or $f(nT)$; the latter definition is used in this book. If Tables 3.11-1 and 3.12-1 are used to find the inverse transform, every t in the left-hand column should be replaced by nT.

Theorem 3.12-4. The Z transform of $f(t/T)$ is independent of the sampling interval T. From the definition of Eq. 3.11-8,

$$\mathscr{Z}[f(t/T)] = \sum_{n=0}^{\infty} f(n)z^{-n}$$

which does not depend upon T. This theorem is useful in finding the inverse transform by partial fractions, particularly if the available tables assume $T = 1$.

Other Properties of the Z Transform

Table 3.12-1 lists the most useful properties of the Z transform. Equations 3.12-1 through 3.12-5 are the basis for the solution of fixed difference equations, as discussed in the next section.

To prove Eq. 3.12-3, for example, let $m = n + 1$, and write

$$\mathscr{Z}[f(t + T)] = \sum_{n=0}^{\infty} f(nT + T)z^{-n} = z\sum_{m=1}^{\infty} f(mT)z^{-m}$$

$$= z\sum_{m=0}^{\infty} f(mT)z^{-m} - zf(0) = zF(z) - zf(0)$$

Equation 3.12-6 offers a method of summing a finite or infinite series. Equations 3.12-7 through 3.12-10 are useful when finding the direct and

Table 3.12-1

Property	Equation Number
$\mathscr{Z}[f_1(t) + f_2(t)] = \mathscr{Z}[f_1(t)] + \mathscr{Z}[f_2(t)] = F_1(z) + F_2(z)$	(3.12-1)
$\mathscr{Z}[af(t)] = aF(z)$	(3.12-2)
$\mathscr{Z}[f(t + T)] = zF(z) - zf(0)$	(3.12-3)
$\mathscr{Z}[f(t + 2T)] = z^2F(z) - z^2f(0) - zf(T)$	(3.12-4)
$\mathscr{Z}[f(t + mT)] = z^mF(z) - z^mf(0) - z^{m-1}f(T) - \cdots - zf(mT - T)$	(3.12-5)
$\mathscr{Z}\left[\sum_{k=0}^{n} f(kT)\right] = \dfrac{z}{z-1} F(z)$	(3.12-6)
$\mathscr{Z}[\epsilon^{-at}f(t)] = F(\epsilon^{aT}z)$	(3.12-7)
$\mathscr{Z}[tf(t)] = -Tz\dfrac{d}{dz} F(z)$	(3.12-8)
$\mathscr{Z}\left[\dfrac{f(t)}{t}\right] = -\dfrac{1}{T}\displaystyle\int \dfrac{F(z)}{z}\, dz$	(3.12-9)
$\mathscr{Z}[f(t - kT)U_{-1}(t - kT)] = z^{-k}F(z)$	(3.12-10)
$\mathscr{Z}[a^tf(t)] = F\left(\dfrac{z}{a^T}\right)$	(3.12-11)
$\mathscr{Z}\left[\sum_{k=0}^{n} f_1(t - kT)f_2(kT)\right] = F_1(z)F_2(z)$	(3.12-12)
$\mathscr{Z}[f_1(t)f_2(t)] = \dfrac{1}{2\pi j}\displaystyle\oint_C \dfrac{F_1(w)F_2(z/w)}{w}\, dw$, where the contour of integration separates the poles of $F_1(w)$ from those of $F_2(z/w)$	(3.12-13)
$\lim_{t\to 0} f(t) = \lim_{z\to\infty} F(z)$, provided that the limit exists	(3.12-14)
$\lim_{t\to\infty} f(t) = \lim_{z\to 1} \left(\dfrac{z-1}{z}\right) F(z)$, provided that $[(z - 1)/z]F(z)$ is analytic on and outside the unit circle	(3.12-15)

inverse transforms of given functions. Equation 3.12-11 shows the effect of a change of scale in the z-plane, while Eqs. 3.12-12 and 3.12-13 are convolution formulas. The latter one is proved and discussed in Section 3.14. Equations 3.12-14 and 3.12-15 are known as the initial and final value theorems, respectively.

Table 3.12-2 lists three properties that further relate the Laplace and Z transforms. $F(s)$ and $F^*(s)$ represent the Laplace transforms of $f(t)$ and

Table 3.12-2

Property	Equation Number
$F(z) = \sum \left[\text{residues of } \dfrac{F(s)}{1 - \epsilon^{sT} z^{-1}} \text{ at the poles of } F(s) \right]$	(3.12-16)
$F^*(s) = \dfrac{1}{T} \sum_{k=-\infty}^{\infty} F\left(s + j\dfrac{2\pi k}{T}\right) = F^*\left(s + j\dfrac{2\pi n}{T}\right)$	(3.12-17)
If $F(s) = F_1(s)F_2^*(s)$, then $F(z) = F_1(z)F_2(z)$	(3.12-18)

$f^*(t)$, respectively. To prove the first property, rewrite Eq. 3.11-1 as

$$f^*(t) = f(t)i(t)$$

where $i(t)$ denotes the impulse train

$$i(t) = \sum_{n=-\infty}^{\infty} U_0(t - nT)$$

Then

$$F^*(s) = \mathscr{L}[f(t)i(t)]$$

which suggests the use of the complex convolution theorem of Eqs. 3.7-7 and 3.7-8. Using the result of Example 3.4-3,

$$I(s) = \frac{1}{1 - \epsilon^{-Ts}}$$

This function has first order poles at

$$\epsilon^{-Ts} = 1$$

$$Ts = \pm j2\pi k$$

$$s = \pm j\frac{2\pi k}{T} \qquad \text{for } k = 0, 1, 2, \ldots$$

Using Eqs. 3.7-7 and 3.7-8,

$$F^*(s) = \frac{1}{2\pi j} \int_C \frac{F(w)}{1 - \epsilon^{-T(s-w)}} \, dw$$

$$= \sum \left[\text{residues of } \frac{F(w)}{1 - \epsilon^{-Ts}\epsilon^{wT}} \text{ at the poles of } F(w) \right] \qquad (3.12\text{-}19)$$

Replacing ϵ^{Ts} by z,

$$F(z) = \sum \left[\text{residues of } \frac{F(w)}{1 - \epsilon^{wT} z^{-1}} \text{ at the poles of } F(w) \right]$$

This is identical with Eq. 3.12-16, if the dummy variable w is changed back to s. The expression provides a method of obtaining $F(z)$ directly in closed form, without having to sum an infinite series. Furthermore, in the analysis of systems, $F(s)$ rather than $f(t)$ may be given. In this case, Eq. 3.12-16 permits the calculation of $F(z)$ without having to find $f(t)$ as an intermediate step.

Example 3.12-1. Find $\mathscr{Z}[t\epsilon^{-at}]$.

$$F(z) = \left[\text{residue of } \frac{1}{(s+a)^2(1-\epsilon^{sT}z^{-1})} \text{ at } s = -a\right]$$

$$= \left[\frac{d}{ds}\left(\frac{1}{1-\epsilon^{sT}z^{-1}}\right)\right]_{s=-a} = \frac{T\epsilon^{-aT}z}{(z-\epsilon^{-aT})^2}$$

The same result follows from Table 3.11-1 and Eq. 3.12-7 or 3.12-8.

Example 3.12-2. Find $F(z)$ corresponding to $F(s) = [s(2s+3)]/[(s+1)^2(s+2)]$.

The residue of $\dfrac{F(s)}{[1-\epsilon^{sT}z^{-1}]}$ at $s = -1$ is

$$\left[\frac{d}{ds}\left\{\frac{s(2s+3)}{(s+2)(1-\epsilon^{sT}z^{-1})}\right\}\right]_{s=-1} = \frac{-T\epsilon^{-T}z}{(z-\epsilon^{-T})^2}$$

The residue at $s = -2$ is

$$\left[\frac{s(2s+3)}{(s+1)^2(1-\epsilon^{sT}z^{-1})}\right]_{s=-2} = \frac{2z}{z-\epsilon^{-2T}}$$

Hence

$$F(z) = \frac{2z}{z-\epsilon^{-2T}} - \frac{T\epsilon^{-T}z}{(z-\epsilon^{-T})^2}$$

The same result can be obtained by finding

$$\mathscr{L}^{-1}\left[\frac{s(2s+3)}{(s+1)^2(s+2)}\right] = 2\epsilon^{-2t} - t\epsilon^{-t}$$

and by looking up the Z transform of the two functions of time in the tables.

The proof of Eq. 3.12-17 follows from a simple modification of Eq. 3.12-19. As shown in Eq. 3.7-8, it is possible to replace the sum of the residues of $F(w)/[1-\epsilon^{-T(s-w)}]$ at the poles of $F(w)$ by minus the sum of the residues of the same function at the poles of $1/[1-\epsilon^{-T(s-w)}]$. These poles are located at

$$w = s \pm j\frac{2\pi k}{T} \quad \text{for } k = 0, 1, 2, \ldots$$

The residue of $F(w)/[1-\epsilon^{-T(s-w)}]$ at one of these first order poles is, by Eq. 3.6-20,

$$\left\{\frac{F(w)}{(d/dw)[1-\epsilon^{-T(s-w)}]}\right\}_{w=s\pm j(2\pi k/T)} = -\frac{1}{T}F\left(s \pm j\frac{2\pi k}{T}\right)$$

Thus

$$F^*(s) = \frac{1}{T}\sum_{k=-\infty}^{\infty} F\left(s + j\frac{2\pi k}{T}\right)$$

which is identical with the first half of Eq. 3.12-17. An alternative proof of this result is suggested in one of the problems at the end of the chapter. If s is replaced by $s + j(2\pi n/T)$,

$$F^*\left(s + j\frac{2\pi n}{T}\right) = \frac{1}{T}\sum_{k=-\infty}^{\infty} F\left(s + j\frac{2\pi(k+n)}{T}\right)$$

The right side may be written

$$\frac{1}{T}\sum_{m=-\infty}^{\infty} F\left(s + j\frac{2\pi m}{T}\right) = F^*(s)$$

which proves the second half of Eq. 3.12-17.†

Although perhaps the most important application of Eq. 3.12-17 is in the proof of other theorems, the equation also gives considerable insight into the sampling of continuous signals. An assumed plot of the frequency spectrum of a continuous signal $f(t)$ is shown in Fig. 3.12-1*a*. It is assumed that the signal is band-limited, so that it contains no frequency components above ω_c radians per second. The frequency spectrum for $F^*(s)$ is given by

$$|F^*(j\omega)| = \frac{1}{T}\left|\sum_{k=-\infty}^{\infty} F\left[j\left(\omega + \frac{2\pi k}{T}\right)\right]\right|$$

and is shown in parts (*b*) and (*c*) of the figure for two different values of the sampling interval T. The plot consists of the spectrum of the continuous signal *repeated every* $2\pi/T$ *radians per second.* If $\pi/T > \omega_c$, as assumed in part (*b*), the original shape of $|F(j\omega)|$ is not destroyed by the

† Equations 3.12-16 and 3.12-17 are given in the form usually used in the literature, but they are not consistent if $f(t)$ has a discontinuity at $t = 0$. Equation 3.12-16 is correct if the value of the function at the origin is defined to be $f(0+)$, while Eq. 3.12-17 assumes that this value is $\frac{1}{2}[f(0-) + f(0+)]$. The inconsistency can be explained by recalling that Eq. 3.7-8 requires that the integral vanish along *DFA* and *DHA* in Fig. 3.7-7, and by carefully examining the integration along these semicircles. Unless otherwise stated, it is customary to use $f(0+)$ as the function's value at $t = 0$. Since $f(0-) = 0$ in the one-sided transform, Eq. 3.12-17 should then read

$$F^*(s) = \tfrac{1}{2}f(0+) + \frac{1}{T}\sum_{k=-\infty}^{\infty} F\left(s + j\frac{2\pi k}{T}\right)$$

In the proof of Eq. 3.12-17 that is suggested in Problem 3.31, the Fourier series for the impulse train assumes that $i(t)$ is an even function of time. The one-sided Laplace transform then includes the effect of only half of the impulse at the origin, so the extra term $\frac{1}{2}f(0+)$ is again needed.

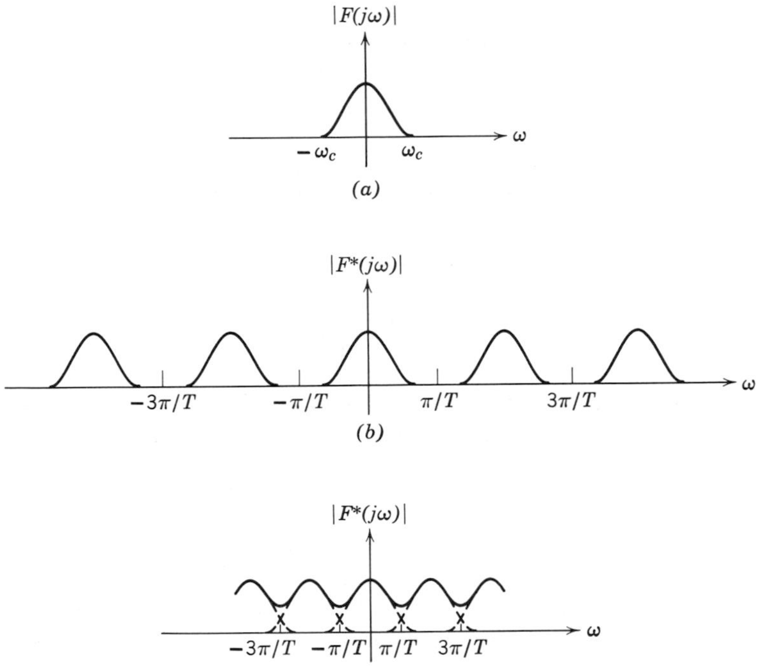

Fig. 3.12-1

sampling process. If, however, $\pi/T < \omega_c$ as in part (c), the original shape of $|F(j\omega)|$ no longer appears in the frequency spectrum of the sampled signal. Thus the continuous signal $f(t)$ can be theoretically reproduced from the sampled signal $f^*(t)$ by linear filtering if and only if $\pi/T > \omega_c$. This statement agrees with Shannon's sampling theorem.[17]

The reconstruction of the continuous signal could be accomplished by an ideal low-pass filter, but approximating an ideal filter requires a large number of components and results in a great time delay. In many sampled-data systems, the circuit following the sampling switch does act as a crude low-pass filter, reducing the high-frequency components and smoothing out the signal.[18]

If the s-plane is divided into horizontal strips as in Fig. 3.12-2, Eq. 3.12-17 indicates that the pole-zero pattern in each strip is the same. Values of $F^*(s)$ at corresponding points in different strips are identical. Thus $F^*(s)$ is uniquely described by its values in the strip $-\pi/T < \omega < \pi/T$.

It is useful to note that the left half of the s-plane maps into the interior

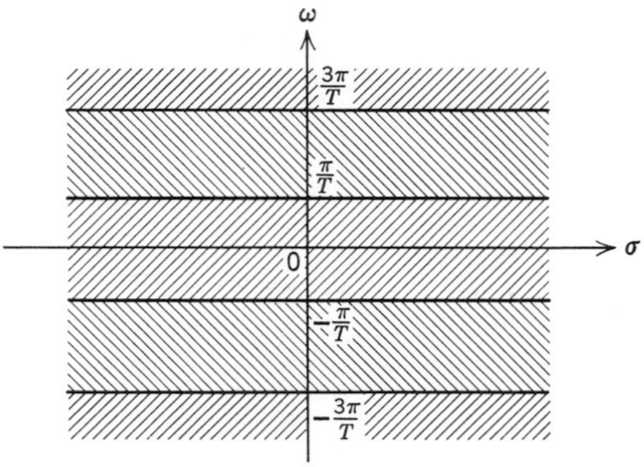

Fig. 3.12-2

of the unit circle in the z-plane, as shown in Fig. 3.12-3. If $s = \sigma + j\omega$ and $z = \rho\epsilon^{j\theta}$, where z has been written in polar form, the transformation $z = \epsilon^{sT}$ gives

$$\rho\epsilon^{j\theta} = \epsilon^{\sigma T}\epsilon^{j\omega T} \tag{3.12-20}$$

Then $\rho = \epsilon^{\sigma T}$ and $\theta = \omega T$. For $\sigma < 0$, $\sigma = 0$, or $\sigma > 0$, the magnitude of z is $\rho < 1$, $\rho = 1$ (the unit circle), or $\rho > 1$, respectively. Note also that, if ω is replaced by $\omega + 2\pi/T$, the value of z is unchanged. Thus *each* horizontal strip in Fig. 3.12-2 maps into the entire z-plane. This is, of course, consistent with the previous conclusion that the values of $F^*(s)$ for $-\pi/T < \omega < \pi/T$ uniquely determine $F^*(s)$ and hence $F(z)$. It also explains why there are a finite number of poles of $F(z)$, despite an infinite number of poles of $F^*(s)$.

Equation 3.12-18, which is used in the next section, follows directly from Eq. 3.12-17. If $F(s) = F_1(s)F_2^*(s)$, then

$$F^*(s) = \frac{1}{T}\sum_{k=-\infty}^{\infty}\left[F_1\left(s + j\frac{2\pi k}{T}\right)F_2^*\left(s + j\frac{2\pi k}{T}\right)\right]$$

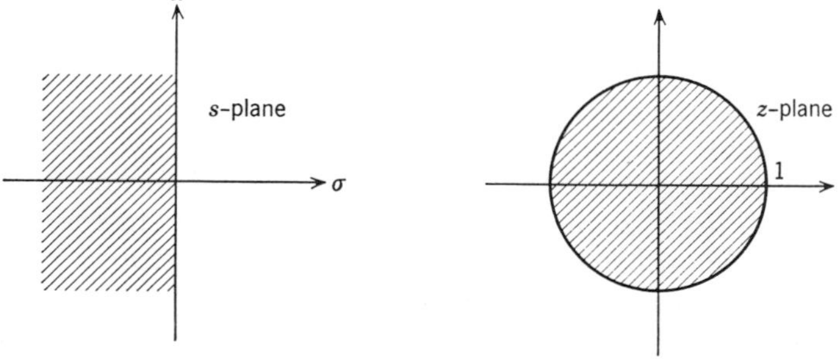

Fig. 3.12-3

Since

$$F_2^*\left(s + j\,\frac{2\pi k}{T}\right) = F_2^*(s)$$

for all integral values of k, it follows that

$$F^*(s) = F_2^*(s)\,\frac{1}{T}\sum_{k=-\infty}^{\infty} F_1\left(s + j\,\frac{2\pi k}{T}\right) = F_2^*(s)F_1^*(s)$$

or

$$F(z) = F_2(z)F_1(z)$$

The Inverse Transform

There are three common ways of finding $f(nT)$, or equivalently $f^*(t)$, when $F(z)$ is given: by infinite series, partial fraction expansion, and an inversion integral. As shown in Examples 3.11-1 and 3.11-2, when the two-sided Z transform is used, the region of convergence must be specified in order to uniquely determine $f(t)$ at the sampling instants. This ambiguity is best resolved by the third of the three inversion methods; thus the initial discussion is restricted to the one-sided transform.

The first method reconstructs the infinite series appearing in the definition of Eq. 3.11-8, repeated below.

$$F(z) = f(0) + f(T)z^{-1} + \cdots + f(nT)z^{-n} + \cdots \tag{3.12-21}$$

The values of $f(nT)$ can then be determined by inspection. Equation 3.12-21 can be regarded as a Taylor series written about the point at infinity. If $\phi(z)$ denotes the function formed from $F(z)$ by replacing z by $1/z$,

$$\phi(z) = f(0) + f(T)z + \cdots + f(nT)z^n + \cdots$$

is a Taylor series about the origin. From Eq. 3.6-12,

$$f(nT) = \left(\frac{1}{n!}\right)\frac{d^n\phi(0)}{dz}$$

As discussed in Section 3.6, however, it is usually easier to find the coefficients by long division. If $\phi(z)$ is the quotient of two polynomials, they should both be arranged in ascending order to obtain an infinite series of the proper form. When dealing directly with $F(z)$, therefore, both the numerator and denominator must be written in ascending powers of (z^{-1}).

Example 3.12-3. $F(z) = \dfrac{2z}{(z-2)(z-1)^2} = \dfrac{2z^{-2}}{1 - 4z^{-1} + 5z^{-2} - 2z^{-3}}$. Determine $f(nT)$.

By long division,

$$F(z) = 2z^{-2} + 8z^{-3} + 22z^{-4} + \cdots$$

Thus

$$\begin{aligned} f(0) &= 0 \\ f(T) &= 0 \\ f(2T) &= 2 \\ f(3T) &= 8 \\ f(4T) &= 22 \\ &\cdots\cdots\cdots \end{aligned}$$

This method is often easier than any other when $f(nT)$ is desired for only a few values of n. A disadvantage is that it does not give a general expression for the nth term in closed form.

A second method is based upon the partial fraction expansion of $F(z)/z$ and the use of a short table of transforms, such as Table 3.11-1. $F(z)$ itself is not expanded in partial fractions, because the functions of z appearing in the table have the factor z in their numerators.

Example 3.12-4. $F(z) = \dfrac{2z}{(z-2)(z-1)^2}$. Find $f(nT)$.

Using the techniques of Section 3.4,

$$\frac{F(z)}{z} = \frac{2}{(z-2)(z-1)^2} = \frac{2}{z-2} - \frac{2}{(z-1)^2} - \frac{2}{z-1}$$

$$F(z) = 2\left[\frac{z}{z-2} - \frac{z}{(z-1)^2} - \frac{z}{z-1}\right]$$

Assuming that $T = 1$, and using Table 3.11-1,

$$f(n) = 2[2^n - n - 1] \quad \text{for } n \geqslant 0$$

By Theorem 3.12-4, if T is not necessarily unity,

$$f(nT) = 2[2^n - n - 1] \quad \text{for } n \geqslant 0$$

The reader should verify that, for $n = 0, 1, 2, \ldots$, this result checks the answer of the previous example.

Example 3.12-5. $F(z) = \dfrac{z(z^2 - 2z - 1)}{(z^2+1)^2}$. Find $f(nT)$.

The partial fraction expansion is

$$\frac{F(z)}{z} = \frac{A}{(z-j)^2} + \frac{B}{(z-j)} + \frac{C}{(z+j)^2} + \frac{D}{(z+j)}$$

where

$$A = \frac{1+j}{2}, \quad C = \frac{1-j}{2}, \quad B = D = 0$$

Then

$$F(z) = \frac{1+j}{2}\,\frac{-j(-jz)}{(-jz-1)^2} + \frac{1-j}{2}\,\frac{j(jz)}{(jz-1)^2}$$

$$= \frac{1-j}{2}\,\frac{\epsilon^{-j\pi/2}z}{(\epsilon^{-j\pi/2}z-1)^2} + \frac{1+j}{2}\,\frac{\epsilon^{j\pi/2}z}{(\epsilon^{j\pi/2}z-1)^2}$$

Using Table 3.11-1, and Eq. 3.12-7,

$$f(nT) = \tfrac{1}{2}[\sqrt{2}\,\epsilon^{-j\pi/4}n\epsilon^{j\pi n/2} + \sqrt{2}\,\epsilon^{j\pi/4}n\epsilon^{-j\pi n/2}]$$

$$= \sqrt{2}\,n\cos\left(\frac{\pi}{2}n - \frac{\pi}{4}\right) \qquad \text{for } n \geqslant 0$$

This example is typical of the manipulations used when $F(z)$ contains complex conjugate poles.

The third method of finding $f(nT)$ is to use the inversion integral, which may be derived in several ways. For generality, the derivation is performed in terms of the two-sided transform. From Section 3.7, the inversion integral for the two-sided Laplace transform is

$$f(t) = \frac{1}{2\pi j}\int_C F(s)\epsilon^{st}\,ds$$

where the contour C is any path from $c - j\infty$ to $c + j\infty$ within the region of convergence $\sigma_{c_1} < \sigma < \sigma_{c_2}$. For convenience, the contour is taken as a straight vertical line. On the basis of the discussion associated with Figs. 3.12-2 and 3.12-3, it is logical to break this contour up into the individual sections $\ldots, -3\pi/T < \omega < -\pi/T,\ -\pi/T < \omega < \pi/T,\ \pi/T < \omega < 3\pi/T, \ldots.$ Also, since the purpose of the derivation is to obtain $f(nT)$ from $F(z)$, t is replaced by nT.

$$f(nT) = \frac{1}{2\pi j}\sum_{k=-\infty}^{\infty}\int_{c+[j(2k-1)\pi]/T}^{c+[j(2k+1)\pi]/T} F(s)\epsilon^{nTs}\,ds \qquad (3.12\text{-}22)$$

The right side of this equation must be rewritten so as to include $F(z)$, or equivalently $F^*(s)$, which is given in terms of $F(s)$ by Eq. 3.12-17. This is accomplished by replacing the dummy variable s by $s + j(2\pi k/T)$ in Eq. 3.12-22. Since $\epsilon^{j2\pi nk} = 1$,

$$f(nT) = \frac{1}{2\pi j}\sum_{k=-\infty}^{\infty}\int_{c-j\pi/T}^{c+j\pi/T} F\left(s + j\frac{2\pi k}{T}\right)\epsilon^{nTs}\,ds$$

Interchanging the order of the summation and integration,

$$f(nT) = \frac{T}{2\pi j}\int_{c-j\pi/T}^{c+j\pi/T}\frac{1}{T}\sum_{k=-\infty}^{\infty} F\left(s + j\frac{2\pi k}{T}\right)\epsilon^{nTs}\,ds$$

$$= \frac{T}{2\pi j}\int_{c-j\pi/T}^{c+j\pi/T} F^*(s)\epsilon^{nTs}\,ds$$

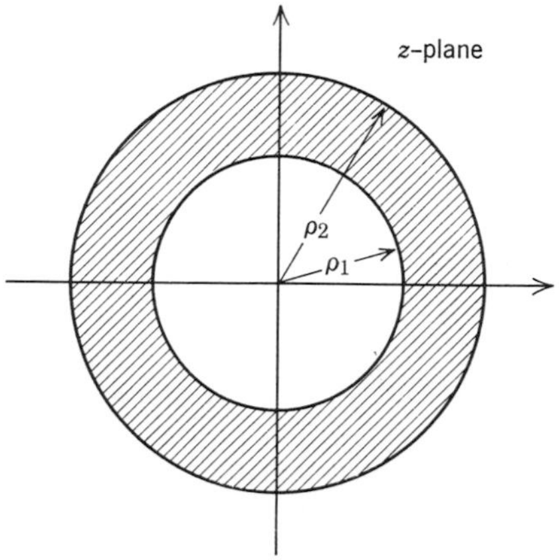

Fig. 3.12-4

Letting $z = \epsilon^{Ts}$ and $dz = Tz\,ds$, and recalling from Eq. 3.12-20 that the vertical line $\sigma = c$ $(-\pi/T < \omega < \pi/T)$ corresponds to a circle of radius ϵ^{cT} in the z-plane,

$$f(nT) = \frac{1}{2\pi j}\oint F(z)z^{n-1}\,dz \tag{3.12-23}$$

Figure 3.12-4 shows two circles of radii $\rho_1 = \exp(T\sigma_{c_1})$ and $\rho_2 = \exp(T\sigma_{c_2})$. The closed contour in Eq. 3.12-23 must be confined to the shaded area, which is the region of convergence for the direct Z transform. $F(z)$ is analytic at all points in this region. By the residue theorem of Eq. 3.6-16,

$$f(nT) = \sum [\text{residues of } F(z)z^{n-1}] \tag{3.12-24}$$

at the poles in the region $|z| < \rho_1$. This equation is perfectly general, and it can be used for both positive and negative values of n.

Example 3.12-6. $F(z) = z[(2 - \epsilon)z - 1]/[(z - \epsilon)(z - 1)^2]$, with the region of convergence $1 < |z| < \epsilon$. Find $f(nT)$.

$$F(z)z^{n-1} = \frac{z^n[(2 - \epsilon)z - 1]}{(z - \epsilon)(z - 1)^2}$$

For $n \geqslant 0$, the only pole inside the contour of integration is at $z = 1$. The residue at this pole is

$$\left[\frac{d}{dz}\left(\frac{z^n[(2 - \epsilon)z - 1]}{z - \epsilon}\right)\right]_{z=1} = n + 1$$

Hence

$$f(nT) = n + 1 \quad \text{for } n \geqslant 0$$

For $n = -1$, $F(z)z^{n-1}$ has an additional pole at $z = 0$. The residue at $z = 1$ is zero, and the residue at $z = 0$ is

$$\left[\frac{(2-\epsilon)z - 1}{(z-\epsilon)(z-1)^2}\right]_{z=0} = \epsilon^{-1}$$

Thus

$$f(-T) = \epsilon^{-1}$$

For $n = -2$, the residue at the pole at $z = 1$ is -1. The residue at the second order pole at the origin is

$$\left[\frac{d}{dz}\frac{(2-\epsilon)z - 1}{(z-\epsilon)(z-1)^2}\right]_{z=0} = 1 + \epsilon^{-2}$$

so $f(-2T) = \epsilon^{-2}$.

The procedure above for calculating $f(nT)$ for negative values of n is unnecessarily tedious. If $F(z)z^{n-1}$ has a second or higher order zero at infinity, then[19]

$$\frac{1}{2\pi j}\oint_C F(z)z^{n-1}\,dz$$
$$= -\sum [\text{residues of } F(z)z^{n-1} \text{ at the poles outside contour } C] \quad (3.12\text{-}25)$$

In the preceding example, the residue at the pole at $z = \epsilon$, the only one outside of the contour of integration, is

$$\left[\frac{z^n[(2-\epsilon)z - 1]}{(z-1)^2}\right]_{z=\epsilon} = -\epsilon^n$$

Thus $f(nT) = \epsilon^n$ for $n \leqslant 0$. It is seen that poles of $F(z)$ in the regions $|z| < \rho_1$ or $|z| > \rho_2$ in Fig. 3.12-4 give rise to components of $f(nT)$ for $n > 0$ and $n < 0$, respectively. In the one-sided Z transform, the discussion of Section 3.4 indicates that $\rho_2 \to \infty$, and $f(nT)$ becomes zero for $n < 0$. The region of convergence is then the entire z-plane outside of the circle of radius ρ_1, which encloses all the poles of $F(z)$.

Example 3.12-7. The one-sided transform is $F(z) = 2z/(z-2)(z-1)^2$. Find $f(nT)$.
The residue of $F(z)z^{n-1}$ at $z = 2$ is

$$\left[\frac{2z^n}{(z-1)^2}\right]_{z=2} = 2(2^n)$$

and at $z = 1$ is

$$\left[\frac{d}{dz}\left(\frac{2z^n}{z-2}\right)\right]_{z=1} = -2(n+1)$$

so

$$f(nT) = 2[2^n - n - 1] \quad \text{for } n \geqslant 0$$

agreeing with the result of Examples 3.12-3 and 3.12-4.

The Significance of Poles and Zeros

The nature of the terms in $f(nT)$ depends upon the location of the poles of $F(z)$. Table 3.12-3, showing the effect of pole position in the one-sided transform, can be constructed from Eq. 3.12-24 or its equivalent. Poles

Table 3.12-3

Location of Poles of $F(z)$ in the z-plane	Corresponding Terms in $f(nT)$
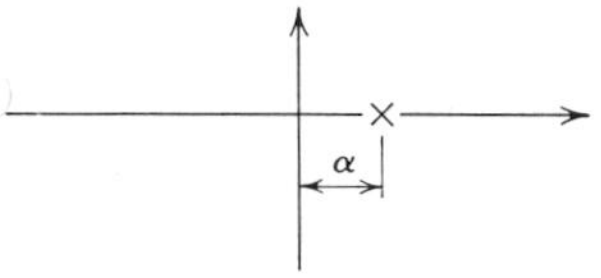	$A\alpha^n$
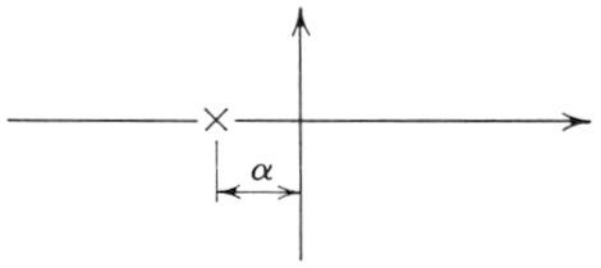	$A(-1)^n\alpha^n$
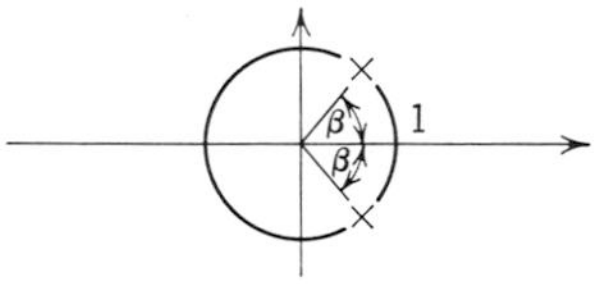	$A \cos (\beta n + \theta)$
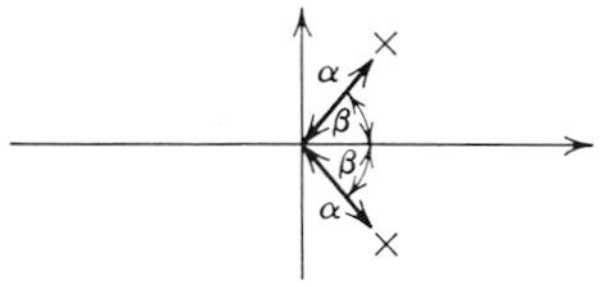	$A\alpha^n \cos (\beta n + \theta)$
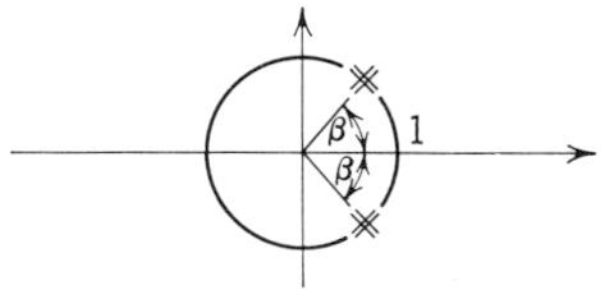	$An \cos (\beta n + \theta_1)$ $+B \cos (\beta n + \theta_2)$

inside or outside the unit circle yield terms that approach zero or infinity, respectively, as t approaches infinity. First order poles on the unit circle (except at $z = 1$) yield sinusoidal terms of constant amplitude, but higher order poles produce terms that increase without limit as t approaches infinity.

When the two-sided transform is used, the form of $f(nT)$ depends upon the region of convergence as well as the location of the poles. If the function of time is known to be bounded, the region of convergence must include the unit circle. In this case, poles inside and outside the unit circle yield components of $f(nT)$ for positive and negative time, respectively.

3.13 APPLICATION OF THE *Z* TRANSFORM TO DISCRETE SYSTEMS

The two basic methods of analyzing fixed continuous systems using the Laplace transform are the transformation of differential equations into algebraic equations and the system function concept. Similar methods exist for the analysis of fixed systems with discrete signals using the Z transform.

The Solution of Difference Equations

Any fixed linear system whose input $v(t)$ and output $y(t)$ are defined only at the discrete instants $t = kT$ can be described by the difference equation

$$a_q y(kT + qT) + a_{q-1} y(kT + qT - T) + \cdots + a_1 y(kT + T) + a_0 y(kT) = b_m v(kT + mT) + \cdots + b_1 v(kT + T) + b_0 v(kT) \qquad (3.13\text{-}1)$$

where the a_i's and b_j's are constants. By the use of Eqs. 3.12-1 through 3.12-5, the difference equation can be transformed into an algebraic equation in z, which can be solved for $Y(z)$. $y(kT)$ can then be found by the inversion methods of the last section. Note that the transformed equation will contain $y(0)$ through $y(qT - T)$. These terms represent the q boundary conditions needed to determine the arbitrary constants in the classical solution.

Example 3.13-1. Find a general expression for $Y(z)$ when

$$a_3 y(k + 3) + a_2 y(k + 2) + a_1 y(k + 1) + a_0 y(k) = b_3 v(k + 3) + b_2 v(k + 2) + b_1 v(k + 1) + b_0 v(k)$$

and when $y(k)$ and $v(k)$ are zero for $k < 0$.

Using Eqs. 3.12-1 through 3.12-5 with $T = 1$, the transformed equation becomes

$$[a_3 z^3 + a_2 z^2 + a_1 z + a_0] Y(z) - z^3[a_3 y(0)] - z^2[a_3 y(1) + a_2 y(0)] - z[a_3 y(2) + a_2 y(1) + a_1 y(0)]$$
$$= [b_3 z^3 + b_2 z^2 + b_1 z + b_0] V(z) - z^3[b_3 v(0)] - z^2[b_3 v(1) + b_2 v(0)] - z[b_3 v(2) + b_2 v(1) + b_1 v(0)]$$

The constants $y(0), v(0), \ldots, y(2), v(2)$ are not directly given. Substituting $k = -3, -2$, and -1 into the original difference equation, however, and noting that $y(k)$ and $v(k)$ are zero for $k < 0$,

$$a_3y(0) = b_3v(0)$$
$$a_3y(1) + a_2y(0) = b_3v(1) + b_2v(0)$$
$$a_3y(2) + a_2y(1) + a_1y(0) = b_3v(2) + b_2v(1) + b_1v(0)$$

The transformed equation then becomes

$$Y(z) = \frac{b_3z^3 + b_2z^2 + b_1z + b_0}{a_3z^3 + a_2z^2 + a_1z + a_0} V(z)$$

By generalizing the result of the last example, it is possible to write down the transformed output by inspection of the difference equation for any nonanticipatory system that is initially at rest. For such systems, $m \leqslant q$ in Eq. 3.13-1, and $y(k)$ and $v(k)$ are zero for $k < 0$. The boundary condition terms diappear, and

$$Y(z) = \frac{b_mz^m + \cdots + b_1z + b_0}{a_qz^q + \cdots + a_1z + a_0} V(z) \tag{3.13-2}$$

This equation is valid for any value of T.

Example 3.13-2. Solve

$$y(k + 2) - 3y(k + 1) + 2y(k) = 2v(k + 1) - 2v(k)$$

when the system is initially at rest, and when

$$v(k) = \begin{cases} k & \text{for } k \geqslant 0 \\ 0 & \text{for } k < 0 \end{cases}$$

From Eq. 3.13-2 and Table 3.11-1,

$$Y(z) = \frac{2(z - 1)}{z^2 - 3z + 2} V(z) = \left[\frac{2}{z - 2}\right]\left[\frac{z}{(z - 1)^2}\right]$$

From Example 3.12-7,

$$y(n) = 2[2^n - n - 1] \quad \text{for } n \geqslant 0$$

which agrees with the answer to Example 1.9-1.

The System Function

Equation 3.13-2 indicates that there is a direct relationship between the transformed input and output of a system initially at rest. The system function (or "transfer function")† is defined as

$$H(z) = \frac{Y(z)}{V(z)} \tag{3.13-3}$$

† $H(z)$ is sometimes called the "pulsed transfer function" or "sampled transfer function" to distinguish it from $H(s)$.

and can be written down directly from the difference equation, as in Eq. 3.13-2. In Section 2.12, it was shown that $H(z)z^k$ is the forced response to $v(kT) = z^k$. From Eqs. 2.12-37 and 2.12-38,

$$H(z) = \frac{b_m z^m + \cdots + b_1 z + b_0}{a_q z^q + \cdots + a_1 z + a_0} = \left[\frac{y_P(kT)}{v(kT)}\right]_{v(kT)=z^k} \qquad (3.13\text{-}4)$$

Equation 3.13-3 enables the output $y(nT)$ to be found by the same three-step procedure used for the Fourier and Laplace transforms. For a discrete system subjected to the unit delta function $v(kT) = \begin{cases} 1 \text{ for } k = 0 \\ 0 \text{ for } k \neq 0 \end{cases}$, $V(z) = 1$, and

$$d(nT) = \mathscr{Z}^{-1}[H(z)] \qquad (3.13\text{-}5)$$

where $d(nT)$ is the response to the unit delta function. Equations 3.13-4 and 3.13-5 provide a convenient way of finding a system's delta response from its difference equation.

Example 3.13-3. If the discrete system of Fig. 1.9-1a is described by the difference equation of Example 3.13-2, find its delta response.

$$d(n) = \mathscr{Z}^{-1}\left[\frac{2}{z-2}\right]$$

This form does not appear in Table 3.11-1, but, by the initial value theorem of Eq. 3.12-14,

$$d(0) = \lim_{z\to\infty} \frac{2}{z-2} = 0$$

By Eq. 3.12-3,

$$d(k+1) = \mathscr{Z}^{-1}\left[\frac{2z}{z-2}\right]$$

This form does appear in Table 3.11-1, hence

$$d(k+1) = 2^{k+1} \quad \text{for } k \geqslant 0$$

or

$$d(k) = 2^k \quad \text{for } k \geqslant 1$$

which agrees with the answer to Example 2.12-6.

For a continuous system preceded by a sampling switch, as in Fig. 3.11-1, the Laplace transformed output is

$$Y(s) = H(s)V^*(s) \qquad (3.13\text{-}6)$$

where $H(s)$ is the transform of the impulse response $h(t)$. By Eq. 3.12-18,

$$Y(z) = V(z)\mathscr{Z}[h(t)] \qquad (3.13\text{-}7)$$

By comparison with Eq. 3.13-3,

$$H(z) = \mathscr{Z}[h(t)] = \mathscr{Z}[h^*(t)] \tag{3.13-8}$$

It is important to note that, while Eq. 3.13-6 can be solved for the output as a continuous function of time, Eq. 3.13-7 yields the output only at the sampling instants. The calculation of the output between sampling instants is deferred until the next section. The determination of $H(z)$ for more complicated configurations is illustrated by the following examples.

Example 3.13-4. Find $H(z) = Y(z)/V(z)$ for the configurations of Fig. 3.13-1. The systems are initially at rest, and the sampling switches operate synchronously.

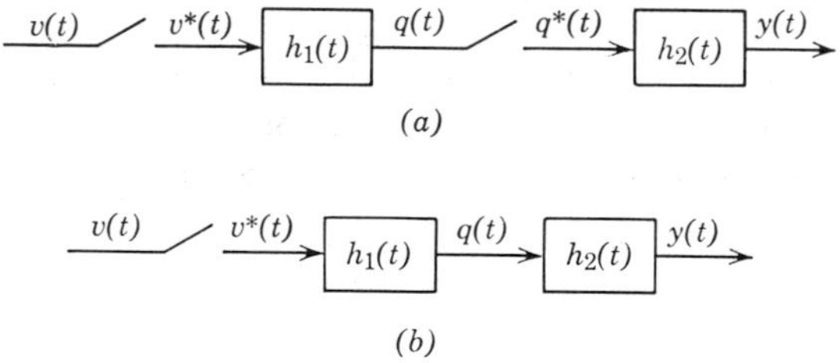

Fig. 3.13-1

In part (a) of the figure, $Q(z) = H_1(z)V(z)$ and $Y(z) = H_2(z)Q(z)$; so $H(z) = H_1(z)H_2(z)$. In part (b), $H(z)$ is the Z transform corresponding to $H_1(s)H_2(s)$. This can be written

$$H(z) = \mathscr{Z}[h(t)]$$

where $h(t) = \mathscr{L}^{-1}[H_1(s)H_2(s)]$, or, by Eq. 3.12-16,

$$H(z) = \sum \left[\text{residues of } \frac{H_1(s)H_2(s)}{1 - \epsilon^{sT}z^{-1}} \text{ at the poles of } H_1(s)H_2(s) \right]$$

The shorthand notation

$$H(z) = H_1H_2(z)$$

is often used, but the reader must remember that $H_1H_2(z) \neq H_1(z)H_2(z)$.

Example 3.13-5. In the previous example, find $y(t)$ for $t = nT$ if $h_1(t) = h_2(t) = t\epsilon^{-t}$ for $t \geqslant 0$, $T = 1$, and $v(t) = \epsilon^{-t}U_{-1}(t)$.

For the configuration of part (a) of Fig. 3.13-1,

$$H_1(z) = H_2(z) = \mathscr{Z}[t\epsilon^{-t}] = \frac{z\epsilon}{(z\epsilon - 1)^2}$$

$$V(z) = \frac{z\epsilon}{z\epsilon - 1}$$

$$Y(z) = \frac{(z\epsilon)^3}{(z\epsilon - 1)^5}$$

By Eqs. 3.12-7 and 3.12-24,

$$y(n) = \epsilon^{-n}\left[\text{residue of } \frac{z^{n+2}}{(z-1)^5} \quad \text{at } z = 1\right]$$

$$= \epsilon^{-n}\left[\frac{1}{4!}\frac{d^4}{dz^4}z^{n+2}\right]_{z=1} = \frac{1}{24}(n+2)(n+1)(n)(n-1)\epsilon^{-n} \quad \text{for } n \geqslant 0$$

which agrees with the answer to Example 1.9-4 at the sampling instants.

For the configuration of part (*b*) of the figure, $h(t) = t^3\epsilon^{-t}/6$ by Example 1.6-2, so

$$H(z) = \mathscr{Z}\left[\frac{t^3\epsilon^{-t}}{6}\right] = \frac{(z\epsilon)(z^2\epsilon^2 + 4z\epsilon + 1)}{6(z\epsilon - 1)^4}$$

Then

$$Y(z) = \frac{(z\epsilon)^2(z^2\epsilon^2 + 4z\epsilon + 1)}{6(z\epsilon - 1)^5}$$

and

$$y(n) = \frac{\epsilon^{-n}}{6}\left[\frac{1}{4!}\frac{d^4}{dz^4}(z^{n+3} + 4z^{n+2} + z^{n+1})\right]_{z=1}$$

$$= \frac{n^2(n+1)^2}{24}\epsilon^{-n} \quad \text{for } n \geqslant 0$$

which agrees with the answer to Example 1.9-3.

Example 3.13-6. Figure 3.13-2 shows a common sampled-data system, labeled with Laplace transformed quantities. Find an expression for $Y(z)$.

$$Q(s) = V(s) - G(s)H(s)Q^*(s)$$

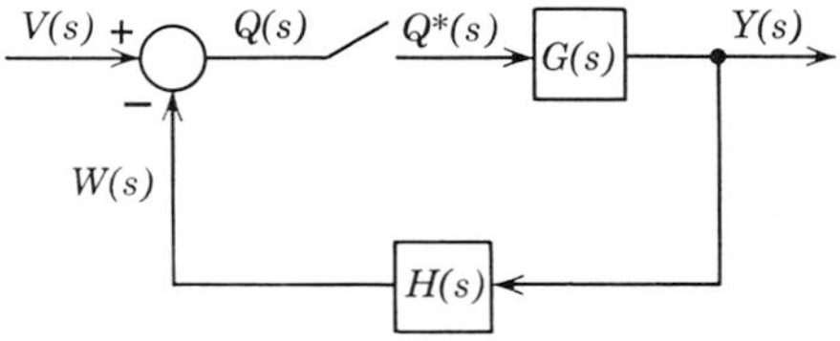

Fig. 3.13-2

Using Eq. 3.12-18,

$$Q(z) = V(z) - GH(z)Q(z)$$

$$Y(z) = G(z)Q(z) = \frac{G(z)V(z)}{1 + GH(z)}$$

where the symbol $GH(z)$ is explained in Example 3.13-4.

The Convolution Summations

Equation 3.12-12, rewritten below, provides a bridge between the time-domain and transform techniques. If $F(z) = F_1(z)F_2(z)$, then

$$f(nT) = \sum_{k=0}^{n} f_1(nT - kT)f_2(kT) \tag{3.13-9}$$

The proof of this convolution theorem follows from the basic definition of the Z transform,

$$F_1(z) = \sum_{n=0}^{\infty} f_1(nT)z^{-n}$$

$$F_2(z) = \sum_{n=0}^{\infty} f_2(nT)z^{-n}$$

The product of the two infinite series is

$$F_1(z)F_2(z) = [f_1(0)f_2(0)] + z^{-1}[f_1(0)f_2(T) + f_1(T)f_2(0)] + \cdots + z^{-n}\left[\sum_{k=0}^{n} f_1(nT - kT)f_2(kT)\right] + \cdots$$

thus proving Eq. 3.13-9.

For nonanticipatory systems initially at rest, this equation can be combined with Eqs. 3.13-3 through 3.13-8. For a discrete system with input $v(kT)$, output $y(kT)$, and delta response $d(kT)$, $Y(z) = V(z)\mathscr{Z}[d(kT)]$. By Eq. 3.13-9,

$$y(nT) = \sum_{k=0}^{n} v(kT)\, d(nT - kT) = \sum_{k=0}^{n} v(nT - kT)\, d(kT) \qquad (3.13\text{-}10)$$

agreeing with Eqs. 1.9-1 and 1.9-2. For the continuous system and sampling switch of Fig. 3.11-1, $Y(z) = V(z)\mathscr{Z}[h(t)]$; hence the output at the sampling instants is

$$y(nT) = \sum_{k=0}^{n} v(kT)h(nT - kT) = \sum_{k=0}^{n} v(nT - kT)h(kT) \qquad (3.13\text{-}11)$$

agreeing with Eqs. 1.9-5 and 1.9-6.

The Free and the Forced Response

Consider the output

$$y(nT) = \mathscr{Z}^{-1}[V(z)H(z)] \qquad (3.13\text{-}12)$$

As in Section 3.8, the poles of $V(z)$ and $H(z)$ give rise to the forced (or particular) and free (or complementary) response, respectively. A system is unstable if its free response increases without limit as t approaches infinity. Table 3.12-3 indicates that a system is unstable if and only if $H(z)$ has poles outside the unit circle, or multiple order poles on the unit circle. From Eq. 3.13-4, the poles of $H(z)$ are the roots of the characteristic equation

$$a_q z^q + \cdots + a_1 z + a_0 = 0$$

Hence the above conclusions about stability agree with those of Section 2.12.

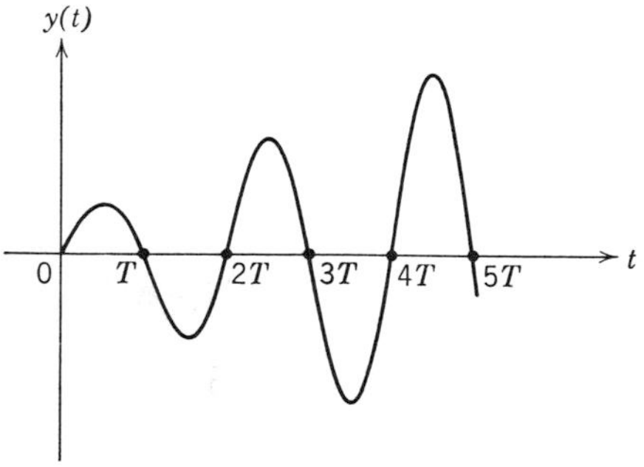

Fig. 3.13-3

A systematic procedure for determining whether or not a polynomial contains roots on or outside the unit circle does exist.[20] Alternatively, the bilinear transformation $z = (1 + w)/(1 - w)$ can be used to map the interior of the unit circle into the left half of the w-plane. Routh's criterion can then be used to determine stability. The references also discuss the application of graphical techniques in the z-plane for stability analysis.[21]

It is theoretically possible for the free response to contain growing oscillations, as in Fig. 3.13-3, even though it remains bounded at the sampling instants. This situation rarely occurs, however.[22]

Time-Varying Systems

The analysis of time-varying systems by the Laplace transform in Section 3.9 has its parallel when using the Z transform. The time-varying system function is defined as

$$H(nT, z) = \mathscr{Z}[d_*(nT, kT)] \tag{3.13-13}$$

for a discrete system, where the transformation is with respect to kT, and where $d_*(nT, kT)$ is the response at time nT to a unit delta function occurring kT seconds earlier. For the configuration of Fig. 3.11-1, the system function is

$$H(nT, z) = \mathscr{Z}[h_*(nT, kT)] \tag{3.13-14}$$

Then

$$y(nT) = \mathscr{Z}^{-1}[V(z)H(nT, z)] \tag{3.13-15}$$

The principal difficulty with this scheme is finding $H(nT, z)$. If, however, the delta or impulse response is known, this can be done by Eq. 3.13-13 or 3.13-14.

3.14 THE MODIFIED Z TRANSFORM

The use of the Z transform in the analysis of Fig. 3.11-1 yields the output $y(t)$ only at the sampling instants $t = nT$. Although a number of methods have been suggested for obtaining the output between sampling instants, the most useful one seems to be the modified Z transform.

The calculation of $\mathscr{Z}[f(t)]$ is based upon samples of $f(t)$ at the instants $t = nT$. In order for the Z transform to contain information about the function of time at other instants, $f(t)$ should be sampled at $t = (n + \gamma)T$, where γ is a real, independent parameter which may assume any value between zero and one. Parts (a) and (b) of Fig. 3.14-1 show an arbitrary function $f(t)$ and a train of unit impulses $i(t)$. Recall that the product

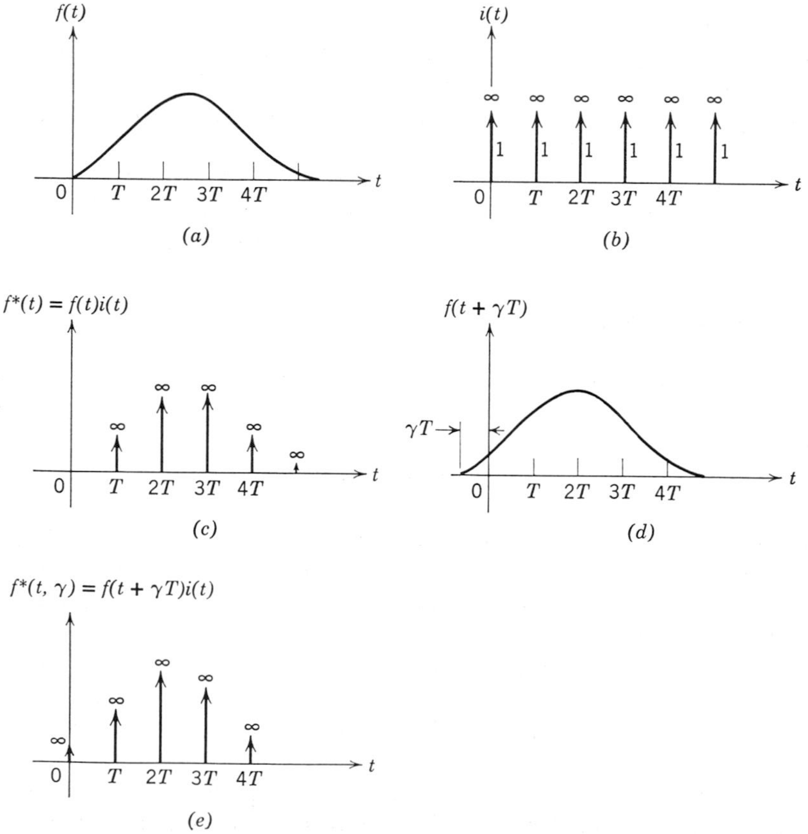

Fig. 3.14-1

$f(t)i(t)$ gives

$$f^*(t) = \sum_{n=-\infty}^{\infty} f(nT)U_0(t - nT) \tag{3.14-1}$$

In order to sample $f(t)$ at the instants $t = (n + \gamma)T$, the plot of $i(t)$ can be moved forward γT seconds. Alternatively, the plot of $f(t)$ can be moved backwards γT seconds, as in part (*d*) of the figure. The product $f(t + \gamma T)i(t)$ gives

$$f^*(t, \gamma) = \sum_{n=-\infty}^{\infty} f(nT + \gamma T)U_0(t - nT) \tag{3.14-2}$$

The Laplace and Z transforms of Eq. 3.14-2 are

$$\mathscr{L}[f^*(t, \gamma)] = \sum_{n=0}^{\infty} f(nT + \gamma T)\epsilon^{-nTs} \tag{3.14-3}$$

$$\mathscr{Z}[f^*(t, \gamma)] = \mathscr{Z}[f(t + \gamma T)] = \mathscr{Z}_\gamma[f(t)]$$

$$= F(z, \gamma) = \sum_{n=0}^{\infty} f(nT + \gamma T)z^{-n} \tag{3.14-4}$$

The various nomenclatures introduced in Eq. 3.14-4 all stand for the Z transform of $f(t + \gamma T)$, otherwise known as the *modified Z transform of $f(t)$*.† Note that

$$F(z) = \lim_{\gamma \to 0} F(z, \gamma) \tag{3.14-5}$$

If $f(t)$ is given, the modified Z transform can be expressed as an infinite series directly from Eq. 3.14-4. A better method, however, is to form $f(t + \gamma T)$ and find the ordinary Z transform of this function by standard tables. Tables of modified transforms do exist, but they save only a small amount of effort.[23]

If $F(s)$ is given, $F(z, \gamma)$ can be found, if desired, without the intermediate step of determining $f(t)$. By a proof similar to that used in deriving Eq. 3.12-16,

$$F(z, \gamma) = \sum\left[\text{residues of } \frac{F(s)\epsilon^{\gamma Ts}}{1 - \epsilon^{Ts}z^{-1}} \text{ at the poles of } F(s)\right] \tag{3.14-6}$$

If the ordinary inverse transform of $F(z, \gamma)$ is taken, with γ treated as an independent parameter,

$$\mathscr{Z}^{-1}[F(z, \gamma)] = f(nT + \gamma T) \tag{3.14-7}$$

† Many authors define the modified Z transform as $F(z, m) = z^{-1}\sum_{n=0}^{\infty} f(nT + mT)z^{-n}$. This is the same as the definition of Eq. 3.14-4, except for the multiplying factor of z^{-1}, but it is not quite as convenient to use. Note that $F(z) = \lim_{m \to 0} zF(z, m)$.

For the configuration of Fig. 3.11-1, one would expect that

$$Y(z, \gamma) = V(z)H(z, \gamma) \tag{3.14-8}$$

where

$$H(z, \gamma) = \mathscr{Z}[h(t + \gamma T)] = \sum_{n=0}^{\infty} h(nT + \gamma T)z^{-n} \tag{3.14-9}$$

is the modified Z transform of the impulse response $h(t)$. To prove this result, form the product

$$\begin{aligned} V(z)H(z, \gamma) &= [v(0) + v(T)z^{-1} + v(2T)z^{-2} + \cdots] \\ &\quad \times [h(\gamma T) + h(T + \gamma T)z^{-1} + h(2T + \gamma T)z^{-2} + \cdots] \\ &= [v(0)h(\gamma T)] + z^{-1}[v(0)h(T + \gamma T) + v(T)h(\gamma T)] \\ &\quad + \cdots + z^{-n}\left[\sum_{k=0}^{n} v(kT)h(nT + \gamma T - kT)\right] + \cdots \end{aligned}$$

By Eq. 1.9-3, the coefficient of z^{-n} is $y(nT + \gamma T)$, so the infinite series above is identical with the definition of $Y(z, \gamma)$, thus proving Eq. 3.14-8. Finally,

$$y(nT + \gamma T) = \mathscr{Z}^{-1}[V(z)H(z, \gamma)] \tag{3.14-10}$$

Example 3.14-1. Solve Example 3.13-5 for $y(t, \gamma)$. In Fig. 3.13-1, $h_1(t) = h_2(t) = t\epsilon^{-t}$ for $t \geqslant 0$, $T = 1$, and $v(t) = \epsilon^{-t}U_{-1}(t)$.

For the configuration of part (*a*) of Fig. 3.13-1,

$$Q(z) = V(z)H_1(z)$$

$$Y(z, \gamma) = V(z)H_1(z)H_2(z, \gamma)$$

where

$$V(z) = \frac{z\epsilon}{(z\epsilon - 1)^2}$$

$$H_1(z) = \frac{z\epsilon}{(z\epsilon - 1)^2}$$

$$H_2(z, \gamma) = \mathscr{Z}[(t + \gamma)\epsilon^{-(t+\gamma)}] = \frac{\epsilon^{-\gamma}(z\epsilon)}{(z\epsilon - 1)^2} + \frac{\gamma\epsilon^{-\gamma}(z\epsilon)}{z\epsilon - 1}$$

Using Eqs. 3.12-7 and 3.12-24,

$$\begin{aligned} y(n, \gamma) &= \mathscr{Z}^{-1}\left[\frac{\epsilon^{-\gamma}(z\epsilon)^3}{(z\epsilon - 1)^5} + \frac{\gamma\epsilon^{-\gamma}(z\epsilon)^3}{(z\epsilon - 1)^4}\right] \\ &= \epsilon^{-n}\epsilon^{-\gamma}\left[\frac{1}{4!}\frac{d^4}{dz^4}(z^{n+2}) + \frac{\gamma}{3!}\frac{d^3}{dz^3}(z^{n+2})\right]_{z=1} \\ &= \frac{1}{24}(n + 2)(n + 1)(n)(n - 1 + 4\gamma)\epsilon^{-(n+\gamma)} \qquad \text{for } n \geqslant 0 \end{aligned}$$

As a check, note that, when $\gamma = 0$ or 1, $y(n + \gamma)$ reduces to $y(n)$ or $y(n + 1)$, respectively, in Example 3.13-5. For $0 < t < 1$, $n = 0$ and $t = \gamma$, while for $1 < t < 2$, $n = 1$, and $t = 1 + \gamma$, so

$$y(t) = 0 \qquad \text{for } 0 \leqslant t \leqslant 1$$

$$y(t) = (t - 1)\epsilon^{-t} \quad \text{for } 1 \leqslant t \leqslant 2$$

$$\cdots\cdots\cdots\cdots\cdots\cdots\cdots$$

agreeing with the answer to Example 1.9-4.

For the configuration of part (*b*) of Fig. 3.13-1,

$$Y(z, \gamma) = V(z)H_1H_2(z, \gamma)$$

where

$$H_1H_2(z, \gamma) = \mathscr{Z}\left[\frac{(t + \gamma)^3\epsilon^{-(t+\gamma)}}{6}\right]$$

Then

$$y(n, \gamma) = \frac{\epsilon^{-\gamma}}{6}\mathscr{Z}^{-1}\left[\frac{(z\epsilon)^2(z^2\epsilon^2 + 4z\epsilon + 1)}{(z\epsilon - 1)^5} + \frac{3\gamma(z\epsilon)^2(z\epsilon + 1)}{(z\epsilon - 1)^4} + \frac{3\gamma^2(z\epsilon)^2}{(z\epsilon - 1)^3} + \frac{\gamma^3(z\epsilon)^2}{(z\epsilon - 1)^2}\right]$$

$$= \epsilon^{-(n+\gamma)}\left[\frac{n^2(n + 1)^2}{24} + \frac{\gamma(n + 1)(n)(2n + 1)}{12} + \frac{3\gamma^2(n + 1)n}{12} + \frac{\gamma^3}{6}(n + 1)\right]$$

and

$$y(t) = t^3\frac{\epsilon^{-t}}{6} \qquad \text{for } 0 \leqslant t \leqslant 1$$

$$y(t) = [2t^3 - 3t^2 + 3t - 1]\frac{\epsilon^{-t}}{6} \quad \text{for } 1 \leqslant t \leqslant 2$$

$$\cdots\cdots\cdots\cdots\cdots\cdots\cdots$$

agreeing with the answer to Example 1.9-3.

Example 3.14-2. Find an expression for $Y(z, \gamma)$ for the configuration of Fig. 3.13-2.

As in Example 3.13-6,

$$Q(z) = \frac{V(z)}{1 + GH(z)}$$

Then

$$Y(z, \gamma) = Q(z)G(z, \gamma) = \frac{G(z, \gamma)V(z)}{1 + GH(z)}$$

REFERENCES

1. G. A. Campbell and R. M. Foster, *Fourier Integrals for Practical Applications*, D. Van Nostrand Company, Princeton, New Jersey, 1948.
2. A. Erdélyi (editor), *Tables of Integral Transforms*, Vol. I, McGraw-Hill Book Company, New York, 1954.
3. R. V. Churchill, *Operational Mathematics*, Second Edition, McGraw-Hill Book Company, New York, 1958, Chapters 1 and 6.
4. M. F. Gardner and J. L. Barnes, *Transients in Linear Systems*, John Wiley and Sons, New York, 1942, Chapters V, VI, and VIII.

5. S. Seshu and N. Balabanian, *Linear Network Analysis*, John Wiley and Sons, New York, 1959, Section 5.5.
6. Gardner and Barnes, *op. cit.*, p. 346, Eq. 2.1362.
7. R. V. Churchill, *Complex Variables and Applications*, Second Edition, McGraw-Hill Book Company, New York, 1960.
8. B. van der Pol, and H. Bremmer, *Operational Calculus Based on the Two-Sided Laplace Integral*, Second Edition, Cambridge University Press, Cambridge, 1955, Section VI.11.
9. W. R. LePage, *Complex Variables and the Laplace Transform for Engineers*, McGraw-Hill Book Company, New York, 1961, Chapter 10.
10. S. Goldman, *Transformation Calculus and Electrical Transients*, Prentice-Hall, Englewood Cliffs, N.J., 1949, Section 11.8.
11. J. A. Aseltine, *Transform Method in Linear System Analysis*, McGraw-Hill Book Company, New York, 1958, Chapter 17.
12. L. A. Zadeh, "Frequency Analysis of Variable Networks," *Proc. IRE*, Vol. 38, No. 3, March 1950, pp. 291–299.
13. D. Graham, E. J. Brunelle, Jr., W. Johnson, and H. Passmore, III, "Engineering Analysis Methods for Linear Time Varying Systems," Report ASD-TDR-62-362, Flight Control Laboratory, Wright-Patterson Air Force Base, Ohio, January, 1963, pp. 132–150.
14. L. A. Zadeh, "A General Theory of Linear Signal Transmission Systems," *J. Franklin Inst.*, Vol. 253, April 1952, pp. 293–312.
15. L. A. Zadeh, "Time-Varying Networks I," *IRE Intern. Conv. Record*, Vol. 9, Pt. 4, March 1961, pp. 251–267.
16. Gardner and Barnes, *op. cit.*, Chapter IX.
17. C. E. Shannon, "Communication in the Presence of Noise," *Proc. IRE*, Vol. 37, No. 1, January 1949, pp. 10–21.
18. J. G. Truxal, *Automatic Feedback Control System Synthesis*, McGraw-Hill Book Company, New York, 1955, Section 9.2.
19. W. Kaplan, *Advanced Calculus*, Addison-Wesley Publishing Company, Reading, Mass., 1952, p. 569.
20. M. Marden, *The Geometry of the Zeros of a Polynomial in a Complex Variable*, American Mathematical Society Mathematical Survey, No. III, New York, 1949, p. 152.
21. Truxal, *op. cit.*, Section 9.6.
22. J. R. Ragazzini and G. F. Franklin, *Sampled-Data Control Systems*, McGraw-Hill Book Company, New York, 1958, p. 93.
23. E. I. Jury, *Theory and Application of the z-Transform Method*, John Wiley and Sons, New York, 1964, pp. 289–296.

Problems

3.1 Derive Eqs. 3.2-6 and 3.2-7.

3.2 A periodic waveform is to be approximated by a finite trigonometric series of n terms. Show that the approximation having the least mean square error consists of the first n terms in the Fourier series.

3.3 Find the Fourier series for an infinite train of unit impulses, spaced T seconds apart, as in Fig. 1.8-1f.

3.4 By extending the result of Example 3.2-1, find and sketch the complete output voltage for Fig. 3.2-2*a*. Assume that the circuit contains no initial stored energy. Compare the answer with Example 2.12-3.

3.5 If the system function of a fixed linear system is $H(s) = s/(s^2 + 0.2s + 100)$, and if the input is the periodic waveform shown in Fig. P3.5, find and sketch the Fourier series representing the steady-state output.

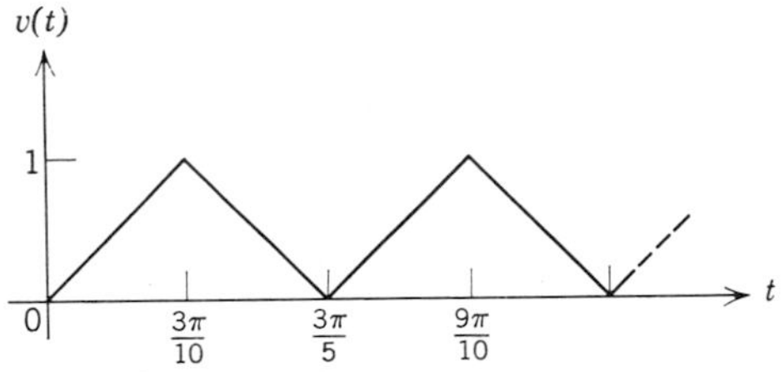

Fig. P3.5

3.6 Find the Fourier transform of $f(t) = (\cos t)U_{-1}(t + L/2)U_{-1}(-t + L/2)$. Sketch the frequency spectrum for $L = 2\pi$, $L = 4\pi$, and $L \to \infty$.

3.7 Prove the Fourier energy theorem:

$$\int_{-\infty}^{\infty} [f(t)]^2\, dt = \frac{1}{2\pi}\int_{-\infty}^{\infty} |G(\omega)|^2\, d\omega$$

3.8 Show that, if $f_2(t) = (d/dt)f_1(t)$, then $G_2(\omega) = j\omega G_1(\omega)$.

3.9 Show that the ideal low-pass filter described by Fig. P3.9 is not a realizable system, by examining the impulse response $h(t)$.

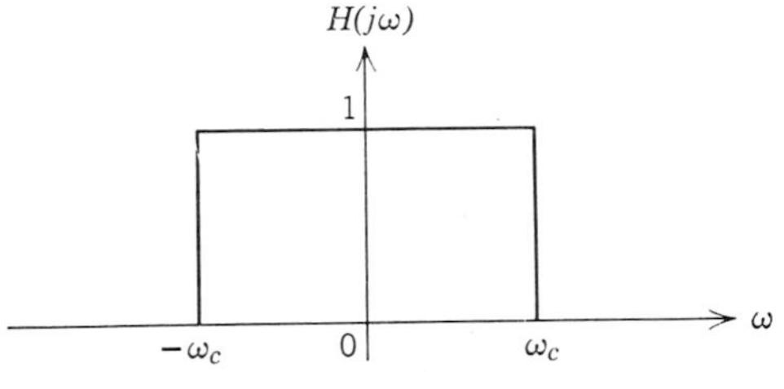

Fig. P3.9

3.10 Verify the entries in Table 3.3-1.

3.11 Prove the properties in Table 3.4-1.

3.12 Find $\mathscr{L}[\cos \beta t]$ by using Eq. 3.4-5 with $n = 2$. Using this result, find $\mathscr{L}[\sin \beta t]$ by Eq. 3.4-6. Then find $\mathscr{L}[t\epsilon^{-t} \sin (t + \pi/4)]$ by Eqs. 3.4-8 and 3.4-9.

3.13 Find the inverse transform of the following functions by partial fraction expansions, and Tables 3.3-1 and 3.4-1.

$$F_1(s) = \frac{1}{(s+1)^2(s^2+4)}$$

$$F_2(s) = \frac{s}{(s+1)^2(s^2+4)}$$

$$F_3(s) = \frac{\epsilon^{-2s}}{(s+1)^2(s^2+4)}$$

$$F_4(s) = \frac{2s^2+1}{s(s^2+2s+5)}$$

3.14 Find $\mathscr{F}^{-1}\left[\dfrac{j\omega}{j\omega^3+\omega^2+2}\right]$ from the tables of Laplace transforms.

3.15 Solve Examples 2.6-6 and 2.6-7 and Problem 2.2 by the Laplace transform.

3.16 A system is described by the differential equation

$$(p^3+3p^2+4p+2)y(t) = (p^2+2p+3)v(t).$$

Find the impulse response $h(t)$, and evaluate $h(0+)$, $\dot{h}(0+)$, and $\ddot{h}(0+)$. If it is possible, find these initial conditions by the initial value theorem of Eq. 3.4-15. Check the results by the methods of Chapter 2.

3.17 The circuit in Fig. P3.17 illustrates the application of the Laplace transform when there is initial stored energy. The circuit is originally operating

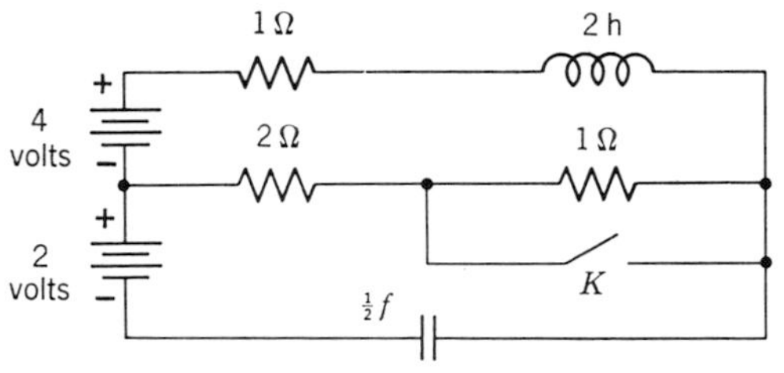

Fig. P3.17

in the steady state with the switch K open. If the switch closes at $t = 0$, shorting out a resistance, find expressions for the currents for $t \geqslant 0$. Represent the energy stored in the inductance and capacitance at $t = 0$ by added sources.

3.18 Write the Laurent series for each of the following functions about the singularity at $s = -1$. The quantity t should be treated as an independent parameter.

(a) $\dfrac{\epsilon^{st}}{s+1}$ (b) $\dfrac{s\epsilon^{st}}{(s+1)^2}$

Find the residue at $s = -1$ from the series, and also by Eqs. 3.6-17 through 3.6-20.

3.19 If $F(s) = \dfrac{b_m s^m + \cdots + b_1 s + b_0}{a_n s^n + \cdots + a_1 s + a_0}$, show that

$$\frac{1}{2\pi j}\oint_C \frac{F'(s)}{F(s)}\,ds = Z - P$$

where the contour C encloses the right half of the s-plane, and where Z and P denote the number of zeros and poles, respectively, of $F(s)$ in the right half-plane. Higher order poles and zeros should be counted according to their multiplicity. This result leads to the Nyquist stability criterion.

3.20 If the one-sided transform is $F(s) = s/[(s + 2)(s - 1)]$, find $f(t)$ by the inversion integral. Repeat for the two-sided transform

$$F(s) = s/[(s + 2)(s - 1)]$$

when the region of convergence is given by $-2 < \sigma < 1$.

3.21 Solve Problem 3.13 by the use of residues.

3.22 Show that the residue and partial fraction methods are essentially identical, when $F(s)$ is the quotient of two polynomials, with the denominator polynomial of higher order.

3.23 Show that $\int (1/s^2)\epsilon^{st}\,ds$ vanishes on contour $DEFGA$ in Fig. 3.7-1 for $t > 0$, and vanishes on DHA for $t < 0$. Generalize your proof so as to verify Theorem 3.7-1.

3.24 Use the Laplace transform to find the output voltage for Fig. 3.2-2*a*. Since $E_0(s)$ has an infinite number of poles, the following procedure, based upon Reference 5, is suggested. First find the transient response by evaluating the proper residue. Next find the complete response for the *first* cycle by neglecting the source voltage for $t > 2$. The steady-state response for the first cycle is found by subtracting the transient response from this result. Since the steady-state response is periodic, it is the same for all cycles. Compare the answer with the solutions to Example 2.12-3 and Problem 3.4.

3.25 Two time-varying systems are characterized by the system functions $H_1(t, s)$ and $H_2(t, s)$, respectively. If the systems are connected in cascade,

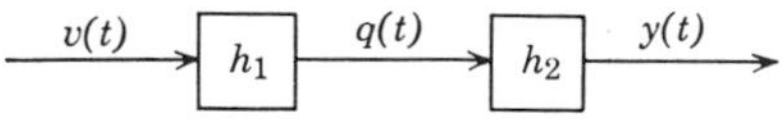

Fig. P3.25

as in Fig. P3.25, show that the overall system function

$$H(t, s) \neq H_1(t, s)H_2(t, s).$$

Find a general expression for H in terms of H_1 and H_2.

3.26 For a system described by the differential equation

$$\frac{dy}{dt} + \frac{1}{t+1}y = \frac{dv}{dt} + v$$

find $h_*(t, \tau)$ and $H(t, s)$. Find the step response using transform techniques, and compare the answer with Problem 1.12.

3.27 Find the system function $H(t, s)$ corresponding to each of the following differential equations.

(a) $(tp^2 + 2p)y = v$

(b) $(t^2p^2 + 4tp + 2)y = (p + 1)v$

Find the response of each system to $v = \epsilon^{-t}U_{-1}(t)$.

3.28 Verify the entries in Table 3.11-1 by Eq. 3.11-6, and also by Eq. 3.12-16.

3.29 Find the two-sided Z transform of the following functions by Eq. 3.11-6, and also by Table 3.11-1. Give the regions of convergence.

(a) $f_1(t) = \epsilon^{-|t|}$ for all t

(b) $f_2(t) = \epsilon^{t}$ for $t < 0$, and $t + 1$ for $t > 0$

3.30 Prove those properties in Table 3.12-1 which are not derived in this chapter.

3.31 Prove the first half of Eq. 3.12-17 by replacing $i(t)$ by its complex Fourier series, before transforming the relationship $f^*(t) = f(t)i(t)$. Problem 3.3 and Eq. 3.4-8 are helpful.

3.32 If $\mathscr{Z}_{II}[f(t)] = z/[(z + 2)(4z^2 + 1)]$, and if $f(t)$ is known to be a bounded function, find $f(nT)$. Calculate the numerical values of $f(nT)$ for $-2 \leqslant n \leqslant 4$ by the inversion integral and by modifying the partial fraction method in view of Eq. 3.11-10.

3.33 Find $f(nT)$ for the following one-sided Z transforms, using Eq. 3.12-21, Eq. 3.12-24, and partial fraction expansions.

(a) $F_1(z) = \dfrac{z}{(z + 2)(4z^2 + 1)}$

(b) $F_2(z) = \dfrac{2z^3 - 2z^2 + z}{(z - 1)(z^2 - z + 1)}$

3.34 Solve Problems 2.14 and 2.15 by the Z transform, again commenting on the stability of the systems.

3.35 In Fig. 3.13-2, $G(s) = 1/(s + 1)$, $H(s) = 1$, and the ideal sampling switch has a period T given by $\epsilon^{-T} = \frac{1}{2}$.

(a) Find the step response $r(nT)$ by the Z transform.

(b) Find the step response $r[(n + \gamma)T]$ by the modified Z transform.

(c) Determine the difference equation which relates $v(t)$ and $y(t)$ at the sampling instants.

3.36 In Fig. 3.13-2, $G(s) = [K(s + 1.04)]/[s(s + 0.692)]$, $H(s) = 1$, and the period of the sampling switch is $T = 1$.
(*a*) By a root locus diagram, determine the values of the real constant K for which the system is stable.
(*b*) Let $v(t) = 0$, but assume that some disturbance produces the signal $q(0) = q(1) = 1$. For $K = 1$, find $y(nT)$.

3.37 A sampling switch is added at the right of $H(s)$ in Fig. 3.13-2, and it operates in synchronism with the original one. Find expressions for $Y(z)$ and $Y(z, \gamma)$.

3.38 A "sampler with a zero order hold" is described by the output-input relationship $y(nT + \gamma T) = v(nT)$ for integral values of n and for $0 \leqslant \gamma < 1$. It samples the input every T seconds and maintains the last sampled value between sampling instants, as in Fig. 1.8-1*e*.

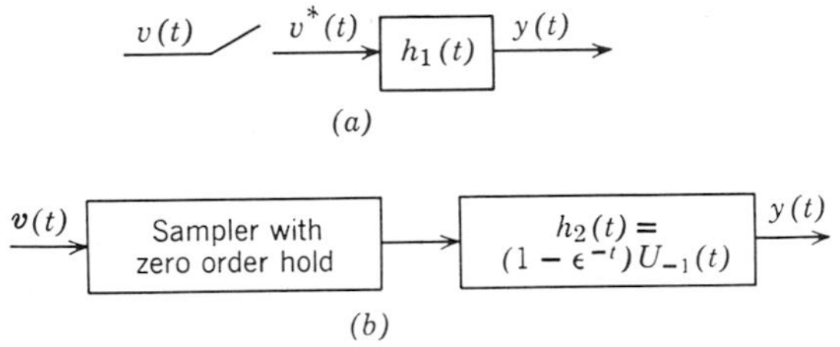

Fig. P3.38

(*a*) If the component is to be represented by the configuration of Fig. P3.38*a*, find $h_1(t)$ and $H_1(s)$ for the element following the ideal sampler.
(*b*) Find $H(z, \gamma)$ for the system in Fig. P3.38*b*.

3.39 Solve Problems 1.15 through 1.17 by the Z transform.

3.40 Formally prove Eq. 3.14-6.

4

Matrices and Linear Spaces

4.1 INTRODUCTION

The purpose of this chapter is to introduce the remaining analytical tools required in the following chapters. Generally, these analytical tools come under the general heading of linear algebra. More specifically, the topics of interest are matrices, determinants, linear vector spaces, linear transformations, characteristic value problems, and functions of a matrix.

The reason for the use of these analytical techniques in the study of systems is due primarily to the great deal of information which is required to describe completely a large-scale system. This information, which may consist of sets of differential or difference equations, can be expressed conveniently in the compact notation of matrices. Once this notation is adopted, the analysis of a system is largely the analysis of the properties of the matrix. The advent of high-speed digital computers has made this approach practical.

The topics chosen for inclusion in this chapter by no means exhaust the detailed knowledge which exists about linear algebra. Rather, the topics were chosen for their special pertinence to the study of control systems. In particular, the topics of vector spaces and linear transformations have special meaning, since the properties of a system may be made more evident by the proper choice of a coordinate system. It then becomes desirable to know the proper linear transformation which yields the desired coordinate system.

The characteristic value problem lies at the heart of the analysis of linear systems by a matrix approach. If a matrix is used to describe the structure of a system, the characteristic values of that matrix describe the normal modes of response of the system. Perhaps the characteristic value problem and the associated topic of functions of a matrix are the most important topics in this chapter.

4.2 BASIC CONCEPTS

The set of linear equations

$$\begin{aligned}
a_{11}x_1 + a_{12}x_2 + \cdots + a_{1n}x_n &= y_1 \\
a_{21}x_1 + a_{22}x_2 + \cdots + a_{2n}x_n &= y_2 \\
\cdots\cdots\cdots\cdots\cdots\cdots\cdots\cdots & \\
a_{m1}x_1 + a_{m2}x_2 + \cdots + a_{mn}x_n &= y_m
\end{aligned}$$

constitutes a set of relationships between the variables $x_1, x_2, \ldots, x_n$ and the variables $y_1, y_2, \ldots, y_m$. This relationship, or linear transformation of the x variables into the y variables, is completely characterized by the ordered array of the coefficients a_{ij}. If this ordered array is denoted by **A**, and written as

$$\mathbf{A} = \begin{bmatrix} a_{11} & a_{12} & \cdots & a_{1n} \\ a_{21} & a_{22} & \cdots & a_{2n} \\ \cdots & \cdots & \cdots & \cdots \\ a_{m1} & a_{m2} & \cdots & a_{mn} \end{bmatrix} \tag{4.2-1}$$

then it will be shown that the set of linear equations can be written as $\mathbf{Ax} = \mathbf{y}$ by a suitable definition of the "product **Ax**." Certainly, this expression is considerably simpler in form than the set of linear equations. This is one of the major reasons for using matrices. A matrix equation, or a set of matrix equations, contains a great deal of information in a compact form. Without the use of this compact notation, the task of analyzing sets of linear equations is quite cumbersome.

Consider then, the rectangular array of ordered elements of Eq. 4.2-1. The typical element in the array may be a real or complex number, or a function of specified variables. A *matrix* is a rectangular array such as shown, but distinguished from simply a rectangular array by the fact that matrices obey certain rules of addition, subtraction, multiplication, and equality. The elements of the matrix $a_{11}, a_{12}, \ldots, a_{ij}$ are written with a double subscript notation. The first subscript indicates the row where the element appears in the array, and the second subscript indicates the column. A matrix is denoted here by a boldface letter **A**, **B**, **a**, **b**, etc., or by writing the general element $[a_{ij}]$ enclosed by square brackets. The columns of the matrix are called column vectors, and the rows of the matrix are called row vectors. A matrix with m rows and n columns is called an

$(m \times n)$ matrix or is said to be of *order* m by n. For a square matrix $(m = n)$ the matrix is of order n.

Principal Types of Matrices

(*a*) *Column matrix.* An $(m \times 1)$ matrix is called a column matrix or a column vector, since it consists of a single column and m rows. It is denoted here by a lower-case, boldface letter, as

$$\mathbf{a} = \begin{bmatrix} a_1 \\ a_2 \\ \cdot \\ \cdot \\ a_m \end{bmatrix}$$

(*b*) *Row matrix.* A matrix containing a single row of elements, such as a $(1 \times n)$ matrix, is called a row matrix or a row vector.

(*c*) *Diagonal matrix.* The principal diagonal of a square matrix consists of the elements a_{ii}. A diagonal matrix is a square matrix all of whose elements which do not lie on the principal diagonal are zero.

$$\text{Diagonal matrix} = \begin{bmatrix} a_{11} & & & \text{All elements zero} \\ & a_{22} & & \\ & & \cdot & \\ & & & \cdot \\ \text{All elements zero} & & \cdot & a_{nn} \end{bmatrix}$$

(*d*) *Unit matrix.* A unit matrix is a diagonal matrix whose principal diagonal elements are equal to unity. The unit matrix here is given the symbol **I**.

$$\mathbf{I} = \begin{bmatrix} 1 & 0 & 0 & \cdots & 0 \\ 0 & 1 & 0 & \cdots & 0 \\ 0 & 0 & 1 & \cdots & 0 \\ \cdot & \cdot & \cdot & \cdots & \cdot \\ 0 & 0 & 0 & \cdots & 1 \end{bmatrix}$$

(*e*) *Null matrix.* A matrix which has all of its elements identically equal to zero is called a zero or null matrix.

(*f*) *Transpose matrix.* Transposing a matrix **A** is the operation whereby the rows and columns are interchanged. The transpose matrix is denoted by $\mathbf{A}^T$. Thus, if $\mathbf{A} = [a_{ij}]$, then $\mathbf{A}^T = [a_{ji}]$, i.e., the element of the ith row, jth column of $\mathbf{A}$ appears in the jth row, ith column of $\mathbf{A}^T$. If $\mathbf{A}$ is an $(m \times n)$ matrix, then $\mathbf{A}^T$ is an $(n \times m)$ matrix.

Special Types of Matrices

(*a*) *Symmetric matrix.* A square matrix, all of whose elements are real, is said to be symmetric if it is equal to its transpose, i.e., if

$$\mathbf{A} = \mathbf{A}^T \quad \text{or} \quad a_{ij} = a_{ji} \quad (i, j = 1, \ldots, n)$$

(*b*) *Skew-symmetric matrix.* A real square matrix is said to be skew-symmetric if

$$\mathbf{A} = -\mathbf{A}^T \quad \text{or} \quad a_{ij} = -a_{ji} \quad (i, j = 1, \ldots, n)$$

This, of course, implies that the elements on the principal diagonal are identically zero.

(*c*) *Conjugate matrix.* If the elements of the matrix $\mathbf{A}$ are complex $(a_{ij} = \alpha_{ij} + j\beta_{ij})$, then the conjugate matrix $\mathbf{B}$ has elements $b_{ij} = \alpha_{ij} - j\beta_{ij}$. This is written in the form $\mathbf{B} = \mathbf{A}^*$.

(*d*) *Associate matrix.* The associate matrix of $\mathbf{A}$ is the transposed conjugate of $\mathbf{A}$, i.e., associate of $\mathbf{A} = (\mathbf{A}^*)^T$.

(*e*) *Real matrix.* If $\mathbf{A} = \mathbf{A}^*$, then $\mathbf{A}$ is a real matrix.

(*f*) *Imaginary matrix.* If $\mathbf{A} = -\mathbf{A}^*$, then $\mathbf{A}$ is pure imaginary.

(*g*) *Hermitian matrix.* If a matrix is equal to its associate matrix, the matrix is said to be Hermitian, i.e., if $\mathbf{A} = (\mathbf{A}^*)^T$, then $\mathbf{A}$ is Hermitian.

(*h*) *Skew-Hermitian matrix.* If $\mathbf{A} = -(\mathbf{A}^*)^T$, then $\mathbf{A}$ is skew-Hermitian.

Elementary Operations

Addition of Matrices. If two matrices $\mathbf{A}$ and $\mathbf{B}$ are both of order $(m \times n)$, where m and n are two given integers not necessarily different, then the sum of these two matrices is the matrix $\mathbf{C}$. The i, jth element of $\mathbf{C} = \mathbf{A} + \mathbf{B}$ is defined by

$$c_{ij} = a_{ij} + b_{ij} \tag{4.2-2}$$

Example 4.2-1. Evaluate the sum of the matrices indicated.

$$\begin{bmatrix} 2 & 4 \\ -3 & 1 \end{bmatrix} + \begin{bmatrix} 2 & -2 \\ 3 & 0 \end{bmatrix} = \begin{bmatrix} 4 & 2 \\ 0 & 1 \end{bmatrix}$$

Addition of matrices is commutative and associative, i.e.,

$$\mathbf{A} + \mathbf{B} = \mathbf{B} + \mathbf{A} \qquad \text{Commutative}$$

$$\mathbf{A} + (\mathbf{B} + \mathbf{C}) = (\mathbf{A} + \mathbf{B}) + \mathbf{C} \qquad \text{Associative}$$

Subtraction of Matrices. The difference of two matrices **A** and **B** both, say, of order (m x n) where m and n are two given integers not necessarily different, is the matrix **D**, where the i,jth element of $\mathbf{D} = \mathbf{A} - \mathbf{B}$ is defined by

$$d_{ij} = a_{ij} - b_{ij} \tag{4.2-3}$$

Example 4.2-2. Evaluate the difference of the matrices indicated.

$$\begin{bmatrix} 2 & 4 \\ -3 & 1 \end{bmatrix} - \begin{bmatrix} 2 & -2 \\ 3 & 0 \end{bmatrix} = \begin{bmatrix} 0 & 6 \\ -6 & 1 \end{bmatrix}$$

Equality of Matrices. Two matrices **A** and **B**, which are of equal order, are *equal* if their corresponding elements are equal. Thus

$$\mathbf{A} = \mathbf{B} \tag{4.2-4}$$

if an only if $a_{ij} = b_{ij}$.

Multiplication of Matrices. The definition of the product of two matrices **A** and **B** comes about in a natural way from the study of linear transformations. Consider then the linear transformation

$$\begin{aligned} y_1 &= a_{11}x_1 + a_{12}x_2 \\ y_2 &= a_{21}x_1 + a_{22}x_2 \end{aligned} \tag{4.2-5}$$

The elements y_1 and y_2 can be considered as components of a vector **y**. Similarly, x_1 and x_2 can be considered as components of a vector **x**. Equation 4.2-5 can then be visualized as the matrix **A** transforming the vector **x** into the vector **y**. This transformation can be written in the form

$$\mathbf{y} = \mathbf{Ax} \tag{4.2-6}$$

where

$$\mathbf{y} = \begin{bmatrix} y_1 \\ y_2 \end{bmatrix}, \quad \mathbf{A} = \begin{bmatrix} a_{11} & a_{12} \\ a_{21} & a_{22} \end{bmatrix} \quad \text{and} \quad \mathbf{x} = \begin{bmatrix} x_1 \\ x_2 \end{bmatrix}$$

Equation 4.2-6 could also have been written

$$y_i = \sum_{j=1}^{2} a_{ij}x_j, \qquad i = 1, 2 \tag{4.2-7}$$

This leads to the definition of the product of an (m x n) matrix by an (n x 1) matrix or column vector as

$$\mathbf{Ax} = \left[\sum_{j=1}^{n} a_{ij}x_j\right] \tag{4.2-8}$$

This product is referred to as the *postmultiplication* of **A** by **x**, or the *premultiplication* of **x** by **A**. It is important that this distinction be made, since in general multiplication is not commutative, i.e., $\mathbf{AB} \neq \mathbf{BA}$.

Assume now that the vector **x** is formed by the linear transformation

$$\begin{aligned} x_1 &= b_{11}z_1 + b_{12}z_2 \\ x_2 &= b_{21}z_2 + b_{22}z_2 \end{aligned} \tag{4.2-9}$$

or

$$x_j = \sum_{k=1}^{2} b_{jk}z_k, \quad j = 1, 2$$

The relationship between the vectors **y** and **z** can be obtained by substituting Eq. 4.2-9 into 4.2-7. The result of this substitution is

$$y_i = \sum_{j=1}^{2} a_{ij} \sum_{k=1}^{2} b_{jk}z_k, \quad i = 1, 2 \tag{4.2-10}$$

Since the order in which the summations are taken can be interchanged,

$$y_i = \sum_{k=1}^{2} \left(\sum_{j=1}^{2} a_{ij}b_{jk} \right) z_k, \quad i = 1, 2 \tag{4.2-11}$$

The transformation from the column vector **z** to the column vector **y** can be written in matrix notation as

$$\mathbf{y} = \mathbf{ABz} \tag{4.2-12}$$

The product **AB** can be viewed as the matrix **C** where

$$\mathbf{C} = \mathbf{AB} \tag{4.2-13}$$

or

$$[c_{ik}] = \left[\sum_{j=1}^{2} a_{ij}b_{jk} \right]$$

A typical element c_{ik} of **C** is the summation shown inside the brackets. The first subscript inside the brackets is the row index of the matrix product, and the last subscript inside the brackets is the column index. In order for the product to be defined, the number of columns of **A** must equal the number of rows of **B**. Proceeding to the general case, the product of two matrices, **A**(m x n) by **B**(n x p), is defined in terms of the typical element of the product **C** as

$$\mathbf{C} = \mathbf{AB} = \left[\sum_{k=1}^{n} a_{ik}b_{kj} \right] \tag{4.2-14}$$

Thus the i, jth element of **C** is the sum of the products of the elements of the ith *row* of **A** and the corresponding elements of the jth *column* of **B**. The resulting matrix **C** is (m x p).

If the number of columns of **A** is equal to the number of rows of **B**, the two matrices are said to be *conformable*, in that the product **AB** exists. For the case where **A** is (m x n) and **B** is (n x m), then both products **AB** and **BA** exist. The product **AB** is (m x m), and the product **BA** is (n x n). They are, of course, generally not equal. Even in the case where $m = n$, and hence both products **AB** and **BA** are (m x m), the two products are not necessarily equal. However, if they are equal, i.e., $\mathbf{AB} = \mathbf{BA}$, the two matrices are said to *commute*.

Example 4.2-3. Evaluate the products of the matrices indicated.

(a)

$$\mathbf{AB} = \begin{bmatrix} 1 & -1 \\ 2 & 2 \end{bmatrix} \begin{bmatrix} 1 & 2 \\ 0 & 1 \end{bmatrix} = \begin{bmatrix} 1 & 1 \\ 2 & 6 \end{bmatrix}$$

(b)

$$\mathbf{BA} = \begin{bmatrix} 1 & 2 \\ 0 & 1 \end{bmatrix} \begin{bmatrix} 1 & -1 \\ 2 & 2 \end{bmatrix} = \begin{bmatrix} 5 & 3 \\ 2 & 2 \end{bmatrix} \neq \mathbf{AB}$$

(c)

$$\begin{bmatrix} 1 & 1 \\ 2 & 2 \end{bmatrix} \begin{bmatrix} -1 & 1 \\ 1 & -1 \end{bmatrix} = \begin{bmatrix} 0 & 0 \\ 0 & 0 \end{bmatrix}$$

Note that $\mathbf{AB} = [0]$ does not imply that $\mathbf{A} = [0]$ or $\mathbf{B} = [0]$.

From the preceding discussion it may be seen that matrix multiplication is associative and distributive, but not in general commutative.

$$\mathbf{A(BC)} = \mathbf{(AB)C} \qquad \text{Associative}$$

$$\left.\begin{aligned} \mathbf{A(B + C)} &= \mathbf{AB + AC} \\ \mathbf{(A + B)C} &= \mathbf{AC + BC} \end{aligned}\right\} \quad \text{Distributive} \tag{4.2-15}$$

Scalar Multiplication. Premultiplication or postmultiplication of a matrix by a scalar multiplier k multiplies each element of the matrix by k. The typical element of the product $k\mathbf{A}$ is ka_{ij}.

Example 4.2-4. Perform the indicated multiplication of the matrix by the scalar 2.

$$2\begin{bmatrix} 1 & -1 \\ 2 & 3 \end{bmatrix} = \begin{bmatrix} 2 & -2 \\ 4 & 6 \end{bmatrix}$$

Multiplication of Transpose Matrices. The product of two transpose matrices, $\mathbf{B}^T$ and $\mathbf{A}^T$, is equal to the transpose of the product of the original two matrices **B** and **A**, taken in reverse order, i.e.,

$$\mathbf{B}^T\mathbf{A}^T = (\mathbf{AB})^T \tag{4.2-16}$$

This is easily shown by taking the transpose of $\mathbf{C} = \mathbf{AB}$. The typical element of the product **AB** is given by

$$c_{ij} = \sum_{k=1}^{n} a_{ik}b_{kj} \qquad \begin{cases} i = 1, \ldots, m \\ j = 1, \ldots, p \end{cases}$$

where it is assumed that **A** is $(m \times n)$ and **B** is $(n \times p)$. The typical element of $\mathbf{C}^T$ is

$$c_{ij}{}^T = c_{ji} = \sum_{k=1}^{n} a_{jk} b_{ki} = \sum_{k=1}^{n} b_{ik}{}^T a_{kj}{}^T$$

Therefore $\mathbf{C}^T = (\mathbf{AB})^T = \mathbf{B}^T\mathbf{A}^T$. The transpose of **AB** is equal to the product of the transposed matrices taken in reverse order.

Multiplication by a Diagonal Matrix. Postmultiplication of a matrix **A** by the diagonal matrix **D** is equivalent to an operation on the *columns* of **A**. *Premultiplication* of a matrix **A** by the diagonal matrix **D** is an operation on the *rows* of **A**. Obviously, premultiplication or postmultiplication by the unit matrix **I** leaves the matrix unchanged, i.e.,

$$\mathbf{IA} = \mathbf{AI} = \mathbf{A}$$

Example 4.2-5. Evaluate the given matrix products.

Postmultiplication:
$$\begin{bmatrix} a_{11} & a_{12} \\ a_{21} & a_{22} \end{bmatrix} \begin{bmatrix} d_{11} & 0 \\ 0 & d_{22} \end{bmatrix} = \begin{bmatrix} a_{11}d_{11} & a_{12}d_{22} \\ a_{21}d_{11} & a_{22}d_{22} \end{bmatrix}$$

Premultiplication:
$$\begin{bmatrix} d_{11} & 0 \\ 0 & d_{22} \end{bmatrix} \begin{bmatrix} a_{11} & a_{12} \\ a_{21} & a_{22} \end{bmatrix} = \begin{bmatrix} a_{11}d_{11} & a_{12}d_{11} \\ a_{21}d_{22} & a_{22}d_{22} \end{bmatrix}$$

Products of Partitioned Matrices. It is sometimes convenient to construct a matrix from elements which are matrices, or to reduce a given matrix to another matrix whose elements are submatrices of the original matrix. For example, by drawing the vertical and horizontal dotted lines in matrix **A**, which is of order (3 x 3), a (2 x 2) matrix can be written using the submatrices $\mathbf{A}_1$, $\mathbf{A}_2$, $\mathbf{A}_3$, and $\mathbf{A}_4$.

$$\mathbf{A} = \left[\begin{array}{c:cc} a_{11} & a_{12} & a_{13} \\ \hdashline a_{21} & a_{22} & a_{23} \\ a_{31} & a_{32} & a_{33} \end{array}\right] = \begin{bmatrix} \mathbf{A}_1 & \mathbf{A}_3 \\ \mathbf{A}_2 & \mathbf{A}_4 \end{bmatrix}$$

where

$$\mathbf{A}_1 = [a_{11}] \quad \mathbf{A}_2 = \begin{bmatrix} a_{21} \\ a_{31} \end{bmatrix} \quad \mathbf{A}_3 = [a_{12} \quad a_{13}] \quad \mathbf{A}_4 = \begin{bmatrix} a_{22} & a_{23} \\ a_{32} & a_{33} \end{bmatrix}$$

Assume that the matrix **B**, which is also of order (3 x 3) is partitioned in the same manner, yielding

$$\mathbf{B} = \left[\begin{array}{c:cc} b_{11} & b_{12} & b_{13} \\ \hdashline b_{21} & b_{22} & b_{23} \\ b_{31} & b_{32} & b_{33} \end{array}\right] = \begin{bmatrix} \mathbf{B}_1 & \mathbf{B}_3 \\ \mathbf{B}_2 & \mathbf{B}_4 \end{bmatrix}$$

Clearly, the sum of matrices **A** and **B** can be expressed in terms of the submatrices as

$$\mathbf{A} + \mathbf{B} = \begin{bmatrix} \mathbf{A}_1 + \mathbf{B}_1 & \mathbf{A}_3 + \mathbf{B}_3 \\ \mathbf{A}_2 + \mathbf{B}_2 & \mathbf{A}_4 + \mathbf{B}_4 \end{bmatrix}$$

The product of **A** and **B** can also be expressed in terms of the submatrices as

$$\mathbf{AB} = \begin{bmatrix} (\mathbf{A}_1\mathbf{B}_1 + \mathbf{A}_3\mathbf{B}_2) & (\mathbf{A}_1\mathbf{B}_3 + \mathbf{A}_3\mathbf{B}_4) \\ (\mathbf{A}_2\mathbf{B}_1 + \mathbf{A}_4\mathbf{B}_2) & (\mathbf{A}_2\mathbf{B}_3 + \mathbf{A}_4\mathbf{B}_4) \end{bmatrix}$$

In general, the product of two matrices can be expressed in terms of the submatrices only if the partitioning produces submatrices which are conformable. The grouping of columns in **A** must be equal to the grouping of rows in **B**. If this condition is satisfied (it is assumed that **A** and **B** are originally conformable), then the submatrices formed by the partitioning may be treated as ordinary elements.

Example 4.2-6. Evaluate the indicated matrix product by means of partitioning.

$$\mathbf{AB} = \left[\begin{array}{cc:c} 0 & 2 & 3 \\ 1 & -1 & 0 \end{array}\right] \left[\begin{array}{ccc} 1 & 0 & 3 \\ -1 & 2 & 1 \\ \hdashline 0 & 1 & -1 \end{array}\right]$$

$$= \begin{bmatrix} 0 & 2 \\ 1 & -1 \end{bmatrix} \begin{bmatrix} 1 & 0 & 3 \\ -1 & 2 & 1 \end{bmatrix} + \begin{bmatrix} 3 \\ 0 \end{bmatrix} [0 \quad 1 \quad -1]$$

$$= \begin{bmatrix} -2 & 4 & 2 \\ 2 & -2 & 2 \end{bmatrix} + \begin{bmatrix} 0 & 3 & -3 \\ 0 & 0 & 0 \end{bmatrix} = \begin{bmatrix} -2 & 7 & -1 \\ 2 & -2 & 2 \end{bmatrix}$$

Differentiation of a Matrix. The usual ideas of differentiation and integration associated with scalar variables carry over to the differentiation and integration of matrices and matrix products, provided that the original order of the factors involved is preserved. Let $\mathbf{A}(t)$ be an $(m \times n)$ matrix whose elements $a_{ij}(t)$ are differentiable functions of the scalar variable t. The derivative of $\mathbf{A}(t)$ with respect to the variable t is defined as

$$\frac{d}{dt}[\mathbf{A}(t)] = \dot{\mathbf{A}}(t) = \begin{bmatrix} \frac{da_{11}(t)}{dt} & \frac{da_{12}(t)}{dt} & \cdots & \frac{da_{1n}(t)}{dt} \\ \frac{da_{21}(t)}{dt} & \frac{da_{22}(t)}{dt} & \cdots & \frac{da_{2n}(t)}{dt} \\ \cdots & \cdots & \cdots & \cdots \\ \frac{da_{m1}(t)}{dt} & \frac{da_{m2}(t)}{dt} & \cdots & \frac{da_{mn}(t)}{dt} \end{bmatrix} \quad (4.2\text{-}17)$$

From this definition it is evident that the derivative of the sum of two matrices is the sum of the derivatives of the matrices, or

$$\frac{d}{dt}[\mathbf{A}(t) + \mathbf{B}(t)] = \dot{\mathbf{A}}(t) + \dot{\mathbf{B}}(t) \tag{4.2-18}$$

The derivative of a matrix product is formed in the same manner as the derivative of a scalar product, with the exception that the order of the product must be preserved. Thus, as typical examples,

$$\frac{d}{dt}[\mathbf{A}(t)\mathbf{B}(t)] = \dot{\mathbf{A}}(t)\mathbf{B}(t) + \mathbf{A}(t)\dot{\mathbf{B}}(t) \tag{4.2-19}$$

and

$$\frac{d}{dt}[A^3(t)] = \dot{\mathbf{A}}(t)\mathbf{A}^2(t) + \mathbf{A}(t)\dot{\mathbf{A}}(t)\mathbf{A}(t) + \mathbf{A}^2(t)\dot{\mathbf{A}}(t) \tag{4.2-20}$$

Integration of a Matrix. Similar to the definition of the derivative of a matrix, the integral of a matrix is defined as the matrix of the integrals of the elements of the matrix. Thus

$$\int \mathbf{A}(t)\,dt = \begin{bmatrix} \int a_{11}(t)\,dt & \int a_{12}(t)\,dt & \cdots & \int a_{1n}(t)\,dt \\ \int a_{21}(t)\,dt & \int a_{22}(t)\,dt & \cdots & \int a_{2n}(t)\,dt \\ \cdots & \cdots & \cdots & \cdots \\ \int a_{m1}(t)\,dt & \int a_{m2}(t)\,dt & \cdots & \int a_{mn}(t)\,dt \end{bmatrix} \tag{4.2-21}$$

The operator notation $\mathbf{Q} = \int(\ \ )\,dt$ is commonly used to signify the integral of a matrix. When the superscript t and the subscript t_0 are affixed to $\mathbf{Q}$, they indicate the upper and lower limits of the integration. Thus

$$\mathbf{Q}_{t_0}^{t}(\ \ ) = \int_{t_0}^{t} (\ \ )\,dt \tag{4.2-22}$$

Example 4.2-7.

Find $\mathbf{Q}_0^t(\mathbf{A})$, if $A = \begin{bmatrix} t & 1 \\ 1 & t \end{bmatrix}$.

$$\mathbf{Q}_0^t(\mathbf{A}) = \begin{bmatrix} \int_0^t t\,dt & \int_0^t dt \\ \int_0^t dt & \int_0^t t\,dt \end{bmatrix} = \begin{bmatrix} \frac{t^2}{2} & t \\ t & \frac{t^2}{2} \end{bmatrix}$$

4.3 DETERMINANTS

The theory of determinants is also useful when dealing with the solution of simultaneous linear algebraic equations. Determinant notation simplifies the solution of these equations, reducing the solution to a set of rules of procedure. As an example, the set of equations

$$\begin{aligned} a_{11}x_1 + a_{12}x_2 &= y_1 \\ a_{21}x_1 + a_{22}x_2 &= y_2 \end{aligned} \tag{4.3-1}$$

can be solved by finding an expression for x_1 from the first equation, and then substituting this expression into the second equation. The result of performing this operation is the solution

$$x_1 = \frac{y_1a_{22} - y_2a_{12}}{a_{11}a_{22} - a_{12}a_{21}} \qquad x_2 = \frac{y_2a_{11} - y_1a_{21}}{a_{11}a_{22} - a_{12}a_{21}}$$

This solution assumes that the denominator $(a_{11}a_{22} - a_{12}a_{21})$ is not zero.

Equation 4.3-1 can be written in matrix notation as

$$\mathbf{Ax} = \mathbf{y} \tag{4.3-2}$$

where

$$\mathbf{A} = \begin{bmatrix} a_{11} & a_{12} \\ a_{21} & a_{22} \end{bmatrix}, \quad \mathbf{x} = \begin{bmatrix} x_1 \\ x_2 \end{bmatrix} \quad \text{and} \quad \mathbf{y} = \begin{bmatrix} y_1 \\ y_2 \end{bmatrix}$$

The *determinant* of the matrix **A** is written as

$$|\mathbf{A}| = \begin{vmatrix} a_{11} & a_{12} \\ a_{21} & a_{22} \end{vmatrix}$$

and the *value* of the determinant is defined to be $(a_{11}a_{22} - a_{12}a_{21})$. A determinant is only defined for a square array of elements, and the *order* of the determinant is equal to the number of rows or columns of elements. Thus this determinant is called a *second order* determinant. If the determinants

$$|\mathbf{A}_1| = \begin{vmatrix} y_1 & a_{12} \\ y_2 & a_{22} \end{vmatrix} = (y_1a_{22} - y_2a_{12})$$

$$|\mathbf{A}_2| = \begin{vmatrix} a_{11} & y_1 \\ a_{21} & y_2 \end{vmatrix} = (y_2a_{11} - y_1a_{21})$$

are formed, then the solution to Eq. 4.3-1 can be expressed in terms of these determinants as

$$x_1 = \frac{|\mathbf{A}_1|}{|\mathbf{A}|} \qquad x_2 = \frac{|\mathbf{A}_2|}{|\mathbf{A}|}$$

provided $|\mathbf{A}| \neq 0$.

The definition of the value of the determinant of any square matrix **A** follows. The determinant of the (n x n) square matrix **A**, written as $|\mathbf{A}|$, has a value which is the algebraic sum of all possible products of n elements which contain one and only one element from each row and column, where each product is either positive or negative depending upon the following rule: Arrange the possible products in ascending order with respect to the first subscript, e.g., $a_{13}a_{22}a_{31} \cdots$. Define an *inversion* as the occurrence of a greater integer before a smaller one. The sign of the product is positive if the number of inversions of the second subscript is even; otherwise it is negative. For example, the sequence 321 has three inversions: 3 before 2, 3 before 1, and 2 before 1.

Example 4.3-1. Evaluate the determinant

$$|\mathbf{A}| = \begin{vmatrix} a_{11} & a_{12} & a_{13} \\ a_{21} & a_{22} & a_{23} \\ a_{31} & a_{32} & a_{33} \end{vmatrix}$$

The possible products and their signs are

Possible Products	Number of Inversions	Sign
$a_{11}a_{22}a_{33}$	0	+
$a_{11}a_{23}a_{32}$	1	−
$a_{12}a_{21}a_{33}$	1	−
$a_{12}a_{23}a_{31}$	2	+
$a_{13}a_{21}a_{32}$	2	+
$a_{13}a_{22}a_{31}$	3	−

The determinant of **A** is then

$$|\mathbf{A}| = a_{11}a_{22}a_{33} + a_{12}a_{23}a_{31} + a_{13}a_{21}a_{32} - (a_{11}a_{23}a_{32} + a_{12}a_{21}a_{33} + a_{13}a_{22}a_{31})$$

For a determinant of *order n*, there are $n!$ such products.

From the preceding definition, the following properties of a determinant can be established:

1. The value of the determinant is unity if all the elements on the *principal diagonal* ($a_{11}, a_{22}, \ldots, a_{nn}$) are unity and all other elements are zero.

2. The value of the determinant is zero if all the elements of any row (or column) are zero, or if the corresponding elements of any two rows (or two columns) are equal or have a common ratio.

3. The value of the determinant is unchanged if the rows and columns are interchanged.

4. The sign of the determinant is reversed if any two rows (or two columns) are interchanged.

5. The value of the determinant is multiplied by a constant k if all the elements of any row (or column) are multiplied by k.

6. The value of the determinant is unchanged if k times the elements of any row (or column) are added to the corresponding elements of another row (or column).

Minors and Cofactors

Minors. If the ith row and jth column of the determinant $|\mathbf{A}|$ are deleted, the remaining $n - 1$ rows and $n - 1$ columns form a determinant $|\mathbf{M}_{ij}|$. This determinant is called the *minor* of element a_{ij}. For example, if the determinant $\mathbf{A}$ is given by

$$\begin{vmatrix} a_{11} & a_{12} & a_{13} \\ a_{21} & a_{22} & a_{23} \\ a_{31} & a_{32} & a_{33} \end{vmatrix}$$

the minor of element a_{12} is

$$|\mathbf{M}_{12}| = \begin{vmatrix} a_{21} & a_{23} \\ a_{31} & a_{33} \end{vmatrix}$$

A minor of $|\mathbf{A}|$, whose diagonal elements are also diagonal elements of $|\mathbf{A}|$, is called a *principal minor* of $|\mathbf{A}|$. These minors are of particular importance, as is seen later.

Cofactors. The cofactor of the element a_{ij} is equal to the minor of a_{ij}, with the sign $(-1)^{i+j}$ affixed to it. Thus the cofactor of a_{ij}, written C_{ij}, is defined as

$$C_{ij} = (-1)^{i+j}\,|\mathbf{M}_{ij}| \qquad (4.3\text{-}3)$$

The cofactor C_{12} of the previous example is

$$C_{12} = (-1)\begin{vmatrix} a_{21} & a_{23} \\ a_{31} & a_{33} \end{vmatrix}$$

Example 4.3-2. Evaluate the minors and cofactors of the third order determinant shown in Example 4.3-1.

$$|\mathbf{M}_{11}| = \begin{vmatrix} a_{22} & a_{23} \\ a_{32} & a_{33} \end{vmatrix} = a_{22}a_{33} - a_{23}a_{32} = C_{11}$$

$$|\mathbf{M}_{12}| = \begin{vmatrix} a_{21} & a_{23} \\ a_{31} & a_{33} \end{vmatrix} = a_{21}a_{33} - a_{23}a_{31} = -C_{12}$$

$$|\mathbf{M}_{13}| = \begin{vmatrix} a_{21} & a_{22} \\ a_{31} & a_{32} \end{vmatrix} = a_{21}a_{32} - a_{22}a_{31} = C_{13}$$

$$|\mathbf{M}_{21}| = \begin{vmatrix} a_{12} & a_{13} \\ a_{32} & a_{33} \end{vmatrix} = a_{12}a_{33} - a_{13}a_{32} = -C_{21}$$

$$|\mathbf{M}_{22}| = \begin{vmatrix} a_{11} & a_{13} \\ a_{31} & a_{33} \end{vmatrix} = a_{11}a_{33} - a_{13}a_{31} = C_{22}$$

$$|\mathbf{M}_{23}| = \begin{vmatrix} a_{11} & a_{12} \\ a_{31} & a_{32} \end{vmatrix} = a_{11}a_{32} - a_{12}a_{31} = -C_{23}$$

$$|\mathbf{M}_{31}| = \begin{vmatrix} a_{12} & a_{13} \\ a_{22} & a_{23} \end{vmatrix} = a_{12}a_{23} - a_{13}a_{22} = C_{31}$$

$$|\mathbf{M}_{32}| = \begin{vmatrix} a_{11} & a_{13} \\ a_{21} & a_{23} \end{vmatrix} = a_{11}a_{23} - a_{13}a_{21} = -C_{32}$$

$$|\mathbf{M}_{33}| = \begin{vmatrix} a_{11} & a_{12} \\ a_{21} & a_{22} \end{vmatrix} = a_{11}a_{22} - a_{12}a_{21} = C_{33}$$

Laplace Expansion of a Determinant

If the results of Examples 4.3-1 and 4.3-2 are compared, it is seen that the determinant of the third order matrix **A** can be expressed in terms of the elements of a single row or column and their respective cofactors. Thus

$$\begin{aligned}
|\mathbf{A}| &= a_{11}C_{11} + a_{12}C_{12} + a_{13}C_{13} \\
&= a_{21}C_{21} + a_{22}C_{22} + a_{23}C_{23} \\
&= a_{31}C_{31} + a_{32}C_{32} + a_{33}C_{33} \\
&= a_{11}C_{11} + a_{21}C_{21} + a_{31}C_{31} \\
&= a_{12}C_{12} + a_{22}C_{22} + a_{32}C_{32} \\
&= a_{13}C_{13} + a_{23}C_{23} + a_{33}C_{33}
\end{aligned}$$

The Laplace expansion formula for the determinant of any (n x n) matrix **A** is a direct generalization of the preceding observation. This formula states that the determinant of a square matrix **A** is given by the sum of the products of the elements of any single row or column and their respective cofactors. Thus

$$|\mathbf{A}| = \sum_{i=1}^{n} a_{ij}C_{ij} \qquad j = 1, \text{ or } 2, \ldots, \text{ or } n \text{ (column expansion)}$$
$$|\mathbf{A}| = \sum_{j=1}^{n} a_{ij}C_{ij} \qquad i = 1, \text{ or } 2, \ldots, \text{ or } n \text{ (row expansion)} \tag{4.3-4}$$

Example 4.3-3. Find the determinant of **A**, where

$$\mathbf{A} = \begin{bmatrix} 0 & -1 & 1 \\ 1 & 2 & 0 \\ 2 & 0 & 2 \end{bmatrix}$$

Using the formula established in Example 4.3-1,

$$|\mathbf{A}| = [(0)(2)(2) + (-1)(0)(2) + (1)(1)(0)] - [(0)(0)(0) + (-1)(1)(2) + (1)(2)(2)] = -2$$

Using the Laplace expansion along the first row,

$$\begin{aligned} |M_{11}| &= 4 \qquad & C_{11} &= 4 \\ |M_{12}| &= 2 \qquad & C_{12} &= -2 \\ |M_{13}| &= -4 \qquad & C_{13} &= -4 \end{aligned}$$

$$|\mathbf{A}| = (0)(4) + (-1)(-2) + (1)(-4) = -2$$

Using the Laplace expansion along the second column,

$$\begin{aligned} |M_{12}| &= 2 \qquad & C_{12} &= -2 \\ |M_{22}| &= -2 \qquad & C_{22} &= -2 \\ |M_{32}| &= -1 \qquad & C_{32} &= 1 \end{aligned}$$

$$|\mathbf{A}| = (-1)(-2) + (2)(-2) + (0)(1) = -2$$

Using the six fundamental properties of a determinant, the value of the determinant can be found by reducing the determinant to a diagonal determinant. The value of the determinant is then the product of the diagonal elements. The following steps show how this method is applied.

Step 1. Multiply the third row by $(-\frac{1}{2})$ and add this to the first row.

$$|\mathbf{A}| = \begin{vmatrix} -1 & -1 & 0 \\ 1 & 2 & 0 \\ 2 & 0 & 2 \end{vmatrix}$$

Step 2. Add the new first row to the second row.

$$|\mathbf{A}| = \begin{vmatrix} -1 & -1 & 0 \\ 0 & 1 & 0 \\ 2 & 0 & 2 \end{vmatrix}$$

Step 3. Subtract the third column from the first column.

$$|\mathbf{A}| = \begin{vmatrix} -1 & -1 & 0 \\ 0 & 1 & 0 \\ 0 & 0 & 2 \end{vmatrix}$$

Step 4. Subtract the first column from the second column.

$$|\mathbf{A}| = \begin{vmatrix} -1 & 0 & 0 \\ 0 & 1 & 0 \\ 0 & 0 & 2 \end{vmatrix}$$

The value of the determinant is then

$$|\mathbf{A}| = a_{11}a_{22}a_{33} = (-1)(1)(2) = -2$$

The second property of a determinant states that the value of a determinant is zero if the corresponding elements of any two rows (or two columns) are equal or have a common ratio. In terms of the Laplace expansion of a determinant this can be formulated as

$$\sum_{j=1}^{n} a_{kj}C_{ij} = 0 \qquad k \neq i \qquad \text{(elements in the } i\text{th row replaced by elements in the } k\text{th row)}$$
$$\sum_{j=1}^{n} a_{jk}C_{ji} = 0 \qquad k \neq i \qquad \text{(elements in the } i\text{th column replaced by elements in the } k\text{th column)} \tag{4.3-5}$$

Using the Kronecker delta notation,

$$\delta_{ik} = \begin{cases} 0 & i \neq k \\ 1 & i = k \end{cases}$$

Eqs. 4.3-4 and 4.3-5 can be combined into the more useful relationship

$$\sum_{j=1}^{n} a_{kj}C_{ij} = |\mathbf{A}|\,\delta_{ik}$$
$$\sum_{j=1}^{n} a_{jk}C_{ji} = |\mathbf{A}|\,\delta_{ik} \tag{4.3-6}$$

This relationship is of importance in the derivation of Cramer's rule, as is seen in Section 4.6.

Pivotal Condensation[1,2]

Chio's method of pivotal condensation is a convenient procedure for evaluating a given determinant by computation of a set of second order determinants. The method is as follows.

Choose any element a_{ij} in the determinant as the pivot term. Select any element a_{ik} which is in the same row as a_{ij}, and any element a_{qj} which is in the same column as a_{ij}. The elements a_{qk}, a_{qj}, a_{ik}, and a_{ij} are then used to form a second order determinant, with the elements kept in their proper order. Form all such second order determinants with the pivot term as one of the elements. The original determinant can now be expressed as an $n-1$ order determinant using the second order determinants as elements, and $1/a_{ij}^{n-2}$ as a multiplying factor. The position of the new elements can be found by subtracting one from each of the subscripts of the element in the second order determinant that lies on a diagonal with the pivot term, if this element lies below the pivot term; the subscripts are unchanged if this diagonal element lies above the pivot term.

By repeating this procedure, the value of a determinant of high order can be computed by successively reducing the order of the determinant by one.

Example 4.3-4. Write the given determinant as a determinant of second order determinants by means of pivotal condensation.

$$\begin{vmatrix} a_{11} & a_{12} & a_{13} \\ a_{21} & a_{22} & a_{23} \\ a_{31} & a_{32} & a_{33} \end{vmatrix} = \frac{1}{a_{11}} \begin{Vmatrix} \begin{vmatrix} a_{11} & a_{12} \\ a_{21} & a_{22} \end{vmatrix} & \begin{vmatrix} a_{11} & a_{13} \\ a_{21} & a_{23} \end{vmatrix} \\ \begin{vmatrix} a_{11} & a_{12} \\ a_{31} & a_{32} \end{vmatrix} & \begin{vmatrix} a_{11} & a_{13} \\ a_{31} & a_{33} \end{vmatrix} \end{Vmatrix}$$

Example 4.3-5. Evaluate

$$|\mathbf{A}| = \begin{vmatrix} 1 & 3 & 4 & 2 \\ 4 & 5 & 6 & 1 \\ 3 & 5 & 8 & 9 \\ 4 & 6 & 2 & 5 \end{vmatrix}$$

using pivotal condensation.

Using the term a_{11} as the pivot,

$$|\mathbf{A}| = \begin{Vmatrix} \begin{vmatrix} 1 & 3 \\ 4 & 5 \end{vmatrix} & \begin{vmatrix} 1 & 4 \\ 4 & 6 \end{vmatrix} & \begin{vmatrix} 1 & 2 \\ 4 & 1 \end{vmatrix} \\ \begin{vmatrix} 1 & 3 \\ 3 & 5 \end{vmatrix} & \begin{vmatrix} 1 & 4 \\ 3 & 8 \end{vmatrix} & \begin{vmatrix} 1 & 2 \\ 3 & 9 \end{vmatrix} \\ \begin{vmatrix} 1 & 3 \\ 4 & 6 \end{vmatrix} & \begin{vmatrix} 1 & 4 \\ 4 & 2 \end{vmatrix} & \begin{vmatrix} 1 & 2 \\ 4 & 5 \end{vmatrix} \end{Vmatrix} = \begin{vmatrix} -7 & -10 & -7 \\ -4 & -4 & 3 \\ -6 & -14 & -3 \end{vmatrix}$$

Again using the new a_{11} as the pivot term, after multiplying the first and second columns by -1,

$$|\mathbf{A}| = \frac{1}{7} \begin{Vmatrix} \begin{vmatrix} 7 & 10 \\ 4 & 4 \end{vmatrix} & \begin{vmatrix} 7 & -7 \\ 4 & 3 \end{vmatrix} \\ \begin{vmatrix} 7 & 10 \\ 6 & 14 \end{vmatrix} & \begin{vmatrix} 7 & -7 \\ 6 & -3 \end{vmatrix} \end{Vmatrix} = \frac{1}{7} \begin{vmatrix} -12 & 49 \\ 38 & 21 \end{vmatrix} = -302$$

Note that each step required only the computation of a set of second order determinants.

To show how the method of pivotal condensation was obtained, a proof is given for a third order determinant. The extension to high order order determinants is then evident. Let

$$|\mathbf{A}| = \begin{vmatrix} a_{11} & a_{12} & a_{13} \\ a_{21} & a_{22} & a_{23} \\ a_{31} & a_{32} & a_{33} \end{vmatrix}$$

Multiply all the rows of the determinant, except the row containing the pivot term, by the value of the pivot term. This multiplies the determinant by the pivot term raised to the $n - 1$ power. To keep the value of the determinant unchanged, the resulting determinant is divided by the pivot term raised to the $n - 1$ power. Using a_{11} as the pivot term, the result of performing this step is

$$|\mathbf{A}| = \frac{1}{a_{11}^{2}} \begin{vmatrix} a_{11} & a_{12} & a_{13} \\ a_{21}a_{11} & a_{22}a_{11} & a_{23}a_{11} \\ a_{31}a_{11} & a_{32}a_{11} & a_{33}a_{11} \end{vmatrix}$$

The first row is then multiplied by the term directly beneath the pivot term (in the original determinant) and is then subtracted from the row containing that term.

$$|\mathbf{A}| = \frac{1}{a_{11}^{2}} \begin{vmatrix} a_{11} & a_{12} & a_{13} \\ 0 & (a_{22}a_{11} - a_{21}a_{12}) & (a_{23}a_{11} - a_{21}a_{13}) \\ a_{31}a_{11} & a_{32}a_{11} & a_{33}a_{11} \end{vmatrix}$$

This process is repeated until all the terms in the column of the pivot term are zero except for the pivot term, i.e.,

$$|\mathbf{A}| = \frac{1}{a_{11}^{2}} \begin{vmatrix} a_{11} & a_{12} & a_{13} \\ 0 & (a_{22}a_{11} - a_{21}a_{12}) & (a_{23}a_{11} - a_{21}a_{13}) \\ 0 & (a_{32}a_{11} - a_{12}a_{31}) & (a_{33}a_{11} - a_{13}a_{31}) \end{vmatrix}$$

The resulting determinant is then seen to be equal to a determinant of order $n - 1$ divided by the pivot term raised to the $(n - 2)$ power.

$$|\mathbf{A}| = \frac{1}{a_{11}} \begin{vmatrix} \begin{vmatrix} a_{11} & a_{12} \\ a_{21} & a_{22} \end{vmatrix} & \begin{vmatrix} a_{11} & a_{13} \\ a_{21} & a_{23} \end{vmatrix} \\ \begin{vmatrix} a_{11} & a_{12} \\ a_{31} & a_{32} \end{vmatrix} & \begin{vmatrix} a_{11} & a_{13} \\ a_{31} & a_{33} \end{vmatrix} \end{vmatrix}$$

The method of pivotal condensation is quite useful in finding the solution to a set of linear nonhomogeneous equations and in the analysis of circuits. In the latter application, the method of pivotal condensation is a rapid procedure for eliminating unnecessary nodes from an admittance matrix formulation of a circuit.

It is interesting to note that the evaluation of an nth order determinant by the Laplace expansion rule generally requires $(n!)(n-1)$ multiplications. The method of pivotal condensation requires $(n^3/2 + n^2 - n/3)$ multiplications; hence this method is a more efficient procedure. For example, if $n = 6$, then the Laplace expansion requires 3600 multiplications. The method of pivotal condensation requires only 106 multiplications.

Product of Determinants

It may be shown that the determinant of the product of two square matrices $\mathbf{A}$ and $\mathbf{B}$ of order n is the product of the determinants $|\mathbf{A}|$ and $|\mathbf{B}|$. That is, if $\mathbf{C} = \mathbf{AB}$, then

$$|\mathbf{C}| = |\mathbf{A}| \cdot |\mathbf{B}| \tag{4.3-7}$$

Example 4.3-6. Evaluate the product of the given determinants.

$$\begin{vmatrix} a_{11} & a_{12} \\ a_{21} & a_{22} \end{vmatrix} \cdot \begin{vmatrix} b_{11} & b_{12} \\ b_{21} & b_{22} \end{vmatrix} = \begin{vmatrix} (a_{11}b_{11} + a_{12}b_{21}) & (a_{11}b_{12} + a_{12}b_{22}) \\ (a_{21}b_{11} + a_{22}b_{21}) & (a_{21}b_{12} + a_{22}b_{22}) \end{vmatrix}$$

Derivative of a Determinant

As a consequence of Eq. 4.3-4 (Laplace expansion formula), the derivative of a determinant with respect to one of its elements is equal to the cofactor of that element, i.e.,

$$\frac{\partial}{\partial a_{ij}} |\mathbf{A}| = \frac{\partial}{\partial a_{ij}} \left(\sum_{i=1}^{n} a_{ij} C_{ij} \right) = C_{ij} \tag{4.3-8}$$

If the elements are functions of a parameter, then the derivative of the determinant with respect to the parameter is the sum of n $(n \times n)$ determinants, in which the ith $(i = 1, 2, \ldots, n)$ is the original determinant except that each element in the ith row (or column) is replaced by its derivative. For example, if the determinant is given by

$$|\mathbf{A}| = \begin{vmatrix} a_{11}(x) & a_{12}(x) \\ a_{21}(x) & a_{22}(x) \end{vmatrix} = a_{11}(x)a_{22}(x) - a_{21}(x)a_{12}(x)$$

then the derivative with respect to x is given by

$$\frac{d}{dx}|\mathbf{A}| = a_{11}(x)\frac{d}{dx}[a_{22}(x)] + a_{22}(x)\frac{d}{dx}[a_{11}(x)] - a_{21}(x)\frac{d}{dx}[a_{12}(x)] - a_{12}(x)\frac{d}{dx}[a_{21}(x)]$$

$$= \begin{vmatrix} a_{11}(x) & \frac{d}{dx}[a_{12}(x)] \\ a_{21}(x) & \frac{d}{dx}[a_{22}(x)] \end{vmatrix} + \begin{vmatrix} \frac{d}{dx}[a_{11}(x)] & a_{12}(x) \\ \frac{d}{dx}[a_{21}(x)] & a_{22}(x) \end{vmatrix}$$

4.4 THE ADJOINT AND INVERSE MATRICES

If $\mathbf{A}$ is a square matrix and C_{ij} is the cofactor of a_{ij}, then the matrix formed by the cofactors C_{ji} is defined as the *adjoint matrix* of $\mathbf{A}$, i.e.,

$$\textbf{Adj A} = [C_{ji}] \tag{4.4-1}$$

Thus the adjoint matrix is the transpose of the matrix formed by replacing the elements a_{ij} by their cofactors.

Example 4.4-1. Find the adjoint matrix of

$$\mathbf{A} = \begin{bmatrix} 1 & 0 & -2 \\ 2 & 3 & 0 \\ 1 & 2 & -1 \end{bmatrix}$$

$$C_{11} = -3 \qquad C_{12} = 2 \qquad C_{13} = 1$$
$$C_{21} = -4 \qquad C_{22} = 1 \qquad C_{23} = -2$$
$$C_{31} = 6 \qquad C_{32} = -4 \qquad C_{33} = 3$$

Therefore

$$\textbf{Adj A} = [C_{ji}] = \begin{bmatrix} -3 & -4 & 6 \\ 2 & 1 & -4 \\ 1 & -2 & 3 \end{bmatrix}$$

Equation 4.3-6 indicates that

$$[a_{kj}][C_{ij}]^T = |\mathbf{A}|\,\mathbf{I} \tag{4.4-2}$$

Therefore the product of the matrix $[a_{ij}]$ and the adjoint matrix $[C_{ji}]$ is equal to

$$[a_{ij}][C_{ji}] = |\mathbf{A}|\,\mathbf{I} \tag{4.4-3}$$

which is a diagonal matrix with all its elements equal to the determinant of the coefficient matrix $\mathbf{A}$. If both sides of Eq. 4.4-3 are divided by $|\mathbf{A}|$, provided $|\mathbf{A}| \neq 0$,

$$\mathbf{A}\,\frac{\mathbf{Adj\ A}}{|\mathbf{A}|} = \mathbf{I} \tag{4.4-4}$$

From this equation it seems natural to define the matrix $(\mathbf{Adj\ A})/|\mathbf{A}|$ as the *inverse* or reciprocal of **A**, such that

$$\mathbf{A}\mathbf{A}^{-1} = \mathbf{I} \tag{4.4-5}$$

where

$$\mathbf{A}^{-1} = \frac{\mathbf{Adj\ A}}{|\mathbf{A}|}, \quad (|\mathbf{A}| \neq 0) \tag{4.4-6}$$

It is evident that $\mathbf{A}^{-1}\mathbf{A} = \mathbf{I}$, i.e., that a matrix and its inverse commute. Only square matrices can possess inverses.

Example 4.4-2. Find the inverse of **A** and verify that $\mathbf{A}\mathbf{A}^{-1} = \mathbf{I}$, if

$$\mathbf{A} = \begin{bmatrix} 2 & 3 \\ 1 & 2 \end{bmatrix}, \quad |\mathbf{A}| = 1, \quad \mathbf{A}^{-1} = \frac{\mathbf{Adj\ A}}{|\mathbf{A}|} = \begin{bmatrix} 2 & -3 \\ -1 & 2 \end{bmatrix}$$

To show that $\mathbf{A}\mathbf{A}^{-1} = \mathbf{I}$,

$$\mathbf{A}\,\frac{\mathbf{Adj\ A}}{|\mathbf{A}|} = \begin{bmatrix} 2 & 3 \\ 1 & 2 \end{bmatrix}\begin{bmatrix} 2 & -3 \\ -1 & 2 \end{bmatrix} = \begin{bmatrix} 1 & 0 \\ 0 & 1 \end{bmatrix} = \mathbf{I}$$

Products of Inverse Matrices

The products of a string of inverse matrices obey the same rules of transposition as do the products of transpose matrices. To show this, consider the product $\mathbf{C} = \mathbf{AB}$. Premultiply both sides of the equation by $\mathbf{B}^{-1}\mathbf{A}^{-1}$ and postmultiply both sides by $\mathbf{C}^{-1}$. This results in the relationship

$$\mathbf{B}^{-1}\mathbf{A}^{-1} = \mathbf{C}^{-1} \tag{4.4-7}$$

Example 4.4-3. Find the inverse matrix of the product **AB** where

$$\mathbf{A} = \begin{bmatrix} 1 & -2 \\ 0 & 2 \end{bmatrix} \qquad \mathbf{B} = \begin{bmatrix} 1 & -1 \\ 2 & 1 \end{bmatrix}$$

The product **AB** is

$$\mathbf{AB} = \begin{bmatrix} -3 & -3 \\ 4 & 2 \end{bmatrix}$$

The inverse matrix $(\mathbf{AB})^{-1}$ is given by

$$(\mathbf{AB})^{-1} = \frac{\mathbf{Adj\ (AB)}}{|\mathbf{AB}|} = \frac{\begin{bmatrix} 2 & 3 \\ -4 & -3 \end{bmatrix}}{6}$$

Alternatively, the separate inverses $\mathbf{A}^{-1}$ and $\mathbf{B}^{-1}$ are given by

$$\mathbf{A}^{-1} = \frac{\begin{bmatrix} 2 & 2 \\ 0 & 1 \end{bmatrix}}{2} \qquad \mathbf{B}^{-1} = \frac{\begin{bmatrix} 1 & 1 \\ -2 & 1 \end{bmatrix}}{3}$$

The product $\mathbf{B}^{-1}\mathbf{A}^{-1}$ is given by

$$\mathbf{B}^{-1}\mathbf{A}^{-1} = \frac{\begin{bmatrix} 1 & 1 \\ -2 & 1 \end{bmatrix}}{3} \frac{\begin{bmatrix} 2 & 2 \\ 0 & 1 \end{bmatrix}}{2} = \frac{\begin{bmatrix} 2 & 3 \\ -4 & -3 \end{bmatrix}}{6}$$

Thus $(\mathbf{AB})^{-1} = \mathbf{B}^{-1}\mathbf{A}^{-1}$, which agrees with Eq. 4.4-7.

Derivative of the Inverse Matrix

For a t-value at which $\mathbf{A}(t)$ is differentiable and possesses an inverse, the derivative of $\mathbf{A}^{-1}(t)$ is given by

$$\frac{d}{dt}[\mathbf{A}^{-1}(t)] = -\mathbf{A}^{-1}(t)\dot{\mathbf{A}}(t)\mathbf{A}^{-1}(t) \tag{4.4-8}$$

This relationship can be derived by considering

$$\frac{d}{dt}[\mathbf{A}^{-1}(t)\mathbf{A}(t)] = \frac{d\mathbf{I}}{dt} = [0]$$

This is

$$\mathbf{A}^{-1}(t)\dot{\mathbf{A}}(t) + \frac{d}{dt}[\mathbf{A}^{-1}(t)]\mathbf{A}(t) = [0]$$

Hence

$$\frac{d}{dt}[\mathbf{A}^{-1}(t)] = -\mathbf{A}^{-1}(t)\dot{\mathbf{A}}(t)\mathbf{A}^{-1}(t)$$

Note that this is not the same as $(d\mathbf{A}/dt)^{-1}$.

Example 4.4-4. Find the derivative of the inverse of $\mathbf{A}$, where

$$\mathbf{A}(t) = \begin{bmatrix} 0 & 1 \\ -1 & -t \end{bmatrix}$$

The inverse of $\mathbf{A}$ is

$$\mathbf{A}^{-1}(t) = \begin{bmatrix} -t & -1 \\ 1 & 0 \end{bmatrix}$$

Clearly,

$$\frac{d}{dt}[\mathbf{A}^{-1}(t)] = \begin{bmatrix} -1 & 0 \\ 0 & 0 \end{bmatrix}$$

Checking this result with Eq. 4.4-8,

$$\frac{d}{dt}[\mathbf{A}^{-1}(t)] = -\begin{bmatrix} -t & -1 \\ 1 & 0 \end{bmatrix}\begin{bmatrix} 0 & 0 \\ 0 & -1 \end{bmatrix}\begin{bmatrix} -t & -1 \\ 1 & 0 \end{bmatrix} = \begin{bmatrix} -1 & 0 \\ 0 & 0 \end{bmatrix}$$

Some Special Inverse Matrices

Involutary matrix. If $\mathbf{AA} = \mathbf{I}$ (a matrix is its own inverse), the matrix is said to be *involutary.*

Orthogonal matrix. If $\mathbf{A}^{-1} = \mathbf{A}^T$, the matrix $\mathbf{A}$ is said to be an *orthogonal matrix.*

Unitary matrix. If $\mathbf{A} = \{(\mathbf{A}^*)^T\}^{-1}$, then $\mathbf{A}$ is *unitary.* ($\mathbf{A}$ is equal to the inverse of the associate matrix of $\mathbf{A}$.)

4.5 VECTORS AND LINEAR VECTOR SPACES

The rows and columns of a matrix are called row *vectors* and column *vectors*, respectively, in Section 4.2. This is simply an extension of the more familiar concept of vectors in two- or three-dimensional spaces to an n-dimensional space. When n is greater than three, the geometrical visualization becomes obscure, but the terminology associated with the familiar coordinate systems is still quite useful. For example, the coordinate system having the unit vectors

$$\hat{\mathbf{i}}_1 = \begin{bmatrix} 1 \\ 0 \\ 0 \\ 0 \\ \cdot \\ \cdot \\ \cdot \\ 0 \end{bmatrix}, \quad \hat{\mathbf{i}}_2 = \begin{bmatrix} 0 \\ 1 \\ 0 \\ 0 \\ \cdot \\ \cdot \\ \cdot \\ 0 \end{bmatrix}, \ldots, \hat{\mathbf{i}}_n = \begin{bmatrix} 0 \\ 0 \\ 0 \\ 0 \\ \cdot \\ \cdot \\ \cdot \\ 0 \\ 1 \end{bmatrix}$$

can be thought of as an n-dimensional system with mutually orthogonal coordinate axes.

Scalar Products

The scalar product (or inner product) of two vectors $\mathbf{x}$ and $\mathbf{y}$ is written as $\langle \mathbf{x}, \mathbf{y} \rangle$ and is defined as

$$\langle \mathbf{x}, \mathbf{y} \rangle = (\mathbf{x}^*)^T\mathbf{y} = x_1{}^*y_1 + x_2{}^*y_2 + \cdots + x_n{}^*y_n$$
$$= \mathbf{y}^T\mathbf{x}^* \neq \mathbf{x}^T\mathbf{y}^* \quad (4.5\text{-}1)$$

Note that the complex conjugates of the components of $\mathbf{x}$ are indicated, since these components may be complex, in general. For the case where both $\mathbf{x}$ and $\mathbf{y}$ are real, Eq. 4.5-1 reduces to the more familiar form

$$\langle \mathbf{x}, \mathbf{y} \rangle = \sum_{i=1}^{n} x_i y_i = x_1 y_1 + x_2 y_2 + \cdots + x_n y_n$$

The scalar product can then (for real $\mathbf{x}$ and $\mathbf{y}$) be written as

$$\langle \mathbf{x}, \mathbf{y} \rangle = \mathbf{x}^T \mathbf{y} = \mathbf{y}^T \mathbf{x} = \langle \mathbf{y}, \mathbf{x} \rangle \tag{4.5-2}$$

Example 4.5-1. Find the scalar product $\langle \mathbf{x}, \mathbf{y} \rangle$ of the vectors

$$\mathbf{x} = \begin{bmatrix} 1 + j \\ 2 \end{bmatrix} \qquad \mathbf{y} = \begin{bmatrix} 1 + j \\ 2j \end{bmatrix}$$

The scalar product $(\mathbf{x}^*)^T \mathbf{y}$ is equal to $\langle \mathbf{x}, \mathbf{y} \rangle = (1 - j)(1 + j) + 2(2j) = 2 + 4j$. This is not equal to, but is the conjugate of, $\mathbf{x}^T \mathbf{y}^* = (1 + j)(1 - j) + 2(-2j) = 2 - 4j$.

Outer Product

If the (n x 1) column vector $\mathbf{x}$ is denoted by $\mathbf{x}\rangle$ and the (1 x m) *row* vector $(\mathbf{y}^*)^T$ is denoted by $\langle\mathbf{y}$, then the *outer product* $\mathbf{x}\rangle\langle\mathbf{y}$ is the (n x m) matrix.†

$$\mathbf{x}\rangle\langle\mathbf{y} = \mathbf{x}(\mathbf{y}^*)^T = \begin{bmatrix} x_1 y_1^* & x_1 y_2^* & \cdots & x_1 y_m^* \\ x_2 y_1^* & x_2 y_2^* & \cdots & x_2 y_m^* \\ \cdots & \cdots & \cdots & \cdots \\ x_n y_1^* & x_n y_2^* & \cdots & x_n y_m^* \end{bmatrix} \tag{4.5-3}$$

Orthogonal Vectors

Two vectors $\mathbf{x}$ and $\mathbf{y}$ are said to be *orthogonal* if their scalar product $\langle \mathbf{x}, \mathbf{y} \rangle$ is equal to zero, i.e.,

$$\langle \mathbf{x}, \mathbf{y} \rangle = 0 \tag{4.5-4}$$

Length of a Vector

The length of a vector $\mathbf{x}$, denoted by $\|\mathbf{x}\|$ and sometimes called the *norm* of $\mathbf{x}$, is defined as the square root of the scalar product of $\mathbf{x}$ and $\mathbf{x}$, i.e.,

$$\|\mathbf{x}\| = \sqrt{\langle \mathbf{x}, \mathbf{x} \rangle} = \sqrt{x_1^* x_1 + x_2^* x_2 + \cdots + x_n^* x_n} \tag{4.5-5}$$

As a consequence of this definition, it can be shown that

$$\|\mathbf{x} + \mathbf{y}\| \le \|\mathbf{x}\| + \|\mathbf{y}\| \qquad \text{(Triangle inequality)} \tag{4.5-6}$$

and

$$|\langle \mathbf{x}, \mathbf{y} \rangle| \le \|\mathbf{x}\| \cdot \|\mathbf{y}\| \qquad \text{(Schwarz inequality)} \tag{4.5-7}$$

† The outer product is often called the *dyadic product*.

Unit Vectors

A vector is said to be a *unit vector* if its length is unity, so that $\langle \hat{\mathbf{x}}, \hat{\mathbf{x}} \rangle = 1$. A unit vector can be obtained from the vector $\mathbf{x}$ by dividing each component of the vector $\mathbf{x}$ by its length. Thus

$$\hat{\mathbf{x}} = \frac{\mathbf{x}}{\sqrt{\langle \mathbf{x}, \mathbf{x} \rangle}} \tag{4.5-8}$$

Example 4.5-2. Find the unit vector $\hat{\mathbf{x}}$ corresponding to the vector $\mathbf{x}$, where

$$\mathbf{x} = \begin{bmatrix} 1 + j2 \\ 2 - j \end{bmatrix}$$

The scalar product $\langle \mathbf{x}, \mathbf{x} \rangle = 10$. Therefore

$$\hat{\mathbf{x}} = \begin{bmatrix} \dfrac{1 + j2}{\sqrt{10}} \\ \\ \dfrac{2 - j}{\sqrt{10}} \end{bmatrix}$$

Linear Independence

A set of m vectors $\mathbf{x}_i$, having n components $x_{1i}, x_{2i}, \ldots, x_{ni}$, is said to be *linearly independent* provided that no set of constants $k_1, k_2, \ldots, k_m$ exists (at least one k_i must be nonzero) such that

$$k_1\mathbf{x}_1 + k_2\mathbf{x}_2 + \cdots k_m\mathbf{x}_m = \mathbf{0} \tag{4.5-9}$$

On the basis of the concept of linear independence of vectors, several important definitions can be stated as follows:

Singular matrix. A square matrix is said to be *singular* if the rows or columns are not linearly independent. In this case $|\mathbf{A}| = 0$. If $|\mathbf{A}| \neq 0$, then the matrix is said to be *nonsingular*. Hence only nonsingular matrices possess inverses.

Degeneracy.† If the rows (columns) of a singular matrix are linearly related by a single relationship, the matrix is said to be *simply degenerate*, or of degeneracy 1. If more than a single relationship connects the rows (columns), the matrix is said to be *multiply degenerate.* If there are q such relationships, the matrix is of degeneracy q.

Rank. The rank r of a matrix $\mathbf{A}$ is the largest square array in $\mathbf{A}$ with a nonvanishing determinant. If the order of a square matrix is n, then

$$r = n - q \tag{4.5-10}$$

Clearly, an $(n \times n)$ matrix has rank $r < n$, only if the matrix is singular.

† The term *nullity* is often used in place of *degeneracy*.

Sylvester's law of degeneracy. The degeneracy of the product of two matrices is at least as great as the degeneracy of either matrix, and at most as great as the sum of the degeneracies of either matrix.

A clear illustration of Sylvester's law is given by the product of two diagonal matrices **A** and **B**,

$$\mathbf{AB} = \begin{bmatrix} a_1b_1 & & & & & & \\ & a_2b_2 & & & & & \\ & & a_3b_3 & & & & \\ & & & \cdot & & & \\ & & & & \cdot & & \\ & & & & & \cdot & \\ & & & & & & a_nb_n \end{bmatrix}$$

where

$$\mathbf{A} = [a_i\delta_{ij}] \qquad \mathbf{B} = [b_i\delta_{ij}]$$

Assume that **A** has m zero elements and **B** has q zero elements, where $q \leqslant m$. If the zero elements of **A** do not occur where the zero elements of **B** are located, then the total number of zeros in the product **AB** is $m + q$. If the zero elements of **B** are located where the zero elements of **A** occur, then the total number of zero elements in the product **AB** is m.

Returning to Eq. 4.5-9 and linear independence, the latter can be expressed in terms of the rank of the matrix formed by the elements of the m vectors $\mathbf{x}_1, \mathbf{x}_2, \ldots, \mathbf{x}_m$. This matrix is

$$\mathbf{A} = \begin{bmatrix} x_{11} & x_{12} & \cdots & x_{1m} \\ x_{21} & x_{22} & \cdots & x_{2m} \\ \cdots & \cdots & \cdots & \cdots \\ x_{n1} & x_{n2} & \cdots & x_{nm} \end{bmatrix} \qquad m \leqslant n$$

If the rank of the matrix **A** associated with these m vectors is less than m, i.e., $r < m$, then there are only r vectors of the set which are linearly independent. The remaining $m - r$ vectors can be expressed as a linear combination of these r vectors. Therefore a necessary and sufficient condition for the vectors to be linearly independent is that the rank of **A** be equal to m.

Gramian

The Gramian of a set of vectors is formed by assuming that a relationship such as Eq. 4.5-9 does exist. By successively taking the scalar products of $\mathbf{x}_i$ and both sides of Eq. 4.5-9, the set of equations

$$\begin{aligned} k_1\langle \mathbf{x}_1, \mathbf{x}_1\rangle + k_2\langle \mathbf{x}_1, \mathbf{x}_2\rangle + \cdots + k_m\langle \mathbf{x}_1, \mathbf{x}_m\rangle &= 0 \\ k_1\langle \mathbf{x}_2, \mathbf{x}_1\rangle + k_2\langle \mathbf{x}_2, \mathbf{x}_2\rangle + \cdots + k_m\langle \mathbf{x}_2, \mathbf{x}_m\rangle &= 0 \\ \cdots\cdots\cdots\cdots\cdots\cdots\cdots\cdots\cdots\cdots \\ k_1\langle \mathbf{x}_m, \mathbf{x}_1\rangle + k_2\langle \mathbf{x}_m, \mathbf{x}_2\rangle + \cdots + k_m\langle \mathbf{x}_m, \mathbf{x}_m\rangle &= 0 \end{aligned}$$

is obtained. As shown in the next section, this set of homogeneous equations possesses a nontrivial solution for the k_i, only if the determinant of the coefficient matrix $[\langle \mathbf{x}_i, \mathbf{x}_j\rangle]$ vanishes. This determinant is called the *Gramian* or Gram determinant, and it is

$$G = \begin{vmatrix} \langle \mathbf{x}_1, \mathbf{x}_1\rangle & \langle \mathbf{x}_1, \mathbf{x}_2\rangle & \cdots & \langle \mathbf{x}_1, \mathbf{x}_m\rangle \\ \langle \mathbf{x}_2, \mathbf{x}_1\rangle & \langle \mathbf{x}_2, \mathbf{x}_2\rangle & \cdots & \langle \mathbf{x}_2, \mathbf{x}_m\rangle \\ \cdots & \cdots & \cdots & \cdots \\ \langle \mathbf{x}_m, \mathbf{x}_1\rangle & \langle \mathbf{x}_m, \mathbf{x}_2\rangle & \cdots & \langle \mathbf{x}_m, \mathbf{x}_m\rangle \end{vmatrix} \tag{4.5-11}$$

Therefore a set of vectors is linearly dependent if and only if the Gramian of the set of vectors is equal to zero. Note that, if the vectors are orthogonal, the Gramian is a diagonal determinant.

Linear Vector Space[3]

A linear vector space S consists of a set of vectors such that

(*a*) the sum of any two vectors in S is also a vector in the set;
(*b*) every scalar multiple of a vector in the set is a vector in the set; and
(*c*) the rules for forming sums of vectors, and products of vectors with scalars have the following properties:

(1) $\mathbf{x} + \mathbf{y} = \mathbf{y} + \mathbf{x}$ for any $\mathbf{x}$ and $\mathbf{y}$ in S;
(2) $(\mathbf{x} + \mathbf{y}) + \mathbf{z} = \mathbf{x} + (\mathbf{y} + \mathbf{z})$ for any $\mathbf{x}$, $\mathbf{y}$, and $\mathbf{z}$ in S;
(3) there exists a vector equal to $\mathbf{0}$ in S, such that $\mathbf{x} + \mathbf{0} = \mathbf{x}$ for any $\mathbf{x}$ in S;
(4) for any $\mathbf{x}$ in S, there exists a vector $\mathbf{y}$ in S such that $\mathbf{x} + \mathbf{y} = \mathbf{0}$;
(5) $1\mathbf{x} = \mathbf{x}$ for any $\mathbf{x}$ in S;
(6) $a(b\mathbf{x}) = (ab)\mathbf{x}$ for any scalar a and b, and any $\mathbf{x}$ in S;
(7) $(a + b)\mathbf{x} = a\mathbf{x} + b\mathbf{x}$ for any scalar a and b, and any $\mathbf{x}$ in S;
(8) $a(\mathbf{x} + \mathbf{y}) = a\mathbf{x} + a\mathbf{y}$ for any scalar a, and any $\mathbf{x}$ and $\mathbf{y}$ in S.

The most common example of a linear vector space is the set of all vectors contained within a three-dimensional Euclidian space. For example, all the forces acting on a space vehicle constitute a vector space, where the forces are described as vectors in the particular coordinate system chosen.

If a set of vectors $\mathbf{x}_1, \mathbf{x}_2, \ldots, \mathbf{x}_m$ are contained in the space S, then the set of all vectors $\mathbf{y}$ which are linear combinations of these vectors, i.e.,

$$\mathbf{y} = k_1\mathbf{x}_1 + k_2\mathbf{x}_2 + \cdots + k_m\mathbf{x}_m \tag{4.5-12}$$

is a vector space. The *dimension* of the space is the maximum number of linearly independent vectors in the space.

If only r of the $\mathbf{x}_i$ vectors in Eq. 4.5-12 are linearly independent, the dimension of the space that can be generated by these vectors is equal to r, the rank of the set of the $\mathbf{x}_i$ vectors. For example, consider the vectors

$$\mathbf{x}_1 = \begin{bmatrix} 1 \\ 1 \\ 0 \end{bmatrix} \qquad \mathbf{x}_2 = \begin{bmatrix} 0 \\ 1 \\ 1 \end{bmatrix} \qquad \mathbf{x}_3 = \begin{bmatrix} 1 \\ 2 \\ 1 \end{bmatrix}$$

These three vectors are linearly dependent, as $\mathbf{x}_1 + \mathbf{x}_2 = \mathbf{x}_3$. A vector $\mathbf{y}$, which is the linear combination of these three vectors,

$$\mathbf{y} = k_1\mathbf{x}_1 + k_2\mathbf{x}_2 + k_3\mathbf{x}_3 = (k_1 + k_3)\mathbf{x}_1 + (k_2 + k_3)\mathbf{x}_2$$

cannot have all three of its components specified independently. Only two of the components of $\mathbf{y}$ can be specified independently. The third component must be a function of the other two. The dimension of the space generated by these particular $\mathbf{x}_1$, $\mathbf{x}_2$, and $\mathbf{x}_3$ is equal to two.

In an n-dimensional space, the n components of any vector $\mathbf{y}$ can be specified independently if the vector $\mathbf{y}$ is generated by a set of vectors of rank n. In that case, the set of n linearly independent vectors is said to "span the space." These n linearly independent vectors can also be used to form a basis in the space. A *basis* of a space is a set of vectors such that any vector in the space is a *unique* linear combination of the basis vectors. A basis is, in essence, a coordinate system.

When the components of a vector $\mathbf{y}$ are given, it is necessary to indicate the basis or coordinate system with respect to which the components are specified. For example, the location of a point in a three-dimensional space could be specified with respect to a rectangular coordinate system, a cyclindrical coordinate system, a spherical coordinate system, etc. The statement that $\mathbf{y}$ has the components 1, 0, 2, i.e.,

$$\mathbf{y} = \begin{bmatrix} 1 \\ 0 \\ 2 \end{bmatrix}$$

is meaningless, unless the basis is also specified.

Example 4.5-3. Show that the vectors

$$\mathbf{x}_1 = \begin{bmatrix} 1 \\ 1 \\ 1 \end{bmatrix}, \quad \mathbf{x}_2 = \begin{bmatrix} 1 \\ 2 \\ 3 \end{bmatrix}, \quad \mathbf{x}_3 = \begin{bmatrix} 1 \\ 3 \\ 2 \end{bmatrix}$$

specified in terms of the orthogonal basis (1, 0, 0), (0, 1, 0), (0, 0, 1) span a three-dimensional space. Specify any vector $\mathbf{y}$ of this space in terms of the vectors $\mathbf{x}_1$, $\mathbf{x}_2$, $\mathbf{x}_3$, and in terms of the basis vectors.

Since the Gramian of these three vectors is

$$G = \begin{vmatrix} 3 & 6 & 6 \\ 6 & 14 & 13 \\ 6 & 13 & 14 \end{vmatrix} \neq 0$$

the three vectors are linearly independent, and therefore they *span* the three-dimensional space. Thus any vector $\mathbf{y}$ in the three-dimensional space can be expressed as a linear combination of these three vectors as $\mathbf{y} = k_1\mathbf{x}_1 + k_2\mathbf{x}_2 + k_3\mathbf{x}_3$.

If the vectors $\mathbf{x}_1$, $\mathbf{x}_2$, and $\mathbf{x}_3$ are used as a basis for the three-dimensional space, then, relative to this basis, the vector $\mathbf{y}$ is

$$\mathbf{y} = \begin{bmatrix} k_1 \\ k_2 \\ k_3 \end{bmatrix}$$

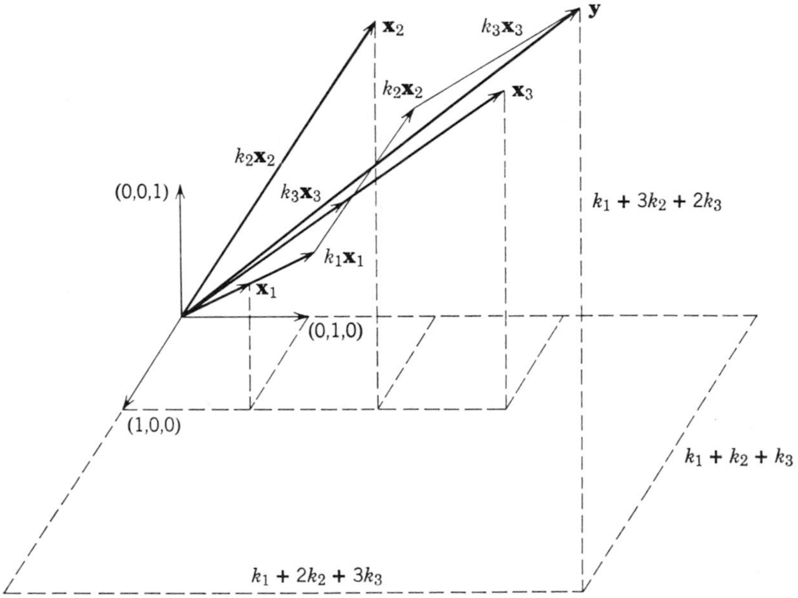

Fig. 4.5-1

Relative to the particular orthogonal system of coordinates (1, 0, 0), (0, 1, 0), and (0, 0, 1), the vector $\mathbf{y}$ is

$$\mathbf{y} = \begin{bmatrix} (k_1 + k_2 + k_3) \\ (k_1 + 2k_2 + 3k_3) \\ (k_1 + 3k_2 + 2k_3) \end{bmatrix}$$

These vectors are illustrated in Fig. 4.5-1 for a specific $\mathbf{y}$.

Gram-Schmidt Orthogonalization of a Vector Set

If a set of m linearly independent vectors is given, an *orthogonal* set of m linearly independent vectors can be found, expressed in terms of the original set of vectors. If the length of each vector in the orthogonal set is unity, the set is said to be an *orthonormal* set.

Assume that the set $\mathbf{x}_1, \mathbf{x}_2, \ldots, \mathbf{x}_m$ is given, and that an orthonormal set $\hat{\mathbf{y}}_1, \hat{\mathbf{y}}_2, \ldots, \hat{\mathbf{y}}_m$ is desired. The procedure used to obtain this set is as follows: Select any one of the $\mathbf{x}_i$ vectors as the first vector in the $\mathbf{y}_i$ set. For example, select $\mathbf{y}_1 = \mathbf{x}_1$. Select another vector $\mathbf{x}_2$ from the original set. Let $\mathbf{y}_2 = \mathbf{x}_2 - k\mathbf{y}_1$, where k is to be chosen such that $\mathbf{y}_2$ is orthogonal to $\mathbf{y}_1$, i.e., such that $\langle \mathbf{y}_1, \mathbf{y}_2 \rangle = \langle \mathbf{y}_1, \mathbf{x}_2 \rangle - k\langle \mathbf{y}_1, \mathbf{y}_1 \rangle = 0$. Therefore

$$k = \frac{\langle \mathbf{y}_1, \mathbf{x}_2 \rangle}{\langle \mathbf{y}_1, \mathbf{y}_1 \rangle}$$

Note that $k = 0$ if and only if $\langle \mathbf{y}_1, \mathbf{x}_2 \rangle = 0$, i.e., $\mathbf{x}_2$ is orthogonal to $\mathbf{y}_1$, in which case take $\mathbf{y}_2 = \mathbf{x}_2$. Generally,

$$\mathbf{y}_2 = \mathbf{x}_2 - \frac{\langle \mathbf{y}_1, \mathbf{x}_2 \rangle}{\langle \mathbf{y}_1, \mathbf{y}_1 \rangle} \mathbf{y}_1$$

The geometrical idea behind the process is that any vector, in particular $\mathbf{x}_2$, may be decomposed into two components, one of which is parallel to $\mathbf{y}_1$ and the other perpendicular to $\mathbf{y}_1$. To obtain the latter, which is called $\mathbf{y}_2$ here, the component of $\mathbf{x}_2$ in the direction of $\mathbf{y}_1$ is subtracted from $\mathbf{x}_2$.

In a similar fashion, $\mathbf{y}_3$ is written as $\mathbf{y}_3 = \mathbf{x}_3 - k_2\mathbf{y}_2 - k_1\mathbf{y}_1$, i.e., as the component of $\mathbf{x}_3$ perpendicular to the plane defined by $\mathbf{y}_1$ and $\mathbf{y}_2$. Analytically, if the vector $\mathbf{y}_3$ is to be orthogonal to both $\mathbf{y}_2$ and $\mathbf{y}_1$, this leads to the equations

$$\langle \mathbf{y}_1, \mathbf{x}_3 \rangle = k_2\langle \mathbf{y}_1, \mathbf{y}_2 \rangle + k_1\langle \mathbf{y}_1, \mathbf{y}_1 \rangle = k_1\langle \mathbf{y}_1, \mathbf{y}_1 \rangle$$

$$\langle \mathbf{y}_2, \mathbf{x}_3 \rangle = k_2\langle \mathbf{y}_2, \mathbf{y}_2 \rangle + k_1\langle \mathbf{y}_2, \mathbf{y}_1 \rangle = k_2\langle \mathbf{y}_2, \mathbf{y}_2 \rangle$$

or

$$\mathbf{y}_3 = \mathbf{x}_3 - \frac{\langle \mathbf{y}_2, \mathbf{x}_3 \rangle}{\langle \mathbf{y}_2, \mathbf{y}_2 \rangle} \mathbf{y}_2 - \frac{\langle \mathbf{y}_1, \mathbf{x}_3 \rangle}{\langle \mathbf{y}_1, \mathbf{y}_1 \rangle} \mathbf{y}_1$$

This procedure leads to the jth member of the set, $\mathbf{y}_j$, as $\mathbf{y}_1 = \mathbf{x}_1$;

$$\mathbf{y}_j = \mathbf{x}_j - \sum_{i=1}^{j-1} \frac{\langle \mathbf{y}_i, \mathbf{x}_j \rangle}{\langle \mathbf{y}_i, \mathbf{y}_i \rangle} \mathbf{y}_i \quad (j = 2, 3, \ldots, m)$$

The $\mathbf{y}_i$'s now form an orthogonal set, and the unit vectors

$$\hat{\mathbf{y}}_i = \frac{\mathbf{y}_i}{\|\mathbf{y}_i\|} \tag{4.5-13}$$

form an orthonormal set.

Example 4.5-4. Using the set of vectors of Example 4.5-3, form an orthonormal set $\hat{\mathbf{y}}_i$.

Choosing $\mathbf{y}_1 = \mathbf{x}_1$,

$$\mathbf{y}_1 = \begin{bmatrix} 1 \\ 1 \\ 1 \end{bmatrix} \quad \text{and} \quad \hat{\mathbf{y}}_1 = \frac{1}{\sqrt{3}} \begin{bmatrix} 1 \\ 1 \\ 1 \end{bmatrix}$$

Then, since $\langle \mathbf{y}_1, \mathbf{x}_2 \rangle / \langle \mathbf{y}_1, \mathbf{y}_1 \rangle = 2$,

$$\mathbf{y}_2 = \begin{bmatrix} 1 \\ 2 \\ 3 \end{bmatrix} - 2 \begin{bmatrix} 1 \\ 1 \\ 1 \end{bmatrix} = \begin{bmatrix} -1 \\ 0 \\ 1 \end{bmatrix} \quad \text{and} \quad \hat{\mathbf{y}}_2 = \frac{1}{\sqrt{2}} \begin{bmatrix} -1 \\ 0 \\ 1 \end{bmatrix}$$

Similarly,

$$\mathbf{y}_3 = \begin{bmatrix} 1 \\ 3 \\ 2 \end{bmatrix} - \tfrac{1}{2} \begin{bmatrix} -1 \\ 0 \\ 1 \end{bmatrix} - 2 \begin{bmatrix} 1 \\ 1 \\ 1 \end{bmatrix} = \begin{bmatrix} -\frac{1}{2} \\ 1 \\ -\frac{1}{2} \end{bmatrix}$$

and

$$\hat{\mathbf{y}}_3 = \frac{1}{\sqrt{6}} \begin{bmatrix} -1 \\ 2 \\ -1 \end{bmatrix}$$

The orthonormal basis is then

$$\hat{\mathbf{y}}_1 = \frac{1}{\sqrt{3}} \begin{bmatrix} 1 \\ 1 \\ 1 \end{bmatrix}, \quad \hat{\mathbf{y}}_2 = \frac{1}{\sqrt{2}} \begin{bmatrix} -1 \\ 0 \\ 1 \end{bmatrix}, \quad \hat{\mathbf{y}}_3 = \frac{1}{\sqrt{6}} \begin{bmatrix} -1 \\ 2 \\ -1 \end{bmatrix}$$

Reciprocal Basis

The solution for the k_i's of Eq. 4.5-12 can be simplified by defining a set of vectors $\mathbf{r}_i$ such that

$$\langle \mathbf{r}_i, \mathbf{x}_j \rangle = \delta_{ij} \quad (i, j = 1, 2, \ldots, m) \tag{4.5-14}$$

Given a basis for a space, it is not difficult to show that there always exists a set of vectors such that a set of relations of the form of Eq. 4.5-14 is satisfied. The vectors $\mathbf{r}_1, \mathbf{r}_2, \ldots, \mathbf{r}_m$ are linearly independent and thus span the m-dimensional space spanned by the basis vectors $\mathbf{x}_i$. Therefore they constitute a basis for the space. Owing to the relationship (Eq. 4.5-14) between this basis and the basis formed by the $\mathbf{x}_i$ vectors, the basis formed by the $\mathbf{r}_i$ vectors is called a *reciprocal basis.*

The principal use of the reciprocal basis is in finding the constants k_i of Eq. 4.5-12. If the scalar products of both sides of Eq. 4.5-12 are taken with $\mathbf{r}_1$, the result is

$$\langle \mathbf{r}_1, \mathbf{y} \rangle = k_1\langle \mathbf{r}_1, \mathbf{x}_1 \rangle + k_2\langle \mathbf{r}_1, \mathbf{x}_2 \rangle + \cdots + k_m\langle \mathbf{r}_1, \mathbf{x}_m \rangle$$

However, by the definition of the reciprocal basis, the scalar products $\langle \mathbf{r}_1, \mathbf{x}_2 \rangle, \ldots, \langle \mathbf{r}_1, \mathbf{x}_m \rangle$ vanish so that $k_1 = \langle \mathbf{r}_1, \mathbf{y} \rangle$, and in general

$$k_i = \langle \mathbf{r}_i, \mathbf{y} \rangle \tag{4.5-15}$$

Equation 4.5-12 can then be written in the form

$$\mathbf{y} = \sum_{i=1}^{m} \langle \mathbf{r}_i, \mathbf{y} \rangle \mathbf{x}_i \tag{4.5-16}$$

The scalar product of $\langle \mathbf{r}_i, \mathbf{y} \rangle$ is the component of the vector $\mathbf{y}$ in the direction of the vector $\mathbf{x}_i$. This equation proves to be extremely useful when the mode characterization of systems is discussed in the next chapter.

4.6 SOLUTIONS OF LINEAR EQUATIONS

Cramer's Rule

Assume that the nonhomogeneous system of n equations

$$\begin{aligned} a_{11}x_1 + a_{12}x_2 + \cdots + a_{1n}x_n &= y_1 \\ a_{21}x_1 + a_{22}x_2 + \cdots + a_{2n}x_n &= y_2 \\ \cdots\cdots\cdots\cdots&\cdots\cdots \\ a_{n1}x_1 + a_{n2}x_2 + \cdots + a_{nn}x_n &= y_n \end{aligned} \tag{4.6-1}$$

is given. The a_{ij} and y_i are a known set of numbers. The n unknowns x_i are to be determined. Note that the number of equations is equal to the number of unknowns. By the use of determinants or matrices, a systematic procedure exists by which the unknowns generally can be determined.

The set of equations, Eq. 4.6-1, can be written in the compact forms,

$$\sum_{j=1}^{n} a_{ij}x_j = y_i, \quad i = 1, 2, \ldots, n \quad \text{or} \quad \mathbf{A}\mathbf{x} = \mathbf{y} \tag{4.6-2}$$

where

$$\mathbf{x} = \begin{bmatrix} x_1 \\ x_2 \\ \cdot \\ \cdot \\ \cdot \\ x_n \end{bmatrix} \quad \text{and} \quad \mathbf{y} = \begin{bmatrix} y_1 \\ y_2 \\ \cdot \\ \cdot \\ \cdot \\ y_n \end{bmatrix}$$

By multiplying both sides of Eq. 4.6-2 by the cofactor C_{ik} and summing over all values of i between 1 and n, it follows that

$$\sum_{j=1}^{n}\left(\sum_{i=1}^{n} C_{ik}a_{ij}\right)x_j = \sum_{i=1}^{n} C_{ik}y_1, \quad k = 1, 2, \ldots, n \qquad (4.6\text{-}3)$$

or

$$(\textbf{Adj A})\mathbf{Ax} = (\textbf{Adj A})\mathbf{y}$$

However, the term in parentheses in Eq. 4.6-3 is equal to zero, except when $j = k$ (see Eq. 4.3-6). When $j = k$, this term is equal to the determinant of **A**. Therefore

$$|\mathbf{A}|\, x_k = \sum_{i=1}^{n} C_{ik}y_1, \quad k = 1, 2, \ldots, n \quad \text{or} \quad |\mathbf{A}|\,\mathbf{x} = (\textbf{Adj A})\mathbf{y} \quad (4.6\text{-}4)$$

If the determinant $|\mathbf{A}|$ is not identically zero, the n equations are linearly independent, i.e., none of the equations can be written as a linear combination of the other equations. Then the unknown x_k is uniquely given by

$$x_k = \frac{\sum_{i=1}^{n} C_{ik}y_i}{|\mathbf{A}|}, \quad k = 1, 2, \ldots, n \quad \text{or} \quad \mathbf{x} = \frac{(\textbf{Adj A})}{|\mathbf{A}|}\mathbf{y} = \mathbf{A}^{-1}\mathbf{y} \quad (4.6\text{-}5)$$

The numerator of the expression for x_k is simply the determinant of **A** with the kth column replaced by the column formed by the right side of Eq. 4.6-1. Thus Cramer's rule for a solution by determinants can be stated as follows:

For a set of n linear algebraic equations in n unknowns $x_1, x_2, \ldots, x_n$, a solution for the unknowns exists if **A** is a *nonsingular* matrix. The value of a given variable, x_k, for example, is the quotient of two determinants. The denominator of the quotient is the determinant of the coefficient matrix. The numerator of the quotient is the determinant of the coefficient matrix with the kth column replaced by the column consisting of the right-hand members of the set of n equations.

Example 4.6-1. Determine the solution for the two simultaneous equations

$$x_1 + 2x_2 = 0$$
$$x_1 - 2x_2 = 2$$

using Cramer's rule.

The coefficient matrix $\mathbf{A}$ is given by

$$\mathbf{A} = \begin{bmatrix} 1 & 2 \\ 1 & -2 \end{bmatrix}$$

The determinant of $\mathbf{A}$ is equal to -4. Replacing the first column of $\mathbf{A}$ by the right-hand members of the simultaneous equations, the first unknown x_1 is

$$x_1 = \frac{\begin{vmatrix} 0 & 2 \\ 2 & -2 \end{vmatrix}}{-4} = 1$$

Replacing the second column of $\mathbf{A}$ by the right-hand members of the simultaneous equations, the second unknown x_2 is

$$x_2 = \frac{\begin{vmatrix} 1 & 0 \\ 1 & 2 \end{vmatrix}}{-4} = -\tfrac{1}{2}$$

Since

$$\mathbf{y} = \begin{bmatrix} 0 \\ 2 \end{bmatrix}$$

and $\mathbf{A}^{-1}$ is easily determined to be

$$\mathbf{A}^{-1} = \begin{bmatrix} \frac{1}{2} & \frac{1}{2} \\ \frac{1}{4} & -\frac{1}{4} \end{bmatrix}$$

the x's are given by matrix methods as

$$\begin{bmatrix} x_1 \\ x_2 \end{bmatrix} = \begin{bmatrix} \frac{1}{2} & \frac{1}{2} \\ \frac{1}{4} & -\frac{1}{4} \end{bmatrix} \begin{bmatrix} 0 \\ 2 \end{bmatrix} = \begin{bmatrix} 1 \\ -\frac{1}{2} \end{bmatrix}$$

This solution, which agrees with that determined by Cramer's rule, can be verified by inspection.

Homogeneous System of Linear Equations

If the right-hand members of Eq. 4.6-1 are zero, the set of equations is said to be *homogeneous*. In this case, the numerator of Eq. 4.6-5 vanishes. Consequently, if the determinant $|\mathbf{A}|$ does not vanish, the set of equations has only the trivial solution $\mathbf{x} = \mathbf{0}$, i.e., $x_1 = 0$, $x_2 = 0, \ldots, x_n = 0$. If the determinant $|\mathbf{A}|$ does vanish, two or more rows or columns of $\mathbf{A}$ are linearly related. Then a q-parameter family of solutions can be obtained, where q is the degeneracy of $\mathbf{A}$.

Assume that the rank of the coefficient matrix is r. The values of r variables can then be expressed in terms of the other $q = n - r$ variables by the following procedure:

1. Omit $q = n - r$ equations, such that the coefficient matrix of the r unknowns has a nonvanishing determinant.

2. Form r equations with the r unknowns on the left side of the equation, and the remaining $q = n - r$ unknowns on the right side.
3. Solve for the r unknowns in terms of the $q = n - r$ unknowns by the use of Cramer's rule.

These steps yield q *independent* solutions, as can be seen by writing Eq. 4.6-2 for the homogeneous case as

$$\mathbf{A}\mathbf{x} = [\mathbf{a}_1 \ \mathbf{a}_2 \cdots \mathbf{a}_n]\mathbf{x} = x_1\mathbf{a}_1 + x_2\mathbf{a}_2 + \cdots + x_n\mathbf{a}_n = \mathbf{0}$$

where $\mathbf{a}_1, \mathbf{a}_2, \ldots, \mathbf{a}_n$ denote the n columns of $\mathbf{A}$. If $\mathbf{A}$ is of degeneracy q, then q linear dependency relations of the form

$$k_1\mathbf{a}_1 + k_2\mathbf{a}_2 + \cdots + k_n\mathbf{a}_n = \mathbf{0}$$

can be determined. Hence $\mathbf{A}\mathbf{x} = \mathbf{0}$ has q independent solutions.

In order to view the situation geometrically, let $\mathbf{b}_1, \mathbf{b}_2, \ldots, \mathbf{b}_n$ denote vectors which are the transposed rows of $\mathbf{A}$. Thus $\mathbf{A} = [\mathbf{b}_1 \ \mathbf{b}_2 \cdots \mathbf{b}_n]^T$, and the homogeneous equations can be written as $[\mathbf{b}_1 \ \mathbf{b}_2 \cdots \mathbf{b}_n]^T\mathbf{x} = \mathbf{0}$. This is the same as

$$\begin{aligned} \langle \mathbf{b}_1, \mathbf{x} \rangle &= 0 \\ \langle \mathbf{b}_2, \mathbf{x} \rangle &= 0 \\ \cdots&\cdots\cdots \\ \langle \mathbf{b}_n, \mathbf{x} \rangle &= 0 \end{aligned} \tag{4.6-6}$$

If $\mathbf{x}$ is to satisfy Eq. 4.6-6, and hence $\mathbf{A}\mathbf{x} = \mathbf{0}$, it must be simultaneously orthogonal to all the vectors $\mathbf{b}_1, \mathbf{b}_2, \ldots, \mathbf{b}_n$. But if the rank of $\mathbf{A}$ is n, then the $\mathbf{b}_i$'s are linearly independent and utilize all the available n dimensions. No vector can be found which is orthogonal to all the $\mathbf{b}$'s. However, if $\mathbf{A}$ is of rank $n - 1$, there is one linear dependency relationship between the $\mathbf{b}$'s. Thus one of the n dimensions of the linear vector space is not occupied and is available for $\mathbf{x}$. Similarly, if the rank of $\mathbf{A}$ is $n - 2$, two dimensions are available. Hence there are two linearly independent vectors which satisfy Eq. 4.6-6. In general, if the degeneracy of $\mathbf{A}$ is q, so that the rank of $\mathbf{A}$ is $n - q$, then q dimensions of the linear vector space are available for $\mathbf{x}$. The number of linearly independent solutions of the homogeneous system of linear equations is equal to q.

The preceding discussion was directed toward the case in which the number of equations and number of unknowns are equal. If the number of unknowns, n, exceeds the number of equations, m, the rank of $\mathbf{A}$ is less than n. Nontrivial solutions always exist for this case. If $m > n$, nontrivial solutions exist only if the rank of $\mathbf{A}$ is less than n. In both cases, the number of linearly independent solutions is $n - r$, where r, assumed less than n, is the rank of $\mathbf{A}$. These solutions can be found by the three steps indicated above.

The solution vectors $\mathbf{x}$ generally are not orthogonal to one another. However, the Gram-Schmidt orthogonalization procedure can be used to orthogonalize the solution vectors.

Example 4.6-2. Solve

$$\begin{aligned} 4x_1 + 2x_2 + x_3 + 3x_4 &= 0 \\ 6x_1 + 3x_2 + x_3 + 4x_4 &= 0 \\ 2x_1 + x_2 \qquad\quad + x_4 &= 0 \end{aligned}$$

The rank of the coefficient matrix is two, i.e., the highest order array having a non-vanishing determinant is obtained by omitting the first and third columns and the second row. Consequently, a set of linearly independent equations is

$$\begin{aligned} 2x_2 + 3x_4 &= -4x_1 - x_3 \\ x_2 + x_4 &= -2x_1 \end{aligned}$$

Using Cramer's rule, $x_2 = x_3 - 2x_1$ and $x_4 = -x_3$.

Since x_1 and x_3 are arbitrary, infinitely many sets of solutions exist. However, only $q = 2$ of these are independent. For example, let $x_1 = x_3 = 1$. Then $x_2 = x_4 = -1$. Thus a solution vector is

$$\mathbf{x}_1 = \begin{bmatrix} 1 \\ -1 \\ 1 \\ -1 \end{bmatrix}$$

where $\mathbf{x}_1$ denotes the first solution vector. Similarly, $x_1 = 1$, $x_3 = 2$ leads to a second solution vector

$$\mathbf{x}_2 = \begin{bmatrix} 1 \\ 0 \\ 2 \\ -2 \end{bmatrix}$$

Since the Gramian

$$G = \begin{vmatrix} 4 & 5 \\ 5 & 9 \end{vmatrix} \neq 0$$

$\mathbf{x}_1$ and $\mathbf{x}_2$ are linearly independent. A further choice for x_1 and x_3 leads to a vector $\mathbf{x}$ which is not linearly independent of $\mathbf{x}_1$ and $\mathbf{x}_2$. For example, $x_1 = 2$, $x_3 = 3$ yields a vector which is the sum of $\mathbf{x}_1$ and $\mathbf{x}_2$.

Since $\langle \mathbf{x}_1, \mathbf{x}_2 \rangle = 5 \neq 0$, $\mathbf{x}_1$ and $\mathbf{x}_2$ are not mutually orthogonal. Using the Gram-Schmidt orthogonalization procedure,

$$\mathbf{x}_1' = \begin{bmatrix} 1 \\ -1 \\ 1 \\ -1 \end{bmatrix} \quad \text{and} \quad \mathbf{x}_2' = \tfrac{1}{4} \begin{bmatrix} -1 \\ 5 \\ 3 \\ -3 \end{bmatrix}$$

are determined. $\mathbf{x}_1'$ and $\mathbf{x}_2'$ are mutually orthogonal.

Normalization of $\mathbf{x}_1'$ by $\|\mathbf{x}_1'\| = 2$ and of $\mathbf{x}_2'$ by $\|\mathbf{x}_2'\| = \sqrt{11}/2$ yields the orthonormal solutions

$$\hat{\mathbf{x}}_1' = \tfrac{1}{2}\begin{bmatrix} 1 \\ -1 \\ 1 \\ -1 \end{bmatrix} \quad \text{and} \quad \hat{\mathbf{x}}_2' = \frac{1}{2\sqrt{11}}\begin{bmatrix} -1 \\ 5 \\ 3 \\ -3 \end{bmatrix}$$

The fact that $\mathbf{x}_1, \mathbf{x}_2, \mathbf{x}_1', \mathbf{x}_2', \hat{\mathbf{x}}_1'$ and $\hat{\mathbf{x}}_2'$ all are solutions to the original equations can be determined by substitution.

A particularly useful special case of homogeneous systems is one in which the rank of the coefficient matrix is equal to $n - 1$. Under this condition one independent solution can be found, since $q = 1$. It can be shown that the unknowns are proportional to the cofactors of their coefficients in any row of the coefficient matrix.† Thus a solution is given by

$$x_j = kC_{ij} \qquad (j = 1, 2, \ldots, n) \tag{4.6-7}$$

where k is an arbitrary scalar and i takes any of the values $1, 2, \ldots, n$.

Example 4.6-3. Solve

$$\begin{aligned} x_1 - 2x_2 + x_3 - x_4 &= 0 \\ x_1 + x_2 - 3x_3 \qquad &= 0 \\ 2x_1 + x_2 \qquad + x_4 &= 0 \\ x_2 + x_3 + x_4 &= 0 \end{aligned}$$

The rank of the coefficient matrix is equal to 3. Therefore the unknowns are proportional to the cofactors of any row of the coefficient matrix. Calculating C_{4j},

$$C_{41} = -\begin{vmatrix} -2 & 1 & -1 \\ 1 & -3 & 0 \\ 1 & 0 & 1 \end{vmatrix} = -2$$

$$C_{42} = \begin{vmatrix} 1 & 1 & -1 \\ 1 & -3 & 0 \\ 2 & 0 & 1 \end{vmatrix} = -10$$

$$C_{43} = -\begin{vmatrix} 1 & -2 & -1 \\ 1 & 1 & 0 \\ 2 & 1 & 1 \end{vmatrix} = -4$$

$$C_{44} = \begin{vmatrix} 1 & -2 & 1 \\ 1 & 1 & -3 \\ 2 & 1 & 0 \end{vmatrix} = 14$$

† Compare Eq. 4.3-6, $\sum_{j=1}^{n} a_{kj}C_{ij} = |\mathbf{A}|\delta_{ik}$, and Eq. 4.6-2, $\sum_{j=1}^{n} a_{ij}x_j = y_i$, under the conditions that $|\mathbf{A}| = y_i = 0$, $i = 1, 2, \ldots, n$.

The solution to this set of equations is

$$x_1 = -k, \quad x_2 = -5k, \quad x_3 = -2k, \quad x_4 = 7k$$

These equations comprise one independent solution and all scalar multiples of it.

Dependent Nonhomogeneous Case

The general procedure for the homogeneous case can be utilized for the nonhomogeneous case in which the number of equations is less than the number of unknowns. Such a case is

$$\begin{aligned} a_{11}x_1 + a_{12}x_2 + \cdots + a_{1n}x_n &= y_1 \\ a_{21}x_1 + a_{22}x_2 + \cdots + a_{2n}x_n &= y_2 \\ \cdots\cdots\cdots\cdots\cdots\cdots\cdots\cdots \\ a_{m1}x_1 + a_{m2}x_2 + \cdots + a_{mn}x_n &= y_m \end{aligned} \tag{4.6-8}$$

where $m < n$. These equations can be written in the homogeneous form

$$\begin{aligned} a_{11}x_1 + a_{12}x_2 + \cdots + a_{1n}x_n + y_1x_{n+1} &= 0 \\ a_{21}x_1 + a_{22}x_2 + \cdots + a_{2n}x_n + y_2x_{n+1} &= 0 \\ \cdots\cdots\cdots\cdots\cdots\cdots\cdots\cdots\cdots\cdots \\ a_{m1}x_1 + a_{m2}x_2 + \cdots + a_{mn}x_n + y_mx_{n+1} &= 0 \end{aligned} \tag{4.6-9}$$

Equation 4.6-9 may be solved using the procedure outlined for homogeneous systems. If x_{n+1} is then set equal to -1, the solution to Eq. 4.6-8 can be obtained. If it turns out that Eq. 4.6-9 has a solution only when $x_{n+1} = 0$, then Eq. 4.6-8 has no solution. Consequently, Eq. 4.6-8 has a solution only when the coefficient matrix of Eq. 4.6-8 and the coefficient matrix of Eq. 4.6-9, called the *augmented matrix*, have equal rank.

Example 4.6-4. Solve

$$\begin{aligned} x_1 + 2x_2 - 3x_3 - 4x_4 &= 6 \\ x_1 + 3x_2 + x_3 - 2x_4 &= 4 \\ 2x_1 + 5x_2 - 2x_3 - 5x_4 &= 10 \end{aligned}$$

The coefficient matrix and the augmented matrix are of rank 3; therefore the set of equations possesses a solution. The square array formed by eliminating the third column has a nonvanishing determinant. The set of equations can be written as

$$\begin{aligned} x_1 + 2x_2 - 4x_4 &= 6 + 3x_3 \\ x_1 + 3x_2 - 2x_4 &= 4 - x_3 \\ 2x_1 + 5x_2 - 5x_4 &= 10 + 2x_3 \end{aligned}$$

The determinant of the new coefficient matrix is equal to unity. Therefore

$$x_1 = \begin{vmatrix} (6 + 3x_3) & 2 & -4 \\ (4 - x_3) & 3 & -2 \\ (10 + 2x_3) & 5 & -5 \end{vmatrix} = 10 + 11x_3$$

$$x_2 = \begin{vmatrix} 1 & (6 + 3x_3) & -4 \\ 1 & (4 - x_3) & -2 \\ 2 & (10 + 2x_3) & -5 \end{vmatrix} = -2 - 4x_3$$

and

$$x_4 = \begin{vmatrix} 1 & 2 & (6 + 3x_3) \\ 1 & 3 & (4 - x_3) \\ 2 & 5 & (10 + 2x_3) \end{vmatrix} = 0$$

Since x_3 is arbitrary, the complete solution is then

$$x_1 = 10 + 11c$$
$$x_2 = -2 - 4c$$
$$x_3 = c$$
$$x_4 = 0$$

where c is an arbitrary constant.

Example 4.6-5. Solve the equations

$$a_{11}x_1 + a_{12}x_2 = y_1$$
$$a_{21}x_1 + a_{22}x_2 = y_2$$

for the following cases:

(*a*) nonhomogeneous, i.e., $\mathbf{y} \neq \mathbf{0}$;
(*b*) homogeneous, i.e., $\mathbf{y} = \mathbf{0}$;
(*c*) dependent nonhomogeneous, i.e., $a_{21} = a_{22} = y_2 = 0$.

Assuming that $a_{12} \neq 0 \neq a_{22}$, the equations can be written in the form

$$x_2 = -\frac{a_{11}}{a_{12}}x_1 + \frac{y_1}{a_{12}}$$

$$x_2 = -\frac{a_{21}}{a_{22}}x_1 + \frac{y_2}{a_{22}}$$

These equations can be represented in an x_1x_2-plane as in Fig. 4.6-1. As indicated in the following, such a representation aids in clarifying the points of this section.

(*a*) *Nonhomogeneous.* Two possibilities exist for this case. If $|\mathbf{A}| = 0$, then $a_{11}a_{22} - a_{12}a_{21} = 0$, or

$$-\frac{a_{11}}{a_{12}} = -\frac{a_{21}}{a_{22}}$$

Thus the slopes of the two lines of Fig. 4.6-1 are equal, i.e., the lines are parallel. Consequently the lines do not intersect, and no solution vector exists.

If $|\mathbf{A}| \neq 0$, the lines intersect at one and only one point, given by the solution

$$x_1 = \frac{y_1a_{22} - y_2a_{12}}{|\mathbf{A}|}, \quad x_2 = \frac{y_2a_{11} - y_1a_{21}}{|\mathbf{A}|}$$

These are the components of the solution vector.

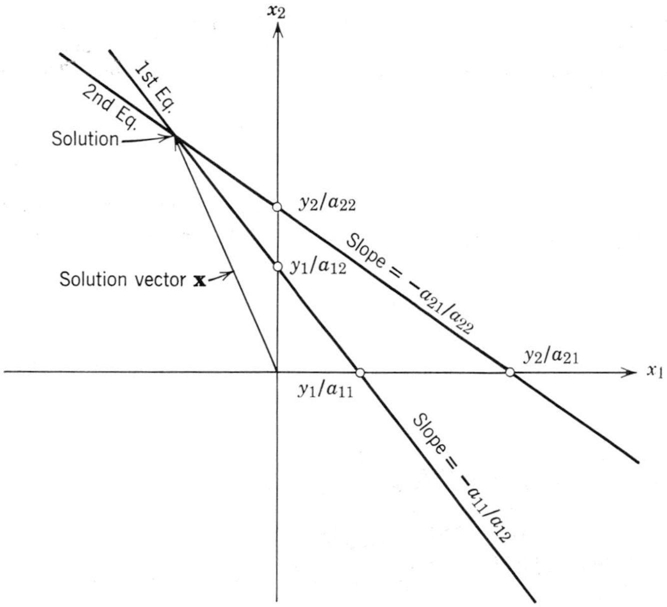

Fig. 4.6-1

(*b*) *Homogeneous.* Again two possibilities exist. Since $y_1 = y_2 = 0$, both lines pass through the origin. If the slopes are unequal, i.e., if $|\mathbf{A}| \neq 0$, then there is one solution. It is the trivial one, $x_1 = x_2 = 0$.

If $|\mathbf{A}| = 0$ and $q = 1$, then both lines have the same slope. Thus the lines coalesce, and any point on the lines is a solution. Since the equation of the coalesced lines is $a_{11}x_1 + a_{12}x_2 = 0 = a_{21}x_1 + a_{22}x_2$, the one independent solution is given by

$$\frac{x_2}{x_1} = -\frac{a_{11}}{a_{12}} = -\frac{a_{21}}{a_{22}}$$

In terms of the preceding discussion, $n = 2$, but $r = 1$. Hence omit one of the original equations (either one), and the solution is

$$x_2 = -\frac{a_{11}}{a_{12}}x_1$$

or

$$x_2 = -\frac{a_{21}}{a_{22}}x_1$$

which are equal.

Note that

$$\mathbf{b}_1 = \begin{bmatrix} a_{11} \\ a_{12} \end{bmatrix} \quad \text{and} \quad \mathbf{b}_2 = \begin{bmatrix} a_{21} \\ a_{22} \end{bmatrix}$$

They are linearly related and hence parallel in the plane. The solution **x** is orthogonal to both $\mathbf{b}_1$ and $\mathbf{b}_2$.

(*c*) *Dependent nonhomogeneous*. The equation is $a_{11}x_1 + a_{12}x_2 = y_1$. This can be written in homogeneous form as $a_{11}x_1 + a_{12}x_2 + x_3y_1 = 0$. In order to solve as a homogeneous equation, the form $a_{12}x_2 = -a_{11}x_1 - x_3y_1$ is written. The solution is

$$x_2 = -\frac{a_{11}}{a_{12}}x_1 - x_3\frac{y_1}{a_{12}}$$

This is the equation of a plane in three dimensions, and any point on the plane is a solution. In particular, along the line in the plane for which $x_3 = -1$, the solution is

$$x_2 = -\frac{a_{11}}{a_{12}}x_1 + \frac{y_1}{a_{12}}$$

Any point on this line is a solution. For this simple case, of course, this equation is simply the original equation in rewritten form.

4.7 CHARACTERISTIC VALUES AND CHARACTERISTIC VECTORS†

The topic of characteristic values and characteristic vectors is an extremely important one, as the dynamic behavior of a linear system is dependent upon the characteristic values of the system. It is important that the reader understand this topic, as much of system analysis and synthesis depends upon the solution of characteristic value problems.

Consider the vector-matrix equation

$$\mathbf{y} = \mathbf{A}\mathbf{x} \tag{4.7-1}$$

where $\mathbf{y}$ and $\mathbf{x}$ are column vectors, and $\mathbf{A}$ is a square $(n \times n)$ matrix. This equation can be viewed as a transformation of the vector $\mathbf{x}$ into the vector $\mathbf{y}$. The question arises whether there exists a vector $\mathbf{x}$, such that the transformation $\mathbf{A}$ produces a vector $\mathbf{y}$, which has the same direction in vector space as the vector $\mathbf{x}$. If such a vector $\mathbf{x}$ exists, then $\mathbf{y}$ is proportional to $\mathbf{x}$, or

$$\mathbf{y} = \mathbf{A}\mathbf{x} = \lambda\mathbf{x} \tag{4.7-2}$$

where λ is a scalar of proportionality. This is known as a characteristic value problem, and a value of λ, e.g., λ_i, for which Eq. 4.7-2 has a solution $\mathbf{x}_i \neq \mathbf{0}$, is called a *characteristic value* of $\mathbf{A}$. The corresponding vector solution $\mathbf{x}_i \neq \mathbf{0}$ is called a *characteristic vector* of $\mathbf{A}$ associated with the characteristic value λ_i.

† The terms *eigenvalues* and *eigenvectors* are frequently used in place of characteristic values and characteristic vectors, respectively. The terms *latent roots* and *latent vectors* are also used.

Equation 4.7-2 can be written in the form of a homogeneous set of linear equations as

$$\begin{aligned}(a_{11} - \lambda)x_1 + a_{12}x_2 + \cdots + a_{1n}x_n &= 0\\ a_{21}x_1 + (a_{22} - \lambda)x_2 + \cdots + a_{2n}x_n &= 0\\ \cdots\cdots\cdots\cdots\cdots\cdots\cdots\cdots&\cdots\\ a_{n1}x_1 + a_{n2}x_2 + \cdots + (a_{nn} - \lambda)x_n &= 0\end{aligned}$$

or

$$[\lambda \mathbf{I} - \mathbf{A}]\mathbf{x} = \mathbf{0} \tag{4.7-3}$$

This system of homogeneous equations has a nontrivial solution if, and only if, the determinant of the coefficients vanishes, i.e., if

$$|\lambda \mathbf{I} - \mathbf{A}| = 0 \tag{4.7-4}$$

The Characteristic Equation

The nth order polynomial in λ, given by Eq. 4.7-4, is called the *characteristic equation* corresponding to the matrix **A**. The general form for the equation is

$$P(\lambda) = \lambda^n + a_1\lambda^{n-1} + a_2\lambda^{n-2} + \cdots + a_{n-1}\lambda + a_n = 0 \tag{4.7-5}$$

The roots of the characteristic equation are precisely the characteristic values of **A**. When all the characteristic values of **A** are different, **A** is said to have *distinct* roots. When a characteristic value occurs m times, the characteristic value is said to be a repeated root of order m. When a characteristic root is of the form $\alpha + j\beta$, the root is said to be *complex*. Complex roots must occur in conjugate pairs, assuming that the elements of **A** are real.

The coefficients a_1 and a_n of the characteristic equation, Eq. 4.7-5, are of special interest. If λ is set equal to zero, then

$$P(0) = |-\mathbf{A}| = a_n \tag{4.7-6}$$

Therefore

$$a_n = (-1)^n\,|\mathbf{A}| \tag{4.7-7}$$

If the polynomial $P(\lambda)$ is written in the factored form (assuming the characteristic values are distinct)

$$P(\lambda) = (\lambda - \lambda_1)(\lambda - \lambda_2)\cdots(\lambda - \lambda_n)$$

and λ is again set equal to zero, it follows that

$$P(0) = (-1)^n\,|\mathbf{A}| = (-1)^n(\lambda_1\lambda_2\cdots\lambda_n)$$

or

$$\lambda_1\lambda_2\lambda_3\cdots\lambda_n = |\mathbf{A}| \tag{4.7-8}$$

The product of the characteristic values is equal to the determinant of **A**. Note that, if any of the characteristic values is equal to zero, **A** is singular.

By expanding the factored form of the characteristic equation, the coefficients of the various powers of λ can be obtained in terms of the characteristic values. For example, the coefficient of λ^{n-1} is

$$a_1 = (\lambda_1 + \lambda_2 + \cdots + \lambda_n)$$

If the determinant $|\lambda\mathbf{I} - \mathbf{A}|$ is also expanded, it is found that the coefficient of λ^{n-1} is equal to the negative sum of the diagonal elements of **A**, i.e.,

$$a_1 = -(\lambda_1 + \lambda_2 + \cdots + \lambda_n) = -(a_{11} + a_{22} + \cdots + a_{nn}) \quad (4.7\text{-}9)$$

Thus the sum of the diagonal elements of a square matrix is equal to the sum of its characteristic values. Because of its importance, the sum of the diagonal elements of a matrix is given a title, namely, *trace* of the matrix. Hence the above can be written as

$$\lambda_1 + \lambda_2 + \cdots + \lambda_n = \text{trace of } \mathbf{A} = \text{Tr}\,[a_{ij}] = a_{11} + a_{22} + \cdots + a_{nn} \quad (4.7\text{-}10)$$

If the trace of $\mathbf{A}^k$ (**A** multiplied by itself k times) is given the symbol T_k, then a useful recursive formula for expressing the coefficients of the characteristic equation in terms of the various T_k can be written as

$$\begin{aligned}
a_1 &= -T_1 \\
a_2 &= -\tfrac{1}{2}(a_1T_1 + T_2) \\
a_3 &= -\tfrac{1}{3}(a_2T_1 + a_1T_2 + T_3) \\
&\cdots\cdots\cdots\cdots\cdots\cdots \\
a_n &= -\frac{1}{n}(a_{n-1}T_1 + a_{n-2}T_2 + \cdots + a_1T_{n-1} + T_n)
\end{aligned} \quad (4.7\text{-}11)$$

This formula provides an alternative means to Eq. 4.7-4 for determining the characteristic equation.† In particular, it is more helpful if a computer is to be used to determine the coefficients in the characteristic equation.

Example 4.7-1. Find the characteristic equation of **A**, where

$$\mathbf{A} = \begin{bmatrix} 2 & -2 & 3 \\ 1 & 1 & 1 \\ 1 & 3 & -1 \end{bmatrix}$$

Using Bôcher's formula,

$$a_1 = -T_1 = -(2 + 1 - 1) = -2$$

† This formula is called Bôcher's formula. See Reference 4.

The matrix product of **A** with itself yields

$$\mathbf{A}^2 = \begin{bmatrix} 5 & 3 & 1 \\ 4 & 2 & 3 \\ 4 & -2 & 7 \end{bmatrix}$$

Thus $a_2 = -\frac{1}{2}(a_1T_1 + T_2) = -5$. Similarly,

$$\mathbf{A}^3 = \begin{bmatrix} 14 & -4 & 17 \\ 13 & 3 & 11 \\ 13 & 11 & 3 \end{bmatrix}$$

so that $a_3 = -\frac{1}{3}(a_2T_1 + a_1T_2 + T_3) = 6$. Thus the characteristic polynomial is

$$\lambda^3 - 2\lambda^2 - 5\lambda + 6 = (\lambda - 1)(\lambda + 2)(\lambda - 3) = 0$$

The characteristic values are $\lambda_1 = 1$, $\lambda_2 = -2$, and $\lambda_3 = 3$.

Modal Matrix

For each of the n characteristic values $\lambda_i (i = 1, 2, \ldots, n)$ of **A**, a solution of Eq. 4.7-3 for **x** can be obtained, provided that the roots of Eq. 4.7-4 are distinct. The vectors $\mathbf{x}_i$, which are the solutions of

$$[\lambda_i\mathbf{I} - \mathbf{A}]\mathbf{x}_i = \mathbf{0} \quad (i = 1, 2, \ldots, n) \tag{4.7-12}$$

are the characteristic vectors of **A**. Since Eq. 4.7-12 is homogeneous, $k_i\mathbf{x}_i$, where k_i is any scalar, is also a solution. Thus only the directions of each of the $\mathbf{x}_i$ are uniquely determined by Eq. 4.7-12. The matrix formed by the column vectors $k_i\mathbf{x}_i$ is called the *modal matrix*.†

For this case of distinct characteristic values, columns of the modal matrix can be taken to be equal, or proportional, to any nonzero column of **Adj** $[\lambda_i\mathbf{I} - \mathbf{A}]$. This is based upon the fact that $[\lambda_i\mathbf{I} - \mathbf{A}]$ is of rank $n - 1$. The rank must be less than n, because of Eq. 4.7-4. However, it cannot be less than $n - 1$, because all the $(n - 1)$ rowed minors of $|\lambda_i\mathbf{I} - \mathbf{A}|$ would also be zero. This would in turn require (see Eq. 4.3-8)

$$\frac{d}{d\lambda}\left\{|\lambda\mathbf{I} - \mathbf{A}|\right\}\Big|_{\lambda=\lambda_i} = 0$$

indicating that λ_i is a repeated root of Eq. 4.7-4. But this has been ruled out by the assumption that the characteristic values are distinct. Thus $[\lambda_i\mathbf{I} - \mathbf{A}]$ is of rank $n - 1$, and application of Eqs. 4.6-7 and 4.4-1 then shows that the columns of the modal matrix are proportional to *any*

† The term "mode" is used because, as later chapters show, the modes of dynamic behavior of a linear system can be expressed in terms of motion along the characteristic vectors.

nonzero column of **Adj** $[\lambda_i \mathbf{I} - \mathbf{A}]$. Since the columns of **Adj** $[\lambda_i \mathbf{I} - \mathbf{A}]$ are linearly related for a given λ_i, each choice of λ_i specifies only one column of the modal matrix.

Example 4.7-2. Find the characteristic values and a model matrix corresponding to the matrix **A**, where

$$\mathbf{A} = \begin{bmatrix} 2 & -2 & 3 \\ 1 & 1 & 1 \\ 1 & 3 & -1 \end{bmatrix}$$

The characteristic equation is, from Eq. 4.7-4,

$$\begin{vmatrix} \lambda - 2 & 2 & -3 \\ -1 & \lambda - 1 & -1 \\ -1 & -3 & \lambda + 1 \end{vmatrix} = 0 = \lambda^3 - 2\lambda^2 - 5\lambda + 6 = (\lambda - 1)(\lambda + 2)(\lambda - 3)$$

The characteristic values are $\lambda_1 = 1$, $\lambda_2 = -2$, and $\lambda_3 = 3$.

The adjoint matrix **Adj** $[\lambda \mathbf{I} - \mathbf{A}]$ is

$$\begin{bmatrix} (\lambda^2 - 4) & (-2\lambda + 7) & (3\lambda - 5) \\ (\lambda + 2) & (\lambda^2 - \lambda - 5) & (\lambda + 1) \\ (\lambda + 2) & (3\lambda - 8) & (\lambda^2 - 3\lambda + 4) \end{bmatrix}$$

For $\lambda_1 = 1$, the adjoint matrix is

$$\begin{bmatrix} -3 & 5 & -2 \\ 3 & -5 & 2 \\ 3 & -5 & 2 \end{bmatrix}$$

For $\lambda = -2$, the adjoint matrix is

$$\begin{bmatrix} 0 & 11 & -11 \\ 0 & 1 & -1 \\ 0 & -14 & 14 \end{bmatrix}$$

For $\lambda = 3$, the adjoint matrix is

$$\begin{bmatrix} 5 & 1 & 4 \\ 5 & 1 & 4 \\ 5 & 1 & 4 \end{bmatrix}$$

Since the characteristic vectors are uniquely determined only in direction, these vectors can be multiplied by any scalar and still satisfy Eq. 4.7-12. Consequently, a modal matrix is

$$\mathbf{M} = \begin{bmatrix} -1 & 11 & 1 \\ 1 & 1 & 1 \\ 1 & -14 & 1 \end{bmatrix}$$

Each column of the modal matrix is a characteristic vector which spans a one-dimensional vector space. The three columns of the modal matrix form a basis in the corresponding three-dimensional space.

The preceding discussion considered the modal matrix when the characteristic values of **A** are distinct. For the case in which therc is a repeated characteristic value and **A** is nonsymmetric, the determination of the number of independent modal columns is not quite as clear. The reason for the ambiguity is that there is no unique correspondence between the order of a repeated root of the characteristic equation and the degeneracy of the corresponding characteristic matrix $[\lambda_i\mathbf{I} - \mathbf{A}]$.

If λ_i is a repeated root of order p, the degeneracy of the characteristic matrix cannot be greater than p, and the dimension of the associated vector space spanned by the corresponding $\mathbf{x}_i$ is not greater than p. The problem arises when there is a repeated root λ_i of order p, and the degeneracy q of $[\lambda_i\mathbf{I} - \mathbf{A}]$ is less than p. Only $q < p$ linearly independent solutions to Eq. 4.7-12 can be found. The dimension of the associated vector space for the $\mathbf{x}_i$ is less than p, and p linearly independent characteristic vectors corresponding to λ_i cannot be obtained. Only when the $(n \times n)$ matrix **A** is *symmetric* is the degeneracy of $[\lambda_i\mathbf{I} - \mathbf{A}]$ definitely equal to p for a p-fold root, so that p linearly independent characteristic vectors can be found.

For the case in which the degeneracy of $[\lambda_i\mathbf{I} - \mathbf{A}]$ is equal to one (simple degeneracy), the modal column can be chosen to be proportional to any nonzero column of **Adj** $[\lambda_i\mathbf{I} - \mathbf{A}]$. This is the only column that can be obtained for the set of p equal roots. For the case where the degeneracy of $[\lambda_i\mathbf{I} - \mathbf{A}]$ is equal to $q > 1$, **Adj** $[\lambda_i\mathbf{I} - \mathbf{A}]$ and all its derivatives up to and including

$$\frac{d^{q-2}}{d\lambda^{q-2}} \{\mathbf{Adj}\ [\lambda\mathbf{I} - \mathbf{A}\}|_{\lambda=\lambda_i}$$

are null matrices.† The q linearly distinct solutions for the modal columns can be obtained from the columns of differentiated adjoint matrices which are non-null. For example, if $q = p$ (so-called "full degeneracy"), then the p linearly distinct modal columns can be obtained from the nonzero columns of

$$\frac{d^{p-1}}{d\lambda^{p-1}} \{\mathbf{Adj}\ [\lambda\mathbf{I} - \mathbf{A}]\}|_{\lambda=\lambda_i}$$

This matrix (called a derived or differentiated adjoint), being of rank p, can be placed into the form of the matrix product **CB**, where the p columns of the rectangular matrix **C** and the p rows of the rectangular matrix **B** are linearly independent. The p columns of **C** can be selected for the required modal columns.[6]

† The proof of this can be found in Reference 5.

Example 4.7-3. Find the characteristic vectors of the given **A** matrices.

Repeated characteristic roots and simple degeneracy.

$$\mathbf{A} = \begin{bmatrix} 0 & 1 & 0 \\ 0 & 0 & 1 \\ 1 & -3 & 3 \end{bmatrix}$$

The roots of the characteristic equation $|\lambda\mathbf{I} - \mathbf{A}| = 0$ are equal to 1, 1, 1, a triple unity root. However, the degeneracy of the characteristic matrix

$$[\lambda\mathbf{I} - \mathbf{A}]\,|_{\lambda=1} = \begin{bmatrix} 1 & -1 & 0 \\ 0 & 1 & -1 \\ -1 & 3 & -2 \end{bmatrix}$$

is equal to one. Therefore there is only one modal column corresponding to the triple root. Since

$$\mathbf{Adj}\,[\lambda_i\mathbf{I} - \mathbf{A}]\,|_{\lambda_1=1} = \begin{bmatrix} 1 & -2 & 1 \\ 1 & -2 & 1 \\ 1 & -2 & 1 \end{bmatrix}$$

the only linearly independent characteristic vector is

$$\mathbf{x}_1 = \begin{bmatrix} 1 \\ 1 \\ 1 \end{bmatrix}$$

It can span only a one-dimensional space. The determination of two additional vectors for this case is considered in the next section.

Repeated characteristic roots and multiple degeneracy.

$$\mathbf{A} = \begin{bmatrix} 2 & 1 & 1 \\ 1 & 2 & 1 \\ 0 & 0 & 1 \end{bmatrix} \qquad [\lambda\mathbf{I} - \mathbf{A}] = \begin{bmatrix} \lambda - 2 & -1 & -1 \\ -1 & \lambda - 2 & -1 \\ 0 & 0 & \lambda - 1 \end{bmatrix}$$

The characteristic values of this matrix are 1, 1, 3, a double root at $\lambda = 1$ and a single root at $\lambda = 3$. The degeneracy of $[\lambda_i\mathbf{I} - \mathbf{A}]$ for $\lambda_i = 1$ is equal to two. Since the characteristic matrix has "full degeneracy," it is possible to obtain a linearly independent vector solution for each of the repeated roots.

The adjoint matrix is

$$\mathbf{Adj}\,[\lambda\mathbf{I} - \mathbf{A}] = \begin{bmatrix} (\lambda - 2)(\lambda - 1) & (\lambda - 1) & (\lambda - 1) \\ (\lambda - 1) & (\lambda - 2)(\lambda - 1) & (\lambda - 1) \\ 0 & 0 & (\lambda - 1)(\lambda - 3) \end{bmatrix}$$

Substitution of $\lambda_i = 1$ in any column of the adjoint matrix yields a null column. The first derivative of the adjoint matrix is

$$\frac{d}{d\lambda}\{\mathbf{Adj}\,[\lambda\mathbf{I} - \mathbf{A}]\} = \begin{bmatrix} 2\lambda - 3 & 1 & 1 \\ 1 & 2\lambda - 3 & 1 \\ 0 & 0 & 2\lambda - 4 \end{bmatrix}$$

Evaluating this matrix at $\lambda = 1$ yields

$$\begin{bmatrix} -1 & 1 & 1 \\ 1 & -1 & 1 \\ 0 & 0 & -2 \end{bmatrix} = \begin{bmatrix} -1 & 1 \\ 1 & 1 \\ 0 & -2 \end{bmatrix} \begin{bmatrix} 1 & -1 & 0 \\ 0 & 0 & 1 \end{bmatrix}$$

Hence the two modal columns corresponding to the repeated root at $\lambda = 1$ are given by

$$\mathbf{x}_1 = \begin{bmatrix} -1 \\ 1 \\ 0 \end{bmatrix} \qquad \mathbf{x}_2 = \begin{bmatrix} 1 \\ 1 \\ -2 \end{bmatrix}$$

The characteristic vector or modal column for $\lambda_i = 3$ can be chosen to be proportional to any nonzero column of the matrix $\mathbf{Adj}\,[\lambda_i\mathbf{I} - \mathbf{A}]$. Thus

$$\mathbf{x}_3 = \begin{bmatrix} 1 \\ 1 \\ 0 \end{bmatrix}$$

The complete modal matrix is then

$$\mathbf{M} = \begin{bmatrix} -1 & 1 & 1 \\ 1 & 1 & 1 \\ 0 & -2 & 0 \end{bmatrix}$$

Symmetric Matrices

The case where the matrix $\mathbf{A}$ is real and symmetric occurs so frequently when dealing with linear circuits that special attention should be paid to the form of the characteristic values and characteristic vectors associated with such a matrix.

A fundamental property of real symmetric matrices is that the characteristic values of a real symmetric matrix must be real. This can be shown by assuming that the characteristic values are complex. They must then occur in complex conjugate pairs, as must the characteristic vectors. Thus $\lambda\mathbf{x} = \mathbf{A}\mathbf{x}$ and $\lambda^*\mathbf{x}^* = \mathbf{A}\mathbf{x}^*$. Premultiplying the first equation by $\mathbf{x}^{*T}$ and postmultiplying the transposed form of the second equation by $\mathbf{x}$ gives $\lambda\langle\mathbf{x}, \mathbf{x}\rangle = \langle\mathbf{x}, \mathbf{A}\mathbf{x}\rangle$ and $\lambda^*\langle\mathbf{x}, \mathbf{x}\rangle = \langle\mathbf{x}, \mathbf{A}^T\mathbf{x}\rangle$. Noting that, for a symmetric matrix $\mathbf{A} = \mathbf{A}^T$, these expressions require

$$(\lambda - \lambda^*)\,\langle\mathbf{x}, \mathbf{x}\rangle = 0.$$

Since $\langle\mathbf{x}, \mathbf{x}\rangle \neq 0$, $\lambda = \lambda^*$ and hence the characteristic values are real.

A second property of real symmetric matrices is that the characteristic vectors form an orthogonal set. Let λ_1 and λ_2 be two distinct characteristic values with associated characteristic vectors $\mathbf{x}_1$ and $\mathbf{x}_2$. Then $\lambda_1\mathbf{x}_1 = \mathbf{A}\mathbf{x}_1$ and $\lambda_2\mathbf{x}_2 = \mathbf{A}\mathbf{x}_2$. If the first of these relationships is transposed and postmultiplied by $\mathbf{x}_2$, then

$$\lambda_1\mathbf{x}_1{}^T\mathbf{x}_2 = \mathbf{x}_1{}^T\mathbf{A}^T\mathbf{x}_2 = \mathbf{x}_1{}^T\mathbf{A}\mathbf{x}_2 \qquad (4.7\text{-}13)$$

Premultiplying the second relationship by $\mathbf{x}_1^T$ yields

$$\lambda_2 \mathbf{x}_1^T \mathbf{x}_2 = \mathbf{x}_1^T \mathbf{A} \mathbf{x}_2$$

If Eq. 4.7-13 is subtracted from this expression, then

$$(\lambda_2 - \lambda_1)\langle \mathbf{x}_1, \mathbf{x}_2 \rangle = 0 \tag{4.7-14}$$

Since it was assumed that $\lambda_1 \neq \lambda_2$, it follows that

$$\langle \mathbf{x}_1, \mathbf{x}_2 \rangle = 0 \tag{4.7-15}$$

Equation 4.7-15 shows that the characteristic vectors of a symmetric matrix form an orthogonal set. If the matrix has n distinct characteristic values, these vectors form an orthogonal basis in the n-dimensional space. Furthermore, it can be shown that, if a symmetric matrix has a characteristic value λ_i of order p, then associated with this characteristic value are p linearly independent characteristic vectors. As was shown earlier, this is not generally true for a nonsymmetric matrix.

In a similar manner it can be shown that the characteristic values of a Hermitian matrix are real, and that the characteristic vectors are orthogonal, such that

$$\langle \mathbf{x}_1, \mathbf{x}_2 \rangle = \mathbf{x}_1^{*T} \mathbf{x}_2 = 0$$

Example 4.7-4. Show that the characteristic vectors of the symmetric matrix $\mathbf{A}$ are orthogonal, where

$$\mathbf{A} = \begin{bmatrix} -3 & 2 \\ 2 & 0 \end{bmatrix}$$

The characteristic equation is $\lambda^2 + 3\lambda - 4 = 0$, and the characteristic values are $\lambda_1 = 1$ and $\lambda_2 = -4$. The adjoint matrix is

$$\mathbf{Adj}\,[\lambda \mathbf{I} - \mathbf{A}] = \begin{bmatrix} \lambda & 2 \\ 2 & \lambda + 3 \end{bmatrix}$$

A suitable set of characteristic vectors is

$$\mathbf{x}_1 = \begin{bmatrix} 1 \\ 2 \end{bmatrix} \quad \text{and} \quad \mathbf{x}_2 = \begin{bmatrix} 2 \\ -1 \end{bmatrix}$$

Clearly, $\langle \mathbf{x}_1, \mathbf{x}_2 \rangle = 0$

Diagonalizing a Square Matrix

Consider the case in which the modal matrix $\mathbf{M}$ is *nonsingular*, such that its inverse $\mathbf{M}^{-1}$ exists.† Under this restriction, the solution to Eq.

† This is always the case if the characteristic values of $\mathbf{A}$ are distinct, or in the case of a symmetric $\mathbf{A}$ matrix. See Reference 7.

4.7-12 can be combined to form the single equation

$$\begin{bmatrix} \lambda_1 x_{11} & \lambda_2 x_{12} & \cdots & \lambda_n x_{1n} \\ \lambda_1 x_{21} & \lambda_2 x_{22} & \cdots & \lambda_n x_{2n} \\ \cdots & \cdots & \cdots & \cdots \\ \lambda_1 x_{n1} & \lambda_2 x_{n2} & \cdots & \lambda_n x_{nn} \end{bmatrix} = \begin{bmatrix} a_{11} & a_{12} & \cdots & a_{1n} \\ a_{21} & a_{22} & \cdots & a_{2n} \\ \cdots & \cdots & \cdots & \cdots \\ a_{n1} & a_{n2} & \cdots & a_{nn} \end{bmatrix} \begin{bmatrix} x_{11} & x_{12} & \cdots & x_{1n} \\ x_{21} & x_{22} & \cdots & x_{2n} \\ \cdots & \cdots & \cdots & \cdots \\ x_{n1} & x_{n2} & \cdots & x_{nn} \end{bmatrix} \tag{4.7-16}$$

or

$$\mathbf{M\Lambda} = \mathbf{AM}$$

where

$$\mathbf{\Lambda} = \begin{bmatrix} \lambda_1 & & & & \\ & \lambda_2 & & & \\ & & \cdot & & \\ & & & \cdot & \\ & & & & \cdot \\ & & & & & \lambda_n \end{bmatrix}$$

is a diagonal matrix composed of the characteristic values $\lambda_1, \lambda_2, \ldots, \lambda_n$. Since $\mathbf{M}^{-1}$ exists, the diagonal matrix $\mathbf{\Lambda}$ can be found by premultiplying both sides of Eq. 4.7-16 by $\mathbf{M}^{-1}$, yielding

$$\mathbf{\Lambda} = \mathbf{M}^{-1}\mathbf{AM} \tag{4.7-17}$$

Higher powers of $\mathbf{A}$ can be diagonalized in the same manner. For example,

$$\mathbf{\Lambda}^2 = (\mathbf{M}^{-1}\mathbf{AM})(\mathbf{M}^{-1}\mathbf{AM}) = \mathbf{M}^{-1}\mathbf{A}^2\mathbf{M} \tag{4.7-18}$$

A transformation of the type $\mathbf{B} = \mathbf{Q}^{-1}\mathbf{AQ}$, where $\mathbf{A}$ and $\mathbf{B}$ are square matrices and $\mathbf{Q}$ is a nonsingular square matrix, is called a *collineatory* or *similarity* transformation. More is said about this type of transformation in the next section, when transformations are discussed.

The significance of this transformation is evident when the set of linear equations $\mathbf{y} = \mathbf{Ax}$ of Eq. 4.6-2 is examined. If the transformation $\mathbf{x} = \mathbf{Mq}$ is made, then Eq. 4.6-2 can be written as

$$\mathbf{y} = \mathbf{AMq} \tag{4.7-19}$$

Premultiplying both sides of Eq. 4.7-19 by $\mathbf{M}^{-1}$ gives

$$\mathbf{M}^{-1}\mathbf{y} = \mathbf{M}^{-1}\mathbf{AMq} = \mathbf{\Lambda q} \tag{4.7-20}$$

Letting $\mathbf{z} = \mathbf{M}^{-1}\mathbf{y}$ yields

$$\mathbf{z} = \mathbf{\Lambda q} \tag{4.7-21}$$

or

$$\begin{aligned} z_1 &= \lambda_1 q_1 \\ z_2 &= \lambda_2 q_2 \\ &\cdots\cdots \\ z_n &= \lambda_n q_n \end{aligned}$$

In terms of the new coordinate system $q_1, q_2, q_3, \ldots, q_n$, the set of equations described by Eq. 4.7-21 is uncoupled. Note that the q_i coordinates are in the direction of the characteristic vectors. These coordinates are called the *normal coordinates* of the system. By transforming to the normal coordinates, the characteristic values, and hence the modes of the system, are isolated. It is for this reason that **M** is called the modal matrix. Recognition of the characteristic vectors as uncoupling the coordinates forms the core of the mode interpretation of linear systems.

The columns of the modal matrix **M** form a basis, and the rows of $\mathbf{M}^{-1}$ form a reciprocal basis, in the original space. If the columns of the modal matrix are called $\mathbf{u}_1, \mathbf{u}_2, \ldots, \mathbf{u}_n$, and the reciprocal basis is denoted by $\mathbf{r}_1, \mathbf{r}_2, \ldots, \mathbf{r}_n$, then any vector $\mathbf{y}$ can be expressed as

$$\mathbf{y} = \langle \mathbf{r}_1, \mathbf{y}\rangle \mathbf{u}_1 + \langle \mathbf{r}_2, \mathbf{y}\rangle \mathbf{u}_2 + \cdots + \langle \mathbf{r}_n, \mathbf{y}\rangle \mathbf{u}_n \tag{4.7-22}$$

To illustrate the equivalence of this form and the use of the normal coordinates, consider

$$\mathbf{y} = \mathbf{M}\mathbf{M}^{-1}\mathbf{y} = [\mathbf{u}_1 \quad \mathbf{u}_2 \quad \cdots \quad \mathbf{u}_n]\mathbf{M}^{-1}\mathbf{y}$$

$$= [\mathbf{u}_1 \quad \mathbf{u}_2 \quad \cdots \quad \mathbf{u}_n] \begin{bmatrix} z_1 \\ z_2 \\ \cdot \\ \cdot \\ \cdot \\ z_n \end{bmatrix} = z_1\mathbf{u}_1 + z_2\mathbf{u}_2 + \cdots + z_n\mathbf{u}_n$$

If this expression is compared with Eq. 4.7-22, it is evident that $z_i = \langle \mathbf{r}_i, \mathbf{y}\rangle$ and therefore the rows of $\mathbf{M}^{-1}$ form a reciprocal basis.

This point could have been made by recognizing the fact that, if the modal columns $\mathbf{u}_1, \mathbf{u}_2, \ldots, \mathbf{u}_n$ constitute a proper basis, then, in view of the definition of the reciprocal basis $\langle \mathbf{r}_i, \mathbf{u}_j\rangle = \delta_{ij}$, the rows of $\mathbf{M}^{-1}$ must constitute the reciprocal basis, i.e., $\mathbf{M}\mathbf{M}^{-1} = \mathbf{I}$. However, the details of the preceding observation are instructive, as the reader will find both forms in the literature. Some authors prefer expressing a vector in terms of scalar products, while others use the normal coordinate form. The reader should be aware of the fact that both forms produce identical results. The scalar product form is a little more geometric in interpretation, possibly giving the reader an intuitive or physical feeling for the results. The normal coordinate form is written slightly more compactly and is generally used for proofs and derivations. These points will be made evident when the dynamic behavior of linear systems is discussed in Chapter 5.

Example 4.7-5. For each of the cases below, show that the product $\mathbf{M}^{-1}\mathbf{A}\mathbf{M}$ is a diagonal matrix with its elements equal to the characteristic values.

Distinct roots (see Example 4.7-2).

$$\mathbf{A} = \begin{bmatrix} 2 & -2 & 3 \\ 1 & 1 & 1 \\ 1 & 3 & -1 \end{bmatrix} \qquad \mathbf{M} = \begin{bmatrix} -1 & 11 & 1 \\ 1 & 1 & 1 \\ 1 & -14 & 1 \end{bmatrix} \qquad \mathbf{M}^{-1} = \frac{1}{30}\begin{bmatrix} -15 & 25 & -10 \\ 0 & 2 & -2 \\ 15 & 3 & 12 \end{bmatrix}$$

Consequently,

$$\mathbf{M}^{-1}\mathbf{A}\mathbf{M} = \begin{bmatrix} 1 & 0 & 0 \\ 0 & -2 & 0 \\ 0 & 0 & 3 \end{bmatrix}$$

Repeated roots and multiple degeneracy (see Example 4.7-3).

$$\mathbf{A} = \begin{bmatrix} 2 & 1 & 1 \\ 1 & 2 & 1 \\ 0 & 0 & 1 \end{bmatrix} \qquad \mathbf{M} = \begin{bmatrix} -1 & 1 & 1 \\ 1 & 1 & 1 \\ 0 & -2 & 0 \end{bmatrix} \qquad \mathbf{M}^{-1} = \tfrac{1}{2}\begin{bmatrix} -1 & 1 & 0 \\ 0 & 0 & -1 \\ 1 & 1 & 1 \end{bmatrix}$$

$$\mathbf{M}^{-1}\mathbf{A}\mathbf{M} = \begin{bmatrix} 1 & 0 & 0 \\ 0 & 1 & 0 \\ 0 & 0 & 3 \end{bmatrix}$$

This procedure yields a diagonal form when the degeneracy of the characteristic matrix $[\lambda_i \mathbf{I} - \mathbf{A}]$ is equal to the order of the root. If it is not, a similar procedure yields a more general Jordan form discussed in the next section.

Example 4.7-6. For the linear transformation given, show that the columns of $\mathbf{M}$ form a basis and the rows of $\mathbf{M}^{-1}$ form a reciprocal basis. Express $\mathbf{y}$ in terms of the characteristic vectors.

$$\mathbf{y} = \begin{bmatrix} 0 & 1 \\ -2 & -3 \end{bmatrix}\mathbf{x}$$

The characteristic equation $|\lambda\mathbf{I} - \mathbf{A}| = 0$ is then

$$\lambda^2 + 3\lambda + 2 = 0$$

The roots of this equation are $\lambda_1 = -1$ and $\lambda_2 = -2$. The adjoint matrix is

$$\mathbf{Adj}\,[\lambda\mathbf{I} - \mathbf{A}] = \begin{bmatrix} \lambda + 3 & 1 \\ -2 & \lambda \end{bmatrix}$$

The characteristic vectors are

$$\mathbf{x}_1 = \begin{bmatrix} 1 \\ -1 \end{bmatrix} \quad \text{and} \quad \mathbf{x}_2 = \begin{bmatrix} 1 \\ -2 \end{bmatrix}$$

A normalized set of characteristic vectors are

$$\mathbf{u}_1 = \begin{bmatrix} 1/\sqrt{2} \\ -1/\sqrt{2} \end{bmatrix} \quad \text{and } \mathbf{u}_2 = \begin{bmatrix} 1/\sqrt{5} \\ -2/\sqrt{5} \end{bmatrix}$$

The modal matrix is

$$\mathbf{M} = \begin{bmatrix} 1/\sqrt{2} & 1/\sqrt{5} \\ -1/\sqrt{2} & -2/\sqrt{5} \end{bmatrix}$$

The inverse modal matrix is

$$\mathbf{M}^{-1} = \begin{bmatrix} 2\sqrt{2} & \sqrt{2} \\ -\sqrt{5} & -\sqrt{5} \end{bmatrix}$$

Note that the rows of $\mathbf{M}^{-1}$ form a reciprocal basis

$$\mathbf{r}_1 = \begin{bmatrix} 2\sqrt{2} \\ \sqrt{2} \end{bmatrix} \quad \text{and} \quad \mathbf{r}_2 = \begin{bmatrix} -\sqrt{5} \\ -\sqrt{5} \end{bmatrix}$$

such that

$$\langle \mathbf{r}_1, \mathbf{u}_1 \rangle = 1 \qquad \langle \mathbf{r}_2, \mathbf{u}_1 \rangle = 0$$

$$\langle \mathbf{r}_1, \mathbf{u}_2 \rangle = 0 \qquad \langle \mathbf{r}_2, \mathbf{u}_2 \rangle = 1$$

The vector $\mathbf{y}$, expressed in terms of the vector $\mathbf{x}$, is

$$\mathbf{y} = \begin{bmatrix} x_2 \\ -2x_1 - 3x_2 \end{bmatrix}$$

If the characteristic vectors $\mathbf{u}_i$ are used as a basis, then the vector $\mathbf{y}$ can be expressed as

$$\begin{aligned} \mathbf{y} &= \langle \mathbf{r}_1, \mathbf{y} \rangle \mathbf{u}_1 + \langle \mathbf{r}_2, \mathbf{y} \rangle \mathbf{u}_2 = \mathbf{M}\mathbf{M}^{-1}\mathbf{y} = [\mathbf{u}_1 \quad \mathbf{u}_2]\mathbf{M}^{-1}\mathbf{y} \\ &= (2\sqrt{2}\,y_1 + \sqrt{2}\,y_2)\mathbf{u}_1 + (-\sqrt{5}\,y_1 - \sqrt{5}\,y_2)\mathbf{u}_2 \\ &= (-2\sqrt{2}\,x_1 - \sqrt{2}\,x_2)\mathbf{u}_1 + (2\sqrt{5}\,x_1 + 2\sqrt{5}\,x_2)\mathbf{u}_2 \end{aligned}$$

Note that substitution for $\mathbf{u}_1$ and $\mathbf{u}_2$ yields

$$y_1 = (-2\sqrt{2}\,x_1 - \sqrt{2}\,x_2)(1/\sqrt{2}) + (2\sqrt{5}\,x_1 + 2\sqrt{5}\,x_2)(1/\sqrt{5}) = x_2$$

$$y_2 = (-2\sqrt{2}\,x_1 - \sqrt{2}\,x_2)(-1/\sqrt{2}) + (2\sqrt{5}\,x_1 + 2\sqrt{5}\,x_2)(-2/\sqrt{5}) = -2x_1 - 3x_2$$

4.8 TRANSFORMATIONS ON MATRICES

Before discussing the subject of transformed matrices, certain *elementary* operations on a matrix are considered. These operations are:

1. Interchange of any two rows (columns).
2. Addition to a row (column) of a multiple of another row (column).
3. Multiplication of a row (column) by a nonzero constant.

Performing any one of these elementary operations on a square matrix is equivalent to premultiplication or postmultiplication of the matrix by a nonsingular matrix, such that the rank of the transformed matrix is equal to the rank of the original matrix.

Operation 1. This operation is equivalent to renumbering the rows (columns), and therefore cannot change the rank of the matrix. Let $\mathbf{Q}_1$ denote the $(n \times n)$ unit matrix with the ith and jth rows interchanged, and let $\mathbf{A}$ be any $(n \times n)$ matrix. Clearly, premultiplication of $\mathbf{A}$ by $\mathbf{Q}_1$ produces a matrix with the ith and jth rows interchanged. Postmultiplication of $\mathbf{A}$ by $\mathbf{Q}_1$ produces a matrix with the ith and jth columns interchanged. For example,

$$\begin{bmatrix} 1 & 0 & 0 \\ 0 & 0 & 1 \\ 0 & 1 & 0 \end{bmatrix}\begin{bmatrix} a_{11} & a_{12} & a_{13} \\ a_{21} & a_{22} & a_{23} \\ a_{31} & a_{32} & a_{33} \end{bmatrix} = \begin{bmatrix} a_{11} & a_{12} & a_{13} \\ a_{31} & a_{32} & a_{33} \\ a_{21} & a_{22} & a_{23} \end{bmatrix}$$

$$\begin{bmatrix} a_{11} & a_{12} & a_{13} \\ a_{21} & a_{22} & a_{23} \\ a_{31} & a_{32} & a_{33} \end{bmatrix}\begin{bmatrix} 1 & 0 & 0 \\ 0 & 0 & 1 \\ 0 & 1 & 0 \end{bmatrix} = \begin{bmatrix} a_{11} & a_{13} & a_{12} \\ a_{21} & a_{23} & a_{22} \\ a_{31} & a_{33} & a_{32} \end{bmatrix}$$

Operation 2. If k times the jth row is added to the ith row of $\mathbf{A}$, this is equivalent to premultiplication of $\mathbf{A}$ by $\mathbf{Q}_2$, where $\mathbf{Q}_2$ is the $(n \times n)$ unit matrix with the element k in the ith row and jth column $(i \neq j)$. In a similar fashion, the addition of k times the ith column to the jth column is equivalent to postmultiplication of $\mathbf{A}$ by $\mathbf{Q}_2$. For example,

$$\begin{bmatrix} 1 & 0 & 0 \\ 0 & 1 & 0 \\ 0 & k & 1 \end{bmatrix}\begin{bmatrix} a_{11} & a_{12} & a_{13} \\ a_{21} & a_{22} & a_{23} \\ a_{31} & a_{32} & a_{33} \end{bmatrix} = \begin{bmatrix} a_{11} & a_{12} & a_{13} \\ a_{21} & a_{22} & a_{23} \\ a_{31} + ka_{21} & a_{32} + ka_{22} & a_{33} + ka_{23} \end{bmatrix}$$

$$\begin{bmatrix} a_{11} & a_{12} & a_{13} \\ a_{21} & a_{22} & a_{23} \\ a_{31} & a_{32} & a_{33} \end{bmatrix}\begin{bmatrix} 1 & 0 & 0 \\ 0 & 1 & 0 \\ 0 & k & 1 \end{bmatrix} = \begin{bmatrix} a_{11} & a_{12} + ka_{13} & a_{13} \\ a_{21} & a_{22} + ka_{23} & a_{23} \\ a_{31} & a_{32} + ka_{33} & a_{33} \end{bmatrix}$$

Operation 3. If the ith row of **A** is multiplied by a constant k, this is equivalent to premultiplying **A** by $\mathbf{Q}_3$, where $\mathbf{Q}_3$ is the (n x n) unit matrix with k substituted for the ith element on the principal diagonal. Similarly, multiplication of the ith column of **A** by a constant k is equivalent to postmultiplication of **A** by $\mathbf{Q}_3$. For example,

$$\begin{bmatrix} 1 & 0 & 0 \\ 0 & 1 & 0 \\ 0 & 0 & k \end{bmatrix} \begin{bmatrix} a_{11} & a_{12} & a_{13} \\ a_{21} & a_{22} & a_{23} \\ a_{31} & a_{32} & a_{33} \end{bmatrix} = \begin{bmatrix} a_{11} & a_{12} & a_{13} \\ a_{21} & a_{22} & a_{23} \\ ka_{31} & ka_{32} & ka_{33} \end{bmatrix}$$

$$\begin{bmatrix} a_{11} & a_{12} & a_{13} \\ a_{21} & a_{22} & a_{23} \\ a_{31} & a_{32} & a_{33} \end{bmatrix} \begin{bmatrix} 1 & 0 & 0 \\ 0 & 1 & 0 \\ 0 & 0 & k \end{bmatrix} = \begin{bmatrix} a_{11} & a_{12} & ka_{13} \\ a_{21} & a_{22} & ka_{23} \\ a_{31} & a_{32} & ka_{33} \end{bmatrix}$$

The determinant of $\mathbf{Q}_3$ is k, so that, if the transformed matrix is to maintain the same rank as **A**, the constant k must be nonzero.

Equivalent Matrices

Two matrices **A** and **B** are said to be *equivalent* if one matrix can be obtained from the other matrix by a sequence of the elementary operations. Since any sequence of these operations on the rows of **A** may be performed by premultiplication of **A** by a corresponding sequence of matrices $\mathbf{P}_i$ $(i = 1, 2, 3, \ldots)$ which are all nonsingular, such a sequence of operations on the rows of **A** corresponds to premultiplication of **A** by a nonsingular matrix **P**; and similarly any sequence of operations on the columns of **A** is equivalent to postmultiplication of **A** by a nonsingular matrix **Q**. Consequently, matrix **B** can be said to be equivalent to matrix **A** if and only if two nonsingular matrices **P** and **Q** exist such that

$$\mathbf{B} = \mathbf{PAQ} \tag{4.8-1}$$

Two matrices which are equivalent have the same rank. Conversely, it may be shown that two (m x n) matrices are equivalent if and only if they have the same rank.

Normal Form

Any matrix **A** of rank $r > 0$ can be reduced to an equivalent matrix of the form

$$\mathbf{I}_r, \quad \left[\begin{array}{c:c} \mathbf{I}_r & \mathbf{0} \\ \hdashline \mathbf{0} & \mathbf{0} \end{array}\right], \quad [\mathbf{I}_r \;\vdots\; \mathbf{0}], \quad \text{or} \quad \begin{bmatrix} \mathbf{I}_r \\ \mathbf{0} \end{bmatrix}$$

where $\mathbf{I}_r$ is the $(r \times r)$ unit matrix. These forms are called *normal* forms, or *canonical* forms. Note that, if the nonsingular matrix **A** can be reduced to a unit matrix by a sequence of operations on the rows of **A**, then $\mathbf{P} = \mathbf{A}^{-1}$, $\mathbf{Q} = \mathbf{I}$. This is another method of finding $\mathbf{A}^{-1}$. In general, if **A** is reduced to a unit matrix by a sequence of elementary operations, then

$$\mathbf{A} = \mathbf{P}^{-1}\mathbf{PAQQ}^{-1} = \mathbf{P}^{-1}\mathbf{IQ}^{-1} = \mathbf{P}^{-1}\mathbf{Q}^{-1} \tag{4.8-2}$$

Thus every nonsingular matrix can be expressed as a product of elementary matrices.

Example 4.8-1. Reduce the following matrices to the *normal* form.

Nonsingular matrix.

$$A = \begin{bmatrix} 0 & 1 \\ -2 & -3 \end{bmatrix}$$

Step 1. Interchange rows 2 and 1.

$$\begin{bmatrix} 0 & 1 \\ 1 & 0 \end{bmatrix}\begin{bmatrix} 0 & 1 \\ -2 & -3 \end{bmatrix} = \begin{bmatrix} -2 & -3 \\ 0 & 1 \end{bmatrix}$$

Step 2. Add 3 times the second row to the first row.

$$\begin{bmatrix} 1 & 3 \\ 0 & 1 \end{bmatrix}\begin{bmatrix} -2 & -3 \\ 0 & 1 \end{bmatrix} = \begin{bmatrix} -2 & 0 \\ 0 & 1 \end{bmatrix}$$

Step 3. Multiply the first row by $-\frac{1}{2}$.

$$\begin{bmatrix} -\frac{1}{2} & 0 \\ 0 & 1 \end{bmatrix}\begin{bmatrix} -2 & 0 \\ 0 & 1 \end{bmatrix} = \begin{bmatrix} 1 & 0 \\ 0 & 1 \end{bmatrix}$$

This reduction was performed using only row operations, in order to illustrate how the inverse of a matrix may be obtained. The sequence of elementary operations was

$$\mathbf{P} = \begin{bmatrix} -\frac{1}{2} & 0 \\ 0 & 1 \end{bmatrix}\begin{bmatrix} 1 & 3 \\ 0 & 1 \end{bmatrix}\begin{bmatrix} 0 & 1 \\ 1 & 0 \end{bmatrix} = \begin{bmatrix} -\frac{3}{2} & -\frac{1}{2} \\ 1 & 0 \end{bmatrix}$$

The product **PA** is

$$\begin{bmatrix} -\frac{3}{2} & -\frac{1}{2} \\ 1 & 0 \end{bmatrix}\begin{bmatrix} 0 & 1 \\ -2 & -3 \end{bmatrix} = \begin{bmatrix} 1 & 0 \\ 0 & 1 \end{bmatrix}$$

Therefore $\mathbf{P} = \mathbf{A}^{-1}$.

Singular matrix.

$$\mathbf{A} = \begin{bmatrix} 1 & 2 & -3 \\ -1 & 2 & -1 \\ -1 & -3 & 4 \end{bmatrix}$$

Step 1. Add the first row to the third row.

$$\begin{bmatrix} 1 & 0 & 0 \\ 0 & 1 & 0 \\ 1 & 0 & 1 \end{bmatrix}\begin{bmatrix} 1 & 2 & -3 \\ -1 & 2 & -1 \\ -1 & -3 & 4 \end{bmatrix} = \begin{bmatrix} 1 & 2 & -3 \\ -1 & 2 & -1 \\ 0 & -1 & 1 \end{bmatrix}$$

Step 2. Add the first row to the second row.

$$\begin{bmatrix} 1 & 0 & 0 \\ 1 & 1 & 0 \\ 0 & 0 & 1 \end{bmatrix} \begin{bmatrix} 1 & 2 & -3 \\ -1 & 2 & -1 \\ 0 & -1 & 1 \end{bmatrix} = \begin{bmatrix} 1 & 2 & -3 \\ 0 & 4 & -4 \\ 0 & -1 & 1 \end{bmatrix}$$

Step 3. Add the second column to the third column.

$$\begin{bmatrix} 1 & 2 & -3 \\ 0 & 4 & -4 \\ 0 & -1 & 1 \end{bmatrix} \begin{bmatrix} 1 & 0 & 0 \\ 0 & 1 & 1 \\ 0 & 0 & 1 \end{bmatrix} = \begin{bmatrix} 1 & 2 & -1 \\ 0 & 4 & 0 \\ 0 & -1 & 0 \end{bmatrix}$$

Step 4. Add the first column to the third column.

$$\begin{bmatrix} 1 & 2 & -1 \\ 0 & 4 & 0 \\ 0 & -1 & 0 \end{bmatrix} \begin{bmatrix} 1 & 0 & 1 \\ 0 & 1 & 0 \\ 0 & 0 & 1 \end{bmatrix} = \begin{bmatrix} 1 & 2 & 0 \\ 0 & 4 & 0 \\ 0 & -1 & 0 \end{bmatrix}$$

Step 5. Add (-2) times the first column to the second column.

$$\begin{bmatrix} 1 & 2 & 0 \\ 0 & 4 & 0 \\ 0 & -1 & 0 \end{bmatrix} \begin{bmatrix} 1 & -2 & 0 \\ 0 & 1 & 0 \\ 0 & 0 & 1 \end{bmatrix} = \begin{bmatrix} 1 & 0 & 0 \\ 0 & 4 & 0 \\ 0 & -1 & 0 \end{bmatrix}$$

Step 6. Multiply the second row by $\frac{1}{4}$.

$$\begin{bmatrix} 1 & 0 & 0 \\ 0 & \frac{1}{4} & 0 \\ 0 & 0 & 1 \end{bmatrix} \begin{bmatrix} 1 & 0 & 0 \\ 0 & 4 & 0 \\ 0 & -1 & 0 \end{bmatrix} = \begin{bmatrix} 1 & 0 & 0 \\ 0 & 1 & 0 \\ 0 & -1 & 0 \end{bmatrix}$$

Step 7. Add the second row to the third row.

$$\begin{bmatrix} 1 & 0 & 0 \\ 0 & 1 & 0 \\ 0 & 1 & 1 \end{bmatrix} \begin{bmatrix} 1 & 0 & 0 \\ 0 & 1 & 0 \\ 0 & -1 & 0 \end{bmatrix} = \begin{bmatrix} 1 & 0 & 0 \\ 0 & 1 & 0 \\ 0 & 0 & 0 \end{bmatrix} = \left[\begin{array}{cc|c} \multicolumn{2}{c|}{\mathbf{I}_2} & \begin{matrix} 0 \\ 0 \end{matrix} \\ \hline 0 & 0 & 0 \end{array}\right]$$

The elementary matrices **P** and **Q** are

$$\mathbf{P} = \begin{bmatrix} 1 & 0 & 0 \\ 0 & 1 & 0 \\ 0 & 1 & 1 \end{bmatrix} \begin{bmatrix} 1 & 0 & 0 \\ 0 & \frac{1}{4} & 0 \\ 0 & 0 & 1 \end{bmatrix} \begin{bmatrix} 1 & 0 & 0 \\ 1 & 1 & 0 \\ 0 & 0 & 1 \end{bmatrix} \begin{bmatrix} 1 & 0 & 0 \\ 0 & 1 & 0 \\ 1 & 0 & 1 \end{bmatrix} = \frac{1}{4} \begin{bmatrix} 4 & 0 & 0 \\ 1 & 1 & 0 \\ 5 & 1 & 4 \end{bmatrix}$$

$$\mathbf{Q} = \begin{bmatrix} 1 & 0 & 0 \\ 0 & 1 & 1 \\ 0 & 0 & 1 \end{bmatrix} \begin{bmatrix} 1 & 0 & 1 \\ 0 & 1 & 0 \\ 0 & 0 & 1 \end{bmatrix} \begin{bmatrix} 1 & -2 & 0 \\ 0 & 1 & 0 \\ 0 & 0 & 1 \end{bmatrix} = \begin{bmatrix} 1 & -2 & 1 \\ 0 & 1 & 1 \\ 0 & 0 & 1 \end{bmatrix}$$

The transformation **PAQ** yields

$$\mathbf{PAQ} = \frac{1}{4} \begin{bmatrix} 4 & 0 & 0 \\ 1 & 1 & 0 \\ 5 & 1 & 4 \end{bmatrix} \begin{bmatrix} 1 & 2 & -3 \\ -1 & 2 & -1 \\ -1 & -3 & 4 \end{bmatrix} \begin{bmatrix} 1 & -2 & 1 \\ 0 & 1 & 1 \\ 0 & 0 & 1 \end{bmatrix} = \begin{bmatrix} 1 & 0 & 0 \\ 0 & 1 & 0 \\ 0 & 0 & 0 \end{bmatrix}$$

which is the normal form

$$\left[\begin{array}{c|c} \mathbf{I}_2 & \mathbf{0} \\ \hline \mathbf{0} & \mathbf{0} \end{array}\right]$$

The transformation $\mathbf{B} = \mathbf{PAQ}$ is the most general kind of matrix transformation. Other transformations are defined in terms of the relationship between $\mathbf{P}$ and $\mathbf{Q}$. Specifically, these transformations are

Collineatory (Similarity):

$$\mathbf{B} = \mathbf{Q}^{-1}\mathbf{AQ}, \quad \text{or} \quad \mathbf{P} = \mathbf{Q}^{-1}$$

Orthogonal:

$$\mathbf{B} = \mathbf{Q}^T\mathbf{AQ} = \mathbf{Q}^{-1}\mathbf{AQ}, \quad \text{or} \quad \mathbf{P} = \mathbf{Q}^T = \mathbf{Q}^{-1}$$

Congruent:

$$\mathbf{B} = \mathbf{Q}^T\mathbf{AQ}, \quad \text{or} \quad \mathbf{P} = \mathbf{Q}^T$$

If $\mathbf{A}$ is a Hermitian matrix, the following transformations are defined:

Conjunctive:

$$\mathbf{B} = \mathbf{Q}^{*T}\mathbf{AQ}, \quad \text{or} \quad \mathbf{P} = \mathbf{Q}^{*T}$$

Unitary:

$$\mathbf{B} = \mathbf{Q}^{*T}\mathbf{AQ} = \mathbf{Q}^{-1}\mathbf{AQ}, \quad \text{or} \quad \mathbf{P} = \mathbf{Q}^{*T} = \mathbf{Q}^{-1}$$

Collineatory (Similarity) Transformation

Consider the linear transformation

$$\mathbf{y} = \mathbf{Ax} \tag{4.8-3}$$

where both $\mathbf{x}$ and $\mathbf{y}$ are defined in an n-dimensional space in terms of a basis $\mathbf{z}_i$. Assume now that the basis is to be changed to the set of vectors $\mathbf{w}_i$. This is a generalized problem in coordinate transformations. Let $\mathbf{x}'$ and $\mathbf{y}'$ form the coordinates of $\mathbf{x}$ and $\mathbf{y}$, respectively, in the new basis. Since $\mathbf{z}_i$ and $\mathbf{w}_i$ form two bases in the n-dimensional space, there must exist a nonsingular $\mathbf{P}$ such that

$$\begin{aligned} \mathbf{x}' &= \mathbf{Px} & \mathbf{x} &= \mathbf{P}^{-1}\mathbf{x}' \\ \mathbf{y}' &= \mathbf{Py} & \mathbf{y} &= \mathbf{P}^{-1}\mathbf{y}' \end{aligned} \tag{4.8-4}$$

The relationship between $\mathbf{y}'$ and $\mathbf{x}'$ in the new coordinate system is to be found. To obtain this relationship, premultiply both sides of Eq. 4.8-3 by $\mathbf{P}$, forming $\mathbf{Py} = \mathbf{PAx}$. Using Eq. 4.8-4,

$$\mathbf{y}' = \mathbf{PAP}^{-1}\mathbf{x}' \tag{4.8-5}$$

or

$$\mathbf{y}' = \mathbf{Bx}' \quad \text{where} \quad \mathbf{B} = \mathbf{Q}^{-1}\mathbf{AQ}, \mathbf{P} = \mathbf{Q}^{-1} \tag{4.8-6}$$

The matrix $\mathbf{B}$, relating $\mathbf{x}'$ and $\mathbf{y}'$ in the new coordinate system, is obtained from $\mathbf{A}$ by a *similarity* transformation.

Similarity transformations have the extremely important property that the characteristic values are invariant under such a transformation. To show this, let $\mathbf{B} = \mathbf{PAQ}$. Then

$$\mathbf{B} - \lambda\mathbf{I} = \mathbf{PAQ} - \lambda\mathbf{I} = \mathbf{P}(\mathbf{A} - \lambda\mathbf{P}^{-1}\mathbf{Q}^{-1})\mathbf{Q}$$

The corresponding determinants are

$$|\mathbf{B} - \lambda\mathbf{I}| = |\mathbf{P}| \cdot |\mathbf{Q}| \cdot |\mathbf{A} - \lambda\mathbf{P}^{-1}\mathbf{Q}^{-1}|$$

Since $\mathbf{P} = \mathbf{Q}^{-1}$, the product of the determinants $|\mathbf{P}|$ and $|\mathbf{Q}|$ is equal to unity. It follows that

$$|\mathbf{B} - \lambda\mathbf{I}| = |\mathbf{A} - \lambda\mathbf{I}| \tag{4.8-7}$$

Therefore the characteristic values are invariant under a similarity transformation. If $\mathbf{x}_i'$ is a characteristic vector corresponding to the characteristic value λ_i of $\mathbf{Q}^{-1}\mathbf{AQ}$, then $\mathbf{x}_i = \mathbf{Qx}_i'$ must be a characteristic vector of $\mathbf{A}$ corresponding to the same characteristic value λ_i of $\mathbf{A}$.

Orthogonal Transformation

Consider the linear transformation of Eq. 4.8-4,

$$\mathbf{x} = \mathbf{Qx}', \quad \mathbf{P}^{-1} = \mathbf{Q}$$

where the vector $\mathbf{x}$ is defined in terms of an orthogonal coordinate system. If the new coordinate system is also orthogonal, then the length of the vector $\mathbf{x}'$ in the new coordinate system must be identical with the length of the vector $\mathbf{x}$ in the original coordinate system. Therefore

$$\langle \mathbf{x}, \mathbf{x} \rangle = \langle \mathbf{x}', \mathbf{x}' \rangle$$

Expressing this relationship in terms of the matrix $\mathbf{Q}$,

$$\mathbf{x}'^T\mathbf{x}' = \mathbf{x}^T\mathbf{x} = \mathbf{x}'^T\mathbf{Q}^T\mathbf{Q}\mathbf{x}'$$

This requires that $\mathbf{Q}^T\mathbf{Q} = \mathbf{I}$, or

$$\mathbf{Q}^T = \mathbf{Q}^{-1} \tag{4.8-8}$$

Therefore, if a transformation from one mutually orthogonal basis to another mutually orthogonal basis is made, the transformation matrix $\mathbf{Q}$ which relates a vector in the new coordinate system to a vector in the original coordinate system must satisfy the relation shown in Eq. 4.8-8. The transformation is then called an *orthogonal* transformation. The matrix $\mathbf{Q}$ is called an *orthogonal* matrix. An orthogonal transformation is a special case of a similarity transformation. Lengths and angles are preserved.

As a consequence of Eq. 4.8-8, it is seen that

$$|\mathbf{Q}^T|\,|\mathbf{Q}| = |\mathbf{Q}|^2 = 1 \quad \text{or} \quad |\mathbf{Q}| = \pm 1 \tag{4.8-9}$$

An interpretation of the minus sign for the determinant of **Q** is that an orthogonal transformation may be obtained by a rotation and a reflection. Lengths and angles are still preserved if both a rotation and a reflection are involved in an orthogonal transformation. It should also be noted that the cosine of the angle between axis i' and the axes $1, 2, \ldots, n$ are denoted, respectively, by the elements of the **Q** matrix $q_{i1}, q_{i2}, \ldots, q_{in}$. These quantities are called *direction cosines*, or the directions of the new axes with respect to the old axes.

Unitary Transformation

If **A** is a Hermitian matrix ($a_{ji} = a_{ij}{}^*$), then a treatment corresponding to that shown for an orthogonal transformation can be used. For this case, it can be shown that the transformation from one orthonormal basis to another orthonormal basis is accomplished through the use of a unitary transformation of the form $\mathbf{x} = \mathbf{Q}\mathbf{x}'$, where **Q** is a unitary matrix defined by

$$\mathbf{Q}^{*T}\mathbf{Q} = \mathbf{I} \quad \text{or} \quad \mathbf{Q}^{*T} = \mathbf{Q}^{-1} \tag{4.8-10}$$

The determinant of a unitary matrix is ± 1. Thus a unitary transformation preserves lengths in the complex sense.

Congruent Transformation

Two matrices **A** and **B** are called congruent if one is obtained from the other by a sequence of pairs of elementary operations, each pair consisting of an elementary row transformation followed by the same elementary column transformation. As a consequence of the definitions of elementary transformations, it follows that two matrices are congruent if there exists a nonsingular matrix **Q** such that

$$\mathbf{B} = \mathbf{Q}^T\mathbf{A}\mathbf{Q} \tag{4.8-11}$$

Congruency is a special case of equivalence, so that congruent matrices have equal rank. A congruency transformation is a transformation to a new basis such that, if two vectors **x** and **y** are related in the original basis by Eq. 4.8-3, the vectors $\mathbf{x}'$ and $\mathbf{y}'$ in the new basis are related by the equation

$$\mathbf{y}' = \mathbf{B}\mathbf{x}' = \mathbf{Q}^T\mathbf{A}\mathbf{Q}\mathbf{x}' \tag{4.8-12}$$

Transformations of this type are useful when dealing with quadratic forms, which are discussed in the next section.

Conjunctive Transformation

Two (n x n) Hermitian matrices **A** and **B** are conjunctive if one can be obtained from the other by a sequence of pairs of elementary transformations, each pair consisting of a column transformation and the corresponding conjugate row transformation. In view of the definitions of elementary transformations, two (n x n) Hermitian matrices **A** and **B** are conjunctive (Hermitely congruent) if there exists a nonsingular matrix **Q** such that

$$\mathbf{B} = \mathbf{Q}^{*T}\mathbf{A}\mathbf{Q} \tag{4.8-13}$$

A conjunctive transformation is a transformation to a new basis such that, if two vectors **x** and **y** are related in the original basis by Eq. 4.8-3, the vectors $\mathbf{x}'$ and $\mathbf{y}'$ in the new basis are related by the equation

$$\mathbf{y}' = \mathbf{B}\mathbf{x}' = \mathbf{Q}^{*T}\mathbf{A}\mathbf{Q}\mathbf{x}' \tag{4.8-14}$$

Transformation to a Diagonal Form

Quite frequently it is advantageous to transform the coordinate system of a problem to a new coordinate system, such that a linear transformation in the new coordinate system is dependent upon a diagonal matrix. For this reason, the conditions under which the matrix $\mathbf{B} = \mathbf{PAQ}$ can be reduced to a diagonal form are now discussed.

Congruent Transformations

Real Symmetric Matrices. The real symmetric matrix **A** of rank r can be reduced to a congruent diagonal matrix having the canonical form

$$\mathbf{Q}^T\mathbf{A}\mathbf{Q} = \left[\begin{array}{c:c:c} \mathbf{I}_p & \mathbf{0} & \mathbf{0} \\ \hdashline \mathbf{0} & -\mathbf{I}_{r-p} & \mathbf{0} \\ \hdashline \mathbf{0} & \mathbf{0} & \mathbf{0} \end{array}\right] \tag{4.8-15}$$

The integer p is called the *index* of the matrix, and the integer $s = p - (r - p) = 2p - r$ is called the *signature* of the matrix. Two (n x n) real symmetric matrices are congruent if and only if they have the same rank and the same signature or index.

Example 4.8-2. Reduce the given symmetric **A** to the canonical form of Eq. 4.8-15.

$$\mathbf{A} = \begin{bmatrix} 1 & 2 & 2 \\ 2 & 3 & 5 \\ 2 & 5 & 5 \end{bmatrix}$$

Step 1. Subtract the third row from the second row.

$$\begin{bmatrix} 1 & 0 & 0 \\ 0 & 1 & -1 \\ 0 & 0 & 1 \end{bmatrix}\begin{bmatrix} 1 & 2 & 2 \\ 2 & 3 & 5 \\ 2 & 5 & 5 \end{bmatrix} = \begin{bmatrix} 1 & 2 & 2 \\ 0 & -2 & 0 \\ 2 & 5 & 5 \end{bmatrix}$$

Step 2. Subtract the third column from the second column.

$$\begin{bmatrix} 1 & 2 & 2 \\ 0 & -2 & 0 \\ 2 & 5 & 5 \end{bmatrix}\begin{bmatrix} 1 & 0 & 0 \\ 0 & 1 & 0 \\ 0 & -1 & 1 \end{bmatrix} = \begin{bmatrix} 1 & 0 & 2 \\ 0 & -2 & 0 \\ 2 & 0 & 5 \end{bmatrix}$$

Step 3. Subtract twice the first row from the third row.

$$\begin{bmatrix} 1 & 0 & 0 \\ 0 & 1 & 0 \\ -2 & 0 & 1 \end{bmatrix}\begin{bmatrix} 1 & 0 & 2 \\ 0 & -2 & 0 \\ 2 & 0 & 5 \end{bmatrix} = \begin{bmatrix} 1 & 0 & 2 \\ 0 & -2 & 0 \\ 0 & 0 & 1 \end{bmatrix}$$

Step 4. Subtract twice the first column from the third column.

$$\begin{bmatrix} 1 & 0 & 2 \\ 0 & -2 & 0 \\ 0 & 0 & 1 \end{bmatrix}\begin{bmatrix} 1 & 0 & -2 \\ 0 & 1 & 0 \\ 0 & 0 & 1 \end{bmatrix} = \begin{bmatrix} 1 & 0 & 0 \\ 0 & -2 & 0 \\ 0 & 0 & 1 \end{bmatrix}$$

Step 5. Multiply the second row by $1/\sqrt{2}$ and the second column by $1/\sqrt{2}$.

$$\begin{bmatrix} 1 & 0 & 0 \\ 0 & 1/\sqrt{2} & 0 \\ 0 & 0 & 1 \end{bmatrix}\begin{bmatrix} 1 & 0 & 0 \\ 0 & -2 & 0 \\ 0 & 0 & 1 \end{bmatrix}\begin{bmatrix} 1 & 0 & 0 \\ 0 & 1/\sqrt{2} & 0 \\ 0 & 0 & 1 \end{bmatrix} = \begin{bmatrix} 1 & 0 & 0 \\ 0 & -1 & 0 \\ 0 & 0 & 1 \end{bmatrix}$$

Step 6. Interchange the second column with the third column and the second row with the third row.

$$\begin{bmatrix} 1 & 0 & 0 \\ 0 & 0 & 1 \\ 0 & 1 & 0 \end{bmatrix}\begin{bmatrix} 1 & 0 & 0 \\ 0 & -1 & 0 \\ 0 & 0 & 1 \end{bmatrix}\begin{bmatrix} 1 & 0 & 0 \\ 0 & 0 & 1 \\ 0 & 1 & 0 \end{bmatrix} = \begin{bmatrix} 1 & 0 & 0 \\ 0 & 1 & 0 \\ 0 & 0 & -1 \end{bmatrix}$$

In this case the index of the matrix is $p = 2$, the rank is $r = 3$, and the signature is $s = 1$.

$$\mathbf{Q} = \begin{bmatrix} 1 & 0 & 0 \\ 0 & 1 & 0 \\ 0 & -1 & 1 \end{bmatrix}\begin{bmatrix} 1 & 0 & -2 \\ 0 & 1 & 0 \\ 0 & 0 & 1 \end{bmatrix}\begin{bmatrix} 1 & 0 & 0 \\ 0 & 1/\sqrt{2} & 0 \\ 0 & 0 & 1 \end{bmatrix}\begin{bmatrix} 1 & 0 & 0 \\ 0 & 0 & 1 \\ 0 & 1 & 0 \end{bmatrix} = \begin{bmatrix} 1 & -2 & 0 \\ 0 & 0 & 1/\sqrt{2} \\ 0 & 1 & -1/\sqrt{2} \end{bmatrix}$$

Skew-Symmetric Matrices. A skew-symmetric matrix of rank r can be reduced to the canonical form of Eq. 4.8-16 by the congruent transformation $\mathbf{B} = -\mathbf{Q}^T\mathbf{A}\mathbf{Q}$. The result is

$$\mathbf{B} = -\mathbf{Q}^T\mathbf{A}\mathbf{Q} = \text{diag}\,(\mathbf{D}_1, \mathbf{D}_2, \ldots, \mathbf{D}_m, 0, \ldots, 0) \qquad (4.8\text{-}16)$$

where

$$\mathbf{D}_i = \begin{bmatrix} 0 & 1 \\ -1 & 0 \end{bmatrix} \quad \text{and} \quad m = \tfrac{1}{2}\,\text{rank}\,\mathbf{A}$$

Two (n x n) skew-symmetric matrices are congruent if and only if they have the same rank.

Complex Symmetric Matrices. An (n x n) complex symmetric matrix of rank r can be reduced by a congruency transformation to the canonical form

$$\mathbf{Q}^T\mathbf{A}\mathbf{Q} = \begin{bmatrix} \mathbf{I}_r & \mathbf{0} \\ \mathbf{0} & \mathbf{0} \end{bmatrix} \tag{4.8-17}$$

Two (n x n) complex symmetric matrices are congruent if and only if they have equal rank.

Hermitian Matrices. An (n x n) Hermitian matrix of rank r can be reduced to the canonical matrix of Eq. 4.8-18 by a conjunctive transformation.

$$\mathbf{B} = \mathbf{Q}^{*T}\mathbf{A}\mathbf{Q} = \begin{bmatrix} \mathbf{I}_p & \mathbf{0} & \mathbf{0} \\ \mathbf{0} & -\mathbf{I}_{r-p} & \mathbf{0} \\ \mathbf{0} & \mathbf{0} & \mathbf{0} \end{bmatrix} \tag{4.8-18}$$

The index p and signature s are the same as defined for a real symmetric matrix.

Skew-Hermitian Matrices. An (n x n) skew-Hermitian matrix of rank r can be reduced to the canonical matrix of Eq. 4.8-19 by a conjunctive transformation.

$$\mathbf{B} = \mathbf{Q}^{*T}\mathbf{A}\mathbf{Q} = \begin{bmatrix} j\mathbf{I}_p & \mathbf{0} & \mathbf{0} \\ \mathbf{0} & -j\mathbf{I}_{r-p} & \mathbf{0} \\ \mathbf{0} & \mathbf{0} & \mathbf{0} \end{bmatrix} \tag{4.8-19}$$

Similarity Transformations

In the previous section it was shown that a transformation of the type $\mathbf{M}^{-1}\,\mathbf{A}\mathbf{M}$ produces a diagonal matrix $\mathbf{\Lambda}$ if the matrix $\mathbf{A}$ had n distinct characteristic values. For the case where repeated roots are involved, this transformation is still possible provided the matrix $[\lambda\mathbf{I} - \mathbf{A}]$ has full degeneracy. This similarity transformation is also always possible if the matrix $\mathbf{A}$ is real and symmetric, a case which is prevalent in the study of linear circuits. Since the characteristic vectors of a real symmetric (or Hermitian) matrix are mutually orthogonal, there always exists a real orthogonal matrix such that

$$\mathbf{Q}^{-1}\mathbf{A}\mathbf{Q} = \mathbf{Q}^T\mathbf{A}\mathbf{Q} = \text{diag}\,(\lambda_1, \lambda_2, \ldots, \lambda_n) \tag{4.8-20}$$

However, this is not generally the case when nonsymmetric matrices are involved, and most matrices found in the analysis of control systems are nonsymmetric matrices.

When an $(n \times n)$ nonsymmetric matrix has repeated characteristic values, there may be less than n linearly independent characteristic vectors; thus a similarity transformation to a diagonal form may be impossible. However, it can be shown that any square matrix $\mathbf{A}$ can be transformed by means of a similarity transformation to a *Jordan canonical matrix* having the following properties:[8]

1. The diagonal elements of the matrix are the characteristic values of $\mathbf{A}$.
2. All the elements below the principal diagonal are zero.
3. A certain number of unit elements are contained in the superdiagonal (the elements immediately to the right of the principal diagonal) when the adjacent elements in the principal diagonal are equal.

A typical Jordan form is

$$\mathbf{J} = \begin{bmatrix} \lambda_1 & 1 & & & & \\ & \lambda_1 & 1 & & & \\ & & \lambda_1 & & & \\ & & & \lambda_1 & & \\ & & & & \lambda_2 & 1 \\ & & & & & \lambda_2 \end{bmatrix}$$

Note that the "ones" occur in blocks of the form

$$\begin{bmatrix} \lambda_i & 1 & & & & & \\ & \lambda_i & 1 & & & & \\ & & \lambda_i & 1 & & & \\ & & & \ddots & \ddots & & \\ & & & & \ddots & \ddots & \\ & & & & & \lambda_i & 1 \\ & & & & & & \lambda_i \end{bmatrix}$$

These are called Jordan blocks.

The number of Jordan blocks associated with a given characteristic value λ_i in the Jordan form resulting from a similarity transformation of $\mathbf{A}$ is equal to the number of characteristic vectors associated with the characteristic value, i.e., q, the degeneracy of $[\lambda_i \mathbf{I} - \mathbf{A}]$. Unfortunately however, the orders of the Jordan blocks are not easily determined.†

† This is an extremely complicated problem of linear algebra.[9,10]

The result is that it is not clear whether the Jordan form given above or the form

$$\mathbf{J} = \begin{bmatrix} \lambda_1 & 1 & & & & \\ & \lambda_1 & & & & \\ & & \lambda_1 & 1 & & \\ & & & \lambda_1 & & \\ & & & & \lambda_2 & 1 \\ & & & & & \lambda_2 \end{bmatrix}$$

would be the result of the transformation $\mathbf{J} = \mathbf{M}^{-1}\mathbf{AM}$. The number of "ones" associated with a given λ_i is the order of λ_i in the characteristic equation minus the degeneracy of $[\lambda_i\mathbf{I} - \mathbf{A}]$. However, this does not clearly define the situation, since both of the above forms have two "ones" associated with λ_1.

It is useful to know that in the case of full degeneracy no "ones" are present, as was shown in the preceding section. Also, in the case of simple degeneracy ($q = 1$), all the superdiagonal elements are unity. For the cases which fit neither of these categories, the level of the discussion presented here dictates a trial and error determination of **J** and **M** based upon

$$\mathbf{AM} = \mathbf{MJ} \tag{4.8-21}$$

Let the columns of **M** be denoted by $\mathbf{x}_1, \mathbf{x}_2, \ldots, \mathbf{x}_n$. Then there is a Jordan block of order m associated with λ_i if and only if the m linearly independent vectors $\mathbf{x}_1, \mathbf{x}_2, \ldots, \mathbf{x}_m$ satisfy the equations

$$\begin{aligned} \mathbf{Ax}_1 &= \lambda_i\mathbf{x}_1 & &\text{or} & (\lambda_i\mathbf{I} - \mathbf{A})\mathbf{x}_1 &= \mathbf{0} \\ \mathbf{Ax}_2 &= \lambda_i\mathbf{x}_2 + \mathbf{x}_1 & &\text{or} & (\lambda_i\mathbf{I} - \mathbf{A})\mathbf{x}_2 &= -\mathbf{x}_1 \\ &\cdots\cdots & & & &\cdots\cdots \\ \mathbf{Ax}_m &= \lambda_i\mathbf{x}_m + \mathbf{x}_{m-1} & &\text{or} & (\lambda_i\mathbf{I} - \mathbf{A})\mathbf{x}_m &= -\mathbf{x}_{m-1} \end{aligned} \tag{4.8-22}$$

These expression apply for each Jordan block.

Example 4.8-3. Show that the matrix

$$\mathbf{A} = \begin{bmatrix} \lambda_1 & 1 \\ 0 & \lambda_2 \end{bmatrix}$$

cannot be reduced to a diagonal form by means of a similarity transformation, if $\lambda_1 = \lambda_2$.

Let

$$\mathbf{Q} = \begin{bmatrix} q_{11} & q_{12} \\ q_{21} & q_{22} \end{bmatrix} \quad \text{and} \quad \mathbf{Q}^{-1} = \frac{\begin{bmatrix} q_{22} & -q_{12} \\ -q_{21} & q_{11} \end{bmatrix}}{|\mathbf{Q}|}$$

where $|\mathbf{Q}| = q_{11}q_{22} - q_{21}q_{12}$. The similarity transformation $\mathbf{Q}^{-1}\mathbf{A}\mathbf{Q}$ is then

$$\mathbf{Q}^{-1}\mathbf{A}\mathbf{Q} = \frac{\begin{bmatrix} q_{22}(q_{11}\lambda_1 + q_{21}) - q_{12}q_{21}\lambda_2 & q_{22}(q_{12}\lambda_1 + q_{22}) - q_{12}q_{22}\lambda_2 \\ -q_{21}(q_{11}\lambda_1 + q_{21}) + q_{11}q_{21}\lambda_2 & -q_{21}(q_{12}\lambda_1 + q_{22}) + q_{11}q_{22}\lambda_2 \end{bmatrix}}{|\mathbf{Q}|}$$

If the nondiagonal terms are to vanish, then

$$q_{22}q_{12}(\lambda_1 - \lambda_2) + q_{22}^2 = 0$$
$$q_{11}q_{21}(\lambda_1 - \lambda_2) + q_{21}^2 = 0$$

If $\lambda_1 = \lambda_2$, then q_{22} and q_{21} must vanish. However, if these terms vanish, then $|\mathbf{Q}| = 0$, which violates the similarity transformation. Therefore the matrix $\mathbf{A}$ cannot be diagonalized by a similarity transformation if $\lambda_1 = \lambda_2$. Since $\mathbf{A}$ is already a Jordan form, no further transformations are considered.

Example 4.8-4. Reduce the $\mathbf{A}$ matrix of Example 4.7-3 with repeated characteristic roots and simple degeneracy to Jordan form. Determine $\mathbf{M}$.

Since

$$\mathbf{A} = \begin{bmatrix} 0 & 1 & 0 \\ 0 & 0 & 1 \\ 1 & -3 & 3 \end{bmatrix}$$

has a characteristic value $\lambda = 1$ of order three and only simple degeneracy, there is only one linearly independent characteristic vector. Hence the Jordan form contains one Jordan block. Also, the order of λ_i minus the degeneracy indicates two "ones" in the Jordan form. Hence the Jordan form must be

$$\begin{bmatrix} 1 & 1 & 0 \\ 0 & 1 & 1 \\ 0 & 0 & 1 \end{bmatrix}$$

The characteristic vector $\mathbf{x}_1$ is given by the first of Eqs. 4.8-22, which is

$$x_{21} = x_{11}$$
$$x_{31} = x_{21}$$
$$x_{11} - 3x_{21} + 3x_{31} = x_{31}$$

where x_{11}, x_{21}, and x_{31} denote the elements of $\mathbf{x}_1$. These equations yield $x_{11} = x_{21} = x_{31}$. Thus

$$\mathbf{x}_1 = \begin{bmatrix} 1 \\ 1 \\ 1 \end{bmatrix}$$

is a characteristic vector. Note that it must be, and is, the same characteristic vector determined in Example 4.7-3.

Now, considering $\mathbf{x}_2$, the second of Eqs. 4.8-22 gives

$$x_{22} = x_{12} + 1$$
$$x_{32} = x_{22} + 1$$
$$x_{12} - 3x_{22} + 3x_{32} = x_{32} + 1$$

Substitution of the first two equations into the third yields $x_{12} = x_{12}$. Hence x_{12} is arbitrary. Let $x_{12} = 1$. Then

$$\mathbf{x}_2 = \begin{bmatrix} 1 \\ 2 \\ 3 \end{bmatrix}$$

The third of Eqs. 4.8-22 gives

$$\begin{aligned} x_{23} &= x_{13} + 1 \\ x_{33} &= x_{23} + 2 \\ x_{13} - 3x_{23} + 3x_{33} &= x_{33} + 3 \end{aligned}$$

Again, substitution of the first two equations into the third yields an arbitrary component, since it gives $x_{13} = x_{13}$. Hence let $x_{13} = -1$. Then

$$\mathbf{x}_3 = \begin{bmatrix} -1 \\ 0 \\ 2 \end{bmatrix}$$

Thus

$$\mathbf{M} = \begin{bmatrix} 1 & 1 & -1 \\ 1 & 2 & 0 \\ 1 & 3 & 2 \end{bmatrix}$$

As a check,

$$\mathbf{M}^{-1} = \begin{bmatrix} 4 & -5 & 2 \\ -2 & 3 & -1 \\ 1 & -2 & 1 \end{bmatrix}$$

and

$$\mathbf{M}^{-1}\mathbf{A}\mathbf{M} = \mathbf{J} = \begin{bmatrix} 1 & 1 & 0 \\ 0 & 1 & 1 \\ 0 & 0 & 1 \end{bmatrix}$$

Example 4.8-5. Reduce

$$\mathbf{A} = \begin{bmatrix} 1 & 0 & 0 \\ 1 & 1 & 0 \\ 2 & 3 & 2 \end{bmatrix}$$

to Jordan form and determine **M**.

Evaluation of $|\lambda\mathbf{I} - \mathbf{A}|$ yields the characteristic equation $(\lambda - 1)^2(\lambda - 2) = 0$. Considering the degeneracy of $[\lambda\mathbf{I} - \mathbf{A}]$ for $\lambda = 1$,

$$[\mathbf{I} - \mathbf{A}] = \begin{bmatrix} 0 & 0 & 0 \\ -1 & 0 & 0 \\ -2 & -3 & -1 \end{bmatrix}$$

is of degeneracy one. Hence **J** consists of two Jordan blocks. One is a first order block corresponding to $\lambda = 2$. The second is a second order block with a single "one." It is

$$\mathbf{J} = \begin{bmatrix} 2 & 0 & 0 \\ 0 & 1 & 1 \\ 0 & 0 & 1 \end{bmatrix}$$

The characteristic vector $\mathbf{x}_1$ corresponding to $\lambda = 2$ is given by

$$\begin{aligned} x_{11} &= 2x_{11} \\ x_{11} + x_{21} &= 2x_{21} \\ 2x_{11} + 3x_{21} + 2x_{31} &= 2x_{31} \end{aligned}$$

Choosing $x_{31} = 1$, arbitrarily, leads to

$$\mathbf{x}_1 = \begin{bmatrix} 0 \\ 0 \\ 1 \end{bmatrix}$$

Considering $\mathbf{x}_2$ corresponding to $\lambda = 1$,

$$\begin{aligned} x_{12} &= x_{12} \\ x_{12} + x_{22} &= x_{22} \\ 2x_{12} + 3x_{22} + 2x_{32} &= x_{32} \end{aligned}$$

Choosing $x_{22} = 1$, arbitrarily, leads to

$$\mathbf{x}_2 = \begin{bmatrix} 0 \\ 1 \\ -3 \end{bmatrix}$$

In order to determine a second vector corresponding to $\lambda = 1$, Eq. 4.8-22 gives

$$\begin{aligned} x_{13} &= x_{13} \\ x_{13} + x_{23} &= x_{23} + 1 \\ 2x_{13} + 3x_{23} + 2x_{33} &= x_{33} - 3 \end{aligned}$$

Arbitrarily choosing $x_{23} = 0$ gives

$$\mathbf{x}_3 = \begin{bmatrix} 1 \\ 0 \\ -5 \end{bmatrix}$$

Thus

$$\mathbf{M} = \begin{bmatrix} 0 & 0 & 1 \\ 0 & 1 & 0 \\ 1 & -3 & -5 \end{bmatrix}$$

As a check,

$$\mathbf{M}^{-1} = \begin{bmatrix} 5 & 3 & 1 \\ 0 & 1 & 0 \\ 1 & 0 & 0 \end{bmatrix}$$

and

$$\mathbf{M}^{-1}\mathbf{A}\mathbf{M} = \mathbf{J} = \begin{bmatrix} 2 & 0 & 0 \\ 0 & 1 & 1 \\ 0 & 0 & 1 \end{bmatrix}$$

Example 4.8-6. Reduce

$$\mathbf{A} = \begin{bmatrix} 0 & 0 & 1 & 0 \\ 0 & 0 & 0 & 1 \\ 0 & 0 & 0 & 0 \\ 0 & 0 & 0 & 0 \end{bmatrix}$$

to Jordan form, and determine $\mathbf{M}$.

Evaluation of $|\lambda\mathbf{I} - \mathbf{A}|$ yields the characteristic equation $\lambda^4 = 0$. The degeneracy of $[\lambda\mathbf{I} - \mathbf{A}]$ for $\lambda = 0$ is two. Thus there are two Jordan blocks, and $\mathbf{J}$ has two "ones" in the superdiagonal. These requirements are satisfied by either

$$\mathbf{J} = \left[\begin{array}{c|ccc} 0 & 0 & 0 & 0 \\ \hline 0 & 0 & 1 & 0 \\ 0 & 0 & 0 & 1 \\ 0 & 0 & 0 & 0 \end{array}\right] \quad \text{or} \quad \mathbf{J} = \left[\begin{array}{cc|cc} 0 & 1 & 0 & 0 \\ 0 & 0 & 0 & 0 \\ \hline 0 & 0 & 0 & 1 \\ 0 & 0 & 0 & 0 \end{array}\right]$$

Equations 4.8-22 can be used on a trial and error basis to determine the correct form.

Assume that the correct Jordan form is the first one given, i.e., one consisting of (1 x 1) and (3 x 3) Jordan blocks. The first of Eqs. 4.8-22 gives

$$\begin{aligned} x_{31} &= 0 \\ x_{41} &= 0 \\ 0 &= 0 \\ 0 &= 0 \end{aligned}$$

Thus associated with the (1 x 1) Jordan block is

$$\mathbf{x}_1 = \begin{bmatrix} x_{11} \\ x_{21} \\ 0 \\ 0 \end{bmatrix}$$

where x_{11} and x_{21} cannot both be zero, but are otherwise arbitrary.

Now, considering the (3 x 3) Jordan block, the first of Eqs. 4.8-22 gives

$$\begin{aligned} x_{32} &= 0 \\ x_{42} &= 0 \\ 0 &= 0 \\ 0 &= 0 \end{aligned}$$

Hence

$$\mathbf{x}_2 = \begin{bmatrix} x_{12} \\ x_{22} \\ 0 \\ 0 \end{bmatrix}$$

where x_{12} and x_{22} cannot both be zero and must be chosen so that $\mathbf{x}_1$ and $\mathbf{x}_2$ are linearly independent.

The second of Eqs. 4.8-22 gives

$$\begin{aligned} x_{33} &= 0 + x_{12} \\ x_{43} &= 0 + x_{22} \\ 0 &= 0 + 0 \\ 0 &= 0 + 0 \end{aligned}$$

Thus

$$\mathbf{x}_3 = \begin{bmatrix} x_{13} \\ x_{23} \\ x_{12} \\ x_{22} \end{bmatrix}$$

Using the third of Eqs. 4.8-22,

$$x_{34} = 0 + x_{13}$$
$$x_{44} = 0 + x_{23}$$
$$0 = 0 + x_{12}$$
$$0 = 0 + x_{22}$$

The last two expressions violate $\mathbf{x}_2 \neq \mathbf{0}$. Hence the proper Jordan form cannot be the one consisting of (1 x 1) and (3 x 3) Jordan blocks.

Assuming now that the correct Jordan form consists of two (2 x 2) Jordan blocks, the first of Eqs. 4.8-22 yields

$$\mathbf{x}_1 = \begin{bmatrix} x_{11} \\ x_{21} \\ 0 \\ 0 \end{bmatrix}$$

where x_{11} and x_{21} cannot both be zero.

The second of Eqs. 4.8-22 gives

$$x_{32} = 0 + x_{11}$$
$$x_{42} = 0 + x_{21}$$
$$0 = 0 + 0$$
$$0 = 0 + 0$$

Thus

$$\mathbf{x}_2 = \begin{bmatrix} x_{12} \\ x_{22} \\ x_{11} \\ x_{21} \end{bmatrix}$$

where $\mathbf{x}_1$ and $\mathbf{x}_2$ must be linearly independent.

Now considering the vectors associated with the second (2 x 2) Jordan block, the first of Eqs. 4.8-22 gives

$$x_{33} = 0$$
$$x_{43} = 0$$
$$0 = 0$$
$$0 = 0$$

Thus

$$\mathbf{x}_3 = \begin{bmatrix} x_{13} \\ x_{23} \\ 0 \\ 0 \end{bmatrix}$$

where x_{13} and x_{23} must be chosen so that $\mathbf{x}_1$ and $\mathbf{x}_3$ are linearly independent.

The second of Eqs. 4.8-22 gives

$$x_{34} = 0 + x_{13}$$
$$x_{44} = 0 + x_{23}$$
$$0 = 0 + 0$$
$$0 = 0 + 0$$

Thus

$$\mathbf{x}_4 = \begin{bmatrix} x_{14} \\ x_{24} \\ x_{13} \\ x_{23} \end{bmatrix} \quad \text{and} \quad \mathbf{M} = \begin{bmatrix} x_{11} & x_{12} & x_{13} & x_{14} \\ x_{21} & x_{22} & x_{23} & x_{24} \\ 0 & x_{11} & 0 & x_{13} \\ 0 & x_{21} & 0 & x_{23} \end{bmatrix}$$

where the components of $\mathbf{x}_1$, $\mathbf{x}_2$, $\mathbf{x}_3$, and $\mathbf{x}_4$ must be chosen so that these vectors are linearly independent.

As a simple check, let

$$\mathbf{M} = \begin{bmatrix} 1 & 0 & 0 & 0 \\ 0 & 0 & 1 & 0 \\ 0 & 1 & 0 & 0 \\ 0 & 0 & 0 & 1 \end{bmatrix}$$

Then

$$\mathbf{M}^{-1} = \begin{bmatrix} 1 & 0 & 0 & 0 \\ 0 & 0 & 1 & 0 \\ 0 & 1 & 0 & 0 \\ 0 & 0 & 0 & -1 \end{bmatrix}$$

and

$$\mathbf{J} = \mathbf{M}^{-1}\mathbf{A}\mathbf{M} = \left[\begin{array}{cc|cc} 0 & 1 & 0 & 0 \\ 0 & 0 & 0 & 0 \\ \hline 0 & 0 & 0 & 1 \\ 0 & 0 & 0 & 0 \end{array}\right]$$

There are many other canonical forms which can be obtained, the Jordan form being a special case of the more general hypercompanion form. A fairly complete listing of these forms is given in the literature.[11,12] For most of the problems that the reader will face, knowledge of the Jordan form is adequate.

4.9 BILINEAR AND QUADRATIC FORMS

An expression of the form

$$\begin{aligned} B = {} & a_{11}x_1y_1 + a_{12}x_1y_2 + \cdots + a_{1n}x_1y_n \\ & + a_{21}x_2y_1 + a_{22}x_2y_2 + \cdots + a_{2n}x_2y_n \\ & + \cdots\cdots\cdots\cdots\cdots\cdots\cdots\cdots \\ & + a_{n1}x_ny_1 + a_{n2}x_ny_2 + \cdots + a_{nn}x_ny_n \end{aligned}$$

where all components are real, is called a *bilinear* form in the variables x_i, y_j. This form can be written compactly as

$$B = \sum_{i=1}^{n} \sum_{j=1}^{n} a_{ij}x_iy_j \tag{4.9-1}$$

or in matrix form as

$$B = [x_1 \quad x_2 \quad \cdots \quad x_n]\begin{bmatrix} a_{11} & a_{12} & \cdots & a_{1n} \\ a_{21} & a_{22} & \cdots & a_{2n} \\ \cdots & \cdots & \cdots & \cdots \\ a_{n1} & a_{n2} & \cdots & a_{nn} \end{bmatrix}\begin{bmatrix} y_1 \\ y_2 \\ \vdots \\ y_n \end{bmatrix} = \mathbf{x}^T\mathbf{A}\mathbf{y} = \langle \mathbf{x}, \mathbf{A}\mathbf{y} \rangle \tag{4.9-2}$$

The matrix **A** is called the coefficient matrix of the form, and the rank of **A** is called the rank of the form.

If the vector **x** is equal to the vector **y**, then Eq. 4.9-2 becomes

$$Q = \mathbf{x}^T\mathbf{A}\mathbf{x} = \langle \mathbf{x}, \mathbf{A}\mathbf{x} \rangle \tag{4.9-3}$$

Q is called a *quadratic* form in $x_1, x_2, \ldots, x_n$. An alternative expression for Q is the double summation

$$Q = \sum_{i=1}^{n}\sum_{j=1}^{n} a_{ij}x_i x_j \tag{4.9-4}$$

Note that the coefficient for the term $x_i x_j (i \neq j)$ is equal to $(a_{ij} + a_{ji})$. This coefficient would be unchanged if both a_{ij} and a_{ji} are set equal to $\frac{1}{2}(a_{ij} + a_{ji})$. Therefore the matrix **A** can be said to be a *symmetric* matrix without any loss in generality.

If the matrix **A** is a Hermitian matrix, such that $a_{ij}^* = a_{ji}$, then the corresponding Hermitian form is defined as

$$H = \mathbf{x}^{T*}\mathbf{A}\mathbf{x} = \sum_{i=1}^{n}\sum_{j=1}^{n} a_{ij}x_i^* x_j = \langle \mathbf{x}, \mathbf{A}\mathbf{x} \rangle, \quad (a_{ij}^* = a_{ji}) \tag{4.9-5}$$

The theorems which are developed for a real quadratic form have a set of analogous theorems for the case of a Hermitian form. Since the proofs of the analogous theorems require only minor changes from the proofs for the real quadratic case, the latter theorems are stated without proof.

Transformation of Variables

The linear transformation $\mathbf{x} = \mathbf{B}\mathbf{y}$, where **B** is an arbitrary $(n \times n)$ nonsingular matrix, transforms the quadratic form of Eq. 4.9-3 into a quadratic form in the variables $y_1, y_2, \ldots, y_n$. This form is

$$Q = \mathbf{y}^T\mathbf{B}^T\mathbf{A}\mathbf{B}\mathbf{y} \tag{4.9-6}$$

or

$$Q = \mathbf{y}^T\mathbf{C}\mathbf{y} \quad \text{where } \mathbf{C} = \mathbf{B}^T\mathbf{A}\mathbf{B} \tag{4.9-7}$$

Since $\mathbf{C} = \mathbf{B}^T\mathbf{A}\mathbf{B}$ is a congruent transformation, it follows that the rank of the quadratic (Hermitian) form is unchanged under a nonsingular transformation of the variables.

Reduction to Diagonal Form

In many instances, it is desirable to express Q as a linear combination of the squares of the coordinates, with no cross-terms present. The matrix **A** can be reduced to a diagonal matrix by use of the congruent transformation shown in Eq. 4.9-7. A particularly useful transformation occurs when **B** is an orthogonal matrix ($\mathbf{B}^T = \mathbf{B}^{-1}$), such that the transformation is orthogonal. As shown in the preceding section, this can be accomplished for symmetric matrices by choosing **B** to be equal to the normalized modal matrix **M**. Thus the linear transformation $\mathbf{x} = \mathbf{My}$ produces the quadratic form

$$\begin{aligned} Q &= \mathbf{y}^T\mathbf{M}^{-1}\mathbf{AMy} = \mathbf{y}^T\mathbf{\Lambda y} = \langle \mathbf{y}, \mathbf{\Lambda y}\rangle \\ &= \lambda_1 {y_1}^2 + \lambda_2 {y_2}^2 + \cdots + \lambda_n {y_n}^2 \end{aligned} \tag{4.9-8}$$

If the symmetric matrix **A** is of rank $r < n$, the modal matrix can still be formed such that the transformation shown yields a diagonal matrix with the diagonal terms equal to the characteristic values of **A**. For this situation the modal matrix **M** is not unique. There are infinitely many ways in which a set of m orthogonalized characteristic vectors corresponding to a characteristic value of order m can be chosen. Note that, if there is a zero characteristic value of order m, then there are only $n - m$ nonzero terms in the quadratic form. The matrix **A** is then of rank $r = n - m$.

Example 4.9-1. Reduce Q to a linear sum of squares, where $Q = \langle \mathbf{x}, \mathbf{Ax}\rangle$ and

$$\mathbf{A} = \begin{bmatrix} 3 & 1 & 1 \\ 1 & 0 & 2 \\ 1 & 2 & 0 \end{bmatrix}$$

The characteristic values for **A** and associated normalized characteristic vectors are given by:

$$\lambda_1 = 1, \qquad \mathbf{u}_1 = \frac{1}{\sqrt{3}}\begin{bmatrix} 1 \\ -1 \\ -1 \end{bmatrix}$$

$$\lambda_2 = 4, \qquad \mathbf{u}_2 = \frac{1}{\sqrt{6}}\begin{bmatrix} 2 \\ 1 \\ 1 \end{bmatrix}$$

$$\lambda_3 = -2, \quad \mathbf{u}_3 = \frac{1}{\sqrt{2}}\begin{bmatrix} 0 \\ 1 \\ -1 \end{bmatrix}$$

The normalized modal matrix is then

$$\mathbf{M} = \begin{bmatrix} 1/\sqrt{3} & 2/\sqrt{6} & 0 \\ -1/\sqrt{3} & 1/\sqrt{6} & 1/\sqrt{2} \\ -1/\sqrt{3} & 1/\sqrt{6} & -1/\sqrt{2} \end{bmatrix} \quad \text{and} \quad \mathbf{M}^{-1} = \begin{bmatrix} 1/\sqrt{3} & -1/\sqrt{3} & -1/\sqrt{3} \\ 2/\sqrt{6} & 1/\sqrt{6} & 1/\sqrt{6} \\ 0 & 1/\sqrt{2} & -1/\sqrt{2} \end{bmatrix}$$

The transformation

$$\begin{aligned} x_1 &= (1/\sqrt{3})y_1 + (2/\sqrt{6})y_2 & y_1 &= (1/\sqrt{3})x_1 - (1/\sqrt{3})x_2 - (1/\sqrt{3})x_3 \\ x_2 &= -(1/\sqrt{3})y_1 + (1/\sqrt{6})y_2 + (1/\sqrt{2})y_3 & y_2 &= (2/\sqrt{6})x_1 + (1/\sqrt{6})x_2 + (1/\sqrt{6})x_3 \\ x_3 &= -(1/\sqrt{3})y_1 + (1/\sqrt{6})y_2 - (1/\sqrt{2})y_3 & y_3 &= (1/\sqrt{2})x_2 - (1/\sqrt{2})x_3 \end{aligned}$$

leads to the quadratic form

$$Q = \lambda_1 y_1^2 + \lambda_2 y_2^2 + \lambda_3 y_3^2 = y_1^2 + 4y_2^2 - 2y_3^2$$

The reduction to a sum of squares can also be approached by the *Lagrange technique* of repeated completion of the square. This technique is demonstrated as follows:

$$\begin{aligned} Q &= 3x_1^2 + 2x_1x_2 + 2x_1x_3 + 4x_2x_3 \\ &= 3x_1^2 + 2x_1(x_2 + x_3) + \tfrac{1}{3}(x_2 + x_3)^2 + 4x_2x_3 - \tfrac{1}{3}(x_2 + x_3)^2 \\ &= [\sqrt{3}\,x_1 + (1/\sqrt{3})(x_2 + x_3)]^2 - \tfrac{1}{3}(x_2 - 5x_3)^2 + 8x_3^2 \end{aligned}$$

Let

$$\begin{aligned} y_1 &= \sqrt{3}\,x_1 + (1/\sqrt{3})x_2 + (1/\sqrt{3})x_3 \\ y_2 &= x_2 - 5x_3 \\ y_3 &= x_3 \end{aligned}$$

Then

$$Q = y_1^2 - \tfrac{1}{3}y_2^2 + 8y_3^2$$

The matrix **B** which performed the transformation $\mathbf{x} = \mathbf{B}\mathbf{y}$ is the triangular matrix

$$\mathbf{B} = \begin{bmatrix} 1/\sqrt{3} & -\frac{1}{3} & -2 \\ 0 & 1 & 5 \\ 0 & 0 & 1 \end{bmatrix}$$

This is not an orthogonal matrix, and therefore the new coordinate system does not have mutually orthogonal unit vectors. However, the congruent transformation $\mathbf{B}^T\mathbf{A}\mathbf{B}$ does reduce the quadratic form to a sum of squares.

The results of the preceding discussion show that a real quadratic form can be reduced by a real nonsingular transformation to the form shown in Eq. 4.9-8 or, for the case where the congruent transformation is nonorthogonal, to the form

$$Q = \alpha_1 z_1^2 + \alpha_2 z_2^2 + \cdots + \alpha_p z_p^2 - \alpha_{p+1} z_{p+1}^2 - \cdots - \alpha_n z_n^2 \quad (4.9\text{-}9)$$

The number of positive terms p is called the index of the quadratic form.

If the quadratic form is of rank r, then only r terms are present in Eq. 4.9-9. If the nonsingular transformation

$$w_i = \sqrt{\alpha_i}\, z_i \qquad i = (1, 2, \ldots, r)$$
$$w_i = z_i \qquad i = (r + 1, \ldots, n)$$

is introduced, then Eq. 4.9-9 becomes

$$Q = w_1^2 + w_2^2 + \cdots + w_p^2 - w_{p+1}^2 - \cdots - w_r^2 \qquad (4.9\text{-}10)$$

Equation 4.9-10 can be viewed as a direct consequence of a transformation to the canonical form of Eq. 4.8-15.

In a similar manner, a Hermitian form of rank r can be reduced to the diagonal forms

$$Q = \alpha_1 z_1^* z_1 + \alpha_2 z_2^* z_2 + \cdots + \alpha_r z_r^* z_r = w_1^* w_1 + w_2^* w_2$$
$$+ \cdots + w_p^* w_p - w_{p+1}^* w_{p+1} - \cdots - w_r^* w_r \qquad (4.9\text{-}11)$$

The latter form follows from the definition of the canonical matrix of Eq. 4.8-18.

Definite and Semidefinite Forms

The quadratic form $Q = \langle \mathbf{x}, \mathbf{Ax} \rangle$ is said to be *positive definite* if it is non-negative for all real values of $\mathbf{x}$, and is zero only when the vector $\mathbf{x}$ is a null vector. For these conditions to be satisfied, it is clear from Eq. 4.9-10 that $\mathbf{A}$ must be a nonsingular matrix, and that the index (number of positive terms) and rank of the quadratic form must be equal. From Eq. 4.9-8, it is evident that a quadratic form is positive definite if and only if the characteristic values of the nonsingular matrix $\mathbf{A}$ are all positive. Either of these conditions can be used to define a positive definite quadratic form.

If the quadratic form $\langle \mathbf{x}, \mathbf{Ax} \rangle$ is positive definite, the matrix $\mathbf{A}$ is also said to be positive definite. A real symmetric matrix is positive definite if and only if there exists a nonsingular matrix $\mathbf{C}$ such that

$$\mathbf{A} = \mathbf{C}^T\mathbf{C} \qquad (4.9\text{-}12)$$

To show this, let $\mathbf{A}$ be reduced to a unit matrix by a congruent transformation. Hence $\mathbf{A}$ can be written as $\mathbf{A} = \mathbf{B}^T\mathbf{I}_n\mathbf{B}$. Since $\mathbf{I}_n^2 = \mathbf{I}_n$ and $\mathbf{I}_n^T = \mathbf{I}_n$, $\mathbf{A} = (\mathbf{B}^T\mathbf{I}_n^T)(\mathbf{I}_n\mathbf{B})$. Let $\mathbf{C} = \mathbf{I}_n\mathbf{B}$. Then $\mathbf{A} = \mathbf{C}^T\mathbf{C}$. These steps are also valid when **A** is reduced to the canonical form

$$\begin{bmatrix} \mathbf{I}_r & \mathbf{0} \\ \mathbf{0} & \mathbf{0} \end{bmatrix}$$

by a congruent transformation. In the latter case, **C** is of rank r, and **A** is positive semidefinite.

The quadratic form is called *positive semidefinite* if it is non-negative. It can be zero when the vector **x** is not zero. This case arises when **A** is singular. From Eq. 4.7-8 it follows that, if **A** is singular and of rank r, then it must possess $n - r$ characteristic roots equal to zero. There are then $n - r$ terms of Eq. 4.9-8 which are identically zero, even when the associated y_i components are nonzero.

Analogous to the definitions above, the quadratic form may be negative definite and negative semidefinite. The conditions required for these forms, as well as the corresponding Hermitian cases, are listed in Table 4.9-1.

Determination of Positive Definiteness by Use of the Principal Minors

It is advantageous to be able to determine whether or not a quadratic form is positive definite without solving the characteristic value problem, or reducing **A** to a canonical form. This can be done by examination of the leading principal minors of **A**. A principal minor of **A** is a minor of **A** whose diagonal elements are also diagonal elements of **A**. The mth *leading principal minor* of **A**, denoted by Δ_m, is defined as the determinant obtained by deleting the last $n - m$ columns and rows of **A**. Since **A** is symmetric, the leading principal minors are

$$\Delta_1 = a_{11}, \quad \Delta_2 = \begin{vmatrix} a_{11} & a_{12} \\ a_{12} & a_{22} \end{vmatrix}, \quad \Delta_3 = \begin{vmatrix} a_{11} & a_{12} & a_{13} \\ a_{12} & a_{22} & a_{23} \\ a_{13} & a_{23} & a_{33} \end{vmatrix}, \quad \ldots, \quad \Delta_n = |\mathbf{A}| \tag{4.9-13}$$

These leading principal minors of **A** are also called the *discriminants* of the quadratic form.

It is now shown that a real quadratic form is positive definite if and only if *all* the leading principal minors of **A** are positive. The starting point of this proof is Eq. 4.9-12, $\mathbf{A} = \mathbf{C}^T\mathbf{C}$. Without any loss in generality, let

$$\mathbf{C} = \mathbf{\Lambda}^{1/2}\mathbf{D} \tag{4.9-14}$$

Table 4.9-1

	$\|\mathbf{A}\|$	Rank and Index	Characteristic Values	$\mathbf{A}$	Form	Leading Principal Minors
			Positive Definite			
Real quadratic form	$\|\mathbf{A}\| > 0$	$r = p = n$	$\lambda_i > 0$	$\mathbf{A} = \mathbf{C}^T\mathbf{C}$ $\|\mathbf{C}\| > 0$	$Q = y_1^2 + y_2^2 + \cdots + y_n^2$	$\Delta_1, \Delta_2, \cdots, \Delta_n$ all positive
Hermitian form				$\mathbf{A} = \mathbf{C}^{*T}\mathbf{C}$ $\|\mathbf{C}\| > 0$	$Q = z_1^*z_1 + z_2^*z_2 + \cdots + z_n^*z_n$	
			Positive Semidefinite			
Real quadratic form	$\|\mathbf{A}\| = 0$	$r = p < n$	$n - r$ roots equal zero; others positive	$\mathbf{A} = \mathbf{C}^T\mathbf{C}$ $\|\mathbf{C}\| = 0$	$Q = y_1^2 + y_2^2 + \cdots + y_r^2$	$\Delta_1, \Delta_2, \ldots, \Delta_r$ all positive; $\Delta_{r+1}, \ldots, \Delta_{n-1}$ non-negative
Hermitian form				$\mathbf{A} = \mathbf{C}^{*T}\mathbf{C}$ $\|\mathbf{C}\| = 0$	$Q = z_1^*z_1 + z_2^*z_2 + \cdots + z_r^*z_r$	
			Negative Definite			
Real quadratic form	$(-1)^n \|\mathbf{A}\| > 0$	$r = n$ $p = 0$	$\lambda_i < 0$	$\mathbf{A} = -\mathbf{C}^T\mathbf{C}$ $\|\mathbf{C}\| > 0$	$Q = -y_1^2 - y_2^2 - \cdots - y_n^2$	$-\Delta_1, \Delta_2, \ldots, (-1)^n\Delta_n$ all positive
Hermitian form				$A = -\mathbf{C}^{*T}\mathbf{C}$ $\|\mathbf{C}\| > 0$	$Q = -z_1^*z_1 - z_2^*z_2 - \cdots - z_n^*z_n$	
			Negative Semidefinite			
Real quadratic form	$\|\mathbf{A}\| = 0$	$r < n$ $p = 0$	$n - r$ roots equal zero; others negative	$\mathbf{A} = -\mathbf{C}^T\mathbf{C}$ $\|\mathbf{C}\| = 0$	$Q = -y_1^2 - y_2^2 - \cdots - y_r^2$	$-\Delta_1, \Delta_2, \ldots, (-1)^r\Delta_r$ all positive; $(-1)^r\Delta_{r+1}, \ldots, (-1)^n\Delta_{n-1}$ nonpositive
Hermitian form				$\mathbf{A} = -\mathbf{C}^{*T}\mathbf{C}$ $\|\mathbf{C}\| = 0$	$Q = -z_1^*z_1 - z_2^*z_2 - \cdots - z_r^*z_r$	

where $\boldsymbol{\Lambda}$ is a diagonal matrix whose diagonal elements are $\lambda_1, \lambda_2, \ldots, \lambda_n$, and $\mathbf{D}$ is a triangular matrix of the form

$$\mathbf{D} = \begin{bmatrix} 1 & d_{12} & d_{13} & \cdots & d_{1n} \\ 0 & 1 & d_{23} & \cdots & d_{2n} \\ 0 & 0 & 1 & \cdots & d_{3n} \\ \cdots & \cdots & \cdots & \cdots & \cdots \\ 0 & 0 & 0 & \cdots & 1 \end{bmatrix}$$

Substitution of Eq. 4.9-14 into Eq. 4.9-12 yields

$$\mathbf{A} = \mathbf{D}^T\boldsymbol{\Lambda}\mathbf{D}$$

$$= \begin{bmatrix} 1 & 0 & 0 & \cdots & 0 \\ d_{12} & 1 & 0 & \cdots & 0 \\ d_{13} & d_{23} & 1 & \cdots & 0 \\ \cdots & \cdots & \cdots & \cdots & \cdots \\ d_{1n} & d_{2n} & d_{3n} & \cdots & 1 \end{bmatrix} \begin{bmatrix} \lambda_1 & 0 & 0 & \cdots & 0 \\ 0 & \lambda_2 & 0 & \cdots & 0 \\ 0 & 0 & \lambda_3 & \cdots & 0 \\ \cdots & \cdots & \cdots & \cdots & \cdots \\ 0 & 0 & 0 & \cdots & \lambda_n \end{bmatrix}$$

$$\times \begin{bmatrix} 1 & d_{12} & d_{13} & \cdots & d_{1n} \\ 0 & 1 & d_{23} & \cdots & d_{2n} \\ 0 & 0 & 1 & \cdots & d_{3n} \\ \cdots & \cdots & \cdots & \cdots & \cdots \\ 0 & 0 & 0 & \cdots & 1 \end{bmatrix} \tag{4.9-15}$$

If a new variable $\mathbf{y}$ is defined by the linear transformation

$$\mathbf{y} = \mathbf{D}\mathbf{x} \tag{4.9-16}$$

then the real quadratic form $Q = \langle \mathbf{x}, \mathbf{A}\mathbf{x} \rangle$ can be expressed as

$$Q = \langle \mathbf{y}, \boldsymbol{\Lambda}\mathbf{y} \rangle = \lambda_1 y_1^2 + \lambda_2 y_2^2 + \cdots + \lambda_n y_n^2 \tag{4.9-17}$$

The real quadratic form Q is positive definite if and only if all the diagonal elements $\lambda_1, \lambda_2, \ldots, \lambda_n$ are positive. However, since $\mathbf{D}$ has been chosen to be a triangular matrix with unity diagonal elements, the determinant of $\mathbf{D}$ and $\mathbf{D}^T$ is equal to unity. Hence

$$|\mathbf{A}| = |\mathbf{D}^T|\,|\boldsymbol{\Lambda}|\,|\mathbf{D}| = |\boldsymbol{\Lambda}| = \lambda_1\lambda_2\cdots\lambda_n \tag{4.9-18}$$

If the variable x_n is set equal to zero, then the variable y_n is also zero. The quadratic form obtained by setting $x_n = 0$ is then

$$Q^1 = \lambda_1 y_1^2 + \lambda_2 y_2^2 + \cdots + \lambda_{n-1} y_{n-1}^2$$

By a similar argument, the discriminant of this quadratic form Δ_{n-1} is equal to

$$\Delta_{n-1} = \lambda_1\lambda_2\cdots\lambda_{n-1} \tag{4.9-19}$$

In general, the discriminant Δ_k, obtained by setting $x_{k+1} = x_{k+2} = \cdots = x_n = 0$ is

$$\Delta_k = \lambda_1 \lambda_2 \cdots \lambda_k \tag{4.9-20}$$

Solving for the elements $\lambda_1, \lambda_2, \ldots, \lambda_n$,

$$\begin{aligned} \lambda_1 &= \Delta_1 \\ \lambda_2 &= \frac{\Delta_2}{\Delta_1} \\ \lambda_3 &= \frac{\Delta_3}{\Delta_2} \\ &\cdots\cdots \\ \lambda_n &= \frac{\Delta_n}{\Delta_{n-1}} \end{aligned} \tag{4.9-21}$$

Clearly, if all the elements λ_i are to be positive, then all the *leading principal minors* $\Delta_1, \Delta_2, \ldots, \Delta_n$ must be positive. Therefore a real quadratic form $Q = \langle \mathbf{x}, \mathbf{Ax} \rangle$ is positive definite if and only if all the leading principal minors of $\mathbf{A}$ are positive.

For the case of Hermitian forms, a set of analogous statements can be made. A summary of the useful statements regarding real quadratic and Hermitian forms is given in Table 4.9-1. The statements about the negative definite forms can be proved by requiring $(-\mathbf{A})$ to be the matrix of a positive definite form.

4.10 MATRIX POLYNOMIALS, INFINITE SERIES, AND FUNCTIONS OF A MATRIX

In this section some of the basic ideas regarding matrix polynomials and infinite series are developed. With the introduction of a few modifications, the theorems regarding matrix polynomials and infinite series are directly analogous to the theorems of scalar variables. In addition, some important theorems regarding functions of a matrix are presented. These theorems are essential to the solution of linear vector-matrix differential equations.

Powers of Matrices

The matrix product $\mathbf{AAA} \cdots \mathbf{A}$, where $\mathbf{A}$ is a square matrix of order n, can be written as $\mathbf{A}^k$, where k is the number of factors involved in the product. The multiplication of powers of a matrix follows the usual rules for scalar algebra. The matrix $\mathbf{A}^0$ is defined as the unit matrix of order n.

$$\begin{aligned} \mathbf{A}^k\mathbf{A}^m &= \mathbf{A}^{k+m} \\ (\mathbf{A}^k)^m &= \mathbf{A}^{km} \\ \mathbf{A}^0 &= \mathbf{I}_n \end{aligned} \tag{4.10-1}$$

If the power to which the matrix is to be raised is negative, the same rules apply if the matrix is nonsingular, so that its inverse exists. That is,

$$(\mathbf{A}^{-1})^m = \mathbf{A}^{-m} \tag{4.10-2}$$

A set of similar rules applies in the case where a fractional power of a matrix is to be computed. Thus, if $\mathbf{A}^m = \mathbf{B}$, where $\mathbf{A}$ is a square matrix, then $\mathbf{B}^{1/m}$ is an mth root of $\mathbf{A}$. The number of mth roots of a matrix depends upon the nature of the matrix, there being no general rule as to how many mth roots the matrix $\mathbf{A}$ possesses.

Example 4.10-1. Find the square root of $\mathbf{A}$, where

$$\mathbf{A} = \begin{bmatrix} a_{11} & a_{12} \\ a_{21} & a_{22} \end{bmatrix}$$

Let

$$\mathbf{B}^2 = \begin{bmatrix} b_{11} & b_{12} \\ b_{21} & b_{22} \end{bmatrix}^2 = \mathbf{A}$$

Then

$$\begin{bmatrix} b_{11}^2 + b_{12}b_{21} & b_{12}(b_{11} + b_{22}) \\ b_{21}(b_{11} + b_{22}) & b_{12}b_{21} + b_{22}^2 \end{bmatrix} = \begin{bmatrix} a_{11} & a_{12} \\ a_{21} & a_{22} \end{bmatrix}$$

Equating like terms yields

$$\begin{aligned} b_{11}^2 + b_{12}b_{21} &= a_{11} & b_{12}(b_{11} + b_{22}) &= a_{12} \\ b_{21}(b_{11} + b_{22}) &= a_{21} & b_{12}b_{21} + b_{22}^2 &= a_{22} \end{aligned}$$

This is a set of four nonlinear simultaneous equations which has no general solution. A pair of numerical examples illustrate the ambiguity involved.

Let

$$\mathbf{A} = \begin{bmatrix} 4 & 1 \\ 0 & 1 \end{bmatrix}$$

Then $b_{21} = 0$, $b_{11} = \pm 2$, $b_{22} = \pm 1$, and $b_{12} = \pm\frac{1}{3}$. The square root of $\mathbf{A}$ is then

$$\mathbf{B} = \pm \begin{bmatrix} 2 & \frac{1}{3} \\ 0 & 1 \end{bmatrix}$$

As a second example, let

$$\mathbf{A} = \begin{bmatrix} 4 & 0 \\ 0 & 4 \end{bmatrix}$$

One possible answer to this problem is

$$\mathbf{B} = \pm \begin{bmatrix} 2 & 0 \\ 0 & 2 \end{bmatrix}$$

However, this is not the only solution. Another solution is

$$\mathbf{B} = \begin{bmatrix} b_{11} & b_{12} \\ b_{21} & -b_{11} \end{bmatrix}$$

with $b_{11}^2 + b_{12}b_{21} = 4$. Therefore there are an infinite number of square roots of **A**.

Matrix Polynomials

Consider a polynomial of order n, where the argument of the polynomial is the scalar variable x, i.e.,

$$N(x) = p_n x^n + p_{n-1}x^{n-1} + \cdots + p_1 x + p_0 \tag{4.10-3}$$

If the scalar variable x is replaced by the $(n \times n)$ square matrix **A**, then the corresponding matrix polynomial is defined by

$$\mathbf{N}(\mathbf{A}) = p_n \mathbf{A}^n + p_{n-1}\mathbf{A}^{n-1} + \cdots + p_1 \mathbf{A} + p_0 \mathbf{I}_n \tag{4.10-4}$$

Note that the last term is multiplied by the nth order unit matrix $\mathbf{I}_n$.

Example 4.10-2. Let $N(x) = 3x^2 + 2x + 1$ and

$$\mathbf{A} = \begin{bmatrix} 2 & 1 \\ 0 & 2 \end{bmatrix}$$

Determine $\mathbf{N}(\mathbf{A})$.

$$\mathbf{N}(\mathbf{A}) = 3\mathbf{A}^2 + 2\mathbf{A} + \mathbf{I}$$

$$= 3\begin{bmatrix} 4 & 4 \\ 0 & 4 \end{bmatrix} + 2\begin{bmatrix} 2 & 1 \\ 0 & 2 \end{bmatrix} + \begin{bmatrix} 1 & 0 \\ 0 & 1 \end{bmatrix} = \begin{bmatrix} 17 & 14 \\ 0 & 17 \end{bmatrix}$$

Factorization of a Matrix Polynomial

The polynomial $N(x)$ may be written in the factored form

$$N(x) = p_n(x - \lambda_1)(x - \lambda_2)\cdots(x - \lambda_n)$$

where $\lambda_1, \lambda_2, \ldots, \lambda_n$ are the roots of the polynomial $N(x) = 0$ and are all assumed to be distinct. Similarly, the factored form of a matrix polynomial is

$$\mathbf{N}(\mathbf{A}) = p_n(\mathbf{A} - \lambda_1\mathbf{I})(\mathbf{A} - \lambda_2\mathbf{I})\cdots(\mathbf{A} - \lambda_n\mathbf{I}) \tag{4.10-5}$$

This form is used later to prove Sylvester's theorem.

Infinite Series of Matrices

Consider the infinite series in the scalar variable x,

$$S(x) = a_0 + a_1 x + a_2 x^2 + \cdots = \sum_{k=0}^{\infty} a_k x^k$$

If the argument of the infinite series is replaced by the square nth order matrix $\mathbf{A}$, then the infinite series of $\mathbf{A}$ can be written as

$$\mathbf{S}(\mathbf{A}) = a_0\mathbf{I}_n + a_1\mathbf{A} + a_2\mathbf{A}^2 + \cdots = \sum_{k=0}^{\infty} a_k\mathbf{A}^k \tag{4.10-6}$$

It may be shown that this series converges as k approaches infinity if all the corresponding scalar series $S(\lambda_i)$, $i = 1, 2, \ldots, n$, converge, where the λ_i's are the characteristic values of $\mathbf{A}$. Because the topic of convergence of matrix series is in general rather involved, a detailed discussion of it is omitted.

Some of the more important infinite series of matrices are as follows:

Geometric series:

$$\mathbf{G}(\mathbf{A}) = \mathbf{I} + a\mathbf{A} + a^2\mathbf{A}^2 + \cdots = \sum_{k=0}^{\infty} a^k\mathbf{A}^k \tag{4.10-7}$$

Exponential function:

$$\begin{aligned} \epsilon^{\mathbf{A}} &= \exp \mathbf{A} = \mathbf{I} + \mathbf{A} + \frac{\mathbf{A}^2}{2} + \frac{\mathbf{A}^3}{3!} + \cdots + \frac{\mathbf{A}^k}{k!} + \cdots \\ \epsilon^{-\mathbf{A}} &= \exp(-\mathbf{A}) = \mathbf{I} - \mathbf{A} + \frac{\mathbf{A}^2}{2!} - \frac{\mathbf{A}^3}{3!} + \cdots + \frac{(-1)^k\mathbf{A}^k}{k!} + \cdots \end{aligned} \tag{4.10-8}$$

It can be shown that this series is absolutely and uniformly convergent.[13] Although the scalar multiplication $\epsilon^x\epsilon^y$ can be written as either $\epsilon^x\epsilon^y$ or $\epsilon^y\epsilon^x$, the corresponding matrix product $\epsilon^{\mathbf{A}}\epsilon^{\mathbf{B}}$ cannot be written as $\epsilon^{\mathbf{B}}\epsilon^{\mathbf{A}}$ unless $\mathbf{A}$ and $\mathbf{B}$ commute. Then

$$\epsilon^{\mathbf{A}}\epsilon^{\mathbf{B}} = \epsilon^{\mathbf{B}}\epsilon^{\mathbf{A}} = \epsilon^{\mathbf{A}+\mathbf{B}}, \quad \text{if} \quad \mathbf{AB} = \mathbf{BA} \tag{4.10-9}$$

Clearly, this condition is satisfied if $\mathbf{B} = \mathbf{A}$, or $\mathbf{B} = -\mathbf{A}$. If $\mathbf{B} = -\mathbf{A}$, then Eq. 4.10-9 becomes

$$\epsilon^{\mathbf{A}}\epsilon^{-\mathbf{A}} = \epsilon^{(\mathbf{A}-\mathbf{A})} = \mathbf{I} \tag{4.10-10}$$

From Eq. 4.10-10 it can be concluded that $(\epsilon^{\mathbf{A}})^{-1} = \epsilon^{-\mathbf{A}}$, or that $\epsilon^{-\mathbf{A}}$ is the inverse matrix of $\epsilon^{\mathbf{A}}$.

Sine function:

$$\sin \mathbf{A} = \mathbf{A} - \frac{\mathbf{A}^3}{3!} + \frac{\mathbf{A}^5}{5!} - \cdots = \frac{\exp[j\mathbf{A}] - \exp[-j\mathbf{A}]}{2j} \tag{4.10-11}$$

Cosine function:

$$\cos \mathbf{A} = \mathbf{I} - \frac{\mathbf{A}^2}{2!} + \frac{\mathbf{A}^4}{4!} - \cdots = \frac{\exp[j\mathbf{A}] + \exp[-j\mathbf{A}]}{2} \tag{4.10-12}$$

where the complex exponential is defined by setting $\mathbf{A}$ equal to $j\mathbf{A}$ in Eq. 4.10-8.

$$\exp(j\mathbf{A}) = \left(\mathbf{I} - \frac{\mathbf{A}^2}{2!} + \frac{\mathbf{A}^4}{4!} - \cdots\right) + j\left(\mathbf{A} - \frac{\mathbf{A}^3}{3!} + \frac{\mathbf{A}^5}{5!} - \cdots\right)$$

$$= \cos\mathbf{A} + j\sin\mathbf{A} \tag{4.10-13}$$

$$\exp(-j\mathbf{A}) = \cos\mathbf{A} - j\sin\mathbf{A}$$

Hyperbolic sine:

$$\sinh\mathbf{A} = \mathbf{A} + \frac{\mathbf{A}^3}{3!} + \frac{\mathbf{A}^5}{5!} + \cdots = \frac{\exp\mathbf{A} - \exp(-\mathbf{A})}{2} \tag{4.10-14}$$

Hyperbolic cosine:

$$\cosh\mathbf{A} = \mathbf{I} + \frac{\mathbf{A}^2}{2!} + \frac{\mathbf{A}^4}{4!} + \cdots = \frac{\exp\mathbf{A} + \exp(-\mathbf{A})}{2} \tag{4.10-15}$$

Trigonometric Identities

The matrix trigonometric identities, which are analogous to the corresponding scalar trigonometric identities, can be established by use of the preceding definitions of the matrix trigonometric functions. For example, the identity

$$\cosh^2\mathbf{A} - \sinh^2\mathbf{A} = \mathbf{I} \tag{4.10-16}$$

can be derived from

$$\cosh^2\mathbf{A} = \left(\frac{\exp\mathbf{A} + \exp(-\mathbf{A})}{2}\right)\left(\frac{\exp\mathbf{A} + \exp(-\mathbf{A})}{2}\right)$$

$$= \frac{\exp 2\mathbf{A} + \exp(-2\mathbf{A})}{4} + \frac{\mathbf{I}}{2}$$

$$\sinh^2\mathbf{A} = \left(\frac{\exp\mathbf{A} - \exp(-\mathbf{A})}{2}\right)\left(\frac{\exp\mathbf{A} - \exp(-\mathbf{A})}{2}\right)$$

$$= \frac{\exp 2\mathbf{A} + \exp(-2\mathbf{A})}{4} - \frac{\mathbf{I}}{2}$$

The (2 x 2) real matrix analogous to the scalar $j = \sqrt{-1}$ is defined by

$$\mathbf{J}_0 = \begin{bmatrix} 0 & -1 \\ 1 & 0 \end{bmatrix} \tag{4.10-17}$$

Note that $\mathbf{J}_0^2 = -\mathbf{I}$, $\mathbf{J}_0^3 = -\mathbf{J}_0$, $\mathbf{J}_0^4 = \mathbf{I}, \ldots$, etc. This matrix is useful in finding certain trigonometric identities. For example, if $\mathbf{A} = a\mathbf{J}_0$, then Eq. 4.10-11 can be written as

$$\sin(a\mathbf{J}_0) = a\mathbf{J}_0 + \frac{a^3\mathbf{J}_0}{3!} + \frac{a^5}{5!}\mathbf{J}_0 + \cdots = \mathbf{J}_0\sinh a \tag{4.10-18}$$

Similarly,

$$\cos(a\mathbf{J}_0) = \mathbf{I}\cosh a \tag{4.10-19}$$

Cayley-Hamilton Theorem

The generalization of Eq. 4.7-18 to $\mathbf{A}^p = \mathbf{M}\mathbf{\Lambda}^p\mathbf{M}^{-1}$ produces an interesting and useful relationship. If $N(\lambda)$ is a polynomial in λ of the form

$$N(\lambda) = \lambda^n + c_1\lambda^{n-1} + \cdots + c_{n-1}\lambda + c_n$$

then this generalization shows that the polynomial using $\mathbf{A}$ as the variable is

$$\mathbf{N}(\mathbf{A}) = \mathbf{A}^n + c_1\mathbf{A}^{n-1} + \cdots + c_{n-1}\mathbf{A} + c_n\mathbf{I} = \mathbf{M}\mathbf{N}(\mathbf{\Lambda})\mathbf{M}^{-1}$$

$$= \mathbf{M}\begin{bmatrix} N(\lambda_1) & & & & \\ & N(\lambda_2) & & & \\ & & \cdot & & \\ & & & \cdot & \\ & & & & \cdot \\ & & & & N(\lambda_n) \end{bmatrix}\mathbf{M}^{-1} \qquad (4.10\text{-}20)$$

where $\lambda_1, \lambda_2, \ldots, \lambda_n$ are the zeros of the polynomial $N(\lambda)$.

If the polynomial chosen is the characteristic polynomial, i.e., if $N(\lambda) = P(\lambda)$, then $N(\lambda_1) = N(\lambda_2) = \cdots = N(\lambda_n) = 0$. It follows that

$$\mathbf{P}(\mathbf{A}) = [0] \quad \text{where} \quad P(\lambda) = |\lambda\mathbf{I} - \mathbf{A}| \qquad (4.10\text{-}21)$$

This statement is known as the *Cayley-Hamilton theorem*. The theorem states that "the matrix **A** satisfies its own characteristic equation." The preceding proof is based on the assumption that **A** has distinct characteristic roots. However, it can be shown that this theorem holds true for any *square* matrix.[14] This theorem is of considerable importance when calculating various functions of the matrix **A**.

Example 4.10-3. The matrix

$$\mathbf{A} = \begin{bmatrix} 0 & 1 \\ -2 & -3 \end{bmatrix}$$

has the characteristic polynomial

$$P(\lambda) = \lambda^2 + 3\lambda + 2$$

Show that $\mathbf{P}(\mathbf{A}) = [0]$.

Substituting **A** for the variable λ gives

$$\mathbf{P}(\mathbf{A}) = \mathbf{A}^2 + 3\mathbf{A} + 2\mathbf{I}$$

$$= \begin{bmatrix} -2 & -3 \\ 6 & 7 \end{bmatrix} + 3\begin{bmatrix} 0 & 1 \\ -2 & -3 \end{bmatrix} + 2\begin{bmatrix} 1 & 0 \\ 0 & 1 \end{bmatrix} = \begin{bmatrix} 0 & 0 \\ 0 & 0 \end{bmatrix}$$

Therefore $\mathbf{P}(\mathbf{A}) = \text{null matrix} = [0]$.

Example 4.10-4. Find $\mathbf{A}^{-1}$ for the matrix of the preceding example by using the Cayley-Hamilton theorem.

Since **A** satisfies its own characteristic equation,

$$\mathbf{A}^2 + 3\mathbf{A} + 2\mathbf{I} = [0]$$

or

$$\mathbf{A} + 3\mathbf{I} + 2\mathbf{A}^{-1} = [0]$$

Therefore

$$\mathbf{A}^{-1} = -\tfrac{1}{2}\mathbf{A} - \tfrac{3}{2}\mathbf{I} = \begin{bmatrix} -\frac{3}{2} & -\frac{1}{2} \\ 1 & 0 \end{bmatrix}$$

This is often a convenient way of computing the inverse of a matrix.

Reduction of Polynomials

By means of the Cayley-Hamilton theorem, it is possible to reduce any polynomial of the nth order matrix **A** to a linear combination of **I**, **A**, $\mathbf{A}^2, \ldots, \mathbf{A}^{n-1}$, or a polynomial whose highest degree in **A** is $n - 1$. This is best shown by example.

Example 4.10-5. Find $\mathbf{N}(\mathbf{A}) = \mathbf{A}^4 + \mathbf{A}^3 + \mathbf{A}^2 + \mathbf{A} + \mathbf{I}$ if

$$\mathbf{A} = \begin{bmatrix} 0 & 1 \\ -2 & -3 \end{bmatrix}$$

The characteristic equation of **A** is $\lambda^2 + 3\lambda + 2 = 0$. Therefore

$$\mathbf{A}^2 + 3\mathbf{A} + 2\mathbf{I} = [0] \quad \text{or} \quad \mathbf{A}^2 = -3\mathbf{A} - 2\mathbf{I}$$

Consequently,

$$\mathbf{A}^4 = 9\mathbf{A}^2 + 12\mathbf{A} + 4\mathbf{I} = 9(-3\mathbf{A} - 2\mathbf{I}) + 12\mathbf{A} + 4\mathbf{I} = -15\mathbf{A} - 14\mathbf{I}$$

Similarly,

$$\mathbf{A}^3 = -3\mathbf{A}^2 - 2\mathbf{A} = -3(-3\mathbf{A} - 2\mathbf{I}) - 2\mathbf{A} = 7\mathbf{A} + 6\mathbf{I}$$

Hence

$$\mathbf{N}(\mathbf{A}) = (-15\mathbf{A} - 14\mathbf{I}) + (7\mathbf{A} + 6\mathbf{I}) + (-3\mathbf{A} - 2\mathbf{I}) + \mathbf{A} + \mathbf{I}$$

$$= -10\mathbf{A} - 9\mathbf{I} = \begin{bmatrix} -9 & -10 \\ 20 & 21 \end{bmatrix}$$

This is a polynomial of first degree in **A**.

Sylvester's Theorem†

Sylvester's theorem is a useful method for obtaining a function of a matrix, if the function can be expressed as a matrix polynomial. The following is a statement of Sylvester's theorem, valid when **A** possesses n distinct roots:

If **N**(**A**) is a matrix polynomial in **A**, and if the square matrix **A** possesses

† The proof of this theorem closely follows the proof given in Reference 15.

n distinct characteristic values, the polynomial in $\mathbf{A}$ can be written as

$$\mathbf{N(A)} = \sum_{i=1}^{n} N(\lambda_i)\mathbf{Z}_0(\lambda_i) \tag{4.10-22}$$

where

$$\mathbf{Z}_0(\lambda_i) = \frac{\prod_{\substack{j=1\\ j\neq i}}^{n} (\mathbf{A} - \lambda_j\mathbf{I})}{\prod_{\substack{j=1\\ j\neq i}}^{n} (\lambda_i - \lambda_j)}$$

From the Cayley-Hamilton theorem, it is known that any matrix polynomial $\mathbf{N(A)}$ can be represented by a polynomial in $\mathbf{A}$, whose highest degree is $n - 1$. Thus $\mathbf{N(A)}$ can be written as

$$\mathbf{N(A)} = a_1\mathbf{A}^{n-1} + a_2\mathbf{A}^{n-2} + \cdots + a_{n-1}\mathbf{A} + a_n\mathbf{I} \tag{4.10-23}$$

In order to prove Sylvester's theorem, Eq. 4.10-23 is written in the form

$$\begin{aligned}\mathbf{N(A)} = {} & c_1[(\mathbf{A} - \lambda_2\mathbf{I})(\mathbf{A} - \lambda_3\mathbf{I}) \cdots (\mathbf{A} - \lambda_n\mathbf{I})] \\ & + c_2[(\mathbf{A} - \lambda_1\mathbf{I})(\mathbf{A} - \lambda_3\mathbf{I}) \cdots (\mathbf{A} - \lambda_n\mathbf{I})] \\ & + \cdots\cdots\cdots\cdots\cdots\cdots \\ & + c_k \prod_{\substack{j=1\\ j\neq k}}^{n} (\mathbf{A} - \lambda_j\mathbf{I}) \\ & + \cdots\cdots\cdots\cdots\cdots\cdots \\ & + c_n[(\mathbf{A} - \lambda_1\mathbf{I})(\mathbf{A} - \lambda_2\mathbf{I}) \cdots (\mathbf{A} - \lambda_{n-1}\mathbf{I})]\end{aligned}$$

or

$$\mathbf{N(A)} = \sum_{k=1}^{n} c_k \prod_{\substack{j=1\\ j\neq k}}^{n} (\mathbf{A} - \lambda_j\mathbf{I}) \tag{4.10-24}$$

Since there is one factor missing from each of the product terms, $\mathbf{N(A)}$ is clearly a polynomial of degree $n - 1$, with n arbitrary constants of combination. If the characteristic vectors of $\mathbf{A}$ are denoted by $\mathbf{u}_1, \mathbf{u}_2, \ldots, \mathbf{u}_n$, then postmultiplying Eq. 4.10-24 by $\mathbf{u}_i$ yields the relation

$$\mathbf{N(A)u}_i = \left[\sum_{k=1}^{n} c_k \prod_{\substack{j=1\\ j\neq k}}^{n} (\mathbf{A} - \lambda_j\mathbf{I})\right]\mathbf{u}_i \tag{4.10-25}$$

However, since $\mathbf{Au}_i = \lambda_i\mathbf{u}_i$ or $(\mathbf{A} - \lambda_i\mathbf{I})\mathbf{u}_i = \mathbf{0}$, all the terms except the ith are zero. The ith term is not zero, since it does not contain the factor $(\mathbf{A} - \lambda_i\mathbf{I})$. Therefore

$$\mathbf{N(A)u}_i = \left[c_i \prod_{\substack{j=1\\ j\neq 1}}^{n} (\mathbf{A} - \lambda_j\mathbf{I})\right]\mathbf{u}_i = \left[c_i \prod_{\substack{j=1\\ j\neq 1}}^{n} (\lambda_i - \lambda_j)\right]\mathbf{u}_i \tag{4.10-26}$$

If the characteristic values of $\mathbf{A}$ are distinct, then $\mathbf{N(A)u}_i = N(\lambda_i)\mathbf{u}_i$. Therefore

$$c_i = \frac{N(\lambda_i)}{\prod_{\substack{j=1 \\ j \neq 1}}^{n} (\lambda_i - \lambda_j)}$$

Consequently,

$$\mathbf{N(A)} = \sum_{i=1}^{n} N(\lambda_i) \frac{\prod_{\substack{j=1 \\ i \neq j}}^{n} (\mathbf{A} - \lambda_j \mathbf{I})}{\prod_{\substack{j=1 \\ i \neq j}}^{n} (\lambda_i - \lambda_j)}$$

This concludes the proof of Sylvester's theorem.

Example 4.10-6. Calculate $\epsilon^{\mathbf{A}}$, using Sylvester's theorem, for the $\mathbf{A}$ matrix of Example 4.10-5.

Since $\epsilon^{\mathbf{A}}$ can be expressed as a convergent series in $\mathbf{A}$, Sylvester's theorem can be used directly on $\epsilon^{\mathbf{A}}$, rather than on the infinite series representation of $\mathbf{A}$. Certainly, if the infinite series for $\epsilon^{\mathbf{A}}$ converges, then $\epsilon^{\mathbf{A}}$ can be determined by Sylvester's theorem. Therefore

$$\epsilon^{\mathbf{A}} = \sum_{i=1}^{2} \epsilon^{\lambda_i} \mathbf{Z}_0(\lambda_i)$$

where

$$\mathbf{Z}_0(\lambda_1) = \frac{\mathbf{A} - \lambda_2 \mathbf{I}}{\lambda_1 - \lambda_2} \quad \text{and} \quad \mathbf{Z}_0(\lambda_2) = \frac{\mathbf{A} - \lambda_1 \mathbf{I}}{\lambda_2 - \lambda_1}$$

Since

$$\mathbf{A} = \begin{bmatrix} 0 & 1 \\ -2 & -3 \end{bmatrix}$$

and $\lambda_1 = -1$ and $\lambda_2 = -2$, it follows that

$$\mathbf{Z}_0(\lambda_1) = \begin{bmatrix} 2 & 1 \\ -2 & -1 \end{bmatrix} \qquad \mathbf{Z}_0(\lambda_2) = \begin{bmatrix} -1 & -1 \\ 2 & 2 \end{bmatrix}$$

Consequently,

$$\epsilon^{\mathbf{A}} = \begin{bmatrix} 2\epsilon^{-1} - \epsilon^{-2} & \epsilon^{-1} - \epsilon^{-2} \\ -2(\epsilon^{-1} - \epsilon^{-2}) & -(\epsilon^{-1} - 2\epsilon^{-2}) \end{bmatrix}$$

Example 4.10-7. Calculate $\mathbf{A}^k$, using Sylvester's theorem, for the $\mathbf{A}$ matrix of the previous example.

$$\mathbf{A}^k = \sum_{i=1}^{2} (\lambda_i)^k \mathbf{Z}_0(\lambda_i)$$

Using $\mathbf{Z}_0(\lambda_1)$ and $\mathbf{Z}_0(\lambda_2)$ as determined in the previous example,

$$\mathbf{A}^k = \begin{bmatrix} 2(-1)^k - (-2)^k & (-1)^k - (-2)^k \\ -2(-1)^k + 2(-2)^k & -(-1)^k + 2(-2)^k \end{bmatrix}$$

Since it can be shown that[16]

$$\frac{\prod_{j\neq i}(\mathbf{A}-\lambda_j\mathbf{I})}{\prod_{j\neq i}(\lambda_i-\lambda_j)} = \frac{\mathbf{Adj}\,[\lambda\mathbf{I}-\mathbf{A}]}{dP(\lambda)/d\lambda}\bigg|_{\lambda=\lambda_i} \tag{4.10-27}$$

where $P(\lambda)$ is the characteristic polynomial of **A**, Eq. 4.10-22 can also be written as

$$\mathbf{N(A)} = \sum_{i=1}^{n} \frac{N(\lambda_i)\,\mathbf{Adj}\,[\lambda_i\mathbf{I}-\mathbf{A}]}{dP(\lambda)/d\lambda\,|_{\lambda=\lambda_i}} \tag{4.10-28}$$

Sylvester's Theorem—Confluent Form

When the matrix **A** contains repeated characteristic values, Eq. 4.10-28 must be modified. The modified form of Eq. 4.10-28 is called the confluent form of Sylvester's theorem.† Assume a characteristic value of order s. The contribution to **N(A)** from the ith root λ_i can be shown to be

$$\frac{1}{(s-1)!}\left\{\frac{d^{s-1}}{d\lambda^{s-1}}\left[\frac{N(\lambda)\,\mathbf{Adj}\,(\lambda\mathbf{I}-\mathbf{A})}{\prod_{\substack{j=1\\ j\neq i}}(\lambda-\lambda_j)^s}\right]\right\}_{\lambda=\lambda_i} \tag{4.10-29}$$

The sum of the contributions of all the roots with different values is then **N(A)**. Hence

$$\mathbf{N(A)} = \sum_i \frac{1}{(s-1)!}\left\{\frac{d^{s-1}}{d\lambda^{s-1}}\left[\frac{N(\lambda)\,\mathbf{Adj}\,(\lambda\mathbf{I}-\mathbf{A})}{\prod_{\substack{j=1\\ j\neq i}}(\lambda-\lambda_j)^s}\right]\right\}_{\lambda=\lambda_i} \tag{4.10-30}$$

where the summation is taken over all the roots, with repeated roots taken only once.

Equation 4.10-30 is the confluent form of Sylvester's theorem. A typical term of the summation, corresponding to a multiple root λ_i, can be expanded into the form

$$\frac{1}{(s-1)!}\left\{\frac{d^{s-1}}{d\lambda^{s-1}}\left[\frac{N(\lambda)\,\mathbf{Adj}\,(\lambda\mathbf{I}-\mathbf{A})}{\prod_{\substack{j=1\\ j\neq i}}(\lambda-\lambda_j)^s}\right]\right\}_{\lambda=\lambda_i} = \sum_{k=1}^{s} \frac{N^{(k-1)}(\lambda_i)\mathbf{Z}_{s-k}(\lambda_i)}{(k-1)!} \tag{4.10-31}$$

where

$$N^k(\lambda_i) = \frac{d^k N(\lambda)}{d\lambda^k}\bigg|_{\lambda=\lambda_i}$$

† A proof of the theorem for this form can be found in Reference 17.

and

$$\mathbf{Z}_k(\lambda_i) = \frac{1}{k!}\left\{\frac{d^k}{d\lambda^k}\left[\frac{\mathbf{Adj}\,(\lambda\mathbf{I} - \mathbf{A})}{\prod_{\substack{j=1\\ j\neq i}} (\lambda - \lambda_j)^s}\right]\right\}_{\lambda=\lambda_i}$$

Example 4.10-8. Find the general form for any matrix function of **A**, where the matrix function can be expressed as a matrix polynomial in

$$\mathbf{A} = \begin{bmatrix} 0 & 1 & 3 \\ 6 & 0 & 2 \\ -5 & 2 & 4 \end{bmatrix}$$

Evaluation of

$$|\lambda\mathbf{I} - \mathbf{A}| = P(\lambda) = \lambda^3 - 4\lambda^2 + 5\lambda - 2 = (\lambda - 1)^2(\lambda - 2)$$

shows that the characteristic equation has a double root at $\lambda = 1$, and a single root at $\lambda = 2$. The contribution to $\mathbf{N}(\mathbf{A})$ from the single root is

$$\frac{N(2)\,\mathbf{Adj}\,(\lambda\mathbf{I} - \mathbf{A})\,|_{\lambda=2}}{(2-1)^2}$$

Since

$$\mathbf{Adj}\,[\lambda\mathbf{I} - \mathbf{A}] = \begin{bmatrix} (\lambda^2 - 4\lambda - 4) & (\lambda + 2) & (3\lambda + 2) \\ (6\lambda - 34) & (\lambda^2 - 4\lambda + 15) & (2\lambda + 18) \\ (12 - 5\lambda) & (2\lambda - 5) & (\lambda^2 - 6) \end{bmatrix}$$

and

$$\mathbf{Adj}\,[\lambda\mathbf{I} - \mathbf{A}]_{\lambda=2} = \begin{bmatrix} -8 & 4 & 8 \\ -22 & 11 & 22 \\ 2 & -1 & -2 \end{bmatrix}$$

the contribution to $\mathbf{N}(\mathbf{A})$ from the single root at $\lambda = 2$ is then

$$\mathbf{N}(2)\begin{bmatrix} -8 & 4 & 8 \\ -22 & 11 & 22 \\ 2 & -1 & -2 \end{bmatrix}$$

The contribution to $\mathbf{N}(\mathbf{A})$ from the double root at $\lambda = 1$ is

$$\frac{d}{d\lambda}\left[\frac{N(\lambda)\,\mathbf{Adj}\,[\lambda\mathbf{I} - \mathbf{A}]}{\lambda - 2}\right]_{\lambda=1}$$

or, using Eq. 4.10-31,

$$\mathbf{Z}_1(1) = \frac{d}{d\lambda}\left[\frac{\mathbf{Adj}\,(\lambda\mathbf{I} - \mathbf{A})}{\lambda - 2}\right]_{\lambda=1} = \left[-\mathbf{Adj}\,(\lambda\mathbf{I} - \mathbf{A}) - \frac{d}{d\lambda}\mathbf{Adj}\,(\lambda\mathbf{I} - \mathbf{A})\right]_{\lambda=1}$$

$$\mathbf{Z}_0(1) = -\mathbf{Adj}\,[\lambda\mathbf{I} - \mathbf{A}]_{\lambda=1}$$

Since

$$\frac{d}{d\lambda}\{\mathbf{Adj}\,[\lambda\mathbf{I} - \mathbf{A}]\} = \begin{bmatrix} 2\lambda - 4 & 1 & 3 \\ 6 & 2\lambda - 4 & 2 \\ -5 & 2 & 2\lambda \end{bmatrix}$$

$$\mathbf{Z}_1(1) = \begin{bmatrix} 7 & -3 & -5 \\ 28 & -12 & -20 \\ -7 & 3 & 5 \end{bmatrix} + \begin{bmatrix} 2 & -1 & -3 \\ -6 & 2 & -2 \\ 5 & -2 & -2 \end{bmatrix} = \begin{bmatrix} 9 & 4 & -8 \\ 22 & -10 & -22 \\ -2 & 1 & 3 \end{bmatrix}$$

$$\mathbf{Z}_0(1) = \begin{bmatrix} 7 & -3 & -5 \\ 28 & -12 & -20 \\ -7 & 3 & 5 \end{bmatrix}$$

Therefore the contribution to $\mathbf{N(A)}$ from the double root at $\lambda = 1$ is given by

$$N(1)\mathbf{Z}_1(1) + \frac{d}{d\lambda}N(\lambda)\bigg|_{\lambda=1}\mathbf{Z}_0(1)$$

The sum of these contributions is then

$$\mathbf{N(A)} = N(2)\begin{bmatrix} -8 & 4 & 8 \\ -22 & 11 & 22 \\ 2 & -1 & -2 \end{bmatrix} + N(1)\begin{bmatrix} 9 & -4 & -8 \\ 22 & -10 & -22 \\ -2 & 1 & 3 \end{bmatrix}$$

$$+ \frac{dN(\lambda)}{d\lambda}\bigg|_{\lambda=1}\begin{bmatrix} 7 & -3 & -5 \\ 28 & -12 & -20 \\ -7 & 3 & 5 \end{bmatrix}$$

As a particular example, let $\mathbf{N(A)} = \epsilon^{\mathbf{A}t}$, or $N(\lambda) = \epsilon^{\lambda t}$. Then

$$N(2) = \epsilon^{2t}, \quad N(1) = \epsilon^{t}, \quad \text{and} \quad \frac{dN(\lambda)}{d\lambda}\bigg|_{\lambda=1} = t\epsilon^{t}$$

and

$$\mathbf{N(A)} = \epsilon^{\mathbf{A}t}$$

$$= \begin{bmatrix} (9\epsilon^t + 7t\epsilon^t - 8\epsilon^{2t}) & (-4\epsilon^t - 3t\epsilon^t + 4\epsilon^{2t}) & (-8\epsilon^t - 5t\epsilon^t + 8\epsilon^{2t}) \\ (22\epsilon^t + 28t\epsilon^t - 22\epsilon^{2t}) & (-10\epsilon^t - 12t\epsilon^t + 11\epsilon^{2t}) & (-22\epsilon^t - 20t\epsilon^t + 22\epsilon^{2t}) \\ (-2\epsilon^t - 7t\epsilon^t + 2\epsilon^{2t}) & (\epsilon^t + 3t\epsilon^t - \epsilon^{2t}) & (3\epsilon^t + 5t\epsilon^t - 2\epsilon^{2t}) \end{bmatrix}$$

Cayley-Hamilton Technique

An alternative, and often simpler, procedure for evaluating a function of a matrix is obtained by making use of the Cayley-Hamilton theorem. First, consider the case where $\mathbf{N(A)}$ is a matrix polynomial which is of higher degree than the order of $\mathbf{A}$. If $N(\lambda)$ is divided by the characteristic polynomial of $\mathbf{A}$, then

$$\frac{N(\lambda)}{P(\lambda)} = Q(\lambda) + \frac{R(\lambda)}{P(\lambda)} \tag{4.10-32}$$

where $R(\lambda)$ is the remainder. Then, if Eq. 4.10-32 is multiplied by $P(\lambda)$, the result is

$$N(\lambda) = Q(\lambda)P(\lambda) + R(\lambda) \tag{4.10-33}$$

Now, if $P(\lambda) = 0$, Eq. 4.10-33 becomes

$$N(\lambda) = R(\lambda) \tag{4.10-34}$$

Correspondingly, since $\mathbf{P}(\mathbf{A}) = [0]$ by the Cayley-Hamilton theorem, the matrix function $\mathbf{N}(\mathbf{A})$ is then equal to $\mathbf{R}(\mathbf{A})$.

Example 4.10-9. Solve the problem of Example 4.10-5 using the Cayley-Hamilton technique.

$$\mathbf{N}(\mathbf{A}) = \mathbf{A}^4 + \mathbf{A}^3 + \mathbf{A}^2 + \mathbf{A} + \mathbf{I}$$

or

$$N(\lambda) = \lambda^4 + \lambda^3 + \lambda^2 + \lambda + 1$$

The characteristic polynomial of

$$\mathbf{A} = \begin{bmatrix} 0 & 1 \\ -2 & -3 \end{bmatrix}$$

is $\lambda^2 + 3\lambda + 2 = 0$. Dividing $N(\lambda)$ by $P(\lambda)$,

$$\frac{\lambda^4 + \lambda^3 + \lambda^2 + \lambda + 1}{\lambda^2 + 3\lambda + 2} = \lambda^2 - 2\lambda + 5 + \frac{(-10\lambda - 9)}{\lambda^2 + 3\lambda + 2}$$

The remainder $R(\lambda)$ is then $R(\lambda) = -10\lambda - 9$. Hence $\mathbf{N}(\mathbf{A}) = \mathbf{R}(\mathbf{A}) = -10\mathbf{A} - 9\mathbf{I}$, which is the same result obtained in Example 4.10-5.

The preceding technique is valid only for the case in which $\mathbf{N}(\mathbf{A})$ is a polynomial function of $\mathbf{A}$. When $\mathbf{F}(\mathbf{A})$ is desired, where $F(\lambda)$ is an analytic function of λ, in a region about the origin, an extension of the previous method can be used. If $F(\lambda)$ is an analytic function in a region, it can be expressed by an infinite power series in λ, which converges in the region of analyticity. Therefore the function $\mathbf{F}(\mathbf{A})$ can be expressed as a polynomial in $\mathbf{A}$ of degree $n - 1$. Consequently, the remainder $R(\lambda)$ of Eq. 4.10-33 must be a polynomial of degree $n - 1$. It follows that, if $Q(\lambda)$ is an analytic function of λ in that region,

$$F(\lambda) = Q(\lambda)P(\lambda) + R(\lambda) \tag{4.10-35}$$

where $P(\lambda)$ is the characteristic polynomial of $\mathbf{A}$, and $R(\lambda)$ is a polynomial of the form

$$R(\lambda) = \alpha_0 + \alpha_1\lambda + \alpha_2\lambda^2 + \cdots + \alpha_{n-1}\lambda^{n-1} \tag{4.10-36}$$

The coefficients $\alpha_0, \alpha_1, \ldots, \alpha_{n-1}$ can be obtained by successively substituting $\lambda_1, \lambda_2, \ldots, \lambda_n$ into Eq. 4.10-35. Since $P(\lambda_i) = 0$, the equations

$$\begin{aligned} F(\lambda_1) &= R(\lambda_1) \\ F(\lambda_2) &= R(\lambda_2) \\ &\cdots\cdots\cdots \\ F(\lambda_n) &= R(\lambda_n) \end{aligned} \tag{4.10-37}$$

are obtained. Equation 4.10-37 describes a set of n linear equations in n unknowns. Therefore a unique solution can be obtained for all the coefficients of the polynomial $R(\lambda)$.

It now remains to show that

$$Q(\lambda) = \frac{F(\lambda) - R(\lambda)}{P(\lambda)} \tag{4.10-38}$$

is an analytic function of λ. Since the zeros of the denominator of $Q(\lambda)$ are also zeros of its numerator, the function $Q(\lambda)$ is analytic in the region of analyticity of $F(\lambda)$. Therefore Eq. 4.10-35 is valid for all values of λ in the region of analyticity of $F(\lambda)$. Consequently $\mathbf{A}$ may be substituted for the variable λ, if the region of analyticity includes all the characteristic values of $\mathbf{A}$. This substitution yields

$$\mathbf{F(A)} = \mathbf{Q(A)P(A)} + \mathbf{R(A)} \tag{4.10-39}$$

Since $\mathbf{P(A)}$ is identically zero by the Cayley-Hamilton theorem, it follows that

$$\mathbf{F(A)} = \mathbf{R(A)} \tag{4.10-40}$$

which is the desired result.

Before proceeding to some examples of this technique, the problem of repeated characteristic roots should be investigated. Obviously, if $\mathbf{A}$ possesses a characteristic value λ_i of order s, only one linear independent equation can be obtained by substituting λ_i into Eq. 4.10-35. The remaining $s - 1$ linear equations, which must be obtained in order to solve for the α_i's, can be found by differentiating both sides of Eq. 4.10-35. Therefore, if $\mathbf{A}$ has a characteristic value of order s, a set of linear equations of the form

$$\left.\frac{d^kF(\lambda)}{d\lambda^k}\right|_{\lambda=\lambda_i} = \left.\frac{d^kR(\lambda)}{d\lambda^k}\right|_{\lambda=\lambda_i}, \quad k = 0, 1, \ldots, s - 1 \tag{4.10-41}$$

must be obtained in order to find a unique solution for the coefficients of the polynomial of Eq. 4.10-36.

Example 4.10-10. Find $\epsilon^{\mathbf{A}t}$, using as $\mathbf{A}$ the matrix of Example 4.10-9.

$$\mathbf{A} = \begin{bmatrix} 0 & 1 \\ -2 & -3 \end{bmatrix}, \quad \lambda_1 = -1, \quad \lambda_2 = -2$$

Since $\mathbf{A}$ is a second order matrix, the polynomial $R(\lambda)$ is of first order, i.e.,

$$R(\lambda) = \alpha_0 + \alpha_1\lambda$$

Therefore the two linear equations obtained by substituting λ_1 and λ_2 into Eq. 4.10-35 are

$$\begin{aligned} F(\lambda_1) &= R(\lambda_1) & F(\lambda_2) &= R(\lambda_2) \\ \epsilon^{\lambda_1 t} &= \alpha_0 + \alpha_1\lambda_1 & \epsilon^{\lambda_2 t} &= \alpha_0 + \alpha_1\lambda_2 \\ \epsilon^{-t} &= \alpha_0 - \alpha_1 & \epsilon^{-2t} &= \alpha_0 - 2\alpha_1 \end{aligned}$$

Solving for α_0 and α_1,

$$\alpha_0 = 2\epsilon^{-t} - \epsilon^{-2t}$$
$$\alpha_1 = \epsilon^{-t} - \epsilon^{-2t}$$

Hence

$$\mathbf{F}(\mathbf{A}) = \epsilon^{\mathbf{A}t} = \alpha_0\mathbf{I} + \alpha_1\mathbf{A} = \begin{bmatrix} \alpha_0 & 0 \\ 0 & \alpha_0 \end{bmatrix} + \begin{bmatrix} 0 & \alpha_1 \\ -2\alpha_1 & -3\alpha_1 \end{bmatrix}$$
$$= \begin{bmatrix} 2\epsilon^{-t} - \epsilon^{-2t} & \epsilon^{-t} - \epsilon^{-2t} \\ -2(\epsilon^{-t} - \epsilon^{-2t}) & -(\epsilon^{-t} - 2\epsilon^{-2t}) \end{bmatrix}$$

Example 4.10-11. Determine $\epsilon^{\mathbf{A}t}$, where $\mathbf{A}$ is the matrix used in Example 4.10-8.

This matrix has a double root at $\lambda = 1$ and a single root at $\lambda = 2$. Since this is a third order matrix, the polynomial $R(\lambda)$ is $R(\lambda) = \alpha_0 + \alpha_1\lambda + \alpha_2\lambda^2$. The three equations for the α_i's are given by $F(\lambda_1) = R(\lambda_1)$ and

$$\left.\frac{dF(\lambda)}{d\lambda}\right|_{\lambda=\lambda_1} = \left.\frac{dR(\lambda)}{d\lambda}\right|_{\lambda=\lambda_1}$$

where $\lambda_1 = 1$, and by $F(\lambda_2) = R(\lambda_2)$, where $\lambda_2 = 2$. Thus the α's are specified by

$$\begin{bmatrix} \epsilon^t \\ t\epsilon^t \\ \epsilon^{2t} \end{bmatrix} = \begin{bmatrix} 1 & 1 & 1 \\ 0 & 1 & 2 \\ 1 & 2 & 4 \end{bmatrix} \begin{bmatrix} \alpha_0 \\ \alpha_1 \\ \alpha_2 \end{bmatrix}$$

Solving for the α's,

$$\begin{bmatrix} \alpha_0 \\ \alpha_1 \\ \alpha_2 \end{bmatrix} = \begin{bmatrix} 1 & 1 & 1 \\ 0 & 1 & 2 \\ 1 & 2 & 4 \end{bmatrix}^{-1} \begin{bmatrix} \epsilon^t \\ t\epsilon^t \\ \epsilon^{2t} \end{bmatrix}$$

It is instructive to solve this set of three simultaneous equations by using the Cayley-Hamilton theorem to find the inverse of the coefficient matrix. The characteristic polynomial of the coefficient matrix is $\lambda^3 - 6\lambda^2 + 4\lambda - 1 = 0$. Hence

$$\mathbf{C}^3 - 6\mathbf{C}^2 + 4\mathbf{C} - \mathbf{I} = [0] \quad \text{where } \mathbf{C} = \begin{bmatrix} 1 & 1 & 1 \\ 0 & 1 & 2 \\ 1 & 2 & 4 \end{bmatrix}$$

Then

$$\mathbf{C}^{-1} = \mathbf{C}^2 - 6\mathbf{C} + 4\mathbf{I} = \begin{bmatrix} 0 & -2 & 1 \\ 2 & 3 & -2 \\ -1 & -1 & 1 \end{bmatrix}$$

Therefore

$$\alpha_0 = -2t\epsilon^t + \epsilon^{2t}$$
$$\alpha_1 = 2\epsilon^t + 3t\epsilon^t - 2\epsilon^{2t}$$
$$\alpha_2 = -\epsilon^t - t\epsilon^t + \epsilon^{2t}$$

Hence

$$\epsilon^{\mathbf{A}t} = \alpha_0\mathbf{I} + \alpha_1\mathbf{A} + \alpha_0\mathbf{A}^2$$

where

$$\mathbf{A} = \begin{bmatrix} 0 & 1 & 3 \\ 6 & 0 & 2 \\ -5 & 2 & 4 \end{bmatrix} \quad \text{and} \quad \mathbf{A}^2 = \begin{bmatrix} -9 & 6 & 14 \\ -10 & 10 & 26 \\ -8 & 3 & 5 \end{bmatrix}$$

The result is

$$\epsilon^{\mathbf{A}t} = \begin{bmatrix} (9\epsilon^t + 7t\epsilon^t - 8\epsilon^{2t}) & (-4\epsilon^t - 3t\epsilon^t + 4\epsilon^{2t}) & (-8\epsilon^t - 5t\epsilon^t + 8\epsilon^{2t}) \\ (22c^t + 28t\epsilon^t - 22\epsilon^{2t}) & (-10\epsilon^t - 12t\epsilon^t + 11\epsilon^{2t}) & (-22\epsilon^t - 20t\epsilon^t + 22\epsilon^{2t}) \\ (-2\epsilon^t - 7t\epsilon^t + 2\epsilon^{2t}) & (\epsilon^t + 3t\epsilon^t - \epsilon^{2t}) & (3\epsilon^t + 5t\epsilon^t - 2\epsilon^{2t}) \end{bmatrix}$$

This matrix checks with the result obtained in Example 4.10-8 using Sylvester's theorem. A considerable amount of labor is involved in either method, but this is usually the case when dealing with a matrix of order higher than two. Generally, the Cayley-Hamilton technique requires much less labor than the use of Sylvester's theorem.

Example 4.10-12. Generalize the preceding discussion on the Cayley-Hamilton technique so that any analytic function of **A** can be generated. Assume that **A** has distinct characteristic values.

The n equations which determine the coefficients α_i can be written in matrix form as

$$\begin{bmatrix} F(\lambda_1) \\ F(\lambda_2) \\ \cdots \\ F(\lambda_n) \end{bmatrix} = \begin{bmatrix} 1 & \lambda_1 & \lambda_1^2 & \cdots & \lambda_1^{n-1} \\ 1 & \lambda_2 & \lambda_2^2 & \cdots & \lambda_2^{n-2} \\ \cdots & \cdots & \cdots & \cdots & \cdots \\ 1 & \lambda_n & \lambda_n^2 & \cdots & \lambda_n^{n-1} \end{bmatrix} \begin{bmatrix} \alpha_0 \\ \alpha_1 \\ \cdots \\ \alpha_{n-1} \end{bmatrix}$$

or $\mathbf{F}(\lambda) = \mathbf{C}\boldsymbol{\alpha}$, where $\mathbf{C}$ is the coefficient matrix shown. Consequently, $\boldsymbol{\alpha} = \mathbf{C}^{-1}\mathbf{F}(\lambda)$. Therefore

$$\mathbf{F}(\mathbf{A}) = \sum_{i=0}^{n-1} \alpha_i \mathbf{A}^i$$

where the α_i's are determined from the equation $\boldsymbol{\alpha} = \mathbf{C}^{-1}\mathbf{F}(\lambda)$.

An alternative procedure, again assuming distinct roots, is to diagonalize **A** by means of a similarity transformation, and then utilize the generalization of Eq. 4.10-20 to the case of an analytic function. A simple example illustrates this procedure.

In Example 4.7-6, the matrix

$$\mathbf{A} = \begin{bmatrix} 0 & 1 \\ -2 & -3 \end{bmatrix}$$

was analyzed, and the modal matrices **M** and $\mathbf{M}^{-1}$ were found to be

$$\mathbf{M} = \begin{bmatrix} 1/\sqrt{2} & 1/\sqrt{5} \\ -1/\sqrt{2} & -2/\sqrt{5} \end{bmatrix} \qquad \mathbf{M}^{-1} = \begin{bmatrix} 2\sqrt{2} & \sqrt{2} \\ -\sqrt{5} & -\sqrt{5} \end{bmatrix}$$

Hence

$$\mathbf{A} = \mathbf{M}\boldsymbol{\Lambda}\mathbf{M}^{-1} = \mathbf{M} \begin{bmatrix} -1 & 0 \\ 0 & -2 \end{bmatrix} \mathbf{M}^{-1}$$

From the generalization of Eq. 4.10-20, $\mathbf{F}(\mathbf{A}) = \mathbf{M}\mathbf{F}(\boldsymbol{\Lambda})\mathbf{M}^{-1}$. If $\mathbf{F}(\mathbf{A}) = \epsilon^{\mathbf{A}t}$ is desired, then

$$\mathbf{F}(\mathbf{A}) = \mathbf{M} \begin{bmatrix} \epsilon^{-t} & 0 \\ 0 & \epsilon^{-2t} \end{bmatrix} \mathbf{M}^{-1}$$

Performing the indicated matrix multiplications yields

$$\epsilon^{At} = \begin{bmatrix} 2\epsilon^{-t} - \epsilon^{-2t} & \epsilon^{-t} - \epsilon^{-2t} \\ -2(\epsilon^{-t} - \epsilon^{-2t}) & -(\epsilon^{-t} - 2\epsilon^{-2t}) \end{bmatrix}$$

which checks with the result found in Example 4.10-10.

Once the modal matrices are obtained, this is a very convenient procedure for finding an analytic function of a matrix. However, the procedure requires a complete characteristic vector analysis.

4.11 ADDITIONAL MATRIX CALCULUS

The usual ideas of differentiation and integration associated with scalar variables previously were shown to carry over to the differentiation and integration of matrices and matrix products, provided that the original order of the factors involved is preserved. However, because such operations are frequently performed on the exponential function of a matrix and on quadratic forms in later chapters, they are considered specifically at this point.

Differentiation of the Exponential Function

If $\mathbf{A}$ is a constant matrix and t is a scalar variable, then the exponential function $\epsilon^{\mathbf{A}t}$ is defined, similarly to Eq. 4.10-8, as the infinite series

$$\epsilon^{\mathbf{A}t} = \exp(\mathbf{A}t) = \mathbf{I} + \mathbf{A}t + \frac{\mathbf{A}^2t^2}{2!} + \frac{\mathbf{A}^3t^3}{3!} + \cdots \qquad (4.11\text{-}1)$$

This series is absolutely and uniformly convergent for all values of the scalar variable t. The derivative of the exponential function $\epsilon^{\mathbf{A}t}$ with respect to t is then the term by term differentiation of Eq. 4.11-1, or

$$\frac{d}{dt}[\epsilon^{\mathbf{A}t}] = \mathbf{A} + \mathbf{A}^2t + \frac{\mathbf{A}^3t^2}{2!} + \cdots = \mathbf{A}\epsilon^{\mathbf{A}t} = \epsilon^{\mathbf{A}t}\mathbf{A} \qquad (4.11\text{-}2)$$

If the operator notation $p = d/dt$ is used, then it follows that

$$\frac{d^k}{dt^k}[\epsilon^{\mathbf{A}t}] = p^k(\epsilon^{\mathbf{A}t}) = \mathbf{A}^k\epsilon^{\mathbf{A}t} = \epsilon^{\mathbf{A}t}\mathbf{A}^k \qquad (4.11\text{-}3)$$

In general, if $N(p)$ is a polynomial of the differential operator p, then

$$N(p)\epsilon^{\mathbf{A}t} = \mathbf{N}(\mathbf{A})\epsilon^{\mathbf{A}t} = \epsilon^{\mathbf{A}t}\mathbf{N}(\mathbf{A}) \qquad (4.11\text{-}4)$$

Often the situation arises where the polynomial operator $N(p)$ must operate on the matrix product $\epsilon^{\mathbf{A}t}\mathbf{B}(t)$. It is assumed that the product $\mathbf{AB}(t)$ exists, but that the product $\mathbf{B}(t)\mathbf{A}$ does not. In this case,

$$p[\epsilon^{\mathbf{A}t}\mathbf{B}(t)] = \epsilon^{\mathbf{A}t}\dot{\mathbf{B}}(t) + \epsilon^{\mathbf{A}t}\mathbf{AB}(t) = \epsilon^{\mathbf{A}t}[p\mathbf{I} + \mathbf{A}]\mathbf{B}(t)$$

$$p^2[\epsilon^{\mathbf{A}t}\mathbf{B}(t)] = \epsilon^{\mathbf{A}t}\ddot{\mathbf{B}}(t) + 2\epsilon^{\mathbf{A}t}\mathbf{A}\dot{\mathbf{B}}(t) + \epsilon^{\mathbf{A}t}\mathbf{A}^2\mathbf{B}(t) = \epsilon^{\mathbf{A}t}(p\mathbf{I} + \mathbf{A})^2\mathbf{B}(t)$$

In general,

$$p^k[\epsilon^{\mathbf{A}t}\mathbf{B}(t)] = \epsilon^{\mathbf{A}t}(p\mathbf{I} + \mathbf{A})^k\mathbf{B}(t) \tag{4.11-5}$$

Consequently,

$$N(p)[\epsilon^{\mathbf{A}t}\mathbf{B}(t)] = \epsilon^{\mathbf{A}t}\mathbf{N}(p\mathbf{I} + \mathbf{A})\mathbf{B}(t) \tag{4.11-6}$$

Integration of the Exponential Function

The integral of the exponential function $\epsilon^{\mathbf{A}t}$, where $\mathbf{A}$ is a constant matrix, can be found by integrating the infinite series expression for $\epsilon^{\mathbf{A}t}$, i.e.,

$$\int_0^t \epsilon^{\mathbf{A}t}\,dt = \int_0^t \mathbf{I}\,dt + \int_0^t \mathbf{A}t\,dt + \int_0^t \frac{\mathbf{A}^2t^2}{2!}\,dt + \int_0^t \frac{\mathbf{A}^3t^3}{3!}\,dt + \cdots$$

$$= \mathbf{I}t + \frac{\mathbf{A}t^2}{2!} + \frac{\mathbf{A}^2t^3}{3!} + \frac{\mathbf{A}^3t^4}{4!} + \cdots$$

Hence

$$\mathbf{A}\int_0^t \epsilon^{\mathbf{A}t}\,dt = \epsilon^{\mathbf{A}t} - \mathbf{I}$$

Assuming that $\mathbf{A}$ is nonsingular,

$$\int_0^t \epsilon^{\mathbf{A}t}\,dt = \mathbf{A}^{-1}(\epsilon^{\mathbf{A}t} - \mathbf{I}) = (\epsilon^{\mathbf{A}t} - \mathbf{I})\mathbf{A}^{-1} \tag{4.11-7}$$

Example 4.11-1. Find $\int_0^t \epsilon^{\mathbf{A}t}\,dt$, where $\mathbf{A} = \begin{bmatrix} 0 & 1 \\ -2 & -3 \end{bmatrix}$.

The result of Example 4.10-10 shows that

$$\epsilon^{\mathbf{A}t} = \begin{bmatrix} 2\epsilon^{-t} - \epsilon^{-2t} & \epsilon^{-t} - \epsilon^{-2t} \\ -2(\epsilon^{-t} - \epsilon^{-2t}) & -(\epsilon^{-t} - 2\epsilon^{-2t}) \end{bmatrix}$$

Therefore

$$\int_0^t \epsilon^{\mathbf{A}t}\,dt = \begin{bmatrix} -2\epsilon^{-t} + \dfrac{\epsilon^{-2t}}{2} + \dfrac{3}{2} & -\epsilon^{-t} + \dfrac{\epsilon^{-2t}}{2} + \dfrac{1}{2} \\ 2\epsilon^{-t} - \epsilon^{-2t} - 1 & \epsilon^{-t} - \epsilon^{-2t} \end{bmatrix}$$

This result can be checked by the application of Eq. 4.11-7. Since

$$\mathbf{A}^{-1} = \begin{bmatrix} -\frac{3}{2} & -\frac{1}{2} \\ 1 & 0 \end{bmatrix}$$

$$\mathbf{A}^{-1}\epsilon^{\mathbf{A}t} - \mathbf{A}^{-1} = \begin{bmatrix} -\frac{3}{2} & -\frac{1}{2} \\ 1 & 0 \end{bmatrix} \begin{bmatrix} 2\epsilon^{-t} - \epsilon^{-2t} & \epsilon^{-t} - \epsilon^{-2t} \\ -2\epsilon^{-t} + 2\epsilon^{-2t} & -\epsilon^{-t} + 2\epsilon^{-2t} \end{bmatrix} - \begin{bmatrix} -\frac{3}{2} & -\frac{1}{2} \\ 1 & 0 \end{bmatrix}$$

$$= \begin{bmatrix} -2\epsilon^{-t} + \dfrac{\epsilon^{-2t}}{2} + \dfrac{3}{2} & -\epsilon^{-t} + \dfrac{\epsilon^{-2t}}{2} + \dfrac{1}{2} \\ 2\epsilon^{-t} - \epsilon^{-2t} - 1 & \epsilon^{-t} - \epsilon^{-2t} \end{bmatrix}$$

The solution to a linear time-varying matrix differential equation often depends upon the exponential function

$$\exp\left[\int_0^t \mathbf{A}(\lambda)\, d\lambda\right]$$

This exponential function is defined as the infinite series

$$\exp\left[\int_0^t \mathbf{A}(\lambda)\, d\lambda\right] = \sum_{k=0}^{\infty} \frac{1}{k!}\left[\int_0^t \mathbf{A}(\lambda)\, d\lambda\right]^k \qquad (4.11\text{-}8)$$

Example 4.11-2. Find

$$\exp\left[\int_0^t \mathbf{A}(\lambda)\, d\lambda\right] \quad \text{if } \mathbf{A} = \begin{bmatrix} t & 0 \\ 0 & t \end{bmatrix}$$

The integral of this matrix is

$$\int_0^t \mathbf{A}(\lambda)\, d\lambda \begin{bmatrix} \dfrac{t^2}{2} & 0 \\ 0 & \dfrac{t^2}{2} \end{bmatrix}$$

Consequently,

$$\exp\left[\int_0^t \mathbf{A}(\lambda)\, d\lambda\right] = \sum_{k=0}^{\infty} \frac{1}{k!} \begin{bmatrix} \dfrac{t^2}{2} & 0 \\ 0 & \dfrac{t^2}{2} \end{bmatrix} = \begin{bmatrix} \epsilon^{t^2/2} & 0 \\ 0 & \epsilon^{t^2/2} \end{bmatrix}$$

Differentiation of Quadratic Forms

The stability analysis of dynamical systems often requires the differentiation of a quadratic form. If the quadratic form is $\mathbf{Q} = \langle \mathbf{x}, \mathbf{A}\mathbf{x} \rangle$, where $\mathbf{A}$ is a symmetric matrix, then

$$\mathbf{grad}_{\mathbf{x}}\, Q = 2\mathbf{A}\mathbf{x} \qquad (4.11\text{-}9)$$

where $\mathbf{grad}_\mathbf{x}$ denotes the vector operator

$$\left[\frac{\partial}{\partial x_i}\right] = \begin{bmatrix} \frac{\partial}{\partial x_1} \\ \frac{\partial}{\partial x_2} \\ \cdot \\ \cdot \\ \cdot \\ \frac{\partial}{\partial x_n} \end{bmatrix} \tag{4.11-10}$$

Also, frequently useful is the matrix operator formed by taking the outer product of $\mathbf{grad}_\mathbf{x}$ and $\mathbf{grad}_\mathbf{y}$, yielding

$$\mathbf{grad}_\mathbf{x}\rangle\langle\mathbf{grad}_\mathbf{y} = \begin{bmatrix} \frac{\partial}{\partial x_1} \\ \frac{\partial}{\partial x_2} \\ \cdot \\ \cdot \\ \cdot \\ \frac{\partial}{\partial x_n} \end{bmatrix} \begin{bmatrix} \frac{\partial}{\partial y_1} & \frac{\partial}{\partial y_2} & \cdots & \frac{\partial}{\partial y_m} \end{bmatrix}$$

$$= \begin{bmatrix} \frac{\partial^2}{\partial x_1\,\partial y_1} & \frac{\partial^2}{\partial x_1\,\partial y_2} & \cdots & \frac{\partial^2}{\partial x_1\,\partial y_m} \\ \frac{\partial^2}{\partial x_2\,\partial y_1} & \frac{\partial^2}{\partial x_2\,\partial y_2} & \cdots & \frac{\partial^2}{\partial x_2\,\partial y_m} \\ \cdots & \cdots & \cdots & \cdots \\ \frac{\partial^2}{\partial x_n\,\partial y_1} & \frac{\partial^2}{\partial x_n\,\partial y_2} & \cdots & \frac{\partial^2}{\partial x_n\,\partial y_m} \end{bmatrix} \tag{4.11-11}$$

Example 4.11-3. Evaluate the gradient of $Q = \langle \mathbf{x}, \mathbf{Ax} \rangle$, where

$$\mathbf{A} = \begin{bmatrix} 1 & 3 \\ 3 & 2 \end{bmatrix}$$

Since $Q = x_1^2 + 2x_2^2 + 6x_1x_2$, the gradient of Q with respect to the variables x_1, x_2 is given by

$$\mathbf{grad}_\mathbf{x}\, Q = \begin{bmatrix} \frac{\partial Q}{\partial x_1} \\ \frac{\partial Q}{\partial x_2} \end{bmatrix} = \begin{bmatrix} 2x_1 + 6x_2 \\ 4x_2 + 6x_1 \end{bmatrix}$$

Note that

$$2\mathbf{A}\mathbf{x} = 2\begin{bmatrix}1 & 3\\ 3 & 2\end{bmatrix}\begin{bmatrix}x_1\\ x_2\end{bmatrix} = \begin{bmatrix}2x_1 + 6x_2\\ 4x_2 + 6x_1\end{bmatrix}$$

If both $\mathbf{A}$ and the variables $x_1, x_2, \ldots, x_n$ are functions of time, then the derivative of the quadratic form with respect to t is given by

$$\frac{d}{dt}[Q(t)] = \frac{d}{dt}\langle \mathbf{x}(t), \mathbf{A}(t)\mathbf{x}(t)\rangle$$
$$= \langle \dot{\mathbf{x}}(t), \mathbf{A}(t)\mathbf{x}(t)\rangle + \langle \mathbf{x}(t), \dot{\mathbf{A}}(t)\mathbf{x}(t)\rangle + \langle \mathbf{x}(t), \mathbf{A}(t)\dot{\mathbf{x}}(t)\rangle$$

Since $\mathbf{A}$ is a symmetric matrix,

$$\dot{Q}(t) = 2\langle \dot{\mathbf{x}}(t), \mathbf{A}(t)\mathbf{x}(t)\rangle + \langle \mathbf{x}(t), \dot{\mathbf{A}}(t)\mathbf{x}(t)\rangle$$
$$= \langle \mathbf{grad}_{\mathbf{x}}\, Q, \dot{\mathbf{x}}(t)\rangle + \langle \mathbf{x}(t), \dot{\mathbf{A}}(t)\mathbf{x}(t)\rangle \qquad (4.11\text{-}12)$$

If $\mathbf{A}$ is independent of time, this reduces to

$$\dot{Q}(t) = \langle \mathbf{grad}_{\mathbf{x}}\, Q, \dot{\mathbf{x}}(t)\rangle = 2\langle \mathbf{A}\mathbf{x}(t), \dot{\mathbf{x}}(t)\rangle \qquad (4.11\text{-}13)$$

Example 4.11-4. Find the time derivative of $Q(t) = \langle \mathbf{x}(t), \mathbf{A}(t)\mathbf{x}(t)\rangle$, where

$$\mathbf{A}(t) = \begin{bmatrix}t^2 & 2t\\ 2t & 1\end{bmatrix}$$

Since $Q(t) = t^2x_1^{\,2}(t) + x_2^{\,2}(t) + 4tx_1(t)x_2(t)$, the derivative of $Q(t)$ with respect to t is

$$\dot{Q}(t) = 2tx_1^{\,2}(t) + 2t^2x_1(t)\dot{x}_1(t) + 2x_2(t)\dot{x}_2(t) + 4x_1(t)x_2(t) + 4t\dot{x}_1(t)x_2(t) + 4tx_1(t)\dot{x}_2(t)$$
$$= \dot{x}_1(t)[2t^2x_1(t) + 4tx_2(t)] + \dot{x}_2(t)[2x_2(t) + 4tx_1(t)] + 2tx_1^{\,2}(t) + 4x_1(t)x_2(t)$$

Note that $\dot{Q}(t)$ is the sum of

$$2\langle \dot{\mathbf{x}}(t), \mathbf{A}(t)\mathbf{x}(t)\rangle = 2[\dot{x}_1(t) \quad \dot{x}_2(t)]\begin{bmatrix}t^2 & 2t\\ 2t & 1\end{bmatrix}\begin{bmatrix}x_1(t)\\ x_2(t)\end{bmatrix}$$
$$= \dot{x}_1(t)[2t^2x_1(t) + 4tx_2(t)] + \dot{x}_2(t)[2x_2(t) + 4tx_1(t)]$$

and

$$\langle \mathbf{x}(t), \dot{\mathbf{A}}(t)\mathbf{x}(t)\rangle = [x_1(t) \quad x_2(t)]\begin{bmatrix}2t & 2\\ 2 & 0\end{bmatrix}\begin{bmatrix}x_1(t)\\ x_2(t)\end{bmatrix}$$
$$= 2tx_1^{\,2}(t) + 4x_1(t)x_2(t)$$

4.12 FUNCTION SPACE

This section is devoted to extending the vector concepts of Section 4.5 to the problem of determining a basis in "function space." In the preceding discussion it was shown that any n linearly independent vectors form a basis in n-dimensional space, such that all vectors in that space can be described by the linear combination $c_1\mathbf{x}_1 + c_2\mathbf{x}_2 + \cdots + c_n\mathbf{x}_n$.

Consider now a set of n functions $f_1(t), \ldots, f_n(t)$, which are defined over the interval (a, b), such that no function $f_i(t)$ is a multiple of any of the other $n - 1$ functions over this interval. Certainly, a linear combination $c_1 f_1(t) + c_2 f_2(t) + \cdots + c_n f_n(t)$ does not describe all functions which can be defined over the interval (a, b). However, a basis can be chosen such that any function satisfying regularity conditions to be stated later can be expressed as a linear combination of the members of the basis.

The topic considered here has become important in the area of adaptive and self-optimizing systems. As the systems which are to be controlled become more complex, less is known about their internal mechanisms and parameter interrelationships. The identification and modeling of such systems is an important first step for adaptive or optimum control. The employment of a convenient orthonormal basis of functions to describe the performance of a system has become an increasingly useful approach.

Scalar (Inner) Product

The scalar product of two real-valued functions $f(t)$ and $g(t)$ over the interval (a, b) is defined as†

$$\langle f, g \rangle = \int_a^b f(t) g(t)\, dt \tag{4.12-1}$$

If the functions are complex functions of a real variable t, the definition of the scalar product is modified as

$$\langle f, g \rangle = \int_a^b f^*(t) g(t)\, dt \tag{4.12-2}$$

Norm of a Function

The norm of a real-valued function is analogous to the length of a vector and is defined to be

$$\text{Norm}\, f = \|f\|^{1/2} = \langle f, f \rangle^{1/2} = \left[\int_a^b f^2(t)\, dt \right]^{1/2} \tag{4.12-3}$$

A *normalized* function is a function whose norm is unity. If the norm of a function is zero, then $f(t)$ must also be zero, except for a finite number of points, in the interval (a, b).

† It is assumed that all functions satisfy the Lebesgue condition $\int_a^b |f(t)|^2\, dt < \infty$.

Orthogonal Functions

Two functions $f(t)$ and $g(t)$ are orthogonal over the interval (a, b) if their scalar product vanishes, i.e.,

$$\langle f, g \rangle = 0 \tag{4.12-4}$$

A set of normalized functions $\phi_1(t), \phi_2(t), \ldots$ is said to be an orthonormal set if the members of the set obey the relation

$$\langle \phi_i, \phi_j \rangle = \delta_{ij} \tag{4.12-5}$$

Similar to the approach used in the Gram-Schmidt orthogonalization procedure, a set of n linearly independent functions can be used to derive a suitable orthonormal set of functions.†

Orthogonal Functions as a Basis in Function Space

If the infinite orthonormal set of functions $\phi_1(t), \phi_2(t), \ldots$ are considered coordinate vectors or coordinate functions in a space which has an infinite number of dimensions, then, by analogy with vector space, a function $f(t)$ can be considered to be a vector in this space, and the components of this function in terms of the coordinate functions are given by

$$c_k = \langle f, \phi_k \rangle \tag{4.12-6}$$

These components of the function with respect to the coordinate functions are called the expansion coefficients, or Fourier coefficients, of the function relative to the orthonormal set $\phi_k(t)$.

If the function $f(t)$ is approximated by a linear combination of n orthonormal functions

$$f(t) \doteq \sum_{k=1}^{n} a_k \phi_k(t) \qquad (a < t < b) \tag{4.12-7}$$

then the best approximation in the "least squares" sense is obtained by letting $a_k = c_k$. This can be shown by minimizing $\mathscr{E}$, the square of the norm of the difference between $f(t)$ and $\sum_{k=1}^{n} a_k \phi_k(t)$, i.e.,

$$\mathscr{E} = \left\| f(t) - \sum_{k=1}^{n} a_k \phi_k(t) \right\|^2 = \int_a^b \left[f(t) - \sum_{k=1}^{n} a_k \phi_k(t) \right]^2 dt$$

† A set of functions $f_1, f_2, \ldots, f_n$ are said to be linearly dependent if there exists a relationship $\sum_{i=1}^{n} c_i f_i = 0$ for all values of the scalar argument of the function, where all the c_i's are not zero. If such a relationship does not exist, the functions are said to be linearly independent.

Setting the derivative of $\mathscr{E}$ with respect to a_j equal to zero,

$$\int_a^b f(t)\phi_j(t)\,dt = \int_a^b \left[\phi_j(t)\sum_{k=1}^{n} a_k\phi_k(t)\right] dt$$

Since $\langle f, \phi_j\rangle = c_j$ and $\langle \phi_j, \phi_k\rangle = \delta_{jk}$, it follows that $c_j = a_j$. Therefore the coefficients a_k should be adjusted to the expansion coefficients c_k. This approximation by means of a minimization of the mean square error is known as an approximation "in the mean."

Since $\mathscr{E}$ cannot be negative, it follows that

$$\int_a^b f^2(t)\,dt - \sum_{k=1}^{n} c_k^{\,2} \geqslant 0 \tag{4.12-8}$$

Equation 4.12-8 is known as Bessel's inequality. Since the terms in the approximating series are orthonormal, the addition of orthonormal terms $\phi_{n+1}(t)$, $\phi_{n+2}(t)$, . . . must decrease the mean square error between the function and the approximation. However, even though the summation $\sum_{k=1}^{\infty} c_k^{\,2}$ converges to a positive number which is not greater than $\int_a^b f^2(t)\,dt$, this positive number may not be identically equal to the integral. A good illustration of this point is the fact that a Fourier series for a given function may consist of the mutually orthogonal cosine set if the function is even, or of the mutually orthogonal sine set if the function is odd. If the function is neither odd nor even over the interval in question, however, both the cosine set and the sine set are required to represent the function. Neither set alone generally converges to the function.

A given orthonormal set $\phi_1(t)$, $\phi_2(t)$, . . . , $\phi_n(t)$ is called "complete" if any piecewise continuous function $f(t)$ can be approximated in the mean with an arbitrarily small error by a sufficiently large number of terms, i.e.,

$$\int_a^b \left[f(t) - \sum_{k=1}^{n} c_k\phi_k(t)\right]^2 dt \leqslant \epsilon \quad \text{for a complete set of orthonormal functions} \tag{4.12-9}$$

or

$$\|f\|^2 = \sum_{k=1}^{\infty} c_k^{\,2} \tag{4.12-10}$$

Equation 4.12-10 is known as the "completeness relation." A sufficient condition for the completeness of a set of orthonormal functions is that Eq. 4.12-10 be satisfied for all continuous values of $f(t)$ over the interval (a, b). It is important to note that the completeness relation does not imply that

$$f(t) = \lim_{n\to\infty} \sum_{k=1}^{n} c_k\phi_k(t)$$

Certainly the series converges to $f(t)$ in the mean, such that the mean square error over the interval (a, b) tends to zero. The series does represent $f(t)$ at a given point if $f(t)$ is a continuous function throughout the interval, and if the series converges uniformly in the interval (a, b). However, even when a complete set of orthonormal functions is available, the convergence of the series is a rather involved problem and is not treated here.[18]

Weighted Orthogonal System

A weighting function $w(t)$ may be added to the definition of the orthonormal functions.

$$\int_a^b \phi_i(t)\phi_j(t)w(t)\,dt = \delta_{ij}$$

The weighting function is generally selected to emphasize a region of interest in the overall interval (a, b). The functions $\phi_k(t)$ are then said to be orthonormal relative to the weighting function. The Fourier coefficients of a function $f(t)$ with respect to the weighting function are given by

$$c_k = \int_a^b f(t)\phi_k(t)w(t)\,dt$$

Example 4.12-1. A given function $f(t)$ is to be approximated by a series of orthonormal functions. The orthonormal functions are to be composed of polynomials in t. If the interval of interest is $0 < t < \infty$ and the weighting function is ϵ^{-t}, the orthonormal functions are known as Laguerre polynomials.

The Laguerre polynomials are

$$L_0 = 1$$

$$L_1 = -t + 1$$

$$L_2 = \frac{t^2}{2} - 2t + 1$$

$$\cdots\cdots\cdots\cdots\cdots$$

$$L_n = \frac{1}{n!}\,\epsilon^t \frac{d^n}{dt^n}\,[t^n\epsilon^{-t}]$$

These polynomial functions are orthonormal with respect to the weighting function ϵ^{-t}, since

$$\int_0^\infty L_i(t)L_j(t)\epsilon^{-t}\,dt = \delta_{ij}$$

For a given function $f(t)$, the expansion coefficients c_k are given by

$$c_k = \int_0^\infty f(t)L_k(t)\epsilon^{-t}\,dt$$

If, for example, the function $f(t) = \epsilon^{-t}$ is to be represented by a Laguerre series, the expansion coefficients of the first five terms are

$$c_0 = \tfrac{1}{2}, \quad c_1 = \tfrac{1}{4}, \quad c_2 = \tfrac{1}{8}, \quad c_3 = \tfrac{1}{16}, \quad c_4 = \tfrac{1}{32}$$

Thus

$$f(t) \doteq \sum_{k=0}^{4} c_k L_k = \tfrac{31}{32} - \tfrac{13}{16}t + \tfrac{1}{4}t^2 - \tfrac{1}{32}t^3 + \tfrac{1}{768}t^4$$

The square of the norm of the function relative to the weighting factor is

$$\|f\|_w^2 = \int_0^\infty f^2(t)w(t)\,dt = \int_0^\infty \epsilon^{-3t}\,dt = \tfrac{1}{3}$$

The mean square error of approximation is

$$\mathscr{E} = \|f\|_w^2 - \sum_{k=0}^{n} c_k^2$$

As each orthonormal term is added to the approximation, the mean square error is decreased, as shown in Table 4.12-1.

Table 4.12-1

No. of Terms	$\mathscr{E}$	$\%\mathscr{E}$	Approximation at $t = 2$	Approximation Error at $t = 2$	Maclaurin Series $= 1 - t + \frac{t^2}{2!} - \frac{t^3}{3!} + \frac{t^4}{4!}$ at $t = 2$
1	0.0833	24.99	0.5	+0.3647	+1.0
2	0.0208	6.24	0.25	+0.1147	−1.0
3	0.0052	1.56	0.125	−0.0103	+1.0
4	0.0013	0.39	0.1042	−0.0311	−0.3333
5	0.0003	0.09	0.115	−0.0203	+0.3333

A calculation of $f(t)$ for a given value of t illustrates an interesting point. The approximation for $f(t)$ as each orthonormal term is added is also given in Table 4.12-1 for $t = 2$. The actual value of $f(t)$ at $t = 2$ is $\epsilon^{-2} = 0.1353$. Note that the error at $t = 2$ actually increases as the fourth term is added, and then decreases as the fifth term is added. This is in contrast to the mean square error over the entire interval, which decreases as each term is added. A Maclaurin series expansion for $f(t)$ at $t = 2$ is extremely poor, as shown.

REFERENCES

1. F. Chio, *Mémoire sur les Fonctions Connues sous le Nom de Résultantes ou de Determinants*, Turin, 1853, p. 11.
2. L. A. Pipes, *Matrix Methods for Engineering*, Prentice-Hall Inc., Englewood Cliffs, N.J., 1963, pp. 10–12.
3. B. Friedman, *Principles and Techniques of Applied Mathematics*, John Wiley and Sons, New York, 1956, pp. 2–5.
4. M. Bôcher, *Introduction to Higher Algebra*, The Macmillan Co., New York, p. 296.

5. R. A. Frazer, W. J. Duncan, and A. R. Collar, *Elementary Matrices*, Cambridge, University Press, 1938, p. 62.
6. *Ibid.*, p. 65.
7. E. A. Guillemin, *The Mathematics of Circuit Analysis*, John Wiley and Sons, New York, 1949, pp. 113–114.
8. F. B. Hildebrand, *Methods of Applied Mathematics*, Prentice-Hall Inc., Englewood Cliffs, N.J., 1952, p. 61.
9. L. S. Pontryagin, *Ordinary Differential Equations*, Addison-Wesley Co., Reading Mass., 1962, pp. 95, 291 ff.
10. F. R. Gantmacher, *Applications of the Theory of Matrices*, Interscience Division, John Wiley and Sons, New York, 1959, pp. 301–303.
11. F. Ayres, *Theory and Problems of Matrices*, Schaum Publishing Co., New York, 1962, pp. 188–195, 203–214.
12. H. W. Turnbull and A. C. Aitken, *An Introduction to the Theory of Canonical Matrices*, Blackie and Son, London, 1932.
13. E. A. Coddington and N. Levinson, *Theory of Ordinary Differential Equations*, McGraw-Hill Book Co., New York, 1955, p. 65.
14. Frazer, Duncan, and Collar, *op. cit.*, p. 70.
15. Hildebrand, *op cit.*, pp. 65-67.
16. Frazer, Duncan, and Collar, *op. cit.*, p. 78.
17. *Ibid.*, p. 83.
18. R. Courant and D. Hilbert, *Methods of Mathematical Physics*, Vol. I, Interscience Division, John Wiley and Sons, New York, 1953, Chapters II, V, and VI.

Problems

4.1 Perform the elementary operations $\mathbf{A} + \mathbf{B}$, $\mathbf{A} - \mathbf{B}$, and $\mathbf{AB}$ on the following matrices.

$$\mathbf{A} = \begin{bmatrix} 1 & 2 & -1 \\ 2 & -5 & 1 \\ 4 & 0 & 2 \end{bmatrix} \qquad \mathbf{B} = \begin{bmatrix} 3 & -4 & 1 \\ 1 & 5 & 0 \\ 2 & -2 & 3 \end{bmatrix}$$

$$\mathbf{A} = \begin{bmatrix} a_{11} & a_{12} \\ a_{21} & a_{22} \end{bmatrix} \qquad \mathbf{B} = \begin{bmatrix} b_{11} & 0 \\ 0 & b_{22} \end{bmatrix}$$

$$\mathbf{A} = \begin{bmatrix} 1 & 2 & 3 \\ \\ 3 & 2 & 1 \end{bmatrix} \qquad \mathbf{B} = \begin{bmatrix} 1 \\ 0 \\ 1 \end{bmatrix}$$

4.2 What are the conditions on the elements a_{ij} and b_{ij} of the (2 x 2) matrices $\mathbf{A}$ and $\mathbf{B}$ such that $\mathbf{AB} = \mathbf{BA}$?

4.3 Compute $\mathbf{AB}$, where

$$\mathbf{A} = \left[\begin{array}{cc:cc} 1 & 2 & 1 & 2 \\ 0 & 1 & 2 & 1 \\ \hdashline 0 & 0 & 0 & 1 \\ 0 & 0 & 1 & 0 \end{array}\right] \qquad \mathbf{B} = \left[\begin{array}{cc:cc} 1 & 2 & 1 & 2 \\ 2 & 1 & 0 & 1 \\ \hdashline 0 & 1 & 0 & 0 \\ 1 & 0 & 0 & 0 \end{array}\right]$$

4.4 Under what conditions is $(\mathbf{A} \pm \mathbf{B})^2 = \mathbf{A}^2 + 2\mathbf{AB} + \mathbf{B}^2$?

4.5 Given the matrix equation $\mathbf{AB} = \mathbf{AC}$. Under what conditions is $\mathbf{B} \neq \mathbf{C}$?

4.6 Find $\int_0^t \mathbf{A}(t)\,dt$ and $(d/dt)\mathbf{A}(t)$ for

$$\mathbf{A} = \begin{bmatrix} \cos t & t^2 \\ 1 & \tanh t \end{bmatrix}$$

4.7 Find the value of the following determinant by using only the definition of a determinant.

$$|\mathbf{A}| = \begin{vmatrix} -7 & -4 & 3 & 2 \\ 3 & 2 & -5 & 2 \\ 6 & 4 & 0 & -4 \\ 6 & 4 & 1 & -5 \end{vmatrix}$$

4.8 Find all the minors and cofactors of the determinant given in Problem 4.7. Show that the Laplace expansion of the determinant along any row is equal to the expansion along any column.

4.9 Find the value of the determinant shown in Problem 4.7 by the method of pivotal condensation.

4.10 Show that the product of the determinants $|\mathbf{A}|$, $|\mathbf{B}|$, is equal to the determinant of the product, $|\mathbf{AB}|$, where

$$\mathbf{A} = \begin{bmatrix} 3 & 0 & 2 \\ -2 & -1 & -1 \\ -1 & -3 & 5 \end{bmatrix} \qquad \mathbf{B} = \begin{bmatrix} 1 & -1 & 4 \\ 2 & 3 & 0 \\ 5 & 0 & 2 \end{bmatrix}$$

4.11 Find the derivative d/dx of the determinant

$$|\mathbf{A}| = \begin{vmatrix} x^2 - 1 & x - 1 & 1 \\ x^4 & x^3 & 2x + 5 \\ x + 1 & x^2 & x \end{vmatrix}$$

4.12 Find **Adj A** of

$$\mathbf{A} = \begin{bmatrix} 1 & 2 & 3 \\ 0 & 1 & 2 \\ 0 & 0 & 1 \end{bmatrix} \qquad \mathbf{A} = \begin{bmatrix} 1 & 2 & 1 \\ 2 & 1 & 0 \\ -1 & 0 & 1 \end{bmatrix} \qquad \mathbf{A} = \begin{bmatrix} 1 & 1 & 1 & 0 \\ 2 & 3 & 3 & 2 \\ 1 & 3 & 3 & 2 \\ 4 & 6 & 7 & 4 \end{bmatrix}$$

4.13 Prove that, if $\mathbf{A}$ is of order n and rank $n - 1$, then $\mathbf{Adj\,A}$ is of rank 1.

4.14 Prove that, if $\mathbf{A}$ is of order n and rank $<n - 1$, then $\mathbf{Adj\,A} = \mathbf{0}$.

4.15 Show that $|\mathbf{Adj\,A}| = |\mathbf{A}|^{n-1}$ if $\mathbf{A}$ is of order n.

4.16 Find $\mathbf{A}^{-1}$ for the matrices given in Problem 4.12.

4.17 Repeat Problem 4.16 using partitioned matrices.

4.18 The inverse of a square nonsingular matrix can be found by a technique of pivotal condensation. First, the following array is set down:

$$
\begin{array}{cc|c}
 & & \overbrace{\begin{matrix} & & & & \end{matrix}}^{n} \\
 & \begin{matrix} a_{11} & a_{12} & a_{13} & \cdots & a_{1n} \\ a_{21} & a_{22} & a_{23} & \cdots & a_{2n} \\ \cdots & \cdots & \cdots & \cdots & \cdots \\ a_{n1} & a_{n2} & a_{n3} & \cdots & a_{nn} \end{matrix} & \begin{matrix} -1 & 0 & 0 & \cdots & 0 \\ 0 & -1 & 0 & \cdots & 0 \\ \cdots & \cdots & \cdots & \cdots & \cdots \\ 0 & 0 & 0 & \cdots & -1 \end{matrix} \\
\hline
n\left\{\vphantom{\begin{matrix} 1 \\ 0 \\ \cdots \\ 0 \end{matrix}}\right. & \begin{matrix} 1 & 0 & 0 & \cdots & 0 \\ 0 & 1 & 0 & \cdots & 0 \\ \cdots & \cdots & \cdots & \cdots & \cdots \\ 0 & 0 & 0 & \cdots & 1 \end{matrix} & \begin{matrix} 0 & 0 & 0 & \cdots & 0 \\ 0 & 0 & 0 & \cdots & 0 \\ \cdots & \cdots & \cdots & \cdots & \cdots \\ 0 & 0 & 0 & \cdots & 0 \end{matrix} \\
 & \begin{matrix} 0 & 0 & 0 & \cdots & 0 \end{matrix} & \begin{matrix} 1 & & & & \end{matrix}
\end{array}
$$

Using the principles of pivotal reduction, the array to the left of the broken line is eliminated. The inverse matrix appears in the lower right-hand box, where each element is divided by the element in the $2n + 1$ row and $n + 1$ column. Find the inverse matrix of

$$
\mathbf{A} = \begin{bmatrix} 2 & 2 \\ -1 & 1 \end{bmatrix}
$$

by pivotal reduction and verify the result using $\mathbf{A}^{-1} = \mathbf{Adj\ A}/|\mathbf{A}|$. Why does this technique work?

4.19 Repeat Problem 4.16 using the technique of pivotal reduction.

4.20 Find $\mathbf{A}^{-1}$ and $(d/dt)[\mathbf{A}^{-1}]$ for

$$
\mathbf{A} = \begin{bmatrix} 2t - 2 & t + 2 & -3 \\ 3t - 1 & t & -1 \\ -1 & 4t - 3 & -t + 1 \end{bmatrix}
$$

4.21 If matrix $\mathbf{A}$ is symmetric, show that only $(n/2)(n + 1)$ cofactors need be computed in obtaining $\mathbf{A}^{-1}$.

4.22 An $(m \times n)$ matrix $\mathbf{A}$ is said to have a right inverse $\mathbf{B}$ if $\mathbf{AB} = \mathbf{I}$, and a left inverse $\mathbf{C}$ if $\mathbf{CA} = \mathbf{I}$. Prove that $\mathbf{B}$ exists only if $\mathbf{A}$ is of rank m, and that $\mathbf{C}$ exists only if $\mathbf{A}$ is of rank n.

4.23 Show that $(\mathbf{AB})^{-1} = \mathbf{B}^{-1}\mathbf{A}^{-1}$, where

$$
\mathbf{A} = \begin{bmatrix} 1 & 2 & 3 \\ 2 & 4 & 5 \\ 3 & 5 & 6 \end{bmatrix} \qquad \mathbf{B} = \begin{bmatrix} 2 & 3 & 4 \\ 4 & 3 & 1 \\ 1 & 2 & 4 \end{bmatrix}
$$

4.24 Show that, if the symmetric matrix $\mathbf{A}$ is nonsingular, then the inverse matrix $\mathbf{A}^{-1}$ is also symmetric.

4.25 Find the inner and outer products of the following pairs of vectors:

$$\mathbf{x} = \begin{bmatrix} 1 \\ -1 \\ 2 \end{bmatrix} \quad \text{and} \quad \mathbf{y} = \begin{bmatrix} -1 \\ 1 \\ 1 \end{bmatrix}$$

$$\mathbf{x} = \begin{bmatrix} 1 + j \\ -1 - j \\ 2 + j2 \end{bmatrix} \quad \text{and} \quad \mathbf{y} = \begin{bmatrix} -1 - j \\ 1 + j \\ 1 - j \end{bmatrix}$$

$$\mathbf{x} = \begin{bmatrix} 1 \\ 0 \\ 2 \\ 3 \end{bmatrix} \quad \text{and} \quad \mathbf{y} = \begin{bmatrix} 2 \\ -1 \\ 1 \\ 1 \end{bmatrix}$$

$$\mathbf{x} = \begin{bmatrix} 2 + j3 \\ 1 - j2 \\ 3 \\ 0 \end{bmatrix} \quad \text{and} \quad \mathbf{y} = \begin{bmatrix} 1 + j2 \\ 0 \\ 1 + j \\ 2 \end{bmatrix}$$

4.26 Find the value of α which makes $\|\mathbf{x} - \alpha\mathbf{y}\|$ a minimum. Show that for this value of α the vector $\mathbf{x} - \alpha\mathbf{y}$ is orthogonal to $\mathbf{y}$ and that $\|\mathbf{x} - \alpha\mathbf{y}\|^2 + \|\alpha\mathbf{y}\|^2 = \|\mathbf{x}\|^2$. The vector $\alpha\mathbf{y}$ is called the *projection* of $\mathbf{x}$ on $\mathbf{y}$. Draw a diagram for the case where $\mathbf{x}$ and $\mathbf{y}$ are two-dimensional.

4.27 Find the projection of the vector (1, 1, 1) on the plane

$$x_1 + 2x_2 + 3x_3 = 0.$$

4.28 Assuming a three-dimensional space, prove that the four vectors (1, 0, 0), (0, 1, 0), (0, 0, 1), and (1, 1, 1) form a linearly dependent set, but that any three of these vectors form a linearly independent set.

4.29 Express the vector $\mathbf{y} = (6, 3)$ in terms of the basis vectors $\mathbf{x}_1 = (-1, 2)$, $\mathbf{x}_2 = (1, 3)$. Determine the reciprocal basis.

4.30 Using as a set of basis vectors $\mathbf{x}_1 = (1, 1, 1)$, $\mathbf{x}_2 = (1, 0, 0)$, and $\mathbf{x}_3 = (0, 1, 0)$, find an orthonormal basis by use of the Gram-Schmidt orthogonalization procedure.

4.31 What is the dimension of the vector space spanned by the following sets of vectors?

(*a*) $\mathbf{x}_1 = (1, 2, 2, 1)$, $\mathbf{x}_2 = (1, 0, 0, 1)$, $\mathbf{x}_3 = [3, 4, 4, 3]$

(*b*) $\mathbf{x}_1 = (1, 1, 1)$, $\mathbf{x}_2 = (1, 0, 1)$, $\mathbf{x}_3 = (1, 2, 1)$

(*c*) $\mathbf{x}_1 = (1, 0, 1, 0)$, $\mathbf{x}_2 = (0, 0, 5, 0)$, $\mathbf{x}_3 = (10, 0, 1, 0)$, $\mathbf{x}_4 = (5, 0, 7, 0)$

(*d*) $\mathbf{x}_1 = (1, 0, 0)$, $\mathbf{x}_2 = (0, 1, 0)$, $\mathbf{x}_3 = (0, 0, 1)$, $\mathbf{x}_4 = (1, 1, 1)$

4.32 Find the rank and degeneracy of the following matrices.

$$\mathbf{A} = \begin{bmatrix} 2 & -2 & 3 \\ 10 & -4 & 5 \\ 5 & -4 & 6 \end{bmatrix} \qquad \mathbf{B} = \begin{bmatrix} 7 & 4 & -1 \\ 4 & 7 & -1 \\ -4 & -4 & 4 \end{bmatrix}$$

4.33 Find the degeneracy of the product **AB** where **A** and **B** are the matrices given in Problem 4.32.

4.34 Let S_1 be a subspace of the n-dimensional space S. Show that any vector $\mathbf{x}$ which is not in S_1 can be represented by

$$\mathbf{x} = \mathbf{y} + \mathbf{y}_1$$

where $\mathbf{y}$ is a vector in S_1 and $\mathbf{y}_1 = \mathbf{x} - \mathbf{y}$ is orthogonal to all vectors in S_1. Draw a three-dimensional picture illustrating this theorem. (See Problem 4.26.)

4.35 Show that, if the n-dimensional vector $\mathbf{x}$ is orthogonal to a set of basis vectors of an n-dimensional space, then $\mathbf{x} = \mathbf{0}$.

4.36 Solve the following set of equations using Cramer's Rule.

$$\begin{aligned} x_1 + x_2 + x_3 + x_4 &= 10 \\ 3x_1 + x_2 - x_3 - 4x_4 &= -14 \\ 2x_1 + 2x_2 - x_3 + x_4 &= 7 \\ x_1 + 3x_2 + 4x_3 - x_4 &= 15 \end{aligned}$$

4.37 Find the solution to the equation $\mathbf{y} = \mathbf{A}\mathbf{x}$ using the inverse matrix $\mathbf{A}^{-1}$, where

$$\mathbf{A} = \begin{bmatrix} 1 & 2 & 3 \\ -1 & 0 & 2 \\ 0 & 1 & 0 \end{bmatrix} \qquad \mathbf{x} = \begin{bmatrix} x_1 \\ x_2 \\ x_3 \end{bmatrix} \qquad \mathbf{y} = \begin{bmatrix} 1 \\ 3 \\ 5 \end{bmatrix}$$

4.38 A convenient pivotal reduction scheme for computing the n unknowns in an nth order system is as follows: Add to the coefficient matrix an additional column consisting of the negative of the right-hand members. Add to this array an additional $n + 1$ rows which have unity in the diagonal element and all other terms equal to zero.

$$\begin{array}{cccccc} a_{11} & a_{12} & a_{13} & \cdots & a_{1n} & -y_1 \\ a_{21} & a_{22} & a_{23} & \cdots & a_{2n} & -y_2 \\ \cdots & \cdots & \cdots & \cdots & \cdots & \cdots \\ a_{n1} & a_{n2} & a_{n3} & \cdots & a_{nn} & -y_n \\ \hline 1 & 0 & 0 & \cdots & 0 & 0 \\ 0 & 1 & 0 & \cdots & 0 & 0 \\ \cdots & \cdots & \cdots & \cdots & \cdots & \cdots \\ 0 & 0 & 0 & \cdots & 0 & 1 \end{array} \left.\vphantom{\begin{array}{c}1\\0\\ \cdots \\0\end{array}}\right\} n + 1$$

Choose any element a_{mk} as the pivot term, where m and k are less than or equal to n. Using the method of pivotal condensation, the rows and columns of this array are reduced one at a time until only column $n + 1$ remains. This column consists of $n + 1$ elements $c_1, c_2, \ldots, c_{n+1}$. The solution for the kth unknown x_k is then given by

$$x_k = \frac{c_k}{c_{n+1}}$$

Solve the following set of equations using the method of pivotal condensation.

$$\begin{aligned} x_1 + x_2 + x_3 &= 2 \\ x_1 - x_2 + x_3 &= -2 \\ 2x_1 - x_2 + x_3 &= -1 \end{aligned}$$

Show why this technique works.

4.39 Solve the following sets of linear equations.

(*a*)
$$\begin{aligned} x_1 + x_2 + x_3 + x_4 &= 0 \\ x_1 + x_2 + x_3 - x_4 &= 4 \\ x_1 + x_2 - x_3 + x_4 &= -4 \\ x_1 - x_2 + x_3 + x_4 &= 2 \end{aligned}$$

(*b*)
$$\begin{aligned} x_1 + 2x_2 + 9x_3 &= 0 \\ 2x_1 \qquad + 2x_3 &= 0 \\ 3x_1 - 2x_2 - 5x_3 &= 0 \end{aligned}$$

(*c*)
$$\begin{aligned} 2x_1 \qquad + x_3 - x_4 &= 0 \\ x_1 + 3x_2 + 2x_3 + 4x_4 &= 0 \\ x_1 + x_2 + x_3 + x_4 &= 0 \end{aligned}$$

4.40 Find the characteristic values and characteristic vectors for the following matrices.

$$(a) \begin{bmatrix} 2 & -1 \\ -1 & 3 \end{bmatrix} \quad (b) \begin{bmatrix} 1 & 1 \\ -1 & 1 \end{bmatrix} \quad (c) \begin{bmatrix} \frac{3}{2} & 1 & \frac{3}{2} \\ -\frac{1}{2} & 0 & -\frac{3}{2} \\ \frac{1}{2} & -1 & \frac{1}{2} \end{bmatrix}$$

4.41 Find the characteristic vectors, the modal matrix, and the diagonal form for the matrices

$$(a) \begin{bmatrix} 0 & 1 & 0 \\ 0 & 0 & 1 \\ -2 & -4 & -3 \end{bmatrix} \quad (b) \begin{bmatrix} 2 & -2 & 3 \\ 1 & 1 & 1 \\ 1 & 3 & -1 \end{bmatrix} \quad (c) \begin{bmatrix} 7 & 4 & -1 \\ 4 & 7 & -1 \\ -4 & -4 & 4 \end{bmatrix}$$

4.42 Show that

$$\mathbf{A} = \begin{bmatrix} 2 & 2 & 1 \\ 1 & 3 & 1 \\ 1 & 2 & 2 \end{bmatrix} \quad \text{and} \quad \mathbf{B} = \begin{bmatrix} 2 & 1 & -1 \\ 0 & 2 & -1 \\ -3 & -2 & 3 \end{bmatrix}$$

have the same characteristic values but are not similar matrices.

4.43 If a characteristic equation $p(\lambda_s) = 0$ with a multiple root of order s has degeneracy q, where $q > 1$, prove that the adjoint matrix $\mathbf{Adj}\ [\lambda_s\mathbf{I} - \mathbf{A}]$ and all its derivatives up to and including $(d^{q-2}/d\lambda^{q-2})[\mathbf{Adj}\ (\lambda\mathbf{I} - \mathbf{A})]_{\lambda=\lambda_s}$ are null.

4.44 The matrix

$$\mathbf{A} = \begin{bmatrix} 1 & 0 & 0 \\ 2 & 1 & 0 \\ 1 & 4 & 1 \end{bmatrix}$$

has multiple roots. Show that by adding a perturbation matrix

$$\begin{bmatrix} \epsilon_1 & 0 & 0 \\ 0 & \epsilon_2 & 0 \\ 0 & 0 & \epsilon_3 \end{bmatrix}$$

all the characteristic values of **A** are distinct. Show that as ϵ_i approaches zero all the characteristic vectors collapse into one characteristic vector.

4.45 Find the sum and product of the characteristic values of **A**, where

$$\mathbf{A} = \begin{bmatrix} 1 & 0 & 1 & 3 \\ 0 & 0 & 2 & 0 \\ 0 & 5 & 1 & 2 \\ 5 & 1 & 0 & -2 \end{bmatrix}$$

4.46 If the nth order matrix **A** has a characteristic value λ_i repeated q times, show that the rank of $\lambda_i\mathbf{I} - \mathbf{A}$ is not less than $n - q$, and that the associated invariant vector space (null space) is not greater than q.

4.47 Prove that the characteristic values of a unitary matrix $[\mathbf{A}^{-1} = (\mathbf{A}^T)^*]$ and an orthogonal matrix $(\mathbf{A}^{-1} = \mathbf{A}^T)$ have an absolute value equal to unity.

4.48 Prove that, if λ_i is a nonzero characteristic value of **A**, then $|\mathbf{A}|/\lambda_i$ is a characteristic value of **Adj A**.

4.49 If the matrix **A** has characteristic values $\lambda_1, \ldots, \lambda_n$, show that

(*a*) $\mathbf{A}^m$ has characteristic values $\lambda_1{}^m, \ldots, \lambda_n{}^m$.
(*b*) $\mathbf{A}^{-1}$ has characteristic values $1/\lambda_1, \ldots, 1/\lambda_n$
(*c*) $\mathbf{A}^T$ has characteristic values $\lambda_1, \ldots \lambda_n$.
(*d*) $k\mathbf{A}$ has characteristic values $k\lambda_1, \ldots, k\lambda_n$.

4.50 If $\mathbf{A}$ is a triangular matrix, such that the elements either above or below the principal diagonal are zero, show that the elements of the principal diagonal are the characteristic values of $\mathbf{A}$.

4.51 A vector $\mathbf{x}_k$, for which $(\mathbf{A} - \lambda_i\mathbf{I})^{k-1}\mathbf{x}_k \neq \mathbf{0}$, but for which $(\mathbf{A} - \lambda_i\mathbf{I})^k\mathbf{x}_k = \mathbf{0}$, is called a "generalized characteristic vector" or a characteristic vector of rank k associated with the characteristic value λ_i. These generalized characteristic vectors are useful in determining the modal matrix for systems which have repeated characteristic values.

(*a*) Show that characteristic vectors of different rank are linearly independent.

(*b*) Find the generalized characteristic vectors for

$$\mathbf{A} = \begin{bmatrix} 1 & 0 & 0 \\ 1 & 1 & 0 \\ 2 & 3 & 2 \end{bmatrix}$$

(*c*) Find a modal matrix $\mathbf{M}$ for the matrix of part (*b*).

4.52 The adjoint transformation $\mathbf{A}^*$ of the linear transformation $\mathbf{A}$ is defined as $\langle \mathbf{Ax}, \mathbf{y} \rangle = \langle \mathbf{x}, \mathbf{A}^*\mathbf{y} \rangle$. Show that $(\mathbf{A}^*)^* = \mathbf{A}$.

4.53 Solve the following set of linear equations by using only elementary row transformations.

$$\begin{aligned} x_1 + 2x_2 - x_3 &= 2 \\ 2x_1 + x_2 + x_3 &= 7 \\ 5x_1 - 3x_2 + 3x_3 &= 8 \end{aligned}$$

4.54 Reduce

$$\mathbf{A} = \begin{bmatrix} 0 & 1 & 2 \\ -1 & 0 & -3 \\ -2 & 3 & 0 \end{bmatrix}$$

to the canonical form shown in Eq. 4.8-16.

4.55 Reduce

$$\mathbf{A} = \begin{bmatrix} 6j & j & 1+j \\ j & 2j & -1 \\ -1+j & 1 & j \end{bmatrix}$$

to the canonical form shown in Eq. 4.8-19.

4.56 Reduce

$$\mathbf{A} = \begin{bmatrix} 1 & 3 & 2 \\ 3 & 2 & 5 \\ 2 & 5 & 1 \end{bmatrix}$$

to the canonical form of Eq. 4.8-15.

4.57 *Introduction to Linear Codes:* Suppose that four binary digits $\{c_1, c_2, c_3, c_4\}$, where each c may be either zero or one, are to be stored in a computer memory. However, one or more of the digits may be inadvertently changed during the read or write cycle. Thus, the numbers read out of memory, $\{x_1, x_2, x_3, x_4\}$, may be in error. It is possible to alleviate this situation somewhat by introducing "parity check bits" c_5, c_6, c_7 as specified below. The encoded data $(c_1, \ldots, c_7)$ are then stored in seven bits of memory.

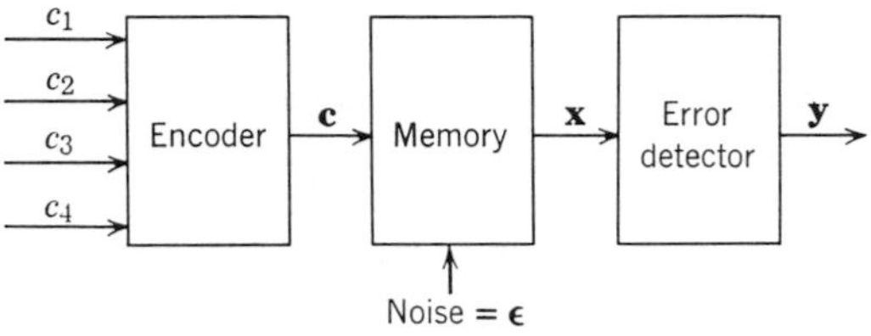

Fig. P4.57

The system shown in Fig. P4.57 illustrates this procedure. c_5, c_6, and c_7 are computed from c_1, c_2, c_3, c_4 in the encoder. The numbers $x_1, \ldots, x_7$ are later read out of the memory and passed through a decoder.

$$\mathbf{c} = \begin{bmatrix} c_1 \\ c_2 \\ \cdot \\ \cdot \\ \cdot \\ c_7 \end{bmatrix} \qquad \boldsymbol{\epsilon} = \begin{bmatrix} \epsilon_1 \\ \epsilon_2 \\ \cdot \\ \cdot \\ \cdot \\ \epsilon_7 \end{bmatrix} \qquad \mathbf{x} = \begin{bmatrix} x_1 \\ x_2 \\ \cdot \\ \cdot \\ \cdot \\ x_7 \end{bmatrix} \qquad y = \begin{bmatrix} y_1 \\ y_2 \\ y_3 \end{bmatrix}$$

$$\mathbf{x} = \mathbf{c} + \boldsymbol{\epsilon}$$

c_5, c_6, c_7 are chosen as follows:

c_5 is chosen so that the parity of (number of ones in) the sequence $s_1 = \{c_2, c_3, c_4, c_5\}$ is even.

c_6 is chosen to make the parity of $s_2 = \{c_1, c_2, c_4, c_6\}$ even.

c_7 is chosen to make the parity of $s_3 = \{x_1, x_3, x_4, x_7\}$ even.

Errors are detected by checking the parity of the sequences corresponding to s_1, s_2, s_3 in the vector $\mathbf{x}$. Define the vector $\mathbf{y}$ such that

$$y_1 = \begin{cases} 1 & \text{if the parity of } s_1 \text{ is odd} \\ 0 & \text{if the parity of } s_1 \text{ is even} \end{cases}$$

and likewise for y_2 and s_2, and y_3 and s_3.

It turns out that, if no more than one error occurs (i.e., $\mathbf{x}$ and $\mathbf{c}$ differ in no more than one position), then it is possible to reconstruct $\mathbf{c}$ merely by computing $\mathbf{y}$. This is called a Hamming code. The problem is formulated in terms of vector spaces and linear transformations. All arithmetic is done modulo 2:

$$\begin{array}{ll} 0+0=0 & 0\cdot 0=0 \\ 0+1=1 & 0\cdot 1=0 \\ 1+0=1 & 1\cdot 0=0 \\ 1+1=0 & 1\cdot 1=1 \end{array}$$

(*a*) Show that the mapping from the **x** vectors to the **y** vectors is a linear transformation T. Find the matrix **A** of this transformation T with respect to the natural basis in V_7 and the natural basis in V_3 (the seven and three-dimensional spaces, respectively).

(*b*) List the eight error vectors $\boldsymbol{\epsilon}$ which correspond to

(1) no error;
(2) an error in the ith place only for $i = 1, 2, \ldots, 7$.

Note: $\mathbf{x} = \mathbf{c} + \boldsymbol{\epsilon}$ defines $\boldsymbol{\epsilon}$

(*c*) Show that the vectors **c** are the null space of **A**. Find the matrix of the form $\begin{bmatrix} \mathbf{I} \\ \mathbf{B} \end{bmatrix}$ whose column vectors are a basis for this null space.

(*d*) If the only possible errors are those listed in (*b*), show that it is always possible to detect which $\boldsymbol{\epsilon}$ has occurred by computing **y**. For each **y** list the corresponding $\boldsymbol{\epsilon}$.

(*e*) In general, any matrix of ones and zeros defines a "linear code." There are other 3 by 7 matrices which define linear codes which have the same error-correcting properties as **A**. How many are there, and what is their relationship to **A**?

(*f*) In general, if a code defined by an (m x n) matrix **A** of a linear transformation T from V_n into V_m is to correct for a possible set of n-dimensional error vectors $\{\mathbf{0}, \boldsymbol{\epsilon}_1, \ldots, \boldsymbol{\epsilon}_p\}$, what condition must be placed upon the transformed vectors $\{T(\boldsymbol{\epsilon}_1), \ldots, T(\boldsymbol{\epsilon}_p)\}$ in V_m? Prove your answer using the linearity property for T.

(*g*) Is it possible to construct a single error-correcting code such as the one above from a (3 x 8) matrix? a (3 x 6) matrix? Tell why, referring to your answer to (*f*).

4.58 If the coefficient matrix **A** of the linear transformation $\mathbf{y} = \mathbf{Ax}$ is singular, then the null space of **A** is the vector space formed by the vectors which satisfy $\mathbf{Ax} = \mathbf{0}$. Find the null space of

$$\mathbf{A} = \begin{bmatrix} 1 & 2 & -2 \\ 1 & 3 & 3 \\ 0 & 1 & 5 \end{bmatrix}$$

4.59 This problem is meant to illustrate the meaning of characteristic vector, and of characteristic value in terms of two-dimensional spaces.

(*a*) Consider the linear transformation which is rotation of vectors by an angle θ as shown in Fig. P4.59*a*.

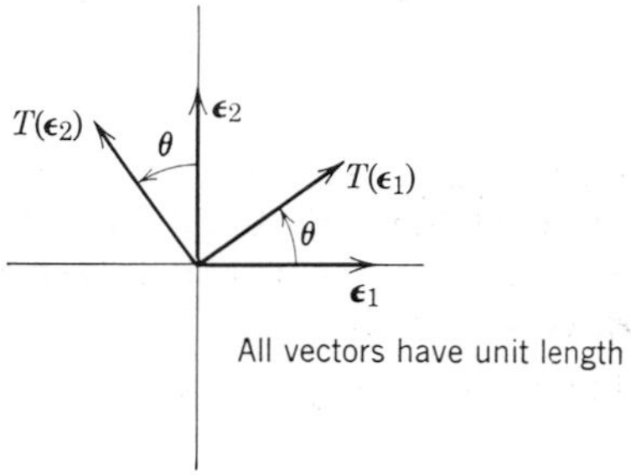

Fig. P4.59*a*

(1) Find the matrix of T with respect to ϵ_1 and ϵ_2.
(2) Find all characteristic vectors, and characteristic values.

(*b*) Consider the linear transformation $T(\epsilon) = \mathbf{A}\epsilon$, where

$$\mathbf{A} = \begin{bmatrix} k_1 & (k_2 - k_1) \\ 0 & k_2 \end{bmatrix}$$

(1) Find the characteristic vectors and characteristic values.
(2) Sketch, in the ϵ_1, ϵ_2 plane, the characteristic vectors ρ_1 and ρ_2 and their transformations $T(\rho_1)$ and $T(\rho_2)$.
(3) Suppose $k_1 = k_2$. Describe the situation
  (i) geometrically;
  (ii) in term of characteristic vectors and characteristic values.

(*c*) To see how it is possible that only one characteristic vector exists, consider the linear transformation obtained as follows: Stand a deck of playing cards on edge so that you are looking at the deck sideways. Draw a vector α on the edge of the deck. Now "skew" the deck as shown in Fig. P4.59*b* and note the "new" vector $\alpha' = T(\alpha)$ obtained. What is the matrix of T with respect to ϵ_1 and ϵ_2?

This transformation has an element of "shear" not present in (a) or (b).

(1) Find the (second order) characteristic vector, and the characteristic value.
(2) Find the Jordan normal form.

(*d*) Consider the transformation which is the result of first performing (*b*) for $k_1 = \sqrt{2}$ and $k_2 = 1/\sqrt{2}$; and then performing (*a*) for $\theta = 45°$.

(1) What is the matrix of this transformation?

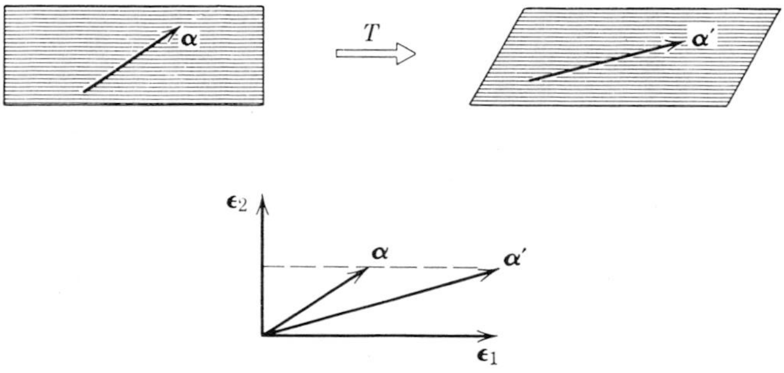

Fig. P4.59*b*

(2) Find the (second order) characteristic vector and the characteristic value, and interpret in terms of (*c*).

4.60 A second order orthogonal matrix must be of the form

$$\mathbf{A}_p = \begin{bmatrix} \cos\theta & \sin\theta \\ -\sin\theta & \cos\theta \end{bmatrix} \quad \text{or} \quad \mathbf{A}_n = \begin{bmatrix} \cos\theta & \sin\theta \\ \sin\theta & -\cos\theta \end{bmatrix}$$

(*a*) Prove that this is true.

(*b*) Consider the linear transformation $\mathbf{y} = \mathbf{A}\mathbf{x}$, where $\mathbf{A} = \mathbf{A}_p$. What can be said about the components of a vector in the original coordinate system with respect to the same components in the transformed coordinate system? How is the relationship between two different vectors in the original coordinate system affected? Draw a picture of this transformation.

(*c*) Repeat (*b*) for $\mathbf{A} = \mathbf{A}_n$.

4.61 Problem 4.59 illustrated the difference between linear transformations whose Jordan forms have ones on the superdiagonal and those which do not. This was done in two dimensions by means of an analog involving a pack of playing cards. In a similar manner (not necessarily involving cards), develop analogies in three dimensions for Jordan forms:

$$(a)\ \begin{bmatrix} 1 & 0 & 0 \\ 0 & 1 & 1 \\ 0 & 0 & 1 \end{bmatrix} \quad \text{and} \quad (b)\ \begin{bmatrix} 1 & 1 & 0 \\ 0 & 1 & 1 \\ 0 & 0 & 1 \end{bmatrix}$$

Make sketches of both (*a*) and (*b*), indicating clearly the directions of *all* characteristic vectors.

4.62 Find the Jordan form and a modal matrix for the following matrices.

$$(a)\begin{bmatrix} 1 & 2 \\ -2 & -3 \end{bmatrix} \qquad (b)\begin{bmatrix} 4 & 2 & 2 & 0 \\ -1 & 0 & -1 & -1 \\ -1 & 0 & 1 & 1 \\ 1 & 2 & 1 & 3 \end{bmatrix}$$

$$(c)\begin{bmatrix} 2 & 0 & 0 & 0 \\ 0 & 1 & 1 & 0 \\ 0 & 1 & 3 & 2 \\ 0 & 1 & -1 & 2 \end{bmatrix} \qquad (d)\begin{bmatrix} 1 & 0 & 0 & 0 \\ 0 & 0 & 1 & 0 \\ 0 & 1 & 0 & 0 \\ 0 & -1 & -1 & -1 \end{bmatrix}$$

$$(e)\begin{bmatrix} 0 & 0 & 0 & 0 \\ 0 & 0 & 0 & 0 \\ 0 & 0 & 0 & 1 \\ 1 & 0 & 0 & 0 \end{bmatrix}$$

• **4.63** Determine if the following matrices are: (1) positive definite, (2) positive semidefinite, (3) negative definite, (4) negative semidefinite

$$(a)\begin{bmatrix} 2 & 1 & 1 \\ 1 & 3 & 0 \\ 1 & 0 & 1 \end{bmatrix} \qquad (b)\begin{bmatrix} \frac{3}{2} & \frac{1}{6} & -\frac{1}{3} \\ \frac{1}{6} & \frac{7}{6} & -\frac{1}{3} \\ -\frac{1}{3} & -\frac{1}{3} & -1 \end{bmatrix}$$

$$(c)\begin{bmatrix} 2 & 1 & 1 \\ 1 & 3 & 2 \\ 1 & 2 & 1 \end{bmatrix} \qquad (d)\begin{bmatrix} -1 & -1 & -1 \\ -1 & -4 & -1 \\ -1 & -1 & -2 \end{bmatrix}$$

$$(e)\begin{bmatrix} 1 & 1 & 1 \\ 1 & 3 & 1 \\ 1 & 1 & 1 \end{bmatrix}$$

4.64 Consider the quadratic equation $\langle \mathbf{x}, \mathbf{A}\mathbf{x} \rangle = 1$, $\mathbf{x}$ real, $\mathbf{A}(2 \times 2)$, real, symmetric with $a_{11} > 0$. The equation defines a curve in the $\mathbf{x}$-plane.

(*a*) Show that, if $\mathbf{A}$ is positive definite, the curve is an ellipse.

(1) What significance have the characteristic vectors in terms of the orientation of the ellipse?

(2) What significance have the characteristic values?

(3) Draw a sketch to illustrate (1) and (2).

(4) What happens if there is only one characteristic value?

(*b*) Show that, if $|\mathbf{A}| < 0$, the curve is a hyperbola. Repeat (1), (2), and (3) above for the hyperbola.

(*c*) What happens if one of the characteristic values is zero?

(*d*) Sketch curves defined by the equation for the following values of **A**. Find characteristic values and characteristic vectors in each case, and indicate these quantities on your sketch.

$$\mathbf{A} = \begin{bmatrix} \frac{5}{2} & \frac{3}{2} \\ \frac{3}{2} & \frac{5}{2} \end{bmatrix} \qquad \mathbf{A} = \begin{bmatrix} \frac{3}{2} & \frac{5}{2} \\ \frac{5}{2} & \frac{3}{2} \end{bmatrix}$$

4.65 Consider the function $Q(\mathbf{x}) = \langle \mathbf{x}, \mathbf{Ax} \rangle$, where **A** is a real, symmetric matrix.

(*a*) Show that $\mathbf{x} = \mathbf{0}$ is a stationary point.

(*b*) Under what condition on **A** is $\mathbf{x} = \mathbf{0}$ a relative maximum? Prove it.

(*c*) Under what condition on **A** is $\mathbf{x} = \mathbf{0}$ a relative minimum? Prove it.

(*d*) If **A** is nonsingular and fits neither of these conditions, then what is $\mathbf{x} = \mathbf{0}$?

(*e*) What happens if **A** is singular with either single or multiple degeneracy?

(*f*) Illustrate your answers to (*b*) through (*e*) by means of sketches of the level curves of (1) to (4) below.

$$(1)\ \mathbf{A} = \begin{bmatrix} 1 & -1 \\ -1 & 1 \end{bmatrix} \qquad (2)\ \mathbf{A} = \begin{bmatrix} -2 & 1 \\ 1 & -1 \end{bmatrix}$$

$$(3)\ \mathbf{A} = \begin{bmatrix} -1 & 1 \\ 1 & 1 \end{bmatrix} \qquad (4)\ \mathbf{A} = \begin{bmatrix} 1 & 1 \\ 1 & 1 \end{bmatrix}$$

• **4.66** For the matrix

$$\mathbf{A} = \begin{bmatrix} 0 & 1 \\ -2 & -3 \end{bmatrix}$$

find (*a*) $\mathbf{A}^{1/2}$, (*b*) cos **A**, (*c*) sinh **A**.

• **4.67** Find $\epsilon^{\mathbf{A}}$ by both the Cayley-Hamilton theorem and Sylvester's theorem for

$$\bullet\ (a)\ \mathbf{A} = \begin{bmatrix} 1 & \frac{1}{2} & \frac{1}{2} \\ -1 & \frac{1}{2} & \frac{3}{2} \\ \frac{3}{2} & \frac{1}{2} & -\frac{1}{2} \end{bmatrix} \qquad (b)\ \mathbf{A} = \begin{bmatrix} -7 & 3 & 6 & 6 \\ -4 & 2 & 4 & 4 \\ -3 & 5 & 0 & -1 \\ -2 & -2 & 4 & 5 \end{bmatrix}$$

$$\bullet\ (c)\ \mathbf{A} = \begin{bmatrix} 0 & 1 & 0 \\ 0 & 0 & 1 \\ 1 & -3 & 3 \end{bmatrix} \qquad (d)\ \mathbf{A} = \begin{bmatrix} 2 & 1 & 0 \\ 1 & 2 & 0 \\ 1 & 1 & 1 \end{bmatrix}$$

4.68 Find $\mathbf{A}^{10}$ where $\mathbf{A}$ is given by

$$(a)\ \mathbf{A} = \begin{bmatrix} 1 & 0 & 0 \\ 0 & 2 & 2 \\ 0 & 2 & 2 \end{bmatrix} \qquad (b)\ \mathbf{A} = \begin{bmatrix} 1 & -\frac{1}{3} & 0 \\ 3 & -1 & 0 \\ 4 & -\frac{4}{3} & -\frac{1}{3} \end{bmatrix}$$

4.69 Equation 4.11-7 defines the integral $\int_0^t \epsilon^{\mathbf{A}t}\, dt$ in terms of $\mathbf{A}^{-1}$. If $\mathbf{A}$ is singular, how can this integration be performed?

4.70 Prove Eq. 4.11-9.

4.71 Prove that $\|f(x) - g(x)\| = \|f(x)\| + \|g(x)\|$ [f and g are defined over the interval (a, b)] only if $f(x)$ and $g(x)$ are orthogonal over the interval (a, b).

4.72 Show that the mean value of a function over a given interval is always less than or equal to the rms value of the function over the same interval.

4.73 Show that Problem 4.72 is a special case of the Schwarz inequality

$$\left| \int_a^b f(x)g(x)\, dx \right| \leqslant \left(\sqrt{\int_a^b f^2(x)\, dx} \right) \left(\sqrt{\int_a^b g^2(x)\, dx} \right)$$

Prove that this equality exists if and only if $g(x) = kf(x)$, where k is a constant.

4.74 Prove that the Laguerre polynomials are orthonormal over the interval $(0, \infty)$.

4.75 One technique which has been proposed for linear process identification using only the input-output operating record of the process is to model the process with a set of orthogonal transfer functions. In this technique, the output of the model of the process is taken as a linear sum of the outputs of the orthogonal transfer functions. The input to the model is the same as the input to the process. See Fig. P4.75. The coefficients

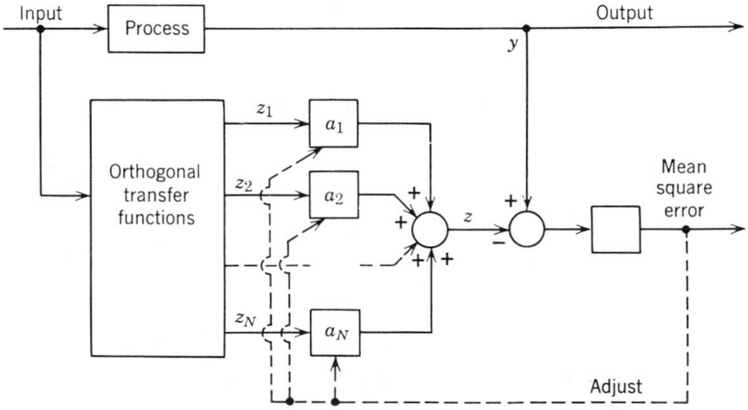

Fig. P4.75

$a_1, a_2, \ldots, a_N$ are adjusted until the mean square difference between the output of the process, $y(t)$, and the output of the model, $z(t)$, is a minimum. The set of orthogonal transfer functions is then said to "model" the process.

(*a*) If

$$z(t) = \sum_{i=1}^{N} a_i z_i(t)$$

show that the optimum settings for the coefficients are given by

$$a_j = \frac{\lim\limits_{T\to\infty} \dfrac{1}{2T}\displaystyle\int_{-T}^{T} y(t)z_j(t)\,dt}{\lim\limits_{T\to\infty} \dfrac{1}{2T}\displaystyle\int_{-T}^{T} z_j^2(t)\,dt} = \frac{\text{time average of } y(t)z_j(t)}{\text{mean square value of } z_j(t)}$$

(*b*) A set of n orthonormal functions is to be constructed for the interval $0 \leqslant t \leqslant \infty$ using as a set of linearly independent transfer functions

$$F_i(s) = \frac{1}{s + s_i} \qquad i = 1, \ldots, n$$

Show that the orthonormal transfer functions (called Kautz orthogonal filters) are given by

$$\Phi_1(s) = \frac{\sqrt{2s_1}}{s + s_1}$$

$$\Phi_n(s) = \sqrt{2s_n}\,\frac{\prod_{k=1}^{n-1}(s - s_k)}{\prod_{k=1}^{n}(s + s_k)}, \quad n \geqslant 2$$

(*c*) If a weighting factor $w(t) = \epsilon^{-2\alpha t}$ is used, show that the orthonormal transfer functions are given by

$$\Phi_1(s) = \frac{\sqrt{2(s_1 + \alpha)}}{s + s_1}$$

$$\Phi_2(s) = \frac{\sqrt{2(s_2 + \alpha)}(s - 2\alpha - s_1)}{(s + s_1)(s + s_2)}$$

$$\Phi_n(s) = \sqrt{2(s_n + \alpha)}\,\frac{\prod_{k=1}^{n}(s - 2\alpha - s_k)}{\prod_{k=1}^{n}(s + s_k)}, \quad n > 2$$

(*d*) If the orthonormal functions are constructed from transfer functions having pairs of complex poles at $s = -\alpha_i \pm j\beta_i$ and a weighting factor

$\epsilon^{-2\alpha t}$, show that the orthonormal functions are

$$\Phi_1(s) = \sqrt{2(\alpha + \alpha_1)}\ \frac{s - \alpha + |s_1|}{(s + \alpha_1)^2 + \beta_1^2}$$

$$\Phi_2(s) = \sqrt{2(\alpha + \alpha_1)}\ \frac{s - \alpha - |s_1|}{(s + \alpha_1)^2 + \beta_1^2}$$

$$\cdots\cdots\cdots\cdots\cdots\cdots\cdots\cdots\cdots$$

$$\Phi_{2i-1} = \sqrt{2(\alpha + \alpha_i)}\ \frac{(s - \alpha + |s_1|)\prod_{k=1}^{i} [(s - 2\alpha - \alpha_k)^2 + \beta_k^2]}{\prod_{k=1}^{i}[(s + \alpha_k)^2 + \beta_k^2]}$$

$$\Phi_{2i} = \sqrt{2(\alpha + \alpha_i)}\ \frac{(s - \alpha - |s_i|)\prod_{k=1}^{i} [(s - 2\alpha - \alpha_k)^2 + \beta_k^2]}{\prod_{k=1}^{i} [(s + \alpha_k)^2 + \beta_k^2]}$$

where $i = 1, 2, \ldots, n/2$ and n = number of complex poles.

$$|s_k|^2 = (\alpha + \alpha_k)^2 + \beta_k^2$$

Hint: Parseval's theorem states that

$$\int_0^\infty f_1(t)f_2(t)\,dt = \frac{1}{2\pi j}\int_{c-j\infty}^{c+j\infty} F_1(s)F_2(-s)\,ds$$

If the weighting factor $w(t) = \epsilon^{-2\alpha t}$ and a set of orthogonal functions $\phi_i(t)$ are used, then

$$\int_0^\infty \phi_i(t)\phi_j(t)\epsilon^{-2\alpha t}\,dt = \frac{1}{2\pi j}\int_{c-j\infty}^{c+j\infty} \Phi_i(s + \alpha)\Phi_j(-s + \alpha)\,ds = \delta_{ij}$$

5

State Variables and Linear Continuous Systems

5.1 INTRODUCTION

In this chapter the matrix techniques and vector space concepts of Chapter 4 are brought to bear upon the problem of multiple input-multiple output systems. Because of the increasing demands for well-designed, complex systems, a search has been made for techniques which are useful in the design of such systems. Not only must these techniques be amenable to computation, but, more importantly, they must also be amenable to conceptual thinking. Most practicing engineers think in intuitive terms, and any proposed technique must enable the engineer to grasp the concept of the technique and relate it to his previous training.

The analysis and synthesis of multiple input-multiple output systems, hereafter called multivariable systems, is a formidable task. The calculation and control of interrelated effects in a multivariable system is a complicated exhausting process and is best done by an electronic computer. However, the engineer who programs the computer must be aware of all the possible techniques to solve his problem. From these reasons, and others, the search for techniques to handle these problems has centered on the "state variable" approach.

From a mathematical viewpoint, the state variable approach is the use of matrix and vector methods to handle the large number of variables which enter into such problems. As such, these are not new methods, but rather they are the rediscovery of existing mathematical techniques. They aid considerably in the solution of linear multivariable problems. More importantly, however, the state variable approach aids conceptual thinking about these problems, and nonlinear system problems as well. Furthermore, it provides a unifying basis for thinking about linear and nonlinear problems. These two topics are frequently treated as somewhat unrelated by the engineer.

5.2 SIMULATION DIAGRAMS

A convenient method of representing the mathematical equations governing a system is a block diagram, similar to the diagram sometimes drawn before simulating a system on an analog computer. Such a diagram consists of blocks, with the function of the simulated element indicated inside the block. The basic elements most frequently utilized are ideal integrators, ideal amplifiers, and ideal adders, as shown in Fig. 5.2-1. The word "ideal" is used because in actual practice such factors as phase shift, sign inversion, and loading must be taken into account before a workable analog computer simulation can be determined.

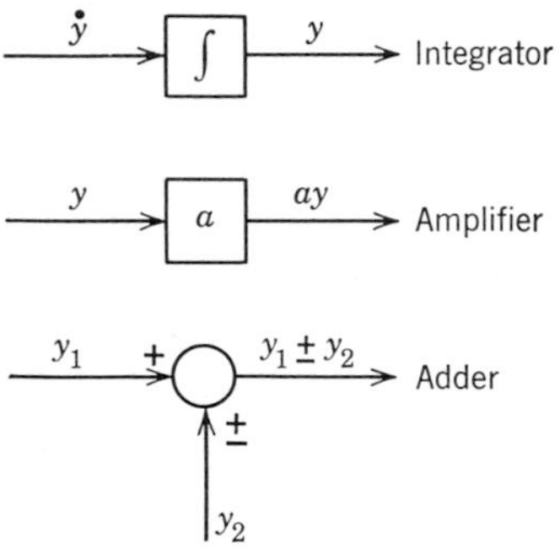

Fig. 5.2-1

The approach used to generate a block diagram of a linear differential equation is to integrate successively the highest derivative to obtain all the lower order derivatives and the dependent variable. The block diagram is completed by satisfying the requirements of the differential equation, i.e., multiplying the derivatives by their respective coefficients and summing these terms to "close the loop." Some illustrations serve to clarify this method.

Example 5.2-1. Draw a block diagram for the system governed by the differential equation

$$\ddot{y} + a\dot{y} + by = v$$

where v is the input, and y the output.

Solving for the highest derivative $\ddot{y}$,

$$\ddot{y} = v - a\dot{y} - by$$

Integrating $\ddot{y}$ twice, both $\dot{y}$ and y are obtained, as shown in Fig. 5.2-2*a*. The loop is then closed by satisfying the requirement of the differential equation

$$\ddot{y} = v - a\dot{y} - by$$

The completed diagram is shown in Fig. 5.2-2*b*.

Figure 5.2-2*b* is essentially the form of an analog computer diagram for this system. However, an analog computer diagram would also have to account for the sign change inherent in the integrators, amplifiers, and adders, the initial conditions of the system, and any time and amplitude scaling factors. Thus Fig. 5.2-2*b* exhibits only the mathematical description of the system with zero initial conditions, and not its practical simulation.

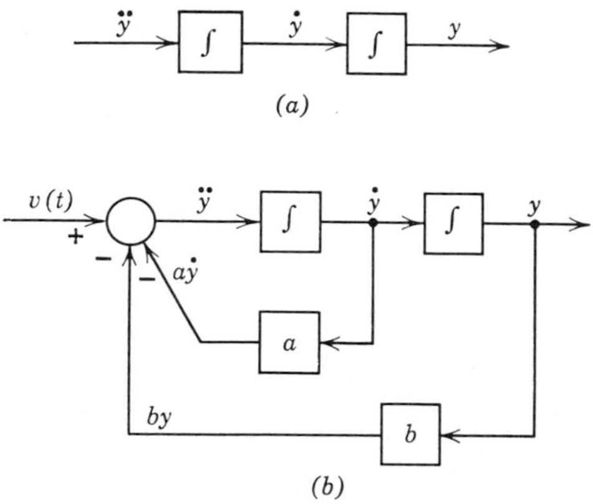

Fig. 5.2-2

Example 5.2-2. Draw a block diagram for the differential equation

$$\ddot{y} + a\dot{y} + by = \dot{v} + v$$

The only change that must be made to the previous block diagram is the addition of a block which provides the $\dot{v}$ term, as shown in Fig. 5.2-3*a*. However, a block containing differentiation is not generally utilized. Differentiators are noise-accentuating devices and therefore are not employed in an analog computer simulation.

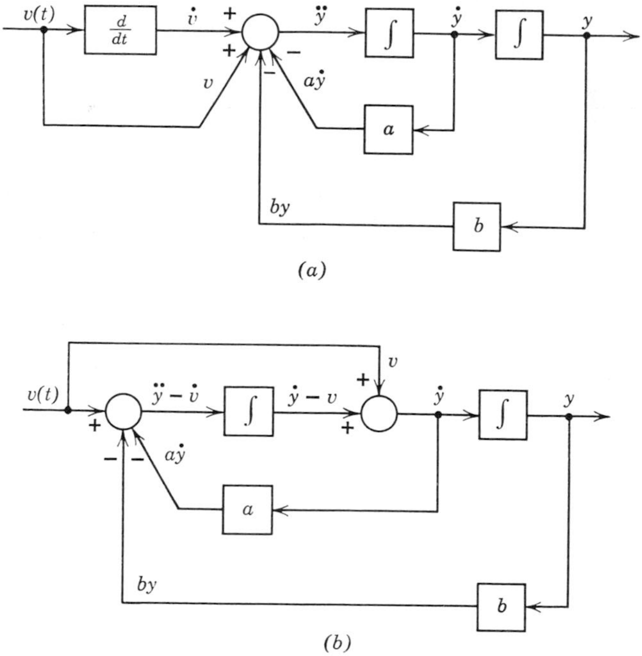

Fig. 5.2-3

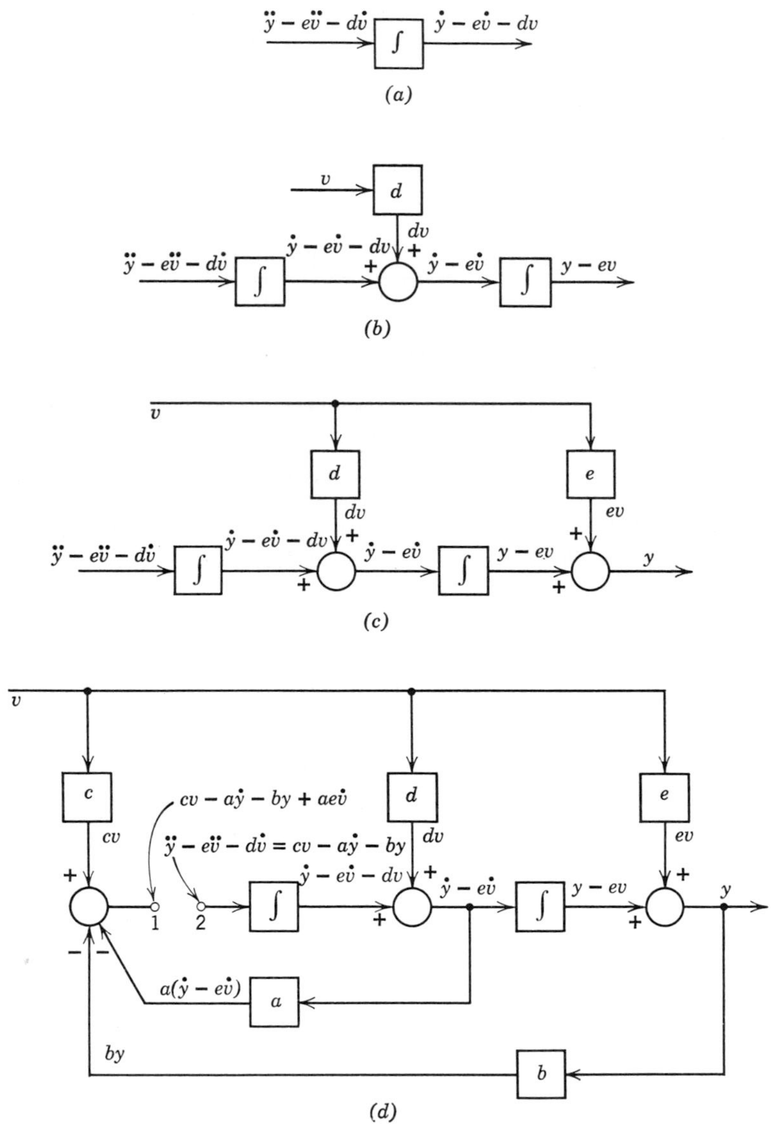

$\ddot{y} - e\ddot{v} - d\dot{v}$
$\dot{y} - e\dot{v} - dv$
(a)
v
d
dv
$\dot{y} - e\dot{v} - dv$
$\dot{y} - e\dot{v}$
$y - ev$
(b)
v
d
e
dv
ev
y
(c)
v
c
d
e
cv
dv
ev
$cv - a\dot{y} - by + ae\dot{v}$
$\ddot{y} - e\ddot{v} - d\dot{v} = cv - a\dot{y} - by$
$\dot{y} - e\dot{v} - dv$
$\dot{y} - e\dot{v}$
$y - ev$
y
1
2
$a(\dot{y} - e\dot{v})$
a
by
b
(d)

Fig. 5.2-4

In order to eliminate the differentiator, assume that the input to the first integrator is $\ddot{y} - \dot{v}$. Thus

$$\ddot{y} - \dot{v} = v - a\dot{y} - by$$

This can be simulated as in Fig. 5.2-3*b*. The output of the first integrator is $\dot{y} - v$. By adding v, $\dot{y}$ is obtained. The diagram is then completed as in the previous example. In essence, the $\dot{v}$ input of Fig. 5.3-2*a* has been shifted to the right of the first integrator, and the differentiation and integration operators cancel.

Example 5.2-3. Draw a block diagram for the system governed by the differential equation

$$\ddot{y} + a\dot{y} + by = e\ddot{v} + d\dot{v} + cv$$

where a, b, c, d, and e are constants.

Using the method of the previous example, the input to the first integrator is assumed to be $\ddot{y} - e\ddot{v} - d\dot{v}$, as shown in Fig. 5.2-4*a*. The term $-d\dot{v}$ can be canceled by adding dv to the output of the first integrator, as in Fig. 5.2-4*b*.

The term $-e\dot{v}$ can be canceled by adding ev at the output of the second integrator. This is shown in Fig. 5.2-4*c*. However, when an attempt is made to close the loop to satisfy the differential equation, it is seen that the derivative $\dot{y}$ does not appear by itself at any point in the diagram. Point 1 in Fig. 5.2-4*d* has the value $cv - a\dot{y} - by + ae\dot{v}$ and does not satisfy the requirements of the differential equation

$$\ddot{y} - e\ddot{v} - d\dot{v} = cv - a\dot{y} - by \tag{5.2-1}$$

An additional term $ae\dot{v}$ is returned to the input of the first integrator. The system can still be simulated in the form shown, however, but with a modification of the blocks containing the multiplying factors c, d, and e.

Assume, then, the form shown in Fig. 5.2-5*a*, where b_0, b_1, and b_2 are the proper multiplying factors. Call the input to the first integrator $\ddot{q}$. The values of other points in Fig. 5.2-5*a* are indicated. The block diagram is a simulation of the differential equation

$$\ddot{q} = -a\dot{q} - bq + (b_0 - ab_1 - bb_2)v - bb_1 \int v\, dt \tag{5.2-2}$$

Recognition that the output of this simulation must be equal to y gives

$$q = y - b_1 \int v\, dt - b_2 v \tag{5.2-3}$$

Then

$$\begin{aligned} \dot{q} &= \dot{y} - b_1 v - b_2 \dot{v} \\ \ddot{q} &= \ddot{y} - b_1 \dot{v} - b_2 \ddot{v} \end{aligned} \tag{5.2-4}$$

Substituting Eqs. 5.2-2 and 5.2-3 into Eq. 5.2-4 yields

$$\begin{aligned} \ddot{y} - b_1\dot{v} - b_2\ddot{v} = -a(\dot{y} - b_1 v - b_2\dot{v}) - b(y - b_1 \int v\, dt - b_2 v) \\ + (b_0 - ab_1 - bb_2)v - bb_1 \int v\, dt \end{aligned} \tag{5.2-5}$$

Collecting terms,

$$\ddot{y} - b_2\ddot{v} - (b_1 + ab_2)\dot{v} = b_0 v - a\dot{y} - by \tag{5.2-6}$$

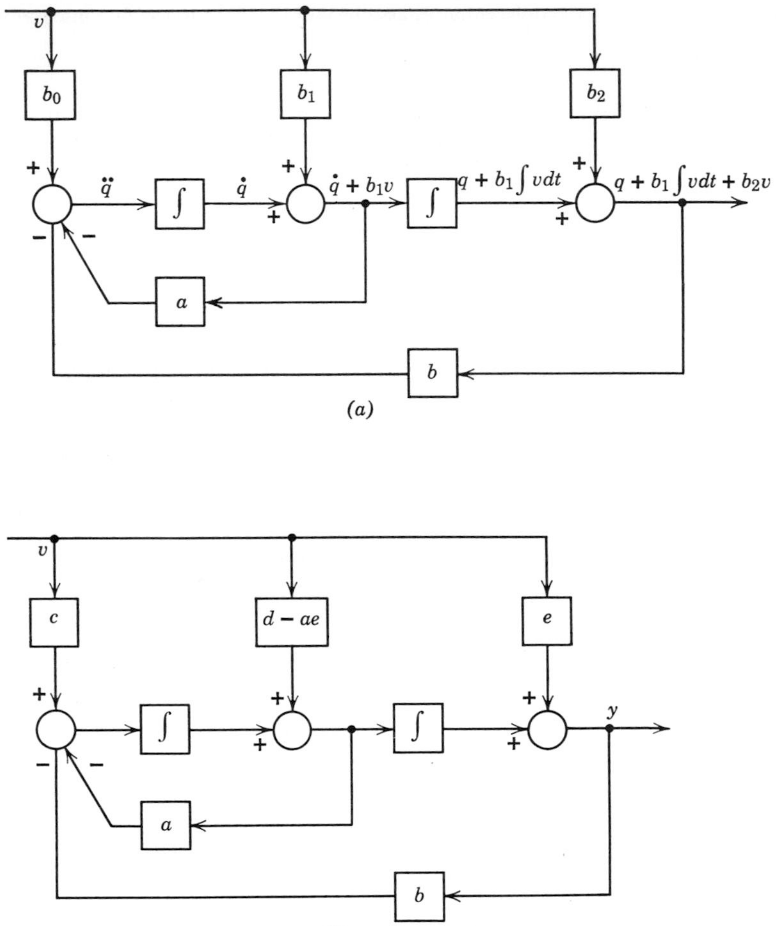

(b)

Fig. 5.2-5

Comparing Eq. 5.2-6 with Eq. 5.2-1, the requirements for b_0, b_1, and b_2 are found by equating like terms. They are

$$\begin{aligned} b_2 &= e \\ b_1 + ab_2 &= d \quad \text{or} \quad b_1 = d - ae \\ b_0 &= c \end{aligned} \tag{5.2-7}$$

The completed diagram is shown in Fig. 5.2-5*b*.

This is a general procedure for linear systems and is utilized in Section 5.5 to derive the standard form of a linear vector matrix differential equation. It is equivalent to canceling the output of the first integrator in Fig. 5.2-4*d* caused by the undesired input $ae\dot{v}$, if the loop were closed. The cancellation is accomplished by modifying the gain of the *d* block. Comparison of Figs. 5.2-4*d* and 5.2-5*b* reveals this viewpoint.

Time-Varying Systems

Time-varying systems can be represented by utilizing blocks with time-varying gains. If derivatives of the input are contained in the differential equation, the procedure indicated in Example 5.2-3 becomes more complicated. It is reserved for a problem at the end of the chapter.

Example 5.2-4. Draw a block diagram for the time-varying system governed by the differential equation $\ddot{y} + a(t)\dot{y} + b(t)y = v$.

Using amplifiers with time-varying gains, this is simulated as shown in Fig. 5.2-6.

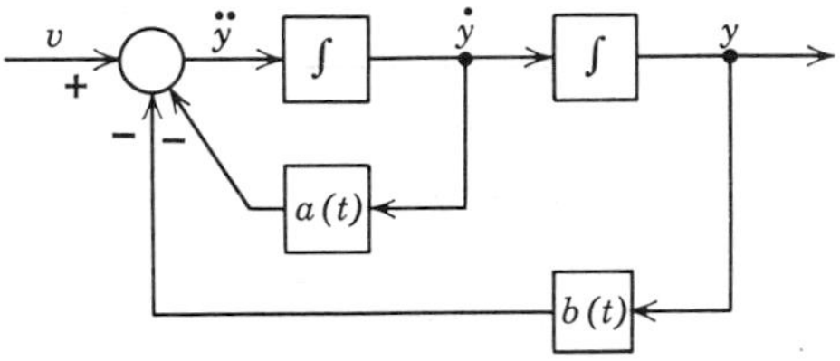

Fig. 5.2-6

Nonlinear Systems

By utilizing blocks known as multipliers and function generators, a diagram can be drawn for systems which are nonlinear. These two blocks are shown in Fig. 5.2-7.

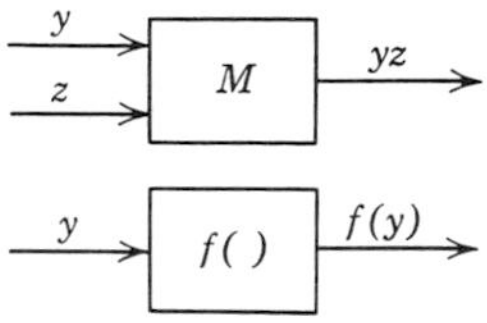

Fig. 5.2-7

Example 5.2-5. Draw the block diagram for the system whose differential equation is

$$\ddot{y} + a\dot{y} + by^3 = v$$

The simulation diagram for this system is shown in Fig. 5.2-8*a*. This system can also be represented as in Fig. 5.2-8*b*. The block indicated by ($)^3$ takes the place of the two multiplier blocks. In this case the function to be generated is $f(y) = y^3$.

Example 5.2-6. Draw a block diagram for

$$\ddot{y} + a\dot{y}y + by^3 = v$$

This is a case where both a function generator and a multiplier must be used. The simulation diagram is shown in Fig. 5.2-9.

Example 5.2-7. If the functional relationship between the input $\dot{y}$ and the output e of a block is defined as $e = d(\dot{y})$, draw a block diagram for the differential equation

$$\ddot{y} + d(\dot{y}) + by = v$$

The nonlinearity is ideal limiting as shown in Fig. 5.2-10.

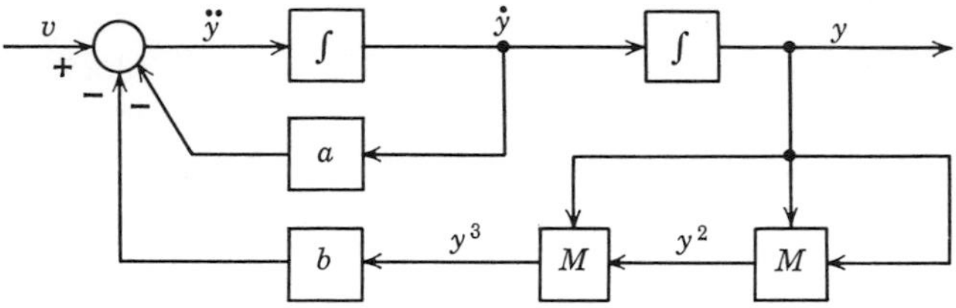

(a)

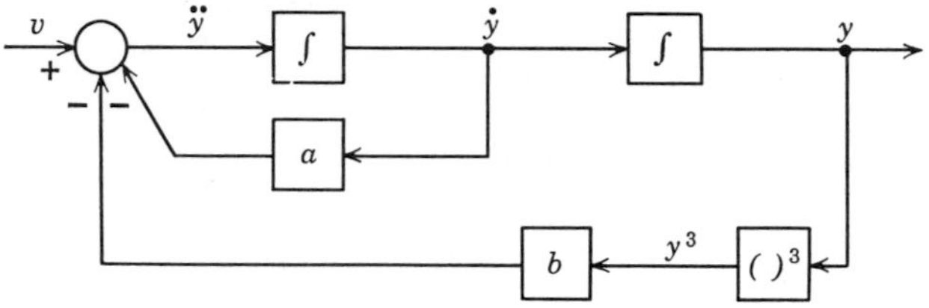

(b)

Fig. 5.2-8

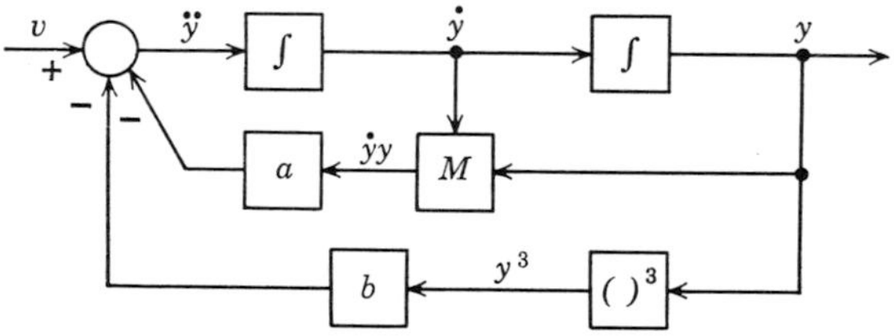

Fig. 5.2-9

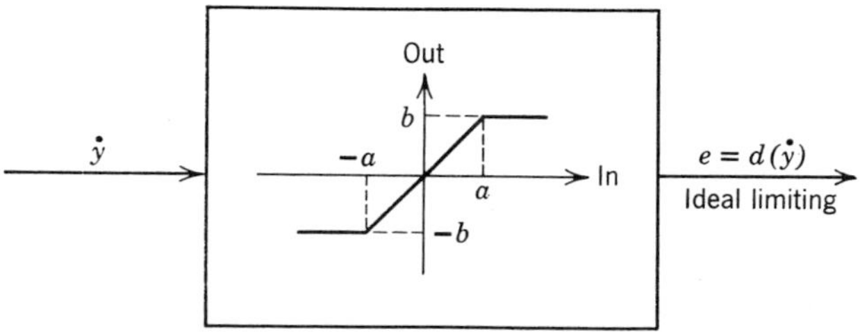

Fig. 5.2-10

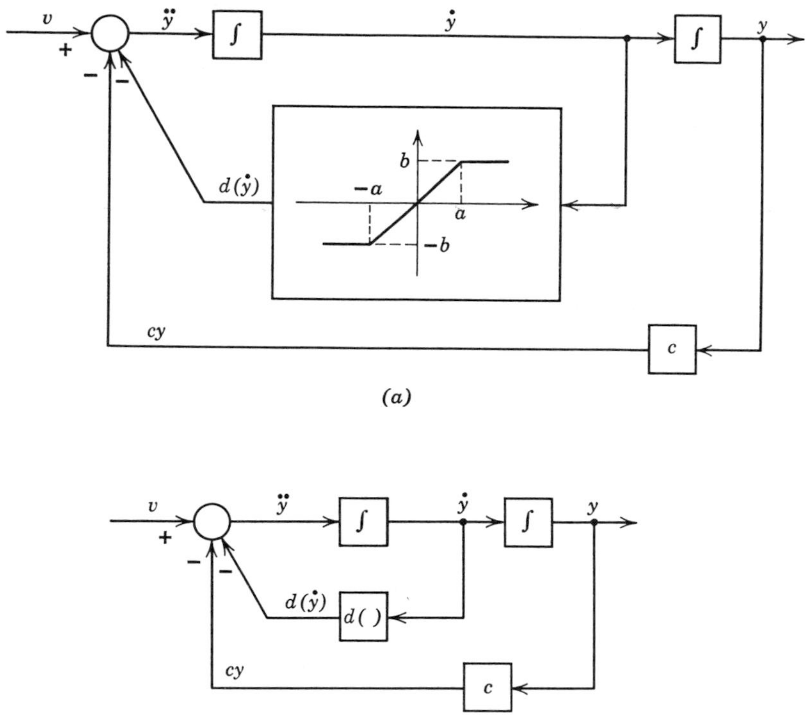

(a)

(b)

Fig. 5.2-11

Often, when the function to be generated cannot be expressed in a convenient mathematical form, the input-output relationship of the desired function is drawn directly in the block. In most cases this is done where the input-output relationship is piecewise linear. This is done in the diagram shown in Fig. 5.2-11*a*, which is more informative than Fig. 5.2-11*b*. Figure 5.2-11*a* indicates at a glance what the nonlinearity is, without having to refer to another defining diagram for $d(\dot{y})$.

Multivariable Systems

For multivariable systems the approach is substantially the same. Examples 5.2-8 and 5.2-9 illustrate two cases.

Example 5.2-8. Diagram the multivariable system

$$\ddot{y}_1 + 3\dot{y}_1 + 2y_2 = v_1$$

$$\ddot{y}_2 + \dot{y}_1 + y_2 = v_2$$

It is obtained by following the principles previously discussed. The simulation diagram is shown in Fig. 5.2-12.

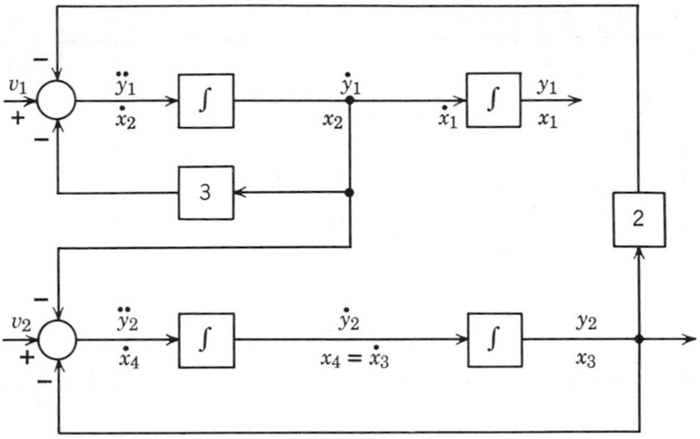

Fig. 5.2-12

Example 5.2-9. Diagram the multivariable system

$$\dot{y}_1 + y_1 = v_1 + 2v_2$$
$$\ddot{y}_2 + 3\dot{y}_2 + 2y_2 = v_1 + v_2 + \dot{v}_2$$

This can be handled by transferring the $\dot{v}_2$ term to the left side of the equation as in Example 5.2-2. The simulation diagram is shown in Fig. 5.2-13.

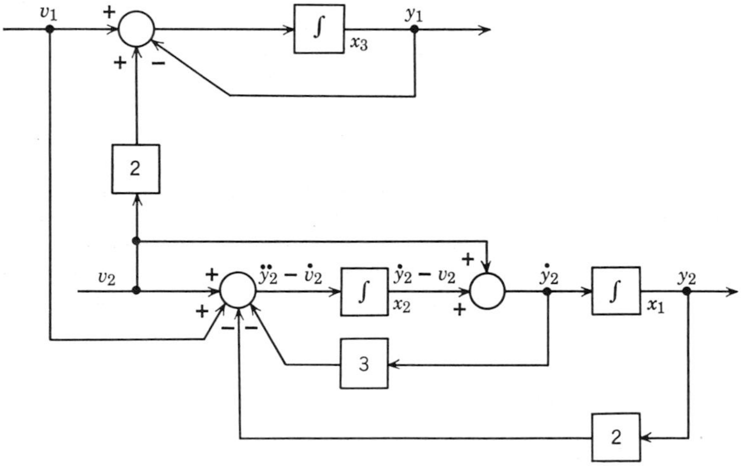

Fig. 5.2-13

5.3 TRANSFER FUNCTION MATRICES

The concept of the transfer function $H(s)$ was introduced for single input-single output fixed linear systems in Section 3.5. For systems which have more than one input or output, transfer functions between various input-output terminals become of interest. The principle of the transfer function is simply extended to cover this more general case. The transfer function $H_{ij}(s)$ is the transfer function between input terminal j and output terminal i and is defined by

$$H_{ij}(s) = \frac{Y_i(s)}{V_j(s)}; \quad V_k(s) = 0, \quad k \neq j \tag{5.3-1}$$

If the elements $H_{ij}(s)$ are ordered into an array, where the first subscript i denotes the row and the second subscript j denotes the column, then this array is called a *transfer function matrix.*

Example 5.3-1. A system with two inputs and two outputs is governed by the set of differential equations

$$\ddot{y}_1 + 5\dot{y}_1 + 6y_1 = \ddot{v}_1 + 3\dot{v}_1 + 4\dot{v}_2 + 8v_2$$
$$\dot{y}_2 + y_2 = \dot{v}_1 + 2\dot{v}_2 + 2v_2$$

Determine the transfer function matrix and draw a block diagram.

Taking the transform of both equations, assuming zero initial conditions, yields

$$(s + 2)(s + 3)Y_1(s) = s(s + 3)V_1(s) + 4(s + 2)V_2(s)$$
$$(s + 1)Y_2(s) = sV_1(s) + 2(s + 1)V_2(s)$$

or

$$Y_1(s) = \frac{s}{s+2}V_1(s) + \frac{4}{s+3}V_2(s)$$

$$Y_2(s) = \frac{s}{s+1}V_1(s) + 2V_2(s)$$

The transfer function matrix $\mathbf{H}(s)$ is then

$$\mathbf{H}(s) = \begin{bmatrix} H_{11}(s) & H_{12}(s) \\ H_{21}(s) & H_{22}(s) \end{bmatrix} = \begin{bmatrix} \dfrac{s}{s+2} & \dfrac{4}{s+3} \\ \dfrac{s}{s+1} & 2 \end{bmatrix}$$

The transfer function diagram is shown in Fig. 5.3-1.

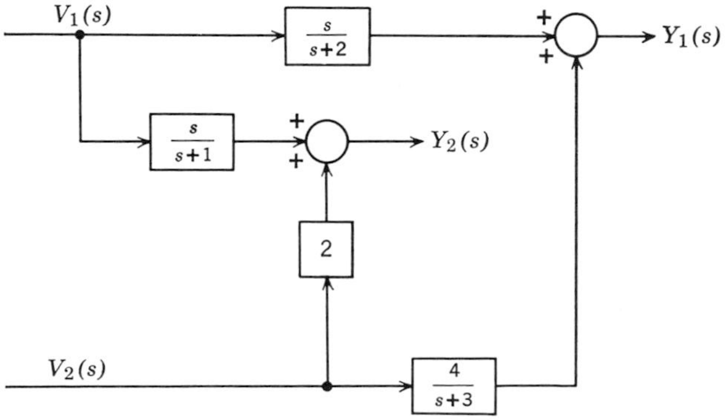

Fig. 5.3-1

Example 5.3-2. A two input-two output system is described by the differential equations

$$\dot{y}_1 + y_2 = v_1 + v_2$$
$$\dot{y}_2 + y_1 = v_2$$

Determine the transfer function matrix and a block diagram.

Transforming the equations assuming zero initial conditions,

$$sY_1(s) + Y_2(s) = V_1(s) + V_2(s)$$
$$sY_2(s) + Y_1(s) = V_2(s)$$

Solving for $Y_1(s)$ and $Y_2(s)$,

$$Y_1(s) = \frac{s}{(s+1)(s-1)} V_1(s) + \frac{1}{s+1} V_2(s)$$

$$Y_2(s) = \frac{-1}{(s+1)(s-1)} V_1(s) + \frac{1}{s+1} V_2(s)$$

Hence

$$\mathbf{H}(s) = \begin{bmatrix} \dfrac{s}{(s+1)(s-1)} & \dfrac{1}{s+1} \\ \dfrac{-1}{(s+1)(s-1)} & \dfrac{1}{s+1} \end{bmatrix}$$

A transfer function diagram for this system is shown in Fig. 5.3-2.

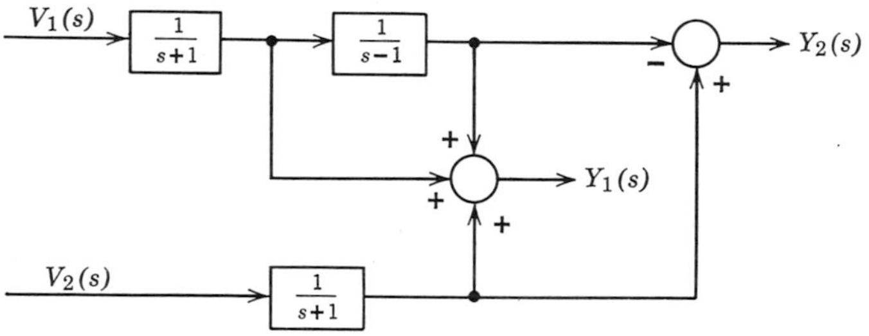

Fig. 5.3-2

It should be noted that the transfer functions $1/(s + a)$ and $s/(s + a)$, commonly shown on diagrams of this sort, can be obtained from a single integrator. This is shown in Fig. 5.3-3. An important point to be stressed here is related to the fact that the transfer function diagram of Fig. 5.3-2 uses three integrators. A conclusion that the system is of

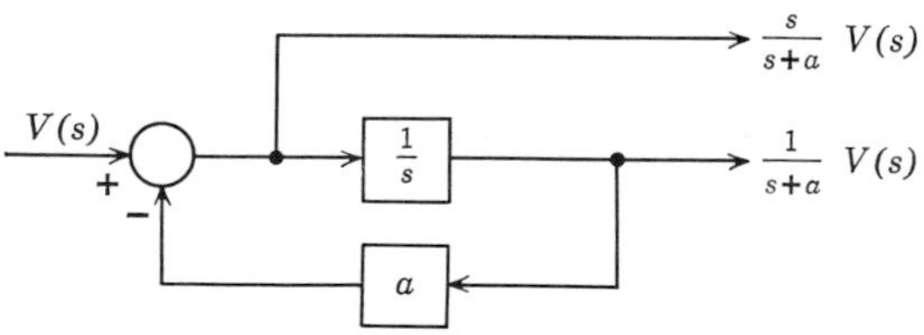

Fig. 5.3-3

third order is incorrect. The original differential equations require only two integrators for simulation, as shown in Fig. 5.3-4. The simulation diagram, obtained from the original differential equations, always shows the correct order of the system. However, the transfer function diagram should be used with the physical problem always in view; otherwise an incorrect conclusion about the order of the system may be made. The

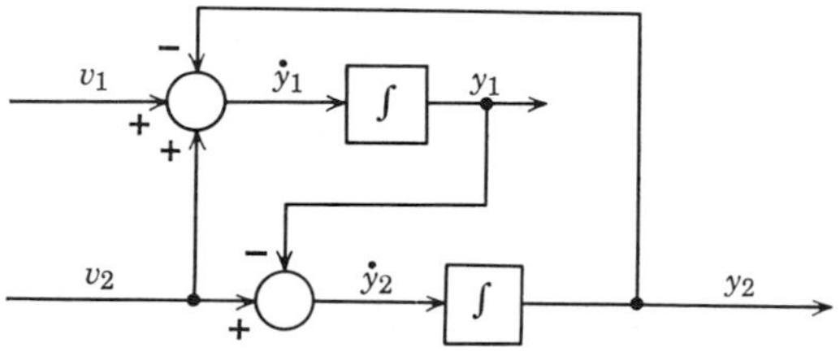

Fig. 5.3-4

transfer function diagram often *masks the physical properties* of a system. This point is covered at greater length in Section 5.7, when the controllability and observability of a multivariable system are discussed.

5.4 THE CONCEPT OF STATE

Consider a single input-single output, linear, electrical network whose structure is known. The input to the network is the time function $v(t)$, and the output of the network is the time function $y(t)$. Since the network is known, complete knowledge of the input $v(t)$ over the time interval $-\infty$ to t is sufficient to determine the output $y(t)$ over the same time interval. However, if the input is known only over the time interval t_0 to t, then the currents through the inductors and the voltages across the capacitors at some time t_1, where $t_0 \leqslant t_1 \leqslant t$ (usually $t_1 = t_0$), must be known in order to determine the output $y(t)$ over the time interval t_0 to t. These currents and voltages constitute the "state" of the network at time t_1. In this sense, the state of the network is related to the memory of the network. For a purely resistive network (zero memory), only the present input is required to determine the present output.

For another example of the state of a system, consider the solution of a linear constant coefficient differential equation for $t \geqslant t_0$. Once the form of the complete solution is obtained in terms of arbitrary constants, these constants can then be determined by the fact that the system must satisfy boundary conditions at time t_0. No other information is required. The boundary conditions can be termed the state of the system at time t_0. Heuristically, the state of a system separates the future from the past, so that the state contains all the relevant information concerning the past history of the system required to determine the response for any input.

The idea of state is a fundamental concept and therefore cannot be defined any more than, for example, the word "set" can be defined in mathematics. The most that can be done is to state the *properties* required of a system whose behavior involves the notion of state. The systems which are considered here are classified as deterministic systems. A *deterministic* system is defined by the following:†

(1) There is a class of time functions $\mathbf{v}(t)$ called admissible input functions;

(2) for *each* time, a set X_t is defined whose elements $\mathbf{x}(t)$ are the possible states at time t; and

(3) to each pair $\mathbf{v}(t)$, $\mathbf{x}(t)$ is assigned at least one time function, called an output function, and for every $t' > t$, a unique element $\mathbf{x}(t')$ contained within $X_{t'}$.

These sets of states and assignment must satisfy the following consistency conditions for a state-determined system:

(1) The admissible input functions must be such that, if $\mathbf{v}_1(t)$ and $\mathbf{v}_2(t)$ are admissible input functions, then

$$\mathbf{v}_3(t) = \mathbf{v}_1(t) \qquad t \leqslant t_0$$

$$\mathbf{v}_3(t) = \mathbf{v}_2(t) \qquad t > t_0$$

is an admissible input function. Figure 5.4-1 illustrates the scalar case.

(2) The manner in which a system reaches a present state does not affect the future output. The present state of a system and the present

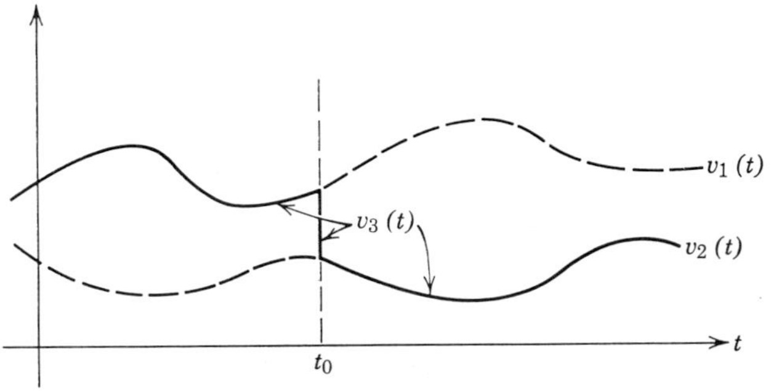

Fig. 5.4-1

† Item 2 allows for the possibility of a time-varying state space, and item 3 involves the concept of a *final* or *terminal* state for each initial state at time t. These ideas differ somewhat from the more general ideas expressed by Zadeh and Desoer,[1] whose concept of state evolves from the characterization of a system by a listing of all observable input-output pairs.

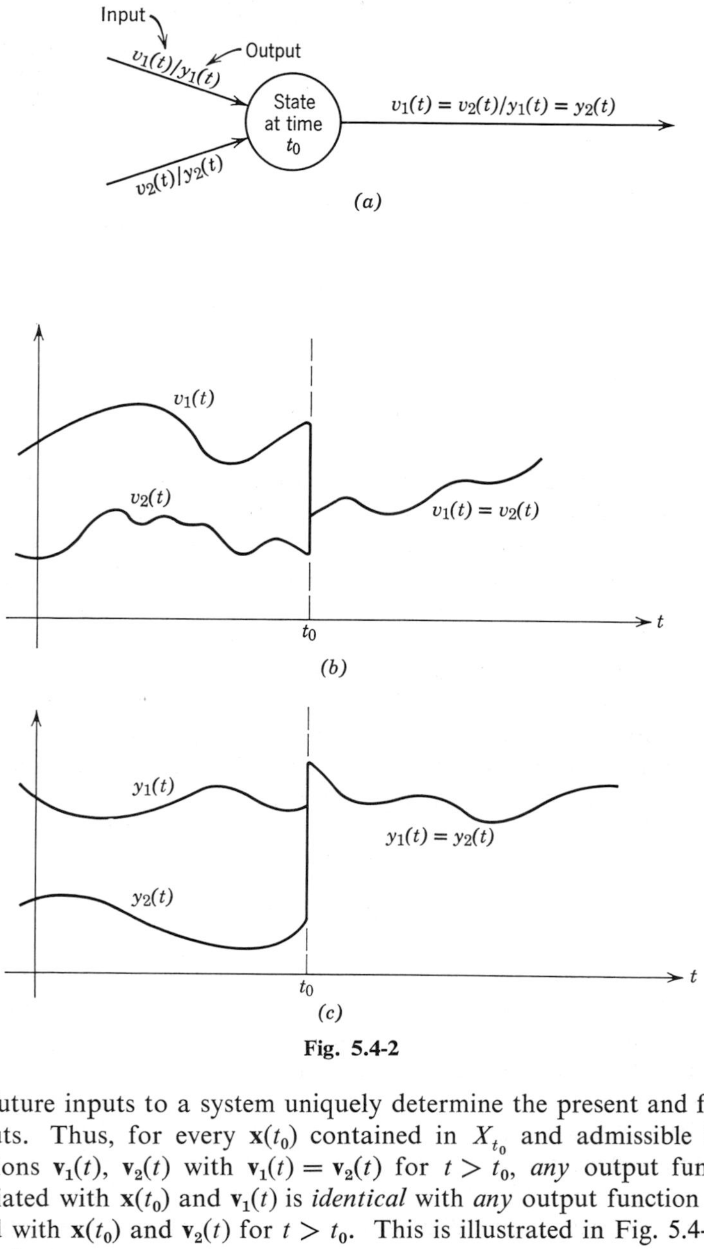

Fig. 5.4-2

and future inputs to a system uniquely determine the present and future outputs. Thus, for every $\mathbf{x}(t_0)$ contained in X_{t_0} and admissible input functions $\mathbf{v}_1(t)$, $\mathbf{v}_2(t)$ with $\mathbf{v}_1(t) = \mathbf{v}_2(t)$ for $t > t_0$, *any* output function associated with $\mathbf{x}(t_0)$ and $\mathbf{v}_1(t)$ is *identical* with *any* output function associated with $\mathbf{x}(t_0)$ and $\mathbf{v}_2(t)$ for $t > t_0$. This is illustrated in Fig. 5.4-2 for the scalar case.

(3) If the initial state of a system and the input $\mathbf{v}(t)$, $t \geqslant t_0$, are given, then the output $\mathbf{y}(t)$, $t \geqslant t_0$, is uniquely determined. Assume that the

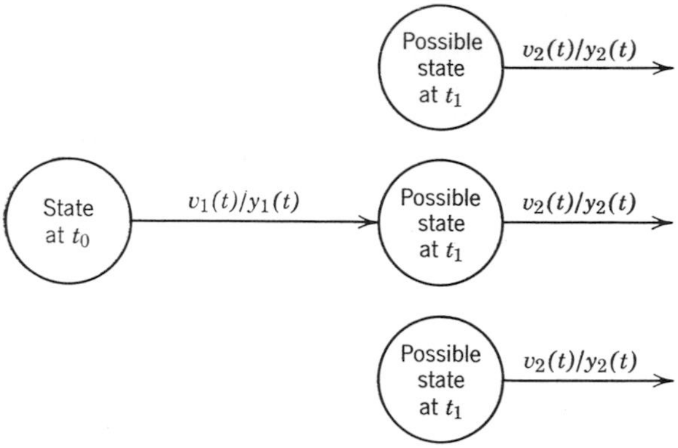

Fig. 5.4-3

known input and output time functions are divided into two time intervals. For the second interval, there may be many initial states for which the given input function over this interval yields the given output function. However, at least one of these possible initial states *must* be the terminal state of the first interval. This is illustrated in Fig. 5.4-3. More formally, for any $\mathbf{x}(t_0)$ contained in X_{t_0} and input function $\mathbf{v}(t)$, let X_1 contained in X_{t_1}, $t_1 > t_0$, be the set of all states for which the output function associated with $\mathbf{v}(t)$ and $\mathbf{x}$ contained in X_1 be the same as the output function associated with $\mathbf{v}(t)$ and $\mathbf{x}(t_0)$ for $t > t_1$. Then $\mathbf{x}(t_1)$ is contained in X_1.

These three consistency conditions can be written as a pair of equations, which are called the *state equations*. They are

$$\mathbf{y}(t_0, t) = \mathbf{g}[\mathbf{x}(t_0); \mathbf{v}(t_0, t)] \tag{5.4-1}$$

$$\mathbf{x}(t) = \mathbf{f}[\mathbf{x}(t_0); \mathbf{v}(t_0, t)] \tag{5.4-2}$$

where both $\mathbf{g}$ and $\mathbf{f}$ are *single-valued* functions. Equation 5.4-1 states that the output $\mathbf{y}$ over the time interval t_0 to t is a single-valued function of the state at the beginning of the interval and the input $\mathbf{v}$ over this time interval. The state at the end of the interval is said, in Eq. 5.4-2, to be a single-valued function of the same argument. These two equations define a *state-determined system.*

For most of what follows, the outputs of the integrators in the simulation diagram are used as the components of the state vector. The state vector is defined in terms of an n-dimensional *state space*, whose coordinates are $x_1, x_2, \ldots, x_n$. The motion of the *tip* of the state vector in state space is called the *trajectory* of the state vector.

Although the outputs of the integrators in the simulation diagram form a natural state vector, these variables may not be physically measurable in a system. In order to control successfully a multivariable system, the control laws must be in terms of measurable information about the system. More is said in this regard later.

5.5 MATRIX REPRESENTATION OF LINEAR STATE EQUATIONS

In the previous section the state equations of a continuous deterministic system were defined to be

$$\begin{aligned} \mathbf{y}(t) &= \mathbf{g}[\mathbf{x}(t_0), \mathbf{v}(t_0, t)] \\ \mathbf{x}(t) &= \mathbf{f}[\mathbf{x}(t_0), \mathbf{v}(t_0, t)] \end{aligned} \tag{5.5-1}$$

If the system can be described by a set of *linear ordinary differential* equations, the state equations can be written as

$$\begin{aligned} \dot{\mathbf{x}}(t) &= \mathbf{A}(t)\mathbf{x}(t) + \mathbf{B}(t)\mathbf{v}(t) \\ \mathbf{y}(t) &= \mathbf{C}(t)\mathbf{x}(t) + \mathbf{D}(t)\mathbf{v}(t) \end{aligned} \tag{5.5-2}$$

where $\mathbf{A}(t)$, $\mathbf{B}(t)$, $\mathbf{C}(t)$, and $\mathbf{D}(t)$ are, in general, time-varying matrices, and

$$\mathbf{x} = \begin{bmatrix} x_1 \\ x_2 \\ \cdot \\ \cdot \\ \cdot \\ x_n \end{bmatrix}, \quad \mathbf{v} = \begin{bmatrix} v_1 \\ v_2 \\ \cdot \\ \cdot \\ \cdot \\ v_m \end{bmatrix}, \quad \mathbf{y} = \begin{bmatrix} y_1 \\ y_2 \\ \cdot \\ \cdot \\ \cdot \\ y_p \end{bmatrix}$$

A general diagram for these equations is shown in Fig. 5.5-1. If the system is fixed, i.e., nontime-varying, then the matrices $\mathbf{A}(t)$, $\mathbf{B}(t)$, $\mathbf{C}(t)$, and $\mathbf{D}(t)$ are constant and can be written simply as **A**, **B**, **C**, and **D**.

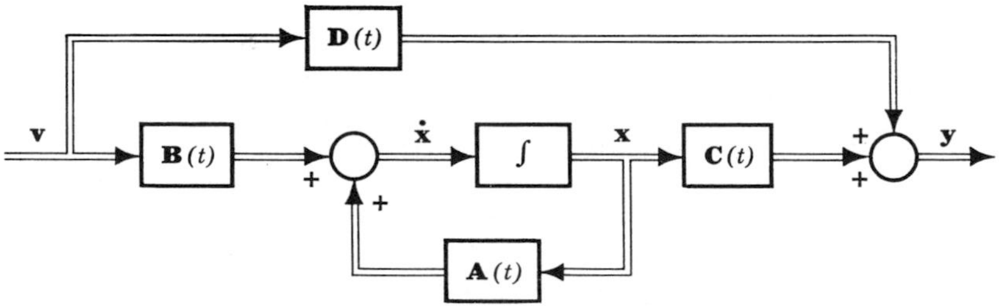

Fig. 5.5-1

Example 5.5-1. Write a set of state equations for the system of Fig. 5.2-12.

For the x's, v's, and y's indicated in Fig. 5.2-12,

$$\begin{bmatrix} \dot{x}_1 \\ \dot{x}_2 \\ \dot{x}_3 \\ \dot{x}_4 \end{bmatrix} = \begin{bmatrix} 0 & 1 & 0 & 0 \\ 0 & -3 & -2 & 0 \\ 0 & 0 & 0 & 1 \\ 0 & -1 & -1 & 0 \end{bmatrix} \begin{bmatrix} x_1 \\ x_2 \\ x_3 \\ x_4 \end{bmatrix} + \begin{bmatrix} 0 & 0 \\ 1 & 0 \\ 0 & 0 \\ 0 & 1 \end{bmatrix} \begin{bmatrix} v_1 \\ v_2 \end{bmatrix}$$

$$\begin{bmatrix} y_1 \\ y_2 \end{bmatrix} = \begin{bmatrix} 1 & 0 & 0 & 0 \\ 0 & 0 & 1 & 0 \end{bmatrix} \begin{bmatrix} x_1 \\ x_2 \\ x_3 \\ x_4 \end{bmatrix} + [0 \quad 0] \begin{bmatrix} v_1 \\ v_2 \end{bmatrix}$$

Thus

$$\mathbf{A} = \begin{bmatrix} 0 & 1 & 0 & 0 \\ 0 & -3 & -2 & 0 \\ 0 & 0 & 0 & 1 \\ 0 & -1 & -1 & 0 \end{bmatrix}, \quad \mathbf{B} = \begin{bmatrix} 0 & 0 \\ 1 & 0 \\ 0 & 0 \\ 0 & 1 \end{bmatrix}, \quad \mathbf{C} = \begin{bmatrix} 1 & 0 & 0 & 0 \\ 0 & 0 & 1 & 0 \end{bmatrix}, \quad \mathbf{D} = [0]$$

Example 5.5-2. Determine a set of state equations for a general nontime-varying linear network consisting of resistors, inductors, and capacitors.

Without loss of generality, the network can be represented as in Fig. 5.5-2*a*. Define the state variables as

$x_1, x_2, \ldots, x_m =$ voltages across the resistive network terminal pairs connected to $C_1, C_2, \ldots, C_m$, respectively; and

$x_{m+1}, x_{m+2}, \ldots, x_n =$ currents into the upper resistive network terminals connected to $L_{m+1}, L_{m+2}, \ldots, L_n$, respectively

Then the network can be described by $\dot{\mathbf{x}} = \mathbf{A}\mathbf{x}$, or

$$\dot{x}_i = \sum_{j=1}^{n} a_{ij} x_j$$

where the a_{ij}'s are to be determined.

Assuming that the network is initially at rest, $x_i = 0$, $i = 1, 2, \ldots, n$. If a unit step of voltage is applied in series with the jth capacitor at $t = 0$, initially all capacitors appear as short circuits and all inductors appear as open circuits. This is true because capacitor voltages and inductor currents cannot change instantaneously in response to a finite signal. The current through the ith capacitor at $t = 0+$ is

$$i_{c_i}(0+) = -C_i \dot{x}_i(0+) = -C_i a_{ij} x_j(0+)$$

This value is the same as the constant current which would flow through the ith terminal pair, if all capacitors were replaced by short circuits and all inductors replaced by open circuits, and a source of 1 volt applied as above. Thus

$$a_{ij} = -\frac{g_{ij}}{C_i} \qquad i, j, = 1, 2, \ldots, m$$

where g_{ij} is the current through the short circuit replacing C_i, due to the unit step voltage source replacing C_j. Note that the resultant network which must be analyzed to determine these a_{ij}'s is a purely resistive network.

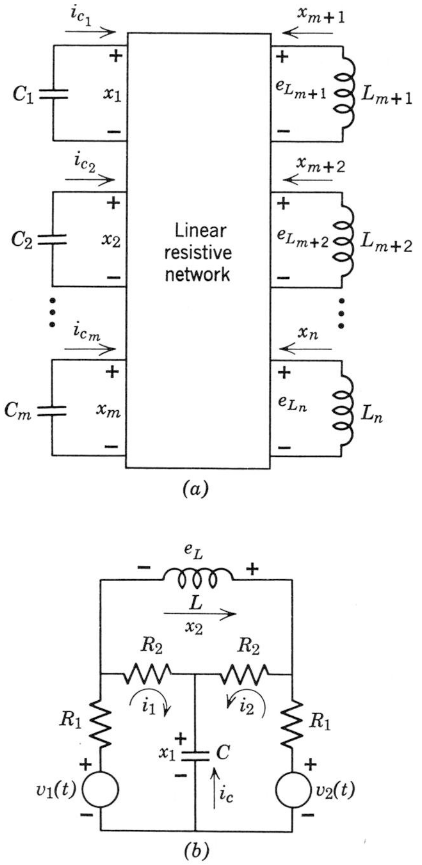

Fig. 5.5-2

For this same step of voltage applied in series with the jth capacitor at $t = 0$, the voltage across the ith inductor is

$$e_{L_i}(0+) = -L_i\dot{x}_i(0+) = -L_i a_{ij} x_j(0+)$$

Thus

$$a_{ij} = -\frac{\alpha_{ij}}{L_i} \qquad \begin{aligned} i &= m+1, m+2, \ldots, n \\ j &= 1, 2, \ldots, m \end{aligned}$$

where α_{ij} is the voltage across the ith open-circuited inductor terminals in response to the 1-volt source replacing C_j. All capacitors and inductors are short- and open-circuited, respectively, so that, again, analysis of the purely resistive network is sufficient.

If instead of the above, a unit step of current is applied in parallel with the jth inductor at $t = 0$, the voltage across the ith inductor at $t = 0+$ is

$$e_{L_i}(0+) = -L_i\dot{x}_i(0+) = -L_i a_{ij} x_j(0+)$$

Thus

$$a_{ij} = -\frac{r_{ij}}{L_i} \qquad i, j = m+1, m+2, \ldots, n$$

where r_{ij} is the voltage across the ith open-circuited terminal pair in response to a unit current source applied to the jth terminal pair.

For the same current applied in parallel with the jth inductor at $t = 0$, the current through the ith capacitor at $t = 0+$ is

$$i_{C_i} = -C_i\dot{x}_i(0+) = -C_i a_{ij} x_j(0+)$$

Thus

$$a_{ij} = -\frac{\beta_{ij}}{C_i} \qquad \begin{array}{l} i = 1, 2, \ldots, m \\ j = m+1, m+2, \ldots, n \end{array}$$

where β_{ij} is the current through the ith short-circuited terminal pair in response to a unit current applied to the jth terminal pair.

By these simple steps, the **A** matrix for a general time-invariant, linear *RLC* network can be found by examination of purely resistive networks. As a specific illustration of the procedure, the reader can confirm that the **A** matrix for the circuit of Fig. 5.5-2*b* is

$$\mathbf{A} = \begin{bmatrix} -\dfrac{2}{(R_1 + R_2)C} & 0 \\ 0 & -\dfrac{2R_1R_2}{(R_1 + R_2)L} \end{bmatrix}$$

Using the description of Eq. 5.5-2, an nth order ordinary differential equation characterizing a system can be rewritten as a set of n first order differential equations in terms of the state variables $\mathbf{x}$.

Example 5.5-3. Determine a state variable representation for the system described by

$$\ddot{y} + a\dot{y} + by = v$$

where $v(t)$ is the input and $y(t)$ is the output.

A simulation diagram for this system is shown in Fig. 5.5-3. Convenient choices for the state variables are the outputs of the integrators, i.e., y and $\dot{y}$. Let $x_1 = y$, $x_2 = \dot{y}$ so that

$$\dot{x}_1 = x_2$$

$$\dot{x}_2 = -bx_1 - ax_2 + v$$

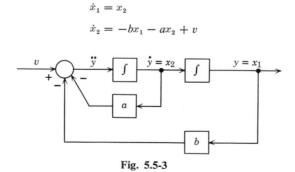

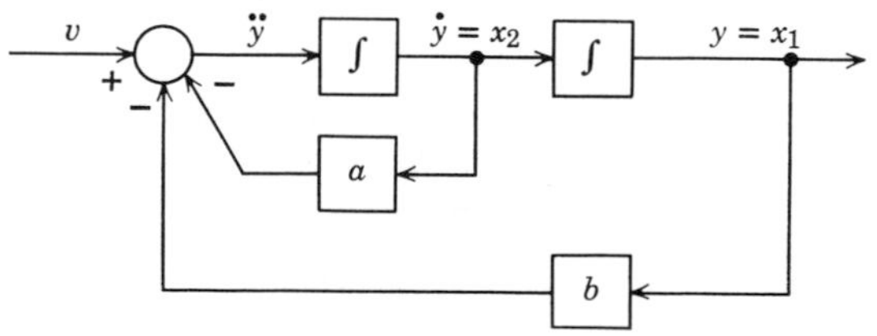

Fig. 5.5-3

The vector matrix representation for this system is

$$\begin{bmatrix} \dot{x}_1 \\ \dot{x}_2 \end{bmatrix} = \begin{bmatrix} 0 & 1 \\ -b & -a \end{bmatrix} \begin{bmatrix} x_1 \\ x_2 \end{bmatrix} + \begin{bmatrix} 0 \\ 1 \end{bmatrix} v$$

$$y = [1 \quad 0] \begin{bmatrix} x_1 \\ x_2 \end{bmatrix} + [0]v$$

Thus

$$\mathbf{A} = \begin{bmatrix} 0 & 1 \\ -b & -a \end{bmatrix}, \quad \mathbf{B} = \begin{bmatrix} 0 \\ 1 \end{bmatrix}, \quad \mathbf{C} = [1 \quad 0], \quad \mathbf{D} = [0]$$

The generalization of the procedure of Example 5.5-3 to the case of an nth order single input-single output, linear, constant coefficient system, described by

$$(p^n + \alpha_{n-1}p^{n-1} + \cdots + \alpha_1 p + \alpha_0)y = (\beta_n p^n + \beta_{n-1}p^{n-1} + \cdots + \beta_1 p + \beta_0)v \quad (5.5\text{-}3)$$

where $p = d/dt$ is now considered. A simulation diagram for this system is shown in Fig. 5.5-4. The outputs of the integrators are chosen as the state variables. The constants a_i and b_i must be determined in terms of the α's and β's in order to relate the diagram to Eq. 5.5-3.

By inspection of the simulation diagram,

$$\begin{aligned} y &= x_1 + b_0 v \\ \dot{x}_k &= x_{k+1} + b_k v, \quad k < n \\ \dot{x}_n &= -(a_0 x_1 + a_1 x_2 + \cdots + a_{n-1}x_n) + b_n v \end{aligned} \quad (5.5\text{-}4)$$

Differentiating y once yields

$$py = \dot{x}_1 + b_0\dot{v}$$

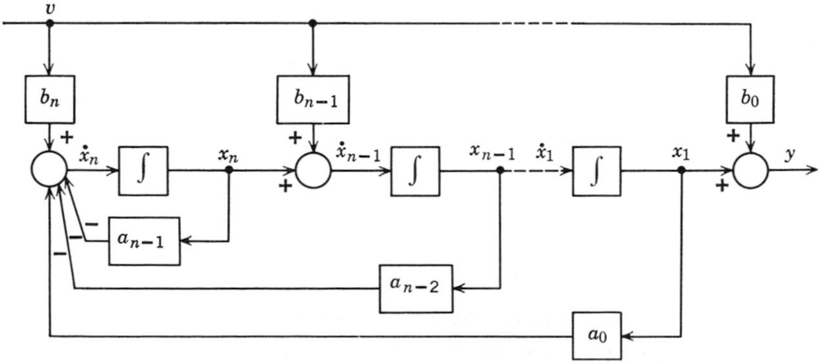

Fig. 5.5-4

Substituting for $\dot{x}_1$ from Eq. 5.5-4 gives

$$py = x_2 + b_1 v + b_0 \dot{v} \tag{5.5-5}$$

Following this procedure, the second and higher derivatives of y are given by

$$\begin{aligned} p^2 y &= \dot{x}_2 + b_1 \dot{v} + b_0 \ddot{v} = x_3 + b_2 v + b_1 \dot{v} + b_0 \ddot{v} \\ &\cdots\cdots\cdots\cdots\cdots\cdots \\ p^{n-1} y &= x_n + b_{n-1} v + b_{n-2} pv + \cdots + b_0 p^{n-1} v \\ p^n y &= -(a_0 x_1 + a_1 x_2 + \cdots + a_{n-1} x_n) + b_n v + b_{n-1} pv + \cdots + b_0 p^n v \end{aligned} \tag{5.5-6}$$

Substituting for $y, py, \ldots, p^{n-1}y$ from Eqs. 5.5-4, 5.5-5, and 5.5-6 into Eq. 5.5-3, and comparing the result with the expression for $p^n y$ as given by Eq. 5.5-6, the expressions for the a_i and b_i are given by

$$a_i = \alpha_i \tag{5.5-7}$$

and

$$\begin{aligned} b_0 &= \beta_n \\ b_1 &= \beta_{n-1} - \alpha_{n-1} b_0 \\ b_2 &= \beta_{n-2} - \alpha_{n-1} b_1 - \alpha_{n-2} b_0 \\ &\cdots\cdots\cdots\cdots \\ b_n &= \beta_0 - \alpha_{n-1} b_{n-1} - \alpha_{n-2} b_{n-2} - \cdots - \alpha_0 b_0 \end{aligned} \tag{5.5-8}$$

Equation 5.5-8 is a convenient, if not explicit, form of an expression for the b_i's. For a given numerical case, the b's can be found by successive substitution. Since Eq. 5.5-8 indicates that the β's can be written as

$$\begin{bmatrix} \beta_n \\ \beta_{n-1} \\ \beta_{n-2} \\ \cdot \\ \cdot \\ \cdot \\ \beta_0 \end{bmatrix} = \begin{bmatrix} 1 & 0 & 0 & \cdots & 0 \\ \alpha_{n-1} & 1 & 0 & \cdots & 0 \\ \alpha_{n-2} & \alpha_{n-1} & 1 & \cdots & 0 \\ \cdot & \cdot & \cdot & \cdot & \cdot \\ \cdot & \cdot & \cdot & \cdot & \cdot \\ \cdot & \cdot & \cdot & \cdot & \cdot \\ \alpha_0 & \alpha_1 & \cdots & \alpha_{n-1} & 1 \end{bmatrix} \begin{bmatrix} b_0 \\ b_1 \\ b_2 \\ \cdot \\ \cdot \\ \cdot \\ b_n \end{bmatrix}$$

the b's can also be determined by premultiplying both sides of this expression by the inverse of the coefficient matrix. As a consequence of these expressions, one form of the matrices **A**, **B**, **C**, and **D** for a system described by Eq. 5.5-3 is given by

$$\mathbf{A} = \begin{bmatrix} 0 & 1 & 0 & \cdots & 0 \\ 0 & 0 & 1 & \cdots & 0 \\ \cdots & \cdots & \cdots & \cdots & \\ & & & & 1 \\ -\alpha_0 & -\alpha_1 & -\alpha_2 & \cdots & -\alpha_{n-1} \end{bmatrix} \tag{5.5-9}$$

$$\begin{bmatrix} \mathbf{D} \\ \mathbf{B} \end{bmatrix} = \begin{bmatrix} b_0 \\ b_1 \\ b_2 \\ \cdot \\ \cdot \\ \cdot \\ b_n \end{bmatrix} = \begin{bmatrix} 1 & 0 & 0 & & \cdots & 0 \\ \alpha_{n-1} & 1 & 0 & & \cdots & 0 \\ \alpha_{n-2} & \alpha_{n-1} & 1 & & \cdots & 0 \\ \cdot & \cdot & \cdot & \cdot & & \cdot \\ \cdot & \cdot & & \cdot & \cdot & \cdot \\ \cdot & \cdot & & \cdot & \cdot & \cdot \\ \alpha_0 & \alpha_1 & \cdots & & \alpha_{n-1} & 1 \end{bmatrix}^{-1} \begin{bmatrix} \beta_n \\ \beta_{n-1} \\ \beta_{n-2} \\ \cdot \\ \cdot \\ \cdot \\ \beta_0 \end{bmatrix}$$

$$\mathbf{C} = [1 \quad 0 \quad 0 \quad \cdots \quad 0]$$
$$\mathbf{D} = b_0 = \beta_n$$

These state equations characterize the system in the so-called *standard form.*

Example 5.5-4. Represent

$$\dddot{y} + 3\ddot{y} + 4\dot{y} + y = 2\dddot{v} + 3\ddot{v} + \dot{v} + 2v$$

in standard form.

From Eq. 5.5-8,

$$\begin{aligned} b_0 &= \beta_n = 2 \\ b_1 &= \beta_{n-1} - \alpha_{n-1}b_0 = 3 - 3(2) = -3 \\ b_2 &= \beta_{n-2} - \alpha_{n-1}b_1 - \alpha_{n-2}b_0 = 1 - 3(-3) - 4(2) = 2 \\ b_3 &= \beta_{n-3} - \alpha_{n-1}b_2 - \alpha_{n-2}b_1 - \alpha_{n-3}b_0 = 2 - 3(2) - 4(-3) - 1(2) = 6 \end{aligned}$$

Thus Eqs. 5.5-2 and 5.5-9 give as a vector matrix representation for this system

$$\begin{bmatrix} \dot{x}_1 \\ \dot{x}_2 \\ \dot{x}_3 \end{bmatrix} = \begin{bmatrix} 0 & 1 & 0 \\ 0 & 0 & 1 \\ -1 & -4 & -3 \end{bmatrix} \begin{bmatrix} x_1 \\ x_2 \\ x_3 \end{bmatrix} + \begin{bmatrix} -3 \\ 2 \\ 6 \end{bmatrix} v$$

$$y = [1 \quad 0 \quad 0] \begin{bmatrix} x_1 \\ x_2 \\ x_3 \end{bmatrix} + 2v$$

The complete simulation diagram is shown in Fig. 5.5-5.

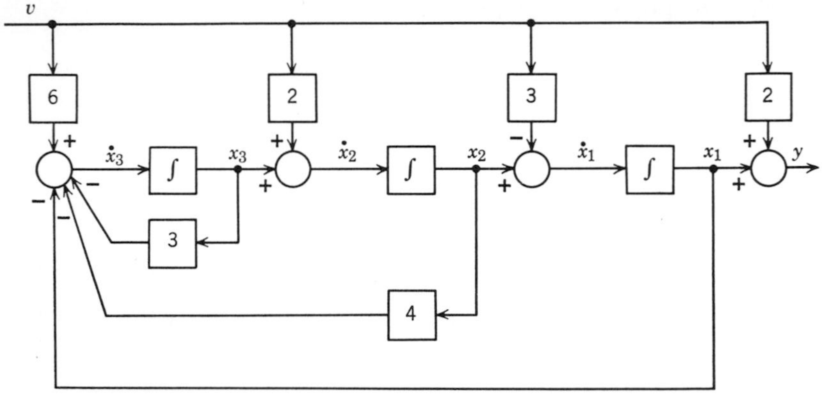

Fig. 5.5-5

State Equations—Partial Fractions Technique

An alternative form for the state equations can be obtained by a partial fraction expansion of the transfer function of the system. Consider a single input-single output system whose transfer function is given by $H(s)$. Assume that the transfer function can be formed as the ratio of two polynomials as $N(s)/D(s)$, where the order of $N(s)$ is at most equal to the order of $D(s)$. If it is assumed that $D(s)$ has only distinct zeros, then $D(s)$ can be written as

$$D(s) = d_n(s - \lambda_1)(s - \lambda_2) \cdots (s - \lambda_n)$$

where λ_i is a characteristic value of the system. The transfer function $H(s)$ can then be expanded into the form

$$H(s) = d_0 + \frac{c_1}{s - \lambda_1} + \frac{c_2}{s - \lambda_2} + \cdots + \frac{c_n}{s - \lambda_n} \tag{5.5-10}$$

where
$$d_0 = \lim_{s \to \infty} H(s)$$

and

$$c_i = (s - \lambda_i)\frac{N(s)}{D(s)}\bigg|_{s=\lambda_i}$$

The output $Y(s) = H(s)V(s)$ can be expressed as

$$Y(s) = d_0V(s) + \sum_{i=1}^{n} \frac{c_i}{s - \lambda_i} V(s) \tag{5.5-11}$$

Consequently, the simulation diagram of Fig. 5.5-6 can be drawn, where the state variable x_i satisfies the first order differential equation

$$\dot{x}_i - \lambda_i x_i = v$$

and

$$y = \sum_{i=1}^{n} c_i x_i + d_0 v$$

The state equations are then

$$\begin{bmatrix} \dot{x}_1 \\ \dot{x}_2 \\ \cdot \\ \cdot \\ \cdot \\ \dot{x}_n \end{bmatrix} = \begin{bmatrix} \lambda_1 & 0 & \cdots & 0 \\ 0 & \lambda_2 & \cdots & 0 \\ \cdots & \cdots & \cdots & \cdots \\ 0 & 0 & \cdots & \lambda_n \end{bmatrix} \begin{bmatrix} x_1 \\ x_2 \\ \cdot \\ \cdot \\ \cdot \\ x_n \end{bmatrix} + \begin{bmatrix} 1 \\ 1 \\ \cdot \\ \cdot \\ \cdot \\ 1 \end{bmatrix} v \tag{5.5-12}$$

$$y = [c_1 \quad c_2 \quad \cdots \quad c_n] \begin{bmatrix} x_1 \\ x_2 \\ \cdot \\ \cdot \\ \cdot \\ x_n \end{bmatrix} + d_0 v$$

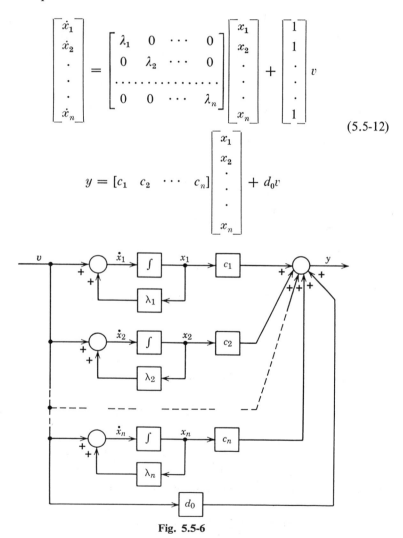

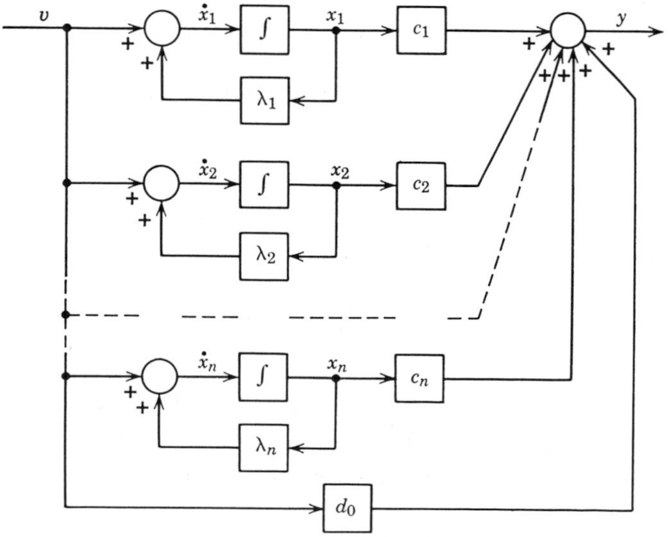

Fig. 5.5-6

or

$$\begin{aligned}\dot{\mathbf{x}} &= \mathbf{\Lambda x} + v\mathbf{1} \\ y &= \langle \mathbf{c}, \mathbf{x} \rangle + d_0 v\end{aligned} \tag{5.5-13}$$

This method of determining a state variable representation of the system leads to state equations in the normal form of Eq. 4.7-21.

Partial Fractions Expansion—Repeated Roots

For the case where the roots of the denominator polynomial $D(s)$ are not distinct, the partial fraction technique leads to a nondiagonal Jordan canonical form. As an example of this, consider the case where $D(s)$ is of the form

$$D(s) = d_n(s - \lambda_1)^k(s - \lambda_2)(s - \lambda_3) \cdots (s - \lambda_n)$$

The partial fraction expansion of $H(s) = N(s)/D(s)$ is then

$$\begin{aligned} H(s) &= d_0 + \frac{c_{11}}{(s - \lambda_1)^k} + \frac{c_{12}}{(s - \lambda_1)^{k-1}} + \cdots + \frac{c_{1k}}{(s - \lambda_1)} \\ &\qquad + \frac{c_2}{s - \lambda_2} + \frac{c_3}{s - \lambda_3} + \cdots + \frac{c_n}{s - \lambda_n} \\ &= d_0 + \sum_{j=1}^{k} \frac{c_{1j}}{(s - \lambda_1)^{(k-j+1)}} + \sum_{i=2}^{n} \frac{c_i}{s - \lambda_i} \end{aligned} \tag{5.5-14}$$

where

$$d_0 = \lim_{s \to \infty} H(s)$$

$$c_{1j} = \frac{1}{(j - 1)!} \frac{d^{j-1}}{ds^{j-1}} \left[(s - \lambda_1)^k \frac{N(s)}{D(s)} \right]_{s=\lambda_1}$$

and

$$c_i = (s - \lambda_i) \frac{N(s)}{D(s)} \bigg|_{s=\lambda_i}$$

The output $Y(s) = H(s)V(s)$ is then

$$Y(s) = d_0 V(s) + \sum_{j=1}^{k} c_{1j} X_j(s) + \sum_{i=2}^{n} c_i X_i(s) \tag{5.5-15}$$

where

$$X_j(s) = \frac{V(s)}{(s - \lambda_1)^{k-j+1}} \quad \text{and} \quad X_i(s) = \frac{V(s)}{s - \lambda_i}$$

However, $X_j(s)$ can be written in terms of $X_{j+1}(s)$ as

$$X_j(s) = \frac{1}{(s - \lambda_1)} X_{j+1}(s) \qquad j = 1, \ldots, k - 1 \tag{5.5-16}$$

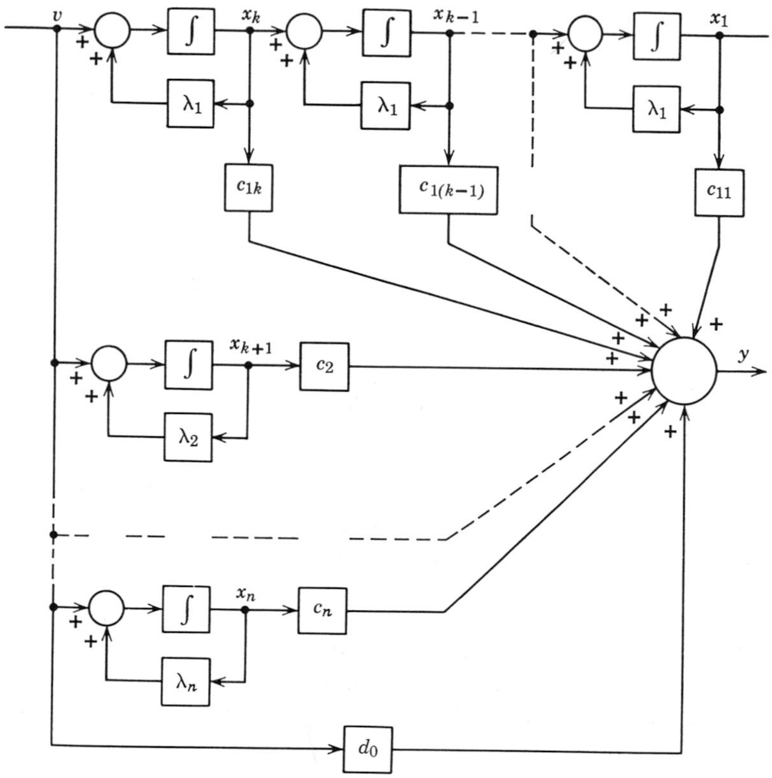

Fig. 5.5-7

As a consequence of Eqs. 5.5-15 and 5.5-16 the simulation diagram of the system can be drawn as shown in Fig. 5.5-7. The state equations of this system are

$$\begin{bmatrix} \dot{x}_1 \\ \dot{x}_2 \\ \cdot \\ \cdot \\ \cdot \\ \dot{x}_k \\ \dot{x}_{k+1} \\ \cdot \\ \cdot \\ \cdot \\ \dot{x}_n \end{bmatrix} = \begin{bmatrix} \lambda_1 & 1 & 0 & \cdots & \cdot & \cdot & \cdot & 0 \\ 0 & \lambda_1 & 1 & \cdots & \cdot & \cdot & \cdot & 0 \\ \cdot & \cdot & \cdot & \cdots & \cdot & \cdot & \cdot & \cdot \\ \cdot & \cdot & \cdot & \cdots & \cdot & \cdot & \cdot & \cdot \\ \cdot & \cdot & \cdot & \cdots & \cdot & \cdot & \cdot & \cdot \\ 0 & 0 & 0 & \cdots & \lambda_1 & 0 & 0 & 0 \\ 0 & 0 & 0 & \cdots & 0 & \lambda_2 & 0 & 0 \\ \cdot & \cdot & \cdot & \cdots & \cdot & \cdot & \cdot & \cdot \\ 0 & 0 & 0 & \cdots & 0 & 0 & 0 & \lambda_n \end{bmatrix} \begin{bmatrix} x_1 \\ x_2 \\ \cdot \\ \cdot \\ \cdot \\ x_k \\ x_{k+1} \\ \cdot \\ \cdot \\ \cdot \\ x_n \end{bmatrix} + \begin{bmatrix} 0 \\ 0 \\ \cdot \\ \cdot \\ 0 \\ 1 \\ 1 \\ \cdot \\ \cdot \\ \cdot \\ 1 \end{bmatrix} v \tag{5.5-17}$$

$$y = [c_{11} \quad c_{12} \quad \cdots \quad c_{1k} \quad c_2 \quad \cdots \quad c_n] \begin{bmatrix} x_1 \\ x_2 \\ \cdot \\ \cdot \\ x_k \\ x_{k+1} \\ \cdot \\ \cdot \\ \cdot \\ x_n \end{bmatrix} + d_0 v$$

or

$$\begin{aligned} \dot{\mathbf{x}} &= \mathbf{J}\mathbf{x} + \mathbf{b}v \\ y &= \langle \mathbf{c}, \mathbf{x} \rangle + d_0 v \end{aligned} \tag{5.5-18}$$

where **J**, **b**, and **c** are defined by the equivalence of Eqs. 5.5-17 and 5.5-18.

The partial fraction technique is appealing when the system has only one input and one output. The result of this approach is an **A** matrix which is either in diagonal form or a nondiagonal Jordan form. This form is particularly convenient when the dynamic behavior of the system is to be found. However, this approach presents certain difficulties. Most of these difficulties can be overcome by the general technique of converting the state equations in *standard form* (if they are available) to the *normal form*.

Normal Form

The method of partial fraction expansion is useful when the state equations of a single input-single output system are to be found. In this case **B**, and **C** are vectors and **D**, **v**, and **y** are scalars. When the system is multivariable, the method of partial fraction expansion can become unwieldy. It then becomes desirable to be able to convert the state equations in standard form to a form where the **A** matrix is either a diagonal matrix or a more general Jordan matrix. This conversion can be accomplished by means of a similarity transformation.

Consider the fixed system defined by the state equations in the standard form

$$\begin{aligned} \dot{\mathbf{x}} &= \mathbf{A}\mathbf{x} + \mathbf{B}\mathbf{v} \\ \mathbf{y} &= \mathbf{C}\mathbf{x} + \mathbf{D}\mathbf{v} \end{aligned} \tag{5.5-19}$$

It is assumed that the characteristic values of **A** are distinct. If the linear transformation $\mathbf{x} = \mathbf{M}\mathbf{q}$ is introduced, where **M** is the modal matrix, then

Eq. 5.5-19 can be written as

$$\begin{aligned} \mathbf{M}\dot{\mathbf{q}} &= \mathbf{AMq} + \mathbf{Bv} \\ \mathbf{y} &= \mathbf{CMq} + \mathbf{Dv} \end{aligned} \qquad (5.5\text{-}20)$$

Premultiplication of the first of these expressions by $\mathbf{M}^{-1}$ yields

$$\dot{\mathbf{q}} = \mathbf{M}^{-1}\mathbf{AMq} + \mathbf{M}^{-1}\mathbf{Bv}$$

Since $\mathbf{M}$ is the modal matrix, the similarity transformation $\mathbf{M}^{-1}\mathbf{AM}$ results in a diagonal Jordan matrix $\mathbf{\Lambda}$. The principal diagonal elements of $\mathbf{\Lambda}$ are the characteristic values $\lambda_1, \lambda_2, \ldots, \lambda_n$. Consequently,

$$\begin{aligned} \dot{\mathbf{q}} &= \mathbf{\Lambda q} + \mathbf{B}_n\mathbf{v} \\ \mathbf{y} &= \mathbf{C}_n\mathbf{q} + \mathbf{D}_n\mathbf{v} \end{aligned} \qquad (5.5\text{-}21)$$

where $\mathbf{\Lambda} = \mathbf{M}^{-1}\mathbf{AM}$, $\mathbf{B}_n = \mathbf{M}^{-1}\mathbf{B}$, $\mathbf{C}_n = \mathbf{CM}$, and $\mathbf{D}_n = \mathbf{D}$. Equation 5.5-21 is known as the *normal form* for the state equations. In this form the differential equations in terms of the state variables $q_1, q_2, \ldots, q_n$ are uncoupled. That is, they are of the form $\dot{q}_i = \lambda_i q_i + f_i$, where f_i is the forcing function applied to the ith state variable. This procedure can be applied when the *standard form* for the state equations is known, and as such it is a more general approach than the partial fraction expansion of the transfer function. It is also shown in later sections that the transfer function approach can lead to serious misunderstanding about the system performance.

In the case of repeated characteristic values, $\mathbf{\Lambda}$ may be replaced by a nondiagonal Jordan matrix. It should be noted however, that the occurrence of repeated roots in the characteristic equation $|\lambda\mathbf{I} - \mathbf{A}| = 0$ does not of necessity imply a nondiagonal Jordan matrix in the normal form. Section 4.8 showed, for example, that, if the degeneracy of the characteristic matrix is equal to the order of the root, then sufficient characteristic vectors can be obtained such that the modal matrix is nonsingular. The matrix $\mathbf{A}$ can then be transformed by the similarity transformation $\mathbf{M}^{-1}\mathbf{AM}$ into a diagonal matrix.

Example 5.5-5. Write the state equations for the dynamics of Fig. 5.5-8*a* in normal form.

For the state variables indicated in Fig. 5.5-8*a*, the standard form of the state equations is

$$\dot{\mathbf{x}} = \begin{bmatrix} 0 & 1 \\ -2 & -3 \end{bmatrix} \mathbf{x} + \begin{bmatrix} 2 \\ 1 \end{bmatrix} v$$

$$y = [1 \quad 0]\mathbf{x} + 2v$$

The modal matrices $\mathbf{M}$ and $\mathbf{M}^{-1}$ associated with the $\mathbf{A}$ matrix are

$$\mathbf{M} = \begin{bmatrix} 1 & 1 \\ -1 & -2 \end{bmatrix} \quad \text{and} \quad \mathbf{M}^{-1} = \begin{bmatrix} 2 & 1 \\ -1 & -1 \end{bmatrix}$$

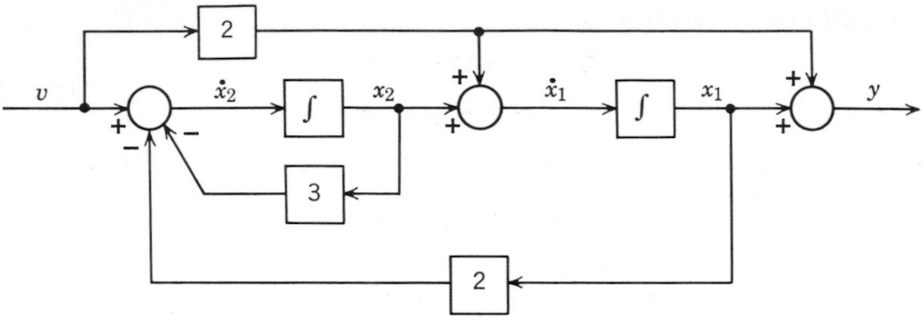

(a)

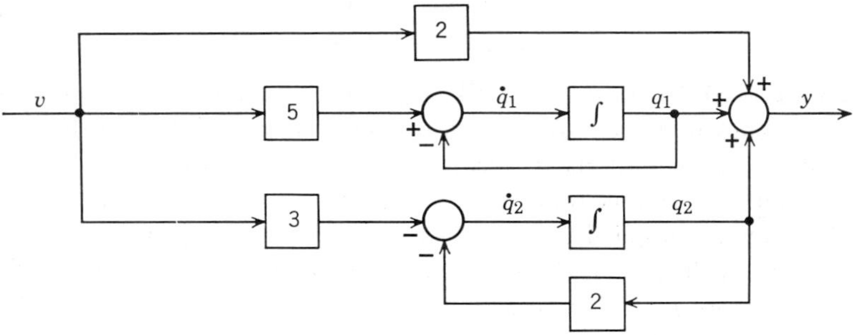

(b)

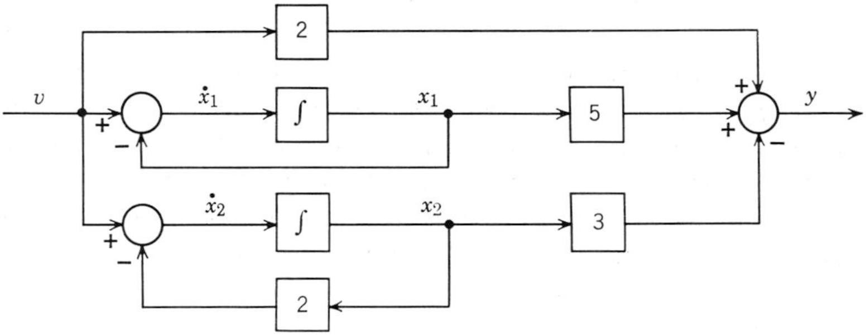

(c)

Fig. 5.5-8

The *normal form* matrices are

$$\mathbf{\Lambda} = \mathbf{M}^{-1}\mathbf{A}\mathbf{M} = \begin{bmatrix} -1 & 0 \\ 0 & -2 \end{bmatrix}$$

$$\mathbf{B}_n = \mathbf{M}^{-1}\mathbf{B} = \begin{bmatrix} 5 \\ -3 \end{bmatrix} = \mathbf{b}_n$$

$$\mathbf{C}_n = \mathbf{C}\mathbf{M} = [1 \quad 1] = \mathbf{c}_n{}^T$$

$$D_n = d_0 = 2$$

The normal form for the state equations is

$$\dot{\mathbf{q}} = \mathbf{\Lambda q} + \mathbf{b}_n v$$
$$y = \langle \mathbf{c}_n, \mathbf{q} \rangle + d_0 v$$

where $\mathbf{q} = \mathbf{M}^{-1}\mathbf{x}$, or $q_1 = 2x_1 + x_2$, $q_2 = -x_1 - x_2$. The block diagram for these equations is shown in Fig. 5.5-8*b*.

The transfer function of this single input-single output system is

$$H(s) = \frac{2s^2 + 8s + 11}{s^2 + 3s + 2} = \frac{2s^2 + 8s + 11}{(s+1)(s+2)}$$

The partial fraction expansion of $H(s)$ is then

$$d_0 = \lim_{s \to \infty} H(s) = 2$$
$$c_1 = H(s)(s+1)\big|_{s=-1} = 5$$
$$c_2 = H(s)(s+2)\big|_{s=-2} = -3$$

Consequently

$$H(s) = 2 + \frac{5}{s+1} - \frac{3}{s+2}$$

The state equations of this system are

$$\dot{\mathbf{x}} = \begin{bmatrix} -1 & 0 \\ 0 & -2 \end{bmatrix} \mathbf{x} + \begin{bmatrix} 1 \\ 1 \end{bmatrix} v$$
$$y = [5 \quad -3]\mathbf{x} + 2v$$

This form is represented in Fig. 5.5-8*c*.

Note that the b_i and c_i of the partial fraction expansion are identical with the c_i and b_i, respectively, of the normal form. This interchange of input and output gain does not affect the validity of either approach. The important factor is the overall gain $b_i c_i$. Certainly the x_i of the partial fraction expansion could be redefined so that the two methods would be identical. However, if this is done, the partial fraction expansion technique for repeated poles of the transfer function does not produce a Jordan matrix. This point is illustrated in Problem 5.13.

Example 5.5-6. Write the state equation for the system of Fig. 5.2-13 in normal form.

For the state variables indicated in Fig. 5.2-13, x_3 is already in normal form. Thus only x_1 and x_2 need be considered. Then

$$\begin{bmatrix} \dot{x}_1 \\ \dot{x}_2 \end{bmatrix} = \begin{bmatrix} 0 & 1 \\ -2 & -3 \end{bmatrix} \begin{bmatrix} x_1 \\ x_2 \end{bmatrix} + \begin{bmatrix} 0 & 1 \\ 1 & -2 \end{bmatrix} \begin{bmatrix} v_1 \\ v_2 \end{bmatrix}$$
$$y_2 = [1 \quad 0] \begin{bmatrix} x_1 \\ x_2 \end{bmatrix}$$

or

$$\mathbf{A} = \begin{bmatrix} 0 & 1 \\ -2 & -3 \end{bmatrix}, \quad \mathbf{B} = \begin{bmatrix} 0 & 1 \\ 1 & -2 \end{bmatrix}, \quad \mathbf{C} = [1 \quad 0], \quad \mathbf{D} = [0]$$

From Example 5.5-5,

$$\mathbf{M} = \begin{bmatrix} 1 & 1 \\ -1 & -2 \end{bmatrix}, \quad \mathbf{M}^{-1} = \begin{bmatrix} 2 & 1 \\ -1 & -1 \end{bmatrix}, \quad \mathbf{\Lambda} = \begin{bmatrix} -1 & 0 \\ 0 & -2 \end{bmatrix}$$

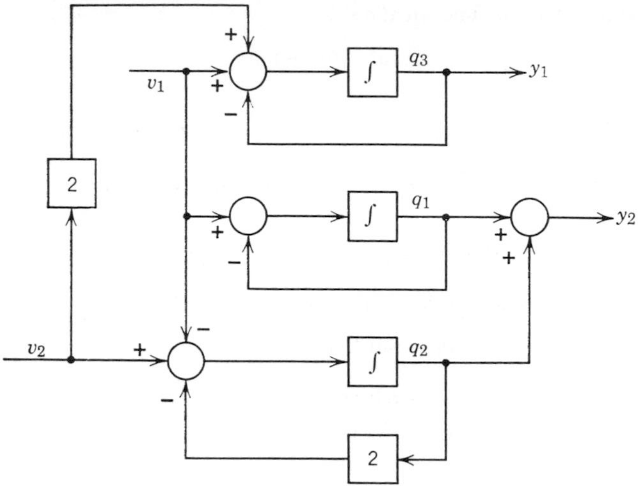

Fig. 5.5-9

Hence

$$\mathbf{B}_n = \mathbf{M}^{-1}\mathbf{B} = \begin{bmatrix} 1 & 0 \\ -1 & 1 \end{bmatrix}, \quad \mathbf{C}_n = \mathbf{CM} = [1 \quad 1]$$

The diagram for these equations is shown in Fig. 5.5-9.

5.6 MODE INTERPRETATION—A GEOMETRICAL CONCEPT[2]

In this section a geometrical interpretation is applied to linear fixed systems. The principal advantages of the approach are its conceptual simplicity, and that it offers the engineer an intuitive "feel" for the nature of solutions of linear systems. The importance of this feel or insight is not to be overlooked.

As an introduction to the mode concept, consider the differential equation

$$\ddot{y} + 3\dot{y} + 2y = f(t)$$

Taking the Laplace transform,

$$Y(s) = \frac{F(s)}{(s+1)(s+2)} + \frac{sy(0) + \dot{y}(0) + 3y(0)}{(s+1)(s+2)}$$

Hence

$$y(t) = \mathscr{L}^{-1}\left[\frac{F(s)}{(s+1)(s+2)}\right] + [2y(0) + \dot{y}(0)]\epsilon^{-t} - [y(0) + \dot{y}(0)]\epsilon^{-2t}$$

If the terms of the response involving ϵ^{-t} and ϵ^{-2t} are called "modes" of the system, then the solution $y(t)$ may or may not exhibit these modes, depending upon the initial conditions and the zeros of $F(s)$. For example, if the initial conditions are such that $2y(0) + \dot{y}(0) = 0$, then the initial condition response does not exhibit the mode ϵ^{-t}. If, in addition, $F(s)$ has a zero at $s = -1$, then the complete solution for $y(t)$ does not exhibit the mode ϵ^{-t}. The mode ϵ^{-t} is completely suppressed. Consequently, if the initial conditions are such that a zero of the response transform occurs at the same point as does a pole of the transfer function of the system, and if a zero of the forcing function transform also occurs there, then the mode corresponding to that pole does not appear in the system response. This pole-zero cancellation, although somewhat obvious, leads to the general concepts which are to be discussed.

Linear Fixed Homogeneous Systems

Assume for simplicity that a system has distinct characteristic roots (eigenvalues) $\lambda_1, \lambda_2, \ldots, \lambda_n$. To each characteristic root λ_i there is a characteristic vector (eigenvector) $\mathbf{u}_i$. Let the characteristic vectors be normalized, so that $|\mathbf{u}_i|^2 = 1$. (This is equivalent to setting the length of each of the characteristic vectors equal to unity.) For the linear unforced system

$$\dot{\mathbf{x}} = \mathbf{A}\mathbf{x} \tag{5.6-1}$$

the characteristic vectors $\mathbf{u}_i$ are defined by

$$\mathbf{A}\mathbf{u}_i = \lambda_i\mathbf{u}_i, \qquad \langle \mathbf{u}_i, \mathbf{u}_i \rangle = 1 \qquad (i = 1, \ldots, n) \tag{5.6-2}$$

Since the characteristic roots are distinct, the characteristic vectors are linearly independent. Therefore $\mathbf{x}(t)$ can be uniquely expressed as a linear combination of these characteristic vectors:

$$\mathbf{x}(t) = \sum_{i=1}^{n} \alpha_i(t)\mathbf{u}_i \tag{5.6-3}$$

The general form for the $\alpha_i(t)$ is given by

$$\alpha_i(t) = c_i\epsilon^{\lambda_i t} \tag{5.6-4}$$

where the c_i's are constants. Thus

$$\mathbf{x}(t) = \sum_{i=1}^{n} c_i\epsilon^{\lambda_i t}\mathbf{u}_i \tag{5.6-5}$$

Evaluating Eq. 5.6-5 at $t = 0$ yields

$$\mathbf{x}(0) = \sum_{i=1}^{n} c_i\mathbf{u}_i \tag{5.6-6}$$

The constants c_i can be determined by utilizing the reciprocal basis $\mathbf{r}_i$ defined by the relation

$$\langle \mathbf{r}_i, \mathbf{u}_j \rangle = \delta_{ij} \qquad (i, j = 1, \ldots, n)$$

By forming the scalar product of both sides of Eq. 5.6-6 with $\mathbf{r}_i$, the constant c_i is found to be

$$c_i = \langle \mathbf{r}_i, \mathbf{x}(0) \rangle \tag{5.6-7}$$

Therefore the initial condition response of a linear fixed system can be written as

$$\mathbf{x}(t) = \sum_{i=1}^{n} \langle \mathbf{r}_i, \mathbf{x}(0) \rangle \epsilon^{\lambda_i t} \mathbf{u}_i \tag{5.6-8}$$

The scalar product $\langle \mathbf{r}_i, \mathbf{x}(0) \rangle$ represents the magnitude of excitation of the ith mode of the system due to the initial conditions. If the initial conditions initially lie along the ith characteristic vector, then only the ith mode is excited. The scalar products $\langle \mathbf{r}_j, \mathbf{x}(0) \rangle$, where $i \neq j$, are identically zero. Therefore, for a linear, fixed, unforced system with distinct characteristic values, the initial condition response is given by a linear weighted sum of the modes $\epsilon^{\lambda_i t} \mathbf{u}_i$, where λ_i is a characteristic root of the $\mathbf{A}$ matrix.

Example 5.6-1. To illustrate the use of the mode expansion technique analyze the system considered at the beginning of this section.

For this system and $x = x_1$, $\dot{x}_1 = x_2$,

$$\mathbf{A} = \begin{bmatrix} 0 & 1 \\ -2 & -3 \end{bmatrix}; \quad \lambda_1 = -1, \quad \lambda_2 = -2$$

The characteristic vectors and a reciprocal basis are

$$\mathbf{u}_1 = \begin{bmatrix} 1/\sqrt{2} \\ -1/\sqrt{2} \end{bmatrix}, \quad \mathbf{u}_2 = \begin{bmatrix} 1/\sqrt{5} \\ -2/\sqrt{5} \end{bmatrix}$$

and

$$\mathbf{r}_1 = \begin{bmatrix} 2\sqrt{2} \\ \sqrt{2} \end{bmatrix}, \quad \mathbf{r}_2 = \begin{bmatrix} -\sqrt{5} \\ -\sqrt{5} \end{bmatrix}$$

Thus the solution for the initial condition response is given by

$$\begin{aligned} \mathbf{x}(t) &= \sum_{i=1}^{2} \langle \mathbf{r}_i, \mathbf{x}(0) \rangle \, \epsilon^{\lambda_i t} \mathbf{u}_i \\ &= [2\sqrt{2}\, y(0) + \sqrt{2}\, \dot{y}(0)]\epsilon^{-t} \mathbf{u}_1 \\ &\quad - [\sqrt{5}\, y(0) + \sqrt{5}\, \dot{y}(0)]\epsilon^{-2t}\, \mathbf{u}_2 \end{aligned}$$

Note that, if the initial condition vector is set equal to one of the characteristic vectors (or any factor times the characteristic vector), then the scalar product $\langle \mathbf{r}_i, \mathbf{x}(0) \rangle$ vanishes for all but the component associated with that characteristic vector. For example, if the initial conditions are such that $y(0) = 1$, $\dot{y}(0) = -1$, then the initial condition vector lies along $\mathbf{u}_1$ and only the mode involving ϵ^{-t} is excited.

Linear, Fixed, Forced Systems

So far, only the homogeneous case case been discussed. For the case in which a forcing function is applied to the system, the same general principles apply if the forcing function is first decomposed among the $\mathbf{u}_i$ vectors as

$$\mathbf{Bv} = \mathbf{f}(t) = \sum_{i=1}^{n} f_i(t)\mathbf{u}_i \tag{5.6-9}$$

where $f_i(t) = \langle \mathbf{r}_i, \mathbf{f}(t) \rangle = \langle \mathbf{r}_i, \mathbf{Bv}(t) \rangle$. Utilizing the convolution integral for linear fixed systems, the general expression for $\mathbf{x}(t)$, assuming distinct characteristic values, is then given by

$$\mathbf{x}(t) = \sum_{i=1}^{n} \langle \mathbf{r}_i, \mathbf{x}(0) \rangle \epsilon^{\lambda_i t}\mathbf{u}_i + \int_0^t \sum_{i=1}^{n} \langle \mathbf{r}_i, \mathbf{Bv}(\tau) \rangle \epsilon^{\lambda_i (t-\tau)}\mathbf{u}_i \, d\tau \tag{5.6-10}$$

The importance of Eq. 5-6.10 lies in that the effect of the forcing function on each mode is considered *independently*. The amount of excitation of the ith mode due to the forcing function is given by

$$\int_0^t \langle \mathbf{r}_i, \mathbf{Bv}(\tau) \rangle \epsilon^{\lambda_i (t-\tau)}\mathbf{u}_i \, d\tau$$

If the forcing function is so selected that it always lies along the direction of one of the characteristic vectors, then only one mode is excited by the forcing function. This situation occurs in circuits that have symmetrical or balanced properties. In these circuits, the modes are referred to as symmetrical or antisymmetrical, such that a "symmetrical" forcing function excites only the symmetrical modes.[3]

Complex Characteristic Roots

If the coefficient matrix possesses a complex characteristic root λ_1, then $\lambda_2 = \lambda_1^*$ is also a complex characteristic root. The characteristic vectors $\mathbf{u}_1$ and $\mathbf{u}_2$ are also complex conjugates, such that $\mathbf{u}_2 = \mathbf{u}_1^*$. The following relationships then exist between the characteristic vectors and the reciprocal basis vectors:

$$\begin{aligned} \lambda_1 &= \alpha_1 + j\beta_1 & \lambda_2 &= \alpha_1 - j\beta_1 \\ \mathbf{u}_1 &= \mathbf{u}_1' + j\mathbf{u}_1'' & \mathbf{u}_2 &= \mathbf{u}_1' - j\mathbf{u}_1'' \\ 2\mathbf{r}_1 &= \mathbf{r}_1' + j\mathbf{r}_1'' & 2\mathbf{r}_2 &= \mathbf{r}_1' - j\mathbf{r}_1'' \end{aligned} \tag{5.6-11}$$

$$\begin{aligned} \langle \mathbf{r}_1', \mathbf{u}_1' \rangle &= 1 & \langle \mathbf{r}_1'', \mathbf{u}_1' \rangle &= 0 \\ \langle \mathbf{r}_1', \mathbf{u}_1'' \rangle &= 0 & \langle \mathbf{r}_1'', \mathbf{u}_1'' \rangle &= 1 \end{aligned} \tag{5.6-12}$$

The normalization condition is given by

$$\langle \mathbf{u}_1', \mathbf{u}_1' \rangle + \langle \mathbf{u}_1'', \mathbf{u}_1'' \rangle = 1 \tag{5.6-13}$$

By utilizing these expressions, the unforced solution for a system with k pairs of complex characteristic roots is given by

$$\begin{aligned} \mathbf{x}(t) = \sum_{i=1}^{k} \epsilon^{\alpha_i t}\{[\langle \mathbf{r}_i', \mathbf{x}(0)\rangle \cos \beta_i t + \langle \mathbf{r}_i'', \mathbf{x}(0)\rangle \sin \beta_i t]\mathbf{u}_i' \\ + [\langle \mathbf{r}_i'', \mathbf{x}(0)\rangle \cos \beta_i t - \langle \mathbf{r}_i', \mathbf{x}(0)\rangle \sin \beta_i t]\mathbf{u}_i''\} \end{aligned} \tag{5.6-14}$$

The amplitude and phase of each oscillatory mode depends on the initial conditions. If the initial conditions are equal to $\mathbf{u}_1'$, then from Eqs. 5.6-12 and 5.6-14 the solution is

$$\mathbf{x}(t) = \epsilon^{\alpha_1 t}[(\cos \beta_1 t)\mathbf{u}_1' - (\sin \beta_1 t)\mathbf{u}_1''] \qquad \text{(Mode 1)} \tag{5.6-15}$$

If the initial conditions are equal to $\mathbf{u}_1''$, then similarly the solution is

$$\mathbf{x}(t) = \epsilon^{\alpha_1 t}[(\sin \beta_1 t)\mathbf{u}_1' + (\cos \beta_1 t)\mathbf{u}_1''] \qquad \text{(Mode 2)} \tag{5.6-16}$$

Note that, for a complex characteristic root, the motion in phase space takes the form of an exponential spiral in the $\mathbf{u}_1'$, $\mathbf{u}_1''$ plane.

Example 5.6-2. Assume that the system under investigation has an **A** matrix given by

$$\mathbf{A} = \begin{bmatrix} 0 & 1 \\ -2 & -2 \end{bmatrix}$$

Determine the initial condition response of each of its modes.

The characteristic roots of this matrix are

$$\lambda_1 = -1 + j$$
$$\lambda_2 = -1 - j$$

The characteristic vectors are found by successively substituting the λ_i's into **Adj** $[\lambda\mathbf{I} - \mathbf{A}]$. Since

$$\mathbf{Adj}\,[\lambda\mathbf{I} - \mathbf{A}] = \begin{bmatrix} \lambda + 2 & 1 \\ -2 & \lambda \end{bmatrix}$$

the normalized characteristic vectors are given by

$$\mathbf{u}_1 = \frac{1}{\sqrt{6}}\begin{bmatrix} 1 + j \\ -2 \end{bmatrix}, \quad \mathbf{u}_2 = \frac{1}{\sqrt{6}}\begin{bmatrix} 1 - j \\ -2 \end{bmatrix}$$

Notice that $\mathbf{u}_2 = \mathbf{u}_1^*$.

The reciprocal basis is found from Eq. 5.6-12 as

$$\mathbf{r}_1' = \sqrt{6}\begin{bmatrix} 0 \\ -\frac{1}{2} \end{bmatrix}, \quad \mathbf{r}_1'' = \sqrt{6}\begin{bmatrix} 1 \\ \frac{1}{2} \end{bmatrix}$$

The complete solution is then

$$\begin{aligned} \mathbf{x}(t) = \epsilon^{-t}\{[\langle \mathbf{r}_1', \mathbf{x}(0)\rangle \cos t + \langle \mathbf{r}_1'', \mathbf{x}(0)\rangle \sin t]\mathbf{u}_1' \\ + [\langle \mathbf{r}_1'', \mathbf{x}(0)\rangle \cos t - \langle \mathbf{r}_1', \mathbf{x}(0)\rangle \sin t]\mathbf{u}_1''\} \end{aligned}$$

If the initial conditions are so chosen that $\mathbf{x}(0) = \mathbf{u}_1'$, then only the mode

$$\mathbf{x}(t) = \epsilon^{-t}[(\cos t)\mathbf{u}_1' - (\sin t)\mathbf{u}_1''] \qquad \text{(Mode 1)}$$

is excited. If the initial conditions are so chosen that $\mathbf{x}(0) = \mathbf{u}_1''$, then the mode

$$\mathbf{x}(t) = \epsilon^{-t}[(\sin t)\mathbf{u}_1' + (\cos t)\mathbf{u}_1''] \qquad \text{(Mode 2)}$$

is excited. The difference between these two modes is the amplitude and phase of the damped oscillations.

The mode expansion is a geometric representation of the diagonalized form of the **A** matrix. This representation shows that, for distinct characteristic roots, the modes of the system are uncoupled and are independently excited by the initial conditions and the input forcing function. The amount of excitation is measured by the scalar product of the reciprocal basis and the vector in question (initial conditions or input forcing function). For systems which do not have distinct characteristic roots, the modes are not necessarily uncoupled. The Jordan canonical form is about the best that can be accomplished in the way of diagonalizing the **A** matrix in this case.

5.7 CONTROLLABILITY AND OBSERVABILITY

The increasing emphasis on linear multivariable control systems has caused a thorough re-examination of the intuitive concepts which have been handed down from the early studies on single input-single output systems. This is due to the large increase in design effort which must be expended in going from the single variable control system to the multivariable system. Most of these early concepts can be extended to multivariable problems by the use of matrix techniques; however, there are some important tools of analysis and synthesis which must be examined with greater care. Transform techniques and transfer functions have been widely used, since they enable the engineer to work with algebraic equations rather than differential equations. However, when single variable subsystems are connected together to form multivariable systems, the transfer functions involved must be examined with care lest some degree of freedom be lost in the combination. In this section, some of the problems connected with multivariable systems are outlined. For a more complete discussion the reader is referred to References 4 to 9.

Consider the set of system equations

$$\begin{aligned} \dot{\mathbf{x}} &= \mathbf{A}\mathbf{x} + \mathbf{B}\mathbf{v} \\ \mathbf{y} &= \mathbf{C}\mathbf{x} + \mathbf{D}\mathbf{v} \end{aligned} \tag{5.7-1}$$

It is assumed that this system has n state variables, m inputs, p outputs, and distinct characteristic values. In terms of the mode expansion technique of Section 5.6, the solution to Eq. 5.7-1 is given by (see Eq. 5.6-10)

$$\mathbf{x}(t) = \sum_{i=0}^{n} \langle \mathbf{r}_i, \mathbf{x}(0)\rangle \epsilon^{\lambda_i t}\mathbf{u}_i + \int_0^t \sum_{i=1}^{n} \langle \mathbf{r}_i, \mathbf{B}\mathbf{v}(\tau)\rangle \epsilon^{\lambda_i(t-\tau)}\mathbf{u}_i \, d\tau \qquad (5.7\text{-}2)$$

If the forcing function is expressed as the sum of its components, i.e.,

$$\mathbf{B}\mathbf{v} = \mathbf{b}_1 v_1 + \mathbf{b}_2 v_2 + \mathbf{b}_3 v_3 + \cdots + \mathbf{b}_m v_m \qquad (5.7\text{-}3)$$

where

$$\mathbf{b}_i = \begin{bmatrix} b_{1i} \\ b_{2i} \\ \cdot \\ \cdot \\ \cdot \\ b_{ni} \end{bmatrix}$$

then the contribution to $\mathbf{x}(t)$ due to the forcing function can be expressed as

$$\int_0^t \sum_{i=1}^{n} \sum_{j=1}^{m} \langle \mathbf{r}_i, \mathbf{b}_j\rangle v_j(\tau) \epsilon^{\lambda_i(t-\tau)}\mathbf{u}_i \, d\tau$$

If the scalar product $\langle \mathbf{r}_i, \mathbf{b}_j\rangle$ is zero for some mode for all j, then the input is not coupled to that mode, and, regardless of what the forcing function is, there is no way for the input to excite or control this mode. Obviously, a criterion for complete *controllability* of a linear time-invariant system is that the scalar products $\langle \mathbf{r}_i, \mathbf{b}_j\rangle$ do not vanish for all j.†

A system which is not *observable* has dynamic modes of behavior which cannot be ascertained from measurements of the available outputs.[8] Considering only the initial condition excitation of the modes, the output $\mathbf{y}$ is given by Eqs. 5.7-1 and 5.7-2 as

$$\mathbf{y} = \mathbf{C} \sum_{i=1}^{n} \langle \mathbf{r}_i, \mathbf{x}(0)\rangle \epsilon^{\lambda_i t}\mathbf{u}_i \qquad (5.7\text{-}4)$$

assuming that $\mathbf{A}$ has distinct λ's. A condition that the ith mode disappear in all the outputs is

$$\langle \mathbf{c}_j, \mathbf{u}_i\rangle = 0 \qquad \text{for all } j \qquad (5.7\text{-}5)$$

† This is a necessary and sufficient condition if $\mathbf{A}$ has distinct λ's. If $\mathbf{A}$ has repeated λ's and is diagonalizable, then this is necessary but not sufficient. In this case or if $\mathbf{A}$ can only be transformed to a more general Jordan form, then one should consider the more basic requirement for controllability, which is that there be an input $\mathbf{v}(t)$, $0 \leqslant t < T < \infty$, such that $\mathbf{x}(0)$ can be forced to $\mathbf{x}(T) = \mathbf{0}$. This leads to the necessary and sufficient condition that the column vectors of the matrix $[\mathbf{B}, \mathbf{AB}, \ldots, \mathbf{A}^{(n-1)}\mathbf{B}]$ span the state space of the system.

where $\mathbf{c}_j$ is a column vector comprised of the elements of the jth row of $\mathbf{C}$. This condition extends to the case where a forcing function is applied.†

If the system equations have distinct λ's and are written in the *normal form*

$$\begin{aligned} \dot{\mathbf{q}} &= \mathbf{\Lambda q} + \mathbf{B}_n\mathbf{v} \\ \mathbf{y} &= \mathbf{C}_n\mathbf{q} + \mathbf{D}_n\mathbf{v} \end{aligned} \tag{5.7-6}$$

then each state variable q_i of the system represents a different mode of behavior. In this form, the conditions for controllability and observability become quite clear. Controllability is a function of the coupling between the inputs and the various modes of the system. A particular mode (or state variable) cannot be controlled if the input is not coupled into this mode. This would be the case if there were a zero *row* in the $\mathbf{B}_n$ matrix.

On the other hand, a particular mode of behavior would not be observed in *any* of the outputs of the system if there were no coupling between that mode and any of the outputs. This would be the case if there were any zero *columns* of the $\mathbf{C}_n$ matrix. Therefore, if the system equations can be written in the normal form with distinct λ's, the conditions for complete controllability and observability are:

Controllability: no *zero* rows of $\mathbf{B}_n$

Observability: no *zero* columns of $\mathbf{C}_n$

Generally speaking, a system can always be divided into four subsystems which display these concepts. This is shown in Fig. 5.7-1. The subsystems have the following characteristics:

System S^*: completely controllable and completely observable

System S^c: completely controllable but unobservable

System S^o: uncontrollable but completely observable

System S^f: uncontrollable and unobservable

The requirement that a system contain only subsystems S^* is that the entire system is controllable and observable. It should be pointed out that disregard of subsystems S^c, S^o, and S^f may lead the system designer into a position where the overall system is unstable, while subsystem S^* is perfectly well-behaved. If, for example, subsystem S^o contains unstable modes, then excitation of these modes by nonzero initial conditions yields

† If $\mathbf{A}$ has repeated λ's, the basic requirement for observability is that, for some $T > 0$ and all initial states $\mathbf{x}(0)$, knowledge of $\mathbf{A}$, $\mathbf{C}$ and $\mathbf{y}(t)$, $0 \leqslant t \leqslant T$, suffices to determine $\mathbf{x}(0)$. This leads to the necessary and sufficient condition for observability that the columns of $[\mathbf{C}^{*T}, \mathbf{A}^{*T}\mathbf{C}^{*T}, \ldots, (\mathbf{A}^{*T})^{(n-1)}\mathbf{C}^{*T}]$ span the system state space.

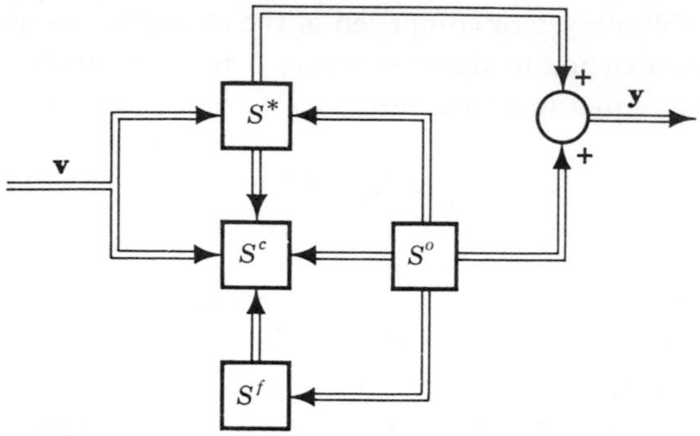

Fig. 5.7-1

an unstable output. Similarly, even though subsystem S^c is unobservable, the required forcing function to subsystem S^* may cause the state variables of S^c to become excessively large, damaging the system. The same type of reasoning holds true for subsystem S^f.

Composite Systems

When a system is comprised of some or all of the components S^*, S^c, S^o, or S^f, the overall system may or may not exhibit the properties of these subsystems, depending upon the connections between the subsystems.

Example 5.7-1.

System 1: $\dot{x}_1 = -x_1 + v_1$ System 2: $\dot{x}_2 = -2x_2 + v_2$

$y_1 = x_1 + v_1$

Now let the output of system 1 be connected to the input of system 2, and let the output of the system be taken as $y = x_2 - y_1$. This is shown in Fig. 5.7-2. Determine the controllability and observability characteristics of the composite system.

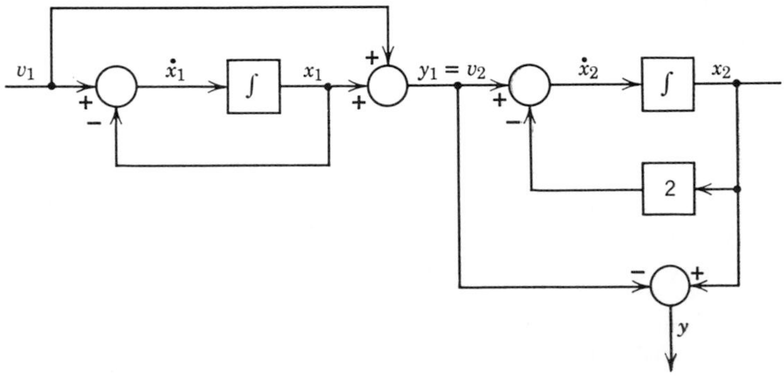

Fig. 5.7-2

The state variable formulation of the composite system is

$$\begin{bmatrix} \dot{x}_1 \\ \dot{x}_2 \end{bmatrix} = \begin{bmatrix} -1 & 0 \\ 1 & -2 \end{bmatrix} \begin{bmatrix} x_1 \\ x_2 \end{bmatrix} + \begin{bmatrix} 1 \\ 1 \end{bmatrix} v_1$$

$$y = [-1 \quad 1] \begin{bmatrix} x_1 \\ x_2 \end{bmatrix} - v_1$$

Using the mode expansion technique, $\lambda_1 = -1$, $\lambda_2 = -2$, and

$$\mathbf{u}_1 = \begin{bmatrix} 1/\sqrt{2} \\ 1/\sqrt{2} \end{bmatrix} \qquad \mathbf{u}_2 = \begin{bmatrix} 0 \\ -1 \end{bmatrix}$$

$$\mathbf{r}_1 = \begin{bmatrix} \sqrt{2} \\ 0 \end{bmatrix} \qquad \mathbf{r}_2 = \begin{bmatrix} 1 \\ -1 \end{bmatrix}$$

For this system,

$$\langle \mathbf{r}_1, \mathbf{b} \rangle = \sqrt{2} \qquad \langle \mathbf{r}_2, \mathbf{b} \rangle = 0$$

$$\langle \mathbf{c}, \mathbf{u}_1 \rangle = 0 \qquad \langle \mathbf{c}, \mathbf{u}_2 \rangle = -1$$

Therefore the mode $\epsilon^{-2t}\mathbf{u}_2$ is not controllable, and the mode $\epsilon^{-t}\mathbf{u}_1$ is not observable. Thus, even though the individual subsystems are both observable and controllable, the overall system is neither observable nor controllable.

An interesting point arises if transfer functions are utilized. The transfer function of the first system is $(s + 2)/(s + 1)$. The transfer function of the second system is $1/(s + 2)$. Therefore the overall transfer function from v_1 to x_2 is

$$\frac{X_2(s)}{V_2(s)} = \frac{1}{s + 1}$$

The overall transfer function is then

$$\frac{Y(s)}{V_1(s)} = -\frac{s + 2}{s + 1} + \frac{1}{s + 1} = -1$$

This confirms the previous analysis that the mode $\epsilon^{-2t}\mathbf{u}_2$ cannot be excited and the mode $\epsilon^{-t}\mathbf{u}_1$ cannot be observed. This analysis by means of transfer functions shows that the use of a transfer function for an overall system may mask some of the modes of the subsystems.

Gilbert has developed a theorem which lists some of the important controllability and observability aspects of a general linear feedback system.[8]

Theorem 5.7-1. Let the linear subsystems S_a and S_b be connected in the feedback arrangement of Fig. 5.7-3. The cascade connection S_aS_b is denoted S_c, and the cascade connection S_bS_a is denoted S_0.

(*a*) The order n of the system is equal to the sum of the orders of S_a and S_b, i.e., $n = n_a + n_b$.

(*b*) A necessary and sufficient condition that the feedback system be controllable (observable) is that $S_c(S_o)$ be controllable (observable).

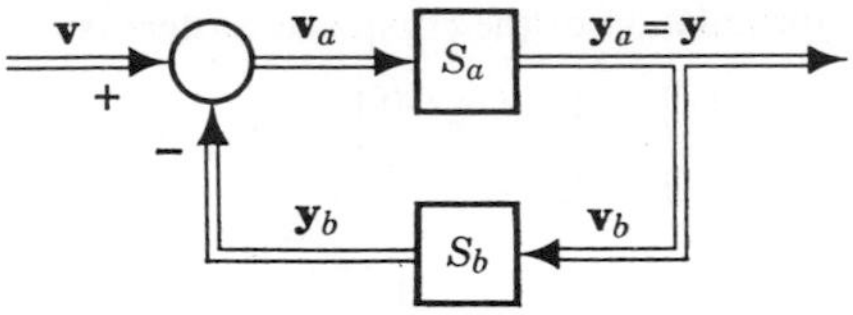

Fig. 5.7-3

(*c*) A necessary but not sufficient condition that the feedback system be controllable (observable) is that both S_a and S_b be controllable (observable).

(*d*) If S_a and S_b are both controllable (observable), any uncontrollable (unobservable) coordinates of the feedback system are uncontrollable (unobservable) coordinates of $S_c(S_o)$ and originate in S_b.

The importance of this theorem lies in the fact that controllability and observability can be determined from the individual open-loop subsystems S_a and S_b. This is a great aid in analysis.

Transfer Functions

The use of transfer functions, or transfer function matrices in the case of multivariable systems, is widespread among practicing engineers, for two reasons. One is that transfer functions allow the use of algebraic equations rather than differential equations. The second is due to the smaller size of the transfer function matrix. For example, if a system has n state variables, $m \leqslant n$ inputs, and $p \leqslant n$ outputs, then the size of the **A** matrix is $(n \times n)$, while the size of the $\mathbf{H}(s)$ matrix is $(m \times p) \leqslant (n \times n)$. In using these matrices however, there are two common errors. The first error is the failure to realize that the transfer function of a composite system may mask the modes of the subsystems. This was illustrated in Example 5.7-1, where the pole $s = -2$ of system 2 was canceled by the zero of system 1. The second error is that, although the elements of the transfer function matrix may indicate all the modes of the subsystems, it may fail to indicate the modes of the composite system. *A transfer function matrix represents only the observable and controllable part of the system, namely* S^*.† A system may consist of subsystems which individually are observable and controllable, but the composite system may be neither observable nor controllable. This was illustrated in Example 5.7-1. Thus the order of the system may be underestimated, and an incorrect conclusion about the stability of the system may result.

† If **A** has repeated characteristic values, this may not even be true. For example, consider such a case in which **A** is diagonalizable.

Example 5.7-2. Determine the transfer function matrix and the order of the system described by

$$\dot{y}_1 + ay_1 = v_1 + v_2$$

$$\ddot{y}_2 + (a + b)\dot{y}_2 + aby_2 = \dot{v}_1 + av_1 + v_2 \qquad a \neq b$$

Transforming these equations for zero initial conditions,

$$Y_1(s) = \frac{V_1(s)}{s + a} + \frac{V_2(s)}{s + a}$$

$$Y_2(s) = \frac{V_1(s)}{s + b} + \frac{V_2(s)}{(s + a)(s + b)}$$

The transfer function matrix is

$$\mathbf{H}(s) = \begin{bmatrix} \dfrac{1}{s + a} & \dfrac{1}{s + a} \\ \dfrac{1}{s + b} & \dfrac{1}{(s + a)(s + b)} \end{bmatrix} = \frac{\begin{bmatrix} 1 & 1 \\ 0 & \dfrac{1}{b - a} \end{bmatrix}}{s + a} + \frac{\begin{bmatrix} 0 & 0 \\ 1 & \dfrac{1}{a - b} \end{bmatrix}}{s + b}$$

The immediate, but erroneous, conclusion is that this matrix represents a second order system. However, if a simulation of this system is made, the minimum number of integrators required is three. This is shown in Fig. 5.7-4. This fact is evident from the original differential equations.

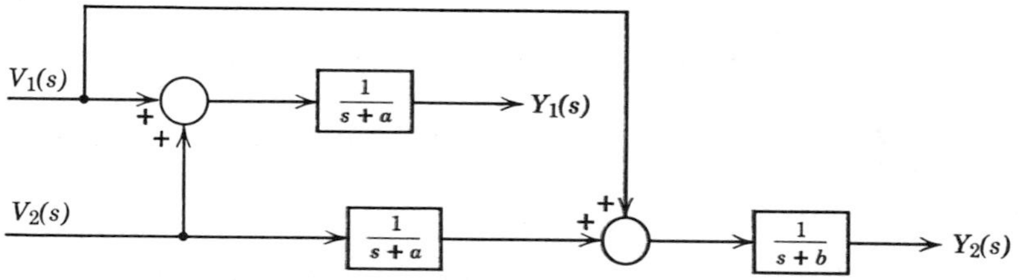

Fig. 5.7-4

To be sure of the order of a system, the use of a theorem by Gilbert always gives the minimum order of the differential equations corresponding to an $\mathbf{H}(s)$ with simple poles.

Theorem 5.7-2. Let $\mathbf{H}(s)$ be expanded into partial fractions as

$$\mathbf{H}(s) = \sum_{i=1}^{m} \frac{\mathbf{K}_i}{s - \lambda_i} + \mathbf{D} \qquad (5.7\text{-}7)$$

where

$$\mathbf{K}_i = \lim_{s \to \lambda_i} (s - \lambda_i)\mathbf{H}(s) \quad \text{and} \quad \mathbf{D} = \lim_{s \to \infty} \mathbf{H}(s)$$

It is assumed that the elements of $\mathbf{H}(s)$ have, at most, m finite simple poles, where $m < \infty$. The rank of the ith pole, r_i, is defined as the rank of the matrix $\mathbf{K}_i$. Then the system cannot be expressed by a set of differential equations of order less than

$$n = \sum_{i=1}^{m} r_i$$

As an example of this theorem consider Example 5.7-2. Let $\lambda_1 = -a$ and $\lambda_2 = -b$. Then $r_1 = 2$, $r_2 = 1$, and $n = r_1 + r_2 = 3$. The minimum order of the system is three.

In conclusion it should be stated that, when dealing with multivariable systems, controllability and observability play an important role in the analysis and synthesis of such systems. When using transform techniques, a firm grasp must be kept on the physical problem, lest the mathematical manipulations obscure the true nature of the system.†

5.8 LINEAR FIXED SYSTEMS—THE STATE TRANSITION MATRIX

The homogeneous differential equation for a linear fixed system is given in vector matrix form by

$$\dot{\mathbf{x}} = \mathbf{A}\mathbf{x} \tag{5.8-1}$$

The matrix $\mathbf{A}$ is a constant (n x n) coefficient matrix, and the vector $\mathbf{x}$ is an (n x 1) column matrix consisting of the n state variables $x_1, x_2, \ldots, x_n$. Similarly to the solution for scalar differential equations, the solution to Eq. 5.8-1 is given by

$$\mathbf{x}(t) = \epsilon^{\mathbf{A}(t-\tau)}\mathbf{x}(\tau) \tag{5.8-2}$$

where the matrix $\epsilon^{\mathbf{A}t}$ is defined by Eq. 4.11-1 to be the infinite series

$$\epsilon^{\mathbf{A}t} = \mathbf{I} + \mathbf{A}t + \frac{\mathbf{A}^2t^2}{2!} + \frac{\mathbf{A}^3t^3}{3!} + \cdots \tag{5.8-3}$$

Equation 5.8-2 can be substituted into Eq. 5.8-1 to verify that it is a solution. Note that, at $t = \tau$, the matrix $\epsilon^{\mathbf{A}(t-\tau)}$ is equal to $\mathbf{I}$, the identity matrix. Therefore the boundary conditions are satisfied, i.e.,

$$\mathbf{x}(\tau) = \epsilon^{\mathbf{A}0}\mathbf{x}(\tau) = \mathbf{I}\mathbf{x}(\tau) = \mathbf{x}(\tau)$$

The matrix $\boldsymbol{\phi}(t) = \epsilon^{\mathbf{A}t}$ is called the *state transition matrix* of the system described by Eq. 5.8-1. Mathematicians prefer the term *fundamental matrix*.‡ The nomenclature *state transition matrix* is more descriptive of

† For complete discussion of this problem, including the case where the characteristic values are not distinct, the reader is referred to References 5, 8, and 10.

‡ For a rigorous discussion of fundamental matrices and their role in the solution of differential equations, see Reference 11 or 12.

the use of this matrix and is generally preferred by the engineering community.

The state transition matrix "describes" the motion of the tip of the state vector from its initial position in state space, and as such it describes the transition of the state of the system. Since the vector $\mathbf{x}(t)$ describes *all* the time functions $x_1(t), x_2(t), \ldots, x_n(t)$, a great deal of information is available in this vector. Necessarily, the computation of the state transition matrix is greater than usually encountered in solving for the dependent variable in a linear differential equation. However, the additional information available enables the system designer to utilize more sophisticated design techniques.

The calculation of the state transition matrix $\boldsymbol{\phi}(t)$ may be performed in several different ways. The principal methods are Sylvester's theorem and the Cayley-Hamilton technique, which were both considered in Section 4.10, and the infinite series method, the frequency-domain method, and the transfer function method. Another method is illustrated in Example 5.9-2 and Problem 5.26.

Infinite Series Method

From the definition of $\epsilon^{\mathbf{A}t}$ (Eq. 5.8-3), the state transition matrix $\boldsymbol{\phi}(t)$ can be calculated by the infinite series

$$\boldsymbol{\phi}(t) = \mathbf{I} + \mathbf{A}t + \frac{\mathbf{A}^2t^2}{2!} + \frac{\mathbf{A}^3t^3}{3!} + \cdots$$

Unless $\mathbf{A}^k$ disappears for some small value of k, this method is the most laborious. Once the summation is performed, the closed form of each series for each element of the $\boldsymbol{\phi}(t)$ matrix must be found. This is generally not a simple task, and, unless the order of $\boldsymbol{\phi}(t)$ is sufficiently low or the analyst is sufficiently clever, the task may be insurmountable. A simple example points out the difficulty involved.

Example 5.8-1. Find the state transition matrix $\boldsymbol{\phi}(t)$ for the matrix

$$\mathbf{A} = \begin{bmatrix} 0 & -2 \\ 1 & -3 \end{bmatrix}$$

The powers of $\mathbf{A}$ can be found by successive multiplication by $\mathbf{A}$, so that

$$\mathbf{A}^2 = \begin{bmatrix} -2 & 6 \\ -3 & 7 \end{bmatrix} \quad \text{and} \quad \mathbf{A}^3 = \begin{bmatrix} 6 & 14 \\ 7 & -15 \end{bmatrix}$$

Therefore, from Eq. 5.8-3, $\boldsymbol{\phi}(t)$ is given by

$$\boldsymbol{\phi}(t) = \begin{bmatrix} 1 & 0 \\ 0 & 1 \end{bmatrix} + \begin{bmatrix} 0 & -2 \\ 1 & -3 \end{bmatrix} t + \begin{bmatrix} -2 & 6 \\ -3 & 7 \end{bmatrix} \frac{t^2}{2!} + \begin{bmatrix} 6 & 14 \\ 7 & -15 \end{bmatrix} \frac{t^3}{3!} + \cdots$$

Collecting terms yields

$$\boldsymbol{\phi}(t) = \begin{bmatrix} 1 - \dfrac{2t^2}{2!} + \dfrac{6t^3}{3!} + \cdots & -2t + \dfrac{6t^2}{2!} + \dfrac{14t^3}{3!} + \cdots \\ t - \dfrac{3t^2}{2!} + \dfrac{7t^3}{3!} + \cdots & 1 - 3t + \dfrac{7t^2}{2!} - \dfrac{15t^3}{3!} + \cdots \end{bmatrix}$$

By recognizing the infinite series for each element (this is the principal drawback of this method),

$$\boldsymbol{\phi}(t) = \begin{bmatrix} 2\epsilon^{-t} - \epsilon^{-2t} & 2(\epsilon^{-2t} - \epsilon^{-t}) \\ \epsilon^{-t} - \epsilon^{-2t} & 2\epsilon^{-2t} - \epsilon^{-t} \end{bmatrix}$$

Frequency-Domain Method

If the Laplace transform of both sides of Eq. 5.8-1 is taken, the result is $s\mathbf{X}(s) - \mathbf{x}(0) = \mathbf{AX}(s)$. Thus

$$\mathbf{X}(s) = [s\mathbf{I} - \mathbf{A}]^{-1}\mathbf{x}(0) \tag{5.8-4}$$

Taking the inverse transform of both sides of Eq. 5.8-4 gives

$$\mathbf{x}(t) = \mathscr{L}^{-1}\{[s\mathbf{I} - \mathbf{A}]^{-1}\}\mathbf{x}(0) \tag{5.8-5}$$

Comparing Eqs. 5.8-5 and 5.8-2, the conclusion is reached that

$$\boldsymbol{\phi}(t) = \epsilon^{\mathbf{A}t} = \mathscr{L}^{-1}[s\mathbf{I} - \mathbf{A}]^{-1} \tag{5.8-6}$$

This method may be the most convenient to use for many problems. The obvious difficulty is finding the inverse of $[s\mathbf{I} - \mathbf{A}]$.

Example 5.8-2. For the same $\mathbf{A}$ matrix as in the previous example, calculate $\boldsymbol{\phi}(t)$ using the frequency-domain approach.

Since

$$[s\mathbf{I} - \mathbf{A}] = \begin{bmatrix} s & 2 \\ -1 & s+3 \end{bmatrix}$$

the inverse matrix $[s\mathbf{I} - \mathbf{A}]^{-1}$ is given by

$$\boldsymbol{\Phi}(s) = [s\mathbf{I} - \mathbf{A}]^{-1} = \frac{\begin{bmatrix} s+3 & -2 \\ 1 & s \end{bmatrix}}{s^2 + 3s + 2}$$

where $\boldsymbol{\Phi}(s)$ is the transform of $\boldsymbol{\phi}(t)$. Taking the inverse transform of $\boldsymbol{\Phi}(s)$ element by element, the expression for $\boldsymbol{\phi}(t)$ is found to be the same as previously computed in Example 5.8-1.

Transfer Function Method

The ith term $x_i(t)$ of Eq. 5.8-2 can be written as the summation

$$x_i(t) = \sum_{j=1}^{n} \phi_{ij}(t)x_j(0) \tag{5.8-7}$$

assuming $\tau = 0$. The element $\phi_{ij}(t)$ can be determined by placing a unit initial condition on state variable x_j and zero initial conditions on all other state variables. The time function $x_i(t)$ is then equal to $\phi_{ij}(t)$. In terms of the simulation diagrams discussed in Section 5.2, this is equivalent to placing a unit initial condition on the *output* of integrator j, and observing the *output* of integrator i. However, placing a unit initial condition on the *output* of an integrator is equivalent to placing a unit impulse on the *input* to the same integrator. Therefore $\phi_{ij}(t)$ is the time response of the *output* of integrator i, when a unit impulse is placed on the *input* to integrator j, and all other integrators have zero initial conditions. Thus $\Phi_{ij}(s)$ can be interpreted as the transfer function from the input to a summer at the *input* of integrator j to the *output* of integrator i. The collection of these transfer functions forms $\mathbf{\Phi}(s) = [s\mathbf{I} - \mathbf{A}]^{-1}$. The state transition matrix in the frequency domain, $\mathbf{\Phi}(s)$, can therefore be determined by an inspection of the simulation diagram.

Example 5.8-3. The simulation diagram for the system described by the $\mathbf{A}$ matrix used for the previous two examples is shown in Fig. 5.8-1a. Determine $\boldsymbol{\phi}(t)$.

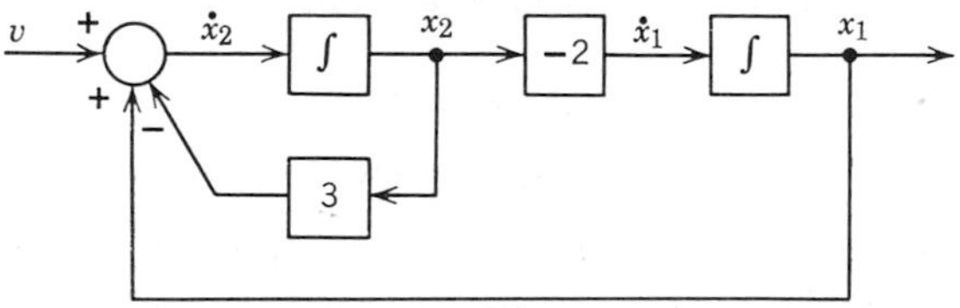

(a)

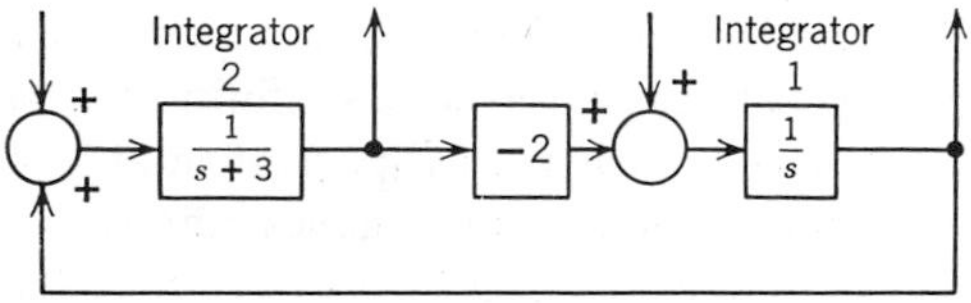

(b)

Fig. 5.8-1

With no loss in generality, the system can be redrawn in the frequency-domain form shown in Fig. 5.8-1*b*. All the transfer functions are then immediately evident.

$$\Phi_{11}(s) = \frac{1/s}{1 + \dfrac{2}{s(s+3)}} = \frac{s+3}{s^2 + 3s + 2}$$

$$\Phi_{12}(s) = \frac{\dfrac{-2}{s(s+3)}}{1 + \dfrac{2}{s(s+3)}} = \frac{-2}{s^2 + 3s + 2}$$

$$\Phi_{21}(s) = \frac{\dfrac{1}{s(s+3)}}{1 + \dfrac{2}{s(s+3)}} = \frac{1}{s^2 + 3s + 2}$$

$$\Phi_{22}(s) = \frac{\dfrac{1}{s+3}}{1 + \dfrac{2}{s(s+3)}} = \frac{s}{s^2 + 3s + 2}$$

Hence

$$\mathbf{\Phi}(s) = \frac{\begin{bmatrix} s+3 & -2 \\ 1 & s \end{bmatrix}}{s^2 + 3s + 2}$$

This matrix checks with $\mathbf{\Phi}(s)$ found in Example 5.8-2 by computation of $[s\mathbf{I} - \mathbf{A}]^{-1}$. The calculation of $\boldsymbol{\phi}(t)$ is then simply the inverse transform of $\mathbf{\Phi}(s)$, element by element.

Linear Fixed Systems—The Complete Solution

The complete set of state variable equations for a linear fixed system is given by

$$\begin{aligned} \dot{\mathbf{x}} &= \mathbf{A}\mathbf{x} + \mathbf{B}\mathbf{v} \\ \mathbf{y} &= \mathbf{C}\mathbf{x} + \mathbf{D}\mathbf{v} \end{aligned} \tag{5.8-8}$$

These equations can be interpreted in the following manner. **A** is the essential matrix of the system, as the structure of this matrix decides the nature of the state transition matrix. The nature of all solutions, whether forced or unforced, depend upon this matrix. Clearly, if all the characteristic values of this matrix have negative real parts, then the solution to Eq. 5.8-1 approaches zero as t approaches infinity. However, if one of the characteristic values has a positive real part, then at least one of the state variables becomes unbounded as t approaches infinity. If the characteristic values which lie along the imaginary axis (zero real part) are simple, $\boldsymbol{\phi}(t)$ does not approach the null matrix as t approaches infinity, but it is bounded.

B is a coupling matrix; the structure of this matrix determines how the input is coupled to the various state variables. **C** is also a coupling matrix, coupling the state variables to the output. Thus the first term of $\mathbf{y}(t)$ in Eq. 5.8-8 represents the coupling of the state variables into the various components of the output vector. **D** again is a coupling matrix, as it directly couples the input vector to the output vector. The structure of this matrix determines how the input forcing functions are distributed among the various outputs. In most physical systems, **D** is a null or zero matrix, so that the term **Dv** is usually zero.

The complete solution for $\mathbf{x}(t)$ and $\mathbf{y}(t)$ in Eq. 5.8-8 can be obtained by a variety of approaches. For purposes of illustration, the method of variation of parameters is used here. The method of adjoint systems is used later to obtain the complete solution for the linear time-varying case.

The homogeneous differential equation $\dot{\mathbf{x}} = \mathbf{A}\mathbf{x}$ has the solution $\mathbf{x}_H(t) = \boldsymbol{\phi}(t-\tau)\mathbf{x}(\tau)$ for $t \geqslant \tau$ from Eq. 5.8-2. Analogous to Eq. 2.6-17, the assumed particular solution is $\mathbf{x}_P(t) = \boldsymbol{\phi}(t-\tau)\mathbf{U}(t)\mathbf{x}(\tau)$. The total solution is

$$\mathbf{x}(t) = \mathbf{x}_H(t) + \mathbf{x}_P(t) = \boldsymbol{\phi}(t-\tau)\mathbf{x}(\tau) + \boldsymbol{\phi}(t-\tau)\mathbf{U}(t)\mathbf{x}(\tau)$$

This is more conveniently written as

$$\mathbf{x}(t) = \boldsymbol{\phi}(t-\tau)[\mathbf{I} + \mathbf{U}(t)]\mathbf{x}(\tau) = \boldsymbol{\phi}(t-\tau)\mathbf{z}(t) \qquad (5.8\text{-}9)$$

where τ is a constant and the vector $\mathbf{z}(t)$ is to be determined.

Substitution of Eq. 5.8-9 into Eq. 5.8-8 yields

$$[\dot{\boldsymbol{\phi}}(t-\tau) - \mathbf{A}\boldsymbol{\phi}(t-\tau)]\mathbf{z}(t) + \boldsymbol{\phi}(t-\tau)\dot{\mathbf{z}}(t) = \mathbf{B}\mathbf{v}(t)$$

However, the first term is a zero matrix, since $\boldsymbol{\phi}(t-\tau)$ is a matrix whose columns are solutions of Eq. 5.8-1. Thus

$$\dot{\mathbf{z}}(t) = \boldsymbol{\phi}^{-1}(t-\tau)\mathbf{B}\mathbf{v}(t) \qquad (5.8\text{-}10)$$

Integrating Eq. 5.8-10 from τ to t is indicated by

$$\mathbf{z}(t) - \mathbf{z}(\tau) = \int_\tau^t \boldsymbol{\phi}^{-1}(\lambda-\tau)\mathbf{B}\mathbf{v}(\lambda)\,d\lambda$$

From Eq. 5.8-9,

$$\boldsymbol{\phi}^{-1}(t-\tau)\mathbf{x}(t) = \boldsymbol{\phi}^{-1}(0)\mathbf{x}(\tau) + \int_\tau^t \boldsymbol{\phi}^{-1}(\lambda-\tau)\mathbf{B}\mathbf{v}(\lambda)\,d\lambda$$

Finally, since

$$\boldsymbol{\phi}(0) = \mathbf{I} \quad \text{and} \quad \boldsymbol{\phi}(t-\tau)\boldsymbol{\phi}^{-1}(\lambda-\tau) = \epsilon^{\mathbf{A}(t-\tau)}\epsilon^{-\mathbf{A}(\lambda-\tau)}$$

$$= \epsilon^{\mathbf{A}(t-\tau)}\epsilon^{\mathbf{A}(\tau-\lambda)} = \epsilon^{\mathbf{A}(t-\lambda)} = \boldsymbol{\phi}(t-\lambda)$$

premultiplication by $\boldsymbol{\phi}(t - \tau)$ yields

$$\mathbf{x}(t) = \boldsymbol{\phi}(t - \tau)\mathbf{x}(\tau) + \int_{\tau}^{t} \boldsymbol{\phi}(t - \lambda)\mathbf{B}\mathbf{v}(\lambda)\, d\lambda \qquad (5.8\text{-}11)$$

The solution for $\mathbf{y}(t)$ follows by substituting Eq. 5.8-11 into Eq. 5.8-8. It is

$$\mathbf{y}(t) = \mathbf{C}\boldsymbol{\phi}(t - \tau)\mathbf{x}(\tau) + \int_{\tau}^{t} \mathbf{C}\boldsymbol{\phi}(t - \lambda)\mathbf{B}\mathbf{v}(\lambda)\, d\lambda + \mathbf{D}\mathbf{v}(t) \qquad (5.8\text{-}12)$$

Equations 5.8-11 and 5.8-12 form the solutions to Eq. 5.8-8. The first term of Eq. 5.8-11 represents the initial condition response of the system state variables, while the second term represents the forced response. Note that the second term of Eq. 5.8-11 is a convolution integral similar to that in Eq. 1.6-2.

Example 5.8-4. Assume that the system whose **A** matrix is given by Example 5.8-1 is subject to a unit step forcing function at $t = 0$ (see Fig. 5.8-1*a*). Find the output $y(t) = x_1(t)$.

For this example,

$$\mathbf{A} = \begin{bmatrix} 0 & -2 \\ 1 & -3 \end{bmatrix}, \quad \boldsymbol{\phi}(t) = \begin{bmatrix} 2\epsilon^{-t} - \epsilon^{-2t} & 2(\epsilon^{-2t} - \epsilon^{-t}) \\ \epsilon^{-t} - \epsilon^{-2t} & 2\epsilon^{-2t} - \epsilon^{-t} \end{bmatrix}$$

$$\mathbf{B} = \begin{bmatrix} 0 \\ 1 \end{bmatrix}, \quad \mathbf{C} = [1 \quad 0], \quad \mathbf{D} = [0]$$

Therefore, from Eq. 5.8-12,

$$\begin{aligned}
\mathbf{y}(t) &= \mathbf{C}\boldsymbol{\phi}(t)\mathbf{x}(0) + \int_0^t \mathbf{C}\boldsymbol{\phi}(t - \lambda)\mathbf{B}U_{-1}(\lambda)\, d\lambda \\
&= [1 \quad 0]\begin{bmatrix} \phi_{11}(t) & \phi_{12}(t) \\ \phi_{21}(t) & \phi_{22}(t) \end{bmatrix}\begin{bmatrix} x_1(0) \\ x_2(0) \end{bmatrix} + \int_0^t [1 \quad 0]\begin{bmatrix} \phi_{11}(t-\lambda) & \phi_{12}(t-\lambda) \\ \phi_{21}(t-\lambda) & \phi_{22}(t-\lambda) \end{bmatrix}\begin{bmatrix} 0 \\ 1 \end{bmatrix} d\lambda \\
&= \phi_{11}(t)x_1(0) + \phi_{12}(t)x_2(0) + \int_0^t \phi_{12}(t - \lambda)\, d\lambda \\
&= (2\epsilon^{-t} - \epsilon^{-2t})x_1(0) + 2(\epsilon^{-2t} - \epsilon^{-t})x_2(0) + \int_0^t 2[\epsilon^{-2(t-\lambda)} - \epsilon^{-(t-\lambda)}]\, d\lambda \\
&= (2\epsilon^{-t} - \epsilon^{-2t})x_1(0) + 2(\epsilon^{-2t} - \epsilon^{-t})x_2(0) + 2\epsilon^{-t} - \epsilon^{-2t} - 1 \qquad t \geqslant 0
\end{aligned}$$

5.9 LINEAR TIME-VARYING SYSTEMS—THE STATE TRANSITION MATRIX

For the case where the **A** matrix is not fixed but varies with time, the homogeneous matrix differential equation of a linear system is

$$\dot{\mathbf{x}} = \mathbf{A}(t)\mathbf{x} \qquad (5.9\text{-}1)$$

It is indeed tempting to return to the scalar case and form an analogy between the solution to the scalar differential equation

$$\dot{x} = a(t)x \tag{5.9-2}$$

considered in Section 2.5, and the solution to Eq. 5.9-1. The solution to Eq. 5.9-2 is

$$x(t) = [\exp b(t)]x(\tau) \tag{5.9-3}$$

where τ is some fixed time instant and

$$b(t) = \int_\tau^t a(\lambda)\, d\lambda$$

The analogous solution for Eq. 5.9-1 would be

$$\mathbf{x}(t) = \left[\exp \int_\tau^t \mathbf{A}(\lambda)\, d\lambda\right] \mathbf{x}(\tau) \tag{5.9-4}$$

However, if Eq. 5.9-4 is substituted into Eq. 5.9-1, it is seen that Eq. 5.9-4 is the correct solution if and only if

$$\frac{d}{dt} \epsilon^{\mathbf{B}(t)} = \frac{dB(t)}{dt} \epsilon^{\mathbf{B}(t)} \tag{5.9-5}$$

where

$$\mathbf{B}(t) = \int_\tau^t \mathbf{A}(\lambda)\, d\lambda$$

Unfortunately, Eq. 5.9-5 is not always valid. In fact, it is seldom valid. Two obvious, but trivial, cases where it is valid are those in which **A** is a constant matrix and those in which **A** is a diagonal matrix. For the former case the solution is known, and for the latter case the state equations are uncoupled, so that Eq. 5.9-3 can be used for each of the x_i's.

It can be shown that the requirement of Eq. 5.9-5 can be written in terms of **A** as the commutativity condition[13]

$$\mathbf{A}(t_1)\mathbf{A}(t_2) = \mathbf{A}(t_2)\mathbf{A}(t_1) \quad \text{for all } t_1 \text{ and } t_2 \tag{5.9-6}$$

If Eq. 5.9-6 is satisfied, then Eq. 5.9-4 is the solution to Eq. 5.9-1 and the state transition matrix is given by

$$\boldsymbol{\phi}(t, \tau) = \exp \int_\tau^t \mathbf{A}(\lambda)\, d\lambda \tag{5.9-7}$$

The Matrizant

For the general case, where the commutativity condition of Eq. 5.9-6 is not satisfied, the state transition matrix is not given by Eq. 5.9-7. However, the solution to Eq. 5.9-1 can be obtained by a method known as the Peano-Baker method of integration.[14] This method follows.

Let the conditions $\mathbf{x}(\tau)$ be given. Then, by integrating Eq. 5.9-1, the integral equation

$$\mathbf{x}(t) = \mathbf{x}(\tau) + \int_{\tau}^{t} \mathbf{A}(\lambda)\mathbf{x}(\lambda)\, d\lambda \tag{5.9-8}$$

is obtained. This equation is sometimes called a vector Volterra integral equation. It can be solved by repeated substitution of the right side of the integral equation into the integral for $\mathbf{x}$. For example, the first iteration is

$$\mathbf{x}(t) = \mathbf{x}(\tau) + \int_{\tau}^{t} \mathbf{A}(\lambda)\left[\mathbf{x}(\tau) + \int_{\tau}^{\lambda} \mathbf{A}(s)\mathbf{x}(s)\, ds\right] d\lambda \tag{5.9-9}$$

The expressions can be simplified somewhat by the introduction of the integral operator $\mathbf{Q}$, where $\mathbf{Q}$ is defined by

$$\mathbf{Q}(\ ) = \int_{\tau}^{t} (\ )\, d\lambda$$

Using this operator notation, Eq. 5.9-8 can be written as

$$\mathbf{x}(t) = \mathbf{x}(\tau) + \mathbf{Q}(\mathbf{A})\mathbf{x}(\lambda) \tag{5.9-10}$$

If the process shown in Eq. 5.9-9 is continued, then $\mathbf{x}(t)$ is obtained as the Neumann series

$$\mathbf{x}(t) = [\mathbf{I} + \mathbf{Q}(\mathbf{A}) + \mathbf{Q}(\mathbf{AQ}(\mathbf{A})) + \mathbf{Q}(\mathbf{AQ}(\mathbf{AQ}(\mathbf{A}))) + \cdots]\mathbf{x}(\tau) \tag{5.9-11}$$

The first term in the parentheses is $\mathbf{I}$, the unit matrix. The second term is the integral of $\mathbf{A}$ between the limits τ and t. The third term is found by premultiplying $\mathbf{Q}(\mathbf{A})$ by $\mathbf{A}$ and then integrating this product between the limits τ and t. The other terms are found in like manner. If the elements of $\mathbf{A}$ remain bounded between the limits of integration, then this series is absolutely and uniformly convergent. This series defines a square matrix $\mathbf{G}(\mathbf{A})$ which is called the *matrizant*.

$$\mathbf{G}(\mathbf{A}) = \mathbf{I} + \mathbf{Q}(\mathbf{A}) + \mathbf{Q}(\mathbf{AQ}(\mathbf{A})) + \mathbf{Q}(\mathbf{AQ}(\mathbf{AQ}(\mathbf{A}))) + \cdots \tag{5.9-12}$$

If both sides of Eq. 5.9-12 are differentiated with respect to t, the fundamental property of the matrizant

$$\frac{d}{dt}\mathbf{G}(\mathbf{A}) = \mathbf{AG}(\mathbf{A}) \tag{5.9-13}$$

is obtained. Therefore $\mathbf{G}(\mathbf{A})$ is indeed the solution to Eq. 5.9-1 and, as such represents the desired state transition matrix for a time-varying system. Thus

$$\boldsymbol{\phi}(t, \tau) = \mathbf{G}(\mathbf{A}) \tag{5.9-14}$$

or

$$\mathbf{x}(t) = \mathbf{G}(\mathbf{A})\mathbf{x}(\tau) = \boldsymbol{\phi}(t, \tau)\mathbf{x}(\tau) \tag{5.9-15}$$

Clearly, if $\mathbf{A}$ is a constant matrix, then

$$\boldsymbol{\phi}(t, \tau) = \mathbf{I} + (t - \tau)\mathbf{A} + \frac{(t-\tau)^2}{2!}\mathbf{A}^2 + \frac{(t-\tau)^3}{3!}\mathbf{A}^3 + \cdots = \epsilon^{\mathbf{A}(t-\tau)}$$

Example 5.9-1. Find $\phi(t, \tau)$ for the first order system $\dot{x} = -tx$.

Obviously, the answer is $\phi(t, \tau) = \exp[(\tau^2 - t^2)/2]$. This could be obtained by direct integration of the given equation, or by recognizing that the commutant condition (Eq. 5.9-6) holds and therefore Eq. 5.9-7 can be applied.

Using Eq. 5.9-12, the matrizant is seen to be

$$G(A) = \phi(t, \tau) = 1 - \left(\frac{t^2 - \tau^2}{2}\right)^2 + \frac{\left(\dfrac{t^2 - \tau^2}{2}\right)^4}{2!} + \cdots$$

Recognizing this as the form for the infinite series ϵ^{-z^2}, the state transition matrix is then given by

$$\phi(t, \tau) = \exp\left[\frac{\tau^2 - t^2}{2}\right]$$

This simple example points out the difficulty in using the matrizant approach. Unless the series (Eq. 5.9-12) converges rapidly, the computation becomes quite lengthy.

An interesting alternative solution was proposed by Kinariwala.[13] The approach is to decompose $\mathbf{A}(t)$ into two matrices, $\mathbf{A}_0(t)$ and $\mathbf{A}_1(t)$, where $\mathbf{A}_0(t)$ satisfies the commutant condition of Eq. 5.9-6. Thus

$$\mathbf{A}(t) = \mathbf{A}_0(t) + \mathbf{A}_1(t) \tag{5.9-16}$$

or

$$\dot{\mathbf{x}} = [\mathbf{A}_0(t) + \mathbf{A}_1(t)]\mathbf{x} \tag{5.9-17}$$

$\mathbf{A}_1(t)$ is interpreted as a perturbation upon $\mathbf{A}_0(t)$.

The unperturbed equation

$$\dot{\mathbf{x}}_0 = \mathbf{A}_0(t)\mathbf{x}_0 \tag{5.9-18}$$

has the solution

$$\mathbf{x}_0(t) = \boldsymbol{\phi}_0(t, \tau)\mathbf{x}(\tau) \tag{5.9-19}$$

where

$$\boldsymbol{\phi}_0(t, \tau) = \exp\left[\int_\tau^t \mathbf{A}_0(\lambda)\, d\lambda\right]$$

The solution to Eq. 5.9-17 is assumed to be of the same form as Eq. 5.9-19, but with successive corrections added to take the perturbations into account.

The perturbations $\mathbf{A}_1(t)\mathbf{x}(t)$ are equivalent to the forcing function term $\mathbf{Bv}$ of the previous section. By direct analogy, using the superposition theorem, the solution for $\mathbf{x}(t)$ is then given by

$$\mathbf{x}(t) = \boldsymbol{\phi}_0(t, \tau)\mathbf{x}(\tau) + \int_\tau^t \boldsymbol{\phi}_0(t, \lambda)\mathbf{A}_1(\lambda)\mathbf{x}(\lambda)\, d\lambda \tag{5.9-20}$$

Again, this is a Volterra equation of the same form as Eq. 5.9-8. Using the same process of iteration as previously described, the Neumann series for $\boldsymbol{\phi}(t, \tau)$ is found to be

$$\boldsymbol{\phi}(t, \tau) = [\mathbf{I} + \mathbf{Q}(\boldsymbol{\phi}_0\mathbf{A}_1) + \mathbf{Q}(\boldsymbol{\phi}_0\mathbf{A}_1\mathbf{Q}(\boldsymbol{\phi}_0\mathbf{A}_1)) + \cdots]\boldsymbol{\phi}_0(t, \tau) \tag{5.9-21}$$

This expression, in conjunction with Eq. 5.9-15, yields $\mathbf{x}(t)$.

If $\mathbf{A}_1(t)$ is relatively small, then only the first few terms of Eq. 5.9-21 are necessary for an adequate approximation. A good first choice for $\mathbf{A}_0(t)$ reduces the number of terms required. However, such a choice is often difficult to determine. Needless to say, the analytical solution of time-varying differential equations is generally quite involved—so involved that the engineer invariably uses a computer to perform the task. In many cases, the state transition matrix can be easily obtained by choosing a set of state variables that are not immediately obvious. It may well be worth the trouble to determine if a set of state variables can be found such that Eq. 5.9-7 can be applied. An example illustrates this point.

Example 5.9-2. The homogeneous differential equation of a system is given by $\ddot{y} - 2t\dot{y} - (2 - t^2)y = 0$. Determine $\boldsymbol{\phi}(t, \tau)$.

The simulation diagram for this system is shown in Fig. 5.9-1*a*. The obvious choice of state variables is $x_1 = y$ and $x_2 = \dot{y}$. The resulting $\mathbf{A}(t)$ matix is

$$\mathbf{A}(t) = \begin{bmatrix} 0 & 1 \\ 2 - t^2 & 2t \end{bmatrix}$$

However, $\mathbf{A}(t_1)\mathbf{A}(t_2) \neq \mathbf{A}(t_2)\mathbf{A}(t_1)$. Therefore Eq. 5.9-7 cannot be applied.

An examination of the differential equation shows that it can be rewritten as

$$\ddot{y} - 2t\dot{y} - (2 - t^2)y = \ddot{y} - \frac{d}{dt}(ty) - y - t(\dot{y} - ty) = 0$$

If the state variables are now chosen to be $z_1 = y$, $z_2 = \dot{y} - ty$, then

$$\dot{z}_2 = \ddot{z}_1 - \frac{d}{dt}(tz_1)$$

Since

$$\ddot{z}_1 - \frac{d}{dt}(tz_1) = z_1 + tz_2$$

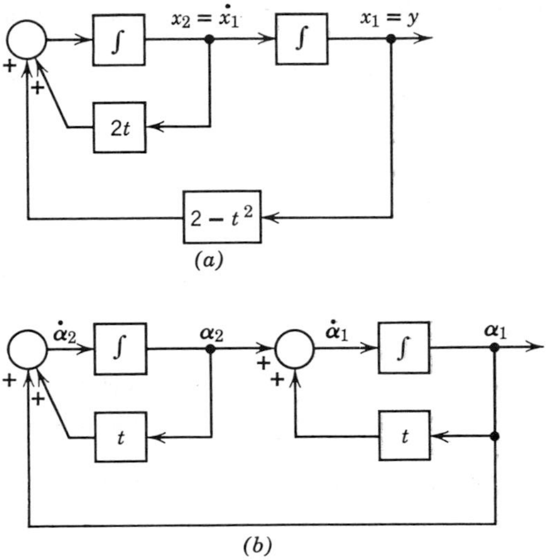

Fig. 5.9-1

then $\dot{z}_2 = z_1 + tz_2$. The expression for $\dot{z}_1 = \dot{y}$ can be obtained from the definition of z_2. Thus $\dot{z}_1 = \dot{y} = z_2 + tz_1$. The matrix differential equation is then

$$\begin{bmatrix} \dot{z}_1 \\ \dot{z}_2 \end{bmatrix} = \begin{bmatrix} t & 1 \\ 1 & t \end{bmatrix} \begin{bmatrix} z_1 \\ z_2 \end{bmatrix}$$

The simulation diagram is shown in Fig. 5.9-1*b*. The new $\mathbf{A}(t)$ matrix

$$\mathbf{A}(t) = \begin{bmatrix} t & 1 \\ 1 & t \end{bmatrix}$$

does obey the commutant condition, i.e., $\mathbf{A}(t_1)\mathbf{A}(t_2) = \mathbf{A}(t_2)\mathbf{A}(t_1)$. Therefore Eq. 5.9-7 can be used.

The state transition matrix $\boldsymbol{\phi}(t, \tau)$ is given by

$$\boldsymbol{\phi}(t, \tau) = \exp\left[\int_\tau^t \mathbf{A}(\sigma)\, d\sigma\right] = \exp \mathbf{B}$$

where

$$\mathbf{B} = \begin{bmatrix} \dfrac{t^2 - \tau^2}{2} & t - \tau \\ t - \tau & \dfrac{t^2 - \tau^2}{2} \end{bmatrix}$$

The matrix $\epsilon^{\mathbf{B}}$ can be found by any one of the techniques previously discussed. However, it is instructive to find $\epsilon^{\mathbf{B}}$ by use of Eq. 4.10-20. Thus

$$\epsilon^{\mathbf{B}} = \mathbf{M}\epsilon^{\mathbf{\Lambda}}\mathbf{M}^{-1}$$

The characteristic values of **B** are $\lambda_{1,2} = [(t^2 - \tau^2)/2] \pm (t - \tau)$, and the modal matrices are

$$\mathbf{M} = (t - \tau)\begin{bmatrix} 1 & 1 \\ 1 & -1 \end{bmatrix}, \quad \mathbf{M}^{-1} = \frac{1}{2(t - \tau)}\begin{bmatrix} 1 & 1 \\ 1 & -1 \end{bmatrix}$$

Therefore

$$\epsilon^{\mathbf{B}} = \tfrac{1}{2}\begin{bmatrix} 1 & 1 \\ 1 & -1 \end{bmatrix}\begin{bmatrix} \exp\left[\dfrac{t^2 - \tau^2}{2} + (t - \tau)\right] & 0 \\ 0 & \exp\left[\dfrac{t^2 - \tau^2}{2} - (t - \tau)\right] \end{bmatrix}\begin{bmatrix} 1 & 1 \\ 1 & -1 \end{bmatrix}$$

and

$$\boldsymbol{\phi}(t, \tau) = \epsilon^{(t^2 - \tau^2)/2}\begin{bmatrix} \cosh (t - \tau) & \sinh (t - \tau) \\ \sinh (t - \tau) & \cosh (t - \tau) \end{bmatrix}$$

The matrix $\boldsymbol{\phi}(t, \tau)$ does satisfy the differential equation $(d/dt)[\boldsymbol{\phi}(t, \tau)] = \mathbf{A}(t)\boldsymbol{\phi}(t, \tau)$, since

$$\frac{d}{dt}\boldsymbol{\phi}(t, \tau) = t\boldsymbol{\phi}(t, \tau) + \begin{bmatrix} 0 & 1 \\ 1 & 0 \end{bmatrix}\boldsymbol{\phi}(t, \tau)$$

and

$$\mathbf{A}(t)\boldsymbol{\phi}(t, \tau) = \begin{bmatrix} t & 1 \\ 1 & t \end{bmatrix}\boldsymbol{\phi}(t, \tau) = t\boldsymbol{\phi}(t, \tau) + \begin{bmatrix} 0 & 1 \\ 1 & 0 \end{bmatrix}\boldsymbol{\phi}(t, \tau)$$

Although almost all the time-varying problems that the control engineer faces must be solved by either an analog or digital computer, it is also true that careful prior inspection of the system can greatly reduce the computation time. The obvious state variables may not be the state variables that should be used to control the system, or used to simulate the system. In many cases, combinations of the obvious state variables may be more valuable. The foregoing example illustrates this point.

The state transition matrix can also be found by using the simulation diagram method of Section 5.8. For a time-varying system, the ith state variable $x_i(t)$ is given by Eq. 5.9-22, similarly to Eq. 5.8-7.

$$x_i(t) = \sum_{j=1}^{n} \phi_{ij}(t, \tau)x_j(\tau) \tag{5.9-22}$$

The term $\phi_{ij}(t, \tau)$ can be found by obtaining the transfer function $\Phi_{ij}(t, s)$ between the *input* to integrator j and the *output* of integrator i. Note that this is a time-varying transfer function (see Section 3.9) and as such may be difficult to evaluate. If the system were actually simulated on an analog computer, a unit initial condition placed on integrator j at time τ_1 would produce the response $\phi_{ij}(t, \tau_1)$ at the output of integrator i. A series of runs and a cross plotting of the results are necessary if $\phi_{ij}(t, \tau)$ for all $\tau \leqslant t$ is to be obtained. This point is discussed in detail in Sections 5.11 and 5.12.

Properties of the State Transition Matrix

Property 1: By definition,

$$\boldsymbol{\phi}(t_0, t_0) = \mathbf{I} \tag{5.9-23}$$

Property 2: The *group property* of a state transition matrix is

$$\boldsymbol{\phi}(t_2, t_0) = \boldsymbol{\phi}(t_2, t_1)\boldsymbol{\phi}(t_1, t_0) \tag{5.9-24}$$

This can be shown from the relations $\mathbf{x}(t_2) = \boldsymbol{\phi}(t_2, t_1)\mathbf{x}(t_1) = \boldsymbol{\phi}(t_2, t_0)\mathbf{x}(t_0)$ and $\mathbf{x}(t_1) = \boldsymbol{\phi}(t_1, t_0)\mathbf{x}(t_0)$. Then $\mathbf{x}(t_2) = \boldsymbol{\phi}(t_2, t_1)\boldsymbol{\phi}(t_1, t_0)\mathbf{x}(t_0)$. Therefore $\boldsymbol{\phi}(t_2, t_0) = \boldsymbol{\phi}(t_2, t_1)\boldsymbol{\phi}(t_1, t_0)$.

Property 3:

$$\boldsymbol{\phi}(t_1, t_2) = \boldsymbol{\phi}^{-1}(t_2, t_1) \tag{5.9-25}$$

This property can be obtained from the expression $\boldsymbol{\phi}^{-1}(t_2, t_1)\boldsymbol{\phi}(t_2, t_1) = \mathbf{I} = \boldsymbol{\phi}(t_1, t_1)$. Since $\boldsymbol{\phi}(t_1, t_2)\boldsymbol{\phi}(t_2, t_1) = \boldsymbol{\phi}(t_1, t_1)$, it follows that $\boldsymbol{\phi}^{-1}(t_2, t_1)\boldsymbol{\phi}(t_2, t_1) = \boldsymbol{\phi}(t_1, t_2)\boldsymbol{\phi}(t_2, t_1)$. Postmultiplying by $\boldsymbol{\phi}^{-1}(t_2, t_1)$ yields $\boldsymbol{\phi}^{-1}(t_2, t_1) = \boldsymbol{\phi}(t_1, t_2)$.

For fixed systems.

Property 4:

$$\boldsymbol{\phi}(t + \tau) = \boldsymbol{\phi}(t)\boldsymbol{\phi}(\tau) \tag{5.9-26}$$

Since $\boldsymbol{\phi}(t) = \epsilon^{\mathbf{A}t}$ for fixed systems, $\boldsymbol{\phi}(t + \tau) = \epsilon^{\mathbf{A}(t+\tau)} = \boldsymbol{\phi}(t)\boldsymbol{\phi}(\tau)$.

Property 5:

$$\boldsymbol{\phi}^{-1}(t) = \boldsymbol{\phi}(-t) \tag{5.9-27}$$

Let $\tau = -t$ in Property 4. Then $\boldsymbol{\phi}(0) = \mathbf{I} = \boldsymbol{\phi}(t)\boldsymbol{\phi}(-t)$ or $\boldsymbol{\phi}(-t) = \boldsymbol{\phi}^{-1}(t)$.

The Inverse State Transition Matrix—The Adjoint System

The inverse state transition matrix $\boldsymbol{\phi}^{-1}(t, \tau)$ plays an important role in the general solution of a time-varying system, in obtaining the impulse response matrix $\mathbf{H}(t, \tau)$, and in the solution of optimal control problems. The usefulness of this matrix is related to Eq. 5.9-25, which states

$$\boldsymbol{\phi}(t, \tau) = \boldsymbol{\phi}^{-1}(\tau, t)$$

The behavior of the system with respect to the variable t is a function of the dynamics of the original system. The behavior of the system with respect to the second variable τ is a function of the dynamics of the system for which $\boldsymbol{\phi}^{-1}(t, \tau)$ is the state transition matrix. This system is known as the *adjoint* to the original system.

If the original system is defined by $\dot{\mathbf{x}} = \mathbf{A}(t)\mathbf{x}$, then the *adjoint* system is defined by

$$\dot{\boldsymbol{\alpha}} = -\boldsymbol{\alpha}\mathbf{A}(t) \tag{5.9-28}$$

where $\boldsymbol{\alpha}$ is a row vector, or

$$\dot{\boldsymbol{\alpha}} = -\mathbf{A}^T(t)\boldsymbol{\alpha} \tag{5.9-29}$$

where $\boldsymbol{\alpha}$ is a column vector. This can be derived from Eq. 4.4-8, which indicates that

$$\frac{d}{dt}[\boldsymbol{\phi}^{-1}(t, \tau)] = \dot{\boldsymbol{\phi}}^{-1}(t, \tau) = -\boldsymbol{\phi}^{-1}(t, \tau)\dot{\boldsymbol{\phi}}(t, \tau)\boldsymbol{\phi}^{-1}(t, \tau)$$

Since $\dot{\boldsymbol{\phi}}(t, \tau) = \mathbf{A}(t)\boldsymbol{\phi}(t, \tau)$, it follows that

$$\dot{\boldsymbol{\phi}}^{-1}(t, \tau) = -\boldsymbol{\phi}^{-1}(t, \tau)\mathbf{A}(t) \tag{5.9-30}$$

Therefore $\boldsymbol{\phi}^{-1}(t, \tau)$ is the state transition matrix for the system whose unforced differential equation is given by

$$\dot{\boldsymbol{\alpha}} = -\boldsymbol{\alpha}\mathbf{A}(t)$$

where $\boldsymbol{\alpha}$ is a row vector. If the transposes of both sides of Eq. 5.9-30 are taken, then†

$$[\dot{\boldsymbol{\phi}}^{-1}(t, \tau)]^T = [\dot{\boldsymbol{\phi}}^T(t, \tau)]^{-1} = -\mathbf{A}^T(t)[\boldsymbol{\phi}^T(t, \tau)]^{-1}$$

Thus $[\boldsymbol{\phi}^T(t, \tau)]^{-1}$ is the state transition matrix for the system whose unforced differential equation is given by

$$\dot{\boldsymbol{\alpha}} = -\mathbf{A}^T(t)\boldsymbol{\alpha} \tag{5.9-31}$$

where $\boldsymbol{\alpha}$ is a column vector.

Example 5.9-3. Compare the state transition matrices of $\dot{x} + tx = 0$ and the adjoint equation.

From Example 5.9-1,

$$x(t) = x(\tau)\exp\left(\frac{\tau^2 - t^2}{2}\right)$$

or

$$\begin{aligned} \phi(t, \tau) &= \exp\left(\frac{\tau^2 - t^2}{2}\right) \qquad && t \geqslant \tau \\ &= 0 && t < \tau \end{aligned}$$

The adjoint differential equation is given by $\dot{\alpha} - t\alpha = 0$. The solution to this equation is given by

$$\alpha(t) = \alpha(\tau)\exp\left(\frac{t^2 - \tau^2}{2}\right)$$

† If $\mathbf{A}(t)$ contains complex elements, the conjugate transpose $[\mathbf{A}^*(t)]^T$ is taken.

or

$$\phi^{-1}(t, \tau) = \exp\left(\frac{t^2 - \tau^2}{2}\right) \qquad t \geqslant \tau$$
$$= 0 \qquad t < \tau$$

If the variables t and τ are interchanged in $\phi^{-1}(t, \tau)$, then

$$\phi^{-1}(\tau, t) = \exp\left(\frac{\tau^2 - t^2}{2}\right) \qquad \tau \geqslant t$$
$$= 0 \qquad \tau < t$$

A comparison of $\phi^{-1}(\tau, t)$ with $\phi(t, \tau)$ shows that the two expressions are identical except that they are valid over different intervals. The physically realizable output $\alpha(t)$ of the adjoint system represents the physically unrealizable portion of the solution $x(t)$. A physically realizable output of a system implies that the observation variable is greater than the application variable.

The Adjoint Operator

The nth order homogeneous differential equation

$$y^{(n)} + a_{n-1}(t)y^{(n-1)} + \cdots + a_0 y = 0 \qquad (5.9\text{-}32)$$

can be written as $L_n y = 0$, where L_n is the linear differential operator defined by

$$L_n = p^n + \sum_{k=0}^{n-1} a_k(t)p^k, \quad p^k = \frac{d^k}{dt^k}$$

and the $a_k(t)$ are real. The linear *adjoint* differential operator is defined as

$$L_n{}^* = (-1)^n p^n + \sum_{k=0}^{n-1} (-1)^k p^k a_k(t) \qquad (5.9\text{-}33)$$

where $p^k a_n(t)$ signifies that p^k operates on the product of $a_k(t)$ and the dependent variable. Consequently, the linear adjoint differential equation $L_n{}^*\alpha = 0$ can be written as

$$(-1)^n p^n \alpha + (-1)^{n-1} p^{n-1}[a_{n-1}(t)\alpha] + \cdots + a_0(t)\alpha = 0$$

If the original differential equation, Eq. 5.9-32, is cast in the form $\dot{\mathbf{x}} = \mathbf{A}(t)\mathbf{x}$, the $\mathbf{A}$ matrix is

$$\mathbf{A} = \begin{bmatrix} 0 & 1 & 0 & \cdots & 0 \\ 0 & 0 & 1 & \cdots & 0 \\ \cdots & \cdots & \cdots & \cdots & \cdots \\ 0 & 0 & 0 & \cdots & 1 \\ -a_0(t) & -a_1(t) & -a_2(t) & \cdots & -a_{n-1}(t) \end{bmatrix} \qquad (5.9\text{-}34)$$

For the adjoint system, $\dot{\boldsymbol{\alpha}} = -\mathbf{A}^T(t)\boldsymbol{\alpha}$ and $-\mathbf{A}^T$ is given by

$$-\mathbf{A}^T = \begin{bmatrix} 0 & 0 & \cdots & 0 & a_0(t) \\ -1 & 0 & \cdots & 0 & a_1(t) \\ 0 & -1 & \cdots & 0 & a_2(t) \\ \cdots & \cdots & \cdots & \cdots & \cdots \\ 0 & 0 & \cdots & -1 & a_{n-1}(t) \end{bmatrix} \tag{5.9-35}$$

To show that the operator notation and the matrix notation are equivalent, the following observations are made on the adjoint matrix formulation.

$$\dot{\alpha}_1 = a_0(t)\alpha_n$$
$$\dot{\alpha}_2 = -\alpha_1 + a_1(t)\alpha_n$$
$$\dot{\alpha}_1 = -\alpha_{i-1} + a_{i-1}(t)\alpha_n$$

Differentiating $\dot{\alpha}_n$,

$$\ddot{\alpha}_n = -\dot{\alpha}_{n-1} + \frac{d}{dt}[a_{n-1}(t)\alpha_n]$$

Substituting for $\dot{\alpha}_{n-1}$,

$$\ddot{\alpha}_n = \alpha_{n-2} - a_{n-2}(t)\alpha_n + \frac{d}{dt}[a_{n-1}(t)\alpha_n]$$

If this process of differentiation and substitution is performed $(n - 1)$ times, it can be shown that

$$(-1)^n p^n \alpha_n + \sum_{k=0}^{n-1}(-1)^k p^k [a_k(t)\alpha_n] = 0 \tag{5.9-36}$$

which is exactly the linear adjoint differential equation.

Alternatively, the adjoint operator can be derived from the definition[15]

$$\langle \boldsymbol{\alpha}, \mathbf{L}\mathbf{x} \rangle = \langle \mathbf{L}^{*T}\boldsymbol{\alpha}, \mathbf{x} \rangle$$

This definition is often useful in formulating existence and uniqueness criteria for the solution of differential equations.

Linear Systems with Periodic Coefficients[16]

Consider the linear system

$$\dot{\mathbf{x}} = \mathbf{A}(t)\mathbf{x} \tag{5.9-37}$$

where $\mathbf{A}(t)$ is a continuous periodic matrix with constant period $T \neq 0$, such that

$$\mathbf{A}(t + T) = \mathbf{A}(t)$$

Because of this relationship, if $\boldsymbol{\phi}(t, \tau)$ is a state transition matrix for Eq. 5.9-37, then so is $\boldsymbol{\phi}(t + T, \tau)$, and there exists a constant nonsingular matrix **C** such that $\boldsymbol{\phi}(t + T, \tau) = \boldsymbol{\phi}(t, \tau)\mathbf{C}$. This can be shown rather easily, since the derivative of this expression yields

$$\dot{\boldsymbol{\phi}}(t + T, \tau) = \dot{\boldsymbol{\phi}}(t, \tau)\mathbf{C} = \mathbf{A}(t)\boldsymbol{\phi}(t, \tau)\mathbf{C} = \mathbf{A}(t)\boldsymbol{\phi}(t + T, \tau)$$

Thus $\boldsymbol{\phi}(t + T, \tau)$ also satisfies Eq. 5.9-37.

It can be shown that there exists a constant matrix **B** (called a logarithm of matrix **C**), such that $\mathbf{C} = \epsilon^{\mathbf{B}T}$. It follows that $\boldsymbol{\phi}(t + T, \tau) = \boldsymbol{\phi}(t, \tau)\epsilon^{\mathbf{B}T}$. $\boldsymbol{\phi}(t, \tau)$ may or may not be periodic. This depends upon certain properties of **B**.

If $\mathbf{P}(t)$ is defined as

$$\mathbf{P}(t) = \boldsymbol{\phi}(t, \tau)\epsilon^{-\mathbf{B}T} \tag{5.9-38}$$

then

$$\begin{aligned} \mathbf{P}(t + T) &= \boldsymbol{\phi}(t + T, \tau)\epsilon^{-\mathbf{B}(t+T)} \\ &= \boldsymbol{\phi}(t, \tau)\epsilon^{\mathbf{B}T}\epsilon^{-\mathbf{B}(t+T)} = \boldsymbol{\phi}(t, \tau)\epsilon^{-\mathbf{B}t} = \mathbf{P}(t) \end{aligned}$$

$\mathbf{P}(t)$ is nonsingular, since $\boldsymbol{\phi}(t, \tau)$ and $\epsilon^{-\mathbf{B}t}$ are nonsingular for $-\infty < t < \infty$. Therefore, for a periodic system, $\mathbf{P}(t)$ is a nonsingular periodic matrix with period T.

It is interesting to note that $\mathbf{P}(t)$ is the solution to the differential equation[17]

$$\dot{\mathbf{P}}(t) = \mathbf{A}(t)\mathbf{P}(t) - \mathbf{P}(t)\mathbf{B} \tag{5.9-39}$$

Thus **B** is such that Eq. 5.9-39 has a periodic solution of period T.

Example 5.9-4. Consider the first order differential equation $\dot{x} - a(t)x = 0$, where $a(t)$ is periodic with period T. Determine the requirements on b of Eq. 5.9-39 for $p(t)$ to be periodic with period T.

Equation 5.9-39 is

$$\dot{p} = [a(t) - b]p$$

The solution to this equation is

$$p(t) = p(0) \exp\left\{\int_0^t [a(\lambda) - b]\, d\lambda\right\}$$

The term $p(t)$ is periodic, with period T, if b is equal to the average value of $a(t)$ over an interval T, i.e.,

$$b = \frac{1}{T}\int_0^T a(\lambda)\, d\lambda$$

Any integral multiple of $2\pi j$ may be added to this value.

Although **B** generally cannot be determined uniquely, since $\mathbf{B}T$ is one of the complex logarithms of **C**, the **A** matrix does determine all the properties associated with **B** that are invariant under a similarity transformation. Specifically, the set of characteristic roots of **C** are uniquely determined by **A**. These characteristic roots ($\lambda_1, \lambda_2, \ldots, \lambda_n$) are called the *multipliers associated with* **A**. These multipliers are all nonzero, since the determinant of **C** is nonzero. The characteristic roots of **B** are called *characteristic exponents.*[18]

Although this discussion is somewhat superficial, its importance lies in the fact that, for a periodic system, the knowledge of the state transition matrix over one period determines the state transition matrix for $-\infty < t < \infty$. For example, if $\boldsymbol{\phi}(t, \tau)$ is known over the t interval $(0 \leqslant t \leqslant T)$, then $\mathbf{C} = \boldsymbol{\phi}^{-1}(0, \tau)\boldsymbol{\phi}(T, \tau)$ and $\mathbf{B} = (\log \mathbf{C})/T$. Since $\mathbf{P}(t)$ is periodic with period T, then $\boldsymbol{\phi}(t, \tau)$ is known for all t.

Example 5.9-5.[19] The equation

$$\ddot{y} + [a^2 + b^2 \operatorname{sgn}(\cos t)]y = 0$$

where

$$\operatorname{sgn} \alpha = \begin{cases} 1, & \alpha > 0 \\ 0, & \alpha = 0 \\ -1, & \alpha < 0 \end{cases}$$

is to be investigated for the two sets of values $a^2 = 1$, $b^2 = 1$, and $a^2 = 4.6$, $b^2 = 5.0$.

This differential equation can be written as

$$\ddot{y} + \omega^2(t)y = 0$$

where the periodic coefficient $\omega^2(t)$ is shown in Fig. 5.9-2. Note that it is piecewise constant. The **A** matrix for this equation is given by $(x_1 = y,\ \dot{x}_1 = x_2)$

$$\mathbf{A} = \begin{bmatrix} 0 & 1 \\ -\omega^2(t) & 0 \end{bmatrix}$$

For each of the intervals where $\omega^2(t)$ is constant, $\boldsymbol{\phi}$ is given by

$$\boldsymbol{\phi} = \mathscr{L}^{-1}[s\mathbf{I} - \mathbf{A}]^{-1} = \begin{bmatrix} \cos \omega t & \dfrac{1}{\omega} \sin \omega t \\ -\omega \sin \omega t & \cos \omega t \end{bmatrix}$$

Since $\omega(t)$ is periodic, $\boldsymbol{\phi}(t, \tau)$ must also be periodic.

Since $\omega^2(t)$ is periodic and piecewise constant, $\boldsymbol{\phi}(3\pi/2, -\pi/2)$ can be found by solving for $\boldsymbol{\phi}(\pi/2, -\pi/2)$ and $\boldsymbol{\phi}(3\pi/2, \pi/2)$. The state transition matix $\boldsymbol{\phi}(3\pi/2, -\pi/2)$ is then $\boldsymbol{\phi}(3\pi/2, -\pi/2) = \boldsymbol{\phi}(3\pi/2, \pi/2)\boldsymbol{\phi}(\pi/2, -\pi/2)$.

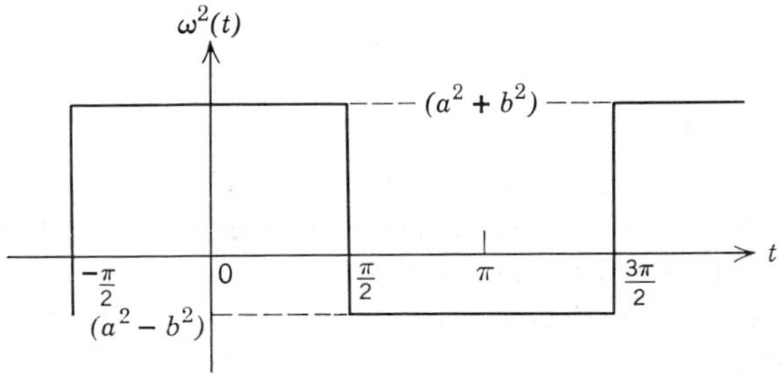

Fig. 5.9-2

For the interval $-\pi/2$ to $\pi/2$, $\omega = \sqrt{2}$ for $a^2 = b^2 = 1$, and $\omega = 3.1$ for $a^2 = 4.6$, $b^2 = 5.0$. Thus

$$\boldsymbol{\phi}\left(\frac{\pi}{2}, -\frac{\pi}{2}\right) = \begin{bmatrix} -0.259 & 0.683 \\ -1.37 & -0.259 \end{bmatrix}_{\omega=\sqrt{2}} = \begin{bmatrix} -0.951 & -0.099 \\ 0.959 & -0.951 \end{bmatrix}_{\omega=3.1}$$

For the interval $\pi/2$ to $3\pi/2$, $\omega = 0$ for $a^2 = b^2 = 1$, and $\omega = j0.634$ for $a^2 = 4.6$, $b^2 = 5.0$. Thus

$$\boldsymbol{\phi}\left(\frac{3\pi}{2}, \frac{\pi}{2}\right) = \begin{bmatrix} 1 & \pi \\ 0 & 1 \end{bmatrix}_{\omega=0} = \begin{bmatrix} 3.73 & 5.66 \\ 2.27 & 3.73 \end{bmatrix}_{\omega=j0.634}$$

For the complete cycle, the state transition matrix is

$$\boldsymbol{\phi}\left(\frac{3\pi}{2}, \frac{-\pi}{2}\right) = \begin{bmatrix} 4.56 & 0.131 \\ -1.37 & -0.259 \end{bmatrix}_{a^2=b^2=1} = \begin{bmatrix} 1.88 & -5.75 \\ 1.41 & -3.78 \end{bmatrix}_{a^2=4.6,\, b^2=5.0}$$

Since $\mathbf{x}[(3\pi/2)k] = \boldsymbol{\phi}^k(3\pi/2, -\pi/2)\mathbf{x}(-\pi/2)$, boundedness of the periodic samples of $\mathbf{x}(t)$ depends upon whether the characteristic values of $\boldsymbol{\phi}$ have an absolute magnitude less than one. Periodicity of $\mathbf{x}(t)$ depends upon whether the characteristic values of $\boldsymbol{\phi}$ have an absolute magnitude equal to one. For $a^2 = b^2 = 1$, the characteristic equation $|\lambda\mathbf{I} - \boldsymbol{\phi}| = 0$ is given by $\lambda^2 - 4.3\lambda - 0.99 = 0$. The roots are $\lambda_{1,2} = 2.15 \pm 2.38$. For $a^2 = 4.6$, $b^2 = 5.0$, the characteristic equation $|\lambda\mathbf{I} - \boldsymbol{\phi}| = 0$ is given by $\lambda^2 + 1.9\lambda + 1.0 = 0$. The roots are $\lambda_{1,2} = -0.95 \pm j0.317$. Therefore, for the first set of constants $(a^2 = b^2 = 1)$, the initial condition response is unstable. For the second set of constants $(a^2 = 4.6, b^2 = 5.0)$, the response is just periodic (within slide rule accuracy) since $|\lambda_i| = 1$. (Note that $\lambda^2 + 2\lambda + 1 = 0$ has a double root at $\lambda = 1$.)

The conclusion to be reached from this example is that a system need not have a periodic solution, even though the coefficients of the governing differential equation are periodic. The solution can be stable, unstable, or periodic.

5.10 LINEAR TIME-VARYING SYSTEMS—THE COMPLETE SOLUTION

Using the results of the previous section, the complete solution to the state equations for a linear time-varying system can be obtained. The state equations for a linear time-varying system are given by

$$\begin{aligned} \dot{\mathbf{x}} &= \mathbf{A}(t)\mathbf{x} + \mathbf{B}(t)\mathbf{v} \\ \mathbf{y} &= \mathbf{C}(t)\mathbf{x} + \mathbf{D}(t)\mathbf{v} \end{aligned} \tag{5.10-1}$$

Equation 5.9-30 states that

$$\dot{\boldsymbol{\phi}}^{-1}(t, \tau) = -\boldsymbol{\phi}^{-1}(t, \tau)\mathbf{A}(t)$$

If the first of Eqs. 5.10-1 is premultiplied by $\boldsymbol{\phi}^{-1}(t, \tau)$ and Eq. 5.9-30 is postmultiplied by $\mathbf{x}$, the resulting equations are

$$\begin{aligned} \boldsymbol{\phi}^{-1}(t, \tau)\dot{\mathbf{x}} &= \boldsymbol{\phi}^{-1}(t, \tau)\mathbf{A}(t)\mathbf{x} + \boldsymbol{\phi}^{-1}(t, \tau)\mathbf{B}(t)\mathbf{v} \\ \dot{\boldsymbol{\phi}}^{-1}(t, \tau)\mathbf{x} &= -\boldsymbol{\phi}^{-1}(t, \tau)\mathbf{A}(t)\mathbf{x} \end{aligned} \tag{5.10-2}$$

Adding these two expressions yields

$$\frac{d}{dt}[\boldsymbol{\phi}^{-1}(t, \tau)\mathbf{x}] = \boldsymbol{\phi}^{-1}(t, \tau)\mathbf{B}(t)\mathbf{v} \tag{5.10-3}$$

Integration of Eq. 5.10-3 from τ to t is indicated by

$$\boldsymbol{\phi}^{-1}(t, \tau)\mathbf{x}(t) - \boldsymbol{\phi}^{-1}(\tau, \tau)\mathbf{x}(\tau) = \int_\tau^t \boldsymbol{\phi}^{-1}(\lambda, \tau)\mathbf{B}(\lambda)\mathbf{v}(\lambda)\, d\lambda \tag{5.10-4}$$

Since $\boldsymbol{\phi}^{-1}(\tau, \tau) = \mathbf{I}$,

$$\mathbf{x}(t) = \boldsymbol{\phi}(t, \tau)\mathbf{x}(\tau) + \boldsymbol{\phi}(t, \tau)\int_\tau^t \boldsymbol{\phi}^{-1}(\lambda, \tau)\mathbf{B}(\lambda)\mathbf{v}(\lambda)\, d\lambda \tag{5.10-5}$$

Using the fact that

$$\boldsymbol{\phi}(t, \tau)\boldsymbol{\phi}^{-1}(\lambda, \tau) = \boldsymbol{\phi}(t, \tau)\boldsymbol{\phi}(\tau, \lambda) = \boldsymbol{\phi}(t, \lambda)$$

$\mathbf{x}(t)$ becomes

$$\mathbf{x}(t) = \boldsymbol{\phi}(t, \tau)\mathbf{x}(\tau) + \int_\tau^t \boldsymbol{\phi}(t, \lambda)\mathbf{B}(\lambda)\mathbf{v}(\lambda)\, d\lambda \tag{5.10-6}$$

The expression for $\mathbf{y}(t)$ is obtained by substituting Eq. 5.10-6 into the second of Eqs. 5.10-1, to yield

$$\mathbf{y}(t) = \mathbf{C}(t)\boldsymbol{\phi}(t, \tau)\mathbf{x}(\tau) + \int_\tau^t \mathbf{C}(t)\boldsymbol{\phi}(t, \lambda)\mathbf{B}(\lambda)\mathbf{v}(\lambda)\, d\lambda + \mathbf{D}(t)\mathbf{v}(t) \tag{5.10-7}$$

Equations 5.10-6 and 5.10-7 represent the complete solution to Eq.5. 10-1. These equations are quite similar to Eqs. 5.8-11 and 5.8-12, respectively.

For a fixed system, the response depends upon $t - \tau$, or the *time difference* between application of cause and observation of effect. The result is the *convolution* integrals of Eqs. 5.8-11 and 5.8-12. The state transition matrix for a fixed system has only one variable, namely the time difference between application of cause and observation of effect. However, for a time-varying system, the solution depends upon both t and τ. For a time-varying system, the cause-effect relationship is varying with time, and therefore *when* the cause is applied and *when* the effect is observed are significant. The state transition matrix for a time-varying system has two variables, one being the time of application of the cause, and the other being the time of observation of the effect. Therefore Eqs. 5.10-6 and 5.10-7 depend upon *superposition* integrals and not upon convolution integrals.

Adjoint Solutions and Integrating Factors

For a scalar differential equation of the form

$$\dot{x} = a(t)x + v(t) \tag{5.10-8}$$

the standard method of solution is by use of the integrating factor $\exp \int [-a(t)]\, dt$, as in Section 2.5. With this method, the solution can be expressed as

$$\int_{\tau}^{t} \frac{d}{d\lambda} [\epsilon^{-\int^{\lambda} a(\alpha)\, d\alpha} x(\lambda)]\, d\lambda = \int_{\tau}^{t} \epsilon^{-\int^{\lambda} a(\alpha)\, d\alpha} v(\lambda)\, d\lambda \qquad (5.10\text{-}9)$$

Performance of the integration on the left side yields

$$\epsilon^{-\int^{t} a(\alpha)\, d\alpha} x(t) - \epsilon^{-\int^{\tau} a(\alpha)\, d\alpha} x(\tau) = \int_{\tau}^{t} \epsilon^{-\int^{\lambda} a(\alpha)\, d\alpha} v(\lambda)\, d\lambda$$

Thus $x(t)$ can be written as

$$x(t) = \epsilon^{\int_{\tau}^{t} a(\alpha)\, d\alpha} x(\tau) + \int_{\tau}^{t} \epsilon^{\int_{\lambda}^{t} a(\alpha)\, d\alpha} v(\lambda)\, d\lambda \qquad (5.10\text{-}10)$$

Notice the similarity between Eqs. 5.10-10 and 5.10-6. $\boldsymbol{\phi}^{-1}$, the state transition matrix for the *adjoint system*, is the *integrating factor* used for the solution of Eq. 5.10-1. This is a fundamental property of the state transition matrix for the adjoint system.

Example 5.10-1. Consider the time-varying system described by the differential equation $\ddot{y} + t\dot{y} + y = v(t)$. The block diagram for this system is shown in Fig. 5.10-1. The $\mathbf{A}$ matrix for the state variables indicated in Fig. 5.10-1 is

$$\mathbf{A}(t) = \begin{bmatrix} 0 & 1 \\ -1 & -t \end{bmatrix}$$

Determine $\boldsymbol{\phi}(t, \tau)$.

$\mathbf{A}(t)$ does not satisfy the commutativity condition, since

$$\mathbf{A}(t_1)\mathbf{A}(t_2) = \begin{bmatrix} -1 & -t_2 \\ t_1 & -1 + t_1 t_2 \end{bmatrix} \neq \mathbf{A}(t_2)\mathbf{A}(t_1) = \begin{bmatrix} -1 & -t_1 \\ t_2 & -1 + t_1 t_2 \end{bmatrix}$$

Therefore the state transition matrix $\boldsymbol{\phi}(t, \tau)$ is not given by Eq. 5.9-7. The state transition matrix could be obtained by use of the matrizant, but the series solution obtained by this method is difficult to express in closed form. However, by use of Eq. 5.9-22, a set of differential equations can be obtained which are solvable, if the integrating factor approach is used.

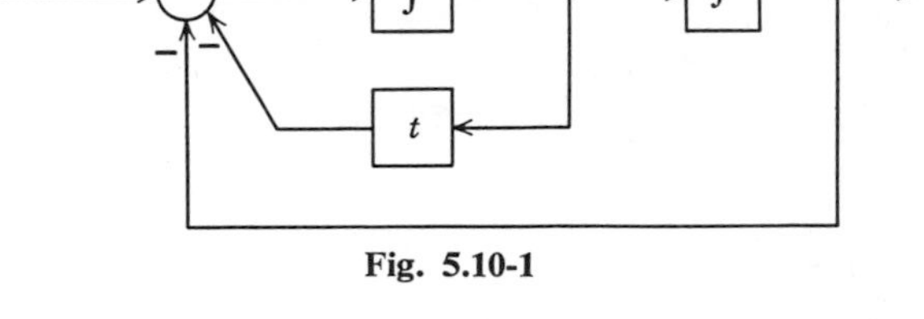

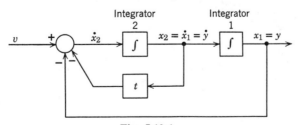

Fig. 5.10-1

The homogeneous form of the given differential equation can be written as

$$\ddot{y} + \frac{d}{dt}(yt) = 0$$

By integrating this equation once, the first order differential equation $\dot{y} = -ty + c_1$ is obtained, where c_1 is a constant of integration. This equation is in the same form as Eq. 5.10-8. The integrating factor is exp $(t^2/2)$. (Note that this integrating factor is the solution to the adjoint differential equation $\dot{\alpha} = t\alpha$.) From Eq. 5.10-10,

$$y(t) = y(\tau)\epsilon^{(\tau^2-t^2)/2} + c_1\epsilon^{-t^2/2}\int_\tau^t \epsilon^{\lambda^2/2}\,d\lambda$$

For the requirements of this example, the expression is best left in this form. The integrated form is quite complicated, as evidenced by

$$y(t) = y(\tau)\,\epsilon^{(\tau^2-t^2)/2} - jc_1\frac{\sqrt{\pi}}{2}\,\epsilon^{-t^2/2}\left[\operatorname{erf}\left(\frac{jt}{\sqrt{2}}\right) - \operatorname{erf}\left(\frac{j\tau}{\sqrt{2}}\right)\right]$$

This illustrates an interesting point. Even when an analytical solution can be obtained for a time-varying differential equation, this solution may not be easily evaluated. In general, the solution for a time-varying differential equation is obtained by a computer simulation, as the analytical solution is either difficult to obtain or difficult to evaluate.

The derivative $\dot{y}(t)$ is

$$\dot{y}(t) = -ty(\tau)\epsilon^{(\tau^2-t^2)/2} + c_1\left[1 - t\int_\tau^t \epsilon^{(\lambda^2-t^2)/2}\,d\lambda\right]$$

The solution for the state transition matrix can now be obtained by application of Eq. 5.9-22. If a unit initial condition is placed on integrator 2, and zero initial conditions on all other integrators, then the time response of the output of integrator 1, $y(t)$, is equal to $\phi_{12}(t, \tau)$. Therefore let $\dot{y}(\tau) = 1$ and $y(\tau) = 0$.

The expression for $\dot{y}(t)$ gives $c_1 = 1$. Therefore

$$\phi_{12}(t, \tau) = \epsilon^{-t^2/2}\int_\tau^t \epsilon^{\lambda^2/2}\,d\lambda = \int_\tau^t \epsilon^{(\lambda^2-t^2)/2}\,d\lambda$$

In a similar manner, the other terms of the state transition matrix are obtained from the following:

$$\begin{aligned}
\phi_{11}(t, \tau) &= y(t) \quad \text{when } \dot{y}(\tau) = 0,\ y(\tau) = 1\\
\phi_{21}(t, \tau) &= \dot{y}(t) \quad \text{when } \dot{y}(\tau) = 0,\ y(\tau) = 1\\
\phi_{22}(t, \tau) &= \dot{y}(t) \quad \text{when } \dot{y}(\tau) = 1,\ y(\tau) = 0
\end{aligned}$$

Thus

$$\phi_{11}(t, \tau) = \epsilon^{(\tau^2-t^2)/2} + \tau\int_\tau^t \epsilon^{(\lambda^2-t^2)/2}\,d\lambda$$

$$\phi_{21}(t, \tau) = \tau - t\left[\epsilon^{(\tau^2-t^2)/2} + \tau\int_\tau^t \epsilon^{(\lambda^2-t^2)/2}\,d\lambda\right]$$

$$\phi_{22}(t, \tau) = 1 - t\int_\tau^t \epsilon^{(\lambda^2-t^2)/2}\,d\lambda$$

The state transition matrix is given by

$$\boldsymbol{\phi}(t, \tau) = \begin{bmatrix} \epsilon^{(\tau^2-t^2)/2} + \tau \int_\tau^t \epsilon^{(\lambda^2-t^2)/2}\, d\lambda & \int_\tau^t \epsilon^{(\lambda^2-t^2)/2}\, d\lambda \\ \tau - t\phi_{11}(t, \tau) & 1 - t\phi_{12}(t, \tau) \end{bmatrix}$$

Another Use of the Adjoint System

The adjoint system is useful for studying the effects of forcing functions and initial conditions on linear combinations of the state variables. Consider an inner product of the adjoint variables $\boldsymbol{\alpha}$ and $\mathbf{Lx}$, where

$$\mathbf{L} = \mathbf{I}\frac{d}{dt} - \mathbf{A}$$

the system operator. This inner product is taken as

$$\langle \boldsymbol{\alpha}, \mathbf{Lx} \rangle = \int_{t_0}^{t} \boldsymbol{\alpha}^T(\lambda)[\dot{\mathbf{x}}(\lambda) - \mathbf{A}(\lambda)\mathbf{x}(\lambda)]\, d\lambda \tag{5.10-11}$$

Integrating the first term by parts yields

$$\langle \boldsymbol{\alpha}, \mathbf{Lx} \rangle = \boldsymbol{\alpha}^T(\lambda)\mathbf{x}(\lambda)\Big|_{t_0}^{t} - \int_{t_0}^{t} [\dot{\boldsymbol{\alpha}}^T(\lambda) + \boldsymbol{\alpha}^T(\lambda)\mathbf{A}(\lambda)]\mathbf{x}(\lambda)\, d\lambda \tag{5.10-12}$$

But

$$[\dot{\boldsymbol{\alpha}}^T(\lambda) + \boldsymbol{\alpha}^T(\lambda)\mathbf{A}(\lambda)] = [\dot{\boldsymbol{\alpha}}(\lambda) + \mathbf{A}^T(\lambda)\boldsymbol{\alpha}(\lambda)]^T = (\mathbf{L}^*\boldsymbol{\alpha})^T = \mathbf{0}^T \tag{5.10-13}$$

from Eq. 5.9-29. Then, from Eqs. 5.10-11, 5.10-12, and 5.10-1,

$$\boldsymbol{\alpha}^T(t)\mathbf{x}(t) = \boldsymbol{\alpha}^T(t_0)\mathbf{x}(t_0) + \int_{t_0}^{t} \boldsymbol{\alpha}^T(\lambda)\mathbf{B}(\lambda)\mathbf{v}(\lambda)\, d\lambda \tag{5.10-14}$$

Note that, for an unforced system, i.e., if $\mathbf{v}(\lambda) = \mathbf{0}$, $\boldsymbol{\alpha}^T\mathbf{x}$ is constant. Equation 5.10-14 can be used to determine linear combinations of the $x_i(t)$'s, once Eq. 5.10-13 has been solved for $\boldsymbol{\alpha}(\lambda)$ corresponding to $t_0 \leqslant \lambda \leqslant t$.

In order to solve Eq. 5.10-13 for $\boldsymbol{\alpha}(\lambda)$, appropriate boundary conditions must be specified on $\boldsymbol{\alpha}(\lambda)$. The boundary conditions which should be specified depend upon the problem to be solved. For example, if the effect of the forcing function and/or initial conditions on $x_1(t)$ is to be determined, appropriate boundary conditions are $\alpha_1(t) = 1$ and $\alpha_2(t) = \alpha_3(t) = \cdots = \alpha_n(t) = 0$. Solution of the adjoint system subject to these boundary conditions, and the subsequent substitution of this solution into Eq. 5.10-14, permit this equation to be solved for the desired result.

Example 5.10-2. A terminal control system described by $\dot{x} = -(1/T)x + m(t)$ starts at $x(0) = 0$ in response to $m(t) = KU_{-1}(t)$. The state of the system at time $t = t_0$ (a fixed value) is desired to be $x^0(t_0) = \epsilon^{(t_1-t_0)/T}$, at which point $m(t)$ is set equal to zero. The system is then to "coast" to $x = 1$ at $t = t_1$ (also a fixed time instant). This is the desired terminal point. When $x(t_0)$ is compared with $x^0(t_0)$, however, it is found that, owing to disturbances, $x(t_0) < x^0(t_0)$. How much longer should $m(t) = K$ be maintained so that $x(t_1) = 1$?

Let Δt be the time duration beyond t_0 for which $m(t) = K$. Then Eq. 5.10-14 becomes

$$\alpha(t_1)x(t_1) = \alpha(t_0)x(t_0) + \int_{t_0}^{t_0+\Delta t} K\alpha(\lambda)\,d\lambda$$

The adjoint equation is

$$\frac{d\alpha}{d\lambda} = \frac{1}{T}\alpha, \quad \alpha(t_1) = 1, \quad t_0 \leqslant \lambda \leqslant t_1$$

The solution for $\alpha(\lambda)$ is

$$\alpha(\lambda) = \epsilon^{(\lambda - t_1)/T}$$

Then

$$x(t_1) = 1 = \epsilon^{(t_0-t_1)/T}x(t_0) + K\epsilon^{-t_1/T}\int_{t_0}^{t_0+\Delta t} \epsilon^{\lambda/T}\,d\lambda$$

Performance of the integration and solution for Δt yield

$$\Delta t = T\ln\left[1 + \frac{\epsilon^{(t_1-t_0)/T} - x(t_0)}{KT}\right] = T\ln\left[1 + \frac{x^0(t_0) - x(t_0)}{KT}\right]$$

As a second illustration of the choice of boundary conditions on $\boldsymbol{\alpha}(\lambda)$, $\alpha_1(t) = \alpha_2(t) = 1$, $\alpha_3(t) = \alpha_4(t) = \cdots = \alpha_n(t) = 0$ enables Eq. 5.10-14 to be solved for $x_1(t) + x_2(t)$. In the time-varying case, of course, this generally must be done by simulation because of the difficulty in analytically solving time-variable differential equations.

With respect to simulation, the adjoint problem with the boundary conditions $\boldsymbol{\alpha}(t)$ may be converted into an initial value problem by the change in variable,

$$\lambda = t - t_1 \tag{5.10-15}$$

Since λ runs forward in "time" from t_0 to t, t_1 runs "backward in time" in the range $0 \leqslant t_1 \leqslant t - t_0$. The boundary conditions on $\boldsymbol{\alpha}(\lambda)$ at $\lambda = t$ correspond to boundary conditions on $\boldsymbol{\alpha}(t_1)$ at $t_1 = 0$. Thus they are initial conditions in the simulation. With the change in independent variable of Eq. 5.10-15, the adjoint equation becomes

$$\frac{d\boldsymbol{\alpha}(t - t_1)}{dt_1} = \mathbf{A}^T(t - t_1)\boldsymbol{\alpha}(t - t_1) \tag{5.10-16}$$

where the change in variable removes a minus sign.

t_1 is time on the simulator, usually an analog computer. The simulator time t_1 runs from $t_1 = 0$ to $t_1 = t - t_0$, as stated above. Thus the simulation method is most useful when t is some fixed, finite value of terminal time T. In this case, Eq. 5.10-16 becomes

$$\frac{d\boldsymbol{\alpha}(T - t_1)}{dt_1} = \mathbf{A}^T(T - t_1)\boldsymbol{\alpha}(T - t_1) \tag{5.10-17}$$

The coefficient matrix of Eq. 5.10-17 is the transposed $\mathbf{A}$ matrix of the original system, with t replaced by $T - t_1$. Thus this system can be simulated by reversing the inputs and outputs of each of the simulation elements of the simulation of $\dot{\mathbf{x}} = \mathbf{A}(t)\mathbf{x}$, and replacing the time t of any time-varying gains by $T - t_1$. This is discussed further in the next two sections.

5.11 IMPULSE RESPONSE MATRICES

The impulse response matrix of a linear system characterizes the dynamic equations of the system in the sense that, given the input vector from $-\infty$ to t and the impulse response matrix $\mathbf{H}(t, \tau)$, the output vector at time t can be uniquely determined. Equation 5.11-1 represents the vector-matrix form of the superposition integral given in Eq. 1.7-2.

$$\mathbf{y}(t) = \int_{-\infty}^{t} \mathbf{H}(t, \lambda)\mathbf{v}(\lambda)\, d\lambda \tag{5.11-1}$$

Equation 5.10-7 can be rewritten in the same form as

$$\mathbf{y}(t) = \int_{-\infty}^{t} \mathbf{C}(t)\boldsymbol{\phi}(t, \lambda)\mathbf{B}(\lambda)\mathbf{v}(\lambda)\, d\lambda + \mathbf{D}(t)\mathbf{v}(t) \tag{5.11-2}$$

or

$$\mathbf{y}(t) = \int_{-\infty}^{t} [\mathbf{C}(t)\boldsymbol{\phi}(t, \lambda)\mathbf{B}(\lambda) + U_0(t - \lambda)\mathbf{D}(t)]\mathbf{v}(\lambda)\, d\lambda \tag{5.11-3}$$

Comparison of Eqs. 5.11-1 and 5.11-3 shows that the impulse response matrix $\mathbf{H}(t, \tau)$ is given by

$$\begin{aligned} \mathbf{H}(t, \tau) &= \mathbf{C}(t)\boldsymbol{\phi}(t, \tau)\mathbf{B}(\tau) + U_0(t - \tau)\mathbf{D}(t), \qquad & t \geqslant \tau \\ &= 0 & t < \tau \end{aligned} \tag{5.11-4}$$

For a fixed system, Eq. 5.11-4 can be written as

$$\begin{aligned} \mathbf{H}(t) &= \mathbf{C}\boldsymbol{\phi}(t)\mathbf{B} + U_0(t)\mathbf{D} \qquad & t \geqslant 0 \\ &= 0 & t < 0 \end{aligned} \tag{5.11-5}$$

Example 5.11-1. Find the impulse response matrix of the *RLC* circuit shown in Fig. 5.5-2*b*. The inputs are $v_1(t)$ and $v_2(t)$, and the outputs are $i_1(t)$ and $i_2(t)$.

From Example 5.5-2, the **A** matrix of this circuit is

$$\mathbf{A} = \begin{bmatrix} -\dfrac{2}{(R_1 + R_2)C} & 0 \\ 0 & \dfrac{2R_1R_2}{(R_1 + R_2)L} \end{bmatrix}$$

Since **A** is diagonal, $\boldsymbol{\phi} = \epsilon^{\mathbf{A}t}$ is easily determined.

Owing to the symmetry of the circuit, the relationships between the elements of the coupling matrices are:

$$\begin{aligned}
b_{11} &= b_{12} && (v_1 \text{ and } v_2 \text{ have the same effect upon } \dot{x}_1) \\
b_{21} &= -b_{22} && (v_1 \text{ and } v_2 \text{ have opposite effects upon } \dot{x}_2) \\
c_{11} &= c_{21} && (x_1 \text{ has the same effect upon } i_1 \text{ and } i_2) \\
c_{22} &= -c_{12} && (x_2 \text{ has opposite effects upon } i_1 \text{ and } i_2) \\
d_{11} &= d_{22} && (v_1 \text{ couples into } i_1 \text{ in the same manner as } v_2 \text{ couples into } i_2) \\
d_{12} &= d_{21} && (v_1 \text{ couples into } i_2 \text{ in the same manner as } v_2 \text{ couples into } i_1)
\end{aligned}$$

The general expression for the impulse response matrix is then

$$\begin{aligned}
\mathbf{H}(t) &= \begin{bmatrix} c_{11} & c_{22} \\ c_{11} & -c_{22} \end{bmatrix} \begin{bmatrix} \phi_{11}(t) & 0 \\ 0 & \phi_{22}(t) \end{bmatrix} \begin{bmatrix} b_{11} & b_{11} \\ -b_{22} & b_{22} \end{bmatrix} + U_0(t) \begin{bmatrix} d_{11} & d_{12} \\ d_{12} & d_{11} \end{bmatrix} \\
&= b_{11}c_{11}\phi_{11}(t) \begin{bmatrix} 1 & 1 \\ 1 & 1 \end{bmatrix} + b_{22}c_{22}\phi_{22}(t) \begin{bmatrix} -1 & 1 \\ 1 & -1 \end{bmatrix} + U_0(t) \begin{bmatrix} d_{11} & d_{12} \\ d_{12} & d_{11} \end{bmatrix}
\end{aligned}$$

From the state equations

$$\dot{\mathbf{x}} = \mathbf{A}\mathbf{x} + \mathbf{B}\mathbf{v}$$

$$\mathbf{y} = \mathbf{C}\mathbf{x} + \mathbf{D}\mathbf{v}$$

it is possible to obtain **B**, **C**, and **D** without resorting to loop equations. The method is similar to that used in Example 5.5-2 to determine the elements of **A**. Note that

$$\dot{x}_1 = a_{11}x_1 + a_{12}x_2 + b_{11}v_1 + b_{12}v_2$$

for all t. Assume that the network is initially at rest ($x_1 = x_2 = v_1 = v_2 = 0$). If v_1 is suddenly made 1 volt by application of a unit step of voltage, $x_1(0+) = x_2(0+) = v_2(0+) = 0$, since the capacitor voltage and inductor current cannot change instantaneously. Thus

$$b_{11} = -\left.\frac{i_c(0+)}{C}\right|_{v_1 = U_{-1}(t)}$$

This is the same as replacing the capacitor by a short circuit and the inductor by an open circuit (the network is now a purely resistive network) and writing

$$b_{11} = -\frac{i_c(t)}{Cv_1(t)} = -\frac{g_{c1}}{C}$$

where g_{c1} is the short-circuit transfer conductance between the input v_1 and the short-circuited capacitor. Thus

$$b_{11} = \frac{1}{(R_1 + R_2)C}$$

In a similar fashion,

$$b_{22} = -\frac{e_L(t)}{Lv_2(t)} = -\frac{\alpha_{L2}}{L}$$

where α_{L2} is the voltage gain between the input v_2 and the open-circuited terminals of the inductor. The capacitor and v_1 are shorted. Therefore

$$b_{22} = -\frac{R_2}{(R_1 + R_2)L}$$

Using similar arguments,

$$c_{11} = \frac{i_1}{x_1}, \quad x_2 = v_1 = v_2 = 0 \quad \text{or} \quad c_{11} = -\frac{1}{R_1 + R_2}$$

$$c_{22} = \frac{i_2}{x_2}, \quad x_1 = v_1 = v_2 = 0 \quad \text{or} \quad c_{22} = -\frac{R_2}{R_1 + R_2}$$

$$d_{11} = \frac{i_1}{v_1}, \quad x_1 = x_2 = v_2 = 0, \quad \text{or} \quad d_{11} = \frac{1}{R_1 + R_2}$$

$$d_{12} = \frac{i_1}{v_2}, \quad x_1 = x_2 = v_1 = 0, \quad \text{or} \quad d_{12} = 0$$

Therefore

$$\mathbf{H}(t) = \frac{-1}{(R_1 + R_2)^2C}\left[\epsilon^{-\frac{2t}{(R_1+R_2)C}}\right]\begin{bmatrix}1 & 1\\1 & 1\end{bmatrix} + \frac{R_2^2}{(R_1 + R_2)^2L}\left[\epsilon^{-\frac{2R_1R_2t}{(R_1+R_2)L}}\right]\begin{bmatrix}-1 & 1\\1 & -1\end{bmatrix} + \frac{U_0(t)}{R_1 + R_2}\begin{bmatrix}1 & 0\\0 & 1\end{bmatrix}$$

Simulation Difficulties

In some control system applications, particularly in missile and space work, it is necessary to determine elements of $\mathbf{H}(t, \tau)$, the impulse response matrix of a time-varying system. One illustration of the use of $\mathbf{H}(t, \tau)$ is in evaluating the mean square outputs of a system in response to random signals.

Consider the case represented in Fig. 5.11-1*a*, where the random signal $\mathbf{v}(t)$ is applied to the system with the impulse response matrix $\mathbf{H}(t, \tau)$ at $t = 0$ by the closing of the switch. Then

$$\mathbf{y}(t) = \int_0^t \mathbf{H}_1(t, \tau)\mathbf{v}(\tau)\, d\tau \tag{5.11-6}$$

Let $\mathbf{Y}$ be the autocorrelation matrix $\overline{\mathbf{y}(t)\mathbf{y}^T(t)}$, where the bar denotes the statistical average. Thus

$$\mathbf{Y} = \begin{bmatrix} \overline{y_1^2} & \overline{y_1y_2} & \cdots & \overline{y_1y_p} \\ \overline{y_2y_1} & \overline{y_2^2} & \cdots & \overline{y_2y_p} \\ \cdots & \cdots & \cdots & \cdots \\ \overline{y_py_1} & \overline{y_py_2} & \cdots & \overline{y_p^2} \end{bmatrix} \tag{5.11-7}$$

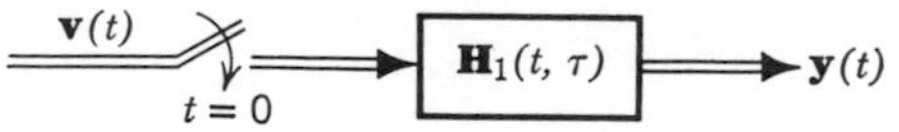

(a)

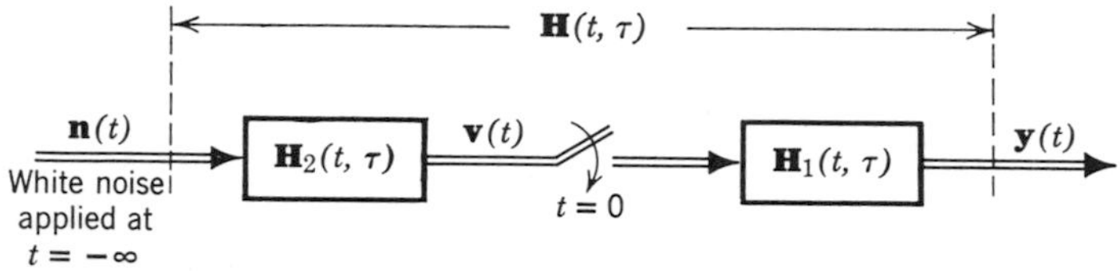

(b)

Fig. 5.11-1

Substitution of Eq. 5.11-6 yields

$$\mathbf{Y} = \overline{\left[\int_0^t \mathbf{H}_1(t, \tau_1)\mathbf{v}(\tau_1)\, d\tau_1\right]\left[\int_0^t \mathbf{v}^T(\tau_2)\mathbf{H}_1^{\,T}(t, \tau_2)\, d\tau_2\right]}$$

Since $\mathbf{H}_1(t, \tau)$ is a deterministic impulse response, $\mathbf{Y}$ can be written as

$$\mathbf{Y} = \int_0^t \mathbf{H}_1(t, \tau_1) \int_0^t \overline{\mathbf{v}(\tau_1)\mathbf{v}^T(\tau_2)}\, \mathbf{H}_1^{\,T}(t, \tau_2)\, d\tau_2\, d\tau_1 \qquad (5.11\text{-}8)$$

Equation 5.11-8 can be used to evaluate the mean square outputs of the system, assuming that $\mathbf{H}_1(t, \tau)$ and the autocorrelation matrix $\overline{\mathbf{v}(\tau_1)\mathbf{v}^T(\tau_2)}$ are known, and that the integrations can readily be performed.

In many cases, Eq. 5.11-8 is too complex to permit analytical evaluation. In the case of time-varying systems for example, it is frequently impossible to determine a closed-form expression for $\mathbf{H}_1(t, \tau)$. In such cases, $\mathbf{Y}$ can be determined by simulation techniques which have the advantage of not requiring random signal generators.[21] These simulation methods are simplest when $\mathbf{Y}$ is to be determined at a fixed instant of time. This is the case considered in the remainder of this discussion.

The basis for making random signal generators unnecessary in the simulation is that Eq. 5.11-8 depends upon the autocorrelation matrix of the random input, rather than upon the random input itself. Thus $\mathbf{Y}$ is the same for Figs. 5.11-1*a* and 5.11-1*b*, if the shaping filter is chosen so that the autocorrelation matrix of $\mathbf{v}$ is the same in both cases. With respect to Fig. 5.11-1*b*,

$$\mathbf{v}(\tau_1) = \int_{-\infty}^{\tau_1} \mathbf{H}_2(\tau_1, \lambda_1)\mathbf{n}(\lambda_1)\, d\lambda_1$$

Then

$$\overline{\mathbf{v}(\tau_1)\mathbf{v}^T(\tau_2)} = \int_{-\infty}^{\tau_1} \mathbf{H}_2(\tau_1, \lambda_1) \int_{-\infty}^{\tau_2} \overline{\mathbf{n}(\lambda_1)\mathbf{n}^T(\lambda_2)}\, \mathbf{H}_2^{\,T}(\tau_2, \lambda_2)\, d\lambda_2\, d\lambda_1$$

Since $\overline{\mathbf{n}(\lambda_1)\mathbf{n}^T(\lambda_2)} = \mathbf{N}(\lambda_1)[U_0(\lambda_1 - \lambda_2)]$ for white noise,

$$\overline{\mathbf{v}(\tau_1)\mathbf{v}^T(\tau_2)} = \int_{-\infty}^{T} \mathbf{H}_2(\tau_1, \lambda_1)\mathbf{N}(\lambda_1)\mathbf{H}_2{}^T(\tau_2, \lambda_1)\, d\lambda_1 \qquad (5.11\text{-}9)$$

after performing the integration with respect to λ_2. Equation 5.11-9 is an integral equation which can be solved for the impulse response matrix of the shaping filter necessary to make Figs. 5.11-1*a* and 5.11-1*b* equivalent with respect to $\mathbf{Y}$. The solution of this equation is frequently easy, particularly in the stationary case where transform techniques can be used.

Assuming that the shaping filter has been appropriately chosen, Eq. 5.11-8 can then be applied to the representation of Fig. 5.11-1*b* by replacing $\overline{\mathbf{v}(\tau_1)\mathbf{v}^T(\tau_2)}$ and $\mathbf{H}_1(t, \tau)$ by $\overline{\mathbf{n}(\tau_1)\mathbf{n}^T(\tau_2)}$ and $\mathbf{H}(t, \tau)$, respectively. $\mathbf{H}(t, \tau)$ is the impulse response matrix of the combination of the shaping filter, the switch, and the system. Since $\overline{\mathbf{n}(\tau_1)\mathbf{n}^T(\tau_2)} = \mathbf{N}(\tau_1)[U_0(\tau_1 - \tau_2)]$, application of Eq. 5.11-8 to the configuration of Fig. 5.11-1*b* yields

$$\mathbf{Y} = \int_{-\infty}^{T} \mathbf{H}(T, \tau)\mathbf{N}(\tau)\mathbf{H}^T(T, \tau)\, d\tau \qquad (5.11\text{-}10)$$

after performing the integration with respect to τ_2, setting $\tau_1 = \tau$, and specializing the result to $t = T$. The lower limit, $\tau = -\infty$, is required because $\mathbf{n}(t)$ is applied at $t = -\infty$.

Equation 5.11-10 indicates that $\mathbf{Y}$ can be determined by applying a deterministic impulse (or its equivalent initial conditions), rather than a noise source, multiplying the appropriate simulator outputs together in pairs, and integrating the results. This assumes that the simulator outputs are the impulse responses of the appropriately chosen shaping filter, the switch, and the system. Note that the switch makes $\mathbf{H}(t, \tau)$ time-varying, even if the shaping filter and system are time-invariant.

Of particular importance is the fact that the variable of integration in Eq. 5.11-10 is τ. Thus the generation of $\mathbf{H}(T, \tau)$ by the simulator should produce $\mathbf{H}(t, \tau)$ for fixed t and variable τ. Unfortunately, straightforward simulation produces $\mathbf{H}(t, \tau)$ for variable t and fixed τ, since t is the response time and τ is the time of application of the impulse. Thus the straightforward simulation would have to be run many times, each for a different τ, and the data replotted for fixed t and variable τ. Multiplication and integration of the replotted data are then required to determine the elements of $\mathbf{Y}$.

To illustrate the bookkeeping task involved in this direct simulation method, assume an m input, p output system. The expression for $y_i(T)$ is,

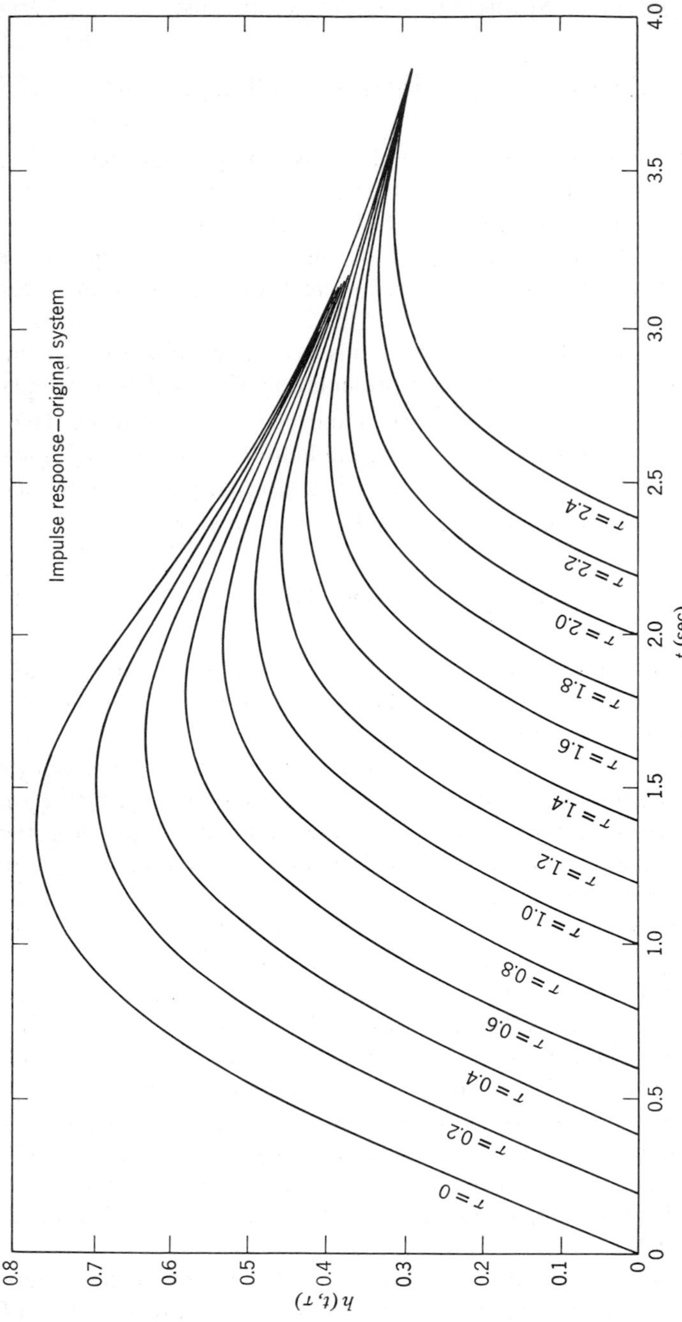

Impulse response—original system
$h(t, \tau)$
0
0.1
0.2
0.3
0.4
0.5
0.6
0.7
0.8
0
0.5
1.0
1.5
2.0
2.5
3.0
3.5
4.0
t (sec)
$\tau = 0$
$\tau = 0.2$
$\tau = 0.4$
$\tau = 0.6$
$\tau = 0.8$
$\tau = 1.0$
$\tau = 1.2$
$\tau = 1.4$
$\tau = 1.6$
$\tau = 1.8$
$\tau = 2.0$
$\tau = 2.2$
$\tau = 2.4$

Fig. 5.11-2a

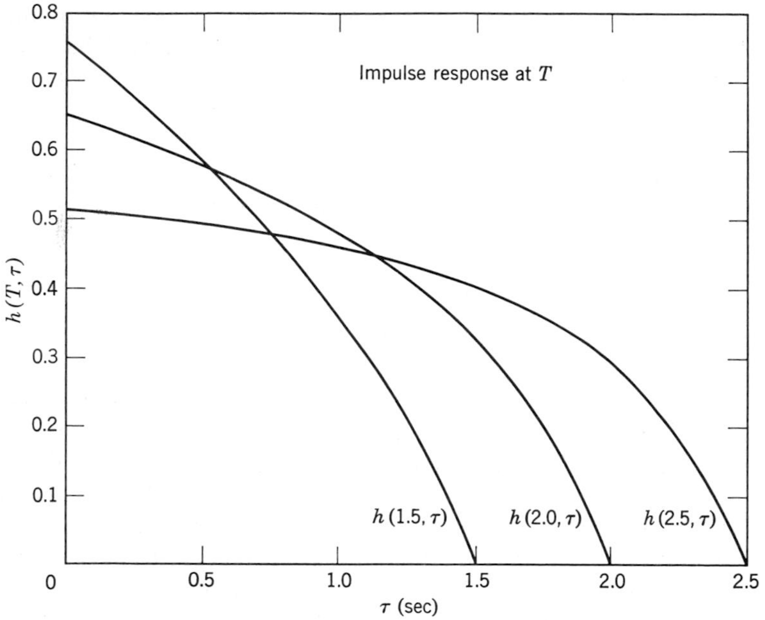

Fig. 5.11-2*b*

from Eq. 5.11-1,

$$y_i(T) = \int_{-\infty}^{T} \sum_{j=1}^{m} h_{ij}(T, \lambda) v_j(\lambda)\, d\lambda, \quad i = 1, 2, \ldots, p \qquad (5.11\text{-}11)$$

If a unit impulse is placed on the jth input at time $\tau \leqslant T$, then

$$y_i(T) = h_{ij}(T, \tau), \quad T \geqslant \tau, \quad i = 1, 2, \ldots, p \qquad (5.11\text{-}12)$$

Consequently, a unit impulse on the jth input produces a *column* of the impulse response matrix for a *fixed* value of τ. The unit impulse must be repeatedly applied to the jth input for various values of τ in order to obtain $h_{ij}(T, \tau)$ for a variable τ.

The impulse response

$$h(t, \tau) = \int_{\tau}^{t} \epsilon^{(\lambda^2 - t^2)/2}\, d\lambda$$

of the system $\ddot{y} + t\dot{y} + y = 0$ is shown in Fig. 5.11-2. Figure 5.11-2*a* shows the results of a set of simulation runs of $h(t, \tau)$ versus t for various values of τ. Figure 5.11-2*b* shows the results of cross-plotting these runs for fixed values of t. This would have to be performed for *each* of the elements $h_{ij}(t, \tau)$. To obtain all the columns of the impulse response matrix, a set of simulation runs must be performed for *each* of the m

inputs. If 10 simulation runs were sufficient to provide enough points to produce a function of τ, then $10m$ simulation runs would have to be performed. This would provide mp sets of curves of the type shown in Fig. 5.11-2*a*. For *each* of the mp sets of curves, a cross-plot would have to be performed to produce mp curves of the type shown in Fig. 5.11-2*b*. In the following section, it is shown that a *single* computer run of a *modified adjoint system* produces a *row* of the impulse response matrix $\mathbf{H}(t, \tau)$ for a fixed value of t and variable τ.

5.12 MODIFIED ADJOINT SYSTEMS[20–22]

The difficulty associated with determining the impulse response matrix $\mathbf{H}(t, \tau)$ for variable τ was indicated in the previous section. It is due to the fact that the response variable is t rather than τ. Since the state transition matrix of the adjoint system is related to $\boldsymbol{\phi}(t, \tau)$, the state transition matrix of the original system, by an interchange of t and τ and a transposition, a transposed adjoint system would seem to answer this difficulty.† Indeed it does, after a change of time axes.

Consider Eq. 5.10-14 with $t = T$ and t_0 equal to $-\infty$. The equation becomes

$$\boldsymbol{\alpha}^T(T)\mathbf{x}(T) = \int_{-\infty}^{T} \boldsymbol{\alpha}^T(\tau)\mathbf{B}(\tau)\mathbf{v}(\tau)\, d\tau$$

since $\mathbf{x}(-\infty) = 0$. Now suppose that $\boldsymbol{\alpha}^T(T)$ is chosen as

$$\boldsymbol{\alpha}^T(T) = [c_{i1}(T) \quad c_{i2}(T) \quad \cdots \quad c_{in}(T)]$$

where the c_{ij}'s are the indicated elements of the system $\mathbf{C}$ matrix. Then

$$y_i(T) = [c_{i1}(T) \quad c_{i2}(T) \quad \cdots \quad c_{in}(T)]\mathbf{x}(T)$$

can be obtained.‡ That is,

$$y_i(T) = \int_{-\infty}^{T} \boldsymbol{\alpha}^T(\tau)\mathbf{B}(\tau)\mathbf{v}(\tau)\, d\tau, \quad \text{if } \boldsymbol{\alpha}(T) = \begin{bmatrix} c_{i1}(T) \\ c_{i2}(T) \\ \cdot \\ \cdot \\ \cdot \\ c_{in}(T) \end{bmatrix}$$

† The term system is used in this section to denote the combination of the shaping filter, switch, and system of Fig. 5.11-1*b*.

‡ The $\mathbf{D}$ matrix is assumed to be a null matrix, since this is true for any practical combinations of shaping filters and systems. Otherwise an infinite $\overline{y^2(T)}$ would result.

Comparing this expression with the ith row of

$$\mathbf{y}(T) = \int_{-\infty}^{T} \mathbf{H}(T, \tau)\mathbf{v}(\tau)\, d\tau$$

it is apparent that

$$\{\mathbf{H}(T, \tau)\}_{i\text{th row}} = \boldsymbol{\alpha}^T(\tau)\mathbf{B}(\tau) \tag{5.12-1}$$

if the adjoint system has the boundary conditions

$$\boldsymbol{\alpha}(T) = \begin{bmatrix} c_{i1}(T) \\ c_{i2}(T) \\ \cdot \\ \cdot \\ \cdot \\ c_{in}(T) \end{bmatrix} \tag{5.12-2}$$

In order to make these boundary conditions initial conditions for the simulation, let $\tau = T - \tau_1$ as was done in Eq. 5.10-15. Then analogous to Eq. 5.10-17,

$$\frac{d\boldsymbol{\alpha}(T - \tau_1)}{d\tau_1} = \mathbf{A}^T(T - \tau_1)\boldsymbol{\alpha}(T - \tau_1) \tag{5.12-3}$$

Also, Eqs. 5.11-10, 5.12-1, and 5.12-2 become

$$\mathbf{Y} = \int_0^{\infty} \mathbf{H}(T, T - \tau_1)\mathbf{N}(T - \tau_1)\mathbf{H}^T(T, T - \tau_1)\, d\tau_1 \tag{5.12-4}$$

$$\{\mathbf{H}(T, T - \tau_1)\}_{i\text{th row}} = \boldsymbol{\alpha}^T(T - \tau_1)\mathbf{B}(T - \tau_1) \tag{5.12-5}$$

$$\boldsymbol{\alpha}(0) = \begin{bmatrix} c_{i1}(T) \\ c_{i2}(T) \\ \cdot \\ \cdot \\ \cdot \\ c_{in}(T) \end{bmatrix} \tag{5.12-6}$$

Remembering from Section 5.10 that Eq. 5.12-3 can be simulated by reversing the inputs and outputs of each of the simulation elements of $\dot{\mathbf{x}} = \mathbf{A}(t)\mathbf{x}$ and replacing the time t of any time-varying gains by $T - t_1$, Eq. 5.12-5 is simulated by multiplying the adjoint signals by $\mathbf{B}(T - t_1)$. Of course, the adjoint system must have the initial conditions of Eq. 5.12-6. A little thought reveals that the initial conditions of Eq. 5.12-6

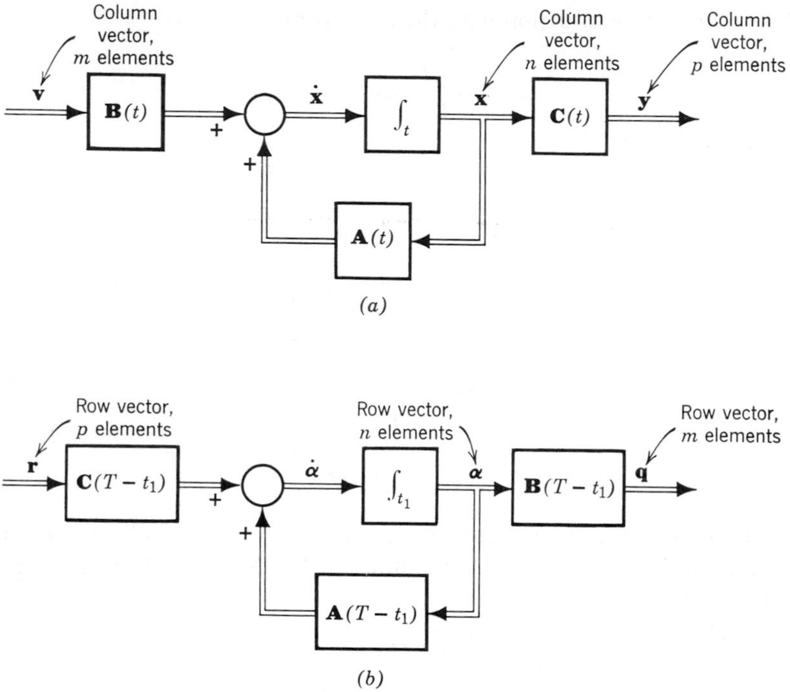

Fig. 5.12-1

can be established by applying a unit impulse at $t_1 = 0$ through gains of

$$\begin{bmatrix} c_{i1}(T - t_1) \\ c_{i2}(T - t_1) \\ \cdot \\ \cdot \\ \cdot \\ c_{in}(T - t_1) \end{bmatrix}$$

Since **B** and **C** are, respectively, the input and output matrices of the original system, the net result of this argument is that the ith row of $\mathbf{H}(T, T - \tau)$ required for Eq. 5.12-4 is generated for variable τ by reversing the inputs and outputs of each of the simulation elements of

$$\dot{\mathbf{x}} = \mathbf{A}(t)\mathbf{x} + \mathbf{B}(t)\mathbf{v}$$
$$\mathbf{y} = \mathbf{C}(t)\mathbf{x}$$

and replacing the time t of any time-varying gains by $T - t_1$. The resulting system is called the *modified adjoint system.* It is compared with the original system in Fig. 5.12-1.

If the original system has m inputs and p outputs, the modified adjoint system has p inputs and m outputs. Each simulation run on the modified adjoint system produces m outputs. These m outputs comprise a row of $\mathbf{H}(T, T - \tau)$, where the running variable of simulation t is equal to τ. The particular row of the matrix is determined by the input on which the unit impulse is placed at the beginning of the simulation run.

Example 5.12-1. For the differential equation $\ddot{y} + t\dot{y} + y = 0$, the curves of Figs. 5.11-2*a* and 5.11-2*b* show the cross-plotting operation required to obtain $h(T, \tau)$. Show that the modified adjoint system produces the desired result in one simulation run.

The impulse response of the system is given by

$$h(t, \tau) = \int_{\tau}^{t} \epsilon^{(\lambda^2 - t^2)/2}\, d\lambda \qquad \tau \leqslant t$$
$$= 0 \qquad \tau > t$$

If T is substituted for t and the change in variable, $\tau = T - \tau_1$, is made, then the impulse response $h(T, T - \tau_1)$ is

$$h(T, T - \tau_1) = \int_{T-\tau_1}^{T} \epsilon^{(\lambda^2 - T^2)/2}\, d\lambda \qquad \tau_1 \geqslant 0$$
$$= 0 \qquad \tau_1 < 0$$

The simulation diagram of the original equation is shown in Fig. 5.12-2*a*. Thus the modified adjoint system has the diagram shown in Fig. 5.12-2*b*. The differential equation corresponding to Fig. 5.12-2*b* is

$$\ddot{\alpha} + (T - t_1)\dot{\alpha} = 0$$

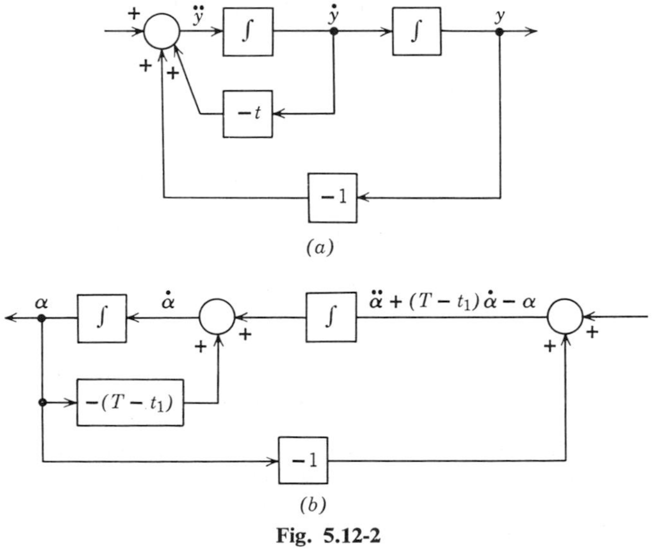

Fig. 5.12-2

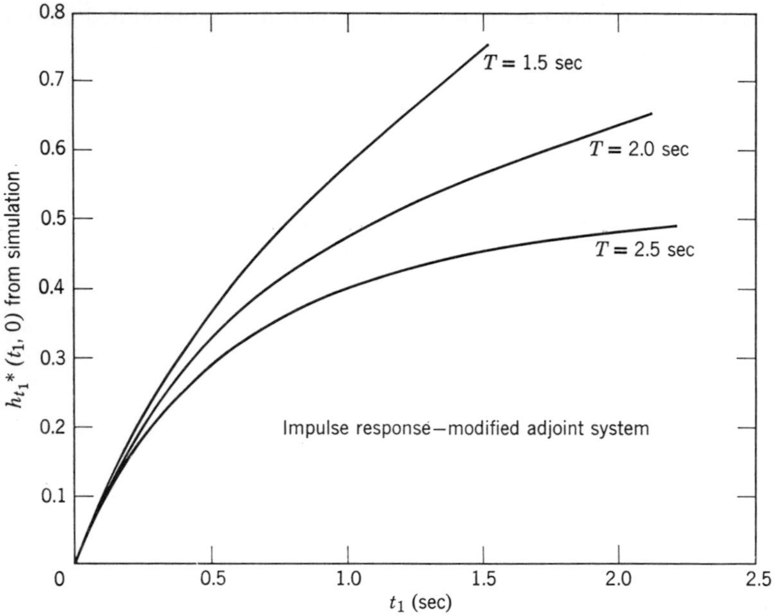

Fig. 5.12-3

The response of the modified adjoint system to an impulse at $t_1 = 0$ is given by the solution of

$$\ddot{\alpha} + (T - t_1)\dot{\alpha} = U_0(t_1)$$

Let $w = \dot{\alpha}$. For $t_1 > 0$,

$$\dot{w} + (T - t_1)w = 0, \quad w(0) = 1$$

Thus

$$w(t_1) = \dot{\alpha} = \exp\left(\frac{t_1^2}{2} - Tt_1\right), \quad t_1 \geqslant 0$$

Then

$$\alpha(t_1) = \int_0^{t_1} \exp\left(\frac{\beta^2}{2} - T\beta\right) d\beta, \quad t_1 \geqslant 0$$

Let $\beta = T - \lambda$. This yields $h^*(t_1, 0)$, the impulse response of the modified adjoint system, as

$$h^*(t_1, 0) = \alpha(t) = \int_{T-t_1}^{T} \epsilon^{(\lambda^2 - T^2)/2}\, d\lambda, \quad t_1 \geqslant 0$$

This is identical with the expression for $h(T, T - \tau_1)$ if t_1 is written for τ_1.

Thus the impulse response of the modified adjoint system *observed over the* t_1 *axis* is exactly the impulse response of the original system, *observed over the* τ_1 *axis*. However, the t_1 axis here is the time axis of *observation* on the computer, while the τ_1 axis is the time axis of *application* of impulses. Therefore a simulation run on the t_1 axis produces exactly the same results as does a cross-plot sketched on the τ_1 axis. This is shown in Fig. 5.12-3, where a series of runs of $h^*(t_1, 0)$ are shown. Note that, if the τ axis in Fig.

5.11-2*b* were changed to $\tau_1 = T - \tau$, these curves would be identical with the curves of Fig. 5.12-3. Therefore the modified adjoint system produces in one simulation run what would take many simulation runs of the original system.

Example 5.12-2. The **A**, **B**, and **C** matrices of a given system are

$$\mathbf{A} = \begin{bmatrix} 0 & 1 \\ a_{21}(t) & a_{22}(t) \end{bmatrix}, \quad \mathbf{B} = \begin{bmatrix} b_1(t) \\ b_2(t) \end{bmatrix}, \quad \mathbf{C} = \begin{bmatrix} c_1(t) & 0 \\ 0 & c_2(t) \end{bmatrix}$$

The simulation diagram for this system is shown in Fig. 5.12-4*a*. Determine the diagram for the modified adjoint system.

Since $\mathbf{H}(t, \tau) = \mathbf{C}(t)\boldsymbol{\phi}(t, \tau)\mathbf{B}(\tau)$, the order of $\mathbf{H}(t, \tau)$ is (2 x 1), or two rows and one column. The application of a unit impulse at time $\tau = \tau_i$ produces an output on both y_1 and y_2, thus giving all the elements of $\mathbf{H}(t, \tau_i)$, since $\mathbf{H}(t, \tau)$ is a single column matrix. However, this is for variable t and fixed $\tau = \tau_i$.

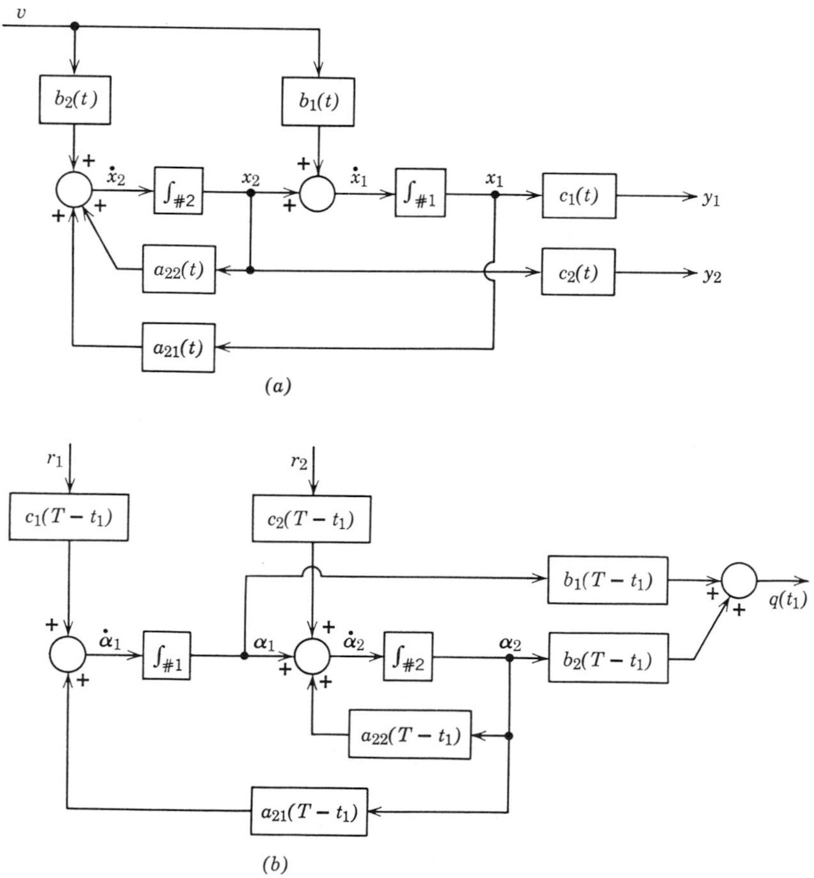

Fig. 5.12-4

If the inputs and outputs of the simulation diagram are reversed and the change in variable $t = T - t_1$ is made, the result is Fig. 5.12-4*b*. Note that at each terminal the number of inputs and of outputs are interchanged. This fact provides a rapid partial check of the correctness of an adjoint simulation diagram. The vector-matrix differential equations for this system are given by

$$[\dot{\alpha}_1 \quad \dot{\alpha}_2] = [\alpha_1 \quad \alpha_2]\begin{bmatrix} 0 & 1 \\ a_{21}(T - t_1) & a_{22}(T - t_1) \end{bmatrix} + [r_1 \quad r_2]\begin{bmatrix} c_1(T - t_1) & 0 \\ 0 & c_2(T - t_1) \end{bmatrix}$$

$$q = [\alpha_1 \quad \alpha_2]\begin{bmatrix} b_1(T - t_1) \\ b_2(T - t_1) \end{bmatrix}$$

or

$$\dot{\boldsymbol{\alpha}}^T = \boldsymbol{\alpha}^T\mathbf{A}(T - t_1) + \mathbf{r}^T\mathbf{C}(T - t_1)$$

$$q = \boldsymbol{\alpha}^T\mathbf{b}(T - t_1)$$

If a unit impulse is placed on *one* of the inputs at time zero, the output $q(t_1)$ represents a *row* (one element in this case) of the original response matrix $\mathbf{H}(T, T - t_1)$. Therefore two simulation runs are required to produce both rows of the impulse response matrix.

REFERENCES

1. L. A. Zadeh, and C. A. Desoer, *Linear System Theory—The State Space Approach*, McGraw-Hill Book Co., New York, 1963, pp. 23–31.
2. *Ibid.*, pp. 311–326.
3. B. Friedland, O. Wing, and R. Ash, *Principles of Linear Networks*, McGraw-Hill Book Co., 1961, pp. 58–64.
4. R. E. Kalman, "Canonical Structure of Linear Dynamical Systems," Proc. Natl. Acad. Sci., Vol. 48, No. 4, April, 1962 pp. 596–600.
5. R. E. Kalman, Y. C. Ho, and K. S. Narendra, "Controllability of Linear Dynamical Systems," *Contrib. Differential Equations*, Vol. I, No. 2, 1962, pp. 189–213.
6. Y. C. Ho, "What Constitutes a Controllable System," *IRE Trans. Auto. Control*, Vol. AC-7, No, 3, April 1962, p. 76.
7. E. B. Lee, "On the Domain of Controllability for Linear Systems," *IRE Trans. Auto. Control*, Vol. AC-8, No. 2, April 1963, pp. 172–173.
8. E. G. Gilbert, "Controllability and Observability in Multivariable Control Systems," *J. Soc. Ind. Appl. Math.—Control Ser.*, Ser. A, Vol. 1, No. 2, 1963, pp. 128–151.
9. I. M. Horowitz, "Synthesis of Linear, Multivariable Feedback Control Systems," *IRE Trans. Auto. Control*, Vol. AC-5, No. 2, June 1960, pp. 94–105.
10. R. E. Kalman, "Mathematical Description of Linear Dynamical Systems," *J. Soc. Ind. Appl. Math.—Control Ser.*, Ser. A, Vol. 1, No. 2, 1963, pp. 152–192.
11. E. A. Coddington, and N. Levinson, *Theory of Ordinary Differential Equations*, McGraw-Hill Book Co., New York, 1955, Chapter 3.
12. R. A. Struble, *Nonlinear Differential Equations*, McGraw-Hill Book Co., New York, 1962, Chapter 4.
13. B. K. Kinariwala, "Analysis of Time Varying Networks," *IRE Inter. Conv. Record*, 1961, Pt. 4, pp. 268–276.
14. Pipes, L. A., *Matrix Methods for Engineering*, Prentice-Hall Inc., Englewood Cliffs, N.J., 1963, pp. 90–92.

15. B. Friedman, *Principles and Techniques of Applied Mathematics*, John Wiley and Sons, New York, 1956, p. 43.
16. Coddington and Levinson, *op. cit.*, p. 65.
17. Stuble, *op. cit.*, p. 64.
18. Coddington and Levinson, *op. cit.*, p. 80.
19. L. A. Pipes, *J. Appl. Phys.*, Vol. 25, pp. 1179–1185.
20. M. Issac, "Deterministic and Stochastic Response of Linear Time-Variable Systems, General Electric, LMED Technical Report, No. R62 EML1, March 1962.
21. J. H. Laning, and R. H. Battin, *Random Processes in Automatic Control*, McGraw-Hill Book Co., New York, 1956, Chapter 6.
22. R. Sussman, "A Method of Solving Linear Problems by Using the Adjoint System," Internal Tech. Memor. No. M-2, Electronics Research Laboratory, University of California, Berkeley, California.

Problems

5.1 Draw the simulation diagram for the following systems.

(*a*) $\ddot{y} + 3\dot{y} + 5y = v$

(*b*) $\ddot{y} + 3\dot{y} + 5y = \ddot{v} + 2\dot{v} + v$

(*c*) $\ddot{y}_1 + 3\dot{y}_1 + 2y_2 = v_1 + 2\dot{v}_2 + 2v_2$

$\ddot{y}_2 + 4\dot{y}_1 + 3y_2 = \ddot{v}_2 + 3\dot{v}_2 + v_1$

5.2 What are the differential equations for the system shown in Fig. P5.2?

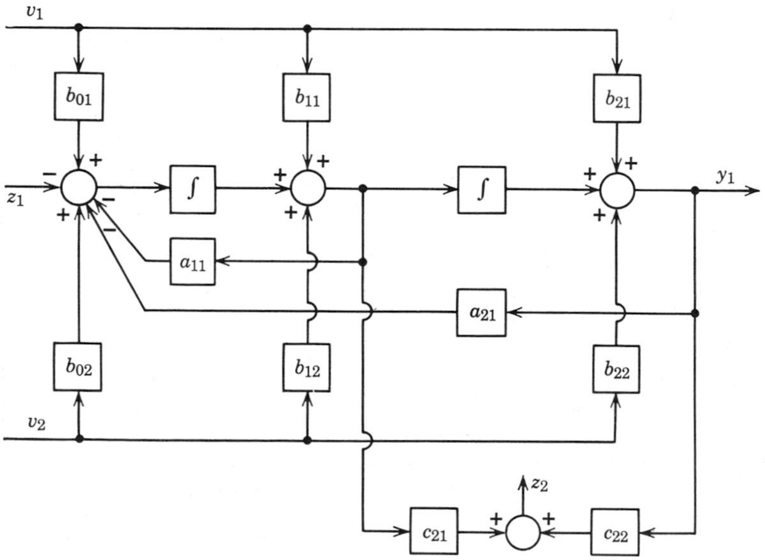

Fig. P5.2 (*Continued on p. 396*)

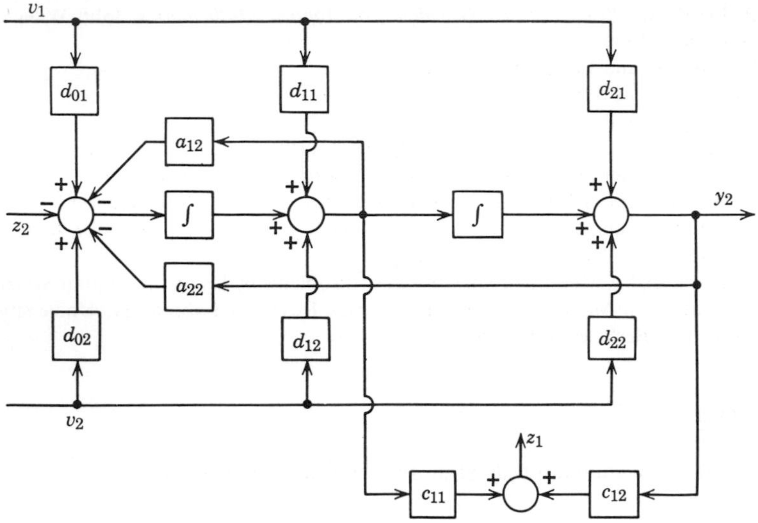

Fig. P5.2 (*Continued*)

5.3 Draw the simulation diagram for the following differential equations.

(*a*) $\ddot{y} + k\dot{y} + (\omega^2 + \epsilon \cos t)y = v$ (Generate cos *t* by simulation.)

(*b*) $\ddot{y} + y^2\dot{y} + 5y = v$

5.4 Show that the simulation in Fig. P5.4 satisfies Eq. 5.5-3.

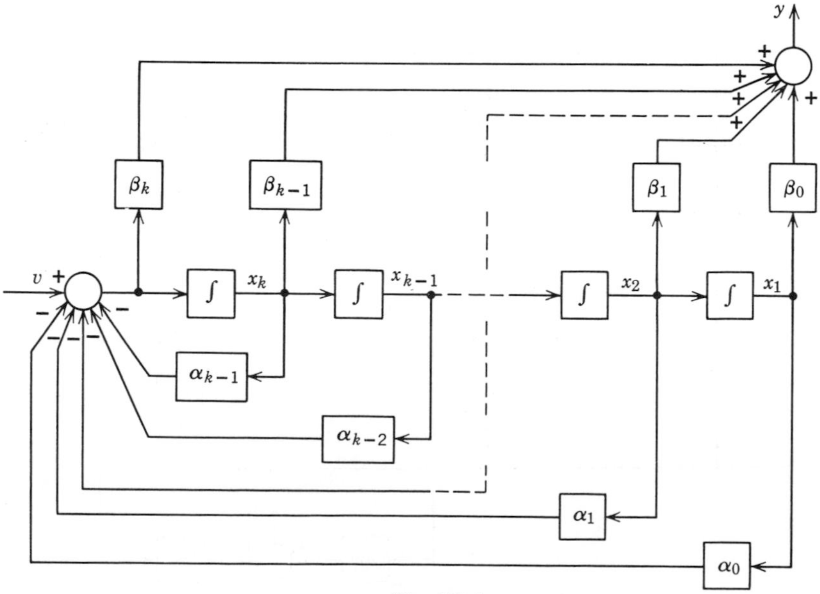

Fig. P5.4

5.5 Find the transfer function matrix and draw the transfer function diagram for the systems described below. Comment on the number of integrators required.

(*a*) $\ddot{y}_1 + 3\dot{y}_1 + 2y_1 = \dot{v}_1 + 2v_1 + \dot{v}_2 + v_2$
$\dot{y}_2 + 2y_2 = -\dot{v}_1 - 2v_1 + v_2$

(*b*) $\dot{y}_1 + y_1 = v_1 + 2v_2$
$\ddot{y}_2 + 3\dot{y}_2 + 2y_2 = \dot{v}_2 + v_2 - \dot{v}_1$

(*c*) $\ddot{y}_1 + 2\dot{y}_2 + y_1 = \dot{v}_1 + v_1 + v_2$
$\ddot{y}_2 + \dot{y}_1 + y_2 = v_2 + v_1$

(*d*) $\ddot{y}_1 + 3\dot{y}_1 + 2y_1 = 3\dot{v}_1 + 4\dot{v}_2 + 8v_2$
$\dot{y}_2 + 3y_2 - 4y_1 - \dot{y}_1 = \dot{v}_1 + 2\dot{v}_2 + 2v_2$

5.6 Write the vector matrix equations for the system shown in Fig. P5.4 in terms of the state variables indicated.

5.7 Write the vector matrix equations for the systems of Problem 5.5.

5.8 Find the vector matrix equations for the following systems using the partial fraction technique. Show that these equations can be determined from the equations obtained by simulation techniques by a coordinate transformation.

(*a*) $\ddot{y} + 3\dot{y} + 2y = v$

(*b*) $\dddot{y} + 4\ddot{y} + 5\dot{y} + 2y = v$

(*c*) $\dddot{y} + 4\ddot{y} + 6\dot{y} + 4y = v$

(*d*) $\ddot{y}_1 - 10\dot{y}_2 + y_1 = v_1$
$\dot{y}_2 + 6y_2 = v_2$

5.9 Find the general expression for the elements of the $\mathbf{A}(t)$ matrix for an *RLC* network if the indicators and capacitors are functions of time.

5.10 Given the system defined by the time-varying differential equation

$$p^n y + \sum_{k=0}^{n-1} \alpha_{n-k}(t) p^k y = \sum_{k=0}^{n} \beta_{n-k}(t) p^k v$$

Show that this system has the state equations

$$\begin{bmatrix} \dot{x}_1 \\ \dot{x}_2 \\ \cdot \\ \cdot \\ \cdot \\ \dot{x}_n \end{bmatrix} = \begin{bmatrix} 0 & 1 & 0 & \cdots & 0 \\ 0 & 0 & 1 & \cdots & 0 \\ \cdot & \cdot & \cdot & \cdots & \cdot \\ \cdot & \cdot & \cdot & \cdots & \cdot \\ \cdot & \cdot & \cdot & \cdots & 1 \\ -\alpha_n & -\alpha_{n-1} & \cdot & \cdots & -\alpha_1 \end{bmatrix} \begin{bmatrix} x_1 \\ x_2 \\ \cdot \\ \cdot \\ \cdot \\ x_n \end{bmatrix} + \begin{bmatrix} b_1 \\ b_2 \\ \cdot \\ \cdot \\ \cdot \\ b_n \end{bmatrix} v$$

$$y = x_1 + b_0 v$$

where

$$b_0(t) = \beta_0(t)$$

$$b_i(t) = \beta_i(t) - \sum_{r=0}^{i-1} \sum_{m=0}^{i-r} \binom{n+m-i}{n-i} \alpha_{i-r-m}(t) p^m b_r(t)$$

5.11 Find the simulation diagram for the system defined by the time-varying differential equation

$$t^2 \frac{d^3y}{dt^2} + (\cos t)\frac{d^2y}{dt^2} + 2\frac{dy}{dt} + (\sin t)y = t^3\frac{d^2v}{dt^2} + (\cos t)\frac{dv}{dt} + v$$

Write the vector matrix equation $\dot{\mathbf{x}} = \mathbf{A}\mathbf{x} + \mathbf{B}\mathbf{v}$.

5.12 Using the partial fractions expansion technique (neglect d_0), but letting

$$Y(s) = \sum_{i=1}^{n} X_i(s)$$

where $x_i(t)$ satisfies the differential equation $\dot{x}_i - \lambda_i x_i = b_i v(t)$, find the corresponding **A**, **B**, and **C**.

5.13 Repeat Problem 5.12 for the case where there is a repeated pole of order k. Show that the ones of the Jordan matrix are replaced by b_i/b_{i-1}, $i = 2, \ldots, k$.

5.14 Find the state transition matrix by each of the five methods presented for the following systems.

(*a*) $\ddot{y} + 4\dot{y} + 3y = 0$

(*b*) $\ddot{y} + 2\dot{y} + y = 0$

(*c*) $\ddot{y} + 2\dot{y} + 2y = 0$

(*d*) $\dddot{y} + 4\ddot{y} + 6\dot{y} + 4y = 0$

5.15 Assume that a unit step is applied to the systems described by Problem 5.14*a*, *b*, and *c*. What is the output $y(t)$?

5.16 Assume that an input, cos ωt, is applied to the system of Problem 5.14*c*. What is the output $y(t)$? For what value of ω does $y(t)$ have its largest peak amplitude? Does $y(t)$ ever become unbounded?

5.17 In the vicinity of the operating point ($N = 1$, $C = 1$, $R = 0$) of a nuclear reactor, it can be shown that the linearized dynamic equations are given by:

$$\dot{x}_1 = -x_1 + x_2 + x_3$$

$$\dot{x}_2 = x_1 - x_2$$

$$\dot{x}_3 = Kx_1$$

where $x_1 = N - 1$, $x_2 = C - 1$, $x_3 = R$, and N = normalized power level, R = normalized reactivity, C = normalized delayed neutron concentration, K = normalized temperature coefficient of reactivity. Find the state transition matrix for the linearized set of equations. For what values of K is this system stable? For a given set of initial conditions, what can be said about the relative magnitudes of the unforced responses $N(t)$, $R(t)$, and $C(t)$?

5.18 Show, by substituting Eq. 5.6-3 into Eq. 5.6-1 and utilizing Eq. 5.6-2 and the fact that the characteristic vectors are independent, that the general form for the $\alpha_i(t)$ in Eq. 5.6-3 is given by $c_i\epsilon^{\lambda_i t}$.

5.19 Show that, by substituting Eq. 5.6-3 into Eq. 5.6-1 and utilizing the reciprocal basis, the result of Problem 5.18 can be obtained.

5.20 For the system defined by the differential equation $\ddot{y} + 5\dot{y} + 6y = f(t)$ carry through the steps outlined in Example 5.6-1. Verify your result by utilizing a Laplace transform method. What should be the forcing function so that only one mode is excited?

5.21 Prove Eq. 5.6-10 by decomposing $\mathbf{f}(t)$ into an infinite sum of unit impulse functions.

5.22 Prove Eq. 5.6-14.

5.23 Assume that the forcing function to a linear fixed system is given by

$$\mathbf{f}(t) = \text{Re}\,[\mathbf{g}\epsilon^{j\omega t}]$$

where $\mathbf{g}$ is a real vector and ω is the angular frequency of the forcing function. Show that the output of the system is given by

$$\mathbf{x}(t) = \text{Re} \sum_{i=1}^{n} \frac{\langle \mathbf{r}_i, \mathbf{g}\rangle}{\lambda_i - j\omega} (\epsilon^{\lambda_i t} - \epsilon^{j\omega t})\mathbf{u}_i$$

If all the λ_i's have negative real parts, show that in the steady state the peak amplitude of oscillation of the ith mode is given by

$$\left| \frac{\langle \mathbf{r}_i, \mathbf{g}\rangle}{j\omega - \lambda_i} \right|$$

for real λ_i's, and is equal to

$$\left| \frac{\langle \mathbf{r}_i, \mathbf{g}\rangle}{j\omega - \lambda_i} \mathbf{u}_i + \frac{\langle \mathbf{r}_i^*, \mathbf{g}\rangle}{j\omega - \lambda_i^*} \mathbf{u}_i^* \right|$$

in the $\mathbf{u}'\mathbf{u}''$ plane for complex λ_i.

5.24 Using the results of Problem 5.23, show that, when λ_i is complex ($\lambda_i = -\alpha_i + j\beta_i$) and $\beta_i \gg \alpha_i$ (high Q), for $\beta_i \approx \omega$ the peak amplitude in the steady state is given by the expression for the peak amplitude when the λ_i's are real and negative. What can be said about the requirements for obtaining the largest steady-state oscillation for a particular mode?

5.25 (*a*) Show that $\epsilon^{\mathbf{A}t}$ can be represented by

$$\sum_{i=1}^{n} \epsilon^{\lambda_i t}\mathbf{u}_i\rangle\langle\mathbf{r}_i$$

if $\mathbf{A}$ has distinct characteristic values. This is called the *spectral representation* of $\epsilon^{\mathbf{A}t}$.

(*b*) Since $\mathbf{x}(t) = \epsilon^{\mathbf{A}t}\,\mathbf{x}(0)$, for an unforced system, show that $\mathbf{x}(t)$ is equal to

$$\sum_{i=1}^{n} \langle \mathbf{r}_i, \mathbf{x}(0)\rangle \epsilon^{\lambda_i t}\mathbf{u}_i$$

5.26 (*a*) Show that the state transition matrix $\epsilon^{\mathbf{A}t}$ can be found by the use of the modal matrix as $\epsilon^{\mathbf{A}t} = \mathbf{M}\epsilon^{\mathbf{\Lambda}t}\mathbf{M}^{-1}$, where $\mathbf{A}$ is a constant matrix. Why is the transformation $\mathbf{x} = \mathbf{M}(t)\mathbf{q}$ not particularly useful for the system $\dot{\mathbf{x}} = \mathbf{A}(t)\mathbf{x}$?

(*b*) Using the result of part (*a*), show that the complete solution for the matrix differential equation $\dot{\mathbf{x}} = \mathbf{A}\mathbf{x} + \mathbf{B}\mathbf{v}$ is given by

$$\mathbf{x}(t) = \int_{-\infty}^{t} \mathbf{M}\epsilon^{\mathbf{\Lambda}(t-\lambda)}\mathbf{M}^{-1}\mathbf{B}\mathbf{v}(\lambda)\,d\lambda$$

(*c*) Using the result of part (*b*) and the fact that the columns of $\mathbf{M}$ are the characteristic vectors $\mathbf{u}_i$ and the rows of $\mathbf{M}^{-1}$ are the reciprocal basis vectors $\mathbf{r}_i$, show that

$$\mathbf{x}(t) = \int_{-\infty}^{t} \sum_{i=1}^{n} \langle \mathbf{r}_i, \mathbf{B}\mathbf{v}(\lambda)\rangle \epsilon^{\lambda_i(t-\lambda)}\mathbf{u}_i\,d\lambda$$

• **5.27** The transform of the state transition matrix of a fixed system is given by $\mathbf{\Phi}(s) = [s\mathbf{I} - \mathbf{A}]^{-1}$. Why is the inverse state transition matrix transform $\mathbf{\Phi}^{-1}(s)$ *not* equal to $[s\mathbf{I} - \mathbf{A}]$?

5.28 For the system shown in Fig. P5.28, solve for the initial condition response $\mathbf{x}(t)$ using the mode expansion technique, and check using the state transition matrix.

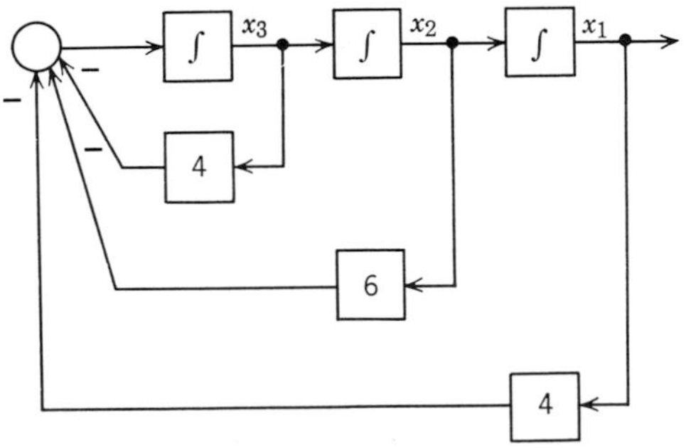

Fig. P5.28

5.29 Consider the constant resistance network shown in Fig. P5.29, where $L(t)/C(t) = R^2$; $L(t), C(t) > 0$. Let $v(t)$ be a voltage source, x_1 be the flux linkages of the inductor ($x_1 = Li_L$), x_2 be the charge across the capacitor ($x_2 = Cv_c$), $y(t)$ be the current to the network. Choose as state variables $x_1' = (x_1 + x_2)/2$, $x_2' = (x_1 - x_2)/2$. Write the differential equations for x_1' and x_2' and show that x_1' is S^c and x_2' is S^o.

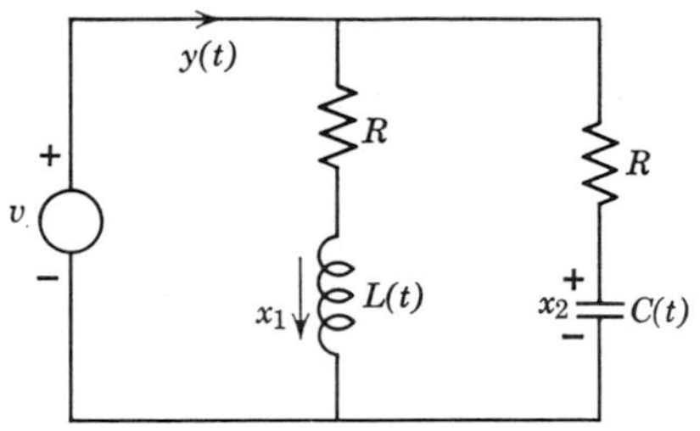

Fig. P5.29

5.30 Consider the following two subsystems, both of which are observable and controllable: $\dot{x}_1 = \alpha_1 x_1 + v(t)$, $\dot{x}_2 = \alpha_2 x_2 + v(t)$. If these two systems are connected in parallel, such that the output $y = x_1 - x_2$, what are the conditions required for the overall system to be both observable and controllable?

5.31 Given the cascade connection shown in Fig. P5.31, show that[8]

(*a*) $n = n_a + n_b$

(*b*) $\lambda_1, \ldots, \lambda_n = \lambda_{1a}, \ldots, \lambda_{na}, \lambda_{1b}, \ldots, \lambda_{nb}$

(*c*) A necessary (but insufficient) condition for the controllability (observability) of S is that both S_a and S_b be controllable (observable).

(*d*) If S_a and S_b are both controllable (observable), any uncontrollable (unobservable) modes of S must originate in $S_b(S_a)$.

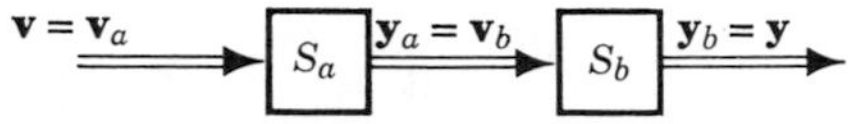

Fig. P5.31

5.32 Given the parallel connection shown in Fig. P5.32, show that[8]

(*a*) $n = n_a + n_b$

(*b*) $\lambda_1, \ldots, \lambda_n = \lambda_{1a}, \ldots, \lambda_{na}, \lambda_{1b}, \ldots, \lambda_{nb}$

(*c*) A necessary and sufficient condition that S be controllable (observable) is that both S_a and S_b be controllable (observable).

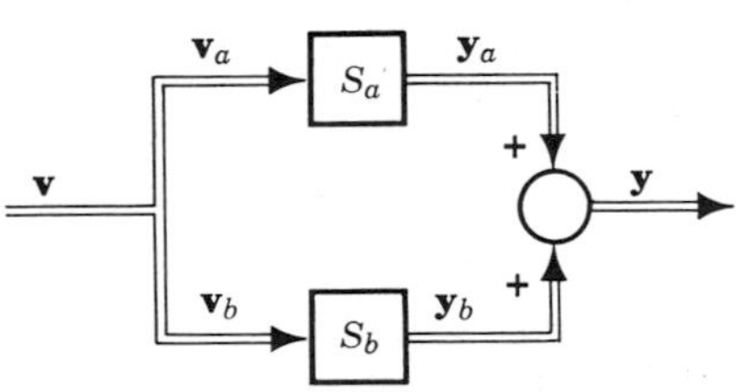

Fig. P5.32

5.33 Show that the transfer function $\mathbf{H}(s)$ of a system S can be written as[8]

$$\mathbf{H}(s) = \sum_{i=1}^{n^*} \frac{\mathbf{K}_i}{s - \lambda_i^*} + \mathbf{D}$$

where the * indicates S^*, and the matrices $\mathbf{K}_i$ have rank one. *Hint:* Start with the expression $\mathbf{H}(s) = \mathbf{C\Phi B} + \mathbf{D}$ where $\mathbf{\Phi} = [s\mathbf{I} - \mathbf{A}]^{-1}$.

5.34 Consider the system

$$\dot{\mathbf{x}} = \begin{bmatrix} -3 & 1 \\ 1 & -3 \end{bmatrix} \mathbf{x} + \begin{bmatrix} 1 & 1 \\ 1 & 1 \end{bmatrix} \mathbf{v}, \quad \mathbf{y} = \begin{bmatrix} 1 & 1 \\ 1 & -1 \end{bmatrix} \mathbf{x}$$

(*a*) Find the response of $\mathbf{x}$ to $\mathbf{x}(0)$, $\mathbf{v}(t) = \mathbf{0}$, in terms of the "modes" of the system. Sketch the trajectories for various initial conditions in both the (q_1, q_2) modal plane, and the x_1, x_2 plane.

(*b*) Comment on observability and controllability.

5.35 Consider the linear fixed system

$$\dot{\mathbf{x}} = \mathbf{Ax} + \mathbf{Bv}$$
$$\mathbf{y} = \mathbf{Cx} + \mathbf{Dv}$$

The characteristic values of $\mathbf{A}$ are not distinct.

(*a*) Suppose that the Jordan form for $\mathbf{A}$ is $\mathbf{M}^{-1}\mathbf{AM} = \mathbf{J} =$ a (2 x 2) diagonal matrix whose diagonal elements are equal; and that $\mathbf{B}$ is (2 x 2), $\mathbf{C}$ is (2 x 2), $\mathbf{D} = [0]$. Draw a block diagram for the system in terms of the new state variables $\mathbf{q} = \mathbf{M}^{-1}\mathbf{x}$. What simplification results if $\mathbf{C}$ is (1 x 2)? Draw a simplified block diagram for

$$\mathbf{A} = \begin{bmatrix} -1 & 0 \\ 0 & -1 \end{bmatrix}, \quad \mathbf{B} = \begin{bmatrix} 1 \\ 1 \end{bmatrix}, \quad \mathbf{C} = [1 \quad -2]$$

(*b*) Suppose that the Jordan form for $\mathbf{A}$ is $\mathbf{M}^{-1}\mathbf{AM} = \mathbf{J} =$ a (2 x 2) matrix with a one on the superdiagonal (and equal diagonal elements). $\mathbf{B}$ is (2 x 1), $\mathbf{C}$ is (1 x 2), $\mathbf{D} = [0]$. Draw a block diagram for the system in terms of new state variables $\mathbf{q} = \mathbf{M}^{-1}\mathbf{x}$.

(*c*) Draw a block diagram for

$$\mathbf{A} = \begin{bmatrix} -1 & 1 \\ 0 & -1 \end{bmatrix}, \quad \mathbf{B} = \begin{bmatrix} 1 \\ 0 \end{bmatrix}, \quad \mathbf{C} = [0 \quad 1]$$

Comment on observability and controllability for this system.

5.36 Consider the unstable system

$$\dot{x} = x + v \qquad |v| \leqslant 1$$
$$y = x$$

(*a*) Find the region of state space for which the system is controllable.

(*b*) Find the minimum time required to return a state in the controllable region to the origin.

5.37 Consider the system

$$\dot{x}_1 = -\tfrac{7}{3}x_1 + \tfrac{1}{3}x_2 + \tfrac{4}{3}x_3 + 2v(t)$$
$$\dot{x}_2 = \tfrac{2}{3}x_1 - \tfrac{8}{3}x_2 - \tfrac{2}{3}x_3$$
$$\dot{x}_3 = x_1 - x_2 - 2x_3 - v(t)$$
$$y(t) = \tfrac{2}{3}x_1 + \tfrac{1}{3}x_2 + \tfrac{1}{3}x_3$$

(*a*) Find the state transition matrix $\boldsymbol{\phi}(t)$.

(*b*) Suppose $v(t) = 0$ and $x_1(0)$, $x_2(0)$, $x_3(0)$ are given. Write $\mathbf{x}(t)$ in terms of the modes of the system.

(*c*) Make a change of state variable $\mathbf{x} = \mathbf{Mq}$ such that the new state variables $\mathbf{q}$ are "uncoupled." Draw a block diagram for the system in terms of the "new" state variables.

(*d*) Comment on observability and controllability, and separate the system into S^*, S^o, S^c, and S^f.

5.38 Consider the system of Fig. P5.38.

(*a*) Write the vector matrix differential equation describing the system, defining the output of the integrators as the state variables.

(*b*) *a* and *b* are constant parameters. Sketch those areas in the *a*, *b* parameter plane for which the system is completely controllable, partially controllable, and completely uncontrollable.

(*c*) Assume that *a* is fixed. Sketch the areas in the *b*, *c* parameter plane for which the system is completely observable, partially observable, and completely unobservable.

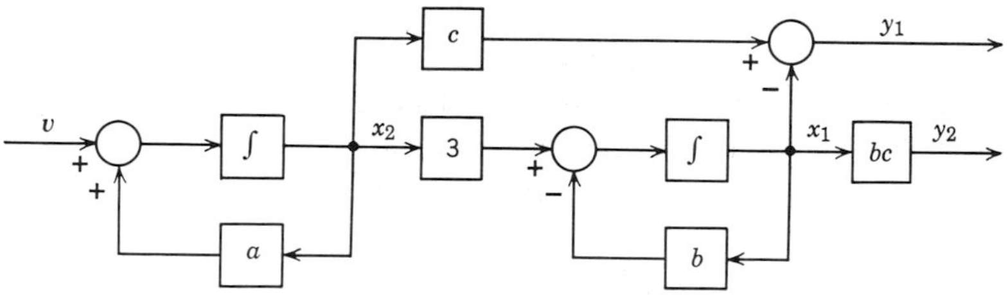

Fig. P5.38

5.39 Find the state transition matrix for the following systems.

(*a*) $\dot{y} + ty = v$

(*b*) $\dot{y} + y/t = v$

(*c*) $\ddot{y} + t\dot{y} + y = v$

(*d*) $t^2\ddot{y} + t\dot{y} + y = v$

(*e*) $\ddot{y} - (1/t)\dot{y} + [1 + (1/t^2)]y = v$

5.40 Prove that if $\boldsymbol{\phi}(t, \tau)$ is a solution to the matrix equation

$$\dot{\mathbf{x}} = \mathbf{A}(t)\mathbf{x}$$

then

$$|\boldsymbol{\phi}(t, \tau)| = |\boldsymbol{\phi}(0, \tau)| \exp \int_0^t \text{trace } \mathbf{A}(\lambda)\, d\lambda$$

From this, what can be concluded about the nonsingularity of $\boldsymbol{\phi}(t, \tau)$ so that its inverse exists? What does this imply about the independence of solutions? *Hint:* Find the Wronskian, $w(t)$, and show that the differential equation for the Wronskian ($w(t) = |\boldsymbol{\phi}(t, \tau)|$) is given by $\dot{w} = (\text{trace } \mathbf{A})w$.

5.41 Find $\boldsymbol{\phi}^{-1}(t, \tau)$ for the systems of Problem 5.39. Show that $\boldsymbol{\phi}^{-1}(t, \tau) = \boldsymbol{\phi}(\tau, t)$. Show that the use of the matrix notation and that of a linear differential adjoint operator are equivalent.

5.42 What physical significance can be attached to the interchange of rows and columns as expressed in Eq. 5.9-29?

5.43 Show that, if $\mathbf{C}$ is a nonsingular matrix and T is a real number $\neq 0$, then there exists a matrix $\mathbf{B}$ such that $\mathbf{C} = \exp \mathbf{B}T$ for a periodic system.

5.44 For the equation $\ddot{y} + p_1(t)\dot{y} + p_2(t)y = 0$ (Hill's equation), where $p_1(t)$ and $p_2(t)$ are continuous functions of time and periodic with period T, a solution which satisfies $y(t + T) = \lambda y(t)$, where λ is a *characteristic multiplier*, is called a normal solution. If $y_1(t)$ and $y_2(t)$ are a given base of real solutions to this equation, show that

$$y_1(t + T) = a_{11}y_1(t) + a_{12}y_2(t)$$
$$y_2(t + T) = a_{21}y_1(t) + a_{22}y_2(t)$$

so that

$$\begin{vmatrix} a_{11} - \lambda & a_{21} \\ a_{12} & a_{22} - \lambda \end{vmatrix} = 0$$

for the system to possess nontrivial solutions.

5.45 With the results of Problem 5.44, show that

$$\lim_{t \to \infty} |y(t)| = \infty, \quad |\lambda| > 1$$
$$\lim_{t \to \infty} |y(t)| = 0, \quad |\lambda| < 1$$

Periodic solutions exist for $\lambda = \pm 1$.

5.46 For the equation $\ddot{y} + p(t)y = 0$, where $p(t)$ is continuous with period T, what are the conditions that this equation admit of periodic solutions?

5.47 A state transition matrix of a system satisfies the equation

$$\boldsymbol{\phi}(t + T, \tau) = \boldsymbol{\phi}(t, \tau)\boldsymbol{\Lambda} \qquad (\boldsymbol{\Lambda} = \mathbf{C} = \epsilon^{\mathbf{B}T})$$

where $\boldsymbol{\Lambda}$ is a diagonal matrix consisting of characteristic multipliers. Show that periodic solutions exist only if the diagonal elements $\lambda_i = \pm 1$. If $\lambda_i = 1$, show that the period of the solution is T. If $\lambda_i = -1$, show that the period of the solution is $2T$.

5.48 Find the impulse response matrix for the systems of Problem 5.5.

5.49 Find the impulse response for the systems of Problem 5.14.

5.50 For the system with the following **A**, **B**, **C**, and **D** matrices, find the impulse response matrix by use of the simulation diagram. Check your solution using Eq. 5.11-5.

$$\mathbf{A} = \begin{bmatrix} -1 & 1 \\ -1 & -1 \end{bmatrix}, \quad \mathbf{B} = \begin{bmatrix} 2 \\ 1 \end{bmatrix}, \quad \mathbf{C} = [1 \quad 1], \quad \mathbf{D} = \begin{bmatrix} 3 \\ 0 \end{bmatrix}$$

5.51 Find the impulse response matrix for the *RLC* circuit shown in Fig. P5.51.

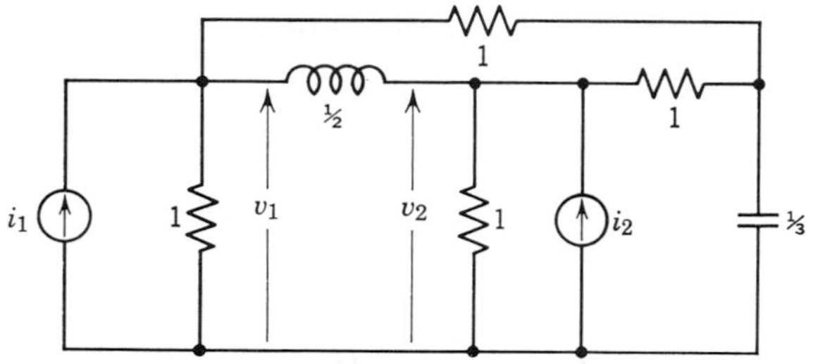

Fig. P5.51

5.52 Find the impulse response matrix $\mathbf{H}(t)$ for the system

$$\dot{\mathbf{x}} = \begin{bmatrix} -1 & 0 & 0 \\ 0 & -3 & 0 \\ 0 & 0 & -5 \end{bmatrix} \mathbf{x} + \begin{bmatrix} 1 & 0 \\ 0 & 1 \\ 1 & 1 \end{bmatrix} \mathbf{v}$$

$$\mathbf{y} = \begin{bmatrix} 1 & 1 & 0 \\ 0 & 1 & 1 \\ 0 & 0 & 1 \end{bmatrix} \mathbf{x}$$

5.53 Obtain Eq. 5.10-7 by the method of "variation of parameters" used in Section 5.8.

5.54 (*a*) Find $\mathbf{H}(s)$ for the following system and determine if the system is controllable and/or observable.

$$\dot{\mathbf{x}} = \begin{bmatrix} 0 & 1 & 0 \\ 5 & 0 & 2 \\ -2 & 0 & -2 \end{bmatrix} \mathbf{x} + \begin{bmatrix} -1 \\ 1 \\ -1 \end{bmatrix} v$$

$$\mathbf{y} = [-2 \quad 1 \quad 0]\,\mathbf{x}$$

(*b*) Show that the intersection of the controllable and observable spaces is the space generated by the transfer function matrix.

5.55 Show that an impulse response matrix $\mathbf{H}(t, \tau)$ is realizable by a finite dimensional dynamical system if and only if there exist two continuous matrices $\mathbf{P}(t)$ and $\mathbf{Q}(\tau)$ such that

$$\mathbf{H}(t, \tau) = \mathbf{P}(t)\mathbf{Q}(\tau) \quad \text{for all } t, \tau$$

Hint: See Section 2.8.

5.56 The impulse response matrix (for a fixed T) of the system

$$\dot{\mathbf{x}} = \mathbf{A}(t)\mathbf{x} + \mathbf{B}(t)\mathbf{v}$$
$$\mathbf{y} = \mathbf{C}(t)\mathbf{x}$$

is given by

$$\begin{aligned} \mathbf{H}(T, \tau) &= \mathbf{C}(T)\boldsymbol{\phi}(T, \tau)\mathbf{B}(\tau) \qquad && T \geqslant \tau \\ &= 0 && T < \tau \end{aligned}$$

• (*a*) Show that the impulse response matrix of the adjoint system

$$\begin{aligned} \dot{\boldsymbol{\alpha}} &= -\boldsymbol{\alpha}\mathbf{A}(t) + \mathbf{v}\mathbf{C}(t) \qquad && (\boldsymbol{\alpha} \text{ and } \mathbf{q} \text{ are row vectors}) \\ \mathbf{q} &= \boldsymbol{\alpha}\mathbf{B}(t) \end{aligned}$$

is given by

$$\begin{aligned} \mathbf{H}^*(t, T) &= \mathbf{C}(T)\boldsymbol{\phi}(T, t)\mathbf{B}(t) \qquad && t \geqslant T \\ &= 0 && t < T \end{aligned}$$

(*b*) If the change in variable $t_1 = T - t$ is made for the adjoint system, and the change in variable $\tau_1 = T - \tau$ is made for the original system, show that

$$\begin{aligned} \mathbf{H}^*(t_1, 0) &= \mathbf{C}(T)\boldsymbol{\phi}(T, T - t_1)\mathbf{B}(T - t_1) \qquad && t_1 \geqslant 0 \\ &= [0] && t_1 < 0 \end{aligned}$$

and

$$\begin{aligned} \mathbf{H}(T, T - \tau_1) &= \mathbf{C}(T)\boldsymbol{\phi}(T, T - \tau_1)\mathbf{B}(T - \tau_1) \qquad && \tau_1 \geqslant 0 \\ &= [0] && \tau_1 < 0 \end{aligned}$$

(*c*) From the result of part *b*, show that observation of the response of the modified adjoint system over the t_1 axis (the running variable of the simulator) is identical with observing the cross-plot at time T of the response of the original system over the τ_1 axis. In addition, show that an impulse placed on one of the inputs to the adjoint system produces a *row* of the desired impulse response matrix.

(*d*) Show that the result of the preceding parts proves that the complete modified adjoint system is obtained by interchanging the inputs and outputs of the original system and making the change in variable $t = T - t_1$.

6

State Variables and Linear Discrete Systems

6.1 INTRODUCTION

The state variable viewpoint is applied to linear discrete time processes in this chapter. Much of the effort is directed toward sampled-data systems. It is also shown by example that the state variable approach is quite useful in dealing with linear sequential systems, which naturally arise out of the theory of coding. Viewed in this fashion, the state variable approach is a unifying concept, as both continuous and discrete systems fall within its general framework.

The theory of linear discrete time systems follows the theory of linear continuous systems closely. Therefore, much of what is said in this chapter is based on the preceding chapters. The similarity between the theory of linear continuous systems and the theory of linear discrete systems is indicated.

6.2 SIMULATION DIAGRAMS

The basic building blocks required to construct a block diagram of a system described by linear difference equations are the adder, the amplifier, and the unit delay. The adder and the amplifier are the same blocks that were used for continuous systems, and the unit delay for difference equations is somewhat analogous to the integrator for differential equations. It is shown in Fig. 6.2-1. The input to the unit delay is $y(kT + T)$, and the corresponding output is $y(kT)$. Thus the input to the unit delay appears at its output one period later, or delayed

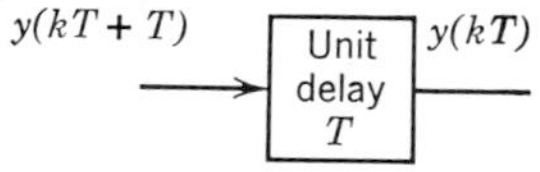

Fig. 6.2-1

by T. It should be noted that unit delays of the order generally required in control systems are quite difficult to obtain in practice, and seldom is a system actually simulated in real time by this method. For sequential circuits, where the variables are either binary in nature or take on discrete values, this type of simulation can be performed in real time.

The approach used to generate a block diagram of a linear difference equation is to assume that the variable $y(nT + kT)$ is available, and then successively pass this variable through unit delays until $y(kT)$ is obtained. The block diagram then is completed by satisfying the requirements of the difference equation, or "closing the loop."

Example 6.2-1. Find the simulation diagram for the system governed by the difference equation $y(kT + 2T) + ay(kT + T) + by(kT) = v(kT)$.

The first step is to solve for $y(kT + 2T)$, as

$$y(kT + 2T) = v(kT) - ay(kT + T) - by(kT)$$

The terms $y(kT + T)$ and $y(kT)$ are obtained as shown in Fig. 6.2-2a. Assuming ideal distortionless delays, a signal which appears at terminal 1 appears at terminal 2 one time period later, and at terminal 3 two time periods later. Similarly, a signal at terminal 2 appears at terminal 3 one time period later. The completed block diagram (Fig. 6.2-2b) is obtained by satisfying the requirements of the difference equation.

If the initial conditions are given in terms of $y(0)$ and $y(T)$, then $y(0)$ is the initial signal at the output of the first delay, and $y(T)$ is the initial output of the second delay. After one time period, $y(T)$ appears at the output of the first delay unit. After two time periods, the output of the first delay is $y(2T) = v(0) - ay(T) - by(0)$.

If a comparison is made between Figs. 5.2-2b and 6.2-2b, it is evident that similar rules hold for constructing block diagrams of difference equations and differential equations. The integrator used in simulating differential equations is analogous to the unit delay used in simulating difference equations.

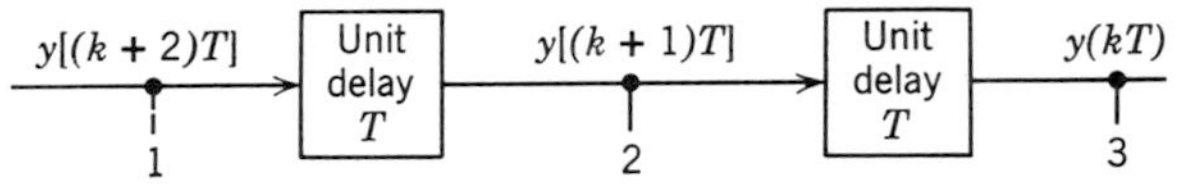

(a)

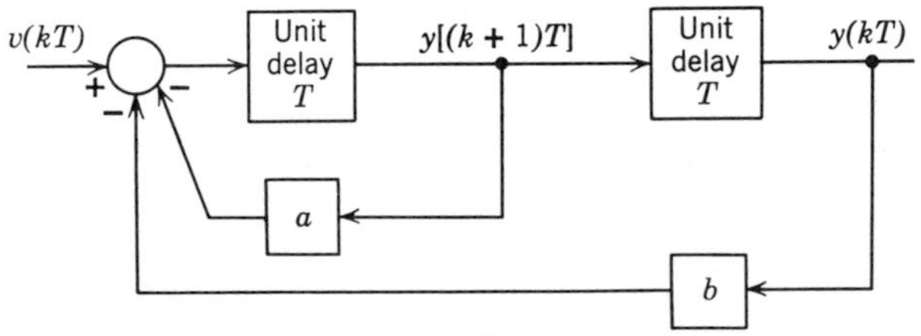

(b)

Fig. 6.2-2

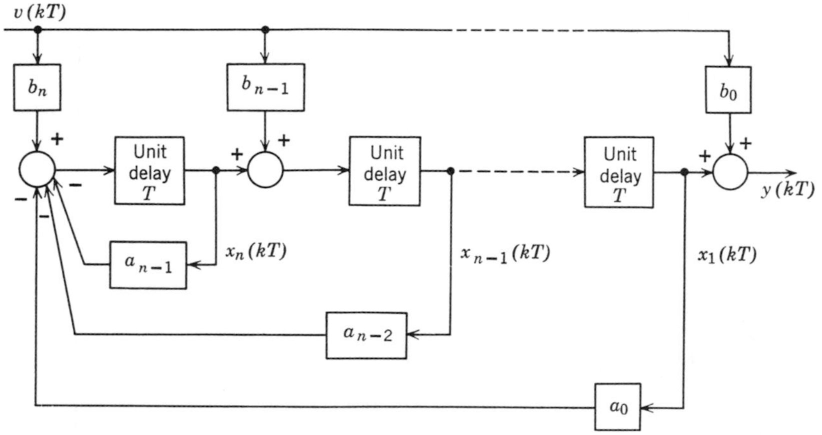

Fig. 6.2-3

Example 6.2-2. Find the simulation diagram for the *n*th order difference equation

$$y(nT + kT) + \alpha_{n-1}y(nT + kT - T) + \cdots + \alpha_0 y(kT)$$
$$= \beta_n v(nT + kT) + \beta_{n-1}v(nT + kT - T) + \cdots + \beta_0 v(kT)$$

The general simulation diagram for this system is shown in Fig. 6.2-3, by analogy to Fig. 5.5-4. The *a*'s and *b*'s of the block diagram are given by Eqs. 5.5-7 and 5.5-8.

For the specific case of the difference equation

$$y(kT + 3T) + 3y(kT + 2T) + 4y(kT + T) + y(kT) = 2v(kT + 3T)$$
$$+ 3v(kT + 2T) + v(kT + T) + 2v(kT)$$
$$b_0 = \beta_n = 2$$
$$b_1 = \beta_{n-1} - \alpha_{n-1}b_0 = -3$$
$$b_2 = \beta_{n-2} - \alpha_{n-1}b_1 - \alpha_{n-2}b_0 = 2$$
$$b_3 = \beta_{n-3} - \alpha_{n-1}b_2 - \alpha_{n-2}b_1 - \alpha_{n-3}b_0 = 6$$

The simulation diagram for this system is shown in Fig. 6.2-4. A comparison of this

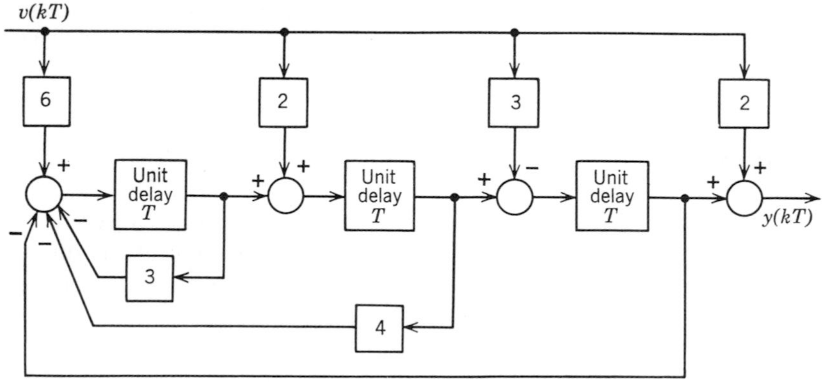

Fig. 6.2-4

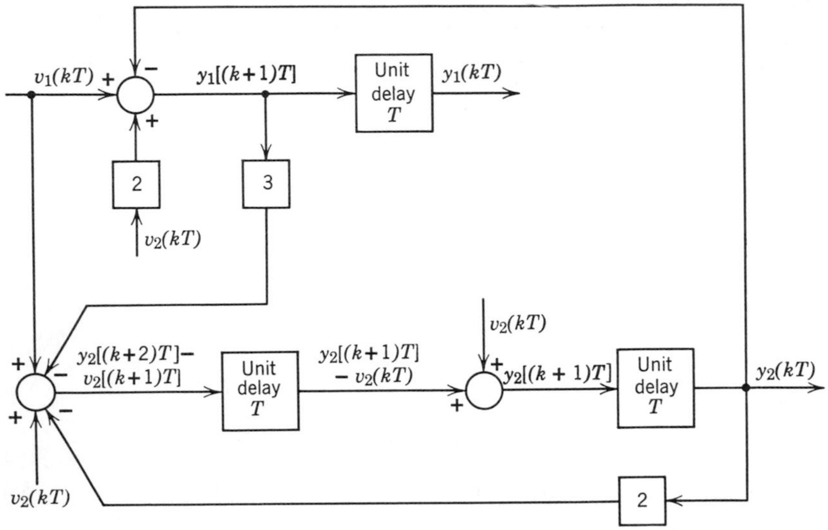

Fig. 6.2-5

diagram with Fig. 5.5-5 shows that the only difference between the two diagrams is that the integrators of Fig. 5.5-5 have been replaced by unit delays.

Example 6.2-3. Find the simulation diagram for the system governed by the difference equations

$$y_1(kT+T) + y_2(kT) = v_1(kT) + 2v_2(kT)$$
$$y_2(kT+2T) + 3y_1(kT+T) + 2y_2(kT) = v_2(kT+T) + v_2(kT) + v_1(kT)$$

These equations can be rewritten as

$$y_1(kT+T) = v_1(kT) + 2v_2(kT) - y_2(kT)$$
$$y_2(kT+2T) - v_2(kT+T) = v_2(kT) + v_1(kT) - 3y_1(kT+T) - 2y_2(kT)$$

Using $y_2(kT+2T) - v_2(kT+T)$ as the input to one delay chain, the block diagram appears as shown in Fig. 6.2-5. The approach is similar to that used for continuous systems.

6.3 TRANSFER FUNCTION MATRICES

The transfer function $H(z)$ for a single input-single output discrete time system is equal to the ratio of the Z transforms of the output and input of the system. For multivariable systems, the transfer function between various input-output terminals is similarly defined. Thus

$$H_{ij}(z) = \frac{Y_i(z)}{V_j(z)}; \quad V_k(z) = 0, \quad k \neq j \tag{6.3-1}$$

where $Y_i(z)$ is the Z transform of the output at terminal i, and $V_j(z)$ is the Z transform of the input at terminal j.

The transfer function matrix is simply the ordered array of these transfer functions, where i denotes the row and j denotes the column in which $H_{ij}(z)$ appears. If the transfer function matrix $\mathbf{H}(z)$ is known, then the output vector transform $\mathbf{Y}(z)$ is given by

$$\mathbf{Y}(z) = \mathbf{H}(z)\mathbf{V}(z) \qquad \text{(assuming zero initial conditions)} \qquad (6.3\text{-}2)$$

where $\mathbf{V}(z)$ is the column matrix of the Z transform of the input vector $\mathbf{v}(kT)$.

Example 6.3-1. Find the transfer function matrix for the two input-two output system described by the difference equations

$$y_1(kT + 3T) + 6y_1(kT + 2T) + 11y_1(kT + T) + 6y_1(kT) = v_1(kT + T) + v_1(kT) + v_2(kT)$$

$$y_2(kT + 2T) + 5y_2(kT + T) + 6y_2(kT) = v_2(kT + T) + v_2(kT)$$

Taking the Z transform of both sides of these equations, assuming zero initial conditions,

$$(z^3 + 6z^2 + 11z + 6)Y_1(z) = (z + 1)V_1(z) + V_2(z)$$

$$(z^2 + 5z + 6)Y_2(z) = (z + 1)V_2(z)$$

Since $z^3 + 6z^2 + 11z + 6 = (z + 1)(z + 2)(z + 3)$ and $z^2 + 5z + 6 = (z + 2)(z + 3)$,

$$H_{11}(z) = \frac{1}{(z + 2)(z + 3)} \qquad H_{12}(z) = \frac{1}{(z + 1)(z + 2)(z + 3)}$$

$$H_{21}(z) = 0 \qquad H_{22}(z) = \frac{z + 1}{(z + 2)(z + 3)}$$

or

$$\mathbf{H}(z) = \begin{bmatrix} \dfrac{1}{(z + 2)(z + 3)} & \dfrac{1}{(z + 1)(z + 2)(z + 3)} \\ 0 & \dfrac{(z + 1)}{(z + 2)(z + 3)} \end{bmatrix}$$

The transfer function block diagram appears in Fig. 6.3-1.

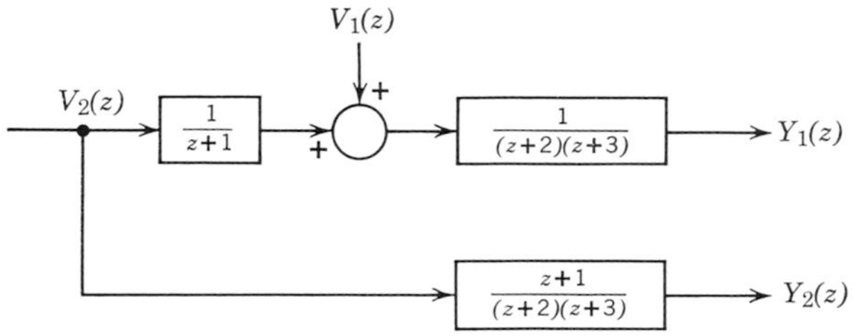

Fig. 6.3-1

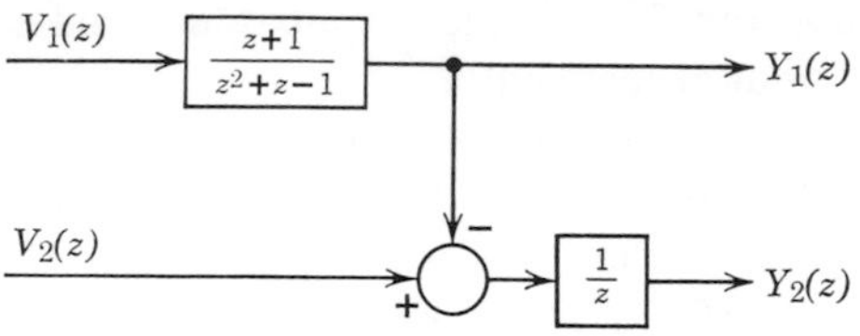

Fig. 6.3-2

Example 6.3-2. Find the transfer function matrix for the system governed by the difference equations

$$y_1(kT + 2T) + y_1(kT + T) + y_2(kT + T) = v_1(kT + T) + v_1(kT) + v_2(kT)$$
$$y_2(kT + T) + y_1(kT) = v_2(kT)$$

Transforming both sides of these equations, assuming zero initial conditions,

$$(z^2 + z)Y_1(z) + zY_2(z) = (z + 1)V_1(z) + V_2(z)$$
$$Y_1(z) + zY_2(z) = V_2(z)$$

Solving for $Y_1(z)$ and $Y_2(z)$,

$$Y_1(z) = \frac{(z + 1)}{z^2 + z - 1}V_1(z) \qquad Y_2(z) = \frac{V_2(z)}{z} - \frac{(z + 1)}{z(z^2 + z - 1)}V_1(z)$$

The transfer function matrix is

$$\mathbf{H}(z) = \begin{bmatrix} \dfrac{z + 1}{z^2 + z - 1} & 0 \\ \dfrac{-(z + 1)}{z(z^2 + z - 1)} & \dfrac{1}{z} \end{bmatrix}$$

The transfer function block diagram is shown in Fig. 6.3-2.

The unit delay, represented by the Laplace transform ϵ^{-sT}, corresponds to $1/z$ in the z domain. Thus the integrator, as represented by $1/s$ for continuous systems, has the unit delay as its analog in the diagram of discrete systems. Hence the transfer functions $1/(z + a)$ and $z/(z + a)$ are obtained in a manner similar to that used for continuous systems (see Fig. 6.3-3.).

The same note of caution that was injected into Section 5.3 should be added here. Inspection of only the transfer function matrix may lead to incorrect conclusions about the order of a system, as the transfer function relates only the controllable and observable aspects of the system. It may

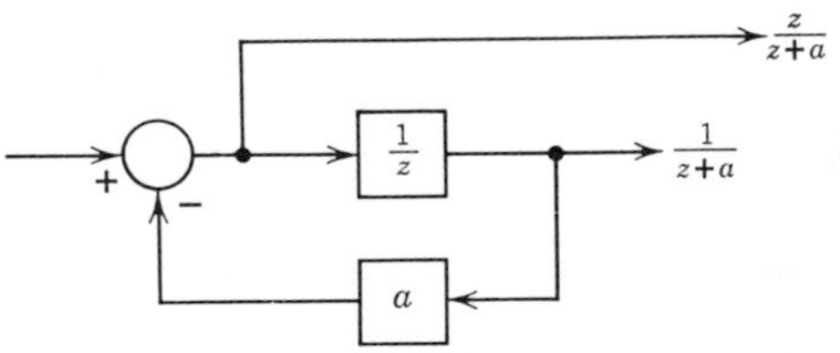

Fig. 6.3-3

mask system properties which would be obtained from the physical system.

6.4 THE CONCEPT OF STATE[1]

The state of a discrete time system can be intuitively defined as the minimum amount of information about the system which is necessary to determine both the output and the future states of the system, if the input function is known. More precisely, a set of variables **x** qualifies as a state vector, if two single-valued functions **f** and **g** can be found such that

$$\begin{aligned} \mathbf{x}(kT + T) &= \mathbf{f}[\mathbf{x}(kT), \mathbf{v}(kT)] \\ \mathbf{y}(kT) &= \mathbf{g}[\mathbf{x}(kT), \mathbf{v}(kT)] \end{aligned} \tag{6.4-1}$$

where $\mathbf{y}(kT)$ = the output vector at time kT, and $\mathbf{v}(kT)$ = the input vector at time kT. It is interesting to note that these requirements for the state of a sequential device were independently determined by Huffman and by Moore while working on related but different problems.[2,3]

In most of the subsequent material, the outputs of the delay elements of the simulation diagram of a system are taken as the state of the system. These outputs provide a convenient and sufficient choice for the state vector.

Example 6.4-1. The capacitor step charger of Fig. 6.4-1 is so designed that the voltage on the capacitor increases in steps each time the input pulse appears.[4] Assume that initially there is no charge on capacitor C_2. The first input pulse charges C_1 through diode D_1 to a voltage V_0. After the input pulse disappears, the charge on C_1 distributes itself between C_1 and C_2 according to the inverse ratio of the capacitances. The second pulse again charges C_1 to a voltage V_0, but the subsequent additional charge which is placed on C_2 is less than that from the first pulse, owing to the previous charge left on C_2. The input pulses are continued, and it is expected that the voltage changes appearing on capacitor C_2 diminish asymptotically. Evaluate these changes and verify that the voltage on C_2 is a state variable.

If $x(k)$ is the voltage on capacitor C_2 after the kth input pulse, then the difference equation for $x(k)$ is given by the principle of conservation of charge as $C_1V_0 + C_2x(k) = (C_1 + C_2)x(k + 1)$ or

$$x(k+1) - \left(\frac{C_2}{C_1 + C_2}\right)x(k) = \frac{C_1V_0}{C_1 + C_2}$$

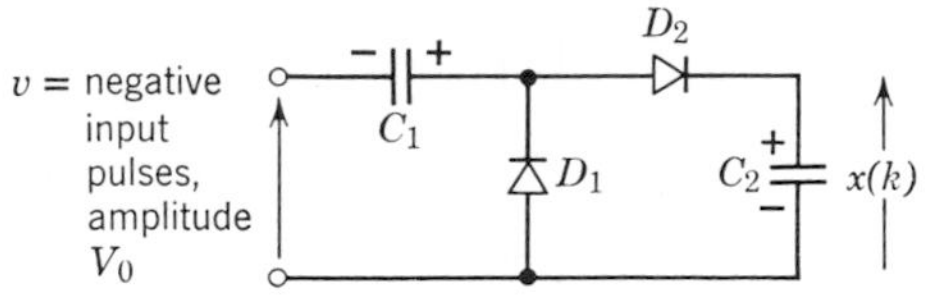

Fig. 6.4-1

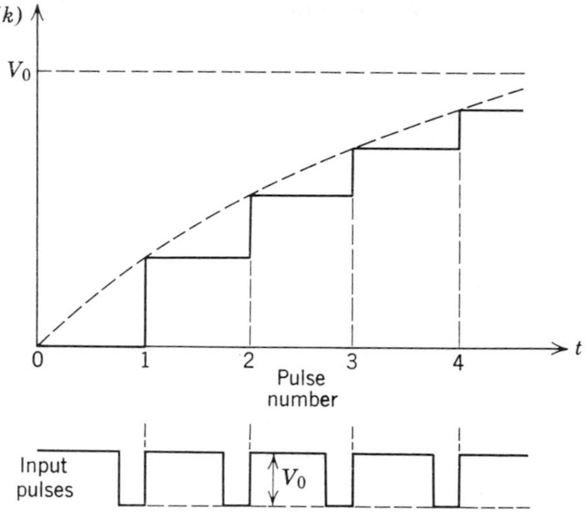

Fig. 6.4-2

Boolean Identities

And	Or
$0 \cdot 0 = 0$	$0 + 0 = 0$
$0 \cdot 1 = 0$	$0 + 1 = 1$
$1 \cdot 0 = 0$	$1 + 0 = 1$
$1 \cdot 1 = 1$	$1 + 1 = 1$
Complement	
$\bar{0} = 1$	$\bar{1} = 0$

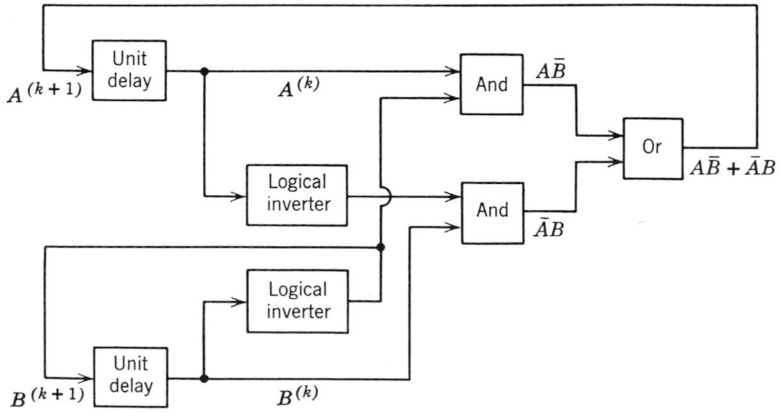

Fig. 6.4-3

Solving this equation in the standard fashion indicated in Section 2.12 yields

$$x(k) = V_0\left[1 - \left(\frac{C_2}{C_1 + C_2}\right)^k\right]$$

Therefore the capacitor voltage increases in discrete steps as shown in Fig. 6.4-2. The step heights are given by

$$x(k + 1) - x(k) = V_0\left(\frac{C_1}{C_1 + C_2}\right)\left(\frac{C_2}{C_1 + C_2}\right)^k$$

The state of this circuit is given by the capacitor voltage $x(k)$. Note that this state takes on only discrete values. The next state of the system, $x(k + 1)$, is uniquely determined by the present state of the system and the present input. The general state equations for this system are given by Eqs. 6.4-1. In this case, both **f** and **g** are linear, single-valued, scalar functions.

Example 6.4-2.[5] A modulo 4 counter is designed so that it cycles through the counts 00, 01, 10, 11, 00, ... The circuit for this device is shown in Fig. 6.4-3, and the reader familiar with sequential circuits can verify that this circuit does indeed cycle through these counts. The outputs of the delay elements represent the state of the system, as well as the outputs of the system. Write the state equations.

The logical equations for this counter are normally written in the form

$$A^{(k+1)T} = (A\bar{B} + \bar{A}B)^{kT}$$
$$B^{(k+1)} = (\bar{B})^{kT}$$

where the superscripts denote time instants and are not exponents. This is an autonomous sequential circuit, and the general state equations for such a circuit are

$$\mathbf{x}[(k + 1)T] = \mathbf{f}[\mathbf{x}(kT)] \qquad \mathbf{x} = \text{state (output of delay elements)}$$
$$\mathbf{y}[(kT)] = \mathbf{g}[\mathbf{x}(kT)] \qquad \mathbf{y} = \text{output (outputs of delay elements)}$$

6.5 MATRIX REPRESENTATION OF LINEAR STATE EQUATIONS

The general form of the state equations for a multivariable discrete time system were given by Eq. 6.4-1 as

$$\begin{aligned} \mathbf{x}[(k + 1)T] &= \mathbf{f}[\mathbf{x}(kT), \mathbf{v}(kT)] \\ \mathbf{y}(kT) &= \mathbf{g}[\mathbf{x}(kT), \mathbf{v}(kT)] \end{aligned} \tag{6.5-1}$$

If the system is linear, then Eq. 6.5-1 can be written as the set of linear vector-matrix difference equations

$$\begin{aligned} \mathbf{x}[(k + 1)T] &= \mathbf{A}(kT)\mathbf{x}(kT) + \mathbf{B}(kT)\mathbf{v}(kT) \\ \mathbf{y}(kT) &= \mathbf{C}(kT)\mathbf{x}(kT) + \mathbf{D}(kT)\mathbf{v}(kT) \end{aligned} \tag{6.5-2}$$

$\mathbf{A}(kT)$, $\mathbf{B}(kT)$, $\mathbf{C}(kT)$, and $\mathbf{D}(kT)$ have been indicated as time-varying matrices. If the system is nontime-varying, these matrices can be written

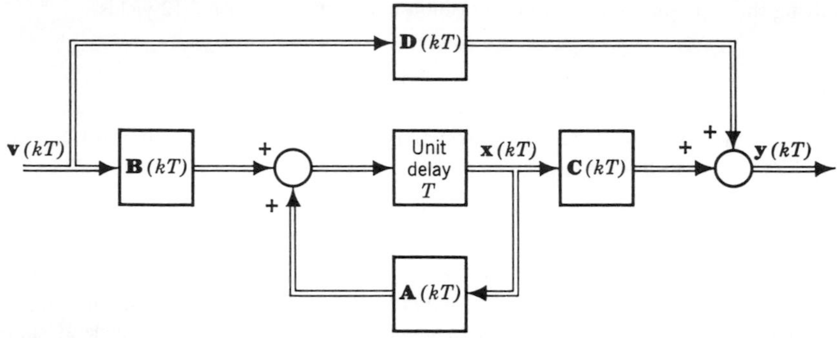

Fig. 6.5-1

as the constant matrices **A**, **B**, **C**, and **D**. The general block diagram, similar to Fig. 5.5-1, is shown in Fig. 6.5-1.

For a system described by a set of nth order difference equations, the form shown in Eq. 6.5-2 can be obtained by writing the given equations as a set of first order difference equations.

Example 6.5-1. Express the second order difference equation

$$y[(k+2)T] + ay[(k+1)T] + by(kT) = v(kT)$$

represented in Fig. 6.2-2*b* as a set of first order difference equations.

Let

$$y(kT) = x_1(kT)$$

$$y[(k+1)T = x_1[(k+1)T] = x_2(kT)$$

Then

$$\begin{bmatrix} x_1(kT+T) \\ x_2(kT+T) \end{bmatrix} = \begin{bmatrix} 0 & 1 \\ -b & -a \end{bmatrix} \begin{bmatrix} x_1(kT) \\ x_2(kT) \end{bmatrix} + \begin{bmatrix} 0 \\ 1 \end{bmatrix} v(kT)$$

$$y(kT) = [1 \quad 0] \begin{bmatrix} x_1(kT) \\ x_2(kT) \end{bmatrix}$$

The variables $x_1(kT)$ and $x_2(kT)$ are the outputs of the delay elements, and they represent the state of the system. These equations are of the general form of Eq. 6.5-2, where

$$\mathbf{A} = \begin{bmatrix} 0 & 1 \\ -b & -a \end{bmatrix}, \quad \mathbf{B} = \begin{bmatrix} 0 \\ 1 \end{bmatrix}, \quad \mathbf{C} = [1 \quad 0], \quad \mathbf{D} = [0]$$

Example 6.5-2. The general form for an nth order difference equation is given in Example 6.2-2. The simulation diagram appears in Fig. 6.2-3. Find the **A**, **B**, **C**, and **D** matrices.

Analogous to the continuous case, the **A**, **B**, **C**, **D** matrices for this system are given by Eq. 5.5-9.

Linear Binary Sequential Networks†

An interesting application of the state variable approach is the analysis of linear binary sequential networks. A linear binary sequential network consists of pure time delays and modulo 2 summing junctions. A modulo 2 summing junction is an exclusive-OR function having the logical equation $f = (x_1 + x_2)\overline{x_1 x_2} = \bar{x}_1 x_2 + x_1 \bar{x}_2$. The output of such a summer is zero if the two inputs x_1 and x_2 are the same, and one if the inputs are different. Although this discussion is limited to modulo 2 networks, the same type of analysis can be used for modulo p networks.[6–8] These linear modular sequential networks have found a limited application in error-correcting codes, computer circuits, and in certain types of radar systems.[9]

For the purpose of illustrating the use of the state variable approach in the analysis of linear binary sequential networks, consider the network of Fig. 6.5-2. This network consists of four unit delays (a four-element shift register) and a single modulo 2 summing junction (a logical exclusive-OR circuit). The equations of this network are

$$\begin{aligned} x_1(k+1) &= x_2(k) \\ x_2(k+1) &= x_3(k) \\ x_3(k+1) &= x_4(k) \\ x_4(k+1) &= x_4(k) \oplus x_3(k) \oplus x_1(k) \end{aligned}$$

In matrix form,

$$\mathbf{x}(k+1) = \begin{bmatrix} 0 & 1 & 0 & 0 \\ 0 & 0 & 1 & 0 \\ 0 & 0 & 0 & 1 \\ 1 & 0 & 1 & 1 \end{bmatrix} \mathbf{x}(k) = \mathbf{A}\mathbf{x}(k) \bmod 2$$

The sequence of states through which this network will pass is $\mathbf{x}(0)$, $\mathbf{A}\mathbf{x}(0)$, $\mathbf{A}^2\mathbf{x}(0)$, ... If $\mathbf{A}$ is nonsingular, then each state $\mathbf{x}(k)$ has a unique preceding state $\mathbf{x}(k-1)$. For a mod 2 network, the determinant $|\mathbf{A}|$ is either one or zero. In this particular example $|\mathbf{A}| = 1$, so that $\mathbf{A}$ is

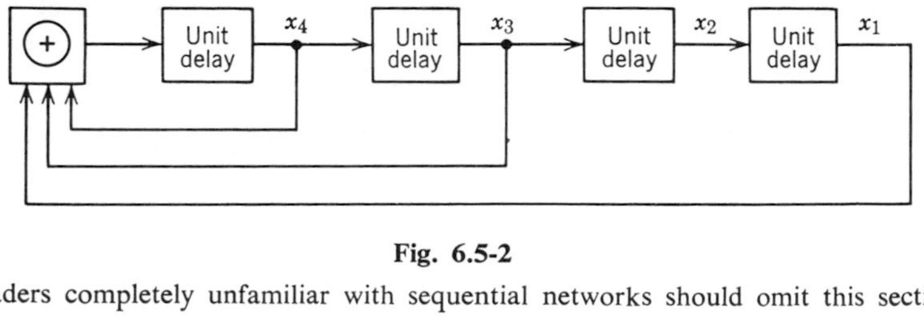

Fig. 6.5-2

† Readers completely unfamiliar with sequential networks should omit this section.

nonsingular. Hence the inverse $\mathbf{A}^{-1}$ does exist.

$$\mathbf{A}^{-1} = \begin{bmatrix} 0 & 1 & 1 & 1 \\ 1 & 0 & 0 & 0 \\ 0 & 1 & 0 & 0 \\ 0 & 0 & 1 & 0 \end{bmatrix}$$

Since there are four state variables, each of which can have the value zero or one, there are $2^4 = 16$ possible states of the system. Therefore the state sequence of an initial state is either periodic with period $P \leq 15$ or goes to an equilibrium state, where the equilibrium state is defined by $\mathbf{x}(k + 1) = \mathbf{x}(k)$. The trivial case where all the state variables are zero is called the *null* state. All other equilibrium states are called *finite* equilibrium states.

Consider the case in which the system reaches an equilibrium state. It is assumed that $\mathbf{A}$ is nonsingular, such that $\mathbf{A}^k$ is not equal to the null matrix. This removes the possibility of a nonzero initial state going to the null state. For the finite equilibrium case $\mathbf{x}(k + 1) = \mathbf{A}\mathbf{x}(k) = \mathbf{x}(k)$. Since $(\mathbf{A} - \mathbf{I})\mathbf{x}(k) = \mathbf{0}$, the existence of an equilibrium requires that the characteristic equation $|\lambda\mathbf{I} - \mathbf{A}| = 0$ have at least one unity root. For the network of Fig. 6.5-2, the characteristic equation is $\lambda^4 + \lambda^3 + \lambda^2 + 1 = 0$ mod 2. This can be factored into $(\lambda + 1)(\lambda^3 + \lambda + 1) = 0$ mod 2. Therefore the characteristic equation has one unity root, and there exist initial states for which this network goes to an equilibrium state. One of these initial states is $\mathbf{x}(0) = (1, 1, 1, 1)$.

When the network has a periodic state sequence of period P, then

$$\mathbf{x}(k + P) = \mathbf{A}^P\mathbf{x}(k) = \mathbf{x}(k) \bmod 2$$

For this condition, $|\lambda\mathbf{I} - \mathbf{A}^P| = 0$ mod 2, must have at least one unity root. The integer P can be found by determining the smallest integer such that $\mathbf{A}^P = \mathbf{I}$ mod 2. If the characteristic polynomial $p(\lambda)$ divides $\lambda^P - 1$ without remainder, then $\lambda^P - 1 = p(\lambda)q(\lambda)$, where $q(\lambda)$ is the quotient polynomial. Therefore $\mathbf{A}^P - \mathbf{I} = \mathbf{P}(\mathbf{A})\mathbf{Q}(\mathbf{A}) = [0]$ mod 2, and $\mathbf{A}^P = \mathbf{I}$ mod 2. Hence the technique to determine the length of the minimum periodic sequence of a network is to determine the smallest integer P such that $p(\lambda)$ divides $\lambda^P - 1$ without remainder. For this particular problem, the length of the minimum periodic sequence is seven. A typical sequence is given by

$$\begin{bmatrix} 0 \\ 0 \\ 0 \\ 1 \end{bmatrix} \rightarrow \begin{bmatrix} 0 \\ 0 \\ 1 \\ 1 \end{bmatrix} \rightarrow \begin{bmatrix} 0 \\ 1 \\ 1 \\ 0 \end{bmatrix} \rightarrow \begin{bmatrix} 1 \\ 1 \\ 0 \\ 1 \end{bmatrix} \rightarrow \begin{bmatrix} 1 \\ 0 \\ 1 \\ 0 \end{bmatrix} \rightarrow \begin{bmatrix} 0 \\ 1 \\ 0 \\ 0 \end{bmatrix} \rightarrow \begin{bmatrix} 1 \\ 0 \\ 0 \\ 0 \end{bmatrix}$$

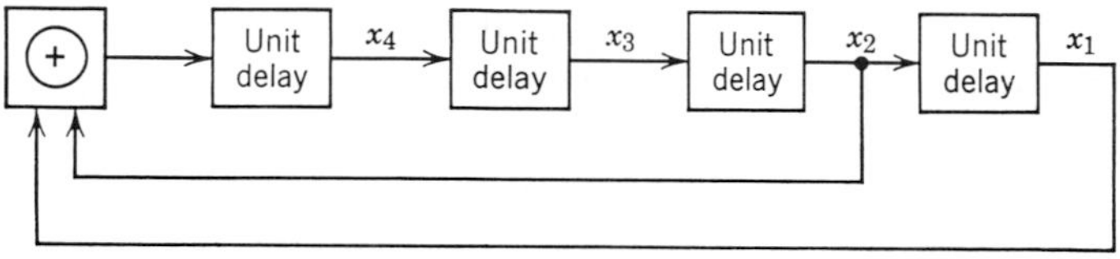

Fig. 6.5-3

The class of networks of order n, whose minimum periodic sequence is equal to $2^n - 1$, are known as *maximal-period* networks. The necessary and sufficient condition for a network to have only maximal-period sequences is that the characteristic polynomial be irreducible and not a divisor for $\lambda^k - 1$, $k < 2^n - 1$. (An irreducible polynomial $f(\lambda)$ is one which cannot be factored into the form $h(\lambda)g(\lambda)$, except for the trivial factorization when $g(\lambda)$ is a constant.) The network of Fig. 6.5-3 has only maximal-period sequences. The equations of this network are

$$\mathbf{x}(k+1) = \begin{bmatrix} 0 & 1 & 0 & 0 \\ 0 & 0 & 1 & 0 \\ 0 & 0 & 0 & 1 \\ 1 & 1 & 0 & 0 \end{bmatrix} \mathbf{x}(k)$$

$$p(\lambda) = |\lambda \mathbf{I} - \mathbf{A}| \bmod 2$$
$$= \lambda^4 + \lambda^3 + 1 = 0 \bmod 2$$

Since $\lambda^4 \oplus \lambda^3 \oplus 1$ is an irreducible polynomial, and since it is a prime factor of $\lambda^{15} - 1$, the minimum length of the periodic sequence is fifteen. The prime factors of $\lambda^{15} - 1$ are

$$\lambda^{15} - 1 = (\lambda^4 + \lambda^3 + 1)(\lambda^4 + \lambda + 1)(\lambda^4 + \lambda^3 + \lambda^2 + \lambda + 1) \times (\lambda^2 + \lambda + 1)(\lambda - 1)$$

Since the cycle structure of such a network is completely determined by the characteristic polynomial (assuming that the network matrix $\mathbf{A}$ is non-singular and that the characteristic polynomial is also the minimum polynomial of the network), it is possible to synthesize these linear sequential networks analogously to ordinary lumped element network synthesis. A set of synthesis procedures is given in Reference 6.

State Equations-Partial Fractions Technique

The partial fractions technique for deriving the state equations for a linear, fixed, discrete time process follows the same procedure as that used for continuous systems. For a single input-output system with transfer function $H(z)$, the transform of the output $Y(z)$ is given by $Y(z) = H(z)V(z)$. If the denominator polynomial of the transfer function $H(z)$ has distinct roots $\lambda_1, \lambda_2, \ldots, \lambda_n$, and if the order of the numerator polynomial of $H(z)$ is less than the order of its denominator polynomial, then $H(z)$ can be written as the partial fraction expansion

$$H(z) = \sum_{i=1}^{n} \frac{c_i}{z - \lambda_i}$$

The output $Y(z)$ is given by

$$Y(z) = \sum_{i=1}^{n} \frac{c_i}{z - \lambda_i} V(z)$$

Therefore $y(kT)$ can be written as a sum of terms of the form

$$y(kT) = \sum_{i=1}^{n} c_i x_i(kT)$$

where the $x_i(kT)$ must satisfy the first order difference equation

$$x_i[(k+1)T] = \lambda_i x_i(kT) + v(kT)$$

The state equations for the system can then be written in the form

$$\begin{bmatrix} x_1[(k+1)T] \\ x_2[(k+1)T] \\ \cdot \\ \cdot \\ \cdot \\ x_n[(k+1)T] \end{bmatrix} = \begin{bmatrix} \lambda_1 & 0 & \cdots & 0 \\ 0 & \lambda_2 & \cdots & 0 \\ \cdots & \cdots & \cdots & \cdots \\ 0 & 0 & \cdots & \lambda_n \end{bmatrix} \begin{bmatrix} x_1(kT) \\ x_2(kT) \\ \cdot \\ \cdot \\ \cdot \\ x_n(kT) \end{bmatrix} + \begin{bmatrix} 1 \\ 1 \\ \cdot \\ \cdot \\ \cdot \\ 1 \end{bmatrix} v(kT)$$

$$[y(kT)] = [c_1 \quad c_2 \quad \cdots \quad c_n] \begin{bmatrix} x_1(kT) \\ x_2(kT) \\ \cdot \\ \cdot \\ \cdot \\ x_n(kT) \end{bmatrix} \tag{6.5-3}$$

or

$$\begin{aligned} \mathbf{x}[(k+1)T] &= \mathbf{\Lambda x}(kT) + \mathbf{B v}(kT) \\ \mathbf{y}(kT) &= \mathbf{C x}(kT) \end{aligned} \tag{6.5-4}$$

where the matrices $\mathbf{\Lambda}$, $\mathbf{B}$, and $\mathbf{C}$ are defined by the equivalence of Eqs. 6.5-3 and 6.5-4. Note that the symbol $\mathbf{\Lambda}$ is used in place of $\mathbf{A}$, to denote that $\mathbf{A}$ is diagonal. Also note $\mathbf{B} = \mathbf{b} = \mathbf{1}$, a vector.

This form of the state equations is of particular importance when dealing with concepts and proofs, since the diagonal form enables one to make concise statements about the properties of the system. This form is particularly convenient when dealing with forcing functions, as the corresponding state transition matrix is also of diagonal form. However, there does arise a computational difficulty, because the transfer function $H(z)$ gives no information about the initial conditions of the system. In fact, to compute the initial conditions on the x_i's in this form, one must find $\mathbf{y}(0), \mathbf{y}(1), \ldots, \mathbf{y}(n-1)$ and then solve a set of simultaneous equations to find the relationships between the boundary conditions on the y's and those on the state variables of Eq. 6.5-4.

Example 6.5-3. Find the **A**, **B**, and **C** matrices for the sampled-data system of Fig. 6.5-4*a*.

With respect to the sampled input and output, the transfer function $H(z)$ of this system is given by

$$\frac{Y(z)}{V(z)} = H(z) = \frac{1}{z-1} - \frac{\epsilon^{-aT}}{z-\epsilon^{-aT}}$$

The output transform $Y(z)$ is then

$$Y(z) = H(z)V(z) = \frac{V(z)}{z-1} - \frac{V(z)\epsilon^{-aT}}{z-\epsilon^{-aT}} = X_1(z) + X_2(z)\epsilon^{-aT}$$

where the difference equations for x_1 and x_2 are given by

$$x_1[(k+1)T] - x_1(kT) = v(kT)$$
$$x_2[(k+1)T] - \epsilon^{-aT}x_2(kT) = v(kT)$$

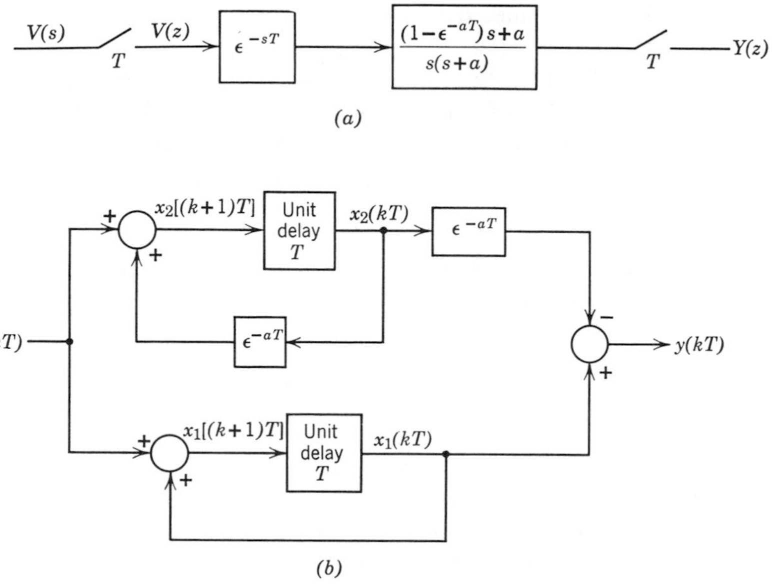

Fig. 6.5-4

The matrix equations for this system are

$$\mathbf{x}(k+1)T = \mathbf{\Lambda x}(kT) + \mathbf{Bv}(kT)$$
$$\mathbf{y}(kT) = \mathbf{Cx}(kT)$$

where

$$\mathbf{\Lambda} = \begin{bmatrix} 1 & 0 \\ 0 & \epsilon^{-aT} \end{bmatrix}, \quad \mathbf{B} = \begin{bmatrix} 1 \\ 1 \end{bmatrix}, \quad \mathbf{C} = [1 \quad -\epsilon^{-aT}]$$

The simulation diagram for this system is shown in Fig. 6.5-4*b*.

For the frequently arising sampled-data case in which the numerator polynomial of $H(z)$ is of the same order as the denominator polynomial of $H(z)$, the **D** matrix is not zero. Since **D** represents a feedthrough term, it can be found by dividing the numerator polynomial by the denominator polynomial, stopping after the first term. The first term represents the feedthrough, or **D** matrix, term. A simple example of this is $H(z) = 1/(1 - z^{-1})$, which represents the z transform corresponding to $1/s$. Using the above criterion for handling this problem,

$$H(z) = \frac{z}{z-1} = 1 + \frac{1}{z-1}$$

This operation is equivalent to

$$\mathbf{D} = D = \lim_{z\to\infty} H(z)$$

The remaining transfer function $H_1(z) = H(z) - D$ can be handled in the same manner as previously described. Actually, once D is found, there is no need to find $H_1(z)$, since $H_1(z)$ has the same poles and the same residues at these poles as does $H(z)$. Therefore, when the order of the numerator polynomial of $H(z)$ is of the same order as the denominator polynomial of $H(z)$ and the poles of $H(z)$ are of first order (distinct roots of the denominator polynomial), the general state equations are

$$\begin{aligned} \mathbf{x}[(k+1)T] &= \mathbf{\Lambda x}(kT) + \mathbf{Bv}(kT) \\ \mathbf{y}(kT) &= \mathbf{Cx}(kT) + \mathbf{Dv}(kT) \end{aligned} \tag{6.5-5}$$

where $\mathbf{\Lambda}$ is the diagonal matrix whose elements are the characteristic roots $\lambda_1, \lambda_2, \ldots, \lambda_n$ of the denominator polynomial of $\mathbf{H}(z)$, and

$\mathbf{B}$ = column matrix whose elements are equal to one

$\mathbf{C}$ = row matrix all of whose elements are equal to $c_i = (z - \lambda_i)H(z)\big|_{z=\lambda_i}$

$\mathbf{D} = d_0$ = single-element matrix $= \lim_{z\to\infty} H(z)$

A general block diagram is shown in Fig. 6.5-5.

For the case where the roots of the denominator polynomial of $H(z)$ are not distinct, the procedure to be followed is the same as that given for the continuous case in Section 5.5. Rather than repeat the procedure here, an illustrative example is given.

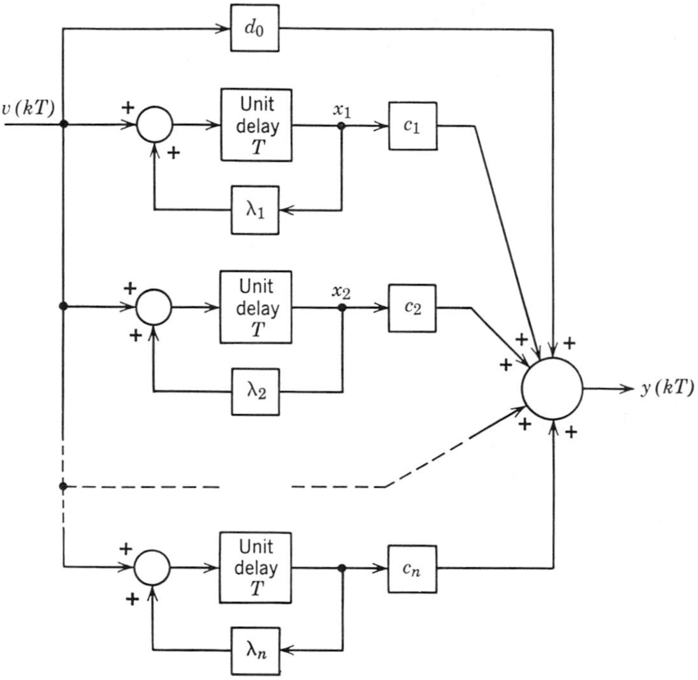

Fig. 6.5-5

Example 6.5-4. Find the **A**, **B**, **C**, and **D** matrices for the system whose transfer function is

$$H(z) = \frac{4z^3 - 12z^2 + 13z - 7}{(z-1)^2(z-2)}$$

This transfer function can be expressed in the partial fraction form

$$H(z) = \frac{c_1}{(z-1)^2} + \frac{c_2}{(z-1)} + \frac{c_3}{(z-2)} + d_0$$

where

$$d_0 = \lim_{z \to \infty} H(z) = 4$$

$$c_1 = (z-1)^2 H(z)\big|_{z=1} = 2$$

$$c_2 = \frac{d}{dz}[(z-1)^2 H(z)]_{z=1} = 1$$

$$c_3 = (z-2)H(z)\big|_{z=2} = 3$$

Therefore the output $Y(z)$ can be written

$$Y(z) = \frac{2V(z)}{(z-1)^2} + \frac{V(z)}{z-1} + \frac{3V(z)}{z-2} + 4V(z) = 2X_1(z) + X_2(z) + 3X_3(z) + 4V(z)$$

Since $X_1(z) = X_2(z)/(z-1)$,

$$x_1[(k+1)T] - x_1(kT) = x_2(kT)$$

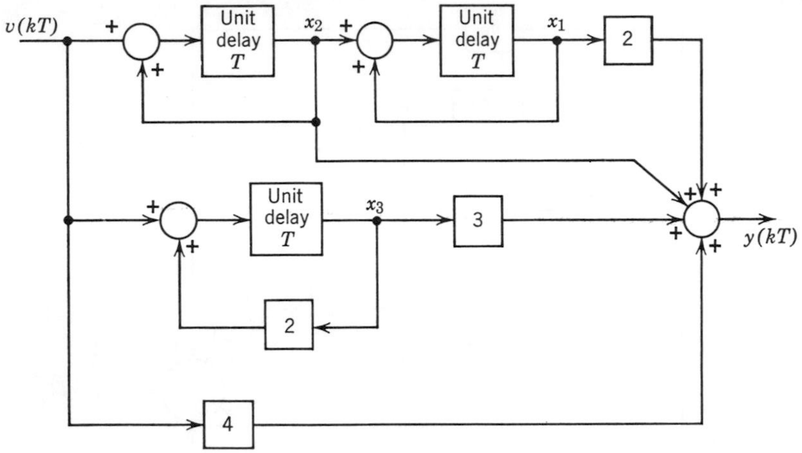

Fig. 6.5-6

The state equations are

$$\mathbf{x}[(k+1)T] = \mathbf{J}\mathbf{x}(kT) + \mathbf{b}v(kT)$$

$$\mathbf{y}(kT) = \mathbf{C}\mathbf{x}(kT) + d_0 v(kT)$$

where

$$\mathbf{J} = \begin{bmatrix} 1 & 1 & 0 \\ 0 & 1 & 0 \\ 0 & 0 & 2 \end{bmatrix}, \quad \mathbf{b} = \begin{bmatrix} 0 \\ 1 \\ 1 \end{bmatrix}, \quad \mathbf{C} = [2 \quad 1 \quad 3], \quad d_0 = 4$$

The simulation diagram for this system is shown in Fig. 6.5-6.

For the multivariable case where there are multiple inputs or outputs, the transfer function matrix $\mathbf{H}(z)$ can be used in a similar manner to find the state equations. However, the approach is not so clear-cut as in the single input-single output case, as there is generally a greater freedom of choice in assigning elements to the **B** and **C** matrices.

Example 6.5-5. Find the **A**, **B**, **C**, and **D** matrices for the system whose transfer function diagram is given in Fig. 6.5-7.

The outputs $Y_1(z)$ and $Y_2(z)$ are given by

$$Y_1(z) = \frac{z}{(z+1)(z+2)} V_1(z) + \frac{1}{z+2} V_2(z)$$

$$Y_2(z) = \frac{z^2}{(z+1)(z+3)} V_1(z) + 4V_2(z)$$

Expanding into partial fractions yields

$$Y_1(z) = -\frac{1}{z+1} V_1(z) + \frac{2}{z+2} V_1(z) + \frac{1}{z+2} V_2(z)$$

$$Y_2(z) = \frac{\frac{1}{2}}{z+1} V_1(z) - \frac{\frac{9}{2}}{z+3} V_2(z) + V_1(z) + 4V_2(z)$$

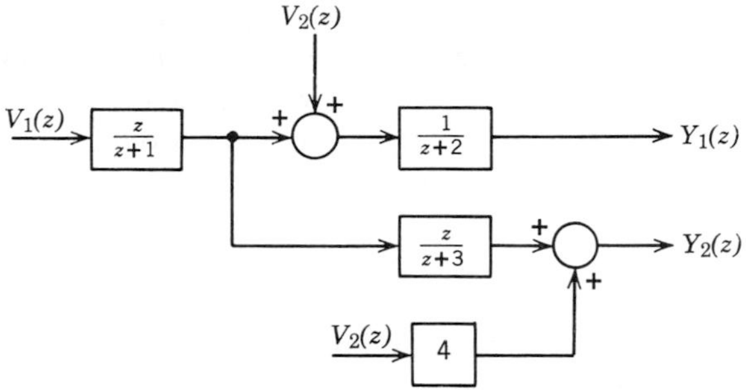

Fig. 6.5-7

Using these equations, Fig. 6.5-8 was drawn. It is readily apparent that the corresponding state equations are

$$\begin{bmatrix} x_1[(k+1)T] \\ x_2[(k+1)T] \\ x_3[(k+1)T] \end{bmatrix} = \begin{bmatrix} -1 & 0 & 0 \\ 0 & -2 & 0 \\ 0 & 0 & -3 \end{bmatrix} \begin{bmatrix} x_1(kT) \\ x_2(kT) \\ x_3(kT) \end{bmatrix} + \begin{bmatrix} 1 & 0 \\ 2 & 1 \\ -\frac{9}{2} & 0 \end{bmatrix} \begin{bmatrix} v_1(kT) \\ v_2(kT) \end{bmatrix}$$

$$\begin{bmatrix} y_1(kT) \\ y_2(kT) \end{bmatrix} = \begin{bmatrix} -1 & 1 & 0 \\ \frac{1}{2} & 0 & 1 \end{bmatrix} \begin{bmatrix} x_1(kT) \\ x_2(kT) \\ x_3(kT) \end{bmatrix} + \begin{bmatrix} 0 & 0 \\ 1 & 4 \end{bmatrix} \begin{bmatrix} v_1(kT) \\ v_2(kT) \end{bmatrix}$$

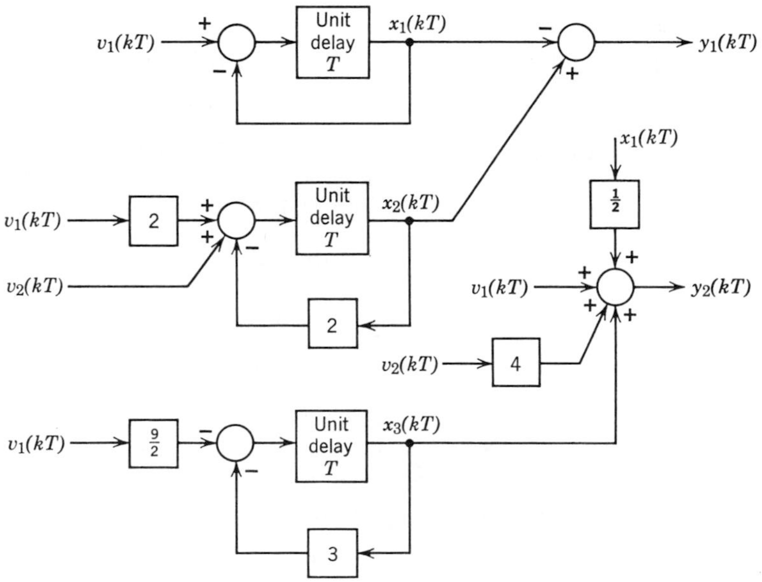

Fig. 6.5-8

Notice that, although the **A** matrix in the preceding example is easily found, the **B** and **C** matrices are by no means unique. Certainly, some of the elements in **B** could be interchanged with some of the elements in **C** and the conditions of the transfer function matrix would still be satisfied. Without any knowledge of the physical properties of the system, either choice is equally valid. As long as all the transfer products $b_{ij}c_{ki}$ remain the same, **B** and **C** can be arranged in any number of ways. The use of the transfer function, or the transfer function matrix, to obtain the state equations of the system is at best a compromise, and it should be used only if the original difference equations are not available.

If the original difference equations are available and **A** can be diagonalized, a diagonal form for **A** can be obtained by use of the modal matrix **M**. Assume then, that the state equations are available in the form of Eq. 6.5-6 below, and that it is desired that a new set of state variables be obtained such that the **A** matrix is a diagonal matrix $\mathbf{\Lambda}$.

$$\begin{aligned} \mathbf{x}[(k+1)T] &= \mathbf{A}\mathbf{x}(kT) + \mathbf{B}\mathbf{v}(kT) \\ \mathbf{y}(kT) &= \mathbf{C}\mathbf{x}(kT) + \mathbf{D}\mathbf{v}(kT) \end{aligned} \tag{6.5-6}$$

Define a new set of state variables **q**, such that

$$\mathbf{x}(kT) = \mathbf{M}\mathbf{q}(kT) \tag{6.5-7}$$

Equation 6.5-6 can be rewritten in terms of these *normal coordinates* as

$$\begin{aligned} \mathbf{M}\mathbf{q}[(k+1)T] &= \mathbf{A}\mathbf{M}\mathbf{q}(kT) + \mathbf{B}\mathbf{v}(kT) \\ \mathbf{y}(kT) &= \mathbf{C}\mathbf{M}\mathbf{q}(kT) + \mathbf{D}\mathbf{v}(kT) \end{aligned}$$

Premultiplying both sides of the first equation by $\mathbf{M}^{-1}$ yields

$$\mathbf{q}[(k+1)T] = \mathbf{M}^{-1}\mathbf{A}\mathbf{M}\mathbf{q}(kT) + \mathbf{M}^{-1}\mathbf{B}\mathbf{v}(kT)$$

However, since $\mathbf{M}^{-1}\mathbf{A}\mathbf{M} = \mathbf{\Lambda}$, vector-matrix state equations in *normal form* are

$$\begin{aligned} \mathbf{q}[(k+1)T] &= \mathbf{\Lambda}\mathbf{q}(kT) + \mathbf{B}_n\mathbf{v}(kT) \\ \mathbf{y}(kT) &= \mathbf{C}_n\mathbf{q}(kT) + \mathbf{D}\mathbf{v}(kT) \end{aligned} \tag{6.5-8}$$

where $\mathbf{B}_n = \mathbf{M}^{-1}\mathbf{B}$, the normal form input matrix, and $\mathbf{C}_n = \mathbf{C}\mathbf{M}$, the normal form output matrix. This is a general procedure, and, since it originates from the difference equations of the system, it is to be preferred over the transfer function matrix approach.

Example 6.5-6. The simulation diagram of Fig. 6.5-9 represents the original system whose transfer function matrix was given in Example 6.5-5. Determine the state equations in normal form.

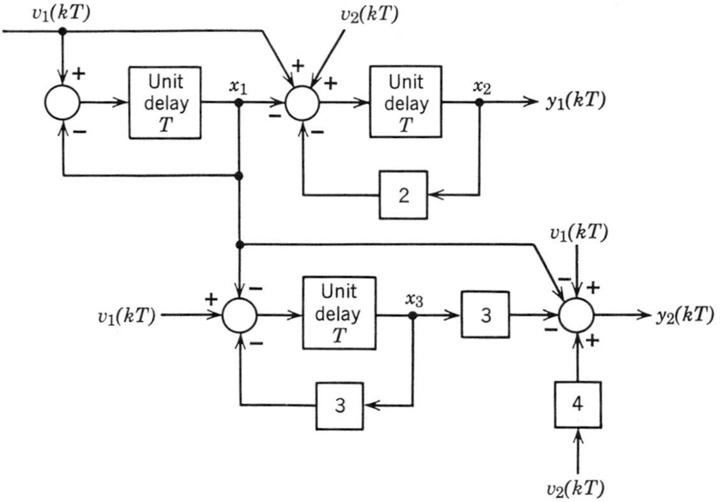

Fig. 6.5-9

The vector-matrix equations for this system are

$$\begin{bmatrix} x_1[(k+1)T] \\ x_2[(k+1)T] \\ x_3[(k+1)T] \end{bmatrix} = \begin{bmatrix} -1 & 0 & 0 \\ -1 & -2 & 0 \\ -1 & 0 & -3 \end{bmatrix} \begin{bmatrix} x_1(kT) \\ x_2(kT) \\ x_3(kT) \end{bmatrix} + \begin{bmatrix} 1 & 0 \\ 1 & 1 \\ 1 & 0 \end{bmatrix} \begin{bmatrix} v_1(kT) \\ v_2(kT) \end{bmatrix}$$

$$\begin{bmatrix} y_1(kT) \\ y_2(kT) \end{bmatrix} \begin{bmatrix} 0 & 1 & 0 \\ -1 & 0 & -3 \end{bmatrix} \begin{bmatrix} x_1(kT) \\ x_2(kT) \\ x_3(kT) \end{bmatrix} \begin{bmatrix} 0 & 0 \\ 1 & 4 \end{bmatrix} \begin{bmatrix} v_1(kT \\ v_2(kT \end{bmatrix}$$

For the $\mathbf{A}$ matrix above, the modal matrix $\mathbf{M}$, and its inverse $\mathbf{M}^{-1}$ are given by

$$\mathbf{M} = \begin{bmatrix} 2 & 0 & 0 \\ -2 & -1 & 0 \\ -1 & 0 & 1 \end{bmatrix}, \quad \mathbf{M}^{-1} = \begin{bmatrix} \frac{1}{2} & 0 & 0 \\ -1 & -1 & 0 \\ \frac{1}{2} & 0 & 1 \end{bmatrix}$$

Substituting into Eq. 6.5-8, the resulting normal form matrices are

$$\mathbf{\Lambda} = \begin{bmatrix} -1 & 0 & 0 \\ 0 & -2 & 0 \\ 0 & 0 & -3 \end{bmatrix}, \quad \mathbf{B}_n = \begin{bmatrix} \frac{1}{2} & 0 \\ -2 & -1 \\ \frac{3}{2} & 0 \end{bmatrix},$$

$$\mathbf{C}_n = \begin{bmatrix} -2 & -1 & 0 \\ 1 & 0 & -3 \end{bmatrix}, \quad \mathbf{D} = \begin{bmatrix} 0 & 0 \\ 1 & 4 \end{bmatrix}$$

A check of all the transfer products $b_{ij}c_{ki}$ shows that this set of matrices represents the same system as that of Example 6.5-5.

The new state variables $\mathbf{q}$ are related to the old state variables $\mathbf{x}$ by the relationship $\mathbf{q} = \mathbf{M}^{-1}\mathbf{x}$, or

$$q_1 = \tfrac{1}{2}x_1 \qquad q_2 = -(x_1 + x_2) \qquad q_3 = \tfrac{1}{2}x_1 + x_3$$

The use of the relationship $\mathbf{q} = \mathbf{M}^{-1}\mathbf{x}$ removes the previous difficulty of finding the initial conditions on the state variables $\mathbf{q}$ in terms of the known system initial conditions.

Finding the modal matrix may involve no more labor than any of the other methods for finding the state transition matrix. In view of the advantages of having a diagonal $\mathbf{A}$ matrix, the normal form is quite desirable. It is also interesting to note that the mode expansion technique, to be considered next, and the normal form produce the same effect of uncoupling the state equations. In this respect, they are identical approaches but are written in different forms. The form in which the system equations are expressed is frequently one of personal preference and familiarity.

6.6 MODE INTERPRETATION

The concept of expanding the response of a linear fixed system into the sum of responses along the characteristic vectors of the $\mathbf{A}$ matrix can also be applied to discrete systems with distinct characteristic values. The development follows directly from the equations in normal form.

From Eq. 6.5-8,

$$\mathbf{q}(T) = \mathbf{\Lambda}\mathbf{q}(0) + \mathbf{B}_n\mathbf{v}(0) \tag{6.6-1}$$

and

$$\mathbf{q}(2T) = \mathbf{\Lambda}\mathbf{q}(T) + \mathbf{B}_n\mathbf{v}(T) \tag{6.6-2}$$

Substitution of Eq. 6.6-1 into Eq. 6.6-2 yields

$$\mathbf{q}(2T) = \mathbf{\Lambda}^2\mathbf{q}(0) + \mathbf{\Lambda}\mathbf{B}_n\mathbf{v}(0) + \mathbf{B}_n\mathbf{v}(T) \tag{6.6-3}$$

Similarly,

$$\mathbf{q}(3T) = \mathbf{\Lambda}\mathbf{q}(2T) + \mathbf{B}_n\mathbf{v}(2T)$$

and Eq. 6.6-3 give

$$\mathbf{q}(3T) = \mathbf{\Lambda}^3\mathbf{q}(0) + \mathbf{\Lambda}^2\mathbf{B}_n\mathbf{v}(0) + \mathbf{\Lambda}\mathbf{B}_n\mathbf{v}(T) + \mathbf{B}_n\mathbf{v}(2T)$$

Continuation of this procedure leads to

$$\mathbf{q}(kT) = \mathbf{\Lambda}^k\mathbf{q}(0) + \sum_{j=0}^{k-1}\mathbf{\Lambda}^{k-j-1}\mathbf{B}_n\mathbf{v}(jT) \tag{6.6-4}$$

Then, since $\mathbf{q} = \mathbf{M}^{-1}\mathbf{x}$ and $\mathbf{B}_n = \mathbf{M}^{-1}\mathbf{B}$,

$$\mathbf{x}(kT) = \mathbf{M}\mathbf{\Lambda}^k\mathbf{M}^{-1}\mathbf{x}(0) + \sum_{j=0}^{k-1}\mathbf{M}\mathbf{\Lambda}^{k-j-1}\mathbf{M}^{-1}\mathbf{B}\mathbf{v}(jT) \tag{6.6-5}$$

Now, recalling from Section 4.7 that the columns of

$$\mathbf{M} = [\mathbf{u}_1 \quad \mathbf{u}_2 \quad \cdots \quad \mathbf{u}_n]$$

form a basis and that the rows of $\mathbf{M}^{-1}$, where

$$\mathbf{M}^{-1} = \begin{bmatrix} \mathbf{r}_1{}^T \\ \mathbf{r}_2{}^T \\ \cdot \\ \cdot \\ \cdot \\ \mathbf{r}_n{}^T \end{bmatrix} \qquad (\mathbf{r}_i = \text{column vectors})$$

form a reciprocal basis, Eq. 6.6-5 becomes

$$\mathbf{x}(kT) = [\mathbf{u}_1 \quad \mathbf{u}_2 \quad \cdots \quad \mathbf{u}_n]\boldsymbol{\Lambda}^k \begin{bmatrix} \mathbf{r}_1{}^T \\ \mathbf{r}_2{}^T \\ \cdot \\ \cdot \\ \cdot \\ \mathbf{r}_n{}^T \end{bmatrix} \mathbf{x}(0)$$

$$+ \sum_{j=0}^{k-1} [\mathbf{u}_1 \quad \mathbf{u}_2 \quad \cdots \quad \mathbf{u}_n]\boldsymbol{\Lambda}^{k-j-1} \begin{bmatrix} \mathbf{r}_1{}^T \\ \mathbf{r}_2{}^T \\ \cdot \\ \cdot \\ \cdot \\ \mathbf{r}_n{}^T \end{bmatrix} \mathbf{B}\mathbf{v}(jT)$$

This can be rewritten as

$$\mathbf{x}(kT) = \sum_{i=1}^{n} \left[\langle \mathbf{r}_i, \mathbf{x}(0)\rangle \lambda_i^k + \sum_{j=0}^{k-1} \langle \mathbf{r}_i, \mathbf{B}\mathbf{v}(jT)\rangle \lambda_i^{k-j-1}\right]\mathbf{u}_i \tag{6.6-6}$$

Equation 6.6-6 is the discrete system analog of Eq. 5.6-10.

The "modes" of the system are given by the terms of Eq. 6.6-6 for $i = 1, 2, \ldots, n$. The mode expansion technique separates or uncouples these modes, so that the response $\mathbf{x}(kT)$ is expressed as a linear weighted sum of the modes. Each mode is directed along the characteristic vector $\mathbf{u}_i$, defined in terms of the state space $x_1, x_2, \ldots, x_n$. For an unforced system, the amount of excitation of each mode is given by the scalar product $\langle \mathbf{r}_i, \mathbf{x}(0)\rangle$. For a forced system, the scalar product $\langle \mathbf{r}_i, \mathbf{B}\mathbf{v}\rangle$ is the amplitude of the forcing function that is coupled to the ith mode.

In effect, the characteristic vectors represent a new coordinate system, such that each mode of the system is directed along one of the coordinate axes. The *normal form* for the state equations performs the same task but expresses it in slightly different form. Thus if the original state equation is

$$\mathbf{x}[(k+1)T] = \mathbf{A}\mathbf{x}(kT) + \mathbf{B}\mathbf{v}(kT)$$

the transformation $\mathbf{q} = \mathbf{M}^{-1}\mathbf{x}$, where $\mathbf{M}$ is the *normalized* modal matrix, results in the new state equation

$$\mathbf{q}[(k+1)T] = \mathbf{\Lambda q}(kT) + \mathbf{M}^{-1}\mathbf{Bv}(kT)$$

In this form the state variables $q_1, q_2, \ldots, q_n$ are uncoupled, since $\mathbf{\Lambda}$ is a diagonal matrix. The state variable q_i is directed along the characteristic vector $\mathbf{u}_i$. The scalar product $\langle \mathbf{r}_i, \mathbf{x}(0)\rangle$ in the mode expansion is simply the ith component of the normal form column vector $\mathbf{q}(0) = \mathbf{M}^{-1}\mathbf{x}(0)$, and the scalar product $\langle \mathbf{r}_i, \mathbf{Bv}\rangle$ is the ith component of the column vector $\mathbf{M}^{-1}\mathbf{Bv}$.

Example 6.6-1. Analyze the system of Fig. 6.6-1 using both the mode expansion technique and the normal form of the state equations.

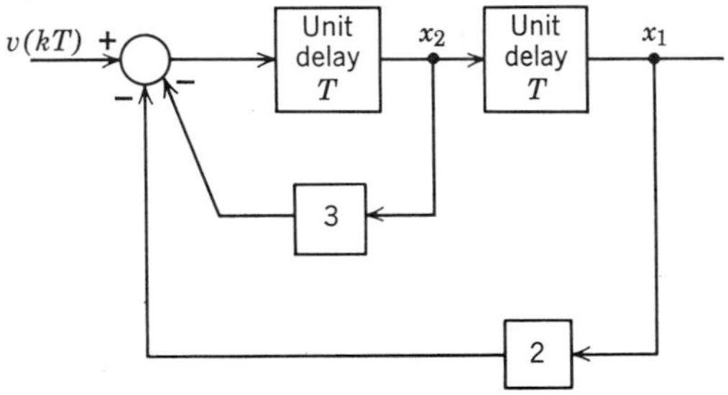

Fig. 6.6-1

The $\mathbf{A}$ matrix for this system is

$$\mathbf{A} = \begin{bmatrix} 0 & 1 \\ -2 & -3 \end{bmatrix}$$

The characteristic roots are $\lambda_1 = -2$, and $\lambda_2 = -1$. The normalized characteristic vectors are

$$\mathbf{u}_1 = \begin{bmatrix} 1/\sqrt{2} \\ -1/\sqrt{2} \end{bmatrix} \qquad \mathbf{u}_2 = \begin{bmatrix} 1/\sqrt{5} \\ -2/\sqrt{5} \end{bmatrix}$$

The reciprocal basis is then

$$\mathbf{r}_1 = \begin{bmatrix} 2\sqrt{2} \\ \sqrt{2} \end{bmatrix} \qquad \mathbf{r}_2 = \begin{bmatrix} -\sqrt{5} \\ -\sqrt{5} \end{bmatrix}$$

The normalized modal matrix $\mathbf{M}$ and its inverse $\mathbf{M}^{-1}$ are given by

$$\mathbf{M} = \begin{bmatrix} 1/\sqrt{2} & 1/\sqrt{5} \\ -1/\sqrt{2} & -2/\sqrt{5} \end{bmatrix} \qquad \mathbf{M}^{-1} = \begin{bmatrix} 2\sqrt{2} & \sqrt{2} \\ -\sqrt{5} & -\sqrt{5} \end{bmatrix}$$

Note that the rows of $\mathbf{M}^{-1}$ are the reciprocal basis vectors $\mathbf{r}_i$.

The scalar products $\langle \mathbf{r}_i, \mathbf{x}(0)\rangle$ are

$$\langle \mathbf{r}_1, \mathbf{x}(0)\rangle = 2\sqrt{2}\,x_1(0) + \sqrt{2}\,x_2(0)$$
$$\langle \mathbf{r}_2, \mathbf{x}(0)\rangle = -\sqrt{5}\,x_1(0) - \sqrt{5}\,x_2(0)$$

The initial conditions on the **q** vector are

$$\mathbf{q}(0) = \mathbf{M}^{-1}\mathbf{x}(0) = \begin{bmatrix} 2\sqrt{2}\,x_1(0) + \sqrt{2}\,x_2(0) \\ -\sqrt{5}\,x_1(0) - \sqrt{5}\,x_2(0) \end{bmatrix}$$

The forcing function $\mathbf{Bv}(kT)$ is the vector **b** times the scalar $v(kT)$, where

$$\mathbf{b} = \begin{bmatrix} 0 \\ 1 \end{bmatrix}$$

Thus the scalar products $\langle \mathbf{r}_i, \mathbf{Bv}(kT)\rangle$ are

$$\langle \mathbf{r}_1, \mathbf{Bv}(kT)\rangle = \sqrt{2}\,v(kT)$$
$$\langle \mathbf{r}_2, \mathbf{Bv}(kT)\rangle = -\sqrt{5}\,v(kT)$$

The forcing function $\mathbf{B}_n\mathbf{v}(kT)$ is

$$B_n\mathbf{v}(kT) = \mathbf{M}^{-1}\mathbf{b}v(kT) = \begin{bmatrix} \sqrt{2} \\ -\sqrt{5} \end{bmatrix} v(kT)$$

The general expression for the time response $\mathbf{x}(kT)$ is

$$\mathbf{x}(kT) = [2\sqrt{2}\,x_1(0) + \sqrt{2}\,x_2(0)](-1)^k\mathbf{u}_1 + \sum_{j=0}^{k-1} \sqrt{2}\,(-1)^{(k-j-1)}v(jT)\mathbf{u}_1$$
$$+ [-\sqrt{5}\,x_1(0) - \sqrt{5}\,x_2(0)](-2)^k\mathbf{u}_2 + \sum_{j=0}^{k-1} (-\sqrt{5})(-2)^{(k-j-1)}v(jT)\mathbf{u}_2$$

The general expression for the time response $\mathbf{q}(kT)$ is, from the above and Eq. 6.6-4,

$$\mathbf{q}(kT) = \begin{bmatrix} (-1)^k q_1(0) + \sum_{j=0}^{k-1} \sqrt{2}\,(-1)^{(k-j-1)}v(jT) \\ (-2)^k q_2(0) + \sum_{j=0}^{k-1} (-\sqrt{5})(-2)^{(k-j-1)}v(jT) \end{bmatrix}$$

where $q_1(0) = 2\sqrt{2}\,x_1(0) + \sqrt{2}\,x_2(0)$ and $q_2(0) = -\sqrt{5}\,x_1(0) - \sqrt{5}\,x_2(0)$.

Obviously both methods are equivalent, the **q** coordinates of the *normal form* being the normalized characteristic vectors of the *mode expansion.*

6.7 CONTROLLABILITY AND OBSERVABILITY

The controllability and observability concepts presented in Section 5.7 carry over directly to the linear discrete system, so there is little need for further discussion on these points. To restate the principal ideas:

Controllability is a function of the coupling between the inputs to the system and the various modes of the system. If the system equations can be written with distinct λ's in the normal form

$$\mathbf{q}[(k+1)T] = \mathbf{\Lambda q}(kT) + \mathbf{B}_n\mathbf{v}(kT)$$
$$\mathbf{y}(kT) = \mathbf{C}_n\mathbf{q}(kT) + \mathbf{Dv}(kT)$$

then all the modes are controllable if there are no zero *rows* of $\mathbf{B}_n$. Stated in terms of the mode expansion method, this means that none of the scalar products $\langle \mathbf{r}_i, \mathbf{Bv}\rangle$ vanishes.

Observability is a function of the coupling between the modes of the system and the output of the system. All the modes of the system are observable if there are no zero *columns* of $\mathbf{C}_n$. Alternatively, this requirement could be stated as: The kth mode is not observable if all the scalar products $\langle \mathbf{c}_i, \mathbf{u}_k\rangle = 0$ for all i's, where the vector $\mathbf{c}_i$ constitutes the ith row of the original $\mathbf{C}$ matrix.

For a sampled-data system there is an additional requirement. If the continuous system has a partial fraction expansion which contains the term $\beta/[(s+a)^2+\beta^2]$, and if the sampling interval $T = \pi/\beta$, then the Z transform of this term,

$$\mathscr{Z}\left[\frac{\beta}{(s+a)^2+\beta^2}\right] = \frac{z^{-1}\epsilon^{-aT}\sin\beta T}{1-2z^{-1}\epsilon^{-aT}\cos\beta T + z^{-2}\epsilon^{-2aT}} = 0$$

The system may even be unstable, with $a < 0$, but this fact could not be inferred from observations of the output. These are called "hidden oscillations," and they occur when the zeros of the oscillation coincide exactly with the time that the system is sampled.[10-12] In this situation the system is neither completely controllable nor completely observable. Therefore the additional requirement for complete controllability and observability of sampled-data systems is that, if a characteristic root of the continuous system is $-a \pm j\beta$, then $T \neq \pi/\beta$.

6.8 THE STATE TRANSITION MATRIX

The state transition matrix for the linear discrete time system is investigated in this section. Similarly to the continuous case, the state transition matrix is the fundamental matrix of Eq. 6.8-1 below, subject to the condition that $\boldsymbol{\phi}[(k_0, k_0)T] = \mathbf{I}$, the unit matrix. Consider, then, the time-varying state difference equation

$$\mathbf{x}[(k+1)T] = \mathbf{A}(kT)\mathbf{x}(kT) \qquad (6.8\text{-}1)$$

If the initial conditions $\mathbf{x}(k_0T)$ are known, then

$$\mathbf{x}[(k_0+1)T] = \mathbf{A}(k_0T)\mathbf{x}(k_0T)$$

Similarly,

$$\mathbf{x}[(k_0 + 2)T] = \mathbf{A}[(k_0 + 1)T]\mathbf{A}(k_0T)\mathbf{x}(k_0T)$$

By a process of iteration, the continued product[13]

$$\mathbf{x}(kT) = \prod_{n=k_0}^{k-1} \mathbf{A}(nT)\mathbf{x}(k_0T) \qquad (k > k_0) \tag{6.8-2}$$

is obtained.

Since the state transition matrix $\boldsymbol{\phi}[(k, k_0)T]$ is defined by the relationship

$$\mathbf{x}(kT) = \boldsymbol{\phi}[(k, k_0)T]\mathbf{x}(k_0T) \tag{6.8-3}$$

then

$$\begin{aligned}\boldsymbol{\phi}[(k, k_0)T] &= \prod_{n=k_0}^{k-1} \mathbf{A}(nT) \qquad (k > k_0) \\ &= \mathbf{I} \qquad (k = k_0)\end{aligned} \tag{6.8-4}$$

This process of obtaining the state transition matrix by iteration is similar to the iterative procedure for computing the matrizant of the analogous continuous system.

For the case where $\mathbf{A}(kT)$ is a constant matrix $\mathbf{A}_0$, the state transition matrix $\boldsymbol{\phi}_0[(k, k_0)T]$ is

$$\boldsymbol{\phi}_0[(k, k_0)T] = \mathbf{A}_0^{(k-k_0)} \qquad \mathbf{A}(kT) = \mathbf{A}_0, \text{ a constant matrix} \tag{6.8-5}$$

This is analogous to the continuous case, where the solution for a fixed system depends only upon time differences; whereas for a time-variable case the solution depends upon both the time of application of cause and the time of observation of effect.

For the time-varying case where $\mathbf{A}(kT)$ can be written as the sum of two matrices $\mathbf{A}_0$ and $\mathbf{A}_1(kT)$, a perturbation technique can be used to obtain the state transition matrix. This procedure is useful if the time-varying matrix $\mathbf{A}_1(kT)$ represents a small perturbation upon the constant matrix $\mathbf{A}_0$. For this case

$$\mathbf{x}[(k + 1)T] = [\mathbf{A}_0 + \mathbf{A}_1(kT)]\mathbf{x}(kT) \tag{6.8-6}$$

This equation can be viewed as a constant system $\mathbf{A}_0$ with a forcing function $\mathbf{v}(kT)$ applied, where $\mathbf{v}(kT) = \mathbf{A}_1(kT)\mathbf{x}(kT)$. Thus

$$\mathbf{x}[(k + 1)T] = \mathbf{A}_0\mathbf{x}(kT) + \mathbf{v}(kT), \qquad \mathbf{v}(kT) = \mathbf{A}_1(kT)\mathbf{x}(kT) \tag{6.8-7}$$

The solution $\mathbf{x}(kT)$ for the system of Eq. 6.8-7 can be determined by the same process of iteration used to give Eqs. 6.6-4 and 6.8-2. Thus

$$\mathbf{x}(kT) = \mathbf{A}_0^{(k-k_0)}\mathbf{x}(k_0T) + \sum_{n=k_0}^{k-1} \mathbf{A}_0^{(k-n-1)}\mathbf{v}(nT) \tag{6.8-8}$$

Substituting $\mathbf{v}(nT) = \mathbf{A}_1(nT)\mathbf{x}(nT)$ into Eq. 6.8-8 yields

$$\mathbf{x}(kT) = \mathbf{A}_0^{(k-k_0)}\mathbf{x}(k_0T) + \sum_{n=k_0}^{k-1} \mathbf{A}_0^{(k-n-1)}\mathbf{A}_1(nT)\mathbf{x}(nT) \tag{6.8-9}$$

This is a summation equation, and it can be solved by the usual methods of iteration.

The first iteration is

$$\mathbf{x}(kT) = \mathbf{A}_0^{(k-k_0)}\mathbf{x}(k_0T) + \sum_{n=k_0}^{k-1} \mathbf{A}_0^{(k-n-1)}\mathbf{A}_1(nT) \times \left[\mathbf{A}_0^{(n-k_0)}\mathbf{x}(k_0T) + \sum_{m=k_0}^{n-1} \mathbf{A}_0^{(n-m-1)}\mathbf{A}_1(mT)\mathbf{x}(mT)\right]$$

Further iterations yield

$$\mathbf{x}(kT) = [\mathbf{I} + \mathbf{S}(\boldsymbol{\phi}_0\mathbf{A}_1) + \mathbf{S}(\boldsymbol{\phi}_0\mathbf{A}_1\mathbf{S}(\boldsymbol{\phi}_0\mathbf{A}_1)) + \mathbf{S}(\boldsymbol{\phi}_0\mathbf{A}_1\mathbf{S}(\boldsymbol{\phi}_0\mathbf{A}_1\mathbf{S}(\boldsymbol{\phi}_0\mathbf{A}_1))) + \cdots]\boldsymbol{\phi}_0\mathbf{x}(k_0T) \tag{6.8-10}$$

where the $\mathbf{S}$ indicates a summation of all terms to the right and $\boldsymbol{\phi}_0$ is given by Eq. 6.8-5. The state transition matrix $\boldsymbol{\phi}(kT, k_0T)$ can then be written as

$$\boldsymbol{\phi}[(k, k_0)T] = [\mathbf{I} + \mathbf{S}(\boldsymbol{\phi}_0\mathbf{A}_1) + \mathbf{S}(\boldsymbol{\phi}_0\mathbf{A}_1\mathbf{S}(\boldsymbol{\phi}_0\mathbf{A}_1)) + \mathbf{S}(\boldsymbol{\phi}_0\mathbf{A}_1\mathbf{S}(\boldsymbol{\phi}_0\mathbf{A}_1\mathbf{S}(\boldsymbol{\phi}_0\mathbf{A}_1))) + \cdots]\boldsymbol{\phi}_0 \tag{6.8-11}$$

which properly reduces to $\boldsymbol{\phi}_0$ for the case when the system is fixed, i.e., $\mathbf{A}_1(kT) = [0]$.

Equation 6.8-11 for discrete systems is analogous to Eq. 5.9-21 for continuous systems. If $\mathbf{A}_1$ represents a small perturbation upon the $\mathbf{A}_0$, then this series is rapidly convergent, and only a few terms are required to find $\boldsymbol{\phi}[(k, k_0)T]$. The advantage of using this form for the state transition matrix is that the general time-varying $\boldsymbol{\phi}$ is expressed in terms of successive corrections upon a constant $\boldsymbol{\phi}$.

Properties of the State Transition Matrix

The state transition matrix for discrete systems has a set of properties which are directly analogous to the properties listed for a continuous system in Section 5.9. Namely,

$$\boldsymbol{\phi}[(k_0, k_0)T] = \mathbf{I} \tag{6.8-12}$$

$$\boldsymbol{\phi}[(k_2, k_1)T]\boldsymbol{\phi}[(k_1, k_0)T] = \boldsymbol{\phi}[(k_2, k_0)T] \tag{6.8-13}$$

$$\boldsymbol{\phi}[(k_1, k_2)T] = \boldsymbol{\phi}^{-1}[(k_2, k_1)T] \tag{6.8-14}$$

For fixed systems,

$$\boldsymbol{\phi}[(k + n)] = \boldsymbol{\phi}(k)\boldsymbol{\phi}(n) \tag{6.8-15}$$

$$\boldsymbol{\phi}(k) = \boldsymbol{\phi}^{-1}(-k) \tag{6.8-16}$$

Computation of $\boldsymbol{\phi}$

In general, the computation of the state transition matrix for the time-varying case is a formidable task. Clearly, for any large value of n, Eq. 6.8-4 becomes most unwieldy. In certain cases, where the difference equations of the system can be handled, an analytical solution can be obtained (see Section 2.13). However, this occurs rarely. Use of a computer is generally the best method to obtain a solution.

For the fixed system, an analytical solution can generally be obtained. Equation 4.10-20 provides one method of computing the state transition matrix. Some others follow.

1. *Cayley-Hamilton Method.* For the discrete time case, the Cayley-Hamilton procedure can be used for computing $\mathbf{A}^k$. Here the $f(\lambda_i)$ to be used is λ_i^k rather than the $\epsilon^{\lambda_i t}$ used for the continuous case.

Example 6.8-1. Compute $\boldsymbol{\phi}(k)$ for the difference equation

$$y[(k+2)T] + 5y[(k+1)T] + 6y(kT) = 0$$

The $\mathbf{A}$ matrix for this system is

$$\mathbf{A} = \begin{bmatrix} 0 & 1 \\ -6 & -5 \end{bmatrix}$$

assuming $x_1(kT) = y(kT)$, $x_2(kT) = y[(k+1)T]$. The characteristic equation $|\lambda\mathbf{I} - \mathbf{A}| = 0$ has two characteristic roots $\lambda_1 = -2$, $\lambda_2 = -3$. Therefore

$$F(\lambda_1) = \lambda_1^k = (-2)^k = \alpha_0 + \alpha_1\lambda_1 = \alpha_0 - 2\alpha_1$$

$$F(\lambda_2) = \lambda_2^k = (-3)^k = \alpha_0 + \alpha_1\lambda_2 = \alpha_0 - 3\alpha_1$$

From these two equations, $\alpha_0 = 3(-2)^k - 2(-3)^k$ and $\alpha_1 = (-2)^k - (-3)^k$. Hence

$$\mathbf{F}(\mathbf{A}) = \mathbf{A}^k = \alpha_0\mathbf{I} + \alpha_1\mathbf{A}$$

or

$$\boldsymbol{\phi}(k) = \begin{bmatrix} 3(-2)^k - 2(-3)^k & (-2)^k - (-3)^k \\ -6[(-2)^k - (-3)^k] & -2(-2)^k + 3(-3)^k \end{bmatrix}$$

Example 6.8-2. Compute $\boldsymbol{\phi}(k)$ for the system whose $\mathbf{A}$ matrix is given by

$$\mathbf{A} = \begin{bmatrix} 0 & 1 \\ -1 & -2 \end{bmatrix}$$

The characteristic equation $|\lambda\mathbf{I} - \mathbf{A}| = 0$ has two characteristic roots located at -1. For this case of repeated roots, the conditions which must be used to obtain the α's are

$$\left.\frac{d^sF(\lambda)}{d\lambda^s}\right|_{\lambda=\lambda_i} = \frac{d^s}{d\lambda^s}\left[\sum_{r=0}^{n-1}\alpha_r\lambda^r\right]_{\lambda=\lambda_i} = \frac{d^s}{d\lambda^s}[\alpha_0 + \alpha_1\lambda]_{\lambda=\lambda_i} \qquad s = 0, 1, 2, \ldots, p_i - 1$$

where p_i is the order of the root. Hence $(-1)^k = \alpha_0 + \alpha_1\lambda_1 = \alpha_0 - \alpha_1$ and $-k(-1)^k = \alpha_1$, or $\alpha_1 = -k(-1)^k$ and $\alpha_0 = (-1)^k(1-k)$. Therefore

$$\boldsymbol{\phi}(k) = \alpha_0\mathbf{I} + \alpha_1\mathbf{A} = \begin{bmatrix} \alpha_0 & \alpha_1 \\ -\alpha_1 & \alpha_0 - 2\alpha_1 \end{bmatrix} = (-1)^k\begin{bmatrix} 1-k & -k \\ k & 1+k \end{bmatrix}$$

Example 6.8-3. It is informative to take up the case in which the **A** matrix may have complex roots. For this reason, determine $\boldsymbol{\phi}(k)$ corresponding to

$$\mathbf{A} = \begin{bmatrix} 0 & 1 \\ -2 & 2 \end{bmatrix}$$

The characteristic values of this matrix are $1 \pm j$. For purposes of evaluating $\mathbf{A}^k$, the polar form $\sqrt{2}\,\epsilon^{\pm j\pi/4}$ is most useful. The computation of $\mathbf{A}^k$ is then

$$F(\lambda_1) = (2)^{k/2}\epsilon^{jk\pi/4} = \alpha_0 + \alpha_1 + j\alpha_1$$

$$F(\lambda_2) = (2)^{k/2}\epsilon^{-jk\pi/4} = \alpha_0 + \alpha_1 - j\alpha_1$$

Adding and subtracting these equations yields $(2)^{k/2}\cos(k\pi/4) = \alpha_0 + \alpha_1$ and $\alpha_1 = (2)^{k/2}\sin(k\pi/4)$, or $\alpha_0 = (2)^{k/2}[\cos(k\pi/4) - \sin(k\pi/4)]$. Since

$$\boldsymbol{\phi}(k) = \alpha_0\mathbf{I} + \alpha_1\mathbf{A} = \begin{bmatrix} \alpha_0 & \alpha_1 \\ -2\alpha_1 & \alpha_0 + 2\alpha_1 \end{bmatrix}$$

then

$$\boldsymbol{\phi}(k) = (2)^{k/2}\begin{bmatrix} \left(\cos\dfrac{k\pi}{4} - \sin\dfrac{k\pi}{4}\right) & \sin\dfrac{k\pi}{4} \\ -2\sin\dfrac{k\pi}{4} & \left(\cos\dfrac{k\pi}{4} + \sin\dfrac{k\pi}{4}\right) \end{bmatrix}$$

2. *Frequency-Domain Method.* The Z transform, analogously to the Laplace transform, can be used to find the state transition matrix of the equation

$$\mathbf{x}[(k+1)T] = \mathbf{A}\mathbf{x}(kT) \tag{6.8-17}$$

Transforming both sides of Eq. 6.8-17, $z\mathbf{X}(z) - z\mathbf{x}(0) = \mathbf{A}\mathbf{X}(z)$, where use has been made of Eq. 3.12-3. Thus

$$\mathbf{X}(z) = (z\mathbf{I} - \mathbf{A})^{-1}z\mathbf{x}(0) \tag{6.8-18}$$

or $\mathbf{X}(z) = \boldsymbol{\Phi}(z)\mathbf{x}(0)$, where

$$\boldsymbol{\Phi}(z) = (z\mathbf{I} - \mathbf{A})^{-1}z \tag{6.8-19}$$

This form is slightly different from the analogous form $\boldsymbol{\Phi}(s) = (s\mathbf{I} - \mathbf{A})^{-1}$ for continuous systems. The state transition matrix is given by the inverse z transform of $\boldsymbol{\Phi}(z)$, or

$$\begin{aligned} \boldsymbol{\phi}(k) &= \mathscr{Z}^{-1}[(z\mathbf{I} - \mathbf{A})^{-1}z] \\ &= \sum \text{residues of } (z\mathbf{I} - \mathbf{A})^{-1}z^k \end{aligned} \tag{6.8-20}$$

Example 6.8-4. Using the **A** matrix of Example 6.8-1, determine $\boldsymbol{\phi}(k)$ by the frequency-domain method.

For this case,

$$(z\mathbf{I} - \mathbf{A}) = \begin{bmatrix} z & -1 \\ 6 & z + 5 \end{bmatrix}$$

so that

$$(z\mathbf{I} - \mathbf{A})^{-1} = \frac{\begin{bmatrix} z + 5 & 1 \\ -6 & z \end{bmatrix}}{(z + 2)(z + 3)}$$

Determining the sum of the residues of $(z\mathbf{I} - \mathbf{A})^{-1}z^k$ gives

$$\boldsymbol{\phi}(k) = \begin{bmatrix} 3(-2)^k - 2(-3)^k & (-2)^k - (-3)^k \\ -6[(-2)^k - (-3)^k] & -2(-2)^k + 3(-3)^k \end{bmatrix}$$

Example 6.8-5. Using the $\mathbf{A}$ matrix of Example 6.8-2, determine $\boldsymbol{\phi}(k)$ by the frequency-domain method.

For this case,

$$(z\mathbf{I} - \mathbf{A}) = \begin{bmatrix} z & -1 \\ 1 & z + 2 \end{bmatrix}$$

so that

$$(z\mathbf{I} - \mathbf{A})^{-1} = \frac{\begin{bmatrix} z + 2 & 1 \\ -1 & z \end{bmatrix}}{(z + 1)^2}$$

Determination of the sum of the residues of $(z\mathbf{I} - \mathbf{A})^{-1}z^k$ yields

$$\boldsymbol{\phi}(k) = (-1)^k \begin{bmatrix} 1 - k & -k \\ k & 1 + k \end{bmatrix}$$

3. *Transfer Function Method.* As for the continuous case, the simulation diagram can be used to obtain the terms $\Phi_{ij}(z)$. The basis for this method is that the solution to Eq. 6.8-17 is

$$\mathbf{x}(kT) = \boldsymbol{\phi}(k)\mathbf{x}(0) \tag{6.8-21}$$

Therefore $x_i(kT)$ is given by

$$x_i(kT) = \sum_{j=1}^{n} \phi_{ij}(k)x_j(0) \tag{6.8-22}$$

If all the state variables except the jth are set equal to zero, and a unit initial condition is placed on x_j, then the response at the ith state variable $x_i(kT)$ represents the term $\phi_{ij}(k)$. Therefore the transfer function from the *output* of the jth delay to the *output* of the ith delay represents the term $\Phi_{ij}(z)$. This is slightly different from the continuous case, where the transfer function was calculated between the *input* to the jth integrator and the *output* of the ith integrator. The difference is due to the fact that the transfer function in the continuous case is expressed as the transform of the *impulse response* (in the continuous case, a unit impulse input establishes a unit initial condition *immediately*), while in the discrete case it is the transform of the *unit initial condition response*. If an analogy is desired, then the transfer function from the *input* of the jth delay to the *output* of the ith delay must be multiplied by z to obtain $\Phi_{ij}(z)$. Since what is generally desired is $(z\mathbf{I} - \mathbf{A})^{-1} = z^{-1}\mathbf{\Phi}(z)$, the transfer function from the *input* of the jth delay to the *output* of the ith delay is perfectly suitable.

Example 6.8-6. The simulation diagram of Fig. 6.8-1a represents the system of Example 6.8-1. Determine $\boldsymbol{\phi}(k)$.

Figure 6.8-1b is the same diagram redrawn for the convenience of computing transfer functions. Since 1/(1—loop transfer function) is equal to $z(z + 5)/(z^2 + 5z + 6)$, the

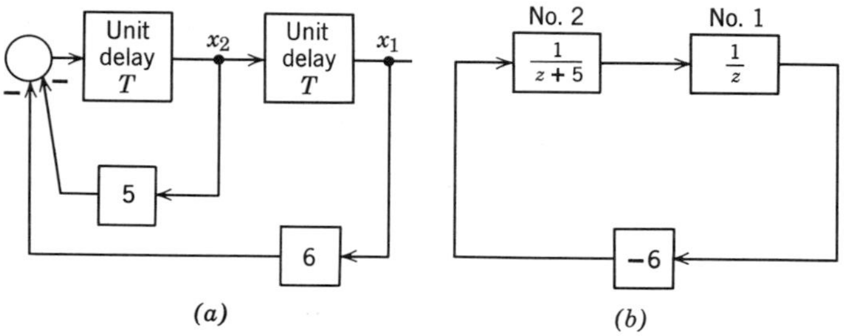

Fig. 6.8-1

various transfer functions $z^{-1}\mathbf{\Phi}_{ij}(z)$ can be obtained by multiplying the forward transfer function from j to i by $z(z + 5)/(z^2 + 5z + 6)$. By performing this operation, the matrix $(z\mathbf{I} - \mathbf{A})^{-1} = z^{-1}\mathbf{\Phi}(z)$ is obtained. Thus

$$[z\mathbf{I} - \mathbf{A}]^{-1} = z^{-1}\mathbf{\Phi}(z) = \frac{\begin{bmatrix} z + 5 & 1 \\ -6 & z \end{bmatrix}}{z^2 + 5z + 6}$$

The inverse transform of $\mathbf{\Phi}(z)$ gives $\boldsymbol{\phi}(k)$ as in Example 6.8-4.

Adjoint System—The State Transition Matrix

Similar to the adjoint of an unforced linear continuous system, there exists an adjoint corresponding to an unforced discrete linear system. The adjoint operator $\mathbf{L}^*$ is defined in terms of the system operator $\mathbf{L}$ by

$$\langle \boldsymbol{\alpha}, \mathbf{L}\mathbf{x} \rangle = \langle \mathbf{L}^*\boldsymbol{\alpha}, \mathbf{x} \rangle \tag{6.8-23}$$

where $\boldsymbol{\alpha}$ is the adjoint vector and the inner product denotes†

$$\langle \mathbf{a}, \mathbf{b} \rangle = \sum_{i=k_0}^{k-1} \mathbf{a}^T[(i + 1)T]\mathbf{b}(iT)$$

† In the case of complex elements, the transpose is replaced by the conjugate transpose.

Considering $\langle \boldsymbol{\alpha}, \mathbf{Lx} \rangle$,

$$\begin{aligned}\langle \boldsymbol{\alpha}, \mathbf{Lx} \rangle &= \sum_{i=k_0}^{k-1} \boldsymbol{\alpha}^T[(i+1)T]\{\mathbf{x}[(i+1)T] - \mathbf{A}(iT)\mathbf{x}(iT)\} \\ &= \boldsymbol{\alpha}^T(kT)\mathbf{x}(kT) + \sum_{i=k_0-1}^{k-2} \boldsymbol{\alpha}^T[(i+1)T]\mathbf{x}(i+1)T] \\ &\qquad - \boldsymbol{\alpha}^T(k_0T)\mathbf{x}(k_0T) - \sum_{i=k_0}^{k-1} \boldsymbol{\alpha}^T[(i+1)T]\mathbf{A}(iT)\mathbf{x}(iT) \\ &= \sum_{i=k_0}^{k-1} \{\boldsymbol{\alpha}^T(iT) - \boldsymbol{\alpha}^T[(i+1)]\mathbf{A}(iT)\}\mathbf{x}(iT) \\ &\qquad + \boldsymbol{\alpha}^T(kT)\mathbf{x}(kT) - \boldsymbol{\alpha}^T(k_0T)\mathbf{x}(k_0T) \end{aligned} \tag{6.8-24}$$

Then by identifying

$$\mathbf{L}^*\boldsymbol{\alpha} = \boldsymbol{\alpha}^T(iT) - \boldsymbol{\alpha}^T[(i+1)T]\mathbf{A}(iT) \tag{6.8-25}$$

so that the adjoint equation is

$$\boldsymbol{\alpha}^T(iT) = \boldsymbol{\alpha}^T[(i+1)T]\mathbf{A}(iT) \tag{6.8-26}$$

the unforced system equation

$$\mathbf{x}[(i+1)T] = \mathbf{A}(iT)\mathbf{x}(iT)$$

can be multiplied by $\boldsymbol{\alpha}^T[(i+1)T]$ and combined with Eq. 6.8-26 to yield

$$\boldsymbol{\alpha}^T[(i+1)T]\mathbf{x}[(i+1)T] = \boldsymbol{\alpha}^T(iT)\mathbf{x}(iT)$$

Then by iteration

$$\boldsymbol{\alpha}^T(kT)\mathbf{x}(kT) = \boldsymbol{\alpha}^T(k_0T)\mathbf{x}(k_0T) \tag{6.8-27}$$

Substitution of Eqs. 6.8-25 and 6.8-27 into Eq. 6.8-24 yields Eq. 6.8-23. Thus Eq. 6.8-26 rewritten as

$$\boldsymbol{\alpha}[(i+1)T] = [\mathbf{A}^{-1}(iT)]^T\boldsymbol{\alpha}(iT) \tag{6.8-28}$$

is the equation for the adjoint system.

Since the system state transition matrix $\boldsymbol{\phi}[(k, k_0)T]$ must satisfy Eq. 6.8-1,

$$\boldsymbol{\phi}[(k+1, k_0)T] = \mathbf{A}(kT)\boldsymbol{\phi}[(k, k_0)T]$$

Taking the inverse and transposing,†

$$\{\boldsymbol{\phi}^{-1}[(k+1, k_0)T]\}^T = [\mathbf{A}^{-1}(kT)]^T\{\boldsymbol{\phi}^{-1}[(k, k_0)T]\}^T \tag{6.8-29}$$

Comparison of Eqs. 6.8-28 and 6.8-29 indicates that the state transition matrix of the adjoint system is

$$\{\boldsymbol{\phi}^{-1}[(k, k_0)T]\}^T = \boldsymbol{\phi}^T[(k_0, k)T] \tag{6.8-30}$$

† If $\mathbf{A}(kT)$ contains complex elements, the conjugate transpose must be taken.

Note that the presentation here is the reverse of that of Chapter 5. There the continuous analogs of Eqs. 6.8-28 and 6.8-30 were defined, and the continuous analog of Eq. 6.8-23 resulted. Here Eqs. 6.8-28 and 6.8-30 are obtained from the definition of the adjoint operator, Eq. 6.8-23.

For the discrete time adjoint system, the state transition matrix

$$\{\boldsymbol{\phi}^{-1}[(k, k_0)T]\}^T$$

can be found by iteration of Eq. 6.8-28. The result is

$$\{\boldsymbol{\phi}^{-1}[(k, k_0)T]\}^T = \prod_{n=k_0}^{k-1} [\mathbf{A}^{-1}(nT)]^T \qquad k > k_0 \tag{6.8-31}$$

The state transition matrix for the original system, for $k_0 < k$, can be found by reverse iteration of Eq. 6.8-1, i.e., Eq. 6.8-1 can be written as $\mathbf{x}(nT) = \mathbf{A}^{-1}(nT)\mathbf{x}[(n + 1)T]$, and this expression can then be iterated from $n = k$ down to $n = k_0$. The result is

$$\mathbf{x}(k_0T) = \left[\prod_{n=k_0}^{k-1} \mathbf{A}^{-1}(nT)\right]\mathbf{x}(kT)$$

so that

$$\boldsymbol{\phi}[(k_0, k)T] = \prod_{n=k_0}^{k-1} \mathbf{A}^{-1}(nT) \qquad k > k_0 \tag{6.8-32}$$

These two equations show the validity of Eq. 6.8-14, namely,

$$\boldsymbol{\phi}^{-1}[(k, k_0)T] = \boldsymbol{\phi}[(k_0, k)T] \qquad k > k_0$$

However, Eq. 6.8-31 was obtained by a forward iteration or running the adjoint system forward in time from k_0T to kT, while Eq. 6.8-32 was obtained by a reverse iteration, or running the original system backward in time from kT to k_0T. This is shown in Fig. 6.8-2.

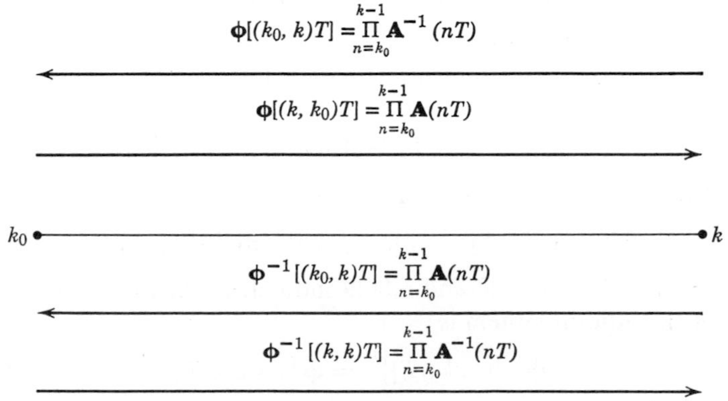

Fig. 6.8-2

Note that the form of the state transition matrix for increasing time (Eq. 6.8-4) is different from the form for the state transition matrix for decreasing time (Eq. 6.8-32). The reason for this difference is that reversing the time direction for a discrete system entails an inverse **A** matrix $\{\mathbf{x}(kT) = \mathbf{A}^{-1}(kT)\mathbf{x}[(k+1)T]\}$. Reversing the time direction for a continuous system simply entails reversing the sign of the **A** matrix $[d/d(-t) = -d/dt]$. A set of alternative forms of the state transition matrix is shown in Table 6.8-1.

Table 6.8-1

Original System		Adjoint System (Transposed)	
Forward Direction		*Forward Direction*	
$\boldsymbol{\phi}[(k, k_0)T] = \prod_{n=k_0}^{k-1} \mathbf{A}(nT)$	$k > k_0$	$\boldsymbol{\phi}^{-1}[(k, k_0)T] = \prod_{n=k_0}^{k-1} \mathbf{A}^{-1}(nT)$	$k > k_0$
$\boldsymbol{\phi}[(k_0, k)T] = \prod_{n=k}^{k_0-1} \mathbf{A}(nT)$	$k_0 > k$	$\boldsymbol{\phi}^{-1}[(k_0, k)T] = \prod_{n=k}^{k_0-1} \mathbf{A}^{-1}(nT)$	$k_0 > k$
Reverse Direction		*Reverse Direction*	
$\boldsymbol{\phi}[(k_0, k)T] = \prod_{n=k_0}^{k-1} \mathbf{A}^{-1}(nT)$	$k > k_0$	$\boldsymbol{\phi}^{-1}[(k_0, k)T] = \prod_{n=k_0}^{k-1} \mathbf{A}(nT)$	$k > k_0$
$\boldsymbol{\phi}[(k, k_0)T] = \prod_{n=k}^{k_0-1} \mathbf{A}^{-1}(nT)$	$k_0 > k$	$\boldsymbol{\phi}^{-1}[(k, k_0)T] = \prod_{n=k}^{k_0-1} \mathbf{A}(nT)$	$k_0 > k$

6.9 THE COMPLETE SOLUTION

The complete solution to the set of state equations

$$
\begin{aligned}
\mathbf{x}[(k+1)T] &= \mathbf{A}(kT)\mathbf{x}(kT) + \mathbf{B}(kT)\mathbf{v}(kT) \\
\mathbf{y}(kT) &= \mathbf{C}(kT)\mathbf{x}(kT) + \mathbf{D}(kT)\mathbf{v}(kT)
\end{aligned}
\tag{6.9-1}
$$

can be found by a process of iteration and induction, similar to the method used to obtain the state transition matrix. However, to illustrate the analogy between discrete and continuous systems, the adjoint system is used in a method similar to the integrating factor method.

The state transition matrix of the adjoint system $\{\boldsymbol{\phi}^{-1}[(k, k_0)T]\}^T$ must satisfy Eq. 6.8-29. Taking the transpose and postmultiplying by $\mathbf{A}(kT)$ yields

$$\boldsymbol{\phi}^{-1}[(k, k_0)T] = \boldsymbol{\phi}^{-1}[(k+1, k_0)T]\mathbf{A}(kT) \tag{6.9-2}$$

If Eq. 6.9-1 for $\mathbf{x}$ is premultiplied by $\boldsymbol{\phi}^{-1}[(k+1, k_0)T]$ and Eq. 6.9-2 is postmultiplied by $\mathbf{x}(kT)$ and the difference between the two equations taken, the result is

$$\boldsymbol{\phi}^{-1}[(k+1, k_0)T]\mathbf{x}[(k+1)T] - \boldsymbol{\phi}^{-1}[(k, k_0)T]\mathbf{x}(kT) = \boldsymbol{\phi}^{-1}[(k+1, k_0)T]\mathbf{B}(kT)\mathbf{v}(kT) \quad (6.9\text{-}3)$$

If k is replaced by m in Eq. 6.9-3 and both sides are then summed from k_0 to $k - 1$, the result is

$$\boldsymbol{\phi}^{-1}[(k, k_0)T]\mathbf{x}(kT) - \boldsymbol{\phi}^{-1}[(k_0, k_0)T]\mathbf{x}(k_0T) = \sum_{m=k_0}^{k-1} \boldsymbol{\phi}^{-1}[(m+1, k_0)T]\mathbf{B}(mT)\mathbf{v}(mT) \quad (6.9\text{-}4)$$

Since $\boldsymbol{\phi}^{-1}[(k_0, k_0)T] = \mathbf{I}$, premultiplication by $\boldsymbol{\phi}[(k, k_0)T]$ gives

$$\mathbf{x}(kT) = \boldsymbol{\phi}[(k, k_0)T]\mathbf{x}(k_0T) + \sum_{m=k_0}^{k-1} \boldsymbol{\phi}[(k, m+1)T]\mathbf{B}(mT)\mathbf{v}(mT) \quad (6.9\text{-}5)$$

The first term on the right side of Eq. 6.9-5 represents the initial condition response of the system, while the second term represents a *superposition* summation of the effects of the forcing function. This equation is analogous to Eq. 5.10-6 for the continuous case.

When the system under investigation is fixed, Eq. 6.9-5 can be written as the sum of an initial condition response and a *convolution* summation:

$$\mathbf{x}(kT) = \boldsymbol{\phi}(k - k_0)\mathbf{x}(k_0T) + \sum_{m=k_0}^{k-1} \boldsymbol{\phi}(k - m - 1)\mathbf{B}\mathbf{v}(mT) \quad (6.9\text{-}6)$$

The corresponding output $\mathbf{y}(kT)$ is obtained by substitution of Eq. 6.9-5 or 6.9-6 into Eq. 6.9-1. Thus

$$\mathbf{y}(kT) = \mathbf{C}(kT)\boldsymbol{\phi}[(k, k_0)T]\mathbf{x}(k_0T) + \sum_{m=k_0}^{k-1} \mathbf{C}(kT)\boldsymbol{\phi}[(k, m+1)T]\mathbf{B}(mT)\mathbf{v}(mT) + \mathbf{D}(kT)\mathbf{v}(kT) \quad (6.9\text{-}7)$$

for a time-varying system, or

$$\mathbf{y}(kT) = \mathbf{C}\boldsymbol{\phi}(k - k_0)\mathbf{x}(k_0T) + \sum_{m=k_0}^{k-1} \mathbf{C}\boldsymbol{\phi}(k - m - 1)\mathbf{B}\mathbf{v}(mT) + \mathbf{D}\mathbf{v}(kT) \quad (6.9\text{-}8)$$

for a fixed system.

For the case in which the system is fixed, it is frequently convenient to use the frequency-domain approach. For this case, $\mathbf{Y}(z)$ can be found by transforming Eq. 6.9-1 directly. The result of this operation is

$$\mathbf{X}(z) = (z\mathbf{I} - \mathbf{A})^{-1}z\mathbf{x}(0) + (z\mathbf{I} - \mathbf{A})^{-1}\mathbf{B}\mathbf{V}(z) \quad (6.9\text{-}9)$$

$$\mathbf{Y}(z) = \mathbf{C}(z\mathbf{I} - \mathbf{A})^{-1}z\mathbf{x}(0) + [\mathbf{C}(z\mathbf{I} - \mathbf{A})^{-1}\mathbf{B} + \mathbf{D}]\mathbf{V}(z) \quad (6.9\text{-}10)$$

An example is now given which illustrates the various approaches that can be taken.

Example 6.9-1. Determine the solution to the difference equation

$$y(k+2) + 5y(k+1) + 6y(k) = 1 \qquad (T = 1)$$

by each of the methods indicated below.

1. *Classical Solution—Time Domain.* Assume $y_H(k) = \beta^k$. Then $(\beta^2 + 5\beta + 6)\beta^k = 0$, or $\beta_1 = -2$ and $\beta_2 = -3$. The homogeneous solution is then $y_H(k) = C_1(-2)^k + C_2(-3)^k$. Assume that $y_P(k) = C_3$. Then $C_3 + 5C_3 + 6C_3 = 1$, or $C_3 = \frac{1}{12}$. The total solution is

$$y(k) = C_1(-2)^k + C_2(-3)^k + \tfrac{1}{12}$$

Using the initial conditions $y(0)$ and $y(1)$,

$$y(0) = C_1 + C_2 + \tfrac{1}{12} \quad \text{and} \quad y(1) = -2C_1 - 3C_2 + \tfrac{1}{12}$$

The constants C_1 and C_2 are

$$C_1 = 3y(0) + y(1) - \tfrac{1}{3}, \quad C_2 = \tfrac{1}{4} - 2y(0) - y(1)$$

Substituting these constants into the total solution gives the complete solution

$$y(k) = [3(-2)^k - 2(-3)^k]y(0) + [(-2)^k - (-3)^k]y(1) + [\tfrac{1}{4}(-3)^k - \tfrac{1}{3}(-2)^k + \tfrac{1}{12}]$$

2. *State Variables Technique—Time Domain.* From Example 6.8-1, the state transition matrix for this system is given by

$$\boldsymbol{\Phi}(k) = \begin{bmatrix} 3(-2)^k - 2(-3)^k & (-2)^k - (-3)^k \\ -6[(-2)^k - (-3)^k] & -2(-2)^k + 3(-3)^k \end{bmatrix}$$

The **B**, **C**, and **D** matrices are

$$\mathbf{B} = \begin{bmatrix} 0 \\ 1 \end{bmatrix}, \quad \mathbf{C} = [1 \quad 0], \quad \mathbf{D} = [0]$$

The output $y(k)$ is given by Eq. 6.9-8 as

$$\mathbf{y}(k) = \mathbf{C}\boldsymbol{\Phi}(k)\mathbf{x}(0) + \sum_{m=0}^{k-1} \mathbf{C}\boldsymbol{\Phi}[(k-m-1)]\mathbf{B}\mathbf{v}(mT) + \mathbf{D}\mathbf{v}(kT)$$

which, for this case, reduces to

$$y(k) = \phi_{11}(k)y(0) + \phi_{12}y(1) + \sum_{m=0}^{k-1} \phi_{12}(k-m-1)$$

The computation of the summation for a forced input may involve some skill in finding a closed-form expression for the resulting series. The summation formula

$$\sum_{m=0}^{k-1} a^{k-m-1} = \frac{1-a^k}{1-a} \qquad \text{(sum of a geometric series)}$$

is of particular use in expressions of this type.[13] Since $\phi_{12}(k-m-1) = (-2)^{k-m-1} - (-3)^{k-m-1}$, the summation in this case is equal to

$$\frac{1-(-2)^k}{1-(-2)} - \frac{1-(-3)^k}{1-(-3)} = \frac{1}{12} + \frac{1}{4}(-3)^k - \frac{1}{3}(-2)^k$$

The complete solution is then

$$y(k) = [3(-2)^k - 2(-3)^k]y(0) + [(-2)^k - (-3)^k]y(1) + [\tfrac{1}{4}(-3)^k - \tfrac{1}{3}(-2)^k + \tfrac{1}{12}]$$

The expression above for the sum of a geometric series is particularly helpful if the matrix summation

$$\sum_{m=0}^{k-1} \mathbf{A}^{k-m-1}$$

is to be found. The Cayley-Hamilton method can be applied where $F(\lambda_i) = (1 - \lambda_i^k)/(1 - \lambda_i)$.

A useful relation in finding the closed-form expression, if one exists, for a summation is the formula for *summation by parts*, analogous to the formula for *integration by parts*. This formula is given by (see Problem 2.12)

$$\sum_{M}^{N} u(k)\,\Delta v(k) = [u(k)v(k)]_M^{N+1} - \sum_{M}^{N} v(k+1)\,\Delta u(k)$$

As an example, the summation

$$\sum_{0}^{N} kr^k \qquad r \neq 1$$

can be found by setting

$$u(k) = k,\ \Delta v(k) = r^k. \text{ Then } \Delta u(k) = 1,$$

$$v(k) = \sum_{n=0}^{k-1} r^n + C_1 = \frac{r^k - 1}{r - 1} + C_1 = \frac{r^k}{r-1} + C_2$$

For convenience, let $C_2 = 0$. Then

$$\sum_{0}^{N} kr^k = \left[\frac{kr^k}{r-1}\right]_0^{N+1} - \frac{1}{r-1}\sum_{0}^{N} r^{k+1}$$

$$= \frac{1}{(r-1)^2}[Nr^{N+2} - (N+1)r^{N+1} + r] \qquad r \neq 1$$

3. *Standard Z Transform Technique*. Taking the Z transform of both sides of the given difference equation, there results

$$z^2[Y(z) - y(0) - z^{-1}y(1)] + 5z[Y(z) - y(0)] + 6Y(z) = \frac{z}{z-1}$$

or

$$Y(z) = \frac{z/(z-1) + z[zy(0) + y(1) + 5y(0)]}{(z+2)(z+3)}$$

The poles of this function are at $z = 1, -2, -3$. Since

$$y(k) = \sum \text{residues of } Y(z)z^{k-1} \text{ at poles of } Y(z)$$

there are three residues to compute.

$$R_1 = (z-1)Y(z)z^{k-1}\Big|_{z=1} = \tfrac{1}{12}$$
$$R_2 = (z+2)Y(z)z^{k-1}\Big|_{z=-2} = [-\tfrac{1}{3} + 3y(0) + y(1)](-2)^k$$
$$R_3 = (z+3)Y(z)z^{k-1}\Big|_{z=-3} = [\tfrac{1}{4} - 2y(0) - y(1)](-3)^k$$

The complete solution is then

$$y(k) = [3(-2)^k - 2(-3)^k]y(0) + [(-2)^k - (-3)^k]y(1) + [\tfrac{1}{4}(-3)^k - \tfrac{1}{3}(-2)^k + \tfrac{1}{12}]$$

4. *State Space Technique—Frequency Domain.* From Eq. 6.9-10, the Z transform of the output $\mathbf{y}(k)$ is

$$\mathbf{Y}(z) = \mathbf{C}(z\mathbf{I} - \mathbf{A})^{-1}z\mathbf{x}(0) + [\mathbf{C}(z\mathbf{I} - \mathbf{A})^{-1}\mathbf{B} + \mathbf{D}]\mathbf{V}(z)$$

From Example 6.8-4, the matrix $(z\mathbf{I} - \mathbf{A})^{-1}$ is given by

$$(z\mathbf{I} - \mathbf{A})^{-1} = \frac{\begin{bmatrix} z+5 & 1 \\ -6 & z \end{bmatrix}}{(z+2)(z+3)}$$

The $\mathbf{B}$, $\mathbf{C}$, and $\mathbf{D}$ matrices were given in part 2 of this example as

$$\mathbf{B} = \begin{bmatrix} 0 \\ 1 \end{bmatrix}, \quad \mathbf{C} = [1 \quad 0], \quad \mathbf{D} = [0]$$

The transform of the output, $Y(z)$, is then

$$Y(z) = \frac{z(z+5)}{(z+2)(z+3)}y(0) + \frac{z}{(z+2)(z+3)}y(1) + \frac{z}{(z-1)(z+2)(z+3)}$$

This z transform corresponds to the $Y(z)$ found in part 3 of this example. Therefore the answer is identical with that given in the three preceding parts of this example.

5. *State Variables Technique—Normal Form.* The modal matrix $\mathbf{M}$ is found by successive substitution of the characteristic values of $\mathbf{A}$ into the matrix $\mathbf{Adj}\,[\lambda\mathbf{I} - \mathbf{A}]$. The characteristic values of $\mathbf{A}$ are $\lambda_1 = -2$, $\lambda_2 = -3$, and $\mathbf{Adj}\,[\lambda\mathbf{I} - \mathbf{A}]$ is given by

$$\mathbf{Adj}\,[\lambda\mathbf{I} - \mathbf{A}] = \begin{bmatrix} \lambda+5 & 1 \\ -6 & \lambda \end{bmatrix}$$

Therefore $\mathbf{M}$ is given by

$$\mathbf{M} = \begin{bmatrix} 1 & 1 \\ -2 & -3 \end{bmatrix} \quad \text{and} \quad \mathbf{M}^{-1} = \begin{bmatrix} 3 & 1 \\ -2 & 1 \end{bmatrix}$$

Since

$$\mathbf{M}^{-1}\mathbf{A}\mathbf{M} = \mathbf{\Lambda} = \begin{bmatrix} -2 & 0 \\ 0 & -3 \end{bmatrix}$$

the transformation $\mathbf{q} = \mathbf{M}^{-1}\mathbf{x}$ leads into the form

$$\mathbf{q}(k+1) = \mathbf{\Lambda}\mathbf{q}(k) + \mathbf{B}_n\mathbf{v}(k) \qquad (T = 1)$$
$$\mathbf{y}(k) = \mathbf{C}_n\mathbf{q}(k) + \mathbf{D}\mathbf{v}(k)$$

where $\mathbf{B}_n = \mathbf{M}^{-1}\mathbf{B}$ and $\mathbf{C}_n = \mathbf{CM}$. Therefore $q_1 = 3x_1 + x_2$, $q_2 = -2x_1 - x_2$, and

$$\mathbf{B}_n = \begin{bmatrix} 1 \\ -1 \end{bmatrix} \quad \text{and} \quad \mathbf{C}_n = [1 \quad 1]$$

The output $\mathbf{y}(k)$ is then

$$\mathbf{y}(k) = \mathbf{C}_n\mathbf{\Lambda}^k\mathbf{q}(0) + \sum_{m=0}^{k-1}\mathbf{C}_n\mathbf{\Lambda}^{k-m-1}\mathbf{B}_n\mathbf{v}(m)$$

from Eq. 6.6-4 and the above. Since $\mathbf{\Lambda}$ is a diagonal matrix, finding $\mathbf{\Lambda}^k$ means simply raising the elements of $\mathbf{\Lambda}$ to the kth power. This is one of the advantages of using this method. It follows that

$$y(k) = (-2)^k q_1(0) + (-3)^k q_2(0) + \sum_{m=0}^{k-1} \{(-2)^{k-m-1} - (-3)^{k-m-1}\}$$

or

$$y(k) = (-2)^k q_1(0) + (-3)^k q_2(0) + [\tfrac{1}{4}(-3)^k - \tfrac{1}{3}(-2)^k + \tfrac{1}{12}]$$

If it is desired that $y(k)$ be expressed in terms of the original initial conditions, then $q_1(0) = 3x_1(0) + x_2(0) = 3y(0) + y(1)$ and $q_2(0) = -2x_1(0) - x_2(0) = -2y(0) - y(1)$ yield

$$y(k) = [3y(0) + y(1)](-2)^k - [2y(0) + y(1)](-3)^k + [\tfrac{1}{4}(-3)^k - \tfrac{1}{3}(-2)^k + \tfrac{1}{12}]$$

6. *Mode Expansion Method.* Since

$$\mathbf{M} = \begin{bmatrix} 1 & 1 \\ -2 & -3 \end{bmatrix} \quad \text{and} \quad \mathbf{M}^{-1} = \begin{bmatrix} 3 & 1 \\ -2 & -1 \end{bmatrix}$$

from part 5, it follows that

$$\mathbf{u}_1 = \begin{bmatrix} 1 \\ -2 \end{bmatrix} \quad \text{and} \quad \mathbf{u}_2 = \begin{bmatrix} 1 \\ -3 \end{bmatrix}$$

form a basis, and

$$\mathbf{r}_1 = \begin{bmatrix} 3 \\ 1 \end{bmatrix} \quad \text{and} \quad \mathbf{r}_2 = \begin{bmatrix} -2 \\ -1 \end{bmatrix}$$

form a reciprocal basis. Thus, from Eq. 6.6-6,

$$y(k) = x_1(k) = [3x_1(0) + x_2(0)](-2)^k + [-2x_1(0) - x_2(0)](-3)^k + \sum_{m=0}^{k-1} [(-2)^{k-m-1} - (-3)^{k-m-1}]$$

This is the same expressions as for $y(k)$ in terms of the ϕ's in part 2. Thus performance of the indicated summations and substitution of $y(0) = x_1(0)$ and $y(1) = x_2(0)$ yields the desired result.

7. *State Variables Technique—Partial Fraction Expansion.* Since

$$H(z) = \frac{1}{z^2 + 5z + 6} = \frac{1}{z+2} - \frac{1}{z+3}$$

then

$$Y(z) = \frac{V(z)}{z+2} - \frac{V(z)}{z+3} = X_1(z) + X_2(z)$$

or $y(k) = x_1(k) + x_2(k)$, where x_1 and x_2 satisfy the first order difference equations

$$x_1(k+1) + 2x_1(k) = v(k) \quad \text{and} \quad x_2(k+1) + 3x_2(k) = -v(k)$$

Therefore the **A**, **B**, **C**, and **D** matrices are

$$\mathbf{A} = \begin{bmatrix} -2 & 0 \\ 0 & -3 \end{bmatrix}, \quad \mathbf{B} = \begin{bmatrix} 1 \\ -1 \end{bmatrix}, \quad \mathbf{C} = [1 \quad 1], \quad \mathbf{D} = [0]$$

Since these are exactly the same matrices which were derived in part 5, $y(k)$ is then

$$y(k) = (-2)^k x_1(0) + (-3)^k x_2(0) + [\tfrac{1}{4}(-3)^k - \tfrac{1}{3}(-2)^k + \tfrac{1}{12}]$$

However, since the transfer function gives no clue to the relationship between **y** and **x**, this relationship must be obtained from examination of $y(k)$. This is necessary because the known initial conditions are in terms of $y(0)$ and $y(1)$. The desired relationships can be found by substituting $k = 0$ and $k = 1$ into $y(k) = x_1(k) + x_2(k)$. This gives $y(0) = x_1(0) + x_2(0)$ and $y(1) = x_1(1) + x_2(1)$. Now, using the equations for $x_1(k + 1)$ and $x_2(k + 1)$ at $k = 0$, these expressions become

$$y(0) = x_1(0) + x_2(0) \quad \text{and} \quad y(1) = -2x_1(0) - 3x_2(0)$$

Solving these two simultaneous equations yields

$$x_1(0) = 3y(0) + y(1) \quad \text{and} \quad x_2(0) = -2y(0) - y(1)$$

which are the same relationships found in part 5 for the q's. Although this procedure is not too difficult for a second order system, it may prove to be laborious for higher order systems with many inputs and outputs.

In looking through these various approaches, there are advantages and disadvantages to each. For single input-output fixed systems, the standard Z transform approach is certainly the easiest to use. For multiple input-output systems, a matrix technique is advisable. Which one of the matrix techniques to use is a different question. Since time-varying systems are almost impossible to solve analytically, the time-domain formulation of a time-varying system is best for computer purposes. For the fixed case, the Z transform of the state equations is quite useful, since numerous Z transform tables are available. The use of the Z transform bypasses some of the difficulties in evaluating the summation forms that are obtained in a time-domain computation. It appears that the normal form is perhaps the best form conceptually, for mathematical proofs, and for time-domain analysis of the unforced system. When the system is subject to a forcing function, the Z transform of the normal form is most convenient.

Stability of Fixed Discrete Systems

For a discrete fixed system, the state transition matrix approaches zero as k approaches infinity, if the characteristic values of **A** are located inside the unit circle. If a characteristic value lies on the unit circle and is of order one, then $\boldsymbol{\phi}(k)$ is bounded as k approaches infinity. For any characteristic values which lie outside the unit circle or for multiple characteristic values which lie on the unit circle, $\boldsymbol{\phi}(k)$ becomes infinite as k approaches infinity. These statements can be proved from the Cayley-Hamilton method of obtaining $\mathbf{A}^k$, which depends upon obtaining certain elements α_m such that

$$\mathbf{A}^k = \sum_{m=0}^{n-1} \alpha_m \mathbf{A}^m$$

where n is the order of the **A** matrix. The α_m are obtained from the equations

$$F(\lambda)\big|_{\lambda=\lambda_i} = \sum_{m=0}^{n-1} \alpha_m \lambda_i^{\,m} = \lambda_i^{\,k}$$

or

$$\frac{d^p}{d\lambda^p} F(\lambda)\bigg|_{\lambda=\lambda_i} = \frac{d^p}{d\lambda^p}\left(\sum_{m=0}^{n-1} \alpha_m \lambda^m\right)\bigg|_{\lambda=\lambda_i} = \frac{d^p}{d\lambda^p}(\lambda^k)\bigg|_{\lambda=\lambda_i}$$

where $p = 0, 1, 2, \ldots, r - 1$ for the case where the characteristic value is of order r. When the characteristic values are distinct, the elements of $\mathbf{A}^k$ contain linear combinations of elements such as $(\lambda_i)^k$. These elements vanish as k approaches infinity if $|\lambda_i| < 1$, become bounded if $|\lambda_i| = 1$, and become unbounded if $|\lambda_i| > 1$. When there are multiple characteristic values, the elements of $\mathbf{A}^k$ contain linear combinations of elements such as $k\lambda_i^{k-1}$. Clearly, these elements are unbounded for $|\lambda_i| \geq 1$ and approach zero as k approaches infinity for $|\lambda_i| < 1$.

6.10 THE UNIT FUNCTION RESPONSE MATRIX

The output $\mathbf{y}(kT)$ of a linear time-varying discrete system can be written in terms of the **A**, **B**, **C**, and **D** matrices as

$$(kT) = \sum_{m=-\infty}^{k-1} \mathbf{C}(kT)\boldsymbol{\phi}[(k, m + 1)T]\mathbf{B}(mT)\mathbf{v}(mT) + \mathbf{D}(kT)\mathbf{v}(kT) \quad (6.10\text{-}1)$$

by setting k_0 equal to $-\infty$ in Eq. 6.9-7. In terms of the unit function response matrix $\mathbf{H}[(k, m)T]$, the output $\mathbf{y}(kT)$ is given by the superposition summation

$$\mathbf{y}(kT) = \sum_{m=-\infty}^{k} \mathbf{H}[(k, m)T]\mathbf{v}(mT) \quad (6.10\text{-}2)$$

A comparison of Eqs. 6.10-1 and 6.10-2 shows that the unit function response matrix is

$$\mathbf{H}[(k, k_0)T] = \begin{cases} \mathbf{C}(kT)\boldsymbol{\phi}[(k, k_0 + 1)T]\mathbf{B}(k_0T) & k \geqslant k_0 + 1 \\ \mathbf{D}(kT) & k = k_0 \end{cases}$$
$$= [0] \qquad k < k_0 \quad (6.10\text{-}3)$$

For a fixed system the unit function response matrix is

$$\mathbf{H}[(k - k_0)T] = \begin{cases} \mathbf{C}\boldsymbol{\phi}[(k - k_0 - 1)]\mathbf{B} & k \geqslant k_0 + 1 \\ \mathbf{D} & k = k_0 \end{cases}$$
$$= [0] \qquad k < k_0 \quad (6.10\text{-}4)$$

In the frequency domain, from Eq. 6.9-10, the unit function response matrix $\mathbf{H}(z)$ for a fixed system is given by

$$\mathbf{H}(z) = \mathbf{C}(z\mathbf{I} - \mathbf{A})^{-1}\mathbf{B} + \mathbf{D} \tag{6.10-5}$$

Example 6.10-1. The system of Fig. 6.10-1 represents a simple sampled-data system. The block $(1 - \epsilon^{-sT})/s$ is commonly called a zero order hold, since it takes the input sample at time kT and provides this value as its output until time $(k + 1)T$. Determine the unit function response matrix $\mathbf{H}(kT)$ assuming $T = 1$.

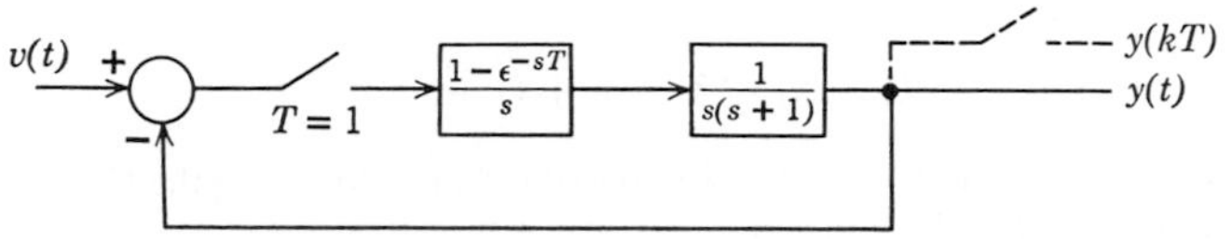

Fig. 6.10-1

The transfer function $G(z)$ of the forward transmission is

$$G(z) = (1 - z^{-1})\mathscr{Z}\left[\frac{1}{s^2(s+1)}\right] = \frac{\epsilon^{-1} - 1}{z - \epsilon^{-1}} + \frac{1}{z - 1}$$

Using this transfer function, the sampled-data system of Fig. 6.10-1 can be redrawn as the discrete time system of Fig. 6.10-2. In this figure, the forward path has been broken

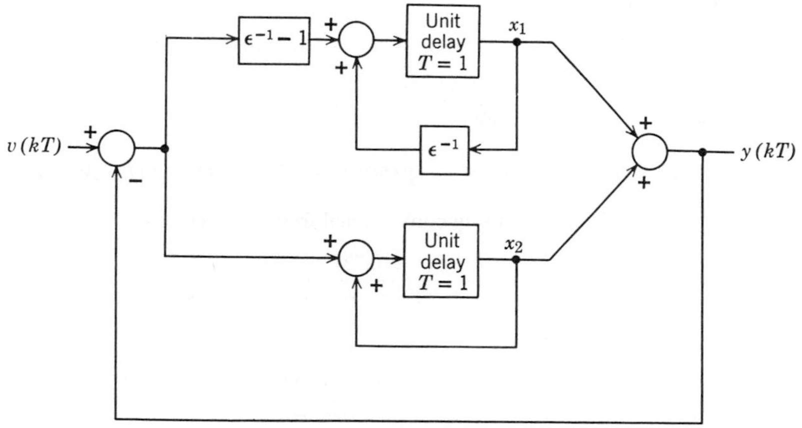

Fig. 6.10-2

up into its partial fraction expansion. The closed-loop system state equations can then be written down by inspection. They are

$$\begin{bmatrix} x_1(k+1) \\ x_2(k+1) \end{bmatrix} = \begin{bmatrix} 1 & 1 - \epsilon^{-1} \\ -1 & 0 \end{bmatrix}\begin{bmatrix} x_1(k) \\ x_2(k) \end{bmatrix} + \begin{bmatrix} \epsilon^{-1} - 1 \\ 1 \end{bmatrix} v(k)$$

$$y(k) = [1 \quad 1]\begin{bmatrix} x_1(k) \\ x_2(k) \end{bmatrix}$$

From Eq. 6.10-4, the unit function matrix $\mathbf{H}(k)$ is

$$\mathbf{H}(k) = [1 \quad 1]\begin{bmatrix} \phi_{11}(k-1) & \phi_{12}(k-1) \\ \phi_{21}(k-1) & \phi_{22}(k-1) \end{bmatrix}\begin{bmatrix} \epsilon^{-1} - 1 \\ 1 \end{bmatrix}$$

Since

$$\boldsymbol{\Phi}(k-1) = \mathbf{A}^{(k-1)} = \alpha_0\mathbf{I} + \alpha_1\mathbf{A} = \begin{bmatrix} \alpha_0 + \alpha_1 & \alpha_1 - \alpha_1\epsilon^{-1} \\ -\alpha_1 & \alpha_0 \end{bmatrix}$$

then $H(k) = \alpha_0\epsilon^{-1} + \alpha_1(1 - \epsilon^{-1})$, where α_0 and α_1 are to be determined. The roots of the characteristic equation

$$|\lambda\mathbf{I} - \mathbf{A}| = \lambda^2 - \lambda + (1 - \epsilon^{-1}) = 0$$

are $\lambda_{1,2} = N\epsilon^{\pm j\theta}$ where $N = 0.795$, $\theta = 0.680$ radian. Applying the Cayley-Hamilton method, where $F(\lambda_i) = \lambda^{k-1}$, the equations

$$N^{(k-1)}\epsilon^{j(k-1)\theta} = \alpha_0 + \alpha_1 N\epsilon^{j\theta}$$

$$N^{(k-1)}\epsilon^{-j(k-1)\theta} = \alpha_0 + \alpha_1 N\epsilon^{-j\theta}$$

are obtained. Solving for α_0 and α_1,

$$\alpha_0 = N^{(k-1)}[\cos(k-1)\theta - \cot\theta\sin(k-1)\theta]$$

$$\alpha_1 = \frac{N^{(k-2)}\sin[(k-1)\theta]}{\sin\theta}$$

The unit function response is then

$$H(k) = \epsilon^{-1}N^{(k-1)}[\cos(k-1)\theta - \cot\theta\sin(k-1)\theta] + \frac{(1-\epsilon^{-1})N^{(k-2)}}{\sin\theta}\sin(k-1)\theta$$

or

$$H(k) = (0.795)^{(k-1)}\{0.368\cos[0.680(k-1)] + 0.724\sin[0.680(k-1)]\}$$

This result can be checked by using conventional feedback methods.

$$H(z) = \frac{G(z)}{1 + G(z)} = \frac{0.368(z + 0.72)}{(z - N\epsilon^{j\theta})(z - N\epsilon^{-j\theta})} \qquad \begin{cases} N = 0.795 \\ \theta = 0.680 \text{ radian} \end{cases}$$

Taking the inverse Z transform,

$$H(k) = 0.368N^{(k-1)}\left[\frac{(N\epsilon^{j\theta} + 0.72)\epsilon^{j(k-1)\theta} - (N\epsilon^{-j\theta} + 0.72)\epsilon^{-j(k-1)\theta}}{N(\epsilon^{j\theta} - \epsilon^{-j\theta})}\right]$$

After some manipulation, this can be written as

$$H(k) = 0.368N^{(k-1)}\left[\cos(k-1)\theta + \left(\cot\theta + \frac{0.72}{N\sin\theta}\right)\sin(k-1)\theta\right]$$
$$= (0.795)^{(k-1)}\{0.368\cos[0.680(k-1)] + 0.724\sin[0.680(k-1)]\}$$

For a time-varying system, the elements of the unit function response can be obtained by simulation in a manner analogous to that used for continuous systems. For an m input-p output system, the ith output $y_i(kT)$ is given by

$$y_i(kT) = \sum_{n=-\infty}^{k} \sum_{j=1}^{m} h_{ij}[(k, n)T]v_j(nT) \tag{6.10-6}$$

The jth *column* of $\mathbf{H}[(k, n)T]$ can be obtained by setting all the inputs except the jth equal to zero, and placing a unit function on the jth input at time nT. The outputs of the system are $h_{ij}[(k, n)T]$ for $i = 1, 2, \ldots, p$ and for fixed j. In order to obtain the complete response $h_{ij}[(k, n)T]$ as a function of both k and n, a set of runs must be performed, each run starting at a different time $n_1T, n_2T, \ldots$. The results of these runs must then be cross-plotted to obtain the variation with respect to nT, the point in time of application. Proceeding to different values of j, these tests must be repeated until all m columns of the unit function response matrix are obtained. This is the same problem that was presented in the last chapter where the impulse response matrix $\mathbf{H}(t, \tau)$ was obtained by a similar cross-plotting procedure. In this discrete case, however, the difference equations can be solved on a digital computer and the necessary cross-plotting also done by the computer. Thus the discrete modified adjoint system is not discussed.

Example 6.10-2. In Section 5.11, the differential equation $\ddot{y} + t\dot{y} + y = 0$ was analyzed, and the impulse response $h(T, \tau)$ was obtained by performing a set of simulation runs for different values of τ and cross-plotting the results for fixed $t = T$. The results of the simulation runs are shown in Fig. 5.11-2*a*, and the results of the cross-plotting are shown in Fig. 5.11-2*b*. Perform the same task, but for the discrete version of the same differential equation. This method is often used to solve a time-varying differential equation numerically by either a desk calculator or a simple digital computer routine.

A discrete version of the equation can be obtained in the following manner. Since

$$\frac{d^2y}{dt^2} = \lim_{h\to 0}\left[\frac{y(t+h) - 2y(t) + y(t-h)}{h^2}\right]$$

$$\frac{dy}{dt} = \lim_{h\to 0}\left[\frac{y(t+h) - y(t)}{h}\right]$$

an approximate solution can be obtained by letting $h = T$ and $t = kT$, and using a "small" value for T. Thus

$$\frac{y(kT+T) - 2y(kT) + y(kT-T)}{T^2} + \frac{kT[y(kT+T) - y(kT)]}{T} + y(kT) = 0$$

or

$$(1 + kT^2)[y(kT+T)] + (T^2 - kT^2 - 2)[y(kT)] + [y(kT-T)] = 0$$

The initial conditions $y(\tau) = 0$, $\dot{y}(\tau) = 1$ are replaced by

$$y(k_0T) = 0$$

$$\frac{y(k_0T+T) - y(k_0T)}{T} = 1$$

or $y(k_0T) = 0$ and $y(k_0T + T) = T$.

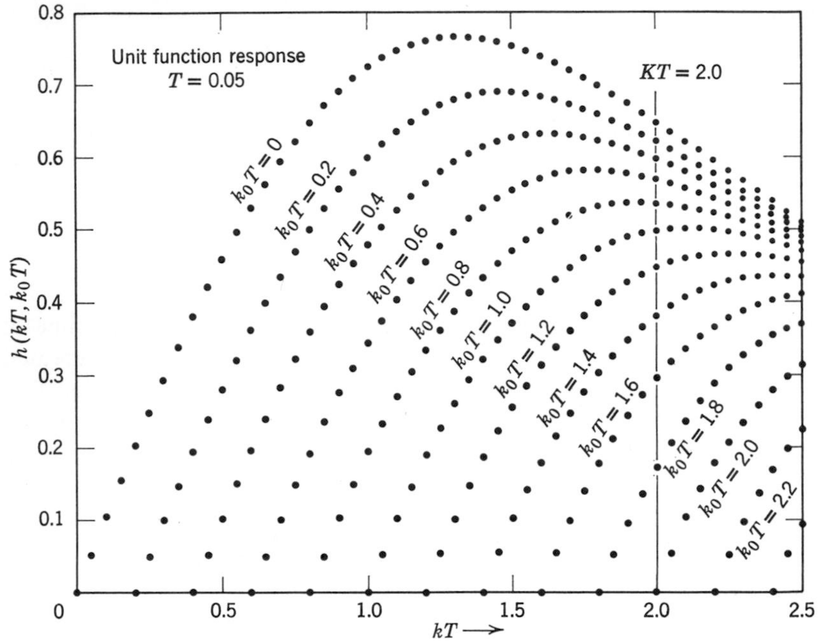

Fig. 6.10-3*a*

Using a value of $T = 0.05$ and $k_0 = 0, 4, 8, 12, \ldots, 48$, the points shown on Fig. 6.10-3*a* are obtained. Obtaining these points is a relatively simple task for a digital computer, and the cross-plotting can also be performed by the computer. A desk calculator can also be used. However, care must be taken to use sufficient accuracy in computing each point, as the round-off errors can build up rapidly. The interested reader can consult any of the many texts written on numerical solution of differential equations.[14,15]

As a comparison of the accuracy that can be obtained by simple numerical methods the results of the continuous system simulation run and the results of the approximate numerical solution for $\tau = k_0 = 0$ are listed in Table 6.10-1. The numerical solution is given to three places, while the simulation run is given to two places, this being the accuracy of reading from the original recording. For this comparison, the discrete simulation reproduces the results of the continuous simulation within the accuracy of the recording of the continuous information.

A crossplot of the points at $KT = 2.0$ is shown in Fig. 6.10-3*b*. The points of Fig. 6.10-3*b* represent $h(2.0, k_0T)$, the unit response of the system as $kT = 2.0$ as a function of the time of application of the unit input. A comparison of the cross-plot obtained by discrete simulation and the cross-plot obtained by the continuous simulation (Fig. 5.11-2*b*) is shown in Table 6.10-2. This comparison shows that the discrete simulation is fairly good, but not within the accuracy of the continuous system. The differences between the continuous and discrete systems are due to the accumulation of round-off error and the basic approximation involved. Because these effects are more noticeable farther out along a simulation run, a cross-plot at $kT = KT$ shows these effects more than a cross-plot at $kT < KT$.

Table 6.10-1

$T = 0.05$

k	$h(k, 0)$	$h(t, 0)$	k	$h(k, 0)$	$h(t, 0)$
0	0	0	22	0.751	0.75
2	0.100	0.10	24	0.766	0.76
4	0.198	0.20	26	0.771	0.77
6	0.291	0.29	28	0.769	0.77
8	0.380	0.38	30	0.760	0.76
10	0.461	0.46	32	0.746	0.75
12	0.534	0.53	34	0.727	0.73
14	0.597	0.60	36	0.705	0.71
16	0.651	0.65	38	0.679	0.68
18	0.695	0.69	40	0.653	0.65
20	0.728	0.72			

Figure 6.10-3*b* can also be obtained by a discrete simulation of the modified adjoint differential equation. The original differential equation is $\ddot{y} + t\dot{y} + y = 0$, and the adjoint differential equation is $\ddot{\alpha} - (d/dt)(t\alpha) + \alpha = 0$ or $\ddot{\alpha} - t\dot{\alpha} = 0$. Making the change in variable $t = T_0 - t_1$, the modified adjoint differential equation is then

$$\frac{d^2\alpha}{dt_1^{\,2}} + (T_0 - t_1)\frac{d\alpha}{dt_1} = 0$$

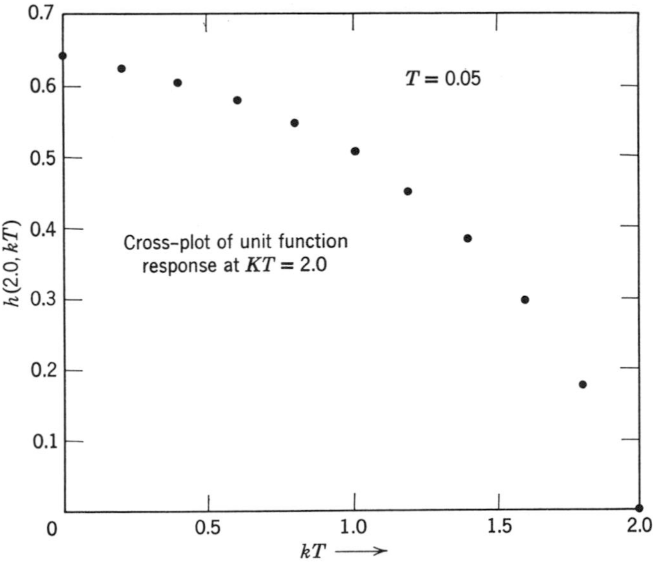

Fig. 6.10-3*b*

Table 6.10-2

	$T = 0.05$	
k	$h(2.0, k_0T)$	$h(2.0, \tau)$
0	0.653	0.65
4	0.628	0.63
8	0.603	0.60
12	0.574	0.56
16	0.541	0.53
20	0.501	0.48
24	0.450	0.43
28	0.385	0.37
32	0.297	0.28
36	0.175	0.17
40	0	0

The discrete simulation of this equation is found by the procedure used to find the discrete simulation of the original differential equation. The resulting difference equation is

$$[1 + (K - k')T^2]y(k'T + T) + [-2 - (K - k')T^2]y(k'T) + y(k'T - T) = 0$$

where $KT = T_0$ is the fixed end time. A comparison of the points obtained by this simulation and those obtained from the continuous system simulation is shown in Table 6.10-3.

Table 6.10-3

	$KT = 1.0, \quad T = 0.1$	
k	$h_{k'}{}^*(k', 0)$	$h_{t_1}{}^*(t_1, 0)$
0	0	0
1	0.100	—
2	0.191	0.18
3	0.276	—
4	0.350	0.33
5	0.420	—
6	0.487	0.47
7	0.550	—
8	0.612	0.60
9	0.673	—
10	0.733	0.72

Transmission Matrices

For single input-output systems, the *transmission matrix* is sometimes used to describe the unit response of the system. The transmission matrix is simply an ordered array of the elements $h(iT, jT)$, where i indicates the row and j indicates the column where the element is located. The general form for the transmission matrix is then

$$\mathbf{H}_T(kT, k_0T) = \begin{bmatrix} h(0,0) & 0 & 0 & \cdots & 0 \\ h(T,0) & h(T,T) & 0 & \cdots & 0 \\ h(2T,0) & h(2T,T) & h(2T,2T) & \cdots & 0 \\ \cdots & \cdots & \cdots & \cdots & \cdots \\ h(mT,0) & h(mT,T) & h(mT,2T) & \cdots & h(mT,mT) \end{bmatrix} \tag{6.10-7}$$

All elements to the right of the element $h(iT, iT)$ are zero, since the system is assumed to be physically realizable, or nonanticipative. If the input $v(kT)$ is ordered into a column vector whose components are $v(0)$, $v(T)$, $v(2T), \ldots, v(mT)$ then the output of the system can be written as

$$\mathbf{y}(kT) = \mathbf{H}_T(kT, k_0T)\mathbf{v}(k_0T) \tag{6.10-8}$$

It is understood that the output $y(kT)$ is ordered into a column vector whose components are $y(0), y(T), y(2T), \ldots, y(mT)$. The ith component of the column vector $\mathbf{y}(kT)$ is then

$$y(iT) = \sum_{k_0=0}^{i} h[(i, k_0)T]v(k_0T) \tag{6.10-9}$$

For a fixed system, the elements of the transmission matrix are $h(iT - jT)$, the argument being the difference between the time of observation and the time of application. Thus

$$\mathbf{H}_T(kT) = \begin{bmatrix} h(0) & 0 & 0 & \cdots & 0 \\ h(T) & h(0) & 0 & \cdots & 0 \\ h(2T) & h(T) & h(0) & \cdots & 0 \\ \cdots & \cdots & \cdots & \cdots & \cdots \\ h(mT) & h(mT-T) & h(mT-2T) & \cdots & h(0) \end{bmatrix} \tag{6.10-10}$$

These matrices are not very useful when dealing with systems with several inputs and outputs, but they have some application when dealing with single input-single output systems or systems comprised of interconnected single input-single output systems.

Example 6.10-3. Repeat Example 1.9-1 using the transmission matrix description. $y(k)$, $k = 0, 1, 2, \ldots$, can be written as

$$\mathbf{y}(k) = \mathbf{D}_T(k)\mathbf{v}(k)$$

where

$$\mathbf{y}(k) = \begin{bmatrix} y(0) \\ y(1) \\ y(2) \\ \cdot \\ \cdot \\ \cdot \end{bmatrix}$$

and from Example 1.9-1,

$$\mathbf{D}_T(k) = \begin{bmatrix} 0 & 0 & \cdot & \cdot & \cdot & \cdot & \cdot \\ 2 & 0 & 0 & \cdot & \cdot & \cdot & \cdot \\ 4 & 2 & 0 & 0 & \cdot & \cdot & \cdot \\ 8 & 4 & 2 & 0 & 0 & \cdots & \\ \cdot & \cdot & \cdot & \cdot & \cdot & & \\ \cdot & \cdot & \cdot & \cdot & & \cdot & \\ \cdot & \cdot & \cdot & \cdot & & & \cdot \end{bmatrix} \quad \text{and} \quad \mathbf{v}(k) = \begin{bmatrix} 0 \\ 1 \\ 2 \\ 3 \\ \cdot \\ \cdot \\ \cdot \\ \cdot \end{bmatrix}$$

Note that by defining

$$\mathbf{V}_T(k) = \begin{bmatrix} v(0) & 0 & \cdot & \cdots & \cdots \\ v(1) & v(0) & 0 & \cdots & \cdots \\ v(2) & v(1) & v(0) & 0 & \cdots \\ \cdot & \cdot & & \cdot & \\ \cdot & \cdot & & & \cdot \\ \cdot & \cdot & & & \cdot \end{bmatrix}$$

and

$$\mathbf{d}(k) = \begin{bmatrix} d(0) \\ d(1) \\ d(2) \\ \cdot \\ \cdot \\ \cdot \end{bmatrix}$$

$\mathbf{y}(k)$ can also be written as $\mathbf{y}(k) = \mathbf{V}_T(k)\mathbf{d}(k)$.

Example 6.10-4. Write an expression for $y(kT)$, $k = 0, 1, 2, \ldots$, for the system of Example 3.13-6.

Defining the indicated transmission matrices and vectors as above,

$$\mathbf{q}(kT) = \mathbf{v}(kT) - \mathbf{GH}_T(kT)\mathbf{q}(kT)$$

or

$$\mathbf{q}(kT) = [\mathbf{I} + \mathbf{GH}_T(kT)]^{-1}\mathbf{v}(kT)$$

Then

$$\mathbf{y}(kT) = \mathbf{G}_T(kT)[\mathbf{I} + \mathbf{GH}_T(kT)]^{-1}\mathbf{v}(kT)$$

6.11 THE METHOD OF LEAST SQUARES

At this point in the presentation of the preceding chapter, the evaluation of the mean square outputs of continuous systems was developed from the viewpoint of the modified adjoint system. As indicated in the previous section, the discrete modified adjoint system has limited value, because a digital computer can be programmed to overcome the analogous difficulty encountered with continuous systems. Thus this discussion of linear discrete systems departs from paralleling the preceding chapter to consider the general topic of least squares. This topic has considerable importance in the areas of communications, control, numerical analysis, prediction, and others, and, upon completion of this section, the reader is encouraged to investigate the continuous analogy to this section.[16,17]

Suppose that the output $y(kT)$ of a linear, stationary, discrete system is given by a weighted sum of the present and a finite number of past values of the input $v[(k-m)T]$, $m = 0, 1, 2, \ldots, M-1$. This can be written in terms of the weighting factors $h(mT)$ as

$$y(kT) = v(kT)h(0) + v[(k-1)T]h(T) + \cdots + v[(k-M-1)T]h[(M-1)T] = \sum_{m=0}^{M-1} v[(k-m)T]h(mT)$$

In order to determine the system-weighting factors, $J \geqslant M$ sets of measurements of $v[(k-m)T]$, $m = 0, 1, \ldots, M-1$, and the corresponding $y(kT)$ could be made. This would yield J equations of the form

$$\begin{aligned} v_{11}h_1 + v_{12}h_2 + \cdots + v_{1M}h_M &= y_1 \\ v_{21}h_1 + v_{22}h_2 + \cdots + v_{2M}h_M &= y_2 \\ \cdots\cdots\cdots\cdots\cdots\cdots&\cdots\cdots \\ v_{J1}h_1 + v_{J2}h_2 + \cdots + v_{JM}h_M &= v_J \end{aligned} \tag{6.11-1}$$

where the subscript on the y's and the first subscript on the v's denote a particular set of measurements. The subscript on the h's and the second subscript on the v's denote the argument $(m + 1)T$. If only M sets of measurements are made, i.e., $J = M$, unique solutions exist for the h_i, but noise and measurement errors generally cause Eq. 6.11-1 to yield incorrect values. Thus, in such an experimental situation, many measurements are usually taken, and more equations than the number of unknowns are found. Hence $J > M$. Now, however, values for the h_i cannot be found which satisfy all the equations. For example, substitution of the values h_1^0, $h_2^0, \ldots, h_m^0$ for the unknown $h_1, h_2, \ldots, h_m$ in the left side of Eq. 6.11-1 might yield $y_1^0, y_2^0, \ldots, y_J^0$ which differ from $y_1, y_2, \ldots, y_J$ by $e_i = y_i^0 - y_i$, $i = 1, 2, \ldots, J$. Faced with this dilemma, one might attempt to determine the h_i in such a way that each of Eqs. 6.11-1 is at least approximately valid, and such that some measure of the total approximation error is as small as possible. For example, the h_i^0 can be determined such that the e_i have the smallest possible mean square deviation, i.e.,

$$I = \sum_{i=1}^{J} e_i^2 = \sum_{i=1}^{J} (y_i^0 - y_i)^2$$

is a minimum. This is known as the *method of least squares.* It is, in essence, an attempt to find the "best" values for the h_i.

The preceding problem can be viewed geometrically. Let the vectors $\mathbf{v}_1, \mathbf{v}_2, \ldots, \mathbf{v}_M$ denote the columns of

$$\mathbf{V} = \begin{bmatrix} v_{11} & v_{12} & \cdots & v_{1M} \\ v_{21} & v_{22} & \cdots & v_{2M} \\ \cdots & \cdots & \cdots & \cdots \\ v_{J1} & v_{J2} & \cdots & v_{JM} \end{bmatrix}$$

Then the vector

$$\mathbf{y}^0 = \begin{bmatrix} y_1^0 \\ y_2^0 \\ \cdot \\ \cdot \\ \cdot \\ y_J^0 \end{bmatrix}$$

is given by $\mathbf{y}^0 = h_1^0\mathbf{v}_1 + h_2^0\mathbf{v}_2 + \cdots + h_M^0\mathbf{v}_M$. The problem becomes one of determining $h_1^0, h_2^0, \ldots, h_M^0$, such that $\|e\|^2 = \|y^0 - y\|^2$ is a minimum, where

$$\mathbf{e} = \begin{bmatrix} e_1 \\ e_2 \\ \cdot \\ \cdot \\ \cdot \\ e_J \end{bmatrix} \quad \text{and} \quad \mathbf{y} = \begin{bmatrix} y_1 \\ y_2 \\ \cdot \\ \cdot \\ \cdot \\ y_J \end{bmatrix}$$

Thus the problem is to determine

$$\mathbf{h}^0 = \begin{bmatrix} h_1^0 \\ h_2^0 \\ \cdot \\ \cdot \\ \cdot \\ h_M^0 \end{bmatrix}$$

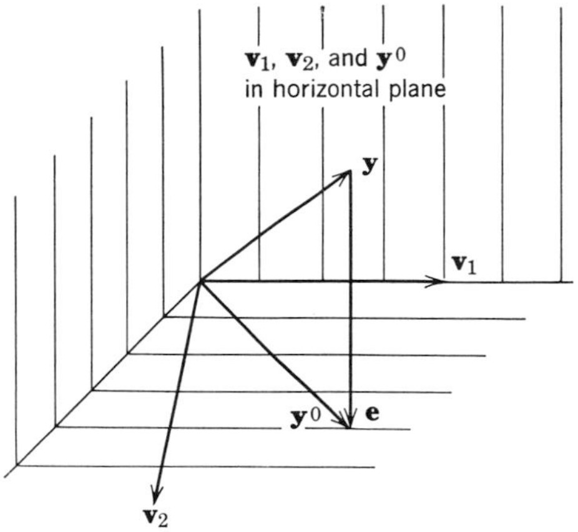

Fig. 6.11-1

such that $\mathbf{y}^0$ has the smallest possible deviation in norm from $\mathbf{y}$. This is represented for the case $M = 2$ in Fig. 6.11-1. In the general case, the set of all linear combinations of $\mathbf{v}_1, \mathbf{v}_2, \ldots, \mathbf{v}_M$ forms a space R^0, and the *orthogonal projection* of $\mathbf{y}$ on R^0 is the vector in R^0 which is the closest to $\mathbf{y}$. This is a simple generalization of what is geometrically apparent for $M = 2$ in Fig. 6.11-1. Thus $\mathbf{h}^0$ is to be chosen so that the linear combination $\mathbf{y}^0 = h_1{}^0\mathbf{v}_1 + h_2{}^0\mathbf{v}_2 + \cdots + h_M{}^0\mathbf{v}_M$ is the orthogonal projection of $\mathbf{y}$ on R^0.

Given a space R^0 and a vector $\mathbf{y}$, which is in general not contained in R^0, $\mathbf{y}$ can always be represented in the form

$$\mathbf{y} = \mathbf{y}^0 - \mathbf{e}$$

where the vector $\mathbf{y}^0$ belongs to R^0 and $\mathbf{e}$ is orthogonal to R^0. This is the geometrical idea behind the Gram-Schmidt orthogonalization procedure of Section 4.5. Taking $\mathbf{v}_1, \mathbf{v}_2, \ldots, \mathbf{v}_M$ as a basis in R^0, $\mathbf{y}^0 = h_1{}^0\mathbf{v}_1 + h_2{}^0\mathbf{v}^2 + \cdots + h_M{}^0\mathbf{v}_M$, where $h_1{}^0, h_2{}^0, \ldots, h_M{}^0$ are to be determined. The vector $\mathbf{e} = \mathbf{y}^0 - \mathbf{y}$ must be orthogonal to R^0, for which is is necessary and sufficient that

$$\langle \mathbf{e}, \mathbf{v}_i \rangle = \langle \mathbf{y}^0 - \mathbf{y}, \mathbf{v}_i \rangle = 0, \qquad i = 1, 2, \ldots, M$$

Substitution for $\mathbf{y}^0$ yields

$$\langle h_1{}^0\mathbf{v}_1 + h_2{}^0\mathbf{v}_2 + \cdots + h_M{}^0\mathbf{v}_M - \mathbf{y}, \mathbf{v}_i \rangle = 0, \quad i = 1, 2, \ldots, M$$

or

$$
\begin{aligned}
h_1^0\langle \mathbf{v}_1, \mathbf{v}_1\rangle + h_2^0\langle \mathbf{v}_2, \mathbf{v}_1\rangle + \cdots + h_M^0\langle \mathbf{v}_M, \mathbf{v}_1\rangle &= \langle \mathbf{y}, \mathbf{v}_1\rangle \\
h_1^0\langle \mathbf{v}_1, \mathbf{v}_2\rangle + h_2^0\langle \mathbf{v}_2, \mathbf{v}_2\rangle + \cdots + h_M^0\langle \mathbf{v}_M, \mathbf{v}_2\rangle &= \langle \mathbf{y}, \mathbf{v}_2\rangle \\
\cdots\cdots\cdots\cdots\cdots\cdots\cdots\cdots\cdots\cdots\cdots\cdots\cdots\cdots & \\
h_1^0\langle \mathbf{v}_1, \mathbf{v}_M\rangle + h_2^0\langle \mathbf{v}_2, \mathbf{v}_M\rangle + \cdots + h_M^0\langle \mathbf{v}_M, \mathbf{v}_M\rangle &= \langle \mathbf{y}, \mathbf{v}_M\rangle
\end{aligned}
\tag{6.11-2}
$$

The determinant of this set of equations is the Gram determinant G defined by Eq. 4.5-11, and it is nonzero as long as $\mathbf{v}_1, \mathbf{v}_2, \ldots, \mathbf{v}_M$ are linearly independent. This is, in essence, an observability condition. Assuming this, the value of $\mathbf{h}^0$, as determined by least squares, is given by

$$
h_i^0 = \frac{1}{G}\begin{bmatrix}
\langle \mathbf{v}_1, \mathbf{v}_1\rangle & \langle \mathbf{v}_2, \mathbf{v}_1\rangle & \cdots & \langle \mathbf{v}_{i-1}, \mathbf{v}_1\rangle & \langle \mathbf{y}, \mathbf{v}_1\rangle & \langle \mathbf{v}_{i+1}, \mathbf{v}_1\rangle & \cdots & \langle \mathbf{v}_M, \mathbf{v}_1\rangle \\
\langle \mathbf{v}_1, \mathbf{v}_2\rangle & \langle \mathbf{v}_2, \mathbf{v}_2\rangle & \cdots & \langle \mathbf{v}_{i-1}, \mathbf{v}_2\rangle & \langle \mathbf{y}, \mathbf{v}_2\rangle & \langle \mathbf{v}_{i+1}, \mathbf{v}_2\rangle & \cdots & \langle \mathbf{v}_M, \mathbf{v}_2\rangle \\
\cdots & \cdots & \cdots & \cdots & \cdots & \cdots & \cdots & \cdots \\
\langle \mathbf{v}_1, \mathbf{v}_M\rangle & \langle \mathbf{v}_2, \mathbf{v}_M\rangle & \cdots & \langle \mathbf{v}_{i-1}, \mathbf{v}_M\rangle & \langle \mathbf{y}, \mathbf{v}_M\rangle & \langle \mathbf{v}_{i+1}, \mathbf{v}_M\rangle & \cdots & \langle \mathbf{v}_M \mathbf{v}_M\rangle
\end{bmatrix}
\tag{6.11-3}
$$

for $i = 1, 2, \ldots, M$.

The corresponding minimum value of the mean square deviation, $I = \|\mathbf{e}\|^2$, can also be determined from geometrical considerations. For $M = 2$, it is the square of the magnitude of the altitude of the parallelepiped determined by $\mathbf{v}_1$, $\mathbf{v}_2$, and $\mathbf{y}$. In general, it is the square of the magnitude of the altitude of the hyperparallelepiped determined by $\mathbf{v}_1, \mathbf{v}_2, \ldots, \mathbf{v}_M, \mathbf{y}$. If V_y is used to denote the volume of this hyperparallelepiped, then $V_y = V\,\|\mathbf{e}\|$, or

$$\|\mathbf{e}\|^2 = \frac{V_y^{\,2}}{V^2}$$

where V is the volume determined by $\mathbf{v}_1, \mathbf{v}_2, \ldots, \mathbf{v}_M$. But $V_y^{\,2}$ is G_y, the Gramian of $\mathbf{v}_1, \mathbf{v}_2, \ldots, \mathbf{v}_M, \mathbf{y}$, and V is G, the Gramian of $\mathbf{v}_1, \mathbf{v}_2, \ldots, \mathbf{v}_M$. This is readily apparent for the case in which the vectors determining the hyperparallelepiped are orthogonal. For the nonorthogonal case, the Gram-Schmidt orthogonalization procedure can be used to arrive at the same conclusion. Thus

$$I_{\min} = \|\mathbf{e}\|^2_{\min} = \frac{G_y}{G}$$

is the minimum value of the mean square deviation.

It is useful to rewrite Eq. 6.11-2 as $\mathbf{V}^T\mathbf{V}\mathbf{h}^0 = \mathbf{V}^T\mathbf{y}$. Then Eq. 6.11-3 is

$$\mathbf{h}^0 = (\mathbf{V}^T\mathbf{V})^{-1}\mathbf{V}^T\mathbf{y} \tag{6.11-4}$$

where the observability condition is that $\mathbf{V}^T\mathbf{V}$ be nonsingular. It is interesting now to assume that one additional measurement y_{J+1} is made, which is supposed to equal $v_{J+1,1}h_1{}^0 + v_{J+1,2}h_2{}^0 + \cdots + v_{J+1,M}h_M{}^0$, and determine a new least squares estimate of $\mathbf{h}^0$ based on these $J + 1$ measurements. Let

$$\mathbf{v} = \begin{bmatrix} v_{J+1,1} \\ v_{J+1,2} \\ \cdot \\ \cdot \\ \cdot \\ v_{J+1,M} \end{bmatrix}$$

Then Eq. 6.11-4 becomes

$$\mathbf{h}^0_{J+1} = \left([\mathbf{V}^T \mid \mathbf{v}] \begin{bmatrix} \mathbf{V} \\ \hline \mathbf{v}^T \end{bmatrix} \right)^{-1} [\mathbf{V}^T \mid \mathbf{v}] \begin{bmatrix} \mathbf{y} \\ \hline \mathbf{y}_{J+1} \end{bmatrix} \qquad (6.11\text{-}5)$$

where the subscript on $\mathbf{h}^0_{J+1}$ indicates that $J + 1$ measurements are used. Equation 6.11-5 provides a means of updating the least squares estimate of $\mathbf{h}^0$. However, it is not satisfactory from a computational viewpoint if the updating is to be continued, because of the matrix inversion required for each new estimate.

In an attempt to avoid repeated matrix inversions, let $\mathbf{P}_J = (\mathbf{V}^T\mathbf{V})^{-1}$. Then define

$$\mathbf{P}_{J+1} = \left([\mathbf{V}^T \mid \mathbf{v}] \begin{bmatrix} \mathbf{V} \\ \hline \mathbf{v}^T \end{bmatrix} \right)^{-1} = [\mathbf{V}^T\mathbf{V} + \mathbf{v}\mathbf{v}^T]^{-1}$$

so that

$$\mathbf{P}_{J+1} = [\mathbf{P}_J{}^{-1} + \mathbf{v}\mathbf{v}^T]^{-1} \qquad (6.11\text{-}6)$$

Direct substitution of

$$\mathbf{P}_{J+1} = \mathbf{P}_J - \mathbf{P}_J\mathbf{v}(\mathbf{v}^T\mathbf{P}_J\mathbf{v} + 1)^{-1}\mathbf{v}^T\mathbf{P}_J \qquad (6.11\text{-}7)$$

into Eq. 6.11-6 indicates that Eq. 6.11-7 is a valid expression for $\mathbf{P}_{J+1}$. Then, denoting the $\mathbf{h}^0$ based on the first J measurements by $\mathbf{h}_J{}^0$, Eq. 6.11-5 can be written, after simplification, as

$$\mathbf{h}^0_{J+1} = \mathbf{h}_J{}^0 + \mathbf{P}_J\mathbf{v}(\mathbf{v}^T\mathbf{P}_J\mathbf{v} + 1)^{-1}(y_{J+1} - \mathbf{v}^T\mathbf{h}_J{}^0) \qquad (6.11\text{-}8)$$

Since $(\mathbf{v}^T\mathbf{P}_J\mathbf{v} + 1)$ is a scalar, repeated matrix inversions are not required to update the estimation of $\mathbf{h}$. The updated $\mathbf{h}^0$ is the former $\mathbf{h}^0$ plus a weighting of the difference between the new value of y and the estimate of y based on J measurements.

REFERENCES

1. J. E. Bertram, "The Concept of State in the Analysis of Discrete Time Control Systems" 1962 Joint Automatic Control Conference, New York University, June 27–29, 1962, Paper No. 11-1.
2. D. A. Huffman, "The Synthesis of Sequential Switching Circuits," *J. Franklin Inst.*, March–April 1954, pp. 161–190, 275–303.

3. E. F. Moore, *Gedanken-Experiments on Sequential Machines*, Automata Studies, Princeton University Press, Princeton, N.J., 1956, pp. 129–153.
4. J. Millman and H. Taub, *Pulse and Digital Circuits*, McGraw-Hill Book Co., New York, 1956, pp. 346–353.
5. M. Phister, *Logical Design of Digital Computers*, John Wiley and Sons, New York, 1958, Chapters 5 and 6.
6. B. Elpas, "The Theory of Autonomous Linear Sequential Networks," Sequential Transducer Issue, *IRE* Trans. *Circuit Theory*, Vol. CT-6, No. 1, March 1959.
7. B. Friedland, "Linear Modular Sequential Circuits," Sequential Transducer Issue, *IRE Trans. Circuit Theory*, Vol. CT-6, No. 1, March 1959.
8. J. Hartmanis, "Linear Multivalued Sequential Coding Networks," Sequential Transducer Issue, *IRE Trans. Circuit Theory*, Vol. CT-6, No. 1, March 1959.
9. D. A. Huffman, "A Linear Circuit Viewpoint on Error-Correcting Codes," *IRE Trans. Infor. Theory*, Vol. IT-2, September 1956, pp. 20–28.
10. R. H. Barker, "The Pulse Transfer Function and Its Application to Sampling Servo Systems," *Proc. IEE*, Vol. 99, IV (1952), pp. 302–317.
11. J. R. Ragazzini and G. F. Franklin, *Sampled Data Control Systems*, McGraw-Hill Book Co., New York, 1958, pp. 93–94, 136, 199, 217–218.
12. R. E. Kalman, Y. C. Ho. and K. S. Narendra, "Controllability of Linear Dynamical Systems", in *Contributions to Differential Equations*, Vol. I, No. 2, John Wiley and Sons, New York, 1963.
13. F. B. Hildebrand, *Methods of Applied Mathematics*, Prentice-Hall, Inc., Englewood Cliffs, N.J., 1952, Chapter 3.
14. T. Fort, *Finite Differences*, Oxford University Press, 1948, pp. 244–245.
15. M. Salvador and M. Baron, *Numerical Methods in Engineering*, Prentice-Hall, Inc., Englewood Cliffs, N.J., 1952.
16. R. E. Kalman and R. S. Bucy, "New Results in Linear Filtering and Prediction Theory," *J. Bas. Eng.* March 1961.
17. Y. C. Ho, "On the Stochastic Approximation Method and Optimal Filtering Theory," *J. Math. Anal. Appl.*, Vol. 6, No. 1, February 1963.

Problems

6.1 Draw the simulation diagrams for the following systems.

(*a*) $y(k+1) - 2y(k)\cosh\alpha + y(k-1) = v(k)$

(*b*) $\nabla^2 y(k) + k\,\nabla y(k) + y(k) = v(k)$

(*c*) $\Delta^2 y(k) + k\,\Delta y(k) + y(k) = v(k)$

(*d*) $y_1(k+2) + 2y_1(k+1) + y_2(k+1) + y_1(k) = v_1(k)$

$y_2(k+2) + 3y_2(k+1) + y_2(k) + 3y_1(k) = v_2(k)$

6.2 Find the transfer function matrix for the system shown in Fig. P6.2.

6.3 A synchronous sequential machine accepts a serial binary coded decimal input. After every fourth pulse, the output of the machine sends out a signal which tells whether the last four inputs formed a correct binary coded decimal number (0 through 9). What is the minimum number of states of this machine?

6.4 If timing pulses $t_1, t_2, t_3, t_4, t_1, t_2, t_3, t_4, \ldots$ are available, what is the answer to Problem 6.3?

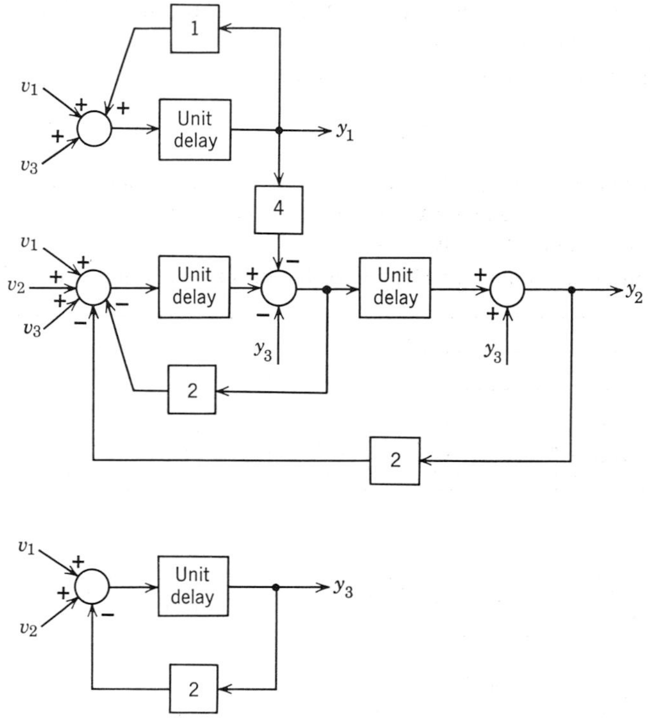

Fig. P6.2

6.5 An electronic lock is designed such that, after the input sequence 101011, the lock is opened. How many states are required in order to build this machine?

6.6 What is a suitable set of state variables for the sampled-data system of Fig. P6.6?

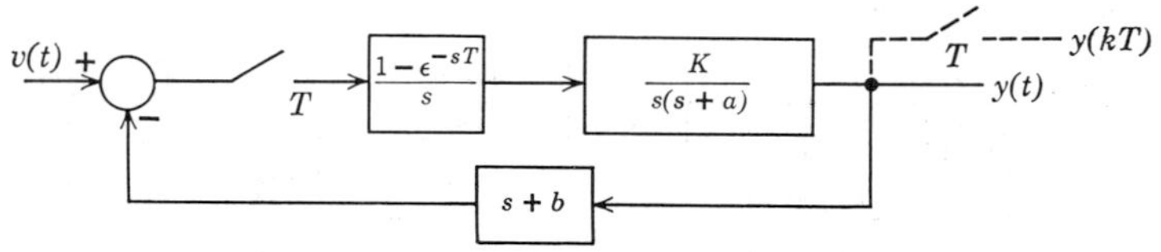

Fig. P6.6

6.7 An n-dimensional **A** matrix can be transformed into a diagonal matrix if the n characteristic roots of **A** are distinct. Similarly, when dealing with a mod p sequential network, the coefficient matrix can be transformed into a diagonal form if the n roots are distinct modulo-p (called incongruent modulo-p). Using Fermat's theorem,

$$\lambda^{p-1} = 1 \bmod p \qquad 0 < \lambda < p - 1$$

Show that, if the matrix $\mathbf{A}^r$ can be diagonalized, then the periods of such a system divide $r(p-1)$.

6.8 The state equation of a linear modular sequential system is given by

$$\mathbf{x}(k+1) = \begin{bmatrix} 3 & 1 \\ 0 & 2 \end{bmatrix} \mathbf{x}(k) \qquad \text{mod } 5$$

(*a*) Show that every state sequence has a period which divides 4.
(*b*) Show that there must be six sequences of period 4, covering 24 different states. Find these sequences.

6.9 Find the minimum cycle length for the linear binary networks of Fig. P6.9.

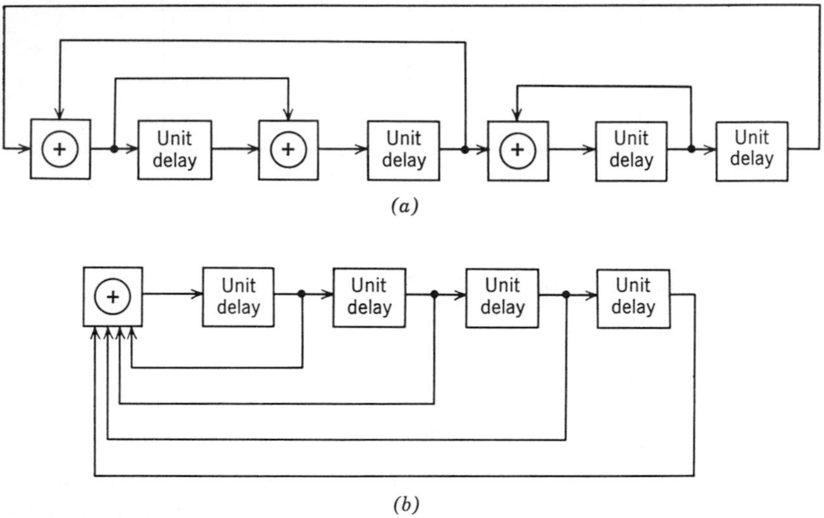

Fig. P6.9

6.10 Set up the matrix equations for the system of Fig. P6.10 in
(*a*) standard **A, B, C, D** form
(*b*) normal form

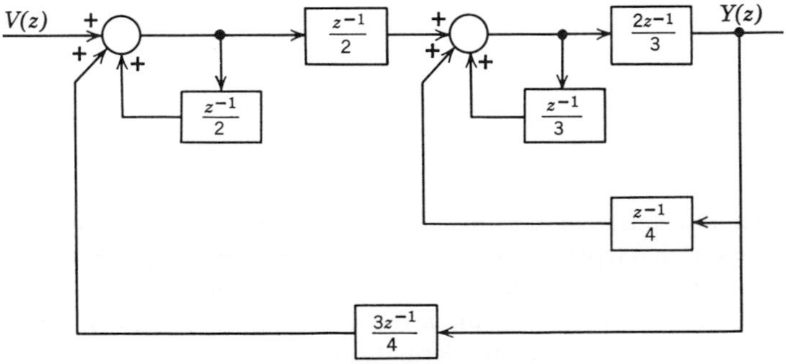

Fig. P6.10

6.11 Using partial fractions expansion, find the matrix equations for the following systems.

(*a*) $H(z) = \dfrac{z + 2}{z^3 + 7z^2 + 14z + 8}$ (*b*) $H(z) = \dfrac{z^2 + 2}{z^2 + 2z + 2}$

(*c*) $H(z) = \dfrac{z^3(z + 3)}{z^4 + 5z^3 + 9z^2 + 7z + 2}$

6.12 For the discrete system shown in Fig. P6.12, (*a*) which modes are controllable? (*b*) which modes are observable?

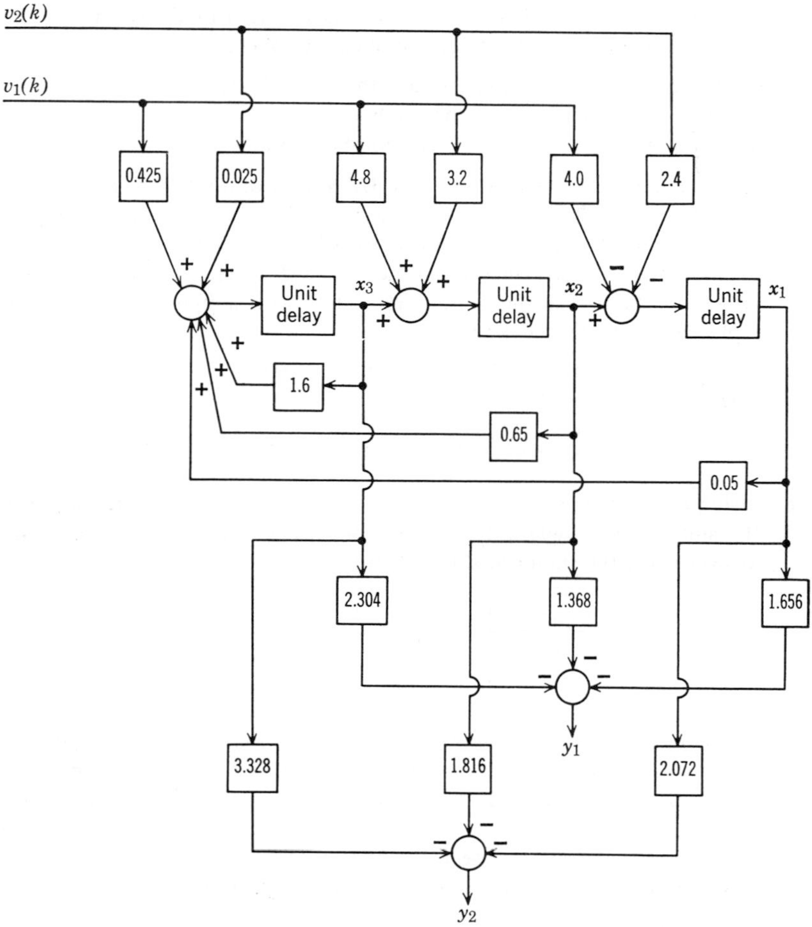

Fig. P6.12

6.13 Given the difference equation

$$y(k + 1) - 3y(k) + 2y(k - 1) = 1 + a^k$$

(*a*) Solve for $y(k)$ by the classical approach.
(*b*) Draw a simulation diagram for the equation.
(*c*) Find $y(k)$ by the mode expansion method.
(*d*) Find $y(k)$ by use of the state transition matrix.
(*e*) Find $y(k)$ by use of the standard Z transform approach.

6.14 Show that the relationships listed in Table 6.8-1 are correct.

6.15 Show that the relationships described in Eq. 6.8-12 through Eq. 6.8-16 are correct.

6.16 The output of an unforced discrete system is the series of numbers 0, 1, 1, 2, 3, 5, 8, . . . , such that each number is the sum of the two preceding numbers. (These are called the Fibonacci numbers.)

(*a*) Find the system which generates these numbers.
(*b*) Find the state transition matrix $\boldsymbol{\phi}(k)$.
(*c*) Find an expression for the kth number $y(k)$.

6.17 The inputs to the system $\dot{\mathbf{x}} = \mathbf{A}\mathbf{x} + \mathbf{B}\mathbf{v}$ consist of a set of piecewise constant signals, such that over the interval $nT \leqslant t \leqslant (n + 1)T$ the inputs are constant. These signals are obtained by sampling the inputs at time nT and holding these values until time $(n + 1)T$, when a new set of samples are taken. Show that the continuous system with sample and hold can be replaced by the discrete system

$$\mathbf{x}[(n + 1)T] = \mathbf{A}_1\mathbf{x}(nT) + \mathbf{B}_1\mathbf{v}(nT)$$

where

$$\mathbf{A}_1 = \epsilon^{\mathbf{A}T} \quad \text{and} \quad \mathbf{B}_1 = \int_0^T \epsilon^{\mathbf{A}\tau}\mathbf{B}\, d\tau$$

6.18 Using the discrete system found in Problem 6.17, solve for the output of the single input-single output system shown in Fig. P6.18. Check your answer using the standard Z transform.

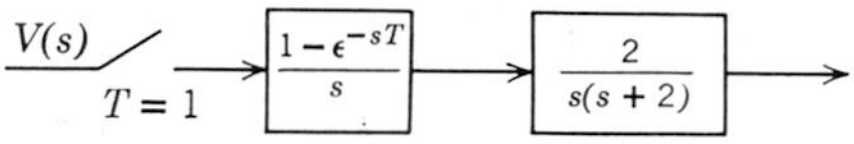

Fig. P6.18

6.19 Consider a network containing any number of switches operating in synchronism. During the first half-cycle $(0^+ \leqslant t \leqslant T_1^-)$, some switches are open and some are closed. In the second half-cycle $(T_1^+ \leqslant t \leqslant T^-)$, the positions of the switches are reversed. At $t = T$, the original condition is restored and the cycle repeated. The system is characterized by $\dot{\mathbf{x}} = \mathbf{A}_1\mathbf{x} + \mathbf{B}_1\mathbf{v}$ $(nT < t < nT + T_1)$ and $\dot{\mathbf{x}} = \mathbf{A}_2\mathbf{x} + \mathbf{B}_2\mathbf{v}$ $(nT + T_1 < t < (n + 1)T$.

(*a*) Find the solutions $\mathbf{x}(nT + T_1^-)$ and $\mathbf{x}[(n + 1)T^-]$.
(*b*) If it is assumed that $\mathbf{x}(nT + T_1^-) = \mathbf{x}(nT + T_1^+)$, $\mathbf{x}(nT^-) = \mathbf{x}(nT^+)$ and $\mathbf{v} =$ constant, find the difference equation for the system

$$\mathbf{x}[(n + 1)T] = \mathbf{A}\mathbf{x}(nT) + \mathbf{B}\mathbf{v}(nT)$$

6.20 For the system described by the difference equation $\nabla^2 y + 2\nabla y + 2y = 1$,
(*a*) Find the matrix formulation in terms of the standard **A**, **B**, **C**, **D** matrices. Solve for $y(k)$.
(*b*) Repeat (*a*) using *normal form* representation.

6.21 The switch in the network of Fig. P6.21 is closed at $t = 0, 2, 4, \ldots$ and opened at $t = 1, 3, 5, \ldots$.

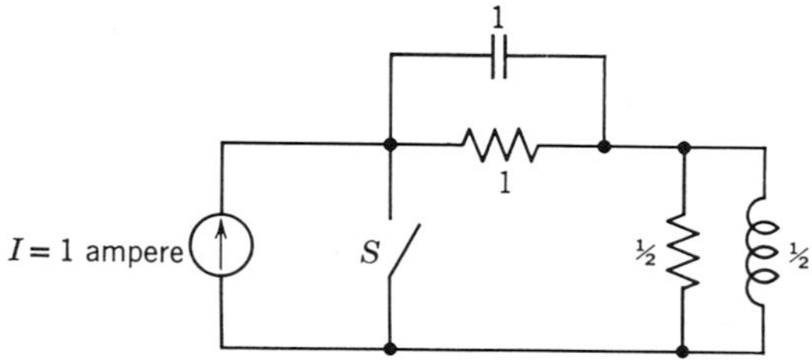

Fig. P6.21

(*a*) Find the difference equation for the inductor current at the end of the kth open-close cycle.
(*b*) Solve the difference equation using the state transition matrix, and draw a sketch of the inductor current as a function of $k/2$.
(*c*) Find the unit response matrix and the transmission matrix of this circuit.

6.22 For the system shown in Fig. P6.22, (*a*) find the state equation formulation; (*b*) solve for $\boldsymbol{\phi}(k)$; (*c*) find the unit response $\mathbf{H}(k)$.

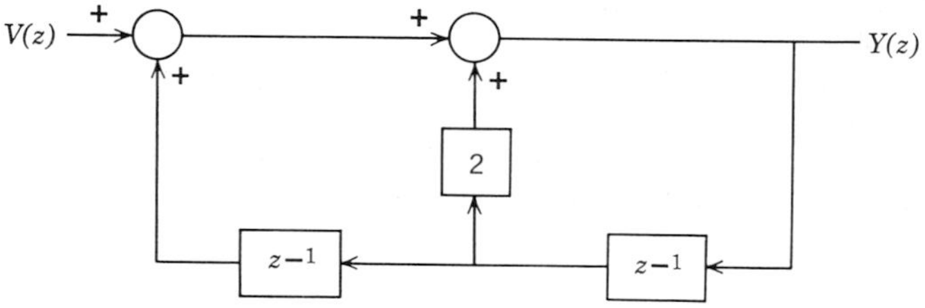

Fig. P6.22

6.23 A system is defined by the following difference equations.

$$x_1(k) - x_2(k-1) + y(k-1) = 0$$
$$x_1(k) - x_2(k+1) = 0$$
$$x_1(k) - 2x_2(k+1) + v(k) = 0$$

Find the unit response $\mathbf{H}(k)$.

6.24 For the system shown in Fig. P6.24*a*:
(*a*) Find the **A** matrix.
(*b*) Find $\boldsymbol{\phi}(k)$. Interpret your answer in terms of signals in the block diagram.

(*c*) Find the transfer function $Y(z)/V(z)$ of the system shown in Fig. P6.24*b*.

(*d*) Find the unit response of this system by dividing out the transfer function. Compare the results of this division with part (*b*).

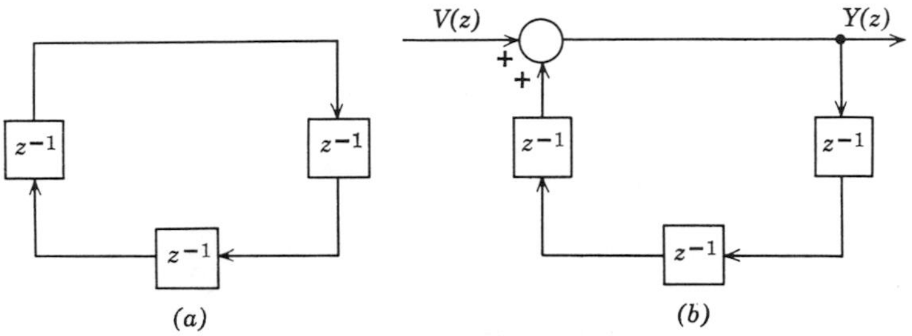

Fig. P6.24

6.25 (*a*) Find the $\boldsymbol{\phi}$ matrix for the system of Problem 6.10 by time domain techniques. (*b*) Repeat (*a*) using Z transform methods.

6.26 Derive the discrete analog of Eq. 5.10-14 and, from the result, the modified adjoint discrete system.

6.27 (*a*) Determine a discrete approximation for the differential equation

$$\dot{y} + \frac{1}{t}y = 0$$

(*b*) Using $T = 0.05$, solve for $y(kT)$ from $k = 0$ to $k = 5$ for $k_0 = 0, 1, 2, 3, 4, 5$.

(*c*) Check the results of part (*b*) by calculating $h(0.25, 0)$ of the original system.

(*d*) Find the discrete modified adjoint of part (*a*).

(*e*) Check the results of part (*b*) by calculating $h_{k'}{}^*(k', 0)$.

7

Introduction to Stability Theory and Lyapunov's Second Method

7.1 INTRODUCTION

The precise definition of stability in the case of a nonlinear system is not simple. The intuitive concept of stability, which for the most part is adequate for time-invariant linear systems, fails when one attempts to extend it to nonlinear systems. A time-invariant linear system is either stable or unstable, depending upon the location of the zeros of the characteristic equation in the s-plane. System stability is independent of the initial conditions or system inputs. This is not true for nonlinear systems.

Whether the response of a nonlinear system is bounded or unbounded may depend upon the initial conditions or the forcing function. Furthermore, a nonlinear system may exhibit oscillations of constant peak value. Should one refer to such a system as being stable or unstable? Intuitively one would say unstable if it is a control system and the oscillations are sufficiently large that the system performance is not satisfactory. A statement such as this is very vague, however, and leads to confusion. In treating nonlinear systems, it is necessary to make careful distinctions between the various intuitive concepts of stability.

The purpose of this chapter is to introduce the reader to the more rigorous definitions of stability and to Lyapunov's methods, and the role that the state variable approach plays. The word "introduce" is carefully chosen, since books have already been exclusively devoted to stability theory and Lyapunov's methods, as the references at the end of the chapter indicate.

7.2 PHASE PLANE CONCEPTS

The response of a physical system may be illustrated by a plot of the system response versus time. It may also be illustrated by using time as a parameter, and plotting the interrelationships of the behavior of the state variables of the system in state space. The latter is a geometrical representation of the system behavior. It is difficult to visualize for systems of third or higher order, but it is particularly useful for second order systems.

For example, consider the linear oscillator illustrated in Fig. 7.2-1 and described by the differential equation

$$\frac{d^2y}{dt^2} + \omega^2 y = 0 \tag{7.2-1}$$

where ω is the frequency of oscillation and is positive. Defining the state variables $x_1(t) = y(t)$, $x_2(t) = \dot{x}_1(t) = \dot{y}(t)$ permits the system description of Eq. 7.2-1 to be written as

$$\dot{x}_2 + \omega^2 x_1 = 0 \tag{7.2-2}$$

Multiplication of Eq. 7.2-2 by x_2 gives

$$0 = x_2\dot{x}_2 + \omega^2 x_1 x_2 = x_2\dot{x}_2 + \omega^2 x_1\dot{x}_1 = \frac{d}{dt}\left[\frac{x_2^{\,2}}{2} + \omega^2\frac{x_1^{\,2}}{2}\right] \tag{7.2-3}$$

Integration of Eq. 7.2-3 yields

$$x_2^{\,2} + \omega^2 x_1^{\,2} = c^2 \tag{7.2-4}$$

where $c =$ constant. Thus, in any solution to Eq. 7.2-1, $y(t) = x_1(t)$ and $\dot{y}(t) = x_2(t)$ are related by Eq. 7.2-4, the equation for an ellipse. Several solutions for various values of c, corresponding to various initial values of x_1 and x_2, are shown in Fig. 7.2-2. A solution in the x_2 versus x_1 plane is called a *trajectory*, and the x_1x_2-plane is called the *phase plane*.

Time is a parameter along any one of the trajectories, and for the particular state variables chosen in this example, i.e., $x_2(t) = \dot{x}_1(t)$, increasing time corresponds to clockwise motion in the x_1x_2-plane, if the axes are as chosen in Fig. 7.2-2. As is illustrated later, however, this need not always

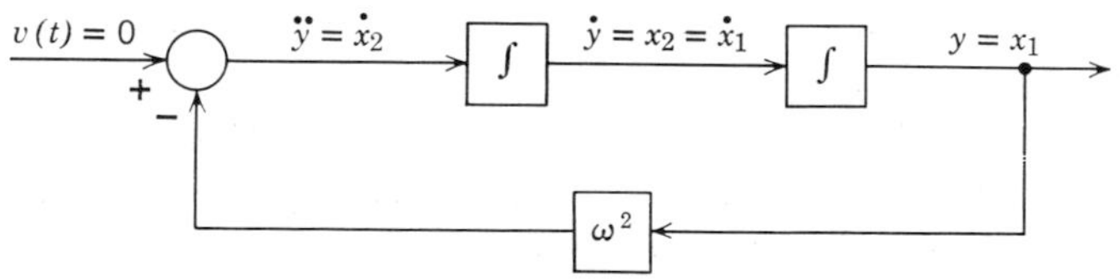

Fig. 7.2-1

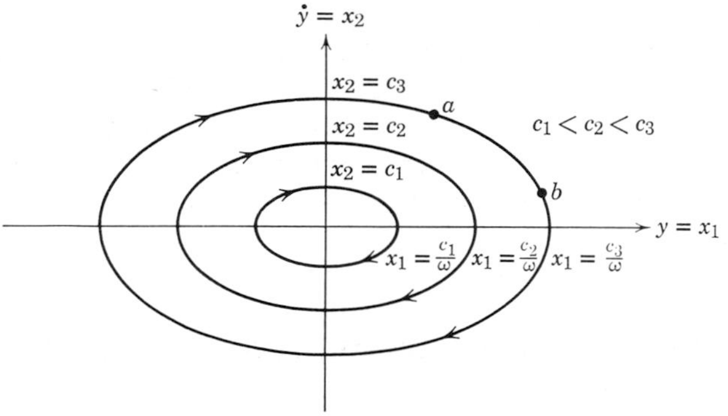

Fig. 7.2-2

be the case. The clockwise motion of the trajectories in this example is evident from the fact that positive x_2, i.e., $\dot{x}_1 > 0$, requires $x_1(t)$ to be increasing, and negative x_2, i.e., $\dot{x}_1 < 0$, requires $x_1(t)$ to be decreasing.

The time necessary for the solution to move from point a to point b on a trajectory can be computed from

$$t_b - t_a = \int_{x_{1a}}^{x_{1b}} dx_1 \bigg/ \frac{dx_1}{dt} = \int_{x_{1a}}^{x_{1b}} \frac{dx_1}{x_2} \tag{7.2-5}$$

For example, let point a be defined by $(0, c_3)$ and point b by $(c_3/\omega, 0)$ on the $c = c_3$ trajectory. Thus a and b are one-fourth of a period apart. Elimination of x_2 from Eq. 7.2-5 by substitution of Eq. 7.2-4 for $c = c_3$ yields

$$t_b - t_a = \frac{T}{4} = \int_0^{c_3/\omega} \frac{dx_1}{\sqrt{c_3^2 - \omega^2 x_1^2}} = \frac{\pi}{2\omega} \tag{7.2-6}$$

where T is the period of oscillation. Equation 7.2-6 is the well-known expression for the period of a linear oscillator, $T = 2\pi/\omega$.

The trajectory corresponding to $c = 0$ is the trivial solution corresponding to $x_1 = x_2 = 0$. This solution is called a *singular solution*, and the point (0, 0) is called a *singular point* or *equilibrium point*. It is a point of equilibrium or rest for the linear oscillator. In the case under consideration it is called a *center*, because all the trajectories form closed paths about it.

The phase plane trajectories of Fig. 7.2-2, corresponding to the behavior of the system described by Eq. 7.2-1, were obtained without explicitly solving Eq. 7.2-1 for $y(t)$. From Fig. 7.2-2 one can see that any solutions to Eq. 7.2-1 are periodic and constant in amplitude, since the trajectories are closed paths. If the trajectories were paths continuously approaching the origin, one would intuitively say that the system is "stable." If the trajectories were paths continuously diverging from the origin, one would intuitively say that the system is "unstable." This representation of the behavior of a second order system and indication of "system stability" without actually solving the system differential equation is an important use of the phase plane approach in connection with nonlinear systems. It is necessary, however, to provide more precise definitions of stability. This is done in Section 7.7.

7.3 SINGULAR POINTS OF LINEAR, SECOND ORDER SYSTEMS

The center singular point observed for the linear oscillator is only one of four possible singular points. The remaining types of singular points can be illustrated by means of the system of Fig. 7.3-1. Notice that, for the case of zero damping ($\xi = 0$), this system is the same as the linear oscillator. Thus, for $\xi = 0$, the origin of the x_1x_2-plane is a center, and the trajectories are ellipses. For $\xi \neq 0$, however, the trajectories are altered significantly.

The differential equation for the system of Fig. 7.3-1 is

$$\frac{d^2y}{dt^2} + 2\xi\omega\frac{dy}{dt} + \omega^2 y = 0 \qquad (7.3\text{-}1)$$

Using the indicated but not unique definition of state variables given in Fig. 7.3-1, this can be written as

$$\begin{aligned} \dot{x}_1 &= x_2 \\ \dot{x}_2 &= -\omega^2 x_1 - 2\xi\omega x_2 \end{aligned} \qquad (7.3\text{-}2)$$

In matrix notation, this is $\dot{\mathbf{x}} = \mathbf{A}\mathbf{x}$, where

$$\mathbf{A} = \begin{bmatrix} 0 & 1 \\ -\omega^2 & -2\xi\omega \end{bmatrix}$$

The characteristic values can be determined as the roots of

$$s^2 + 2\xi\omega s + \omega^2 = 0 \qquad (7.3\text{-}3)$$

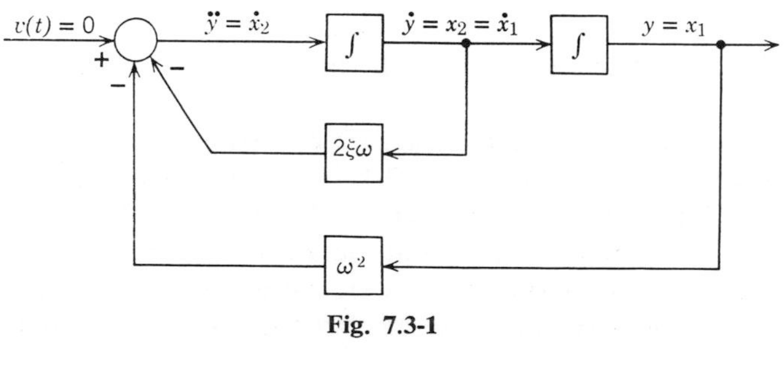

Fig. 7.3-1

or from $|s\mathbf{I} - \mathbf{A}| = 0$. Either approach yields

$$\begin{aligned} s_1 &= -\xi\omega - \omega\sqrt{\xi^2 - 1} \\ s_2 &= -\xi\omega + \omega\sqrt{\xi^2 - 1} \end{aligned} \tag{7.3-4}$$

Thus at this point one could plot the response $y(t)$ for various values of ξ and ω, since the general solution of Eq. 7.3-1 is

$$y(t) = c_1\epsilon^{s_1 t} + c_2\epsilon^{s_2 t}$$

where c_1 and c_2 are arbitrary constants. The purpose here, however, is to illustrate phase plane principles which are useful for nonlinear problems. Hence the nature of the possible singular points is now considered.

Singular points are points of dynamic equilibrium. In essence, they correspond to positions of rest for the system. They may be stable. For example, a pendulum hanging vertically with no kinetic energy is in a stable equilibrium position. Equilibrium points can also be unstable. A pendulum which has its mass at rest directly above its point of support is in equilibrium. In this case, however, the slightest disturbance will set the pendulum into oscillation. Nevertheless, without a disturbance in either case, the pendulum is at rest. Thus the significant feature of a singular or equilibrium point is that all the derivatives characterizing the system behavior are zero. In other words, the derivatives of the state variables are zero. For the linear, second order system of Fig. 7.3-1, the singular point is given by setting $\dot{x}_1 = \dot{x}_2 = 0$. Equation 7.3-2 then reveals that the origin of the x_1x_2-plane is the only singular point, assuming that ω is nonzero.

The slope of the trajectory at any point in the x_1x_2-plane is given by

$$\frac{dx_2}{dx_1} = \frac{\dot{x}_2}{\dot{x}_1} = -\omega\,\frac{[\omega x_1 + 2\xi x_2]}{x_2} \tag{7.3-5}$$

Thus, at a singular point, the trajectory slope is indeterminate. This is true for all singular points and is not merely a result of the particular choice of state variables in this example. Along the x_1-axis, where x_2 is zero, the slope is infinite. This, however, is due to the particular choice of state variables and requires that all trajectories in this example cross the x_1-axis vertically.

At all *ordinary points* (points which are not singular points), Eq. 7.3-5 indicates that the trajectory slope is unique. The Cauchy-Lipschitz theorem then guarantees that only one solution curve passes through a given point in the phase plane.† Thus trajectories cannot cross one another.

† A sufficient condition for this is that $\mathbf{f}(\mathbf{x})$ be a Lipschitz function in a region R containing the point, i.e., for every pair of points $\mathbf{x}$ and $\mathbf{y}$ in R, $\|\mathbf{f}(\mathbf{x}) - \mathbf{f}(\mathbf{y})\| \leqslant K\,\|\mathbf{x} - \mathbf{y}\|$, where K is a positive constant.[1] Note that this is not satisfied where $\mathbf{f}(\mathbf{x})$ is a multivalued function of $\mathbf{x}$. In this case, the x's are not a set of state variables for the system.

In final preparation for investigating the nature of the singular points of the system of Fig. 7.3-1, consider the trajectories which are straight lines passing through the origin. Since they are straight lines passing through the origin, they must be described by $x_2 = mx_1$. Hence $\dot{x}_2 = m\dot{x}_1$, so that the slope of any straight-line trajectories is

$$m = \frac{x_2}{x_1} = \frac{\dot{x}_2}{\dot{x}_1} \tag{7.3-6}$$

Equating $\dot{x}_2/\dot{x}_1$ as given by Eqs. 7.3-5 and 7.3-6 yields

$$m = -\omega\frac{[\omega x_1 + 2\xi x_2]}{x_2} = -\omega\frac{[\omega + 2\xi m]}{m}$$

Thus the equation for the slope of the trajectories which are straight lines is

$$m^2 + 2\xi\omega m + \omega^2 = 0 \tag{7.3-7}$$

Since Eqs. 7.3-7 and 7.3-3 have the same roots, the slopes of any straight-line trajectories are equal to the characteristic values given by Eq. 7.3-4. This result is, of course, dependent upon the choice of the state variables, although the characteristic values themselves are independent of the choice of state variables. These straight-line trajectories correspond to the modes of the response, and could have been obtained from Eq. 5.6-10.

With these preliminary thoughts completed, the nature of the singular points for linear, second order systems is now considered.

Center: $\xi = 0, \ \omega^2 > 0$

This case was investigated in the preceding section. It was found that the origin is a center, and that the trajectories are ellipses about the origin. No straight-line trajectories exist, since there are no real roots to Eq. 7.3-7 for $\xi = 0$, i.e., there are no modes or characteristic vectors corresponding to real characteristic values.

Focus: $|\xi| < 1, \ \omega^2 > 0$

For this case, it is not possible to carry out directly the procedure of the previous section to determine the equation for the trajectories. However, it is again possible to say that there are no straight line trajectories, since Eq. 7.3-7 has no real roots for $|\xi| < 1$.

The substitution of $x_2 = zx_1$ and $dx_2 = z\,dx_1 + x_1\,dz$ permits Eq. 7.3-5 to be written as

$$-\frac{dx_1}{x_1} = \frac{z\,dz}{z^2 + 2\xi\omega z + \omega^2} \tag{7.3-8}$$

Equation 7.3-8 can be integrated to yield an expression for the trajectories in the x_1z-plane. The corresponding result in the $x_1\dot{x}_2$-plane is

$$(x_2 + \xi\omega x_1)^2 + \omega^2(1 - \xi^2)x_1^2 = c^2 \exp\left[\frac{2\xi}{\sqrt{1-\xi^2}} \tan^{-1}\left(\frac{x_2 + \xi\omega x_1}{x_1\omega\sqrt{1-\xi^2}}\right)\right]$$

where c is a constant. This rather formidable expression becomes

$$z_2^2 + z_1^2 = c^2 \exp\left[\frac{2\xi}{\sqrt{1-\xi^2}} \tan^{-1}\frac{z_2}{z_1}\right] \tag{7.3-9}$$

upon introducing the coordinates

$$\begin{bmatrix} z_1 \\ z_2 \end{bmatrix} = \begin{bmatrix} \omega\sqrt{1-\xi^2} & 0 \\ \xi\omega & 1 \end{bmatrix} \begin{bmatrix} x_1 \\ x_2 \end{bmatrix}$$

Using a polar coordinate system in the z_1z_2-plane, let

$$z_1 = r \cos \phi$$
$$z_2 = r \sin \phi$$

In polar coordinates, Eq. 7.3-9 becomes

$$r = c \exp\left[\frac{\xi\phi}{\sqrt{1-\xi^2}}\right]$$

Thus the trajectories in the z_1z_2-plane are logarithmic spirals. If $1 > \xi > 0$, the radius decreases as ϕ becomes more negative (which is the direction of increasing t), corresponding to Fig. 7.3-2*a*. This is the case of a stable focal point. The unstable focal point corresponding to $0 > \xi > -1$ is illustrated in Fig. 7.3-2*b*. As $|\xi|$ is increased from zero toward unity, the rate at which the radius changes with ϕ is increased. This corresponds to moving the system poles, as given by Eq. 7.3-4, away from the $j\omega$ axis of

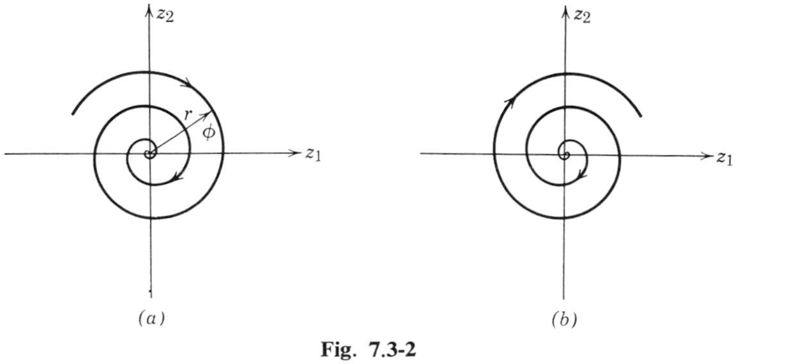

Fig. 7.3-2

the s-plane and agrees with the result one would intuitively expect. The trajectories about the focal point in the x_1x_2-plane are distorted versions of Fig. 7.3-2 and are illustrated in Fig. 7.3-8. Example 5.6-2 is a specific case of a stable focal point.

Node: $|\xi| > 1,\ \omega^2 > 0$

In the situation in which the origin is a nodal point, there are two trajectories which are straight lines. The slopes of these trajectories are given by the roots of Eq. 7.3-7 and are the characteristic values of Eq. 7.3-4. These two trajectories are illustrated in Fig. 7.3-3 for $\xi > 1$. The remaining trajectories about a nodal point in the x_1x_2-plane are more difficult to evaluate.

In order to determine the remaining trajectories, it is useful to introduce the transformation $\mathbf{z} = \mathbf{M}^{-1}\mathbf{x}$ or, more specifically,

$$\begin{bmatrix} z_1 \\ z_2 \end{bmatrix} = \begin{bmatrix} -s_1 & 1 \\ -s_2 & 1 \end{bmatrix} \begin{bmatrix} x_1 \\ x_2 \end{bmatrix} \tag{7.3-10}$$

The z_1- and z_2-axes are the straight-line trajectories shown in Fig. 7.3-3. Thus the z_1z_2-plane is a distortion of the x_1x_2-plane. Using the transformation of Eq. 7.3-10 and Eq. 7.3-4, Eq. 7.3-2 becomes, for the case of a nodal point, the normal form equations

$$\begin{aligned} \dot{z}_1 &= s_2 z_1 \\ \dot{z}_2 &= s_1 z_2 \end{aligned} \tag{7.3-11}$$

Elimination of time as a variable yields

$$\frac{dz_2}{z_2} = \frac{s_1}{s_2}\frac{dz_1}{z_1}$$

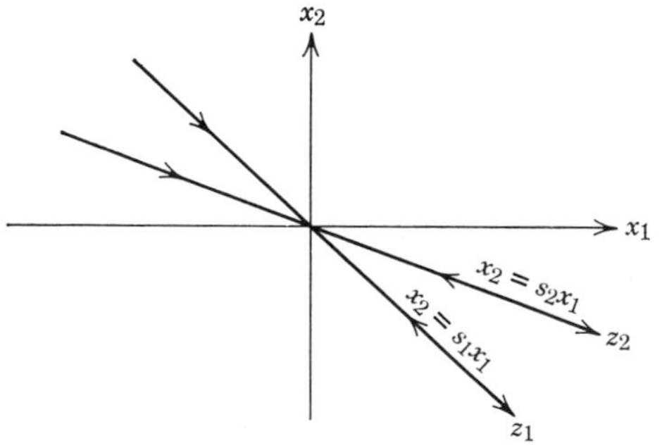

Fig. 7.3-3

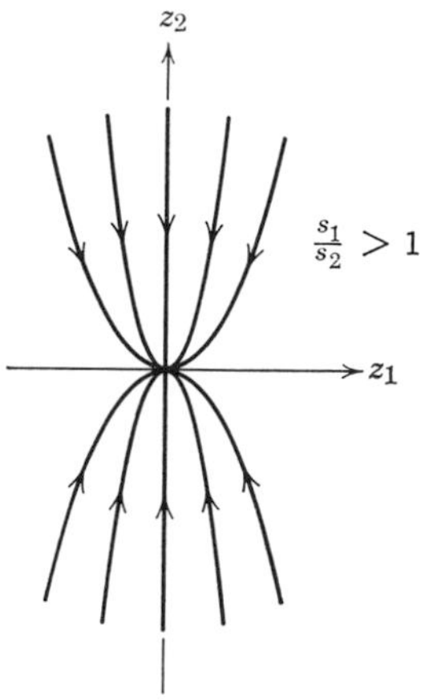

(a)

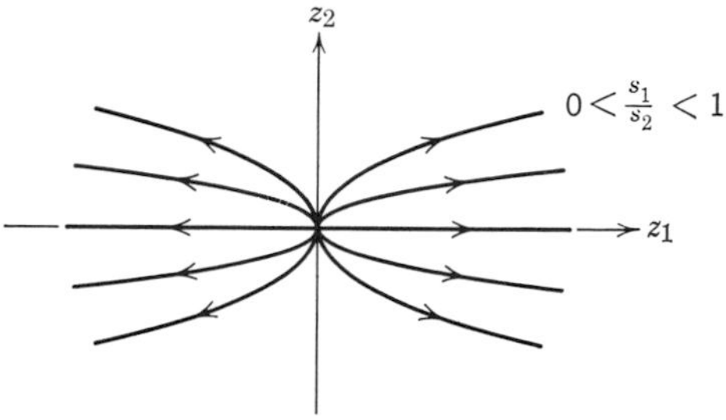

(b)

Fig. 7.3-4

Therefore the trajectories in the z_1z_2-plane are described by

$$z_2 = c(z_1)^{s_1/s_2} \tag{7.3-12}$$

For $\xi > 1, s_1 < s_2 < 0$, so that $s_1/s_2 > 1$. This corresponds to the case of a stable node. The trajectories in the z_1z_2-plane appear as in Fig. 7.3-4*a*. In the case of an unstable node where $\xi < -1$, $s_2 > s_1 > 0$, so that $1 > s_1/s_2 > 0$. The corresponding z_1z_2-trajectories are shown in Fig. 7.3-4*b*. These curves illustrate the fact that the solution curves for a nodal point all (except the curve along the z_2-axis) have the same limiting direction at the nodal point. In returning to the x_1x_2-plane, the trajectories which are not straight lines are distorted as shown in Fig. 7.3-8. Example 5.6-1 is a specific case of stable node.

Saddle Point: $\omega^2 < 0$

If the sign of the gain in the outside feedback loop of the system in Fig. 7.3-1 is changed, the singular point becomes what is known as a saddle point. For this case, the characteristic equation is

$$s^2 + 2\xi\,|\omega|\,s - |\omega^2| = 0$$

The corresponding characteristic values are

$$s_1 = -\xi\,|\omega| - |\omega|\sqrt{\xi^2 + 1}$$
$$s_2 = -\xi\,|\omega| + |\omega|\sqrt{\xi^2 + 1}$$

They correspond to one negative real root and one positive real root. Thus there are two straight-line trajectories as shown in Fig. 7.3-5.

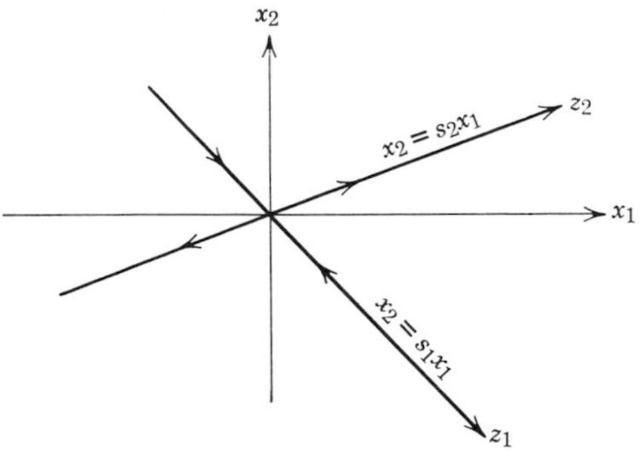

Fig. 7.3-5

In order to determine the remaining trajectories, it is again useful to introduce the transformation of Eq. 7.3-10. The resulting normal form equations are the same as Eq. 7.3-11, so that Eq. 7.3-12 also characterizes the z_1z_2-plane trajectories for a saddle point. In this case, however, if $\xi\,|\omega| > 0$, then $s_1/s_2 < -1$, and, if $\xi\,|\omega| < 0$, then $-1 < s_1/s_2 < 0$. The corresponding z_1z_2-plane trajectories are shown in Fig. 7.3-6*a* and 7.3-6*b*, respectively. The x_1x_2-plane trajectories are shown in Fig. 7.3-8.

Summary

The conditions on Eq. 7.3-1 for the various types of singular points are summarized in Fig. 7.3-7. Not considered are the case for which $|\xi| = 1$ and the case for which $\omega^2 = 0$. The trajectories for these cases are contained in Fig. 7.3-8. The corresponding expressions for the trajectories are given in the problems.

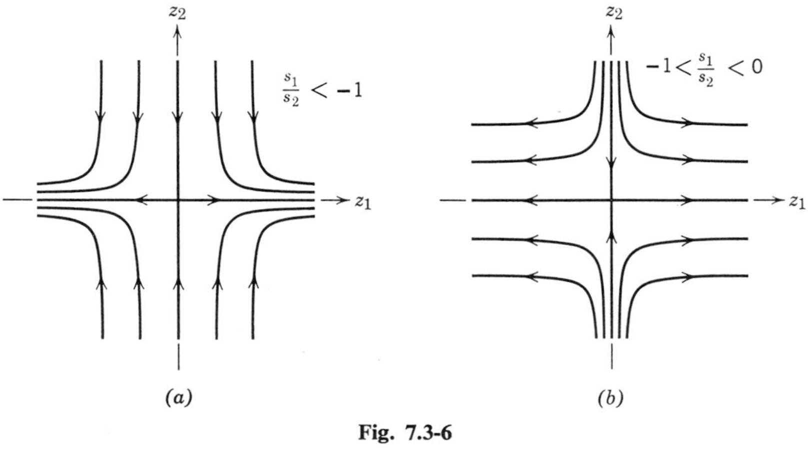

Fig. 7.3-6

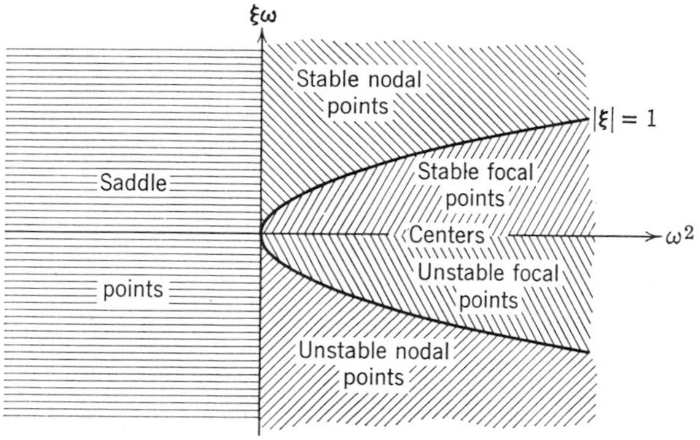

Fig. 7.3-7

The time response of the linear, second order system is easily determined, so that it is possible to compute the phase plane trajectories from the time response. For example, Eqs. 5.6-8 and 5.6-14 could have been utilized. This was not done in this section, because it is not generally possible for nonlinear systems, and the interest in studying the phase plane trajectories for linear systems stems from their relationship to nonlinear cases. In nonlinear cases, past studies often resorted to graphical methods of determining the trajectories, such as the method of isoclines, Lienard's method, and the phase plane delta method.[2–6] In the light of present-day computer simulation capabilities, however, such graphical techniques lose much of their appeal. Except for a brief discussion of the method of isoclines in the next section, the graphical techniques are not pursued further in this book. This is because of the introductory nature of this chapter and because of the belief that the proper use of the phase plane method, in a case where the trajectories cannot be determined analytically, is in conjunction with a computer simulation. The trajectories can be measured from the simulation, and the phase plane approach enables one intelligently to determine proper parameter changes to improve system behavior. For the same reasons, some of the more sophisticated methods for determining the values of time along a trajectory are not considered.[7]

7.4 VARIATIONAL EQUATIONS

As stated in the previous section, there is interest in the phase plane trajectories of the linear, second order system because of their relationship to the trajectories of a nonlinear, second order system. In small neighborhoods about its singular points, a nonlinear system behaves similarly to a linear system. It is possible to utilize this property to determine the stability

$$\ddot{y} + 2\xi\omega\dot{y} + \omega^2 y = 0$$

Singular Point	Eigenvalues	Phase Plane
Center $\xi = 0,\ \omega^2 > 0$	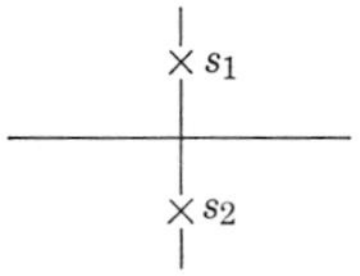	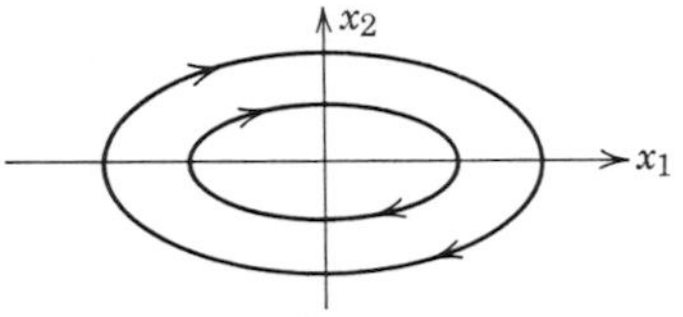
Stable focus $0 < \xi < 1,\ \omega^2 > 0$	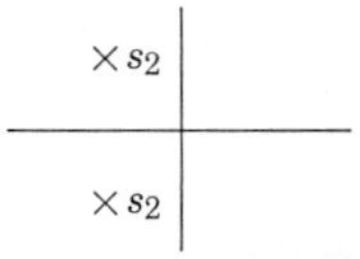	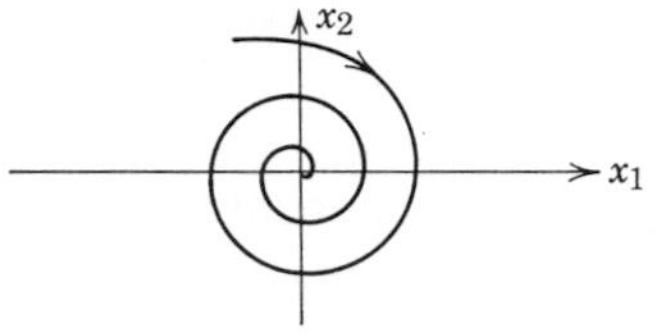
Unstable focus $-1 < \xi < 0,\ \omega^2 > 0$	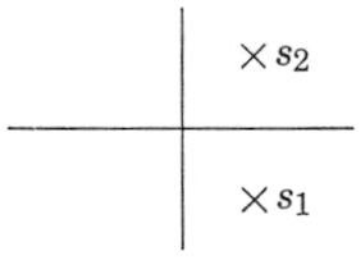	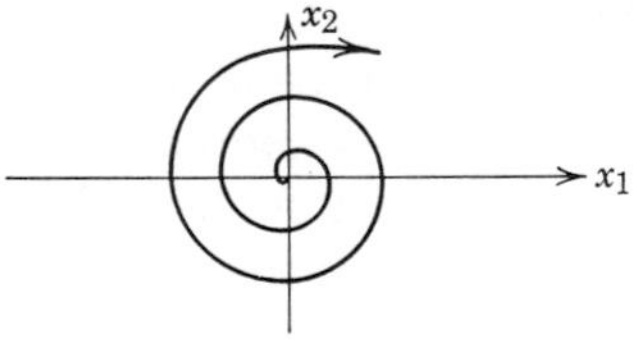
Stable node $1 < \xi,\ \omega^2 > 0$	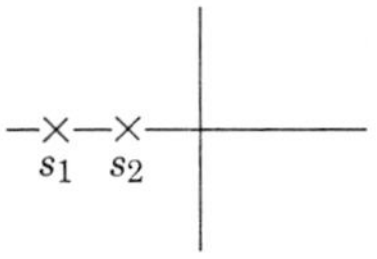	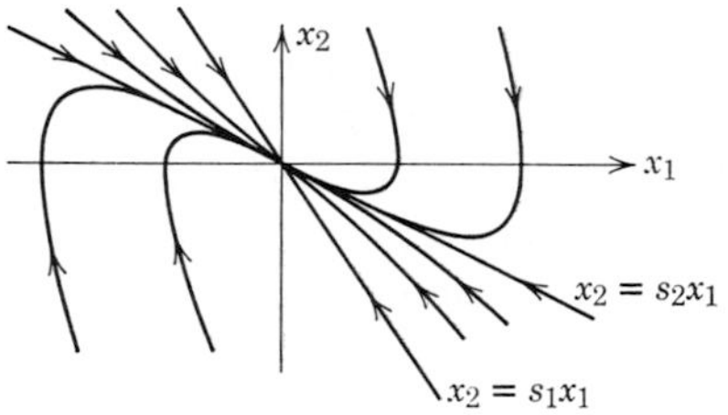
Unstable node $\xi < -1,\ \omega^2 > 0$	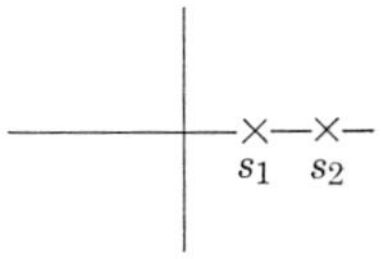	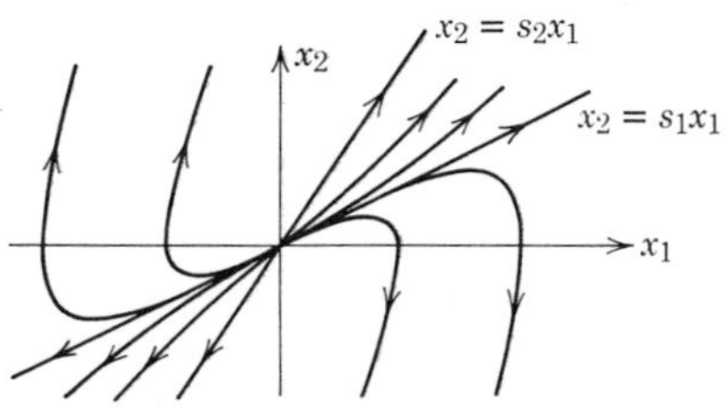

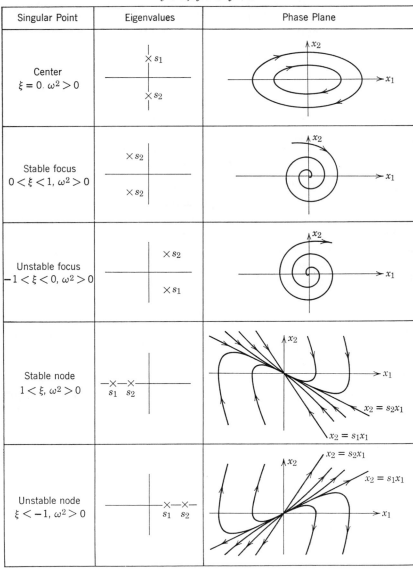

Fig. 7.3-8

$$\ddot{y} + 2\xi\omega\dot{y} + \omega^2 y = 0$$

Singular Point	Eigenvalues	Phase Plane
Saddle point $\omega^2 < 0$ (figure is for $\xi > 0$)	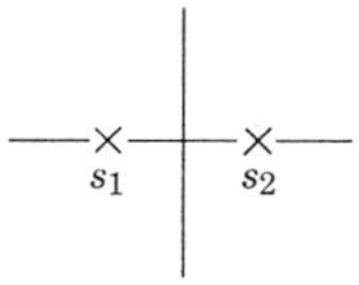	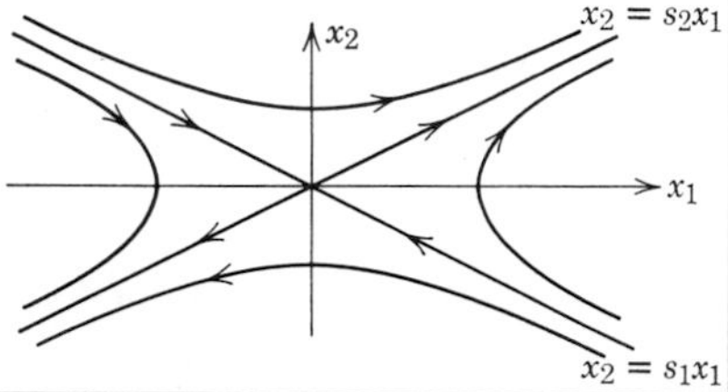
$\xi = 1$ stable	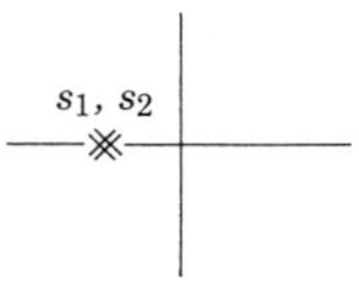	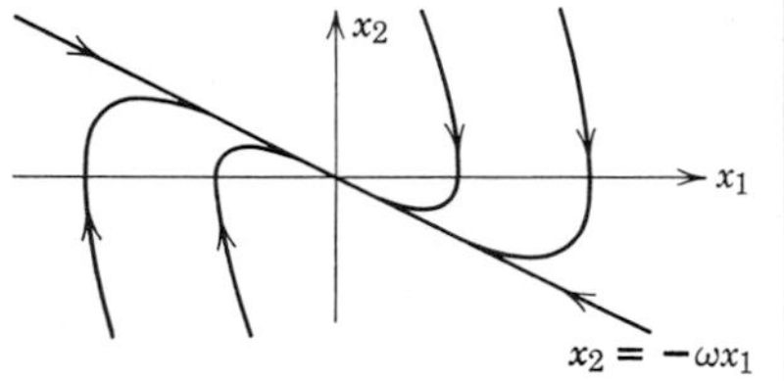
$\xi = -1$ unstable	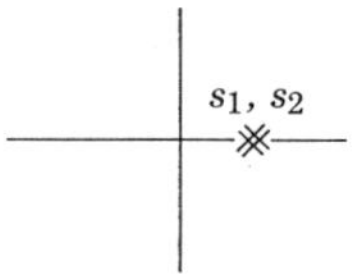	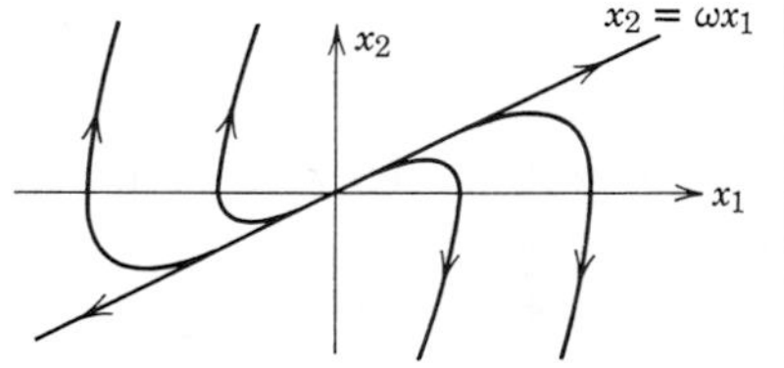
$\omega^2 = 0$ $\xi\omega > 0$	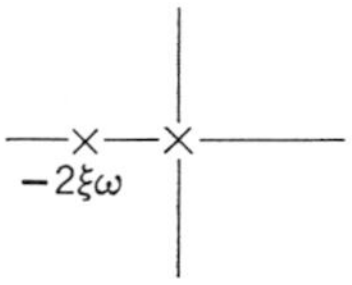	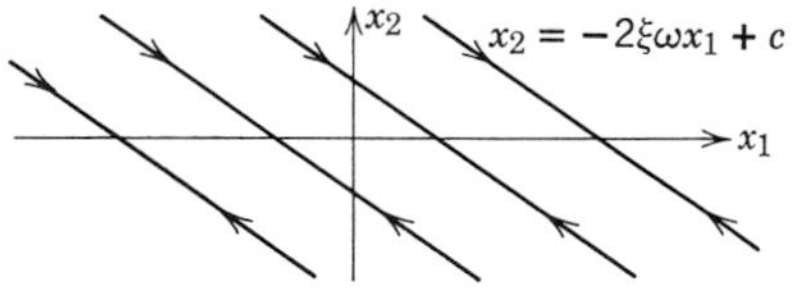
$\omega^2 = 0$ $\xi\omega < 0$	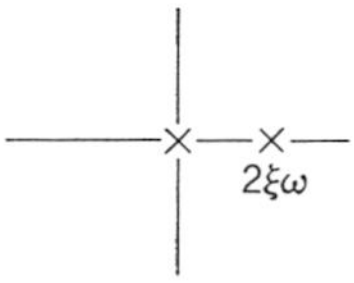	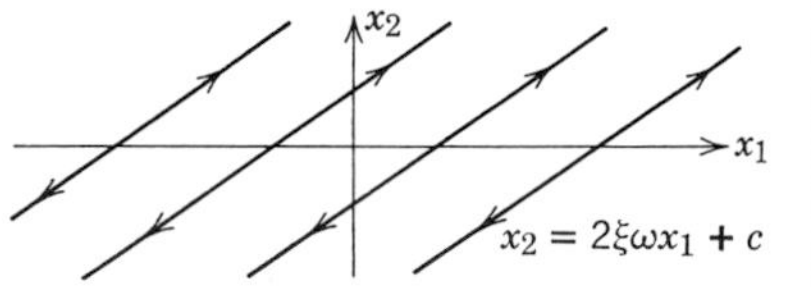

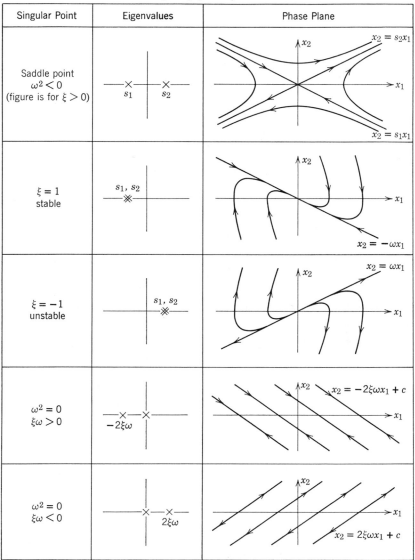

Fig. 7.3-8 (Continued)

of the singular points of a nonlinear system. Stability in larger regions is a subject of later sections.

Consider an nth order dynamical system represented by the state equations

$$\dot{\mathbf{x}} = \mathbf{f}(\mathbf{x}) \tag{7.4-1}$$

This is the *autonomous* case, that is, one in which the independent variable t does not appear explicitly. It corresponds to a system which is both unforced and time-invariant. Equation 7.4-1 is an abbreviated representation of

$$\begin{aligned} \dot{x}_1 &= f_1(x_1, x_2, \ldots, x_n) \\ \dot{x}_2 &= f_2(x_1, x_2, \ldots, x_n) \\ &\cdots\cdots\cdots\cdots \\ \dot{x}_n &= f_n(x_1, x_2, \ldots, x_n) \end{aligned}$$

Thus $\dot{\mathbf{x}}$ and $\mathbf{f}(\mathbf{x})$ are n-dimensional vectors. The singular points of Eq. 7.4-1 are given by $\dot{\mathbf{x}} = \mathbf{0}$. Hence they are the solutions of

$$\mathbf{f}(\mathbf{x}) = \mathbf{0} \tag{7.4-2}$$

In the case of a linear system, $\mathbf{f}(\mathbf{x})$ is a linear function of $\mathbf{x}$. Thus Eq. 7.4-2 has only the solution $\mathbf{x} = \mathbf{0}$, provided that the determinant of the coefficient matrix of the linear system is nonzero. This is true if there are no free integrators. Then only one singular point exists for the corresponding linear system, and it is at the origin of the state space. However, in the nonlinear case, more than one singular point can exist, as is evidenced by the possibility of more than one solution to Eq. 7.4-2. Denoting the ith solution by $\mathbf{x}_{ie}$, Eq. 7.4-2 becomes $\mathbf{f}(\mathbf{x}_{ie}) = \mathbf{0}$, $i = 1, 2, \ldots$. If each of the components of $\mathbf{f}(\mathbf{x})$ can be expanded in a Taylor series about the ith singular point, the result of considering only the linear terms is

$$\frac{d}{dt}(\mathbf{x} - \mathbf{x}_{ie}) = \dot{\mathbf{x}} = \mathbf{J}(\mathbf{x}_{ie})(\mathbf{x} - \mathbf{x}_{ie}) \tag{7.4-3}$$

where $\mathbf{J}(\mathbf{x}_{ie})$ is the Jacobian matrix evaluated at $\mathbf{x} = \mathbf{x}_{ie}$, that is,

$$\mathbf{J}(\mathbf{x}_{ie}) = \begin{bmatrix} \dfrac{\partial f_1}{\partial x_1} & \dfrac{\partial f_1}{\partial x_2} & \cdots & \dfrac{\partial f_1}{\partial x_n} \\ \dfrac{\partial f_2}{\partial x_1} & \dfrac{\partial f_2}{\partial x_2} & \cdots & \dfrac{\partial f_2}{\partial x_n} \\ \cdots & \cdots & \cdots & \cdots \\ \dfrac{\partial f_n}{\partial x_1} & \dfrac{\partial f_n}{\partial x_2} & \cdots & \dfrac{\partial f_n}{\partial x_n} \end{bmatrix}_{\mathbf{x}=\mathbf{x}_{ie}} \tag{7.4-4}$$

In terms of $\mathbf{u} = \mathbf{x} - \mathbf{x}_{ie}$, a variable measured from the singular point, Eq. 7.4-3 are the *variational equations*

$$\dot{\mathbf{u}} = \mathbf{J}(\mathbf{x}_{ie})\mathbf{u} \tag{7.4-5}$$

The variational equations are linear homogeneous differential equations which, for the autonomous case under consideration, have constant coefficients. The stability of Eq. 7.4-5 can thus be determined from the roots of

$$|s\mathbf{I} - \mathbf{J}(\mathbf{x}_{ie})| = 0 \tag{7.4-6}$$

The question then arises as to how the stability of the system of Eq. 7.4-5 relates to the stability of Eq. 7.4-1.

It is possible to state that, if any of the roots of Eq. 7.4-6 has positive real parts, then the ith singular point is unstable. If all the roots of Eq. 7.4-6 have negative real parts, then the ith singular point is stable.[8] (In this case the singular point is actually what is known as asymptotically stable. However, asymptotic stability is not defined until Section 7.7.) These statements about the stable and unstable behavior of the singular points of a nonlinear system are given here without proof. They are most easily proved by using Lyapunov's method, which is introduced in Section 7.8.

For the case in which the variational equations have roots with zero real parts, it is impossible to distinguish between stability and instability of the singular point based on the linear approximation.[9] One can intuitively view this situation as a borderline case, in which the effect of the ignored nonlinear terms can result in either stable or unstable behavior.

Although the preceding statements concerning the stability of the singular points of a nonlinear system, as indicated by the variational equations, are valid regardless of the order of the system, a second order example is chosen so that a phase plane may be used to illustrate the results.

Example 7.4-1. Figure 7.4-1 is a simplified pictorial representation of a relaxation circuit.[10] Its differential equation is

$$-Ke_g = -KRC\frac{dv_L}{dt} = -KRCL\frac{d^2i}{dt^2} = L\frac{di}{dt} + R_Li + v \tag{7.4-7}$$

The negative resistance characteristic is illustrated in Fig. 7.4-2 and represented by

$$v = -r_1i + r_3i^3 \tag{7.4-8}$$

The objective is to determine the behavior of the circuit.

The substitution of Eq. 7.4-8 into Eq. 7.4-7 yields the system equation

$$\frac{d^2i}{dt^2} + 2a\frac{di}{dt} - \frac{b^2}{2}i + c^2i^3 = 0 \tag{7.4-9}$$

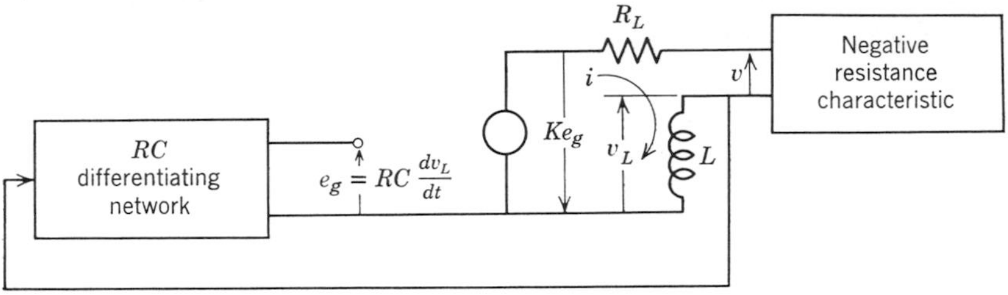

Fig. 7.4-1

where $a = 1/2KRC$, $b^2 = 2(r_1 - R_L)/KRCL$, and $c^2 = r_3/KRCL$. Defining the state variables by $x_1 = i$ and $x_2 = \dot{x}_1$ permits Eq. 7.4-9 to be written as

$$\begin{aligned} \dot{x}_1 &= x_2 \\ \dot{x}_2 &= \frac{b^2}{2} x_1 - c^2 x_1^3 - 2a x_2 \end{aligned} \tag{7.4-10}$$

Equating the right side of Eq. 7.4-10 to zero determines that there are three singular points located at

$$(1)\ x_1 = 0 \quad (2)\ x_1 = b/c\sqrt{2} \quad (3)\ x_1 = -b/c\sqrt{2}$$
$$x_2 = 0 \qquad x_2 = 0 \qquad x_2 = 0$$

The Jacobian matrix of Eq. 7.4-4 evaluated at the first singular point is

$$\mathbf{J}_1 = \begin{bmatrix} 0 & 1 \\ b^2/2 & -2a \end{bmatrix}$$

For this singular point, Eq. 7.4-6 becomes

$$\begin{vmatrix} s & -1 \\ -b^2/2 & s + 2a \end{vmatrix} = s^2 + 2as - \frac{b^2}{2} = 0$$

This corresponds to the case of a saddle point in Fig. 7.3-8, assuming $r_1 > R_L$.

At the second singular point, the Jacobian matrix is

$$\mathbf{J}_2 = \begin{bmatrix} 0 & 0 \\ -b^2 & -2a \end{bmatrix}$$

At this singular point, Eq. 7.4-6 becomes

$$\begin{vmatrix} s & -1 \\ b^2 & s + 2a \end{vmatrix} = s^2 + 2as + b^2 = 0$$

Thus the nature of this singular point depends upon the relative values of a and b, since $\xi = a/b$. Assuming $0 < \xi < 1$, the singular point is a stable focus. The same result is easily obtained for the third singular point. Thus the phase plane behavior of the circuit of Fig. 7.4-1 in the

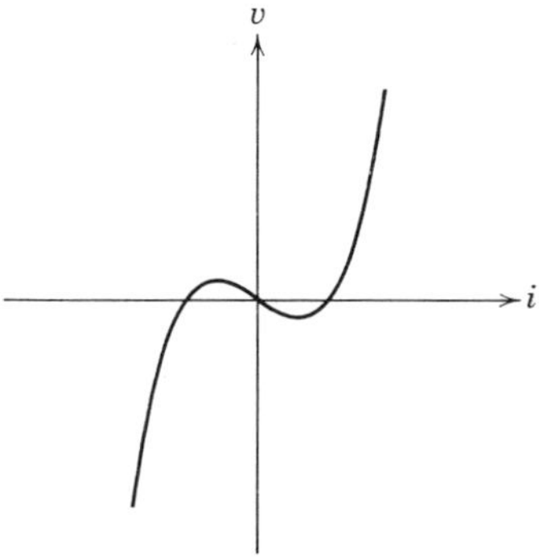

Fig. 7.4-2

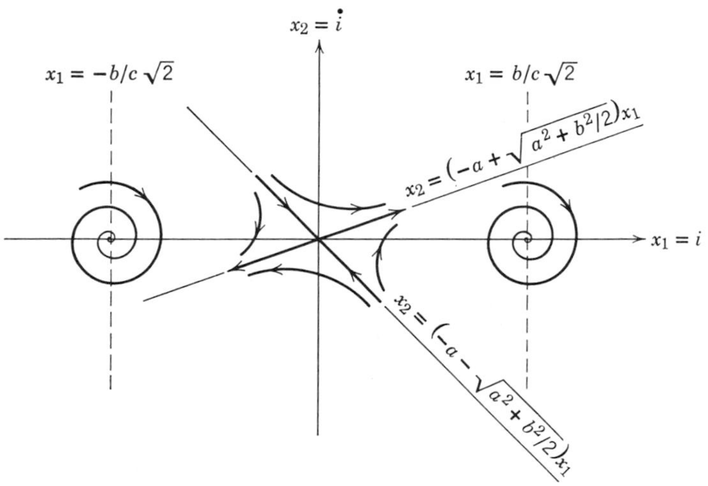

Fig. 7.4-3

neighborhood of its singular points is as shown in Fig. 7.4-3. Notice that this does not indicate the system stability or response for an arbitrary initial condition, but merely indicates the behavior in a small region about each of the singular points. In fact, the trajectories as illustrated are not precisely correct, since the trajectories about a given singular point will be distorted from the curves determined for the linear case by neighboring singular points.

The determination of the trajectories in larger regions of the phase plane is not an easy task. One approach is to use the method of isoclines, which consists in determining curves (for linear systems, they are straight lines) which connect points in the phase plane where the trajectory slopes are equal. One can then sketch through these points with the appropriate slope and estimate the trajectories.

This procedure is illustrated for this example by using Eq. 7.4-10 to write

$$\frac{dx_2}{dx_1} = \frac{\dot{x}_2}{\dot{x}_1} = \frac{(b^2/2)x_1 - c^2x_1^3 - 2ax_2}{x_2} \tag{7.4-11}$$

Equation 7.4-11 gives the slope at any point in the phase plane. Assuming that the slope is a constant k, Eq. 7.4.11 becomes

$$k = \frac{(b^2/2)\, x_1 - c^2x_1^3 - 2ax_2}{x_2}$$

This can be rewritten in the form

$$x_2 = \frac{x_1(b^2/2 - c^2x_1^2)}{k + 2a} \tag{7.4-12}$$

If c were zero (corresponding to a linear system), Eq. 7.4-12 would be the equation of a straight line, and the isoclines could easily be evaluated. For this case, however, it is necessary to pick a value for k and then compute the value of x_2 for various values of x_1. The results are the isoclines shown in Fig. 7.4-4 for the arbitrarily selected case of $a = c = 1, b = 2$. Sketching through each of the isoclines with the appropriate slope permits the trajectories of Fig. 7.4-5 to be determined.

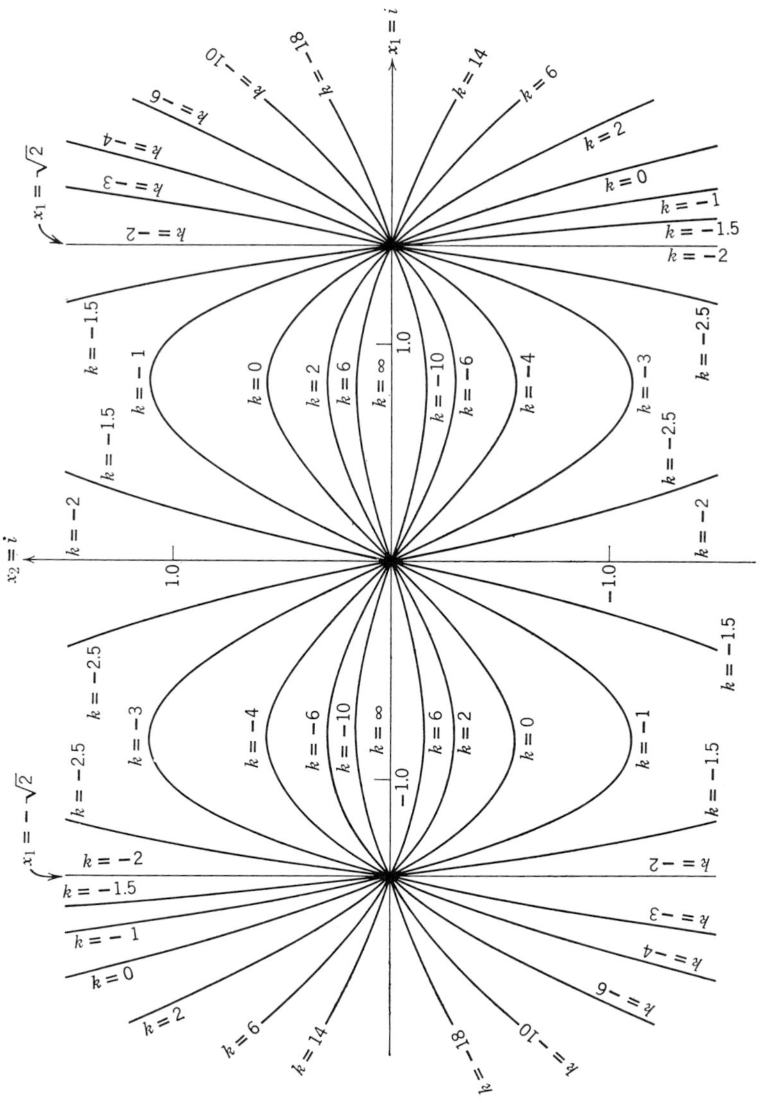

$x_1 = i$
$x_2 = \dot{i}$
$x_1 = \sqrt{2}$
$x_1 = -\sqrt{2}$
1.0
−1.0
$k = 14$
$k = 6$
$k = 2$
$k = 0$
$k = -1$
$k = -1.5$
$k = -2$
$k = -2.5$
$k = -3$
$k = -4$
$k = -6$
$k = -10$
$k = -18$
$k = \infty$

Fig. 7.4-4

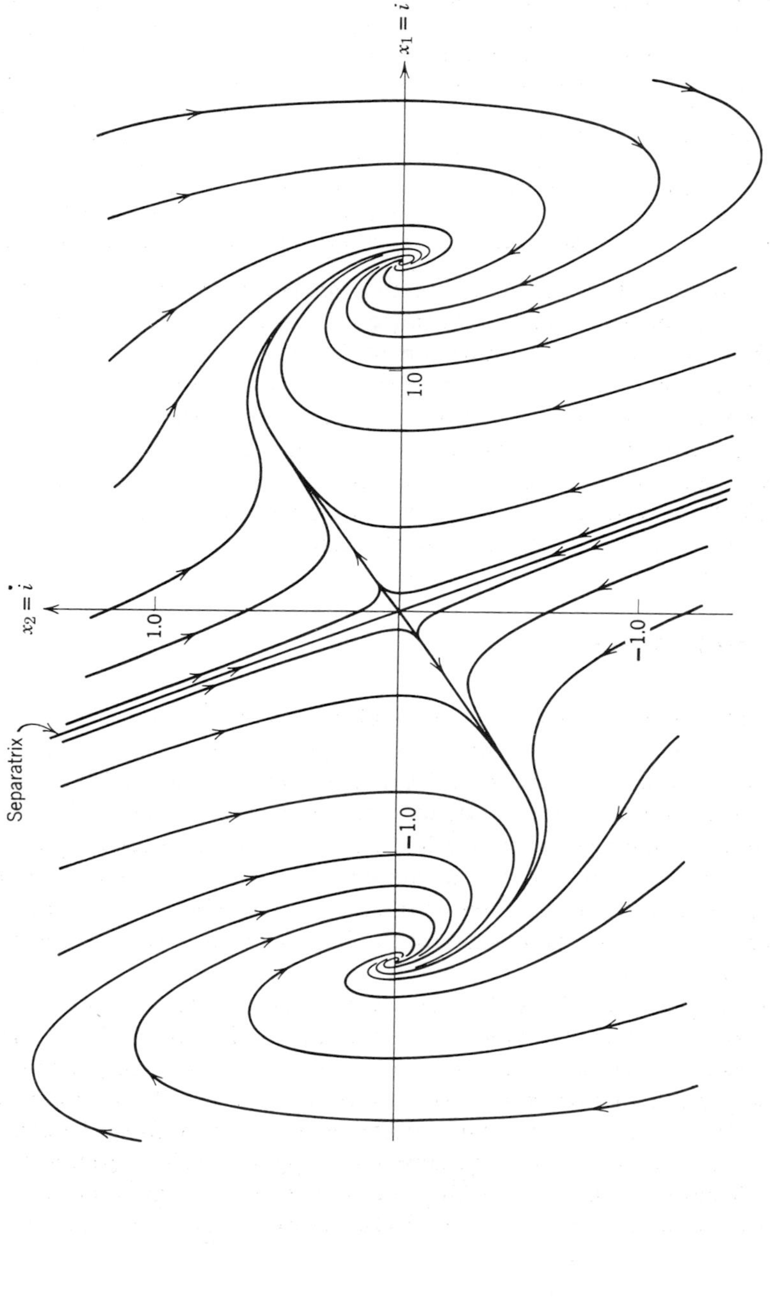

Fig. 7.4-5

The trajectories of Fig. 7.4-5 are useful in that they indicate the system behavior for any of the initial conditions considered therein. They show two stable conditions. Thus the system could be used as a flip-flop. However, for the parameters chosen, the design is poor in the sense that an extremely large triggering signal is required to change the system from one stable condition to the other. This is revealed by the fact that the triggering signal would have to drive the system states across the curve labeled separatrix. A *separatrix* is a trajectory which passes through singular points and divides the phase plane into regions of different character to the trajectories. The separatrix of Fig. 7.4-5 separates the trajectories about the left and right focal points. Since the negative characteristic value of the saddle point is approximately the slope of the separatrix, the phase plane analysis reveals that the size of the triggering signal can be altered by changing a. Adjustment of b would also change the triggering signal requirements, but it has the simultaneous effect of changing the separation of the singular points.

In concluding this section, it is worth emphasizing the proper utilization of the phase plane approach. The trajectories of Fig. 7.4-5 can be obtained much more easily from a simulation of Eq. 7.4-9 than by the method of isoclines utilized here. The proper place for the phase plane approach is in conjunction with the simulation, to indicate the appropriate parameter changes necessary to improve system behavior with respect to the magnitude of the stable states, the size of the triggering signal required, the transient response, etc. All this information is readily available in the equations and figures determined in the example.

7.5 LIMIT CYCLES

Nonlinear systems may have particular trajectories called *limit cycles*, which are *isolated* closed curves in the phase plane corresponding to periodic motions. The linear oscillator considered in Section 7.2 has closed trajectories, but they are not isolated and hence are not limit cycles. The amplitude of the periodic motion in the linear oscillator depends upon the initial conditions, and an infinite number of closed curves exist. A nonlinear system has a fixed number (possibly infinite) of limit cycles, each of which separates the phase plane into two regions where the character of the motion is different.

Limit cycles may be stable, unstable, or semistable. Figure 7.5-1*a* illustrates a stable limit cycle. All near trajectories approach the limit cycle as time approaches infinity. A stable limit cycle corresponds to a stable periodic motion in a physical system. Figure 7.5-1*b* illustrates an unstable limit cycle. The near trajectories move away from the closed curve. In other words, all near trajectories approach the limit cycle as time approaches *minus* infinity. In the case of a semistable limit cycle, for increasing positive time all the near trajectories on one side of the limit cycle

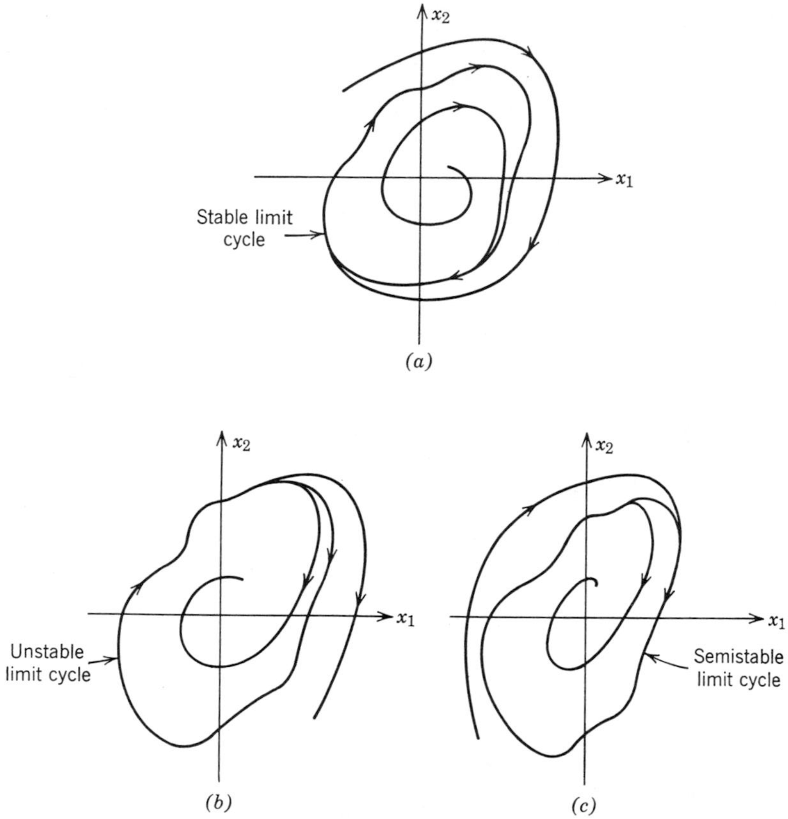

Fig. 7.5-1

approach it, while those on the other side of the limit cycle leave it. Such a case is shown in Fig. 7.5-1*c*.

The existence or nonexistence of limit cycles in the behavior of a system is of fundamental importance to engineers. For example, a control system engineer generally desires systems without limit cycles, although small-amplitude oscillations are sometimes acceptable. On the other hand, an engineer designing an oscillator would definitely want a system with a stable limit cycle. Several theorems are available to guide the engineer in this respect.

Poincaré's Index[12,13]

The index n of a closed curve in the phase plane is given by N, the total number of centers, foci, and nodes enclosed, minus S, the number of saddle points enclosed. That is, $n = N - S$. A *necessary* condition for a closed curve to be a limit cycle is that its index be +1. This criterion is not sufficient, however. As an illustration, consider Example 7.4-1. A large closed curve enclosing the two focal points and the saddle point in Fig. 7.4-5 has an index of +1. Thus a limit cycle conceivably could exist which encloses all the singular points. This is not the case, however, as can be shown by Bendixson's negative criterion.

Bendixson's Negative Criterion

This criterion is sometimes useful for proving that no limit cycle exists in a region of the phase plane. Consider the equations $\dot{x}_1 = f_1(x_1, x_2)$, $\dot{x}_2 = f_2(x_1, x_2)$. The slope of any trajectory is given by

$$\frac{dx_2}{dx_1} = \frac{\dot{x}_2}{\dot{x}_1} = \frac{f_2(x_1, x_2)}{f_1(x_1, x_2)}$$

This can be rewritten as $f_1(x_1, x_2)\, dx_2 - f_2(x_1, x_2)\, dx_1 = 0$. Thus, around a limit cycle,

$$\oint [f_1(x_1, x_2)\, dx_2 - f_2(x_1, x_2)\, dx_1] = 0 \tag{7.5-1}$$

By Gauss' theorem,† Eq. 7.5-1 becomes

$$\iint \left[\frac{\partial f_1}{\partial x_1} + \frac{\partial f_2}{\partial x_2}\right] dx_1\, dx_2 = 0 \tag{7.5-2}$$

If the integrand of Eq. 7.5-2, i.e.,

$$I = \frac{\partial f_1}{\partial x_1} + \frac{\partial f_2}{\partial x_2} \tag{7.5-3}$$

does not change sign or vanish identically within a region of the phase plane, the integral of Eq. 7.5-2 cannot be zero. Since Eq. 7.5-2 applies along a limit cycle, no limit cycle can exist within a region of the phase plane in which I does not change sign or vanish identically.

Example 7.5-1. As an illustration of the use of this theorem, determine I for Example 7.4-1.

Use of Eqs. 7.4-10 and 7.5-3 yields $I = -2a$. Since this does not change sign, nor is it zero anywhere in the x_1x_2-plane, no limit cycle can exist for the flip-flop of Fig. 7.4-1. Thus the circuit of Fig. 7.4-1 cannot oscillate. This is an important property for a flip-flop.

Example 7.5-2. The system of Fig. 7.5-2 is characterized by

$$\begin{aligned} \dot{x}_1 &= x_2 \\ \dot{x}_2 &= -x_1 + 2\xi x_2\left(1 - \frac{{x_2}^2}{3a}\right) \end{aligned} \tag{7.5.4}$$

† See any standard text on advanced calculus or electromagnetic theory.

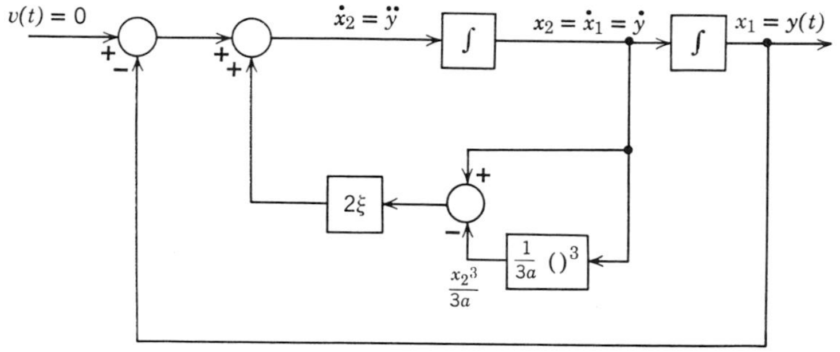

Fig. 7.5-2

Assuming $0 < \xi < 1$, this system has only one singular point, an unstable focus, at $x_1 = x_2 = 0$. Thus its index is $+1$ and a limit cycle could exist. Apply Bendixon's negative criterion.

Evaluation of I given by Eq. 7.5-3 yields

$$I = 2\xi\left[1 - \frac{x_2^{\,2}}{a}\right]$$

Since I does not change sign or identically vanish for $|x_2| < \sqrt{a}$, no limit cycle can exist which is wholly contained within the region specified by $|x_2| < \sqrt{a}$, assuming $a > 0$. A limit cycle which is not wholly contained within this region does exist, however, since Eq. 7.5-4 corresponds to the Rayleigh equation[14] $\ddot{y} - 2\xi\dot{y}\left(1 - \frac{\dot{y}_2}{3a}\right) + y = 0$. Notice, from the expression for I, that no limit cycle can exist for negative values of a.

Poincaré-Bendixson Theorem[15]

The Poincaré-Bendixson theorem states that, if a trajectory remains inside a finite region without approaching any singular points, then the trajectory is either closed or approaches a closed trajectory. Such a closed trajectory need not be a limit cycle, e.g., the closed trajectories about a center.

If a region can be determined for which the theorem applies, then it is of considerable use. Unfortunately, however, the determination of such a region is often very difficult. One approach which is occasionally useful is to choose two concentric circles C_1 and C_2, which define a region R as shown in Fig. 7.5-3. If there are no singular points in R or on C_1 and C_2, and if trajectories enter R through every point of C_1 and C_2, then there is at least one closed trajectory in the region R. The same is true if the trajectories leave R.

Example 7.5-3. Use the concept of the Poincaré-Bendixson theorem to enlarge the known region in which the Rayleigh equation cannot possess a limit cycle.

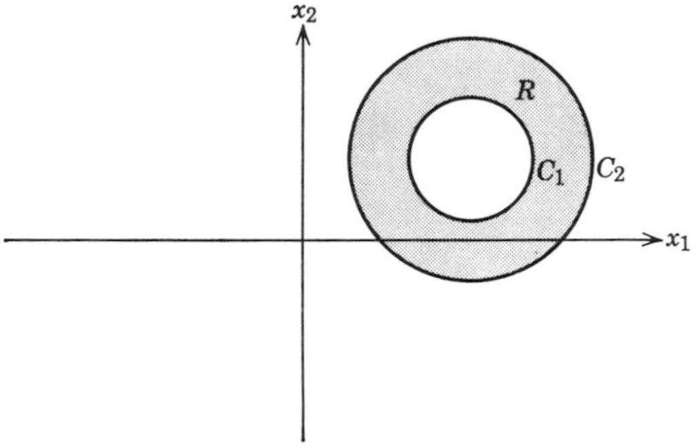

Fig. 7.5-3

From Eq. 7.5-4, the slope of any trajectory is given by

$$\frac{dx_2}{dx_1} = \frac{-x_1 + 2\xi x_2(1 - x_2^2/3a)}{x_2}$$

Let C_1 be a circle of radius r with its center at the origin, so that $x_1^2 + x_2^2 = r^2$. Then the slope of the circle is given by

$$\frac{dx_2}{dx_1} = -\frac{x_1}{\sqrt{r^2 - x_1^2}} = -\frac{x_1}{x_2}$$

The difference between the slope of the trajectory and the slope of the circle at a common point is

$$\delta = 2\xi\left(1 - \frac{x_2^2}{3a}\right)$$

For all $|x_2| < \sqrt{3a}$, $\delta > 0$, so that the slope of the trajectory is more positive than the slope of the circle. This indicates that all trajectories pass out of C_1 for any $r < \sqrt{3a}$. Thus the system of Fig. 7.5-2 has no limit cycles within the region defined by $r \leqslant \sqrt{3a}$. This region is larger than that defined by Bendixson's negative criterion.

In this example, it is not possible to use similar reasoning to determine a circle C_2 for which all trajectories enter, in an attempt to locate the limit cycle. The necessary curve is more complicated.[16]

Poincaré's Successor Function[17,18]

The successor function introduced by Poincaré is useful for proving certain theorems relating to limit cycles. More important to engineers, however, is the fact that the concept can be used to derive conditions on system parameters for which limit cycles can exist. This is particularly true for piecewise linear systems, where expressions for the trajectories in various regions of the phase plane can be determined.

The successor function can be illustrated by means of Fig. 7.5-4. In Fig. 7.5-4*a* is shown a curve C, which is intersected by a trajectory first at

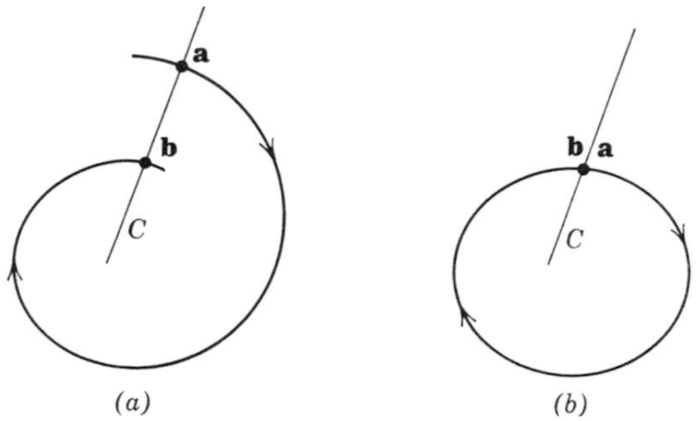

Fig. 7.5-4

point **a**, and next at point **b**. Thus point **b** is the successor to point **a**. If u is a parameter such as arc length on C, points **a** and **b** can be written as functions of u in the form $\mathbf{a} = \boldsymbol{\alpha}(u_a)$, $\mathbf{b} = \boldsymbol{\alpha}(u_b)$. Furthermore, since **a** and **b** are points on a solution curve, they are related by this solution. Thus it is possible to write u_b as a function of u_a in the form $u_b = g(u_a)$. The function $g(u)$ is the successor function. In the case of a limit cycle, points **a** and **b** coincide as shown in Fig. 7.5-4*b*. Such points are given by the solution of $u = g(u)$. The use of the successor function concept to determine conditions on system parameters for which limit cycles can or cannot exist is demonstrated by means of an example.

Example 7.5-4. The system to be considered is the relay control system with rate feedback shown in Fig. 7.5-5*a*. The relay is assumed to have deadband and hysteresis as indicated by the characteristic of Fig. 7.5-5*b*. The relay characteristic has odd symmetry, so that $m(a) = -m(-a)$, where $a = -(x_1 + kx_2)$. The relay output $m(a)$ can have the values 0, +1, or −1, depending upon the value of a and its past history. The system dynamics follow linear relationships for each of these values, so that the system is piecewise linear. That is, the phase plane can be divided into the three regions

$$\begin{aligned} m(a) &= 0, &&\text{Region I} \\ m(a) &= -1, &&\text{Region II} \\ m(a) &= +1, &&\text{Region III} \end{aligned}$$

and the trajectories in each of these regions correspond to those of a linear system. Use the successor function to determine the conditions for a limit cycle.

The trajectories in each of the three regions can be determined from the state equations $\dot{x}_1 = x_2$ and $\dot{x}_2 = -x_2 + m(a)$. The trajectory slopes are given by

$$\frac{dx_2}{dx_1} = \frac{-x_2 + m(a)}{x_2} \tag{7.5-5}$$

In Region I, $m(a)$ is zero, so that Eq. 7.5-5 becomes

$$\frac{dx_2}{dx_1} = -1$$

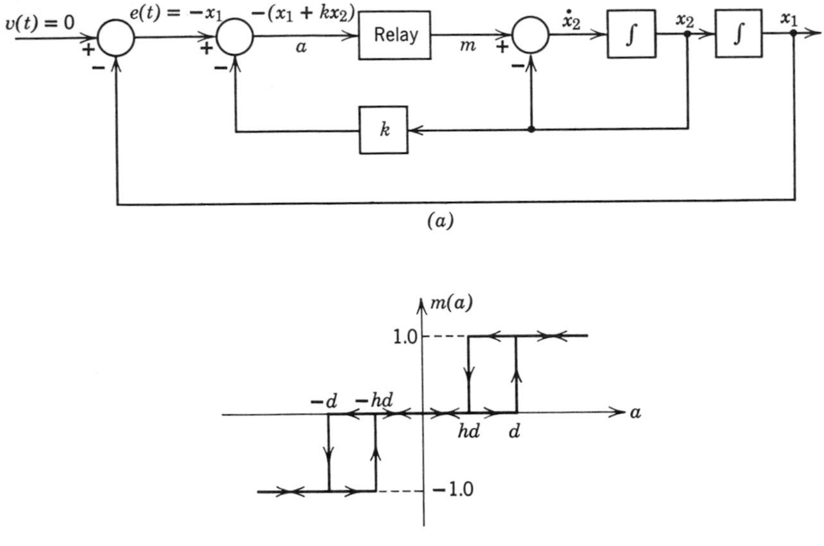

Fig. 7.5-5

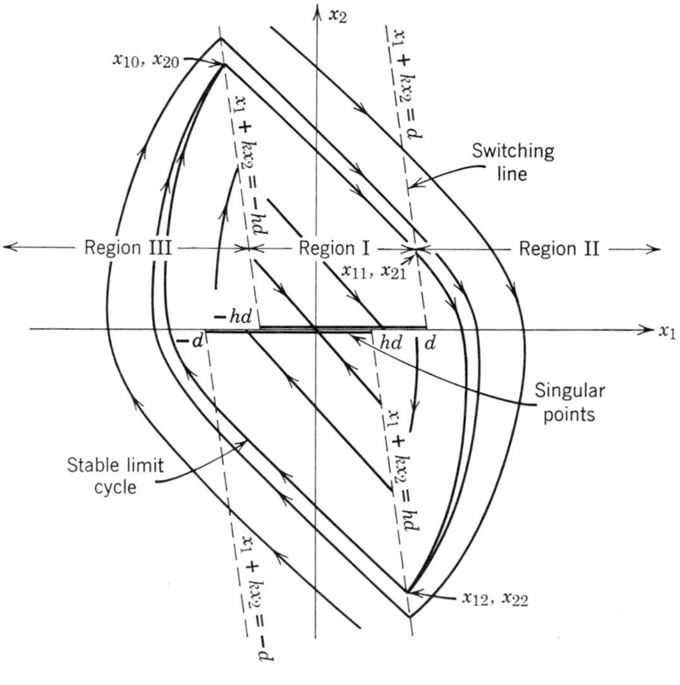

Fig. 7.5-6

This can be integrated directly to give the expression for the trajectories in Region I. They are the straight lines

$$x_2 = -x_1 + c_1, \quad c_1 = \text{constant} \tag{7.5-6}$$

shown in Fig. 7.5-6. It should be observed that there is an infinity of singular points in Region I given by $|x_1| < d$, $x_2 = 0$.

In Region II, Eq. 7.5-5 is

$$\frac{dx_2}{dx_1} = -\frac{1 + x_2}{x_2} \tag{7.5-7}$$

Integration of Eq. 7.5-7 yields

$$x_1 + x_2 = \ln(1 + x_2) + c_2, \quad c_2 = \text{constant} \tag{7.5-8}$$

The trajectories in Region II are shown in Fig. 7.5-6. They are shown only for $x_2 > -1$, since Eq. 7.5-8 is valid only in this range. Since the slope of the trajectories is zero at $x_2 = -1$, as indicated by Eq. 7.5-7, no trajectories can cross $x_2 = -1$. Thus no limit cycle can exist outside the range $|x_2| < 1$. The trajectories in Region III have the same shape as those in Region II, but they are rotated by 180°.

The three regions in Fig. 7.5-6 are divided by the four switching lines as indicated, assuming $k > 0$. These lines are the locus of points at which the relay output switches value. The reader should coordinate these switching lines with the relay characteristic of Fig. 7.5-5*b* and verify that such is the case.

A possible stable limit cycle is shown in Fig. 7.5-6. It is yet to be determined if a limit cycle actually exists. In order to do this, a successor function will be determined. Because of the symmetry which exists in this example, however, the successor function need span only one-half of the possible limit cycle. That is, a successor function will be determined which relates the point $[x_{10} = -(hd + kx_{20}), x_{20}]$ to the point $[x_{12} = hd - kx_{22}, x_{22}]$. If $x_{12} = -x_{10}$ and $x_{22} = -x_{20}$, then a limit cycle exists.

From Eq 7.5-6, it is apparent that the points (x_{10}, x_{20}) and (x_{11}, x_{21}) are related by

$$x_{21} - x_{20} = -(x_{11} - x_{10}) \tag{7.5-9}$$

Also, since the two points are on switching lines, the expressions for these switching lines can be used to eliminate x_{10} and x_{11} from Eq. 7.5-9. That is, since $x_{10} + kx_{20} = -hd$ and $x_{11} + kx_{21} = d$, Eq. 7.5-9 can be written as

$$x_{21} = x_{20} - \frac{d(1 + h)}{1 - k} \tag{7.5-10}$$

Similarly, Eq. 7.5-8 and the appropriate switching line expressions can be used to relate x_{21} and x_{22} by

$$(1 - k)(x_{22} - x_{21}) - d(1 - h) = \ln\left(\frac{1 + x_{22}}{1 + x_{21}}\right) \tag{7.5-11}$$

Equations 7.5-10 and 7.5-11 can be combined to eliminate x_{21} and thereby yield a relationship between x_{20} and x_{22}. After some manipulation, it can be written as

$$(1 + x_{22}) \exp[-x_{22}(1 - k)] = \left[1 + x_{20} - \frac{d(1 + h)}{1 - k}\right] \exp[-x_{20}(1 - k) + 2dh] \tag{7.5-12}$$

Although this expression cannot be solved explicitly for x_{22} in terms of x_{20}, it is in essence a successor function relating these two coordinates. If $x_{22} = -x_{20}$ satisfies Eq. 7.5-12,

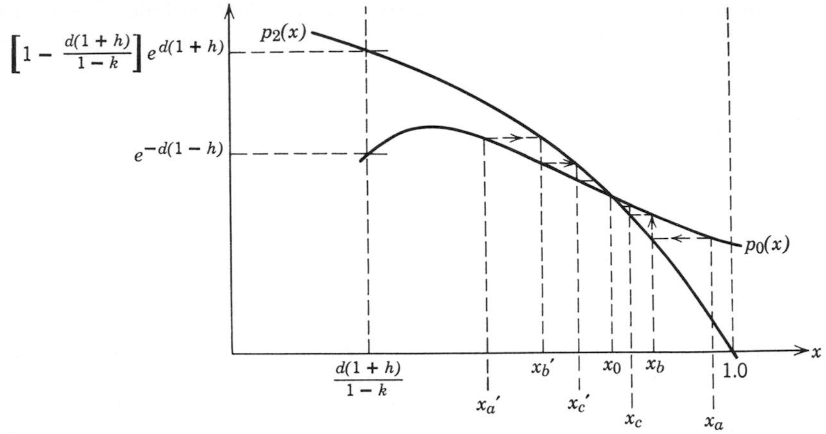

Fig. 7.5-7

there is a limit cycle. This is true since this equality would also necessitate $x_{12} = -x_{10}$, because the two points are on lines of equal slope, symmetrically located about the origin.

In order to determine if any solutions to Eq. 7.5-12 and $x_{22} = -x_{20}$ exist, let

$$p_0(x) = \left[1 + x - \frac{d(1+h)}{1-k}\right] \exp\left[-x(1-k) + 2dh\right]$$

and $p_2(x) = (1 - x) \exp[x(1 - k)]$. If there is a value of x for which $p_0(x) = p_2(x)$, then that value of x is the $x_{20} = -x_{22}$ satisfying Eq. 7.5-12, since $p_0(x)$ and $p_2(-x)$ correspond to the right and left sides of this equation, respectively. Typical curves of $p_0(x)$ and $p_2(x)$ with an intersection, and hence a limit cycle, are shown in Fig. 7.5-7 for x in the range $d(1 + h)/(1 - k) \leqslant x \leqslant 1$. The upper limit was specified previously in connection with Eq. 7.5-8. The lower limit is a consequence of the fact that the Region I portion of any limit cycle trajectory cannot intersect the region of singular points, otherwise the system would come to rest. The initial conditions for which no periodic motions exist are shown in Fig. 7.5-9 in Area I.

From the values of $p_0(x)$ and $p_2(x)$ at $x = d(1 + h)/(1 - k)$, it is apparent that the curves do not intersect unless

$$\epsilon^{-2d} \leqslant \left[1 - \frac{d(1+h)}{1-k}\right] \tag{7.5-13}$$

Thus Eq. 7.5-13 is a necessary condition for a limit cycle. The values of d and h satisfying Eq. 7.5-13 for various values of k are those less than the values on the curves of Fig. 7.5-8. For $k > 0.5$, no limit cycle can exist.

In order to show that a limit cycle exists if Eq. 7.5-13 is satisfied, assume that $x_{20} = x = x_a$ in Fig. 7.5-7. Then $p_0(x_a) = p_2(x_b)$ corresponds to a solution of Eq. 7.5-12, so that the resulting $x_{22} = -x_b$. This is the value of x_{20} for the next half-cycle, so that the new $x_{22} = x_c$. This procedure continues until $x_{20} = x_{22} = x_0$. This is a stable limit cycle, since the same result is achieved starting from x_a'.

For the case illustrated, the phase plane can be divided into three areas as in Fig. 7.5-9. If the initial conditions are such that the initial point is in Area I, the system is

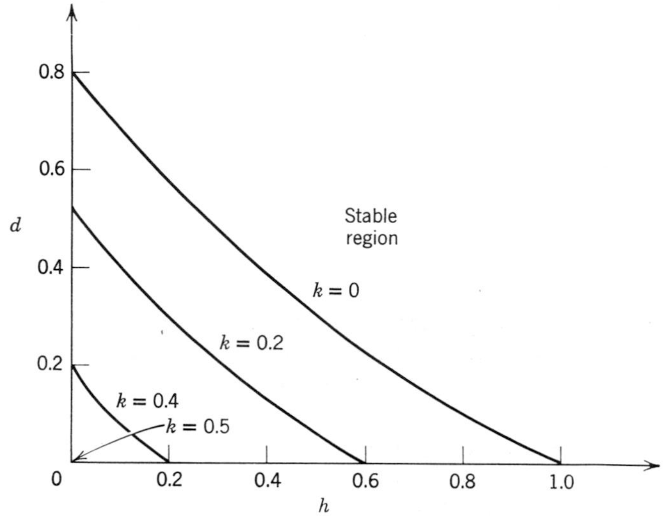

Fig. 7.5-8

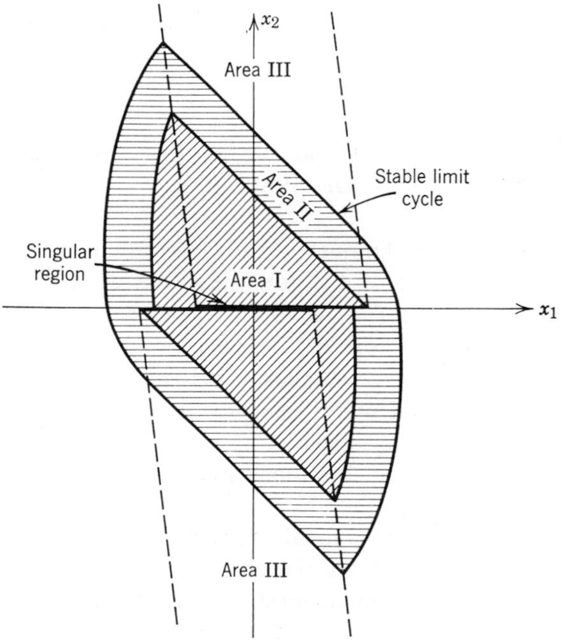

Fig. 7.5-9

stable and the trajectory goes to the singular region. If the initial point is in Area II, the trajectory approaches the stable limit cycle from within. If the initial point is in Area III, the trajectory approaches the stable limit cycle from outside.

This example has demonstrated the use of the successor function for determining conditions under which limit cycles can or cannot exist. Typical results are those of Fig. 7.5-8, which indicate the amount of rate feedback, in terms of the relay deadband and hysteresis, required to prevent the existence of a limit cycle. Figure 7.5-7 illustrates a method of determining the size of the limit cycle if one exists.

Summary

This section has considered limit cycles, and some of the theorems and methods which may be useful in determining their existence. They are, in essence, methods of estimating if a system has periodic motion without explicitly evaluating the system response. Unfortunately, the phase plane approach is limited primarily to unforced, second order systems. Theoretically at least, the concepts may be extended to higher order phase space to treat systems of higher than second order.[5,19] In general, however, the extension has not been an overwhelming success. This is primarily due to the problem of determining, presenting, and visualizing trajectories in phase space. For this reason, most of the theory has not been extended to include higher order systems.

7.6 RATE OF CHANGE OF ENERGY CONCEPT—INTRODUCTION TO THE SECOND METHOD OF LYAPUNOV

Contrasted with the techniques already presented in this chapter, the second or direct method of Lyapunov can be used to determine the stability of the behavior of higher order systems. It applies to systems which may be forced or unforced, linear or nonlinear, stationary or time-varying, and deterministic or stochastic. However, much remains to be done to make the method useful and practical in many cases. For this reason, and because of the introductory nature of this chapter, the discussion which follows is not so all-inclusive.

The philosophy of Lyapunov's second method is the same as most of the stability methods utilized by control engineers, that is, to answer the question of stability without solving the characterizing differential equation. The major limitation of the method is one of determining a suitable Lyapunov function required to indicate stability. If one can be found, then stability is verified. If one cannot be found, however, this does not necessarily indicate instability. It may be a reflection on the experience and ingenuity of the user of the method. This difficulty is gradually being alleviated as more methods for generating possible Lyapunov functions are developed. Although first published in a Russian journal in 1892 and translated into French in 1907, it does not appear that Lyapunov's second method was employed to investigate the stability of the responses of nonlinear control systems until 1944.[20,21] The method was available in the United States by 1949 but did not become widely known to engineers here until 1960.[22,23] Thus extensive experience with the method is yet to be acquired in this country.

As an introduction to the second method of Lyapunov, consider the mechanical configuration of Fig. 7.6-1, in which the unit mass is permitted to move in the y direction only. The spring and friction effects are nonlinear, as represented by the functional force equations $F_{\text{spring}} = -k(y)$ and $F_{\text{friction}} = -b(\dot{y})$. Thus a summation of the forces acting on the mass indicates that the system can be described by the nonlinear differential equation $\ddot{y} + b(\dot{y}) + k(y) = 0$. In the conservative or dissipationless case, $b(\dot{y})$ is zero and the total energy of the system is constant.

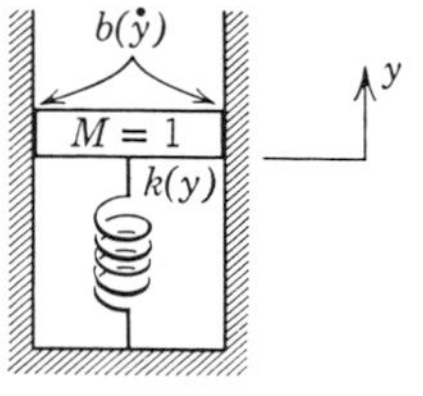

Fig. 7.6-1

Studying first the behavior of this conservative case in the phase plane, let $x_1 = y$ and $x_2 = \dot{x}_1$. Then the system is described by

$$\begin{aligned} \dot{x}_1 &= x_2 \\ \dot{x}_2 &= -k(x_1) \end{aligned} \tag{7.6-1}$$

Assuming $k(0)$ is zero and $k(x_1) \neq 0$ for $x_1 \neq 0$, there is only one equilibrium point, a center, located at the origin of the x_1x_2-plane. The trajectories enclose the center and are given by the solutions of the equation

$$\frac{dx_2}{dx_1} = -\frac{k(x_1)}{x_2} \tag{7.6-2}$$

Equation 7.6-2 can be integrated by separation of variables to yield

$$\frac{x_2^{\,2}}{2} + \int k(x_1)\,dx_1 = c, \quad c = \text{constant} \tag{7.6-3}$$

Thus the trajectories take the form shown in Fig. 7.2-2. The exact shape depends upon $k(y)$.

The total energy of this conservative system is the kinetic $x_2^{\,2}/2$, plus the potential

$$\int k(x_1)\,dx_1$$

Thus the total energy is

$$E(x_1, x_2) = \frac{x_2^{\,2}}{2} + \int k(x_1)\,dx_1 \tag{7.6-4}$$

Equations 7.6-3 and 7.6-4 yield $E(x_1, x_2) = c$. In other words, the x_1x_2 trajectories are contours of constant total energy for this conservative system. Alternatively, the time rate of change of the system energy is zero, as is also shown by

$$\frac{dE}{dt} = x_2\dot{x}_2 + k(x_1)\dot{x}_1 = x_2[\dot{x}_2 + k(x_1)] = 0 \qquad (7.6\text{-}5)$$

where $\dot{x}_2 = -k(x_1)$ from Eq. 7.6-1.

For the case with dissipation $[x_2b(x_2) > 0$, for $x_2 \neq 0]$, the system is described by

$$\begin{aligned} \dot{x}_1 &= x_2 \\ \dot{x}_2 &= -[k(x_1) + b(x_2)] \end{aligned} \qquad (7.6\text{-}6)$$

The trajectories are given by the solution of

$$\frac{dx_2}{dx_1} = -\frac{[k(x_1) + b(x_2)]}{x_2}$$

This expression could be graphically integrated by the method of isoclines to determine solutions.

If only information about the stability of the system is desired, however, a simpler approach is to evaluate the time rate of change of energy. The rate of change of energy, as defined in Eq. 7.6-5, is $dE/dt = x_2[\dot{x}_2 + k(x_1)]$. Using Eq. 7.6-6, the rate of change of energy for the nonconservative case becomes $dE/dt = -x_2b(x_2)$. If $x_2b(x_2)$ is positive for any nonzero x_2, this shows that the energy is always decreasing along any solution of Eq. 7.6-6, except on the line $x_2 = 0$. Therefore, if a trajectory for this case were superimposed on those of Fig. 7.2-2, the result would appear as in Fig. 7.6-2. The motion of the system is from one contour of constant energy to a contour of lesser constant energy, except on the line $x_2 = 0$.

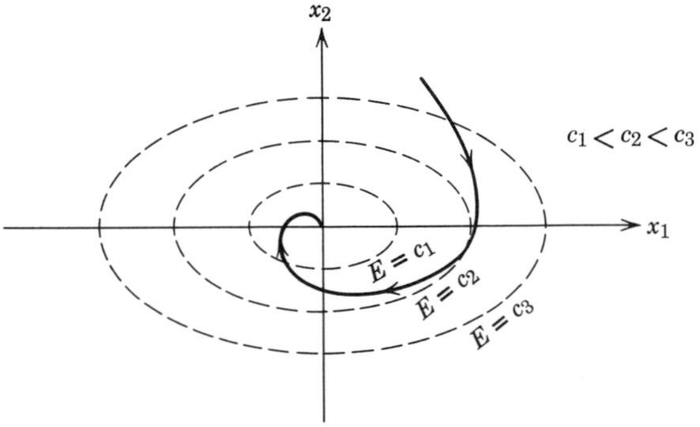

Fig. 7.6-2

On this line, dx_2/dx_1 is infinite if $x_1 \neq 0$. Thus no trajectory can terminate there unless $x_1 = 0$, and hence the system approaches the equilibrium point at the origin.

The preceding discussion is a demonstration of the intuitive reasoning that, if the time rate of change of energy of an isolated physical system is negative for every possible state, except for a single equilibrium state, then the energy will continually decrease until it assumes its minimum value at the equilibrium state.[23] This agrees with one's intuitive concept of stability. A problem arises, however, when one attempts to convert this intuitive concept into a rigorous mathematical technique for determining stability. One of the difficulties is that there is no natural way of defining energy when the system equations are given in purely mathematical form. In order to circumvent this difficulty, the concept of a Lyapunov function is introduced.

A Lyapunov function is a scalar function of the state of the system. As in the preceding discussion, it may sometimes be taken as the total energy of the system. This need not always be the case, however. By means of theorems due to Lyapunov and others, it is possible to specify various types of stability, corresponding to various conditions satisfied by the Lyapunov function. Various types of stability are considered in the next section.

7.7 STABILITY AND ASYMPTOTIC STABILITY

The preceding sections of this chapter have considered the "stability" of autonomous nonlinear systems. In the course of these considerations, several significant points should have occurred to the reader familiar only with stability in connection with linear systems.

First of all, the stability of a *linear* system is a property of the system itself and is not influenced by the states of the system or the input signals. A linear system is stable or unstable in all of state space. Stability of the behavior of a *nonlinear* system, however, as indicated by the system's variational equations, is a concept applicable only in an infinitesimal region about each of the singular points of the system. Furthermore, a nonlinear system may have both stable and unstable singular points and hence may exhibit stable behavior in some regions of state space and unstable behavior in others. Thus, in considering the stability of a nonlinear system, one must associate the concept with a region of state space. As will be seen, this region may extend to all of state space in some cases.

Secondly, excluding the cases of the free integrator and the undamped linear oscillator, if a "stable" linear system is perturbed from an equilibrium point, the system returns to the equilibrium point after an infinite time interval. However, a perturbed nonlinear system may return to the equilibrium point, move to some other stable equilibrium point, exhibit limit cycle oscillations, or have one or more of its states increase without bound. The intuitive concept of stability is not sufficient to characterize these possibilities.

Finally, how are stable, unstable, and semistable limit cycles explained by the intuitive concept of stability? Note that here one is considering the stability of a closed trajectory, not the stability of a system. In the case of a linear system, because stability is a property of the system alone, one often talks about "system stability." In the case of a nonlinear system, however, stability is not a property of the system alone. Hence the proper approach is not to consider stability of the system, but to consider stability of motions or trajectories. That is, the precise concepts of stability relate to deviations or perturbations of the states of a system about some specific motion.

In order to emphasize these concepts, assume that $\mathbf{u}_s(t)$ is a specific solution of

$$\dot{\mathbf{u}} = \mathbf{g}(\mathbf{u}, t) \tag{7.7-1}$$

so that

$$\dot{\mathbf{u}}_s = \mathbf{g}(\mathbf{u}_s, t) \tag{7.7-2}$$

From a precise stability viewpoint, the appropriate question to ask is, "If $\mathbf{u}(t)$ is suddenly perturbed from $\mathbf{u}_s(t)$, how does the resulting solution $\mathbf{u}(t)$ behave with respect to $\mathbf{u}_s(t)$? Does $\mathbf{u}(t)$ remain in the neighborhood of $\mathbf{u}_s(t)$ as time passes, or does it diverge from $\mathbf{u}_s(t)$?" Mathematically this is equivalent to introducing the variable $\mathbf{x} = \mathbf{u} - \mathbf{u}_s$, which measures the deviation of $\mathbf{u}(t)$ from the specific solution $\mathbf{u}_s(t)$. Then Eq. 7.7-1 becomes

$$\dot{\mathbf{u}} = \dot{\mathbf{x}} + \dot{\mathbf{u}}_s = \mathbf{g}[\mathbf{x} + \mathbf{u}_s, t] \tag{7.7-3}$$

Substitution of Eq. 7.7-2 into Eq. 7.7-3 yields the differential equation for the perturbed motion $\mathbf{x}(t)$ as

$$\dot{\mathbf{x}} = \mathbf{g}[\mathbf{x} + \mathbf{u}_s, t] - \mathbf{g}[\mathbf{u}_s, t] \tag{7.7-4}$$

Equation 7.7-4 has an equilibrium point at the origin, since it has the trivial solution $\mathbf{x} = \mathbf{0}$. Thus one can now consider stability of the origin of the equivalent system

$$\dot{\mathbf{x}} = \mathbf{f}(\mathbf{x}, t) \qquad \mathbf{f}(\mathbf{0}, t) = \mathbf{0} \tag{7.7-5}$$

where $\mathbf{f}(\mathbf{x}, t) = \mathbf{g}[\mathbf{x} + \mathbf{u}_s, t] - \mathbf{g}[\mathbf{u}_s, t]$. Unfortunately, Eq. 7.7-5 is often more complicated than the original version, Eq. 7.7-1. Furthermore, the specific solution of Eq. 7.7-1 is required to determine Eq. 7.7-5. However, Eq. 7.7-5 does permit the definitions of stability to be formulated in terms of the stability of an equilibrium state at the origin, and most of the

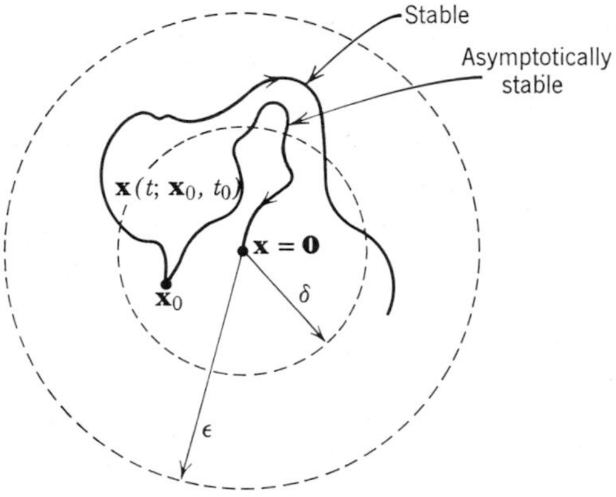

Fig. 7.7-1

stability theorems have been so formulated in recent works. A similar approach is utilized in the definitions which follow.

The equilibrium state of Eq. 7.7-5 is *stable* if, subsequent to a small perturbation from the equilibrium, all motions of the system remain in a correspondingly small neighborhood of the equilibrium. With reference to Fig. 7.7-1, this can be written in precise mathematical terms as follows.[24,25]

The equilibrium of Eq. 7.7-5 is *stable* if, given any $\epsilon > 0$, there exists a $\delta(\epsilon, t_0) > 0$ such that $\|\mathbf{x}_0\| < \delta$ implies $\|\mathbf{x}(t; \mathbf{x}_0, t_0)\| < \epsilon$ for all $t \geqslant t_0$. $\mathbf{x}(t; \mathbf{x}_0, t_0)$ is the response at time t to a sudden perturbation $\mathbf{x}_0 = \mathbf{x}(t_0)$, which exists at time t_0, and $\|\mathbf{x}\|$ is the Euclidean length of the vector $\mathbf{x}$, i.e., $\|\mathbf{x}\|^2 = x_1^2 + x_2^2 + \cdots + x_n^2$. In other words, a bound ϵ on the perturbed response is first specified. If one can then find a bound δ on the size of the perturbation $\mathbf{x}_0$, so that the perturbed response $\mathbf{x}(t; \mathbf{x}_0, t_0)$ due to any $\mathbf{x}_0$ within the bound δ always remains within its bound ϵ, then the equilibrium is said to be stable. The equilibrium is *unstable* if it is not stable.

As indicated previously, this stability concept is a local one. It is sometimes referred to as stability in the small, since one does not know *a priori* how small it may be necessary to choose δ. For example, a stable equilibrium point could be surrounded in the phase plane by an unstable limit cycle. If δ is chosen too large, the perturbed response would not return to the vicinity of the equilibrium point.

A center singular point, characteristic of a linear oscillator, is stable according to the definition above, since one can make the amplitude of the oscillation arbitrarily small by decreasing the initial condition. A stronger type of stability is desired for most control systems (except perhaps for regulators), however. Usually, in that case, one would desire a perturbed system to return to the equilibrium point. This corresponds to asymptotic stability. Such motion is characteristic of stable foci and nodes. Again this is a local concept, defined in the Lyapunov sense as follows.[24,25]

The equilibrium of Eq. 7.7-5, defined to be at the origin, is *asymptotically stable* if, in addition to the equilibrium being stable, there exists a $\delta_0(\epsilon, t_0) > 0$ such that, if $\|\mathbf{x}_0\| < \delta_0$, the solution $\mathbf{x}(t; \mathbf{x}_0, t_0)$ approaches $\mathbf{0}$ as t approaches infinity. Asymptotic stability is also represented in Fig. 7.7-1. It is characterized by the fact that the perturbed response approaches the equilibrium point as time approaches infinity.

There are generalizations of these stability definitions, and in fact many more stability definitions are in the mathematical literature. The generalizations and additional definitions which are useful to the engineer are presented in later sections. The next section considers the application of Lyapunov's second method to investigate the behavior of autonomous continuous systems.

7.8 LYAPUNOV'S SECOND METHOD FOR AUTONOMOUS CONTINUOUS SYSTEMS—LOCAL CONSIDERATIONS

As indicated previously, the stability or asymptotic stability of the equilibrium at the origin of the autonomous system

$$\dot{\mathbf{x}} = \mathbf{f}(\mathbf{x}), \qquad \mathbf{f}(\mathbf{0}) = \mathbf{0} \tag{7.8-1}$$

can often be determined using Lyapunov's second method. If the equilibrium is not at the origin, the method used to arrive at Eq. 7.7-5 can be utilized so that an equilibrium at the origin can be considered. $\mathbf{f}(\mathbf{x})$ is assumed to be continuous, and such that the existence, uniqueness, and continuous dependence of the solutions upon the initial conditions are assured.

Lyapunov's second method requires the utilization of a continuous scalar function of the state variables $V(\mathbf{x})$, in conjunction with Eq. 7.8-1, the state equations. Depending upon the properties of $V(\mathbf{x})$ and its time derivative, instability or various forms of stability of the equilibrium can be proved. Two types of $V(\mathbf{x})$ of particular importance are the *semidefinite* and the *definite* forms.†

The function $V(\mathbf{x})$ is *semidefinite* in a neighborhood about the origin if it is continuous and has continuous first partial derivatives, and if it has the same sign throughout the neighborhood, except points at which it is zero. A $V(\mathbf{x}) \geqslant 0$ is *positive semidefinite*, while a $V(\mathbf{x}) \leqslant 0$ is *negative semidefinite*.

† See also Section 4.9.

The function $V(\mathbf{x})$ is *definite* in a neighborhood about the origin if it is continuous and has continuous first partial derivatives, and if it has the same sign throughout the neighborhood, and is nowhere zero, except possibly at the origin. A $V(\mathbf{x}) > 0$ for $\mathbf{x} \neq \mathbf{0}$ is *positive definite*, while a $V(\mathbf{x}) < 0$ for $\mathbf{x} \neq \mathbf{0}$ is *negative definite.*

These definitions are illustrated by considering examples in an n-dimensional state space where $V(\mathbf{x}) = V(x_1, x_2, \ldots, x_n)$.

Example 7.8-1. $V(\mathbf{x}) = (x_1 + x_2)^2 + x_3^2$, $n = 3$.

This $V(\mathbf{x})$ is positive semidefinite, since it is positive except for $x_1 = x_2 = x_3 = 0$ and $x_1 = -x_2$, $x_3 = 0$.

Example 7.8-2. $V(\mathbf{x}) = x_1^2 + x_2^2$, $n = 2$.

This $V(\mathbf{x})$ is positive definite, since it is positive except at the origin, where it is zero.

Example 7.8-3. $V(\mathbf{x}) = x_1^2 + x_2^2$, $n = 3$.

This $V(\mathbf{x})$ is positive semidefinite, since, for $x_1 = x_2 = 0$, it is zero for any x_3.

Example 7.8-4. $V(\mathbf{x}) = x_1^2 + x_2^2 + \cdots + x_n^2$.

This $V(\mathbf{x})$ is positive definite for arbitrary n.

Example 7.8-5. $V(\mathbf{x}) = -(x_1^2 + x_2^2 + \cdots + x_n^2)$.

This $V(\mathbf{x})$ is negative definite for arbitrary n. The negative of any positive definite function is negative definite.

Example 7.8-6. $V(\mathbf{x}) = \sum_{i,j=1}^{n} q_{ij}x_ix_j = \langle \mathbf{x}, \mathbf{Qx} \rangle$, $q_{ij} = q_{ji}$.

Sylvester's theorem states that the necessary and sufficient conditions for $V(\mathbf{x})$ to be positive definite are that all the successive principal minors of $\mathbf{Q}$ be positive, i.e.,

$$q_{11} > 0, \quad \begin{vmatrix} q_{11} & q_{12} \\ q_{21} & q_{22} \end{vmatrix} > 0, \quad \ldots, \quad \begin{vmatrix} q_{11} & q_{12} & \cdots & q_{1n} \\ q_{21} & q_{22} & \cdots & q_{2n} \\ \cdots & \cdots & \cdots & \cdots \\ q_{n1} & q_{n2} & \cdots & q_{nn} \end{vmatrix} > 0$$

This theorem was proved in Section 4.9.

In addition to the nature of the function $V(\mathbf{x})$, Lyapunov's second method requires consideration of the time derivative of $V(\mathbf{x})$ along the system trajectories. The time derivative of $V(\mathbf{x})$, denoted by W, is

$$W = \dot{V}(\mathbf{x}) = \frac{dV}{dt} = \frac{\partial V}{\partial x_1}\dot{x}_1 + \frac{\partial V}{\partial x_2}\dot{x}_2 + \cdots + \frac{\partial V}{\partial x_n}\dot{x}_n \qquad (7.8\text{-}2)$$

Along any trajectory corresponding to Eq. 7.8-1,

$$\begin{aligned} \dot{x}_1 &= f_1(x_1, x_2, \ldots, x_n) \\ \dot{x}_2 &= f_2(x_1, x_2, \ldots, x_n) \\ &\cdots\cdots\cdots\cdots \\ \dot{x}_n &= f_n(x_1, x_2, \ldots, x_n) \end{aligned}$$

Thus, along any system trajectory, Eq. 7.8-2 is

$$W = \frac{\partial V}{\partial x_1} f_1(\mathbf{x}) + \frac{\partial V}{\partial x_2} f_2(\mathbf{x}) + \cdots + \frac{\partial V}{\partial x_n} f_n(\mathbf{x})$$

In terms of the gradient of $V(\mathbf{x})$, denoted by **grad** $V(\mathbf{x})$, the time derivative of $V(\mathbf{x})$ becomes

$$W = \langle \mathbf{grad}\ V, \mathbf{f} \rangle \tag{7.8-3}$$

The function W is the time derivative of $V(\mathbf{x})$ for the differential equation of Eqs. 7.8-1. Note, however, that the solutions of Eqs. 7.8-1 are not required to evaluate W.

Lyapunov's stability theorem can be written in terms of the functions $V(\mathbf{x})$ and W.

Lyapunov's Stability Theorem.[26–28] Given the differential system of Eqs. 7.8-1, the equilibrium is stable if it is possible to determine a definite $V(\mathbf{x})$, such that $V(\mathbf{0}) = 0$ and W is semidefinite of sign opposite to $V(\mathbf{x})$ or vanishes identically. If $V(\mathbf{x})$ satisfies the requirements of Lyapunov's stability theorem, it is called a *Lyapunov function.*

The requirements of this theorem on the Lyapunov function can be viewed on an intuitive basis by considering the nonlinear mechanical system of Section 7.6. Assume that $V(\mathbf{x})$ is the total energy in the conservative case, as in Eq. 7.6-4. This equation is written below as Eq. 7.8-4.

$$E(x_1, x_2) = V(x_1, x_2) = \frac{x_2^{\,2}}{2} + \int_0^{x_1} k(u)\, du \tag{7.8-4}$$

Note that $V(\mathbf{0}) = 0$, and $V(\mathbf{x})$ is positive definite since any physical spring is such that $k(u)$ has the same sign as u. From Eq. 7.6-5, the time derivative of $V(\mathbf{x})$ is

$$\frac{dE}{dt} = W = x_2[\dot{x}_2 + k(x_1)] \tag{7.8-5}$$

For the conservative case, $\dot{x}_2 = -k(x_1)$ and W vanishes identically. This is the situation in which the origin is a center, and the system moves around a trajectory which is a contour of constant energy. Thus, given a perturbation from the origin, the system remains in the vicinity of the origin and is stable.

For the nonconservative case, Eqs. 7.6-5 and 7.6-6 give $W = -x_2 b(x_2)$. If $x_2 b(x_2) \geqslant 0$, corresponding to a negative semidefinite W, the energy can never increase. The perturbed system remains in the vicinity of the origin and is stable. If $x_2 b(x_2) > 0$, $x_2 \neq 0$, the energy must decrease. The perturbed system returns to the origin and is asymptotically stable. This requirement on W for asymptotic stability is indicated in Lyapunov's asymptotic stability theorem.

Lyapunov's Asymptotic Stability Theorem.[29-31] Given the differential system of Eqs. 7.8-1, the equilibrium is asymptotically stable if it is possible to determine a definite $V(\mathbf{x})$, such that $V(\mathbf{0}) = 0$ and W is definite of sign opposite to $V(\mathbf{x})$.

Example 7.8-7. Figure 7.8-1 represents a second order system with nonlinear feedback. The objective is to determine conditions on a and $g(\;)$ for which the equilibrium is asymptotically stable.

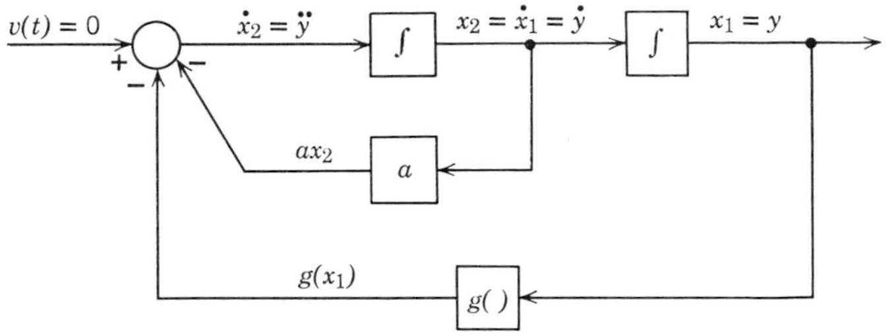

Fig. 7.8-1

The state equations are $\dot{x}_1 = x_2$ and $\dot{x}_2 = -g(x_1) - ax_2$, where it is assumed that $g(x_1) = 0$, only at $x_1 = 0$, so that the only equilibrium is at $x_1 = x_2 = 0$. From the similarity of these equations to those of Eqs. 7.6-6,

$$V(\mathbf{x}) = \frac{x_2^{\,2}}{2} + \int_0^{x_1} g(u)\,du$$

is suggested as a possible Lyapunov function. Note that $V(\mathbf{0}) = 0$, and $V(\mathbf{x})$ for $\mathbf{x} \neq \mathbf{0}$ is positive if the $g(u)$ versus u characteristic is anywhere in the first and third quadrants as shown by the solid curve in Fig. 7.8-2. The latter condition can be expressed as

$$\frac{g(u)}{u} > 0, \quad u \neq 0 \tag{7.8-6}$$

$V(\mathbf{x})$ is positive definite if Eq. 7.8-6 is satisfied.

Now, considering the requirement on the time derivative of $V(\mathbf{x})$, W is given by $W = g(x_1)x_2 - x_2[g(x_1) + ax_2] = -ax_2^{\,2}$. This is negative semidefinite for $a \geqslant 0$, indicating a stable equilibrium.

The equilibrium is asymptotically stable if $a > 0$ and Eq. 7.8-6 is satisfied. This does not come directly from Lyapunov's theorem, since W is not negative definite. However, everywhere that W does not comply with the definition of negative definiteness, i.e., $x_2 = 0$, $x_1 \neq 0$, dx_2/dx_1 is infinite. Hence no trajectory could terminate on the x_1-axis except at the origin.

Note that this equilibrium is still asymptotically stable for the nonlinear characteristic shown dotted in Fig. 7.8-2. $V(\mathbf{x})$ would also be positive definite for this case, since the area under the characteristic is still positive for any u measured from the origin. However, this characteristic introduces other equilibria, which may not be asymptotically stable.

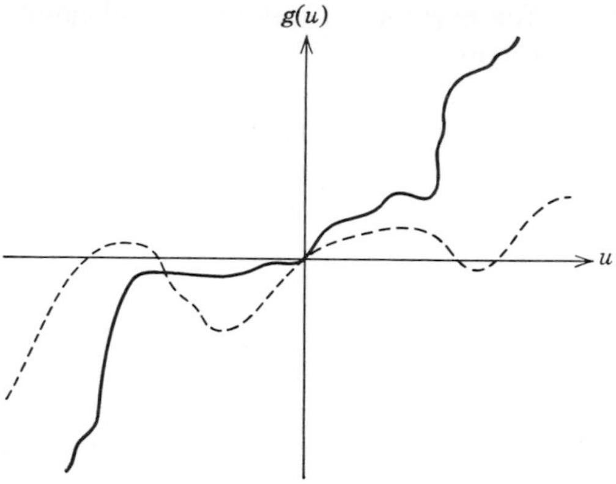

Fig. 7.8-2

Example 7.8-8.[32,33] Barbasin investigated the stability of the third order nonlinear differential equation

$$\frac{d^3y}{dt^3} + a\frac{d^2y}{dt^2} + g\left(\frac{dy}{dt}\right) + h(y) = 0 \tag{7.8-7}$$

The function $g(r)$ is assumed to be continuous, and $h(u)$ to be continuously differentiable. It is also assumed that $g(r) = 0$ and $h(u) = 0$ only at $r = 0$ and $u = 0$, respectively. The state variables are as defined in Fig. 7.8-3, so that the state equations

$$\begin{aligned} \dot{x}_1 &= x_2 \\ \dot{x}_2 &= x_3 \\ \dot{x}_3 &= -h(x_1) - g(x_2) - ax_3 \end{aligned} \tag{7.8-8}$$

indicate a single equilibrium at $x_1 = x_2 = x_3 = 0$. Use Lyapunov's second method to determine the conditions for asymptotic stability.

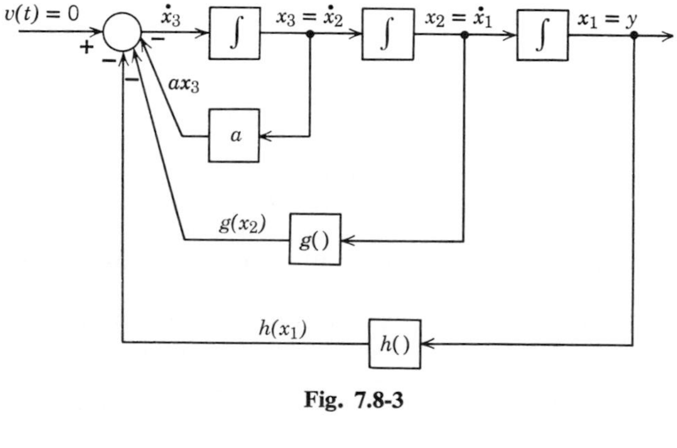

Fig. 7.8-3

A suitable Lyapunov function for this system is†

$$V(\mathbf{x}) = a\int_0^{x_1} h(r)\,dr + \int_0^{x_2} g(r)\,dr + \tfrac{1}{2}(ax_2 + x_3)^2 + x_2h(x_1)$$

Note that $V(\mathbf{0})$ is zero. The time derivative of $V(\mathbf{x})$ is

$$W = ah(x_1)\dot{x}_1 + g(x_2)\dot{x}_2 + (ax_2 + x_3)(a\dot{x}_2 + \dot{x}_3) + x_2\frac{dh(x_1)}{dx_1}\dot{x}_1 + h(x_1)\dot{x}_2$$

Substitution for $\dot{x}_1$, $\dot{x}_2$, and $\dot{x}_3$ as defined by Eq. 7.8-8 yields

$$W = -\left[a\frac{g(x_2)}{x_2} - \frac{dh(x_1)}{dx_1}\right]x_2^2$$

W satisfies the conditions for negative definiteness, except on the line $x_3 = 0$, if

$$a\frac{g(x_2)}{x_2} - \frac{dh(x_1)}{dx_1} > 0, \quad x_2 \neq 0 \tag{7.8-9}$$

Then, using reasoning similar to that of Example 7.8-7, the equilibrium of the system of Fig. 7.8-3 is asymptotically stable if $V(\mathbf{x})$ is positive definite. These requirements on $V(\mathbf{x})$ are now examined.

If a and $h(u)$ are such that

$$\begin{gathered} a > 0 \\ \frac{h(u)}{u} > 0, \quad u \neq 0 \end{gathered} \tag{7.8-10}$$

then the conditions for positive definiteness of the first term of the V-function are satisfied. Furthermore, satisfaction of Eq. 7.8-10, in conjunction with Eq. 7.8-9, yields $g(x_2)/x_2 > 0$ for $x_2 \neq 0$. This means that the second term of the V-function is positive definite if Eqs. 7.8-9 and 7.8-10 are satisfied. Also, the third term of the V-function satisfies the positive definite conditions, with the exception that it is zero for $ax_2 = -x_3$. Even if this is the case, however, $V(\mathbf{x})$ is not zero since the second term is positive.

Thus it remains to consider the last term in the V-function for $x_2 \neq 0$. When $x_2 \neq 0$, it is possible to write the V-function in the form

$$V(\mathbf{x}) = \tfrac{1}{2}(ax_2 + x_3)^2 + \frac{[2G(x_2) + x_2h(x_1)]^2 + 4\displaystyle\int_0^{x_1} h(u)\int_0^{x_2}\left[a\frac{g(r)}{r} - \frac{dh(u)}{du}\right]r\,dr\,du}{4G(x_2)}$$

where

$$G(x_2) = \int_0^{x_2} g(r)\,dr$$

The sum of the first term of $V(\mathbf{x})$ and the first term of the fraction in $V(\mathbf{x})$ is always positive for nonzero x_2. Also, the integration with respect to r is positive because of Eq. 7.8-9. Hence the integration with respect to u is positive. Thus it is now possible to say that $V(\mathbf{x})$ is positive definite and the equilibrium is asymptotically stable if Eqs. 7.8-9 and 7.8-10 are satisfied.

It is interesting to note that the criteria for asymptotic stability of the equilibrium contain two different types of linearizations.[23,34] These are $dh(u)/du$ and $h(u)/u$. These

† This Lyapunov function is derived in Example 7.14-2.

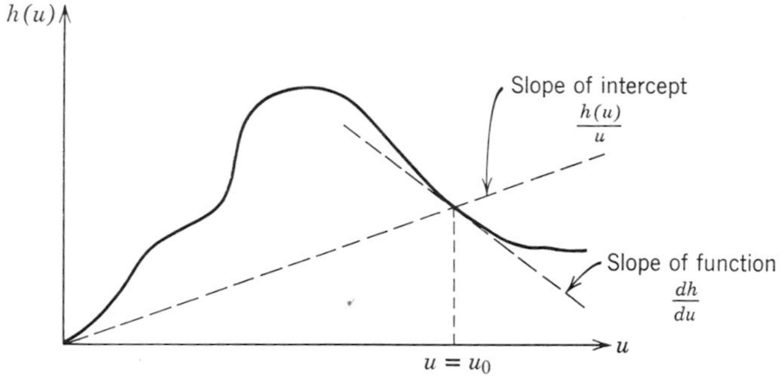

Fig. 7.8-4

linearizations are represented in Fig. 7.8-4 at a point $u = u_0$. The first linearization is the slope of the function, while the second is the slope of the intercept. If $h(u)$ were voltage and u were current for a nonlinear resistance, then dh/du would be the a-c or dynamic resistance, while $h(u)/u$ would be the d-c or static resistance, both evaluated at u_0.

If the dynamic and static linearizations are the same, as they would be for a linear $h(y)$ function, and if $g(\dot{y})$ is linear, then it is possible to replace $h(y)$ by a_0y and $g(\dot{y})$ by $a_1\dot{y}$ in Eq. 7.8-7. It then becomes

$$\frac{d^3y}{dt^3} + a\frac{d^2y}{dt^2} + a_1\frac{dy}{dt} + a_0y = 0 \qquad (7.8\text{-}11)$$

and the stability conditions of Eqs. 7.8-9 and 7.8-10 are

$$\begin{aligned} a &> 0 \\ a_0 &> 0 \\ aa_1 - a_0 &> 0 \end{aligned} \qquad (7.8\text{-}12)$$

These are precisely the Routh-Hurwitz conditions for stability of the linear system described by Eq. 7.8-11. Unfortunately, however, when one attempts to analyze a nonlinear system by introducing some type of linearization, it is usually not evident which type of linearization should be used. This is illustrated by this example, in which both types of linearizations are involved.

In concluding this section, it should be emphasized that the stability or asymptotic stability of the equilibrium at the origin of autonomous systems has been considered. The conclusions about such an equilibrium are local ones, valid only in a small neighborhood of the equilibrium. However, one might feel that the asymptotic stability of the system of Fig. 7.8-3 for the conditions of Eqs. 7.8-9 and 7.8-10 is more than local, since the "equivalent" linear system described by Eq. 7.8-11 is stable everywhere if Eq. 7.8-12 is satisfied. Such is indeed the case, as is verified in the next section.

7.9 ASYMPTOTIC STABILITY IN THE LARGE

The discussion of the preceding section indicated the possibility of an equilibrium being asymptotically stable, regardless of the point $\mathbf{x}_0$ from which the motion originates. In such a case, the equilibrium is said to be *asymptotically stable in the large.* This form of stability is extremely important to the engineer, since it guarantees that all perturbed motions will return to the equilibrium point. Furthermore, this type of stability is related to the effect of disturbances upon system motions, which is considered in the next section.

Theorem 7.9-1.[35,36] If the differential system of Eq. 7.8-1 is asymptotically stable, i.e., if $V(\mathbf{x}) > 0$ and $W < 0$ for all $\mathbf{x} \neq \mathbf{0}$, and if $V(\mathbf{x})$ approaches infinity as $\|\mathbf{x}\|$ approaches infinity, then the origin is asymptotically stable in the large.

The requirement for asymptotic stability in the large, beyond those of asymptotic stability, is that $V(\mathbf{x})$ approach infinity as $\mathbf{x}$ approaches infinity *in any direction.* This requirement can be intuitively comprehended by returning to the nonlinear mechanical system considered in Section 7.6 and again in Section 7.8. The curves of constant energy are given in Eq. 7.8-4 by

$$E(x_1, x_2) = V(x_1, x_2) = \frac{x_2^{\,2}}{2} + \int_0^{x_1} k(u)\, du = c, \quad c = \text{constant} \qquad (7.9\text{-}1)$$

Suppose that $k(u)$ is such that the potential energy term does not become infinite as x_1 approaches infinity, but approaches a finite limit c_0. If $c_0 < c$, the curves of constant V may not all be closed curves as in Fig. 7.6-2, but they could appear as in Fig. 7.9-1. In this case the origin would be asymptotically stable, but not asymptotically stable in the large. It would not be asymptotically stable in the large, since for sufficiently large $\mathbf{x}_0$ it is possible for the system to move from a state of higher energy to a state of lower energy without approaching the equilibrium at the origin. The requirement for asymptotic stability in the large removes this possibility.[37]

Example 7.9-1. Determine the conditions for which the equilibrium of the system of Fig. 7.8-1 is asymptotically stable in the large.

In this case, Eq. 7.8-6 must be replaced by $g(u)/u \geqslant \alpha > 0$, $u \neq 0$. α is a constant. This ensures that $V(\mathbf{x})$ approaches infinity as $\|\mathbf{x}\|$ approaches infinity.

Example 7.9-2. Determine the conditions for which the equilibrium of the system of Fig. 7.8-3 is asymptotically stable in the large.

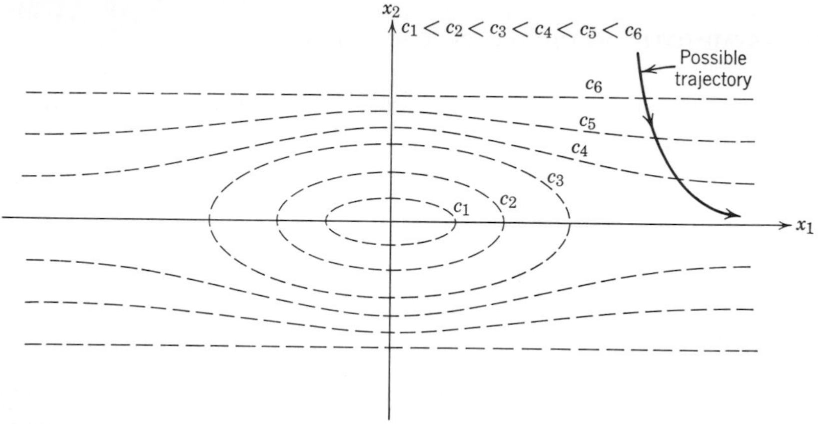

Fig. 7.9-1

The conditions of Eqs. 7.8-9 and 7.8-10 are replaced by

$$a > 0$$

$$\frac{h(u)}{u} \geqslant \alpha_1 > 0, \quad u \neq 0$$

$$a\frac{g(x_2)}{x_2} - \frac{dh(x_1)}{dx_1} \geqslant a\alpha_2 - \alpha_1 > 0, \quad x_2 \neq 0$$

where α_1 and α_2 are constants.

Example 7.9-3.[38] The Euler equations

$$I_x\dot{\omega}_x - (I_y - I_z)\omega_y\omega_z = M_x$$
$$I_y\dot{\omega}_y - (I_z - I_x)\omega_x\omega_z = M_y$$
$$I_z\dot{\omega}_z - (I_x - I_y)\omega_x\omega_y = M_z$$

are the equations of motion of a rigid body written about the principal axes of inertia. I_x, I_y, and I_z denote the moments of inertia about these axes; ω_x, ω_y, and ω_z, the angular velocities; and M_x, M_y, and M_z, the external torques.

Assuming that the rigid body is a space vehicle tumbling in orbit, it is desired to stop the tumbling by applying control torques proportional to the angular velocities. The control torques are $M_x = -k_x\omega_x$, $M_y = -k_y\omega_y$, and $M_z = -k_z\omega_z$. Lyapunov's method can be used to determine the stability of the responses.

Choosing the state variables $x_1 = \omega_x$, $x_2 = \omega_y$, and $x_3 = \omega_z$ permits the system to be described by

$$\dot{\mathbf{x}} = \mathbf{A}(\mathbf{x})\mathbf{x} \tag{7.9-2}$$

where

$$\mathbf{A}(\mathbf{x}) = \begin{bmatrix} -\dfrac{k_x}{I_x} & \dfrac{I_y}{I_x}x_3 & -\dfrac{I_z}{I_x}x_2 \\ -\dfrac{I_x}{I_y}x_3 & -\dfrac{k_y}{I_y} & \dfrac{I_z}{I_y}x_1 \\ \dfrac{I_x}{I_z}x_2 & -\dfrac{I_y}{I_z}x_1 & -\dfrac{k_z}{I_z} \end{bmatrix} \tag{7.9-3}$$

Note that the system has an equilibrium at $\mathbf{x} = \mathbf{0}$. Let $V(\mathbf{x})$ be the positive definite quadratic form of Example 7.8-6, i.e., $V(\mathbf{x}) = \langle \mathbf{x}, \mathbf{Qx} \rangle$, and choose

$$\mathbf{Q} = \begin{bmatrix} I_x^2 & 0 & 0 \\ 0 & I_y^2 & 0 \\ 0 & 0 & I_z^2 \end{bmatrix} \tag{7.9-4}$$

so that $V(\mathbf{x}) = I_x^2 x_1^2 + I_y^2 x_2^2 + I_z^2 x_3^2$, the square of the norm of the total angular momentum. The corresponding time derivative of $V(\mathbf{x})$ is $W = \langle \dot{\mathbf{x}}, \mathbf{Qx} \rangle + \langle \mathbf{x}, \mathbf{Q}\dot{\mathbf{x}} \rangle$. Using Eq. 7.9-2 yields $W = \langle \mathbf{A}(\mathbf{x})\mathbf{x}, \mathbf{Qx} \rangle + \langle \mathbf{x}, \mathbf{QA}(\mathbf{x})\mathbf{x} \rangle$. This can be rewritten as

$$W = \langle \mathbf{x}, [\mathbf{A}^T(\mathbf{x})\mathbf{Q} + \mathbf{QA}(\mathbf{x})]\mathbf{x} \rangle \tag{7.9-5}$$

Equation 7.9-5 is of the form $W = -\langle \mathbf{x}, \mathbf{Px} \rangle$, where

$$-\mathbf{P} = \mathbf{A}^T(\mathbf{x})\mathbf{Q} + \mathbf{QA}(\mathbf{x}) \tag{7.9-6}$$

Thus W is negative definite if $\mathbf{P}$ is positive definite. Then, since $V(\mathbf{x})$ is positive definite, and because of its form must approach infinity as $\mathbf{x}$ approaches infinity, the equilibrium is asymptotically stable in the large, if $\mathbf{P}$ is positive definite. Substitution of Eqs. 7.9-3 and 7.9-4 into Eq. 7.9-6 yields

$$\mathbf{P} = \begin{bmatrix} 2k_x I_x & 0 & 0 \\ 0 & 2k_y I_y & 0 \\ 0 & 0 & 2k_z I_z \end{bmatrix}$$

If each of the k's is positive, $\mathbf{P}$ is positive definite and the equilibrium is asymptotically stable in the large.

7.10 FIRST CANONIC FORM OF LUR'E—ABSOLUTE STABILITY

General stability criteria can be derived which are applicable to classes of system configurations. Several standard configurations have been considered. The most useful of these seems to be the first canonic form of Lur'e, studied by Lur'e, Letov, LaSalle and Lefschetz, and others.[40–43] It is the only standard form considered here.

The first canonic form of Lur'e can be useful for investigating the stability of systems which can be manipulated into the form shown in Fig. 7.10-1*a*. The equivalent form obtained by combining $G_1(s)$, $G_2(s)$, and $H(s)$, so that $G(s) = -G_1(s)G_2(s)H(s)$, is given in Fig. 7.10-1*b*. By expanding $G(s)$ in partial fractions, as in Section 5.5, the system representation takes the form in Fig. 7.10-1*c*. The λ's are the characteristic values or eigenvalues and are assumed to be distinct and negative real.† The b's are the residues of $G(s)$ at its poles. A single free integrator in the linear dynamics is included in the formulation.

† The case in which some of the λ's occur in complex conjugate pairs with negative real parts can be handled in a somewhat similar fashion.

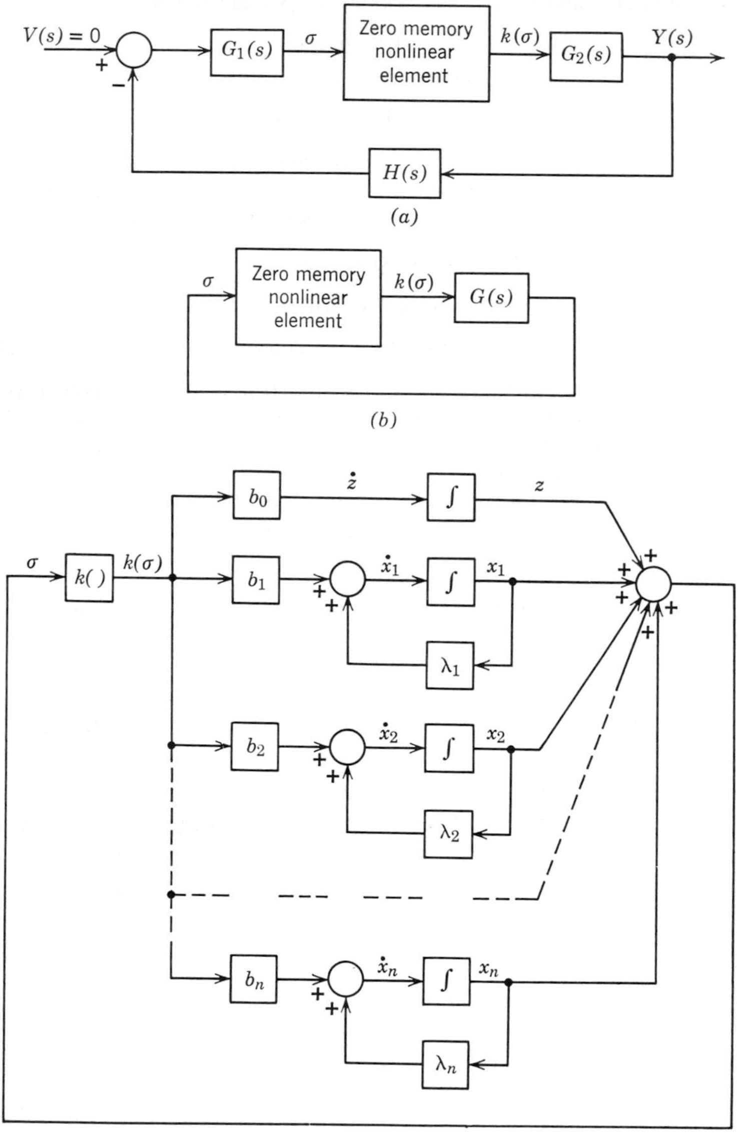
V(s) = 0
+
−
G1(s)
σ
Zero memory nonlinear element
k(σ)
G2(s)
Y(s)
H(s)
(a)
σ
Zero memory nonlinear element
k(σ)
G(s)
(b)
σ
k( )
k(σ)
b0
ż
∫
z
b1
ẋ1
x1
λ1
b2
ẋ2
x2
λ2
bn
ẋn
xn
λn
+
+

(c)

Fig. 7.10-1

Figure 7.10-1c indicates that the systems to be considered are characterized by

$$\begin{aligned} \dot{\mathbf{x}} &= \mathbf{\Lambda x} + k(\sigma)\mathbf{b} \\ \dot{z} &= b_0 k(\sigma) \\ \sigma &= \langle \mathbf{1}, \mathbf{x} \rangle + z \end{aligned} \tag{7.10-1}$$

$\mathbf{\Lambda}$ is a diagonal matrix whose elements are the characteristic values λ_1, $\lambda_2, \ldots, \lambda_n$, and $\mathbf{b}$ and $\mathbf{1}$ are vectors with components $b_1, b_2, \ldots, b_n$ and $1, 1, \ldots, 1$, respectively. Since $\dot{\sigma} = \langle \mathbf{1}, \dot{\mathbf{x}} \rangle + \dot{z}$, it is possible to write $\dot{\sigma}$ as

$$\dot{\sigma} = \langle \boldsymbol{\lambda}, \mathbf{x} \rangle - rk(\sigma) \tag{7.10-2}$$

where the scalar $r = -(b_0 + b_1 + \cdots + b_n)$ and the vector $\boldsymbol{\lambda}$ has components $\lambda_1, \lambda_2, \ldots, \lambda_n$.

A Lyapunov function for this system is

$$V(\mathbf{x}, z, \sigma) = \langle \mathbf{x}, \mathbf{Qx} \rangle + \int_0^\sigma k(\alpha)\, d\alpha$$

where $\mathbf{Q}$ is positive definite and $k(\sigma)$ is restricted to the class for which

$$\begin{aligned} &\frac{k(\sigma)}{\sigma} \geqslant \epsilon > 0, \quad \sigma \neq 0, \quad \epsilon \text{ constant} \\ &k(0) = 0 \end{aligned} \tag{7.10-3}$$

Note that $\sigma = 0$ together with $\mathbf{x} = \mathbf{0}$ requires $z = 0$, so that $V(\mathbf{x}, z, \sigma)$ is positive definite for all $\mathbf{x}$, z, and σ. Furthermore, $V(\mathbf{x}, z, \sigma)$ approaches infinity as $\|\mathbf{x}\|$, σ, or z approaches infinity. The latter condition, i.e., $V(\mathbf{x}, z, \sigma)$ approaches infinity as z approaches infinity, is observed from the fact that, if z approaches infinity in Fig. 7.10-1c, either σ approaches infinity or one of the x's approaches minus infinity.

The time derivative of $V(\mathbf{x}, z, \sigma)$ is $W = \langle \dot{\mathbf{x}}, \mathbf{Qx} \rangle + \langle \mathbf{x}, \mathbf{Q}\dot{\mathbf{x}} \rangle + k(\sigma)\dot{\sigma}$. Substitution of Eqs. 7.10-1 and 7.10-2 yields

$$W = \langle \mathbf{x}, [\mathbf{\Lambda}^T\mathbf{Q} + \mathbf{Q\Lambda}]\mathbf{x} \rangle + k(\sigma)[\langle \mathbf{b}, \mathbf{Qx} \rangle + \langle \mathbf{x}, \mathbf{Qb} \rangle] + k(\sigma)[\langle \boldsymbol{\lambda}, \mathbf{x} \rangle - rk(\sigma)]$$

Similarly to Eq. 7.9-6, it is desirable to choose a positive definite $\mathbf{P}$ such that $-\mathbf{P} = \mathbf{\Lambda}^T\mathbf{Q} + \mathbf{Q\Lambda}$. In this case, however, $\mathbf{\Lambda}$ is diagonal so that $\mathbf{\Lambda}^T = \mathbf{\Lambda}$. Furthermore, since $\mathbf{Q}$ is symmetrical, $\langle \mathbf{b}, \mathbf{Qx} \rangle = \langle \mathbf{x}, \mathbf{Qb} \rangle$. Then

$$W = -\langle \mathbf{x}, \mathbf{Px} \rangle + k(\sigma)[2\mathbf{b}^T\mathbf{Q} + \boldsymbol{\lambda}^T]\mathbf{x} - rk^2(\sigma)$$

W is a quadratic form in $\mathbf{x}$ and $k(\sigma)$ and is negative definite if and only if[44]

$$r > [\mathbf{Qb} + \tfrac{1}{2}\boldsymbol{\lambda}]^T\mathbf{P}^{-1}[\mathbf{Qb} + \tfrac{1}{2}\boldsymbol{\lambda}] \tag{7.10-4}$$

Assume that $\mathbf{Q}$ is given by

$$\mathbf{Q} = \frac{1}{2}\begin{bmatrix} q_{11} & & \cdots & 0 \\ & q_{22} & & \vdots \\ \vdots & & \ddots & \\ 0 & \cdots & & q_{nn} \end{bmatrix}, \quad q_{ii} > 0, \quad i = 1, 2, \ldots, n$$

where all the nondiagonal elements are zero. Then Eq. 7.10-4 becomes

$$r > -S \tag{7.10-5}$$

where S is given by

$$S = \frac{1}{4}\sum_{i=1}^{n} \frac{[b_i + \lambda_i/q_{ii}]^2}{\lambda_i/q_{ii}}$$

This condition, in conjunction with Eq. 7.10-3, is sufficient for asymptotic stability in the large. However, it may be more demanding than necessary. Thus it is desirable to determine the q_{ii} which are the least demanding on r.

S is negative since $\lambda_i < 0$. Thus the least restriction on r is to have $|S|$ a minimum. If $b_i > 0$, choosing q_{ii} such that $\lambda_i/q_{ii} = -b_i$ makes the ith term of S zero. If $b_i < 0$, the ith term of $|S|$ is a minimum for $\lambda_i/q_{ii} = b_i$. The corresponding value of the ith term is b_i. Thus, if the first m of the b_i are positive and the remaining $n - m$ are negative, Eq. 7.10-5 becomes

$$-b_0 > \overbrace{b_1 + b_2 + \cdots + b_m}^{\text{each positive}} \tag{7.10-6}$$

It is readily apparent that Eq. 7.10-6 is applicable only if $(-b_0)$ is more positive than the sum of the positive residues of $G(s)$. Since Eq. 7.10-3, together with Eq. 7.10-6, form sufficient, but not necessary, conditions for asymptotic stability in the large, failure to satisfy these conditions does not necessarily mean that the system is unstable. It may mean that the criterion does not apply in the particular case under consideration.

If Eq. 7.10-6 is satisfied, the equilibrium is asymptotically stable in the large for any nonlinear characteristic satisfying Eq. 7.10-3. Stability of this type, i.e., asymptotic stability in the large for somewhat arbitrary $k(\sigma)$, is called *absolute stability*. Various other absolute stability criteria can be derived by modifying the previous procedure (e.g., by choosing a different $\mathbf{Q}$), or the Lyapunov function.[45,46]

Example 7.10-1. The absolute stability of the equilibrium of the system with $G(s) = -(s + 3)/(s + 2)(s - 1)$ is to be investigated.

Since $G(s)$ does not have a pole at the origin, Eq. 7.10-6 is of no direct use. However it can be utilized by making use of a technique known as *pole-shifting*.[47] The pole-shifting technique can be illustrated by means of Fig. 7.10-2. The feedforward and feedback ϕ's cancel so that the basic system and its conditions for stability remain unchanged.

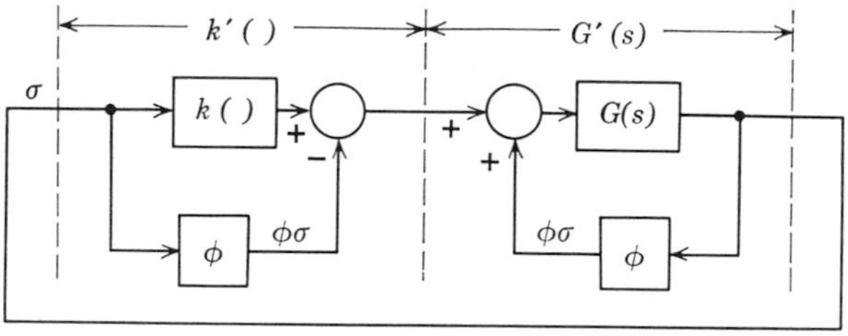

Fig. 7.10-2

However, the feedback around $G(s)$ can be utilized to move the poles of $G'(s)$, since

$$G'(s) = \frac{G(s)}{1 - \phi G(s)}$$

In the example under consideration,

$$G'(s) = \frac{-(s+3)}{s^2 + (1+\phi)s + (3\phi - 2)}$$

so that choosing $\phi = \frac{2}{3}$ results in $G'(s)$ having a pole at the origin. Now Eq. 7.10-6 can be applied.

For $\phi = \frac{2}{3}$,

$$G'(s) = \frac{-(s+3)}{s(s+\frac{5}{3})} = \frac{-\frac{9}{5}}{s} + \frac{\frac{4}{5}}{s+\frac{5}{3}}$$

Thus $b_0 = -\frac{9}{5}$ and $b_1 = \frac{4}{5}$, so that Eq. 7.10-6 indicates absolute stability if

$$\frac{k'(\sigma)}{\sigma} = \left[\frac{k(\sigma) - \frac{2}{3}}{\sigma}\right] \geqslant \epsilon > 0$$

Since $k'(\sigma)$ is restricted to the first and third quadrants by Eq. 7.10-3, $k(\sigma)$ must be restricted to the unshaded area of Fig. 7.10-3. If $k(\sigma)$ is replaced by a linear gain k, the Routh-Hurwitz condition on k for stability is precisely that of Fig. 7.10-3. Aizerman has shown that this is true in general for second order systems.[48]

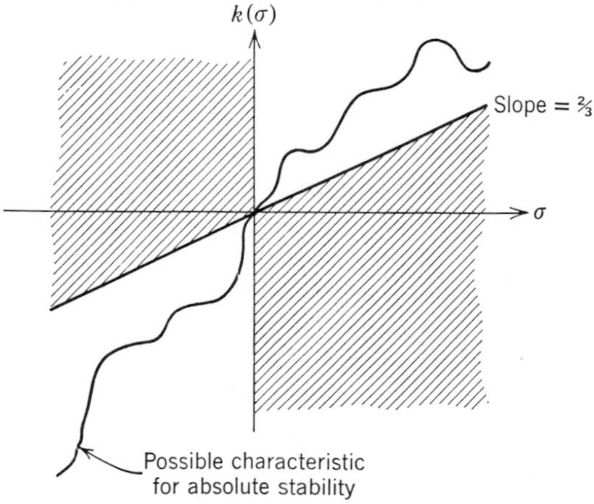

Fig. 7.10-3

7.11 STABILITY WITH DISTURBANCES—PRACTICAL STABILITY

Previous sections considered system motions from the viewpoint of a perturbation suddenly moving the system from its equilibrium state, and the perturbing force then immediately disappearing. Stability is the property of the motion which indicates that the effect of a small perturbation upon the motion is not large. Asymptotic stability further indicates that the effect of a small perturbation disappears with time. Asymptotic stability in the large indicates that the effect of the perturbation tends to disappear, independently of its size. In the physical world of real systems, however, many disturbances are not of the impulsive type. In fact, they are often stochastic in nature. The preceding discussion says nothing about disturbances which are not impulsive.

Total stability is a form of stability relating to constantly acting perturbations. In essence, total stability means that the system motion will remain near the equilibrium if it is not "too far" away initially, and if the perturbations are not "too large." Considering the unperturbed system

$$\dot{\mathbf{x}} = \mathbf{f}(\mathbf{x}), \quad \mathbf{f}(\mathbf{0}) = \mathbf{0} \tag{7.11-1}$$

and denoting the perturbations by $\mathbf{u}(\mathbf{x}, t)$, the perturbed system is

$$\dot{\mathbf{x}} = \mathbf{f}(\mathbf{x}) + \mathbf{u}(\mathbf{x}, t) \tag{7.11-2}$$

Theorem 7.11-1. The equilibrium of Eq. 7.11-1 is *totally stable*, if for every $\epsilon > 0$ there is a $\delta(\epsilon) > 0$ and a $\delta_1(\epsilon) > 0$, such that, if $\|\mathbf{x}_0\| < \delta$ and $\|\mathbf{u}(\mathbf{x}, t)\| < \delta_1$ for all $\mathbf{x}$ and $t \geqslant t_0$, then $\|\mathbf{x}(t; \mathbf{x}_0, t_0)\| < \epsilon$ for all $t > t_0$.[49–53]

Total stability is analogous to stability and asymptotic stability in the sense that one does not know how large $\mathbf{x}_0$ and $\mathbf{u}(\mathbf{x}, t)$ can be in practice. It would be useful to have a form of total stability analogous to asymptotic stability in the large. Such a form is provided by LaSalle and Lefschetz.[53] They refer to total stability as *practical stability* and hence refer to the desired form of stability as *strong practical stability.* This type of stability is the condition under which the equilibrium is totally stable and, in addition, each motion $\mathbf{x}(t; \mathbf{x}_0, t_0)$ of Eq. 7.11-2 for each disturbance $\mathbf{u}(\mathbf{x}, t)$ in a region U lies ultimately in a closed bounded region R containing the equilibrium. The motions are assumed to start at t_0 in a region R_0 contained in R.

Theorem 7.11-2.[53] Let $V(\mathbf{x})$ be a scalar function which for all $\mathbf{x}$ has continuous first partial derivatives and be such that $V(\mathbf{x})$ approaches infinity as $\|\mathbf{x}\|$ approaches infinity. If W, the time derivative of $V(\mathbf{x})$, evaluated for

Eq. 7.11-2 is such that $W \leqslant -\epsilon < 0$ for all $\mathbf{x}$ outside R_0, for all $\mathbf{u}(\mathbf{x}, t)$ in U, and all $t \geqslant t_0$, and if $V(\mathbf{x}) \leqslant V(\mathbf{q})$ for all $\mathbf{x}$ in R_0 and all $\mathbf{q}$ outside R, then Eq. 7.11-1 possesses a strong practical stability.

This theorem provides an indication of the sizes of $\mathbf{x}_0$ and the disturbance for which total stability can be realized.

Example 7.11-1. Returning to Example 7.9-3, the effects of disturbance torques such as solar radiation are to be considered. Thus the external torques are

$$M_x = -k_x\omega_x + u_x$$
$$M_y = -k_y\omega_y + u_y$$
$$M_z = -k_z\omega_z + u_z$$

where the control torques are as previously considered, and the disturbance torques on the x, y, and z axes are u_x, u_y, u_z, respectively.

Using the same $V(\mathbf{x})$ function as in Example 7.9-3, W for the perturbed system is

$$W = -2[I_x x_1(k_x x_1 - u_1) + I_y x_2(k_y x_2 - u_2) + I_z x_3(k_z x_3 - u_3)]$$

where u_1, u_2, and u_3 have been written for u_x, u_y, and u_z, respectively. Let u_{1m}, u_{2m}, and u_{3m} denote the largest absolute values for the disturbances in the region U. Choosing the region R_0 to be the interior of an ellipsoidal surface containing the three points $(u_{1m}/k_1, 0, 0)$, $(0, u_{2m}/k_y, 0)$, and $(0, 0, u_{3m}/k_z)$, then $W \leqslant -\epsilon < 0$ for all $\mathbf{x}$ outside R_0, all $\mathbf{u}(\mathbf{x}, t)$ in U, and all $t \geqslant t_0$. Also, by choosing R to be some region larger than R_0, $V(\mathbf{x}) \leqslant V(\mathbf{q})$ for all $\mathbf{x}$ in R_0 and all $\mathbf{q}$ outside R. Thus the system possesses a strong practical stability. Note that, if the maximum disturbances are increased, the same regions R_0 and R can be maintained by increasing the gains k_x, k_y, and k_z.

7.12 ESTIMATION OF TRANSIENT RESPONSE

If a Lyapunov function is known for an autonomous system with an asymptotically stable equilibrium, then the Lyapunov function can be used to estimate a limit on the transient response.[23] This aspect of Lyapunov's method is best introduced by means of the simple linear system of Fig. 7.12-1. The system speed of response is determined by T, the closed-loop time constant. It is the reciprocal of the loop gain.

The system is characterized by $\dot{x}_1 = -x_1/T$. A suitable $V(\mathbf{x})$ function for stability analysis is $V(x_1) = x_1^2/2$. The rate of change of $V(x_1)$ is $W = dV/dt = x_1\dot{x}_1 = -x_1^2/T$. As expected, $V(x_1)$ and W indicate that the system is asymptotically stable in the large, if T is positive.

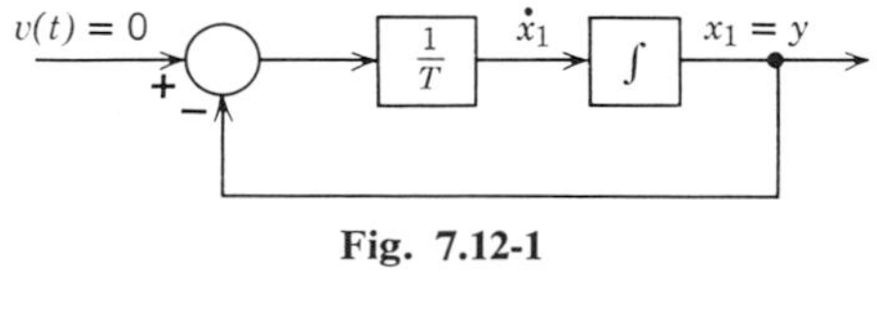

Fig. 7.12-1

The point to be emphasized here is not stability, however. The system speed of response is of interest. A measure of this is the normalized rate at which the Lyapunov function changes. In this case, for example,

$$-\frac{dV/dt}{V} = -\frac{W}{V} = \frac{2}{T} \tag{7.12-1}$$

Integration of Eq. 7.12-1 yields $V = V(x_{1_0})\epsilon^{-2t/T}$ for the variation of the Lyapunov function with time along a motion. Since the Lyapunov function is, or is similar to, depending upon its choice, the system energy, it depends generally upon the squares of the state variables. For that reason, the time constant for the variation in the Lyapunov function is one-half the system time constant. Therefore $-W/V$ is a measure of the system speed of response in this example.

For more general systems, the situation is not as clear as in this simple, linear, first order system. Generally, $-W/V$ changes as the states of the system change. This is illustrated in the following example.

Example 7.12-1. For the system of Example 7.8-7,

$$-\frac{W}{V} = \frac{2a}{1 + (2/x_2^2)\int_0^{x_1} g(u)\,du}, \quad x_2 \neq 0$$

Discuss the effects of the system parameters on the speed of response.

From the viewpoint of speed of response, the designer should choose a as large as possible. One must be careful about drawing conclusions concerning the function $g(\ )$ however, since its effect on $-W/V$ varies with the state variables x_1 and x_2.

7.13 RELAY CONTROLLERS FOR LINEAR DYNAMICS

The concept of the previous section can be used for the design of relay controllers for controlled elements described by

$$\dot{\mathbf{x}} = \mathbf{Ax} + \mathbf{Bm} \tag{7.13-1}$$

where $\mathbf{m}$ is the input to the controlled elements and is restricted to $|m_i(t)| \leqslant M_i$. By determining a Lyapunov function for the unforced system, i.e., $\mathbf{m} = \mathbf{0}$, and then choosing $\mathbf{m}$ to make W as negative as possible, the resulting system removes initial errors rapidly.[23,39]

Lyapunov functions for asymptotically stable in the large linear systems can be determined by choosing $V(\mathbf{x})$ to be the positive definite quadratic form $V(\mathbf{x}) = \langle \mathbf{x}, \mathbf{Qx} \rangle$, where $\mathbf{Q}$ is a positive definite, symmetric matrix which satisfies

$$\mathbf{A}^T\mathbf{Q} + \mathbf{QA} = -\mathbf{P} \tag{7.13-2}$$

and **P** is any symmetric, positive definite matrix. Then

$$W = \langle \dot{\mathbf{x}}, \mathbf{Q}\mathbf{x} \rangle + \langle \mathbf{x}, \mathbf{Q}\dot{\mathbf{x}} \rangle = -\langle \mathbf{x}, \mathbf{P}\mathbf{x} \rangle \qquad (7.13\text{-}3)$$

which is negative definite.

For the system with control, substitution of Eq. 7.13-1 into Eq. 7.13-3 yields $W = -\langle \mathbf{x}, \mathbf{P}\mathbf{x} \rangle + \langle \mathbf{B}\mathbf{m}, \mathbf{Q}\mathbf{x} \rangle + \langle \mathbf{x}, \mathbf{Q}\mathbf{B}\mathbf{m} \rangle$. Since **Q** is symmetric, W can be written as $W = -\langle \mathbf{x}, \mathbf{P}\mathbf{x} \rangle + 2\langle \mathbf{m}, \mathbf{B}^T\mathbf{Q}\mathbf{x} \rangle$. The first term of W is negative definite by choice of **P**. The proper choice of **m** is to make W as negative as possible. Thus it is to set each component of **m** to its maximum magnitude, with a sign opposite that of the corresponding component of $\mathbf{B}^T\mathbf{Q}\mathbf{x}$. Therefore the proper choice of each component of **m** is

$$m_i(t) = -M_i \operatorname{sgn} [\mathbf{B}^T\mathbf{Q}\mathbf{x}]_i$$

This leads to a relay controller operating on a linear combination of the state variables.

Example 7.13-1. The objective is to determine a relay controller for the linear system in which

$$\mathbf{A} = \begin{bmatrix} 0 & 1 \\ -b & -a \end{bmatrix}, \quad \mathbf{B} = \begin{bmatrix} 0 & 0 \\ 0 & 1 \end{bmatrix}, \quad a > 0, \quad b > 0$$

Choose **P** to be the unit matrix. Then the **Q** matrix can be determined as

$$\mathbf{Q} = \begin{bmatrix} \dfrac{a^2 + b(1 + b)}{2ab} & \dfrac{1}{2b} \\ \dfrac{1}{2b} & \dfrac{1 + b}{2ab} \end{bmatrix}$$

$\mathbf{B}^T\mathbf{Q}\mathbf{x}$ can be determined to be

$$\mathbf{B}^T\mathbf{Q}\mathbf{x} = \begin{bmatrix} 0 \\ \dfrac{x_1}{2b} + \dfrac{(1 + b)x_2}{2ab} \end{bmatrix}$$

Thus $m_1 = 0$ and

$$m_2(t) = -M_2 \operatorname{sgn} \left[x_1(t) + \frac{(1 + b)x_2(t)}{a} \right]$$

where the positive constant $1/2b$ has been removed from each of the terms within the brackets, since it does not affect the sign of $m_2(t)$.

The resultant control system is asymptotically stable in the large by design. Furthermore, if u is the maximum value of a disturbance at the point indicated in Fig. 7.13-1, then the system has a strong practical stability since, for $\dot{x}_1 = x_2$ and $\dot{x}_2 = -bx_1 - ax_2 + m_2 + u$, W is given by

$$W = -(x_1^2 + x_2^2) - \left[\frac{x_1}{b} + \left(\frac{1 + b}{ab} \right) x_2 \right] \left\{ M_2 \operatorname{sgn} \left[\frac{x_1}{b} + \left(\frac{1 + b}{ab} \right) x_2 \right] - u \right\}$$

Strong practical stability is a general result for systems of this type.[39]

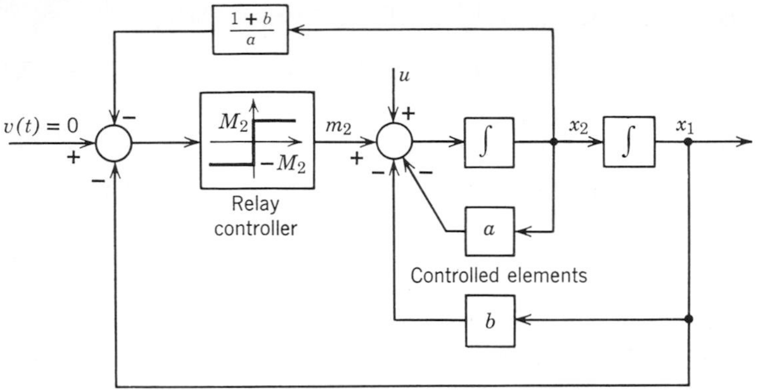

Fig. 7.13-1

It is important to note that the method, as presented, applies only to controlled elements which have feedback around the integrators. If the controlled elements possess free integrations, then the **A** matrix is singular and one cannot satisfy Eq. 7.13-2. This is made evident by letting b approach zero in this example.

For the case in which the controlled elements have a single free integration, the procedure above can be utilized with slight modification. Although presented here for the single input (**m** a scalar) case, the method is readily extended to the multiple input case.

If the state equations describing the linear dynamics are determined using the partial fractions technique of Section 5.5, the result can be put in the form shown in Fig. 7.13-2. The elements comprising $G(s)$ can be characterized by Eq. 7.13-1. In this case, however, m is a scalar, and $\mathbf{B} = \mathbf{b}$, a vector. Thus $\dot{\mathbf{x}} = \mathbf{A}\mathbf{x} + \mathbf{b}m$, where **A** may or may not be diagonal, depending upon how far one carries out the partial fraction expansion. The free integrator can be characterized by $\dot{z} = b_0 m$. Now let $V(\mathbf{x}, z) = \langle \mathbf{x}, \mathbf{Q}\mathbf{x} \rangle + z^2$, which is positive definite, for **Q** as chosen in Eq. 7.13-2. Then

$$W = -\langle \mathbf{x}, \mathbf{P}\mathbf{x} \rangle + 2m[\langle \mathbf{b}, \mathbf{Q}\mathbf{x} \rangle + b_0 z]$$

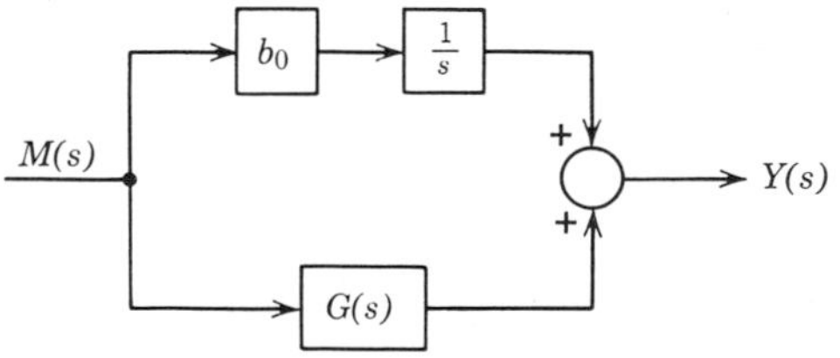

Fig. 7.13-2

Therefore the proper choice of m is

$$m(t) = -M \operatorname{sgn} [\langle \mathbf{b}, \mathbf{Qx} \rangle + b_0 z] \tag{7.13-4}$$

where M is the largest possible value for $m(t)$.

Example 7.13-2. A relay controller is to be designed for the dynamics of Fig. 7.13-3*a*.

$$\frac{Y(s)}{M(s)} = \frac{1}{s(s+1)} = \frac{1}{s} - \frac{1}{s+1}$$

State equations are defined for the system of Fig. 7.13-3*b* as $\dot{x} = -x + m$ and $\dot{z} = m$. Note that the first equation also applies to Fig. 7.13-3*a*. z, defined in Fig. 7.13-3*b*, is $x + y$ of Fig. 7.13-3*a*. Assuming $\mathbf{P} = 2$, then $\mathbf{Q} = 1$. Since $b_0 = 1$ and $\mathbf{b}$ is a scalar

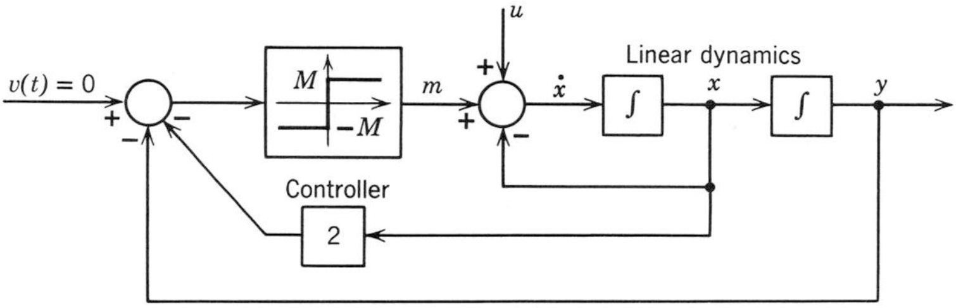

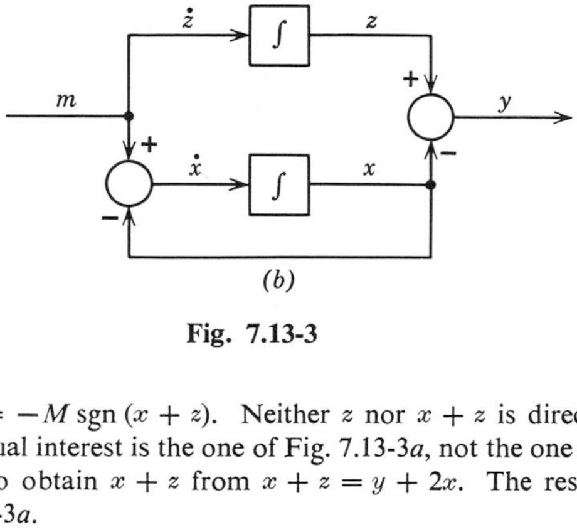

Fig. 7.13-3

equal to unity, $m(t) = -M \operatorname{sgn} (x + z)$. Neither z nor $x + z$ is directly measurable, since the system of actual interest is the one of Fig. 7.13-3*a*, not the one of Fig. 7.13-3*b*. Thus it is necessary to obtain $x + z$ from $x + z = y + 2x$. The resulting system is illustrated in Fig. 7.13-3*a*.

If u is the maximum value of a disturbance input to the linear elements, as shown in Fig. 7.13-3*a*, then since

$$W = -2[x^2 + M(x + z) \operatorname{sgn} (x + z) - u(x + z)]$$

the system has a strong practical stability which can be made as strong as desired by increasing M. This capability, of course, is bounded by the saturation tendencies of the linear dynamics.

7.14 DETERMINATION OF LYAPUNOV FUNCTIONS—VARIABLE GRADIENT METHOD

Lyapunov's second method has been used in preceding sections for the determination of stability and for the design of relay systems. Other than for linear systems and the nonlinear case considered in Problem 7.21, no method has been presented for the determination of Lyapunov functions. Since various theorems guarantee the existence of Lyapunov functions for the stable and asymptotically stable (locally or in the large) cases if $\mathbf{f}(\mathbf{x})$ satisfies the Lipschitz condition† (locally or in the large), it is important to consider possible methods of determining such functions.[54] At least three methods are presented in the literature.[55–60] The variable gradient method is considered here.

Given the existence of a Lyapunov function, its gradient must exist. From $\mathbf{grad}\ V$, $V(\mathbf{x})$ and W may be determined, since, for $\dot{\mathbf{x}} = \mathbf{f}(\mathbf{x})$, Eq. 7.8-3 indicates

$$W = \frac{dV}{dt} = \langle \mathbf{grad}\ V, \mathbf{f} \rangle = \langle \mathbf{grad}\ V, \dot{\mathbf{x}} \rangle \tag{7.14-1}$$

and hence

$$V(\mathbf{x}) = \int_{\mathbf{0}}^{\mathbf{x}} \langle \mathbf{grad}\ V, d\mathbf{x} \rangle \tag{7.14-2}$$

The upper limit of integration does not imply that $V(\mathbf{x})$ is a vector. It indicates that the integral is a line integral to an arbitrary point $\mathbf{x} = (x_1, x_2, \ldots, x_n)$ in state space. For the scalar function $V(\mathbf{x})$ to be obtained uniquely from the line integral of the vector $\mathbf{grad}\ V$, $V(\mathbf{x})$ must be independent of the path of integration. The necessary and sufficient conditions for this are‡

$$\frac{\partial(\mathbf{grad}\ V)_i}{\partial x_j} = \frac{\partial(\mathbf{grad}\ V)_j}{\partial x_i}, \quad i, j = 1, 2, \ldots, n \tag{7.14-3}$$

where $(\mathbf{grad}\ V)_i$ is the component of $\mathbf{grad}\ V$ in the i direction. That is,

$$\mathbf{grad}\ V = \begin{bmatrix} (\mathbf{grad}\ V)_1 \\ (\mathbf{grad}\ V)_2 \\ \cdot \\ \cdot \\ \cdot \\ (\mathbf{grad}\ V)_n \end{bmatrix}$$

† The Lipschitz condition implies continuity of $\mathbf{f}(\mathbf{x})$ in $\mathbf{x}$.
‡ See any standard text on vector calculus.

Since $V(\mathbf{x})$ is independent of the path of integration if Eq. 7.14-3 is satisfied, any path may be chosen. A particularly simple path is the one indicated by

$$\begin{aligned} V(\mathbf{x}) = & \int_0^{x_1} [\mathbf{grad}\ V(u_1, 0, 0, \ldots, 0)]_1\, du_1 \\ & + \int_0^{x_2} [\mathbf{grad}\ V(x_1, u_2, 0, \ldots, 0)]_2\, du_2 + \cdots \\ & + \int_0^{x_n} [\mathbf{grad}\ V(x_1, x_2, x_3, \ldots, x_{n-1}, u_n)]_n\, du_n \end{aligned} \qquad (7.14\text{-}4)$$

This path is utilized in what follows.

Assume $(\mathbf{grad}\ V) = [\alpha]\mathbf{x}$, where

$$[\alpha] = \begin{bmatrix} \alpha_{11}(\mathbf{x}) & \alpha_{12}(\mathbf{x}) & \cdots & \alpha_{1n}(\mathbf{x}) \\ \alpha_{21}(\mathbf{x}) & \alpha_{22}(\mathbf{x}) & \cdots & \alpha_{2n}(\mathbf{x}) \\ \cdots & \cdots & \cdots & \cdots \\ \alpha_{n1}(\mathbf{x}) & \alpha_{n2}(\mathbf{x}) & \cdots & \alpha_{nn}(\mathbf{x}) \end{bmatrix}$$

As the term "variable gradient" implies, **grad** V is assumed to have n undetermined components. The α's consist of a constant and a variable part, which is a function of the state variables. That is, $\alpha_{ij} = \alpha_{ijc} + \alpha_{ijv}(\mathbf{x})$, where the subscripts c and v denote constant and variable, respectively. Without loss of generality, the coefficients α_{iiv} are restricted to be functions of x_i only. Furthermore, to assist in the verification that $V(\mathbf{x}) = \text{constant}$ represents closed surfaces as required (see Section 7.9), $[\alpha]$ is chosen independent of x_n and α_{nn} is set equal to unity. This is a slight loss in generality. With these restrictions, $[\alpha]$ is

$$\begin{bmatrix} \alpha_{11c} + \alpha_{11v}(x_1) & \alpha_{12c} + \alpha_{12v}(x_1, x_2, \ldots x_{n-1}) & \cdots & \alpha_{1nc} + \alpha_{1nv}(x_1, x_2, \ldots x_{n-1}) \\ \alpha_{21c} + \alpha_{21v}(x_1, x_2, \ldots, x_{n-1}) & \alpha_{22c} + \alpha_{22v}(x_2) & \cdots & \alpha_{2nc} + \alpha_{2nv}(x_1, x_2, \ldots, x_{n-1}) \\ \cdots & \cdots & \cdots & \cdots \\ \alpha_{n1c} + \alpha_{n1v}(x_1, x_2, \ldots, x_{n-1}) & \alpha_{n2c} + \alpha_{n2v}(x_1, x_2, \ldots, x_{n-1}) & \cdots & 1 \end{bmatrix}$$

With $[\alpha]$ as defined above, the steps in the variable gradient method of determining a Lyapunov function are:

1. Assume a gradient of the form $\mathbf{grad}\ V = [\alpha]\mathbf{x}$.
2. Determine W from $W = \langle \mathbf{grad}\ V, \mathbf{f} \rangle$.
3. Subject to Eq. 7.14-3, constrain W to be at least negative semidefinite.
4. Determine $V(\mathbf{x})$ from Eq. 7.14-4, and check that it represents closed surfaces.

This procedure is now illustrated by examples.

Example 7.14-1. Determine a Lyapunov function for the system of Example 7.8-7. The system to be considered is second order and is defined by $\dot{x}_1 = x_2$ and $\dot{x}_2 = -g(x_1) - ax_2$. Then **grad** V is

$$\textbf{grad } V = \begin{bmatrix} [\alpha_{11c} + \alpha_{11v}(x_1)]x_1 + [\alpha_{12c} + \alpha_{12v}(x_1)]x_2 \\ [\alpha_{21c} + \alpha_{21v}(x_1)]x_1 + x_2 \end{bmatrix}$$

and

$$W = \left[\alpha_{11c} + \alpha_{11v}(x_1) - a\alpha_{21c} - a\alpha_{21v}(x_1) - \frac{g(x_1)}{x_1}\right]x_1x_2$$
$$- [\alpha_{21c} + \alpha_{21v}(x_1)]x_1 g(x_1) - [a - \alpha_{12c} - \alpha_{12v}(x_1)]x_2^2$$

Equations 7.14-3 yield that

$$\frac{\partial(\textbf{grad } V)_1}{\partial x_2} = \alpha_{12c} + \alpha_{12v}(x_1)$$

equals

$$\frac{\partial(\textbf{grad } V)_2}{\partial x_1} = \alpha_{21c} + \frac{\partial[\alpha_{21v}(x_1)x_1]}{\partial x_1}$$

This requirement can be satisfied and W simplified considerably by choosing $\alpha_{12c} = \alpha_{21c} = \alpha_{12v} = \alpha_{21v} = \alpha_{11c} = 0$ and $\alpha_{11v}(x_1) = g(x_1)/x_1$. Then $W = -ax_2^2$, which is negative semidefinite for $a > 0$.

For these restrictions on the α's, **grad** V is

$$\textbf{grad } V = \begin{bmatrix} g(x_1) \\ x_2 \end{bmatrix}$$

Then, from Eq. 7.14-4, V is given by

$$V(\mathbf{x}) = \int_0^{x_1} g(u_1)\, du_1 + \int_0^{x_2} u_2\, du_2$$

Integration of the second term yields

$$V(\mathbf{x}) = \int_0^{x_1} g(u)\, du + \frac{x_2^2}{2}$$

For $g(u)$ as defined in Example 7.8-7, $V(x)$ is positive definite. The Lyapunov function determined here is precisely the one used in Example 7.8-7.

Example 7.14-2. Determine a Lyapunov function for the system of Example 7.8-8. The system to be considered is third order and is defined by $\dot{x}_1 = x_2$, $\dot{x}_2 = x_3$, and $\dot{x}_3 = -h(x_1) - g(x_2) - ax_3$. For a third order case, **grad** V is

$$\begin{bmatrix} [\alpha_{11c} + \alpha_{11v}(x_1)]x_1 + [\alpha_{12c} + \alpha_{12v}(x_1, x_2)]x_2 + [\alpha_{13c} + \alpha_{13v}(x_1, x_2)]x_3 \\ [\alpha_{21c} + \alpha_{21v}(x_1, x_2)]x_1 + [\alpha_{22c} + \alpha_{22v}(x_2)]x_2 + [\alpha_{23c} + \alpha_{23v}(x_1, x_2)]\, x_3 \\ [\alpha_{31c} + \alpha_{31v}(x_1, x_2)]x_1 + [\alpha_{32c} + \alpha_{32v}(x_1, x_2)]x_2 + x_3 \end{bmatrix}$$

For $i = 1, j = 2$, Eq. 7.14-3 yields

$$\alpha_{12c} + \frac{\partial[\alpha_{12v}(x_1, x_2)x_2]}{\partial x_2} + x_3\frac{\partial[\alpha_{13v}(x_1, x_2)]}{\partial x_2} = \alpha_{12c} + \frac{\partial[\alpha_{21v}(x_1, x_2)x_1]}{\partial x_1} + x_3\frac{\partial[\alpha_{23v}(x_1, x_2)]}{\partial x_1}$$

This can be satisfied simply by $\alpha_{12c} = \alpha_{21c} = \alpha_{13v} = \alpha_{23v} = 0$, $\alpha_{12v} = dh(x_1)/dx_1$, and $\alpha_{21v} = h(x_1)/x_1$. Similarly, for $i = 1, j = 3$, and the α's above, Eq. 7.14-3 yields

$$\alpha_{13c} = \alpha_{31c} + \frac{\partial[\alpha_{31v}(x_1, x_2)x_1]}{\partial x_1} + x_2 \frac{\partial[\alpha_{32v}(x_1, x_2)]}{\partial x_1}$$

This can be satisfied simply by $\alpha_{13c} = \alpha_{31c} = \alpha_{31v} = \alpha_{32v} = 0$. Similarly, for $i = 2$, $j = 3$, and the α's above, Eq. 7.14-3 yields $\alpha_{23c} = \alpha_{32c}$. This can be satisfied simply by $\alpha_{23c} = \alpha_{32c} = a$. For these α's, W is

$$\begin{aligned} W = {} & [\alpha_{11c} + \alpha_{11v}(x_1)]x_1x_2 + \frac{dh(x_1)}{dx_1} x_2^2 \\ & + h(x_1)x_3 + [\alpha_{22c} + \alpha_{22v}(x_2)]x_2x_3 + ax_3^2 \\ & -[ax_2 + x_3][ax_3 + h(x_1) + g(x_2)] \end{aligned}$$

It is now apparent that some of these choices for the α's were determined while observing W.

Now W can be put into a rather simple form by choosing $\alpha_{11c} = 0$, $\alpha_{11v}(x_1) = ah(x_1)/x_1$, $\alpha_{22c} = a^2$, and $\alpha_{22v}(x_2) = g(x_2)/x_2$. Then W becomes

$$W = -\left[a\frac{g(x_2)}{x_2} - \frac{dh(x_1)}{dx_1}\right]x_2^2$$

The corresponding **grad** V is

$$\begin{bmatrix} ah(x_1) + \dfrac{dh(x_1)}{dx_1} x_2 \\ h(x_1) + a^2x_2 + g(x_2) + ax_3 \\ ax_2 + x_3 \end{bmatrix}$$

Then, from Eq. 7.14-4, $V(\mathbf{x})$ is

$$V(\mathbf{x}) = \int_0^{x_1} ah(u_1)\,du_1 + \int_0^{x_2} [h(x_1) + a^2u_2 + g(u_2)]\,du_2 + \int_0^{x_3} (ax_2 + u_3)\,du_3$$

The integrations yield

$$V(\mathbf{x}) = a\int_0^{x_1} h(u)\,du + \int_0^{x_2} g(u)\,du + \tfrac{1}{2}(ax_2 + x_3)^2 + x_2h(x_1)$$

The $V(\mathbf{x})$ determined here is precisely the one utilized in Example 7.8-8. The conditions for asymptotic stability are given there.

From these examples, the usefulness of the variable gradient method should be evident. The only particular difficulty is in the selection of the α's. This choice is somewhat arbitrary, and yet it can affect both the results and the ease in obtaining them.

7.15 NONAUTONOMOUS SYSTEMS

The preceding discussion of Lyapunov's second method has been limited to autonomous cases. If one considers the nonautonomous case

$$\dot{\mathbf{x}} = \mathbf{f}(\mathbf{x}, t), \quad \mathbf{f}(\mathbf{0}, t) = \mathbf{0} \quad \text{for} \quad t \geqslant 0 \qquad (7.15\text{-}1)$$

the stability definitions remain unchanged. However, the definition of a positive definite function must be slightly modified.

The function $V(\mathbf{x}, t)$ is positive definite in a neighborhood about the origin if it is continuous and has continuous partial derivatives with respect to all arguments, $V(\mathbf{0}, t) = 0$ for $t \geqslant 0$, and a positive definite (according to the definition of Section 7.8) function $V_a(\mathbf{x})$ exists such that $V(\mathbf{x}, t) \geqslant V_a(\mathbf{x})$ for all $\mathbf{x}$ in the neighborhood and all $t \geqslant 0$.

Note that now $W = dV/dt$ is given by

$$W = \frac{\partial V}{\partial t} + \frac{\partial V}{\partial x_1} f_1(\mathbf{x}) + \frac{\partial V}{\partial x_2} f_2(\mathbf{x}) + \cdots + \frac{\partial V}{\partial x_n} f_n(\mathbf{x})$$

In terms of the gradient of $V(\mathbf{x}, t)$,

$$W = \frac{\partial V}{\partial t} + \langle \mathbf{grad}\, V, \mathbf{f} \rangle$$

If $W \leqslant 0$ in the neighborhood, then $V(\mathbf{x}, t)$ is a Lyapunov function for Eq. 7.15-1 in the neighborhood.

With this definition of a positive definite $V(\mathbf{x}, t)$, the previous theorems on the stabilities of autonomous systems apply also for nonautonomous systems. In the case of asymptotic stability, $V(\mathbf{x}, t)$ must be dominated by a positive definite $V_b(\mathbf{x})$, i.e., $V(\mathbf{x}, t) \leqslant V_b(\mathbf{x})$. A negative definite $W(\mathbf{x}, t)$ is also defined in terms of some negative definite scalar $W_a(\mathbf{x})$ as $W(\mathbf{x}, t) \leqslant W_a(\mathbf{x})$. Even though the previous theorems apply, the difficulty in determining Lyapunov functions has limited the number of useful applications of Lyapunov's second method in nonautonomous cases.

Example 7.15-1. The stability of the linear time-varying system of Fig. 7.15-1 is to be investigated.

The state equations are $\dot{\mathbf{x}} = \mathbf{A}(t)\mathbf{x}$, where

$$\mathbf{A}(t) = \begin{bmatrix} 0 & \dfrac{1}{\alpha} \\ -\dfrac{1}{\gamma(t)} & -\dfrac{\beta}{\alpha} \end{bmatrix}$$

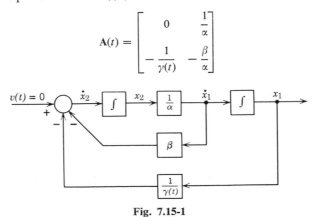

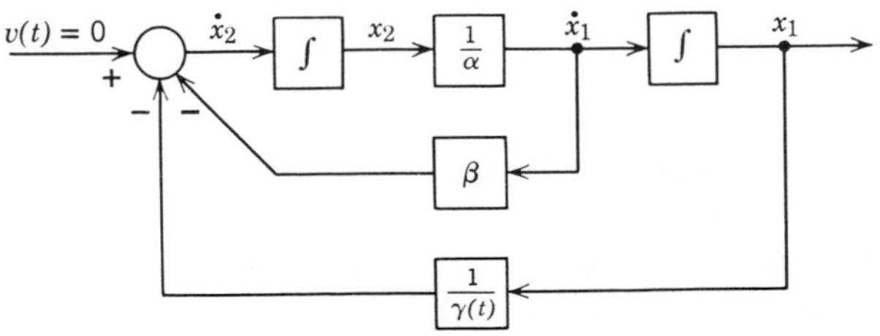

Fig. 7.15-1

Assume $V(\mathbf{x}, t) = \langle \mathbf{x}, \mathbf{Q}(t)\mathbf{x} \rangle$, where

$$\mathbf{Q}(t) = \begin{bmatrix} \beta + \dfrac{2\alpha}{\beta\gamma(t)} & 1 \\ 1 & \dfrac{2}{\beta} \end{bmatrix}$$

Then

$$V(\mathbf{x}, t) = \left[\beta + \frac{2\alpha}{\beta\gamma(t)}\right] x_1^2 + 2x_1x_2 + \frac{2}{\beta} x_2^2$$

Let $V_a(\mathbf{x}) = \epsilon_1 x_1^2 + 2x_1x_2 + \epsilon_2 x_2^2$. $V_a(\mathbf{x})$ is positive definite, and certainly $V(\mathbf{x},t) \geqslant V_a(\mathbf{x})$ if

$$\beta + \frac{2\alpha}{\beta\gamma(t)} \geqslant \epsilon_1 > 1 \quad \text{and} \quad \frac{2}{\beta} > \epsilon_2 > 1$$

Now, considering W,

$$W = \langle \dot{\mathbf{x}}, \mathbf{Q}(t)\mathbf{x} \rangle + \langle \mathbf{x}, \mathbf{Q}(t)\dot{\mathbf{x}} \rangle + \langle \mathbf{x}, \dot{\mathbf{Q}}(t)\mathbf{x} \rangle$$

Substitution of $\mathbf{A}(t)\mathbf{x}$ for $\dot{\mathbf{x}}$ yields $W = -\langle \mathbf{x}, \mathbf{P}(t)\mathbf{x} \rangle$, where $-\mathbf{P}(t) = \mathbf{A}^T(t)\mathbf{Q}(t) + \mathbf{Q}(t)\mathbf{A}(t) + \dot{\mathbf{Q}}(t)$. For this specific example,

$$\mathbf{P} = \begin{bmatrix} \dfrac{2}{\gamma(t)}\left[1 + \dfrac{\alpha\dot{\gamma}(t)}{\beta\gamma(t)}\right] & 0 \\ 0 & \dfrac{2}{\alpha} \end{bmatrix}$$

Then W is negative definite, if $\alpha > 0$ and

$$\frac{1}{\gamma(t)}\left[1 + \frac{\alpha\dot{\gamma}(t)}{\beta\gamma(t)}\right] > 0$$

Sufficient conditions for satisfaction of these inequalities, and that some $V_b(\mathbf{x})$ dominate $V(\mathbf{x}, t)$, and hence asymptotic stability in the large, are $\alpha > \gamma(t) > 0$, $2 > \beta > 0$, and $\dot{\gamma}(t) > 0$.

7.16 EVENTUAL STABILITY

The types of stability discussed in previous sections are concerned with specific motions. The specific motion determines an equilibrium, and the effects of perturbations on the equilibrium are studied.

In some cases, however, an equilibrium does not exist. For example, it is extremely difficult to specify an equilibrium for many types of adaptive control systems in which the system, its environment, and the desired responses are continually changing. This eliminates Lyapunov stability concepts.

For cases of this type, LaSalle and Rath consider *eventual stability*.[61] Essentially eventual stability states that, if a system behaves properly for a sufficiently long time, it can be expected to behave properly in the future.

The origin of $\dot{\mathbf{x}} = \mathbf{f}(\mathbf{x}, t)$ is *eventually stable* if, given $\epsilon > 0$, there exists δ and T such that $\|\mathbf{x}_0\| < \delta$ implies $\|\mathbf{x}(t; \mathbf{x}_0, t_0)\| < \epsilon$ for all $t \geqslant t_0 \geqslant T$.[61]

Lyapunov's second method can be extended to study eventual stability. LaSalle and Rath contribute several theorems to this extension. One of these, and their example relating to adaptive control, are presented here.

Theorem 7.16-1.[61] Assume for the system $\dot{\mathbf{x}} = \mathbf{f}(\mathbf{x}, \mathbf{m}, t), \dot{\mathbf{z}} = \mathbf{g}(\mathbf{x}, \mathbf{z}, t)$ that $\mathbf{f}(\mathbf{x}, \mathbf{z}, t)$ is bounded for bounded $\mathbf{x}$ and $\mathbf{z}$ and all $t \geqslant 0$, and assume also the existence of a scalar function $V(\mathbf{x}, \mathbf{z})$ such that:

1. $V(\mathbf{x}, \mathbf{z})$ is positive definite and has continuous first partial derivatives for all $\mathbf{x}$ and $\mathbf{z}$.
2. $V(\mathbf{x}, \mathbf{z})$ approaches infinity as $\|\mathbf{x}\|^2 + \|\mathbf{z}\|^2$ approaches infinity.
3. $W \leqslant -V_a(\mathbf{x}) + h_1(t)q(\mathbf{x}, \mathbf{z}) + h_2(t)V(\mathbf{x}, \mathbf{z})$, where $V_a(\mathbf{x})$ is continuous and positive definite for all $\mathbf{x}$; $q(\mathbf{x}, \mathbf{z})$ is continuous for all $\mathbf{x}$, $\mathbf{z}$ and

$$\int_0^\infty |h_i(t)|\, dt < \infty, \quad i = 1, 2$$

Under these conditions, the state $\mathbf{x} = \mathbf{z} = \mathbf{0}$ is eventually stable, and, given $r > 0$, there is a T_r such that $\|\mathbf{x}(t_0)\|^2 + \|\mathbf{z}(t_0)\|^2 < r^2$ for some $t_0 \geqslant T_r$ implies that $\mathbf{z}(t)$ is bounded and $\mathbf{x}(t)$ approaches $\mathbf{0}$ as t approaches infinity.

If in addition,

4. For some K and some $0 \leqslant \alpha \leqslant 1$, $|q(\mathbf{x}, \mathbf{z})| \leqslant KV^{-\alpha}(\mathbf{x}, \mathbf{z})$, then all solutions $\mathbf{z}(t; \mathbf{z}_0, t_0)$ are bounded and all $\mathbf{x}(t; \mathbf{x}_0, t_0)$ approach $\mathbf{0}$ as t approaches infinity.

Example 7.16-1.[61,62] It is desired to control the velocity x_1 of a body moving through a viscous fluid, the viscosity of which is incompletely described. The uncontrolled system is $\dot{x}_1 = -\beta(t)x_1$. It is known only that $\beta(t)$ is bounded and $\beta(t)$ approaches β_0, an unknown constant, as t approaches infinity: this implies

$$\int_0^\infty |\beta(t) - \beta_0|\, dt < \infty$$

The desired velocity is $x_1 = k$, a known constant.

If $\beta(t)$ were known, the system described by

$$\dot{x}_1 = -\beta(t)x_1 + [\beta(t) - a]x_1 + ak, \quad a > 0$$

would solve the problem, i.e., the steady-state solution would be k. Since $\beta(t)$ and β_0 are unknown, however, replace $\beta(t)$ in the brackets by an adjustable feedback gain $z_1(t)$. Thus

$$\dot{x}_1 = -\beta(t)x_1 + (z_1 - a)x_1 + ak, \quad a > 0$$

Let $V(x_1, z_1) = \frac{1}{2}[(x_1 - k)^2 + (z_1 - \beta_0)^2]$. Then $W = dV/dt$ is given by

$$W = -a(x_1 - k)^2 + [z_1 - \beta_0][x_1(x_1 - k) + \dot{z}_1] - (\beta - \beta_0)x_1(x_1 - k)$$

For $\dot{z}_1 = -x_1(x_1 - k)$, W becomes

$$W = -a(x_1 - k)^2 - (\beta - \beta_0)x_1(x_1 - k)$$

Denoting $V_a(\mathbf{x}) = a(x_1 - k)^2$, $h_1(t) = \sqrt{2}\ |k| \cdot |\beta(t) - \beta_0|$, $h_2(t) = 2\,|\beta(t) - \beta_0|$, and $q(\mathbf{x}, \mathbf{z}) = \sqrt{V(x_1, z_1)}$, then

$$W \leqslant -V_a(\mathbf{x}) + h_1(t)q(\mathbf{x}, \mathbf{z}) + h_2(t)V(\mathbf{x}, \mathbf{z})$$

Thus the controlled system $\dot{x}_1 = -\beta(t)x_1 + (z_1 - a)x_1 + ak$, $\dot{z}_1 = -x_1(x_1 - k)$ satisfies all the conditions of the theorem, so that the state $x_1 = k$, $z_1 = \beta_0$ is eventually stable. Furthermore, for all initial states of x_1 and z_1, z_1 is bounded and x_1 approaches k as t approaches infinity.

7.17 DISCRETE SYSTEMS

Chapter 6 indicated that the state variable approach is applicable to discrete systems as well as continuous systems. Thus the discussion now turns to the application of the second method of Lyapunov to discrete systems. In the present considerations, there is little difference between this and the case of continuous systems. Hence the discussion is quite brief, indicating only the salient features and indicating the method by means of example.

In the case of discrete systems, the state equations are difference equations of the general form

$$E\mathbf{x}(kT) = \mathbf{x}[(k + 1)T] = \mathbf{f}(\mathbf{x}, kT) \tag{7.17-1}$$

The equilibrium is given by the solutions of $\mathbf{f}(\mathbf{0}, kT) = \mathbf{0}$ for all k, and $\mathbf{f}(\mathbf{x}, kT)$ is assumed to be real and continuous with respect to $\mathbf{x}$ for fixed k.

The fundamental stability definitions, given in preceding sections for continuous systems, are applicable to discrete systems. A Lyapunov function used for the investigation of the stability of an equilibrium of a discrete system need only be defined for the discrete instants of time kT. Similarly, the definiteness requirements need hold only for these discrete instants. Thus, instead of the total derivative of the Lyapunov function, which is of interest for continuous systems, one considers the total difference†

$$W = \Delta V = V\{\mathbf{x}[(k + 1)T], (k + 1)T\} - V[\mathbf{x}(kT), kT] \tag{7.17-2}$$

† W can be divided by T to determine the rate of increase of V; however, this does not change the definiteness of the function.

With these modifications, the previous theorems on the application of Lyapunov's second method to continuous systems may be applied to discrete systems.

For linear, autonomous difference equations, a Lyapunov function can be determined by a procedure analogous to that used in Sections 7.9, 7.10, and 7.13. Let $V(\mathbf{x})$ be the positive definite quadratic form $V(\mathbf{x}) = \langle \mathbf{x}, \mathbf{Qx} \rangle$. Then, corresponding to Eq. 7.17-2, $W = \langle (E\mathbf{x}), \mathbf{Q}(E\mathbf{x}) \rangle - \langle \mathbf{x}, \mathbf{Qx} \rangle$. For linear, autonomous difference equations, Eq. 7.17-1 can be written as $E\mathbf{x} = \mathbf{Ax}$, so that $W = \langle \mathbf{x}, \mathbf{A}^T\mathbf{QAx} \rangle - \langle \mathbf{x}, \mathbf{Qx} \rangle$ or $W = \langle \mathbf{x}, (\mathbf{A}^T\mathbf{QA} - \mathbf{Q})\mathbf{x} \rangle$, where T denotes the transpose and should not be confused with the sampling interval. Let $\mathbf{P}$ be any symmetric, positive definite matrix which is the unique solution of the linear equation $-\mathbf{P} = \mathbf{A}^T\mathbf{QA} - \mathbf{Q}$. Then $W = -\langle \mathbf{x}, \mathbf{Px} \rangle$, which is negative definite. This procedure is successful only if the absolute values of all characteristic values of $\mathbf{A}$ are less than unity.

7.18 APPLICATION TO PULSE WIDTH MODULATED SYSTEMS

Pulse width modulated systems provide an excellent illustration of the application of Lyapunov's second method to nonlinear discrete systems.[63,64] Consider the system represented in Fig. 7.18-1. The pulse width modulator output is

$$m(t) = \{U_{-1}[t - kT] - U_{-1}[t - kT - \delta(k)]\} \operatorname{sgn} \beta(k) \qquad \text{for } kT < t < (k+1)T$$

where $\delta(k) = T \operatorname{sat} |\gamma(k)|$ and

$$\operatorname{sat} |\gamma| = \begin{cases} 1 & \text{for} \quad |\gamma| \geqslant 1 \\ |\gamma| & \text{for} \quad 1 \geqslant |\gamma| \geqslant 0 \end{cases}$$

$\beta(k)$ and $\gamma(k)$ are yet to be determined functions of the state variables of the linear dynamics.

Expanding $G(s)$ in partial fractions, as in Sections 5.5 and 7.10, $\dot{\mathbf{x}} = \mathbf{\Lambda x} + m\mathbf{b}$, where $\mathbf{\Lambda}$ is the diagonal matrix with elements $\lambda_1, \lambda_2, \ldots, \lambda_n$.

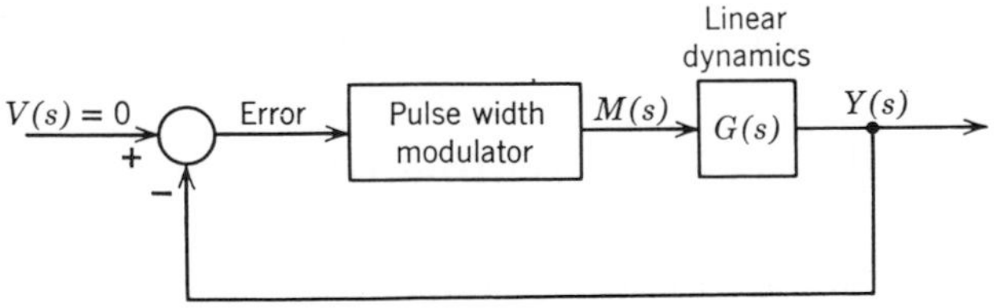

Fig. 7.18-1

The general solution of the state equations is

$$\mathbf{x}(t) = \epsilon^{(t-t_0)\mathbf{\Lambda}}\mathbf{x}(t_0) + \int_{t_0}^{t} m(\tau)\epsilon^{(t-\tau)\mathbf{\Lambda}}\mathbf{b}\, d\tau$$

where t_0 is an arbitrary initial instant. Substitution of the following two sets of values for t_0 and t

$$t_0 = kT \qquad t = kT + \delta(k)$$

$$t_0 = kT + \delta(k) \qquad t = (k+1)T$$

yields

$$\mathbf{x}[kT + \delta(k)] = \epsilon^{\delta(k)\mathbf{\Lambda}}\mathbf{x}(kT) + (\operatorname{sgn}\beta)\int_{kT}^{kT+\delta(k)} \epsilon^{[kT+\delta(k)-\tau]\mathbf{\Lambda}}\mathbf{b}\, d\tau$$

$$\mathbf{x}[(k+1)T] = \epsilon^{[T-\delta(k)]\mathbf{\Lambda}}\mathbf{x}[kT + \delta(k)]$$

Combining these equations gives

$$\mathbf{x}[(k+1)T] = \epsilon^{T\mathbf{\Lambda}}\mathbf{x}(kT) + (\operatorname{sgn}\beta)\epsilon^{[T-\delta(k)]\mathbf{\Lambda}}\int_{kT}^{kT+\delta(k)} \epsilon^{[kT+\delta(k)-\tau]\mathbf{\Lambda}}\mathbf{b}\, d\tau$$

The integration results in

$$\mathbf{x}[(k+1)T] = \epsilon^{T\mathbf{\Lambda}}\mathbf{x}(kT) + (\operatorname{sgn}\beta)\epsilon^{T\mathbf{\Lambda}}\mathbf{\Lambda}^{-1}[\mathbf{I} - \epsilon^{-\delta(k)\mathbf{\Lambda}}]\mathbf{b}$$

where $\mathbf{I}$ is the unit matrix. If $T \ll 1/-\lambda_i$, $i = 1, 2, \ldots, n$, then

$$\frac{1 - \epsilon^{-\delta(k)\lambda_i}}{\lambda_i} \doteq \delta(k)$$

so that

$$\mathbf{x}[(k+1)T] = \epsilon^{T\mathbf{\Lambda}}\mathbf{x}(kT) + (\operatorname{sgn}\beta)\,\delta(k)\epsilon^{T\mathbf{\Lambda}}\mathbf{b}$$

It is assumed that this difference equation satisfactorily characterizes system behavior.

In order to determine the modulator characteristics for stability, choose as a Lyapunov function $V(\mathbf{x}) = \langle\mathbf{x}, \mathbf{x}\rangle$. Then

$$W = -\langle\mathbf{x}, \mathbf{x}\rangle + \langle[\epsilon^{T\mathbf{\Lambda}}\mathbf{x} + (\operatorname{sgn}\beta)\,\delta(k)\epsilon^{T\mathbf{\Lambda}}\mathbf{b}], [\epsilon^{T\mathbf{\Lambda}}\mathbf{x} + (\operatorname{sgn}\beta)\,\delta(k)\epsilon^{T\mathbf{\Lambda}}\mathbf{b}]\rangle$$

which can be rewritten as

$$W = -\langle\mathbf{x}, [\mathbf{I} - \epsilon^{2T\mathbf{\Lambda}}]\mathbf{x}\rangle + 2(\operatorname{sgn}\beta)\,\delta(k)\langle\mathbf{x}, \epsilon^{2T\mathbf{\Lambda}}\mathbf{b}\rangle + \delta^2(k)\langle\mathbf{b}, \epsilon^{2T\mathbf{\Lambda}}\mathbf{b}\rangle$$

The first term is negative definite if $\lambda_i < 0$, $i = 1, 2, \ldots, n$. Considering the second term, let $\operatorname{sgn}\beta = -\operatorname{sgn} < \mathbf{x}, \epsilon^{2T\mathbf{\Lambda}}\mathbf{b}\rangle$. Then

$$W = -\langle\mathbf{x}, [\mathbf{I} - \epsilon^{2T\mathbf{\Lambda}}]\mathbf{x}\rangle - 2\delta(k)\,|\langle\mathbf{x}, \epsilon^{2T\mathbf{\Lambda}}\mathbf{b}\rangle| + \delta^2(k)\langle\mathbf{b}, \epsilon^{2T\mathbf{\Lambda}}\mathbf{b}\rangle$$

W has its most negative value for any $\mathbf{x}$ if

$$\delta(k) = \frac{|\langle\mathbf{x}, \epsilon^{2T\mathbf{\Lambda}}\mathbf{b}\rangle|}{\langle\mathbf{b}, \epsilon^{2T\mathbf{\Lambda}}\mathbf{b}\rangle}$$

Since $\delta(k)$ cannot exceed T, the best choice of $\delta(k)$ is

$$\delta(k) = T \operatorname{sat}\left\{\frac{|\langle \mathbf{x}, \epsilon^{2T\mathbf{\Lambda}}\mathbf{b}\rangle|}{T\langle \mathbf{b}, \epsilon^{2T\mathbf{\Lambda}}\mathbf{b}\rangle}\right\}$$

For stable linear dynamics and the modulator characterized by these definitions, the equilibrium is asymptotically stable in the large.

Example 7.18-1. Determine the characteristics of a pulse width modulator for the linear dynamics of Fig. 7.18-2*a*, where

$$G(s) = \frac{2s + 3}{(s + 1)(s + 2)} = \frac{1}{s + 1} + \frac{1}{s + 2}$$

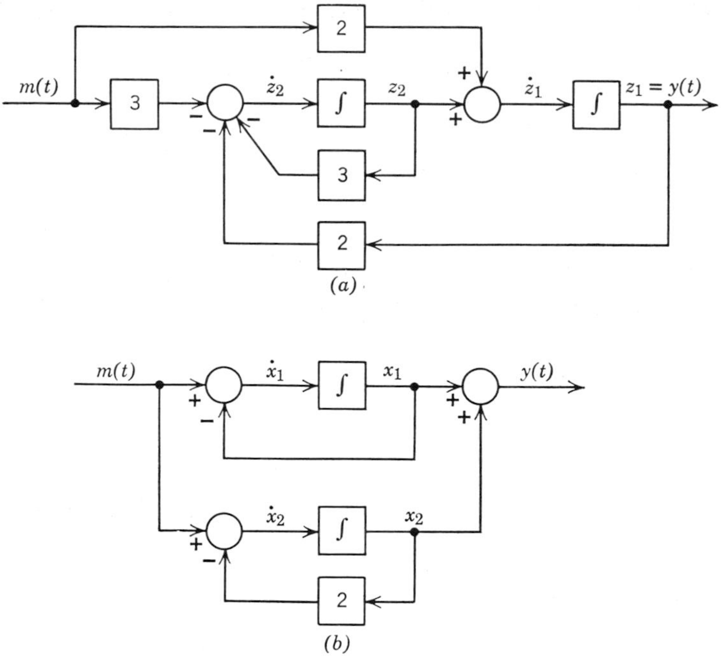

Fig. 7.18-2

For the state variables defined in Fig. 7.18-2*b*, $\langle \mathbf{x}, \epsilon^{2T\mathbf{\Lambda}}\mathbf{b}\rangle = \epsilon^{-2T}(x_1 + \epsilon^{-2T}x_2)$ and $\langle \mathbf{b}, \epsilon^{2T\mathbf{\Lambda}}\mathbf{b}\rangle = \epsilon^{-2T}(1 + \epsilon^{-2T})$. Thus the sign of the modulator output is

$$\operatorname{sgn} \beta = -\operatorname{sgn}[x_1 + \epsilon^{-2T}x_2]$$

and the pulse width is

$$\delta(k) = T \operatorname{sat}\left\{\frac{x_1 + \epsilon^{-2T}x_2}{T(1 + \epsilon^{-2T})}\right\}$$

Note that neither x_1 nor x_1 exists as such in Fig. 7.18-2*a*. However, by the methods of Chapter 5, $x_1 = 2z_1 + z_2$ and $x_2 = -(z_1 + z_2)$. Then, if z_1 and z_2 can be measured by appropriate sensors, the proper switching signals for the pulse width modulator can be obtained from a linear combination of these measurements.

REFERENCES

1. R. A. Struble, *Nonlinear Differential Equations*, McGraw-Hill Book Co., New York, 1962, Chapter 2.
2. A. A. Andronow and C. E. Chaikin, *Theory of Oscillations*, Princeton University Press, Princeton, N.J., 1949.
3. A. Liénard, "Étude des Oscillations Entretenues," *Rev. gén élec.*, Vol. 23, 1928, pp. 901–946.
4. R. N. Buland, "Analysis of Nonlinear Servos by Phase Plane-Delta Method," *J. Franklin Inst.*, Vol. 257, 1954, pp. 37–48.
5. Y. H. Ku, *Analysis and Control of Nonlinear Systems*, The Ronald Press Co., New York, 1958.
6. P. S. Hsia, "A Graphical Analysis for Nonlinear Systems," *Proc. IEE*, Vol. 99, Pt. II, 1952, pp. 120–165.
7. D. Graham and D. McRuer, *Analysis of Nonlinear Control Systems*, John Wiley and Sons, New York, 1961, pp. 304–313.
8. L. S. Pontryagin, *Ordinary Differential Equations*, Addison-Wesley Publishing Co., Reading Mass., 1962, pp. 201–213.
9. Struble, *op. cit.*, p. 177.
10. L. M. Vallese, "A Note on the Analysis of Flip-Flops," Proceedings of the Symposium on Nonlinear Circuit Analysis, Polytechnic Institute of Brooklyn, Vol. VI, 1956, pp. 347–365.
11. L. M. Vallese, "On the Synthesis of Nonlinear Systems," Proceedings of the Symposium on Nonlinear Circuit Analysis, Polytechnic Institute of Brooklyn, Vol. II, 1953, pp. 201–214.
12. Struble, *op. cit.*, pp. 172–177.
13. N. Minorsky, *Nonlinear Oscillations*, D. Van Nostrand Co., Princeton, N.J., 1962, pp. 77–80.
14. A. A. Andronow and C. E. Chaikin, *Theory of Oscillations*, Princeton University Press, Princeton, N.J., 1949, Appendix C, pp. 343–347.
15. N. Minorsky, *op. cit.*, pp. 84–91.
16. J. P. LaSalle, "Relaxation Oscillations," *Quart. Appl. Math.*, Vol. 7, 1949, pp. 1–19.
17. Pontryagin, *op. cit.*, pp. 223–236.
18. N. Minorsky, *op. cit.*, pp. 165–169.
19. S. T. Bow and J. E. Van Ness, "Use of Phase Space in Transient-Stability Studies," *Trans. AIEE*, Vol. 77, Pt. II, 1958, pp. 187–191.
20. A. M. Lyapunov, "Problème général de la stabilité du mouvement," *Ann. fac. sci. Univ. Toulouse*, Vol. 9, 1907, pp. 203–474.
21. A. I. Lur'e and V. N. Postaikow, "Concerning the Stability of Regulating Systems," *Prikl. Math. Mekh.*, Moscow, Vol. 8, 1944, pp. 246–248.
22. Reprint of Reference 20, *Ann. Math. Study* No. 17, 1949, Princeton University Press.
23. R. E. Kalman and J. E. Bertram, "Control System Analysis and Design via the Second Method of Lyapunov: I, Continuous-Time Systems; II, Discrete-Time Systems," *Trans. ASME*, Ser. D, *J. Basic. Eng.*, Vol. 82, 1960, ASME Papers No. 59-NAC-2 and -3, RIAS Monograph M 59-13.
24. S. Lefschetz, *Differential Equations: Geometric Theory*, Interscience Division, John Wiley and Sons, New York, 1957, p. 78.
25. Pontryagin, *op. cit.*, p. 202.

26. J. LaSalle and S. Lefschetz, *Stability by Lyapunov's Direct Method with Applications*, Academic Press, New York, 1961, p. 37–38.
27. Struble, *op. cit.*, p. 161.
28. W. Hahn, *Theory and Application of Liapunov's Direct Method*, Prentice-Hall, Inc., Englewood Cliffs, N.J., 1963, pp. 14–15.
29. *Ibid.*, p. 15.
30. Struble, *op. cit.*, p. 164.
31. LaSalle and Lefschetz, *loc. cit.*
32. Hahn, *op. cit.*, pp. 47–48.
33. E. A. Barbasin, "Stability of the Solution of a Certain Nonlinear Third-Order Equation" (in Russian), *Prikl. Mat. Mekh.*, Vol. 16, 1952, pp. 629–632.
34. W. J. Cunningham, *An Introduction to Lyapunov's Second Method* (AIEE Work Session in Lypanuov's Second Method, edited by L. F. Kazda,) Sept. 1960, p. 30.
35. Hahn, *op. cit.*, p. 15.
36. LaSalle and Lefschetz, *op. cit.*, p. 67.
37. E. A. Barbasin and N. N. Krasovskii, "Concerning the Stability of Motion as a Whole," *Dokl. Akad. Nauk SSSR*, Vol. 36, No. 3, 1953.
38. E. I. Ergin, V. D. Norum, and T. G. Windeknecht, "Techniques for Analysis of Nonlinear Attitude Control Systems for Space Vehicles," Aeronautical Systems Division, Dir./Aeromechanics, Flight Control Lab, Wright-Patterson AFB, Ohio, Rept. No. ASD-TDR-62-208, Vol. II, June 1962.
39. J. P. LaSalle, "Stability and Control," *J. SIAM*, Ser. A on Control, Vol. 1, No. 1, 1962, pp. 3–15.
40. A. I. Lur'e, *Some Nonlinear Problems in the Theory of Automatic Control*, Gostekhizdat (in Russian), 1951; English translation, Her Majesty's Stationery Office, London, 1957.
41. A. M. Letov, *Stability in Nonlinear Control Systems*, Princeton University Press, Princeton, N.J., 1961.
42. LaSalle and Lefschetz, *op. cit.*, pp. 75–105.
43. J. P. LaSalle, "Complete Stability of a Nonlinear Control System," *Proc. Nat. Acad. Sci., U.S.*, Vol. 48, No. 4, April 1962, pp. 600–603.
44. LaSalle and Lefschetz, *op. cit.*, p. 85.
45. J. E. Gibson, *Nonlinear Automatic Control*, McGraw-Hill Book Co., New York, 1963, pp. 324–326.
46. G. Franklin and B. Gragg, "Discussion of Stability Analysis of Nonlinear Control Systems by the Second Method of Liapunov," *Trans. IRE*, AC-7, October 1962, pp. 129–130.
47. Gibson, *op. cit.*, pp. 328–334.
48. M. A. Aizerman, *Lectures on Theory of Automatic Regulation* (in Russian), Moscow, 1958.
49. G. N. Dubošin, "On the Problem of Stability of a Motion under Constantly Acting Perturbations," *Trudy gos. astron. Inst., Sternberg*, 1940.
50. I. I. Vorovich, "On the Stability of Motion with Random Disturbances," *Izv. Akad. Nauk. SSSR, Ser. mat.*, 1956.
51. Hahn, *op. cit.*, p. 107.
52. LaSalle, *loc. cit.*
53. LaSalle and Lefschetz, *op. cit.*, pp. 121–126.
54. Hahn, *op. cit.*, pp. 68–82.
55. D. R. Ingwerson, "A Modified Liapunov Method for Nonlinear Stability Analysis," *Trans. IRE*, AC-6, May 1961.

56. G. P. Szego, "A Contribution to Liapunov's Second Method: Nonlinear Autonomous Systems," Paper No. 61-WA-192, ASME Annual Winter Meeting, November 1961.
57. G. P. Szego, "On a New Partial Differential Equation for the Stability Analysis of Time Invariant Control Systems," *J. SIAM*, Ser. A on Control, Vol. 1, No. 1, 1962. pp. 63–75.
58. Hahn, *op. cit.*, pp. 78–82.
59. S. G. Margolis and W. G. Vogt, "Control Engineering Applications of V. I. Zubov's Construction Procedure for Lyapunov Functions," *Trans. IEEE*, AC-8 No. 2, April 1963, pp. 104–113.
60. D. G. Schultz and J. E. Gibson, "The Variable Gradient Method for Generating Liapunov Functions," *Trans. AIEE*, Vol. 81, Pt. II, 1962, pp. 203–210.
61. J. P. LaSalle and R. J. Rath, "Eventual Stability," Proc. Fourth Joint Automatic Control Conference, 1963, pp. 468–470, 1963 IFAC Conference.
62. E. R. Rang, "Adaptive Controllers Derived by Stability Considerations," Minneapolis-Honeywell MPG Report 1529-TR9, March 1962.
63. W. L. Nelson, "Pulse Width Relay Control in Sampling Systems," ASME Paper No. 60-JAC-4, 1960.
64. T. T. Kadota and H. C. Bourne, "Stability Conditions of Pulse-Width-Modulated Systems through the Second Method of Lyapunov," *Trans. IRE*, AC-6, September 1961, pp. 266–276.

Problems

7.1 Show that the trajectories for the linear, second order system of Fig. 7.3-1 with the outer feedback loop gain equal to zero ($\omega^2 = 0,\ \xi\omega \neq 0$) are given by $x_2 + 2\xi\omega x_1 = c$.

7.2 Show that the trajectories for the linear, second order system of Fig. 7.3-1 with $|\xi| = 1$ are given by

$$x_2 + \xi\omega x_1 = c\epsilon^{\xi\omega x_1/(x_2+\xi\omega x_1)}, \quad \xi = +1 \text{ or } -1$$

7.3 Sketch the phase plane trajectories for each of the cases considered in Fig. 7.3-8 for the system of Fig. 7.3-1 where $v(t)$ now is not zero but is $U_{-1}(t)$, a unit step function.

7.4 Derive the phase plane trajectories of Section 7.3 from the mode interpretation viewpoint of Chapter 5.

7.5 (*a*) Determine the location and nature of any singular points in the example of Section 7.4 for the cases: $a = 0$; $a > b > 0$.

(*b*) Sketch the phase plane portraits for each of the preceding cases. Indicate any separatrices.

(*c*) In which of the cases is the system on oscillator or a flip-flop?

(*d*) On your sketches of part (*b*), indicate the value of $\dot{x}$ required to make the circuit oscillate or flip, as the case may be.

7.6 A one-farad capacitor and a one-henry inductor are connected in series with a tunnel diode negative resistance device characterized by $i = -v + v^3$. Let x_1 be the capacitor voltage and x_2 be the current as shown in Fig. P7.6.

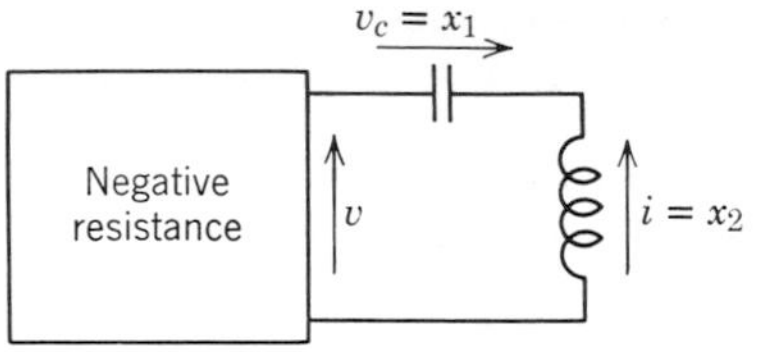

Fig. P7.6

(*a*) Determine the location and nature of any singular points, and sketch the x_1x_2 phase plane portrait. Are x_1 and x_2 a set of state variables for this system?

(*b*) Repeat part (*a*) in a vx_1-plane.

7.7 (*a*) Determine the location and nature of any singular points, and sketch the phase plane portrait for the frictionless pendulum described by $\ddot{\theta} + \sin\theta = 0$. Indicate any separatrices.

(*b*) Repeat part (*a*) for $\ddot{\theta} + 0.1\dot{\theta} + \sin\theta = 0$.

7.8 Determine the location and nature of any singular points, and sketch the phase plane portrait for the nonlinear system described by

$$\ddot{y} - (0.1 - \tfrac{10}{3}\dot{y}^2)\dot{y} + y + y^2 = 0$$

Could the system be used as an oscillator?[11] (Using the Poincaré-Bendixson theorem, prove that a stable limit cycle exists.)

7.9 A feedback system is composed of a relay amplifier driving a motor. The motor speed is sufficiently low for the back emf to be neglected. The block

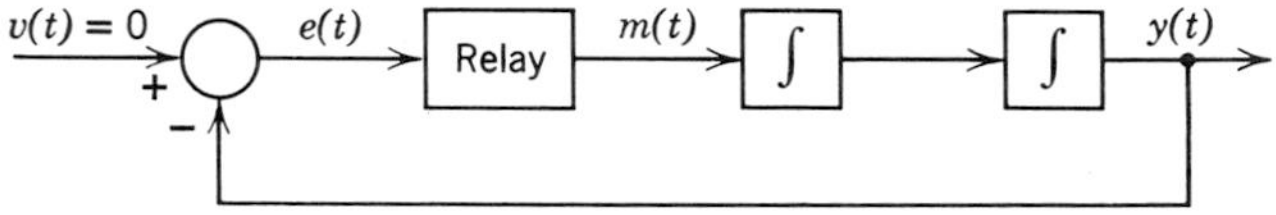

(*a*)

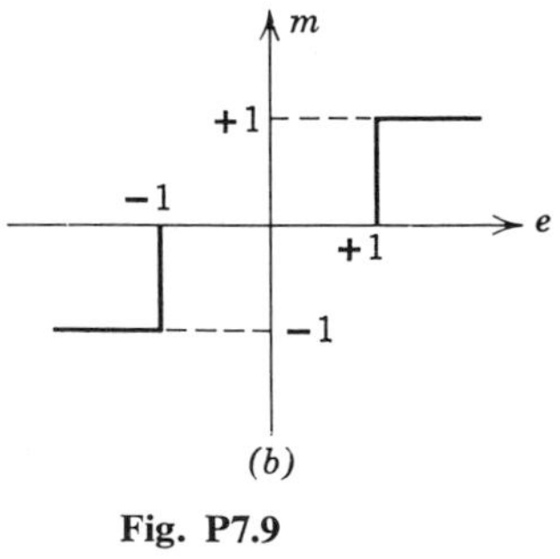

(*b*)

Fig. P7.9

diagram for the system is shown in Fig. P7.9*a*. The relay characteristic is given in Fig. 7.9*b*.

(*a*) Assuming $v(t) = 0$, sketch the phase plane trajectories.

(*b*) If the motion is periodic, what is the period as a function of the maximum velocity during the oscillation?

(*c*) Is a limit cycle present?

7.10 The relay system of Problem 7.9 is modified by the addition of nonlinear feedback. The block diagram for the modified system is given in Fig. P7.10*a*. The nonlinear element has the characteristic shown in Fig. P7.10*b*.

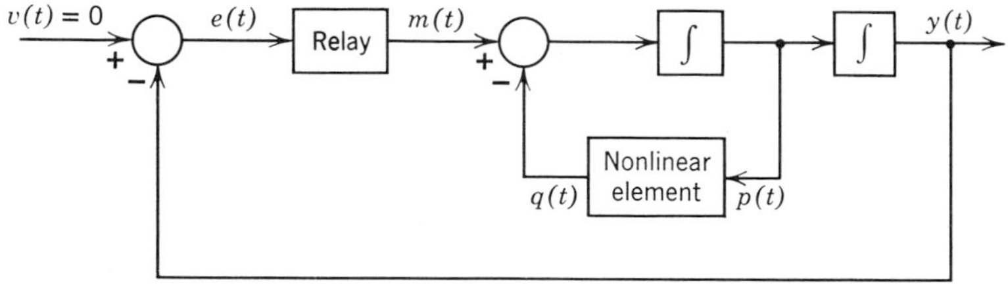

(*a*)

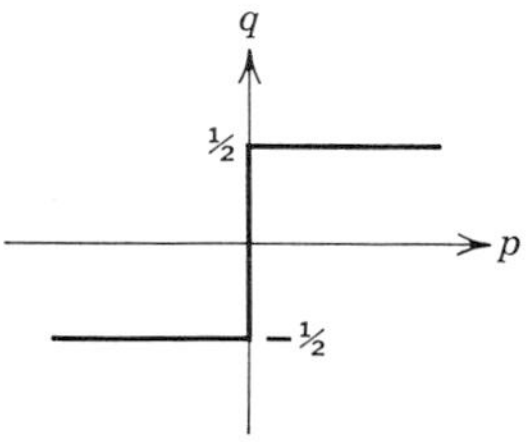

(*b*)

Fig. P7.10

(*a*) Repeat the questions of Problem 7.9 for the modified system.

(*b*) What is the effect of the nonlinear feedback on the system stability?

(*c*) Discuss the investigation of the stability of this system by the describing function method.

7.11 The system of Fig. P7.11*a* has a nonlinear error detection means and deadband in the load. The deadband characteristic is shown in Fig. P7.11*b*.
(*a*) Determine the location and nature of any singular points in an x_1x_2-plane between the limits $|x_1| \leqslant 2\pi + a$.

(*b*) Sketch the phase plane trajectories for this system in the same range, and discuss system stability.

(*c*) Discuss the possibility and difficulty of analyzing this system by the describing function method.

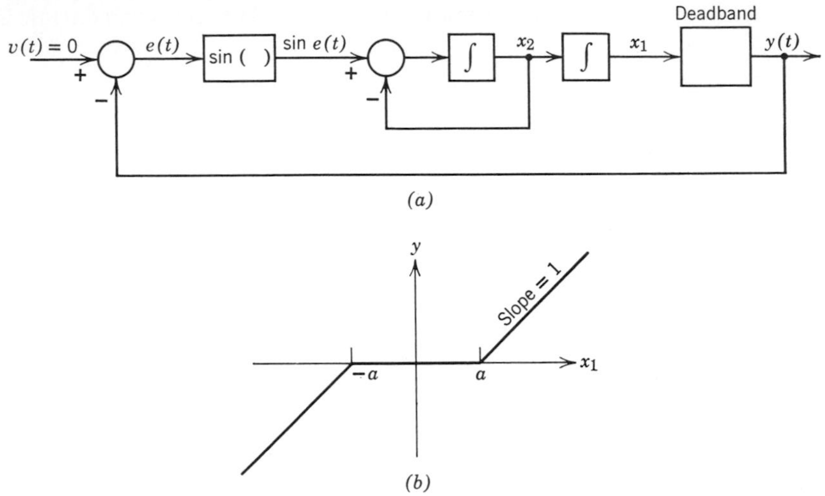

Fig. P7.11

7.12 For the system with coulomb friction illustrated in Fig. P7.12:
(*a*) Determine the differential equation characterizing the system.
(*b*) Sketch the phase portrait, indicating any singular region.
(*c*) Does a limit cycle exist?

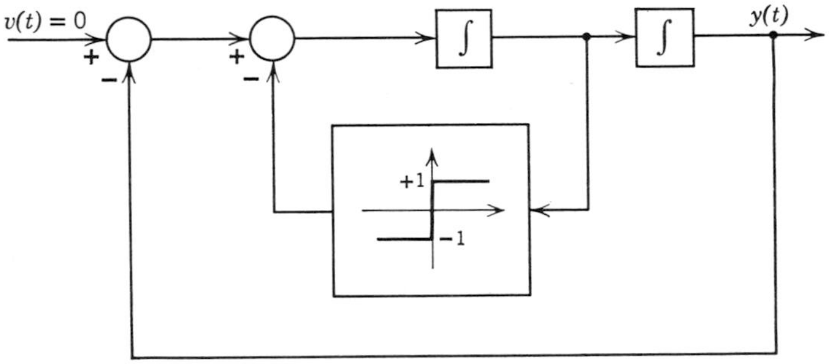

Fig. P7.12

7.13 (*a*) Sketch the trajectories for the system of Fig. 7.5-5 in an $a\dot{a}$-plane.
(*b*) Derive the successor function corresponding to part (*a*), and discuss system stability.

7.14 In the system of Fig. P7.14, $m(t)$ can be $+1$ or -1. The value of $m(t)$ is controlled by the switching logic, so that the transient error and error derivative to a step input (i.e., $v(t) = U_{-1}(t)$) are reduced to zero in the shortest possible time. This corresponds to the least number of changes of sign of $m(t)$ and is the case of optimum switching.

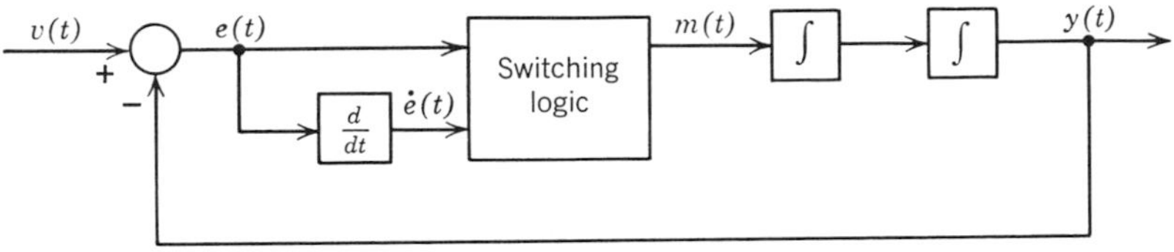

Fig. P7.14

(*a*) Indicate on an $e\dot{e}$-plane the regions where $m(t)$ should be $+1$ and where $m(t)$ should be -1 for optimum switching.

(*b*) Does the system of part (*a*) switch in an optimum fashion for $v(t) = t$?

(*c*) Repeat part (*b*) for $v(t) = t^2/2$.

(*d*) If the switching logic corresponding to part (*a*) has a pure time delay of T_d seconds (i.e., $m(t)$ changes sign T_d seconds after it should according to the optimum switching criterion), determine a relationship in terms of T_d between the $\dot{e}$ when switching should occur, as determined in part (*a*), and the $\dot{e}$ when switching actually occurs.

(*e*) Sketch the switching curves which result with the time delay.

(*f*) Estimate the stability of the system with the time delay.

7.15 (*a*) Sketch the trajectories in an x_1x_2-plane assuming $T > 0$, $\alpha > \beta > 0$ in the system of Fig. P7.15. Where are the singular points?

(*b*) For what values of α, β, and T is the system (i) stable in the large? (ii) unstable in the large? (iii) asymptotically stable in the large? Prove your answer using the successor function.

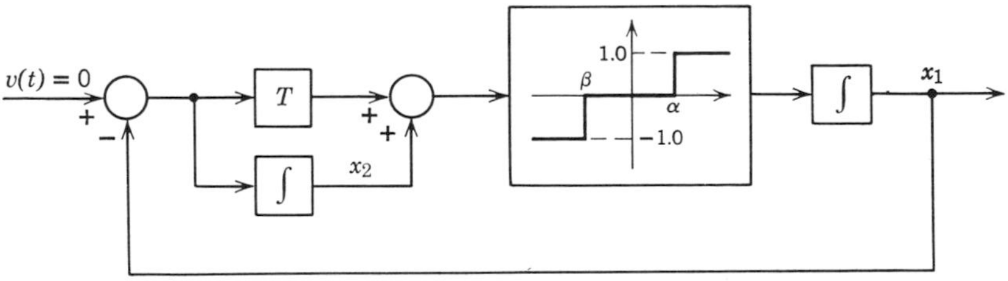

Fig. P7.15

7.16 Let $H = T + V$, where T is the kinetic energy and V is the potential energy, be the Hamiltonian of a system with n degrees of freedom. The system can be described by n generalized coordinates $q_1, q_2, \ldots, q_n$ and n generalized momenta $p_1, p_2, \ldots, p_n$. The equations of motion are $\dot{q} = \partial H/\partial p$ and $p = -\partial H/\partial q$. Let the origin of the **pq** space be an isolated equilibrium. Use Lyapunov's method to determine the conditions under which the equilibrium is stable (Lagrange's theorem).

7.17 Using the Lyapunov function

$$V(\mathbf{x}) = \frac{\omega^2}{2} x_1^2 + \frac{x_2^2}{2}$$

determine the conditions for which the equilibrium of $\dot{x}_1 = x_2$, $\dot{x}_2 = -\omega^2 x_1 - g(x_1, x_2)x_2$ is asymptotically stable in the large.

7.18 (*a*) Show that the system

$$\frac{dx_1}{dt} = -H(x_1) + x_2, \quad \frac{dx_2}{dt} = -g(x_1)$$

where

$$H(x_1) = \int_0^{x_1} h(u)\, du$$

is equivalent to Liénard's equation

$$\frac{d^2x_1}{dt^2} + h(x_1)\frac{dx_1}{dt} + g(x_1) = 0$$

(*b*) Using the V-function

$$V(\mathbf{x}) = \frac{x_2^2}{2} + \int_0^{x_1} g(u)\, du$$

and assuming $g(0) = H(0) = 0$, determine the conditions for asymptotic stability in the large.

7.19 A simplified representation of a single-axis satellite attitude control system with a gravity gradient effect is shown in Fig. P7.19. Using the function

$$V(\mathbf{x}) = \frac{x_1^2}{2} + \frac{x_2^2}{2} + 1 - \cos 2x_1$$

determine any requirements on T for the system to be asymptotically stable in the large. Discuss the practicality of the compensation.

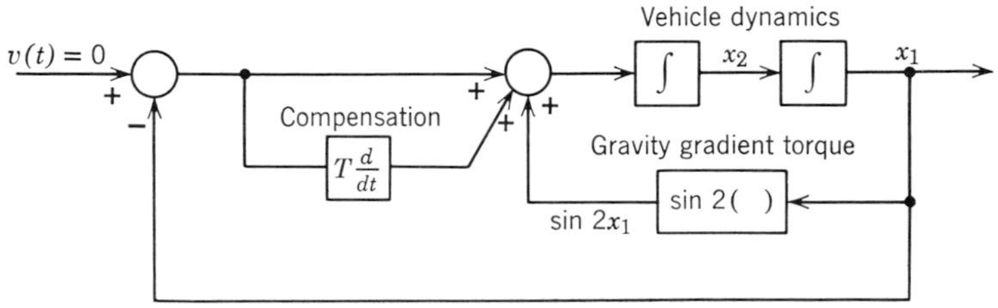

Fig. P7.19

7.20 The *RLC* network of Fig. P7.20 consists of a passive resistance network, and capacitors and inductors, all of which may be nonlinear. Choose as the state variables the charge on each of the capacitors and the flux linkage of each of the inductors. Let the V-function be the stored energy, and

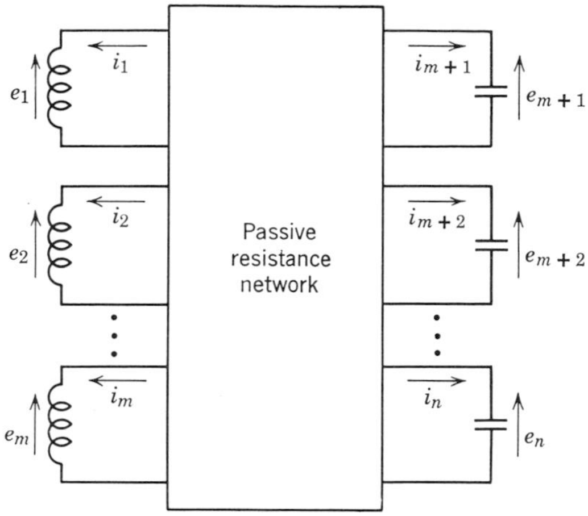

Fig. P7.20

determine the conditions for asymptotic stability in the large. How would hysteresis affect your solution?

7.21 (*a*) Show that

$$f(\mathbf{x}) = \int_0^1 \mathbf{J}(s\mathbf{x})\mathbf{x}\, ds$$

where $\mathbf{J}$ is the Jacobian matrix.

(*b*) Use this result to show that if, for any positive definite quadratic form $V = \langle \mathbf{x}, \mathbf{Q}\mathbf{x} \rangle$, $\mathbf{J}^T(\mathbf{x})\mathbf{Q} + \mathbf{Q}\mathbf{J}(\mathbf{x})$ is negative definite for all $\mathbf{x} \neq 0$, then the equilibrium of $\dot{\mathbf{x}} = \mathbf{f}(\mathbf{x})$, $\mathbf{f}(\mathbf{0}) = \mathbf{0}$ is asymptotically stable in the large.[39]

(*c*) Apply this method to Example 7.9-3.

7.22 (*a*) Investigate the absolute stability of the system in which a zero memory nonlinear controller is in a loop with

$$\text{(i)} \quad G(s) = -\frac{(2s^2 + 4s + 1)}{8s^3 - 11s - 3}$$

$$\text{(ii)} \quad G(s) = -\frac{1}{(s + \alpha)(s + \beta)}$$

(*b*) Derive the equation corresponding to Eq. 7.10-6, but which applies if some of the λ's are complex with negative real parts. Use this result for the cases

$$\text{(i)} \quad G(s) = -\frac{(s^3 + 6s^2 + 12s + 8)}{s(s + 1)(s^2 + 2s + 2)}$$

$$\text{(ii)} \quad G(s) = -\frac{K(s^2 + 1.5s + 1.5)}{s(s^2 + 2s + 2)}$$

7.23 Using a method analogous to that of Section 7.10, let $\mathbf{P} = \mathbf{u}\mathbf{u}^T$, where $\mathbf{u}$ is a vector with components $u_1, u_2, \ldots, u_n$, develop conditions sufficient for *stability in the large* of the system of Fig. 7.10-1*c*. Assume that $b_0 = 0$, and that the controlled elements are second order with both λ's negative real. Can this $\mathbf{P}$ be used to prove asymptotic stability in the large? Why?

7.24 Can an equilibrium be stable without being totally stable? Think of an undamped pendulum. Can an equilibrium be totally stable without being stable?

7.25 (*a*) Assuming $v(t) = 0$, for what values of K is the equilibrium at $x = 0$ of the system of Fig. P7.25 stable?

(*b*) Assuming $v(t) \neq 0$, but is bounded with a maximum absolute value of v_m, determine values of K for which x ultimately lies within a finite value of $x = 0$. How does this finite value vary with v_m and K?

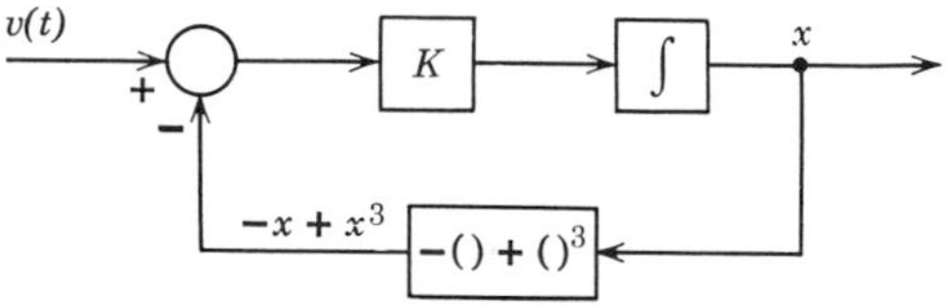

Fig. P7.25

7.26 Discuss the effect of $H(x_1)$ on the speed of response of the system of Problem 7.18.

7.27 Discuss the effect of T on the speed of response of the system of Problem 7.19.

7.28 Use the procedure of Section 7.13 to design a relay controller for the system

$$\mathbf{A} = \begin{bmatrix} 0 & 1 \\ -b & -a \end{bmatrix}, \quad \mathbf{B} = \begin{bmatrix} 1 & 0 \\ 0 & 1 \end{bmatrix}, \quad a > 0, \quad b > 0$$

7.29 Use the procedure of Section 7.13 to design a relay controller for the system

$$\mathbf{A} = \begin{bmatrix} 0 & 1 & 0 \\ 0 & 0 & 1 \\ -c & -b & -a \end{bmatrix}, \quad \mathbf{B} = \begin{bmatrix} 0 & 0 & 0 \\ 0 & 0 & 0 \\ 0 & 0 & 1 \end{bmatrix}$$

where the characteristics of $\mathbf{A}$ are distinct, negative real.

7.30 Extend the procedure of Example 7.13-2 to the case in which m is a vector.

7.31 Use the variable gradient method to determine a Lyapunov function for a system in the form of Fig. 7.10-1*b*, where

$$G(s) = \frac{s + \alpha}{s(s + 1)}$$

What are the conditions on α and $k(\ )$ for which the system is asymptotically stable in the large?

7.32 (*a*) Use the variable gradient method to determine a Lyapunov function for the system of Fig. 7.10-1*b* where

$$G(s) = \frac{s + 1}{s(s + \beta)}$$

(*b*) What are the conditions on β and $k(\ )$ for which the system is asymptotically stable in the large?

(*c*) Resolve your answer to part (*b*) with Aizerman's conjecture (see Example 7.10-1).

7.33 Using the variable gradient method, determine $V(\mathbf{x})$ and W for the system of Example 7.4-1. Assume

$$\alpha_{11c} = -b^2/2, \quad \alpha_{11v}(x_1) = c^2 x_1^{\ 2}, \quad \alpha_{21c} = \alpha_{12c} = \alpha_{21v} = \alpha_{12v} = 0$$

In what regions of the x_1x_2-plane are the $V(\mathbf{x})$ curves closed? What can you say about the behavior of the system?

7.34 Investigate the stability of the system of Example 7.15-1, when α and β are also functions of time.

7.35 Develop the method analogous to Section 7.13 for the case in which $\mathbf{x}$ and $\mathbf{m}$ are discrete signals.

7.36 Investigate the application of Lyapunov's second method to the first canonic form of Lur'e, when σ, $\mathbf{x}$, and z are discrete signals.

7.37 The linear dynamics characterized by $\dot{\mathbf{x}} = \mathbf{A}\mathbf{x} + \mathbf{B}\mathbf{m}$ are such that $|\lambda\mathbf{I} - \mathbf{A}| = 0$ has one root at $\lambda = 0$, negative real roots, and some complex roots with negative real parts. The roots are distinct. These linear dynamics are to be controlled by a controller such that the various components of $\mathbf{m}$ are given by

$$m_i(t) = M_i \operatorname{sgn} \langle \mathbf{x}(k), \boldsymbol{\alpha}_i \rangle$$

for $(k + 1)T > t \geqslant kT$, i.e., m_i can change sign at the discrete time instants kT, $(k + 1)T, \ldots$. Between these discrete instants, m_i is a constant equal to $+M_i$ or $-M_i$. Can the $\boldsymbol{\alpha}_i$ be determined so that the controlled system is asymptotically stable in the large?

8

Introduction to Optimization Theory

8.1 INTRODUCTION

The Second World War provided a great impetus to the development of the feedback control systems area. After a somewhat dormant period in the 1950's, the area again received a strong stimulus. This was caused by the interest in industrial automation and, even more, by the advent of the space age. Modern control system design has become exceedingly complex because of the desire to control large-scale, inherently nonlinear processes which sometimes operate in widely changing environments, and because of extremely stringent specifications on the performance of such systems. Optimization theory appears to offer the control system engineer a means of combating the complexities of modern control system design. It is an excellent example of the usage of linear vector space concepts. Although much research is presently being performed in the optimization area, this chapter is limited to attempting to provide the reader with the basics of optimization theory, and to indicating the nature of some of the difficulties involved with its application.

The philosophy of optimization theory is to design the "best" system. This, of course, implies some criterion or *performance index* for judging what is "best." The determination of a suitable performance index is often a problem in itself. Performance indices are discussed in later sections.

In comparison with more conventional methods for feedback control system design, the advantages of optimization theory include:

1. The design procedure is more direct, because of the inclusion of all the important aspects of performance in a single design index.
2. The best the designer can hope to achieve with respect to the performance index is apparent. Thus the ultimate performance limitations, and

the extent to which these limitations affect a given design problem, are indicated.

3. Inconsistent sets of performance specifications are revealed.
4. Prediction is naturally included in the procedure, because the design index evaluates performance over the future interval of control.
5. The resulting control system is adaptive, if the design index is reformulated and the controller parameters recomputed on-line.
6. Time-varying processes do not cause any added difficulty, assuming that a computer is used to determine the optimum.
7. Nonlinear processes can be treated directly, however, at the expense of increased computational complexity.

The difficulties of optimization theory include:

1. The conversion of prescribed design specifications into a meaningful mathematical performance index is not a straightforward process, and it may involve trial and error.
2. Existing algorithms for the computation of the optimum control signals in nonlinear cases require complex computer programs and, in some cases, a large amount of computer time.
3. Proven techniques for the design of controllers for large regions of state space, rather than merely for small regions about nominal trajectories, are presently unavailable for nonlinear cases.
4. The resulting control system performance is highly sensitive to erroneous assumptions about and/or changes in the values of the parameters of the controlled elements.

Considerable research is presently being devoted to these limitations.

The subject of system optimization had its birth in the optimum linear filter theory of Wiener.[1] This theory was extended to the time-varying case by Booton.[2] Neither of these is directly applicable to control system optimization, however, since the limitations of physical components are not considered. On the basis of Wiener's optimum filter theory, Newton considered the limitations of physical components by introducing constraints on functions of signals in the system.[3] With reference to Fig. 8.1-1*a*, Newton's method can be viewed as determining the transfer function of the optimum compensation for the system. As such, it is necessarily restricted to linear systems. Furthermore, this method neglects the effect of the configuration of the system on its performance.

A departure from the preceding procedure is indicated with respect to Fig. 8.1-1*b*, by seeking the optimum control signals for the controlled elements. The optimum *control law*, i.e., the dependence of these optimum control signals on the state variables of the controlled elements and the

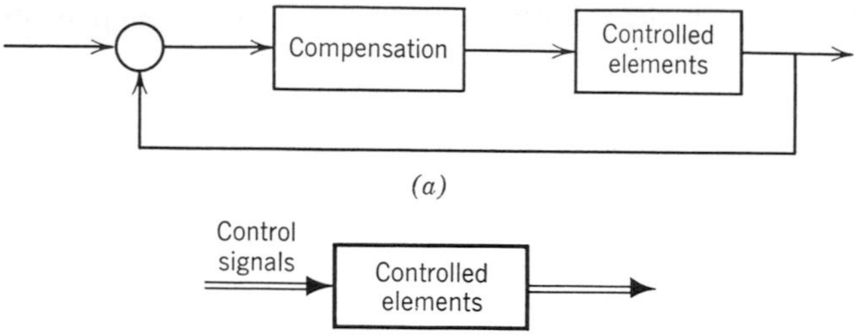

Fig. 8.1-1

desired system behavior, must also be determined in order to realize the system. The optimum control law indicates the optimum system configuration.

The latter approach to system optimization is utilized in the modern procedures developed around the dynamic programing concepts of Bellman, and the extended variational calculus methods of Pontryagin. Numerous others, some of whom are mentioned later in this chapter, have also made many important contributions to optimization theory. Notable among these is Merriam, who has been particularly concerned with making optimization theory of practical value to the control engineer. Much of the material of this chapter has been taken from his writings and those of Ellert, one of his associates.

8.2 DESIGN REQUIREMENTS AND PERFORMANCE INDICES

The primary task of the control engineer is to design practical control systems for physical processes. The application of optimization theory to this problem, as considered here, consists of three fundamental steps. They are:

1. Formulation of mathematical models for both the behavior of the physical process to be controlled and the performance requirements. The mathematical model of the performance requirements is the performance index.
2. Computation of the optimum control signals.
3. Synthesis of a controller to generate the optimum control signals.

This section considers various performance indices and their relationship to the performance requirements.

Control systems must satisfy numerous requirements relating to the performance of the system and its implementation. For example, system performance requirements may include:

1. Desired system response.
2. Desired control effort.
3. Limits on the control effort.
4. Limits on the system response, dictated either by the nature of the system mission or by saturation limits.
5. Desired system response at some future terminal time.
6. Minimization or maximization of some function of a process variable or time.
7. Disturbances, initial conditions, parameter variations, etc., which must be tolerated.
8. Damping ratio.
9. Undamped natural frequency.

Requirements 1 through 7 are objective requirements, since they can be mathematically described for any system. The last two requirements are subjective, however, because they have a precise meaning for linear, second order, time-invariant systems only. Nevertheless, they are useful for approximately characterizing the relative stability and the speed of response of more general feedback systems.

The system implementation requirements may include the specification of

1. Available sensors.
2. Available controller components.
3. System size, weight, cost, and reliability.

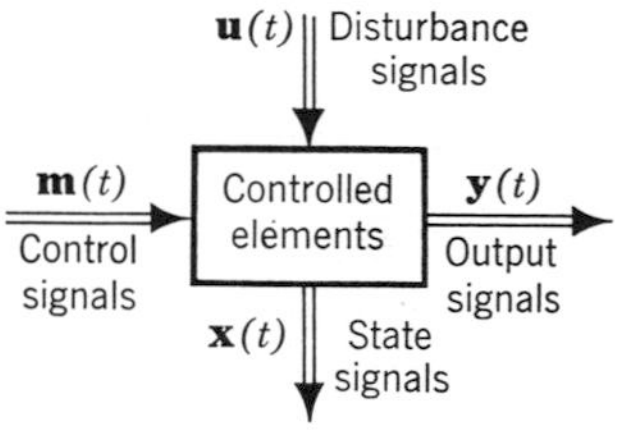

Fig. 8.2-1

Implementation requirements are exceedingly difficult to include directly in any design procedure.

The performance index is a mathematical model of the performance requirements. It is expressed in terms of the inputs, outputs, and state signals of the controlled elements. These are indicated in Fig. 8.2-1. Many performance indices have been proposed in the literature.[4–16] A substantial portion of these are special cases of the performance index

$$I = \int_0^\infty f(t)g[e(t)]\,dt$$

where $f(t)$ is a factor which weights $g[e(t)]$ as a function of time. $g[e(t)]$ is a function of the error $e(t)$. $f(t)$ is usually one of the functions $1, t, t^2, \ldots,$ or t^n, and $g[e(t)]$ is usually $e^2(t)$ or $|e(t)|$. In particular, the integral square error index

$$I = \int_0^\infty e^2(t)\,dt \tag{8.2-1}$$

leads to responses which tend not to be sufficiently damped, because large errors are counted more heavily than small errors. Thus minimization of the index requires that large errors be removed rapidly. However, this performance index is often used because of its analytical tractability.

Performance indices of the form of Eq. 8.2-1 are not suitable when multiple design specifications are encountered, since the error may be only one of these specifications. For this reason, performance indices of more general forms have been proposed. For example, Ellert uses the form[16]

$$I = \int_{t_0}^{t_f} \left\{ \sum_{i=1}^{n} \left(\phi_{ii}(t) \left[\frac{y_i^d(t) - y_i(t)}{l_{y_i}} \right]^2 + \xi_{ii}(t) \left| \frac{y_i^d(t) - y_i(t)}{l_{y_i}} \right|^{\gamma_i} \right) + \sum_{i=1}^{M} \left(\psi_{ii}(t) \left[\frac{m_i^d(t) - m_i(t)}{l_{m_i}} \right]^2 + \beta_{ii}(t) \left| \frac{m_i^d(t) - m_i(t)}{l_{m_i}} \right|^{\mu_i} \right) \right\} dt \tag{8.2-2}$$

where y_i^d and m_i^d are the desired output and control effort, respectively; l_{y_i} and l_{m_i} are related to limits on y_i and m_i, respectively; $\phi_{ii}(t)$, $\xi_{ii}(t)$, $\psi_{ii}(t)$ and $\beta_{ii}(t)$ are time-dependent weighting factors; and γ_i and μ_i are integers. The performance index considers the system behavior during the future time interval $t_0 \leqslant t \leqslant t_f$, where t_f may be a constant, a variable, or infinity.

The *weighting factors* permit the various terms of the performance index to be emphasized or weighted in time, depending upon the relative importance of these terms. The terms raised to the powers γ_i and μ_i are *penalty functions*, which tend to maintain output and control signals within prescribed limits. This is accomplished by heavily weighting these signals, if they exceed their limits.

A unique set of weighting factors and penalty functions to satisfy prescribed design specifications generally does not exist. Furthermore, the selection of these quantities is unfortunately not a straightforward matter. However, the lack of uniqueness of the weighting factors and penalty functions does introduce a flexibility which makes their selection simpler. From an engineering viewpoint, an efficient procedure for selecting weighting factors and penalty functions is needed. As discussed in later sections, Ellert has partially answered this need.

The performance indices above, and many of the specialized indices found in the literature, can be put in the form

$$I = \int_{t_0}^{t_f} q[\mathbf{y}(t), \mathbf{m}(t), t]\,dt \tag{8.2-3}$$

For example, in the flight of a vehicle from one point to another with least fuel consumption, *minimization* of Eq. 8.2-3 is desirable, if $q(\mathbf{y}, \mathbf{m}, t)$ is chosen as the fuel consumption per unit time. In chemical process control, one might seek a *maximum* of Eq. 8.2-3. In the latter case, however, $q[\mathbf{y}, \mathbf{m}, t]$ typically would represent the instantaneous yield of the process. As a final illustration, minimization of the time required for a system to go from one state to another can be accomplished by minimizing Eq. 8.2-3, with $q[\mathbf{y}, \mathbf{m}, t]$ chosen to be a constant. In such a case, constraints would exist on the maximum velocities and accelerations which can be tolerated.

Many more examples of optimization problems could be listed. However, the important aspect of this discussion is that, even though these problems are different, they are all closely related mathematically by the objective of finding a maximum or a minimum of Eq. 8.2-3. Problems of this type can be solved by Pontryagin's method, or by the dynamic programming techniques of Bellman.

8.3 NECESSARY CONDITIONS FOR AN EXTREMUM—VARIATIONAL CALCULUS APPROACH

The problem to be considered is one of determining the control signals $\mathbf{m}(t)$ which minimize (or maximize) the performance index of Eq. 8.2-3. The controlled elements are described by the equations

$$\begin{aligned} \dot{\mathbf{x}} &= \mathbf{f}(\mathbf{x}, \mathbf{m}, t) \\ \mathbf{y} &= \mathbf{g}(\mathbf{x}, t) \end{aligned} \tag{8.3-1}$$

The elements of $\mathbf{f}$ are assumed to be continuous with respect to the elements of $\mathbf{x}$ and $\mathbf{m}$, and continuously differentiable with respect to the elements of $\mathbf{x}$. The controlled elements are assumed to be observable and controllable, i.e., all state variables are measureable, and it is possible to excite every state of the controlled elements. The presentation here is further limited to the special case for which there are no restrictions on the amplitudes of the control signals or state variables. A more general presentation is given by Pontryagin et al.[17]

Before considering minimization (maximization) of a functional, as Eq. 8.2-3, it is worthwhile to consider the more familiar case of minimization (maximization) of a function. All engineers have encountered problems of trying to minimize (maximize) a function of a finite number of independent variables, say $\theta(\mathbf{x})$. Points at which all the first partial derivatives of the function are zero are known as *stationary points*. If the function is a minimum (maximum) at a stationary point, then that point is called an *extremum*.

If the variables of the function are not independent but are subject to equality constraints, e.g., $\mathbf{w}(\mathbf{x}) = \mathbf{0}$, necessary conditions for an extremum can be determined by Lagrange's method of multipliers. This method consists in introducing as many new parameters (*Lagrange multipliers*) $p_1, p_2, \ldots$ (which may be regarded as the components of a vector $\mathbf{p}$) as there are constraint equations, forming the function $\theta_c = \theta(\mathbf{x}) + \langle \mathbf{p}, \mathbf{w} \rangle$ and determining necessary conditions for an extremum from

$$\mathbf{grad}_{\mathbf{x}}\, \theta_c = \mathbf{0} \quad \text{and} \quad \mathbf{grad}_{\mathbf{p}}\, \theta_c = \mathbf{0}.$$

Thus these conditions are

$$\frac{\partial \theta_c}{\partial x_1} = \frac{\partial \theta_c}{\partial x_2} = \cdots = 0$$

$$\frac{\partial \theta_c}{\partial p_1} = w_1(\mathbf{x}) = \frac{\partial \theta_c}{\partial p_2} = w_2(\mathbf{x}) = \cdots = 0$$

Lagrange's method avoids having to solve the constraint equations for the x's and substituting the results into $\theta(\mathbf{x})$. This is accomplished by introducing the above additional restrictions.

The calculus of variations is also concerned with the determination of extrema.† Rather than extrema of functions, however, the object of the calculus of variations is to determine extrema of functionals. Section 1.4 indicates that, if x has a unique value corresponding to each value of t lying in some domain, then $x(t)$ is said to be a function of t for that domain; to each *value* of t, there corresponds a *value* of x. In essence, a *functional* is a function of a function, rather than of a variable. For example, $f[x(t)]$ is a functional if, to each *function* $x(t)$, there corresponds a *value* of f. The performance index I of Eq. 8.2-3 is also a functional.

If the second of Eqs. 8.3-1 is substituted into Eq. 8.2-3, the result can be written as‡

$$I = \int_{t_0}^{t_f} f_0(\mathbf{x}, \mathbf{m}, t)\, dt \tag{8.3-2}$$

Then the problem of determining an extremum of Eq. 8.3-2 for the controlled elements of Eqs. 8.3-1, is one of determining the function $\mathbf{m}(t)$ which makes I an extremum, subject to the n equality constraints $\mathbf{f}(\mathbf{x}, \mathbf{m}, t) - \dot{\mathbf{x}} = \mathbf{0}$. The method of Lagrange multipliers is also useful for

† Reference 18 is a particularly readable presentation of the calculus of variations. Reference 19 provides a higher degree of rigor.

‡ In order to exclude degenerate problems, it is assumed that all state variables contribute to the value of the performance index. This may be due to the state variables appearing explicitly in f_0, or through their effect on other state variables which appear in f_0.

minimizing (maximizing) functionals, subject to functional equality constraints, which is the problem of interest.

Thus the functional

$$I_c = \int_{t_0}^{t_f} (f_0 + \langle \mathbf{p}, \mathbf{f} - \dot{\mathbf{x}} \rangle)\, dt \tag{8.3-3}$$

is formed. The components of **p** are Lagrange multipliers. If the optimum values (i.e., those furnishing the extremum of I) of **x**, **m**, and **p** are denoted by $\mathbf{x}^0$, $\mathbf{m}^0$, and $\mathbf{p}^0$, respectively, then perturbations in these variables from their optimum values are indicated by

$$\begin{aligned} \mathbf{x} &= \mathbf{x}^0 + \mathbf{A}\mathbf{x}^a \\ \mathbf{m} &= \mathbf{m}^0 + \mathbf{B}\mathbf{m}^a \\ \mathbf{p} &= \mathbf{p}^0 + \mathbf{\Gamma}\mathbf{p}^a \end{aligned} \tag{8.3-4}$$

where **A**, **B**, and **Γ** are diagonal matrices with elements α_i, β_i and γ_i, respectively. α_i, β_i, and γ_i are parameters which adjust the amount of perturbation that the quantities x_i^a, m_i^a, and p_i^a introduce into x_i, m_i and p_i, respectively. It is assumed that these perturbations are unrestricted.

From the first of Eqs. 8.3-4, it is apparent that

$$\dot{\mathbf{x}} = \dot{\mathbf{x}}^0 + \mathbf{A}\dot{\mathbf{x}}^a \tag{8.3-5}$$

If Eqs. 8.3-4 and 8.3-5 are substituted into Eq. 8.3-3, I_c has its optimum value I_c^0 for $\boldsymbol{\alpha} = \mathbf{A1} = \mathbf{0}$, $\boldsymbol{\beta} = \mathbf{B1} = \mathbf{0}$, $\boldsymbol{\gamma} = \mathbf{\Gamma 1} = \mathbf{0}$, since **x**, **m**, and **p** then have their optimum values $\mathbf{x}^0$, $\mathbf{m}^0$, and $\mathbf{p}^0$, respectively. Thus Eq. 8.3-3 has a stationary point at $\boldsymbol{\alpha} = \boldsymbol{\beta} = \boldsymbol{\gamma} = \mathbf{0}$, and necessary conditions for the optimum are

$$\begin{aligned} \mathbf{grad}_{\boldsymbol{\alpha}}\, I_c \big|_{\boldsymbol{\alpha}=\boldsymbol{\beta}=\boldsymbol{\gamma}=\mathbf{0}} &= \mathbf{0} \\ \mathbf{grad}_{\boldsymbol{\beta}}\, I_c \big|_{\boldsymbol{\alpha}=\boldsymbol{\beta}=\boldsymbol{\gamma}=\mathbf{0}} &= \mathbf{0} \\ \mathbf{grad}_{\boldsymbol{\gamma}}\, I_c \big|_{\boldsymbol{\alpha}=\boldsymbol{\beta}=\boldsymbol{\gamma}=\mathbf{0}} &= \mathbf{0} \end{aligned} \tag{8.3-6}$$

Application of Eq. 8.3-6 to Eq. 8.3-3, after substitution of Eqs. 8.3-4 and 8.3-5, yields

$$\begin{aligned} \int_{t_0}^{t_f} (\mathbf{X}^a\, \mathbf{grad}_{\mathbf{x}^0} H_c^0 + \dot{\mathbf{X}}^a\, \mathbf{grad}_{\dot{\mathbf{x}}^0} H_c^0)\, dt &= \mathbf{0} \\ \int_{t_0}^{t_f} (\mathbf{M}^a\, \mathbf{grad}_{\mathbf{m}^0} H_c^0)\, dt &= \mathbf{0} \\ \int_{t_0}^{t_f} (\mathbf{P}_a\, \mathbf{grad}_{\mathbf{p}^0} H_c^0)\, dt &= \mathbf{0} \end{aligned} \tag{8.3-7}$$

where $\mathbf{X}^a$, $\mathbf{M}^a$, and $\mathbf{P}^a$ are diagonal matrices whose elements are the elements of $\mathbf{x}^a$, $\mathbf{m}^a$, and $\mathbf{p}^a$, respectively, and H_c^0 is the optimum value of the integrand of Eq. 8.3-3, i.e., $H_c^0 = f_0^0 + \langle \mathbf{p}^0, \mathbf{f}^0 - \dot{\mathbf{x}}^0 \rangle$. Integration by parts of the second term in the first of Eqs. 8.3-7 allows that equation to be written as

$$\int_{t_0}^{t_f} \mathbf{X}^a \left[\mathbf{grad}_{\mathbf{x}^0} H_c^0 - \frac{d}{dt}(\mathbf{grad}_{\dot{\mathbf{x}}^0} H_c^0) \right] dt + \mathbf{X}^a \, \mathbf{grad}_{\dot{\mathbf{x}}^0} H_c^0 \Bigg|_{t=t_0}^{t=t_f} = \mathbf{0}$$

$\mathbf{x}^a$ is an arbitrary perturbation, except at $t = t_0$, and possibly at $t = t_f$. At $t = t_0$, $\mathbf{x}^a = \mathbf{0}$, so that $\mathbf{x}^0(t_0) = \mathbf{x}(t_0)$ in order for the optimum solution to apply to the problem of interest. For a problem with specified terminal conditions $\mathbf{x}(t_f)$, $\mathbf{x}^0(t_f) = \mathbf{x}(t_f)$ and $\mathbf{x}^a(t_f) = \mathbf{0}$. If the terminal conditions on $\mathbf{x}$ are not specified, $\mathbf{x}^a(t_f)$ is arbitrary. Thus the preceding equation requires that

$$\mathbf{grad}_{\mathbf{x}^0} H_c^0 - \frac{d}{dt}(\mathbf{grad}_{\dot{\mathbf{x}}^0} H_c^0) = \mathbf{0} \tag{8.3-8}$$

and either $\mathbf{x}^0(t_f) = \mathbf{x}(t_f)$ or

$$\mathbf{grad}_{\dot{\mathbf{x}}^0} H_c^0 \Big|_{t=t_f} = \mathbf{0} \tag{8.3-9}$$

The last two of Eqs. 8.3-7 are satisfied if

$$\begin{aligned} \mathbf{grad}_{\mathbf{m}^0} H_c^0 &= \mathbf{0} \\ \mathbf{grad}_{\mathbf{p}^0} H_c^0 &= \mathbf{0} \end{aligned} \tag{8.3-10}$$

Equations 8.3-8 and 8.3-10, together with the boundary conditions $\mathbf{x}^0(t_0) = \mathbf{x}(t_0)$, and either $\mathbf{x}^0(t_f) = \mathbf{x}(t_f)$ or Eq. 8.3-9, constitute the first necessary condition for an optimum.

Pontryagin's equations are usually written in a form analogous to Hamilton's equations of analytical mechanics. This can be accomplished by defining H, analogous to the Hamiltonian, as

$$H(\mathbf{x}, \mathbf{m}, \mathbf{p}, t) = \langle \mathbf{p}, \mathbf{f} \rangle \tag{8.3-11}$$

The vector $\mathbf{p}$ in Eq. 8.3-11 differs from the one of Eq. 8.3-3 in that it has a zeroth component equal to unity. Likewise, $\mathbf{f}$ in Eq. 8.3-11 differs from the one of Eq. 8.3-1 in that it has a zeroth component equal to $f_0(\mathbf{x}, \mathbf{m}, t)$ of Eq. 8.3-2. Thus $\mathbf{p}$ and $\mathbf{f}$ are now vectors with $n + 1$ components.

In terms of the optimum H, the first necessary condition for an optimum for the case of unspecified terminal conditions on the state variables can be written as

$$\begin{aligned} \mathbf{grad}_{\mathbf{x}^0} H^0 &= -\dot{\mathbf{p}}^0 \\ \mathbf{grad}_{\mathbf{m}^0} H^0 &= \mathbf{0} \\ \mathbf{grad}_{\mathbf{p}^0} H^0 &= \dot{\mathbf{x}}^0 \end{aligned} \tag{8.3-12}$$

subject to the boundary conditions $\mathbf{x}^0(t_0) = \mathbf{x}(t_0)$ and $\mathbf{p}^0(t_f) = \mathbf{0}$.† For specified terminal conditions on $\mathbf{x}^0$, the latter boundary condition is

† The latter boundary conditions are a special case of the so-called *transversality condition.*

replaced by $\mathbf{x}^0(t_f) = \mathbf{x}(t_f)$.† The $\dot{\mathbf{x}}^0$ equation above is equivalent to Eq. 8.3-2 and the first of Eqs. 8.3-1, and hence is always part of the problem statement. Equations 8.3-12 are called the *Euler* or *Hamilton* equations in canonic form. Their simultaneous solution yields the control signal $\mathbf{m}^0(t)$ which makes I stationary.

In the calculus of variations, a distinction is made between weak and strong maxima and minima. In addition to the Euler equations, two other necessary conditions for a weak maximum or minimum must be satisfied. They are the Legendre condition and the Jacobi condition.[18] The Legendre condition for a weak minimum requires that the matrix (see Eq. 4.11-11) $\mathbf{grad}_{\mathbf{m}^0} > <\mathbf{grad}_{\mathbf{m}^0} H^0$ be positive definite. This is analogous to the requirement that the second derivative be positive in the minimization of a function by the usual techniques of calculus. As indicated in the next section, proper formulation of f_0 guarantees satisfaction of the Legendre condition, if the controlled elements are linear.

The Jacobi condition requires that no conjugate points exist for $t_0 < t \leqslant t_f$. A *conjugate point* is one at which $x_i^a(t_1)$ is restricted, where $t_0 < t_1 \leqslant t_f$.‡ Hence, at such a point, the perturbation in $x_i(t)$ is restricted. If a conjugate point existed, the controller parameters could become unbounded. Thus, to ensure bounded controller parameters, the Jacobi condition must be satisfied. If the controlled elements are linear, proper formulation of f_0 also guarantees satisfaction of the Jacobi condition. This is indicated in Section 8.4.

Coordinate Optimization Interpretation

Pontryagin's formulation of the optimization problem restates the problem as one of optimization of a coordinate. In essence, a zeroth coordinate of $\mathbf{x}$ is introduced as

$$x_0(t) = \int_{t_0}^{t} f_0(\mathbf{x}, \mathbf{m}, \tau)\, d\tau$$

so that $\dot{x}_0(t) = f_0(\mathbf{x}, \mathbf{m}, t)$. Optimization of $x_0(t)$ at $t = t_f$ is optimization of the performance index, since

$$I = x_0(t_f) = \int_{t_0}^{t_f} f_0(\mathbf{x}, \mathbf{m}, t)\, dt \tag{8.3-13}$$

which is Eq. 8.3-2.

For first order controlled elements described by $\dot{x}_1 = f_1(x_1, m_1, t)$ the optimization problem with a specified terminal condition $x_1^0(t_f) = x_1(t_f)$ can be interpreted in the x_0x_1-plane of Fig. 8.3-1. (The generalization to

† Problems in which some, but not all, of the state variables have specified terminal conditions are also considered in the literature.[17]

‡ Conjugate points have very interesting geometrical interpretations.[18,19] However, these are beyond the scope intended here.

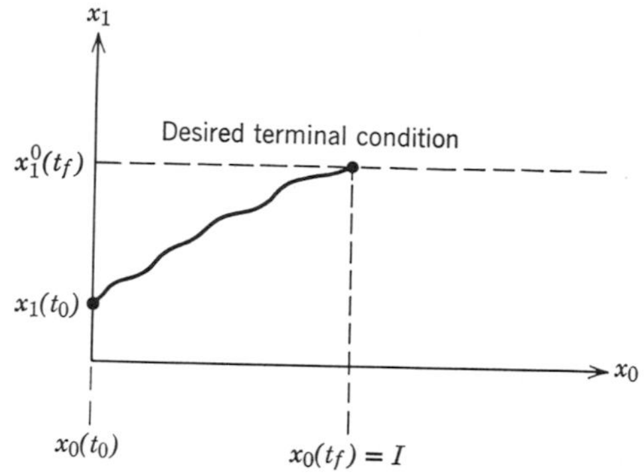

Fig. 8.3-1

higher order controlled elements is straightforward, but difficult to picture.) The desired terminal condition is the line $x_1 = x_1{}^0(t_f)$. For various $m_1(t)$, corresponding values of $I = x_0(t_f)$ can be determined from Eq. 8.3-13. Assuming that minimum I is desired, the optimum control signal $m_1{}^0(t)$ is the one for which $x_0(t_f) = I$ has the smallest coordinate $x_0{}^0(t_f) = I^0$. If I does not depend upon $m_1(t)$, a desirable $m_1{}^0(t)$ is an impulse. Then $x_1(t)$ would be transferred from $x_1(t_0)$ to $x_1{}^0(t_f)$ in zero time, and hence with zero I. However, impulses in $m_1(t)$ could not be realized physically. $m_1{}^0(t)$ must be chosen from a set of *admissible* control signals, which are defined to be bounded, and also continuous for all $t_0 \leqslant t \leqslant t_f$, except possibly at a finite number of t.

In the case of unspecified terminal conditions on $\mathbf{x}^0(t_f)$, all components of $\mathbf{p}^0$ are zero at $t = t_f$ except for the $p_0{}^0(t_f)$ component, which is unity. Thus the problem of minimizing (maximizing) $I = x^0(t_f)$ can be viewed as minimizing (maximizing) $\langle \mathbf{p}, \mathbf{x} \rangle$ at $t = t_f$. In other words, starting from the initial conditions $\mathbf{x}^0(t_0)$, $\mathbf{m}^0(t)$ is to be chosen to move the state of the system (including the x_0 component) as little (much) as possible in the direction of the vector $\mathbf{p}$. But the first and last of Eqs. 8.3-12 are the same as Hamilton's equations of analytical mechanics. H is analogous to the Hamiltonian, or total energy, and $\mathbf{p}$ and $\mathbf{x}$ are analogous to the momenta and generalized coordinates, respectively. Since H is the total energy for moving the state, $\mathbf{x}$, $\mathbf{m}(t)$ should be chosen at each instant of time to minimize (maximize) H. This is indicated by the second of Eqs. 8.3-12.

8.4 LINEAR OPTIMIZATION PROBLEMS

In this section, it is assumed that the controlled elements are described by

$$\dot{\mathbf{x}} = \mathbf{A}(t)\mathbf{x} + \mathbf{B}(t)\mathbf{m} \tag{8.4-1}$$

and the performance index is given by substituting

$$f_0(\mathbf{x}, \mathbf{m}, t) = \tfrac{1}{2}[\langle(\mathbf{x}^d - \mathbf{x}), \mathbf{\Omega}(\mathbf{x}^d - \mathbf{x})\rangle + \langle\mathbf{m}, \mathbf{Z}\mathbf{m}\rangle]$$

into Eq. 8.3-2. $\mathbf{x}^d$ is the desired state behavior, and $\mathbf{\Omega}$ and $\mathbf{Z}$ are symmetric matrices which are possibly time-varying.† The dimensions of $\mathbf{\Omega}$ are less than $(n \times n)$, unless all components of $(\mathbf{x}^d - \mathbf{x})$ are included in f_0.‡ The objective is to determine $\mathbf{x}^0$, $\mathbf{m}^0$ and the dependence of $\mathbf{m}^0$ on $\mathbf{x}^0$ and $\mathbf{x}^d$.

From Eq. 8.3-10,

$$H^0 = \tfrac{1}{2}[\langle\mathbf{x}^d - \mathbf{x}^0, \mathbf{\Omega}(\mathbf{x}^d - \mathbf{x}^0)\rangle + \langle\mathbf{m}^0, \mathbf{Z}\mathbf{m}^0\rangle] + \langle\mathbf{p}^0, \mathbf{A}\mathbf{x}^0 + \mathbf{B}\mathbf{m}^0\rangle$$

From the second of Eqs. 8.3-12, $\mathbf{Z}\mathbf{m}^0 + \mathbf{B}^T\mathbf{p}^0 = \mathbf{0}$ since $\mathbf{Z}$ is symmetric. Then

$$\mathbf{m}^0 = -\mathbf{Z}^{-1}\mathbf{B}^T\mathbf{p}^0 \qquad (8.4\text{-}2)$$

This is an expression for the optimum control signal, but it is in terms of $\mathbf{p}^0$. The control law requires $\mathbf{m}^0(t)$ in terms of $\mathbf{x}^0(t)$.

From the first of Eqs. 8.3-12, $\dot{\mathbf{p}}^0 = -\mathbf{A}^T\mathbf{p}^0 + \mathbf{\Omega}(\mathbf{x}^d - \mathbf{x}^0)$. From Eq. 8.4-1, after substituting Eq. 8.4-2, $\dot{\mathbf{x}}^0 = \mathbf{A}\mathbf{x}^0 - \mathbf{B}\mathbf{Z}^{-1}\mathbf{B}^T\mathbf{p}^0$. The last two equations can be written as

$$\begin{bmatrix} \dot{\mathbf{x}}^0 \\ \dot{\mathbf{p}}^0 \end{bmatrix} = \begin{bmatrix} \mathbf{A} & -\mathbf{B}\mathbf{Z}^{-1}\mathbf{B}^T \\ -\mathbf{\Omega} & -\mathbf{A}^T \end{bmatrix} \begin{bmatrix} \mathbf{x}^0 \\ \mathbf{p}^0 \end{bmatrix} + \mathbf{\Omega}\begin{bmatrix} \mathbf{0} \\ \mathbf{x}^d \end{bmatrix} \qquad (8.4\text{-}3)$$

Equation 8.4-3 represents $2n$ linear, first order differential equations in the $2n$ unknowns $x_1^0, x_2^0, \ldots, x_n^0, p_1^0, p_2^0, \ldots, p_n^0$. They are subject to n boundary conditions at $t = t_0$, i.e., $\mathbf{x}^0(t_0) = \mathbf{x}(t_0)$, and n boundary conditions at $t = t_f$, i.e., either $\mathbf{p}^0(t_f) = \mathbf{0}$ or $\mathbf{x}^0(t_f) = \mathbf{x}(t_f)$, depending upon the nature of the problem. Equation 8.4-3, subject to the preceding boundary conditions, is a *two-point boundary value problem.* Its solution yields the optimum control signal $\mathbf{m}^0(t)$ and the corresponding behavior of the controlled elements $\mathbf{x}^0(t)$ for $t_0 \leqslant t \leqslant t_f$.

Conversion of the Two-Point Boundary Value Problem

For the case under consideration, the two-point boundary value problem can be converted into two one-point boundary value problems. Equation 8.4-3 consists of a set of interrelated linear differential equations for $\mathbf{x}^0$ and

† The Euler equations together with a positive semidefinite $\mathbf{\Omega}$ and a positive definite $\mathbf{Z}$ constitute necessary and sufficient conditions for a minimum of the performance index, for the class of problems considered here. Furthermore, the corresponding linear optimum control system is stable (asymptotically stable if $\mathbf{\Omega}$ is positive definite).[25]

‡ A similar statement holds with respect to $\mathbf{Z}$ in terms of the dimensions of $\mathbf{m}$.

$\mathbf{p}^0$. Thus $\mathbf{x}^0$ and $\mathbf{p}^0$ must be related by a linear transformation. This transformation may be expressed by

$$\mathbf{p}^0 = \mathbf{K}\mathbf{x}^0 - \mathbf{v}^0 \tag{8.4-4}$$

where $\mathbf{K}$ is a square matrix of time-varying gains and $\mathbf{v}$ is a time-varying vector. Substitution of Eq. 8.4-4 into the second of Eqs. 8.4-3 yields

$$\mathbf{K}\dot{\mathbf{x}}^0 + \dot{\mathbf{K}}\mathbf{x}^0 - \dot{\mathbf{v}}^0 = -\mathbf{\Omega}\mathbf{x}^0 - \mathbf{A}^T\mathbf{K}\mathbf{x}^0 + \mathbf{A}^T\mathbf{v}^0 + \mathbf{\Omega}\mathbf{x}^d$$

Then substituting for $\dot{\mathbf{x}}^0$ from the first of Eqs. 8.4-3 and using Eq. 8.4-4 results in

$$(\dot{\mathbf{K}} + \mathbf{KA} + \mathbf{A}^T\mathbf{K} - \mathbf{KBZ}^{-1}\mathbf{B}^T\mathbf{K} + \mathbf{\Omega})\mathbf{x}^0$$
$$= \dot{\mathbf{v}}^0 + (\mathbf{A}^T - \mathbf{BZ}^{-1}\mathbf{B}^T)\mathbf{v}^0 + \mathbf{\Omega}\mathbf{x}^d$$

Since this expression must be valid for all possible $\mathbf{x}$, the conditions are

$$\begin{aligned} \dot{\mathbf{K}} + \mathbf{KA} + \mathbf{A}^T\mathbf{K} - \mathbf{KBZ}^{-1}\mathbf{B}^T\mathbf{K} + \mathbf{\Omega} &= [0] \\ \dot{\mathbf{v}}^0 + (\mathbf{A}^T - \mathbf{KBZ}^{-1}\mathbf{B}^T)\mathbf{v}^0 + \mathbf{\Omega}\mathbf{x}^d &= \mathbf{0} \end{aligned} \tag{8.4-5}$$

The first of Eqs. 8.4-5 is a set of first order nonlinear differential equations of the Riccati type.[20] The second of Eqs. 8.4-5 is a set of linear, time-varying, first order differential equations.† In the case of unspecified terminal conditions on $\mathbf{x}^0$, $\mathbf{p}^0(t_f) = \mathbf{0}$. Thus the boundary conditions on $\mathbf{K}$ and $\mathbf{v}^0$ for this case are that each of the elements of $\mathbf{K}$ and $\mathbf{v}^0$ is zero at $t = t_f$, as indicated by Eq. 8.4-4.

Once $\mathbf{K}$ and $\mathbf{v}^0$ are determined, the control law for the optimum system is given by substituting Eq. 8.4-4 into Eq. 8.4-2 to obtain

$$\mathbf{m}^0 = -\mathbf{Z}^{-1}\mathbf{B}^T(\mathbf{K}\mathbf{x}^0 - \mathbf{v}^0) \tag{8.4-6}$$

Thus, for this case, the control law is linear, and the controller feedback gains $\mathbf{K}$ are independent of the state of the controlled elements. Furthermore, since the control law is independent of the initial conditions of the state variables, the system configuration as defined by Eq. 8.4-6 is optimum for all initial conditions. Merriam, who first noted this property, refers to this as the *optimum configuration*.[14] Figure 8.4-1 illustrates this configuration for the general linear case.

Once $\mathbf{m}^0$ is determined, the response of the optimum system can be obtained from

$$\dot{\mathbf{x}}^0 = (\mathbf{A} - \mathbf{BZ}^{-1}\mathbf{B}^T\mathbf{K})\mathbf{x}^0 + \mathbf{BZ}^{-1}\mathbf{B}^T\mathbf{v}^0 \tag{8.4-7}$$

which results from substituting Eq. 8.4-6 into Eq. 8.4-1. Thus the two-point boundary value problem has been converted into two one-point

† These equations are adjoint to the equations of the closed-loop (controlled) system.

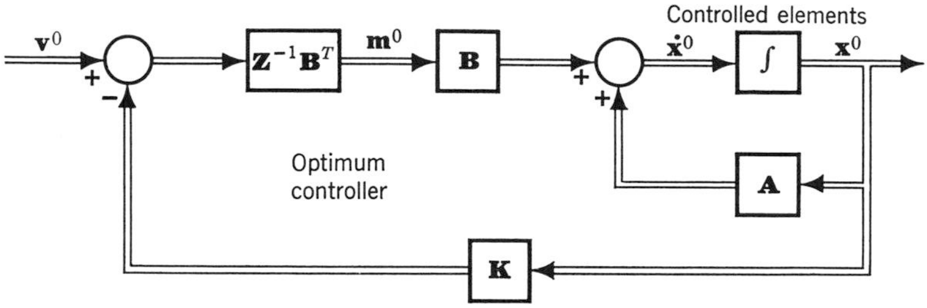

Fig. 8.4-1

boundary value problems. These are the solution of Eq. 8.4-5 backward in time from $t = t_f$ to $t = t_0$, and subsequently solving Eq. 8.4-7 forward in time from $t = t_0$ to $t = t_f$.

In the nonlinear case, where the controlled elements are nonlinear and/or the performance index is nonquadratic, it is not possible to convert the two point boundary value problem in the above manner. Also, the optimum control law is not linear. These aspects then generally demand computer solution of the equations defining the optimum system, as is considered in later sections.

Example 8.4-1. Determine the optimum controller according to the performance index

$$I = \tfrac{1}{2}\int_{t_0}^{t_f} [x_1\omega_{11}x_1 + m_1\zeta_{11}m_1]\, dt$$

for the first order controlled elements described by $\dot{x}_1(t) = a_{11}x_1(t) + b_{11}m_1(t)$. The system is assumed to be a regulator, so that $x_1{}^d = 0$. $x_1(t_f)$ is unspecified.

For $\mathbf{x}^d = \mathbf{0}$, Eq. 8.4-5 indicates $\mathbf{v}^0 = \mathbf{0}$. Therefore $m_1{}^0(t) = -\zeta_{11}{}^{-1}b_{11}k_{11}(t)x_1{}^0(t)$, where $k_{11}(t)$ is given by the solution to

$$\dot{k}_{11} + 2k_{11}a_{11} - \left(\frac{b_{11}{}^2}{\zeta_{11}}\right)k_{11}{}^2 + \omega_{11} = 0, \quad k_{11}(t_f) = 0$$

from Eqs. 8.4-6 and 8.4-5. In order to determine $k_{11}(t)$, let $\tau = t_f - t$ and $k_{11}(t_f - \tau) = k_{11}{}^0(\tau)$. Then

$$\dot{k}_{11}{}^0 = 2a_{11}k_{11}{}^0 - \left(\frac{b_{11}{}^2}{\zeta_{11}}\right)k_{11}{}^{02} + \omega_{11}, \quad k_{11}{}^0(0) = 0$$

Let $k_{11}{}^0(\tau) = \zeta_{11}\dot{z}/b_{11}{}^2z$, which yields

$$\ddot{z} - 2a_{11}\dot{z} - \frac{\omega_{11}b_{11}{}^2}{\zeta_{11}}z = 0$$

This is a linear differential equation with constant coefficients; it can be solved by classical or transform methods. The solution is $z = c_1\epsilon^{\lambda_1\tau} + c_2\epsilon^{\lambda_2\tau}$, where c_1 and c_2 are constants, and $\lambda_1 = a_{11} + \beta$, $\lambda_2 = a_{11} - \beta$, and

$$\beta = \left(a_{11}{}^2 + \omega_{11}\frac{b_{11}{}^2}{\zeta_{11}}\right)^{1/2}$$

Thus

$$k_{11}{}^0(\tau) = \frac{\zeta_{11}(p_1c_1\epsilon^{\beta\tau} + p_2c_2\epsilon^{-\beta\tau})}{b_{11}{}^2(c_1\epsilon^{\beta\tau} + c_2\epsilon^{-\beta\tau})}$$

Since $k_{11}{}^0(0) = 0$, $c_2 = -c_1\lambda_1/\lambda_2$. Then

$$k_{11}{}^0(\tau) = \frac{\omega_{11}\sinh\beta\tau}{\beta\cosh\beta\tau - a_{11}\sinh\beta\tau}$$

Therefore

$$k_{11}(t) = \frac{\omega_{11}\sinh\beta(t_f - t)}{\beta\cosh\beta(t_f - t) - a_{11}\sinh\beta(t_f - t)}$$

and $\mathbf{m}^0(t)$ is given by

$$m_1{}^0(t) = -\frac{\omega_{11}b_{11}}{\zeta_{11}}\left[\frac{\sinh\beta(t_f - t)}{\beta\cosh\beta(t_f - t) - a_{11}\sinh\beta(t_f - t)}\right]x_1{}^0(t)$$

The resultant regulator is shown in Fig. 8.4-2.

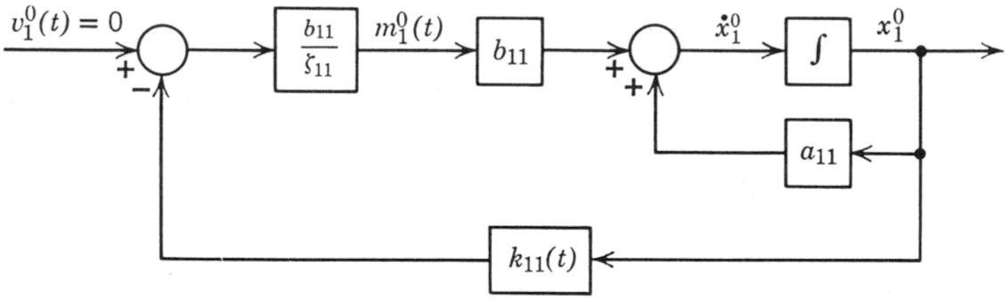

Fig. 8.4-2

If t_f is a constant, the terminal time of the performance index becomes nearer as real time advances, assuming $t < t_f$. In this so-called *shrinking interval* problem, the optimum system is time-varying. If t_f is a fixed time T in the future relative to real time, i.e., $t_f = t + T$, the terminal time of the performance index slides ahead in time as real time advances. This is called a *sliding interval* problem, and, if $\mathbf{x}^d$, $\mathbf{\Omega}$, $\mathbf{Z}$, and the linear controlled elements are time-invariant, the resultant system is stationary. A special case of these is given by infinite t_f. This is the *infinite interval* problem.† If $\mathbf{x}^d$, $\mathbf{\Omega}$, $\mathbf{Z}$, and the controlled elements are time-invariant, the resultant system designed according to an infinite interval performance criterion is stationary. In this example, $m_1{}^0(t)$ becomes

$$m_1{}^0(t) = -\frac{\omega_{11}b_{11}}{\zeta_{11}(\beta - a_{11})}x_1{}^0(t)$$

corresponding to a stationary system.

For the case in which the controlled elements consist of an integrator without feedback, $a_{11} = 0$. Then, for the infinite interval case, $m_1{}^0(t)$ is

$$m_1{}^0(t) = -\left(\frac{\omega_{11}}{\zeta_{11}}\right)^{1/2}x_1{}^0(t)$$

As ω_{11}/ζ_{11} is increased, so that the performance index emphasizes the system error relative to the cost of reducing it, the loop gain increases. Also, the speed of response as

† These names were coined by Merriam.

indicated by

$$x_1^0(t) = x_1^0(t_0) \exp \left[-b_{11} \left(\frac{\omega_{11}}{\zeta_{11}} \right)^{1/2} (t - t_0) \right]$$

increases. This agrees with one's intuition based on conventional feedback control theory.

If the optimum system is stationary, $\dot{\mathbf{K}} = [0]$ and the Riccati equation given in Eq. 8.4-5 reduces to a set of nonlinear algebraic equations defining the elements of $\mathbf{K}$. Even in this special case, however, it generally is not possible to determine $\mathbf{K}$ in closed form for controlled elements above second order. Thus the preceding discussion was presented to indicate that the control law of Eq. 8.4-6 exists for the linear case, rather than to provide a general method of determining it. Since the control law is of the form of Eq. 8.4-6, it is important to choose, as the state variables, variables which can be measured with available sensors.

In the general case, *analytical* determination of the optimum system makes use of direct solution of the two-point boundary value problem, rather than converting it to two one-point boundary value problems.

The Time-Invariant Case

If $\mathbf{A}$, $\mathbf{B}$, $\mathbf{\Omega}$, and $\mathbf{Z}$ are time-invariant, Eq. 8.4-3 can be solved by means of Laplace transforms. Assuming $t_0 = 0$, the transform of Eq. 8.4-3 is

$$\begin{bmatrix} \mathbf{I} & \mathbf{\Phi}(s)\mathbf{B}\mathbf{Z}^{-1}\mathbf{B}^T \\ -\mathbf{\Phi}^T(-s)\mathbf{\Omega} & \mathbf{I} \end{bmatrix} \begin{bmatrix} \mathbf{X}^0(s) \\ \mathbf{P}^0(s) \end{bmatrix} = \begin{bmatrix} \mathbf{\Phi}(s)\mathbf{x}^0(0) \\ -\mathbf{\Phi}^T(-s)[\mathbf{p}^0(0) + \mathbf{\Omega}\mathbf{X}^d(s)] \end{bmatrix}$$

where $\mathbf{\Phi}(s) = (s\mathbf{I} - \mathbf{A})^{-1}$ and $-\mathbf{\Phi}^T(-s) = (s\mathbf{I} + \mathbf{A}^T)^{-1}$, and $\mathbf{X}^0(s)$, $\mathbf{P}^0(s)$ and $\mathbf{X}^d(s)$ are vectors. $\mathbf{\Phi}(s)$ and $-\mathbf{\Phi}(-s)$ are the Laplace transforms of the state transmission matrices for Eq. 8.4-1 and its adjoint, respectively. Since

$$\begin{bmatrix} \mathbf{I} & \boldsymbol{\alpha}_{12} \\ \boldsymbol{\alpha}_{21} & \mathbf{I} \end{bmatrix}^{-1} = \begin{bmatrix} (\mathbf{I} - \boldsymbol{\alpha}_{12}\boldsymbol{\alpha}_{21})^{-1} & -(\mathbf{I} - \boldsymbol{\alpha}_{12}\boldsymbol{\alpha}_{21})^{-1}\boldsymbol{\alpha}_{12} \\ -(\mathbf{I} - \boldsymbol{\alpha}_{21}\boldsymbol{\alpha}_{12})^{-1}\boldsymbol{\alpha}_{21} & (\mathbf{I} - \boldsymbol{\alpha}_{21}\boldsymbol{\alpha}_{12})^{-1} \end{bmatrix}$$

$\mathbf{X}^0(s)$ and $\mathbf{P}^0(s)$ can be written as

$$\begin{aligned} \mathbf{X}^0(s) &= [\mathbf{I} + \mathbf{\Phi}(s)\mathbf{B}\mathbf{Z}^{-1}\mathbf{B}^T\mathbf{\Phi}^T(-s)\mathbf{\Omega}]^{-1} \\ &\quad \times \mathbf{\Phi}(s)\{\mathbf{x}^0(0) + \mathbf{B}\mathbf{Z}^{-1}\mathbf{B}^T\mathbf{\Phi}^T(-s)[\mathbf{p}^0(0) + \mathbf{\Omega}\mathbf{X}^d(s)]\} \\ \mathbf{P}^0(s) &= [\mathbf{I} + \mathbf{\Phi}^T(-s)\mathbf{\Omega}\mathbf{\Phi}(s)\mathbf{B}\mathbf{Z}^{-1}\mathbf{B}^T]^{-1} \\ &\quad \times \mathbf{\Phi}^T(-s)\{\mathbf{\Omega}\mathbf{\Phi}(s)\mathbf{x}^0(0) - [\mathbf{p}^0(0) + \mathbf{\Omega}\mathbf{X}^d(s)]\} \end{aligned} \tag{8.4-8}$$

Equations 8.4-8 and 8.4-2 can be utilized to determine the optimum control law and the response of the optimum system. However, the procedure is less direct than the previous method.

Example 8.4-2. Repeat Example 8.4-1, using the Laplace transform method.
For this case,

$$\mathbf{\Phi}(s) = [s\mathbf{I} - \mathbf{A}]^{-1} = \frac{1}{s - a_{11}}$$

and $\mathbf{X}^d(s) = \mathbf{0}$. Then

$$\begin{aligned} P_1{}^0(s) &= -\left[1 - \frac{\omega_{11}b_{11}{}^2}{\zeta_{11}(s - a_{11})(s + a_{11})}\right]^{-1}\left[\frac{1}{s + a_{11}}\right]\left[\frac{\omega_{11}x_1{}^0(0)}{s - a_{11}} - p_1{}^0(0)\right] \\ &= -\frac{\omega_{11}x_1{}^0(0)}{(s + \beta)(s - \beta)} + \frac{p_1{}^0(0)(s - a_{11})}{(s + \beta)(s - \beta)} \end{aligned}$$

where β is defined in Example 8.4-1. Inverse transformation leads to

$$p_1{}^0(t) = -\frac{\omega_{11}x_1{}^0(0)}{\beta}\sinh\beta t + \frac{p_1{}^0(0)}{\beta}(\beta\cosh\beta t - a_{11}\sinh\beta t)$$

The boundary condition on $p_1{}^0(t)$ is $p_1{}^0(t_f) = 0$. Since the problem is linear, this may be accomplished by adjustment of $p_1{}^0(0)$. Thus

$$p_1{}^0(0) = \frac{\omega_{11}x_1{}^0(0)\sinh\beta t_f}{\beta\cosh\beta t_f - a_{11}\sinh\beta t_f}$$

Then $p_1{}^0(t)$ can be written as

$$p_1{}^0(t) = \frac{\omega_{11}x_1{}^0(0)}{\beta}\left[-\sinh\beta t + \frac{\sinh\beta t_f(\beta\cosh\beta t - a_{11}\sinh\beta t)}{\beta\cosh\beta t_f - a_{11}\sinh\beta t_f}\right]$$

In a similar fashion,

$$x_1{}^0(t) = x_1{}^0(0)\left[\frac{\beta\cosh\beta(t_f - t) - a_{11}\sinh\beta(t_f - t)}{\beta\cosh\beta t_f - a_{11}\sinh\beta t_f}\right]$$

This expression can be solved for $x_1{}^0(0)$, and the result substituted for $x_1{}^0(0)$ in the equation for $p_1{}^0(t)$. This yields

$$p_1{}^0(t) = \left[\frac{\omega_{11}\sinh\beta(t_f - t)}{\beta\cosh\beta(t_f - t) - a_{11}\sinh\beta(t_f - t)}\right]x_1{}^0(t)$$

Then, from Eq. 8.4-2,

$$m_1{}^0(t) = -\frac{\omega_{11}b_{11}}{\zeta_{11}}\left[\frac{\sinh\beta(t_f - t)}{\beta\cosh\beta(t_f - t) - a_{11}\sinh\beta(t_f - t)}\right]x_1{}^0(t)$$

which is the result previously obtained by the more direct method. This result is also illustrated by Fig. 8.4-2.

Example 8.4-3. Determine the optimum system for the controlled elements of Fig. 8.4-3. The performance index is described by

$$\mathbf{\Omega} = \begin{bmatrix} \omega_{11} & 0 \\ 0 & 0 \end{bmatrix} \qquad \mathbf{Z} = \begin{bmatrix} 0 & 0 \\ 0 & \zeta_{22} \end{bmatrix}$$

and t_f is infinite, corresponding to an infinite interval problem. Again a regulator problem is assumed, so that $\mathbf{x}^d(t) = \mathbf{0}$. Also, $\mathbf{x}(t_f)$ is unspecified.

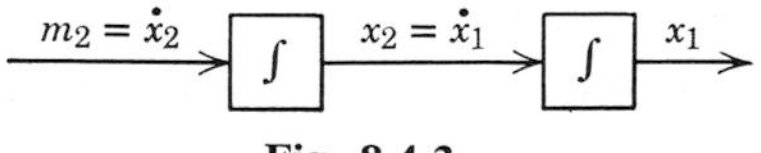

Fig. 8.4-3

Since $\mathbf{Z}$ is singular, the first of Eqs. 8.4-8 cannot be used directly. The factor $\mathbf{Z}^{-1}$ in Eqs. 8.4-8 is due to solving

$$\mathbf{Zm}^0 + \mathbf{B}^T\mathbf{p}^0 = \mathbf{0} \tag{8.4-9}$$

for $\mathbf{m}^0$ and substituting the result into Eq. 8.4-1. In this case, $\mathbf{Z}$ is as given above anc

$$\mathbf{B} = \begin{bmatrix} 0 & 0 \\ 0 & 1 \end{bmatrix}$$

Thus the only information contained in Eq. 8.4-9 is $m_2{}^0 = -(p_2{}^0/\zeta_{22})$. But, by the problem definition, $m_1{}^0 = 0$. Then Eq. 8.4-9 is unchanged if $\mathbf{Z}$ is replaced by

$$\mathbf{Z}_1 = \begin{bmatrix} 1 & 0 \\ 0 & \zeta_{22} \end{bmatrix}$$

Therefore Eqs. 8.4-8 can be used if $\mathbf{Z}^{-1}$ is replaced by $\mathbf{Z}_1{}^{-1}$.† Then since

$$\mathbf{\Phi}(s) = \begin{bmatrix} s^{-1} & s^{-2} \\ 0 & s^{-1} \end{bmatrix}$$

$\mathbf{P}^0(s)$ can be determined to be

$$\mathbf{P}^0(s) = \frac{\omega_{11}\begin{bmatrix} -s^2 & -s \\ s & 1 \end{bmatrix}\mathbf{x}^0(0) - \begin{bmatrix} -s^3 & -\dfrac{\omega_{11}}{\zeta_{22}} \\ s^2 & -s^3 \end{bmatrix}\mathbf{p}^0(0)}{s^4 + (\omega_{11}/\zeta_{22})}$$

where $\mathbf{p}^0(0)$ is to be adjusted so that $\mathbf{p}^0(\infty) = \mathbf{0}$. But $s^4 + (\omega_{11}/\zeta_{22}) = G(s)G(-s)$, where

$$G(s) = s^2 + (2)^{1/2}\left(\frac{\omega_{11}}{\zeta_{22}}\right)^{1/4} s + \left(\frac{\omega_{11}}{\zeta_{22}}\right)^{1/2}$$

This shows that $\mathbf{P}^0(s)$ has two right-half-plane poles and two left half-plane poles, symmetrically located with respect to the origin. In order to have $\mathbf{p}^0(\infty) = \mathbf{0}$, the residue in the right half-plane poles must be zero. A partial fraction expansion of either $P_1{}^0(s)$ or $P_2{}^0(s)$ reveals that the requirements on $\mathbf{p}^0(0)$ for zero residue in each of the right half-plane poles of $\mathbf{P}^0(s)$ are

$$\mathbf{p}^0(0) = \begin{bmatrix} (4\omega_{11}{}^3\zeta_{22})^{1/4} & (\omega_{11}\zeta_{22})^{1/2} \\ (\omega_{11}\zeta_{22})^{1/2} & (4\omega_{11}\zeta_{22}{}^3)^{1/4} \end{bmatrix}\mathbf{x}^0(0)$$

Since

$$\mathbf{X}^0(s) = \frac{\begin{bmatrix} s^3 & s^2 \\ -\dfrac{\omega_{11}}{\zeta_{22}} & s_3 \end{bmatrix}\mathbf{x}^0(0) + \zeta_{22}{}^{-1}\begin{bmatrix} 1 & -s \\ s & -s^2 \end{bmatrix}\mathbf{p}^0(0)}{s^4 + (\omega_{11}/\zeta_{22})}$$

† Note that $\mathbf{Z}_1$ is positive definite and hence the resulting linear optimum system is asymptotically stable, since the other requirements previously given for this are also satisfied.

An obvious alternative to this procedure is to rederive Eq. 8.4-8, but for the case in which $\mathbf{B}$ is a vector.

$\mathbf{x}^0(t)$, for the above values of $\mathbf{p}^0(0)$, is $\mathbf{x}^0(t) = \boldsymbol{\phi}(t)\mathbf{x}^0(0)$, where

$$\boldsymbol{\phi}(t) = \epsilon^{-\alpha t}\begin{bmatrix} (2)^{1/2}\sin\left(\alpha t + \dfrac{\pi}{4}\right) & \alpha^{-1}\sin\alpha t \\ -2\alpha\sin\alpha t & -(2)^{1/2}\sin\left(\alpha t - \dfrac{\pi}{4}\right) \end{bmatrix}$$

and $\alpha = (\omega_{11}/4\zeta_{22})^{1/4}$. Similarly,

$$m_2{}^0(t) = -\zeta_{22}{}^{-1}p_2{}^0(t) = -2\alpha^2\epsilon^{-\alpha t}\left[-x_1{}^0(0)(2)^{1/2}\sin\left(\alpha t - \frac{\pi}{4}\right) - \frac{x_2{}^0(0)}{\alpha}\sin\left(\alpha t + \frac{\pi}{2}\right)\right]$$

Substitution of $\mathbf{x}^0(0) = \boldsymbol{\phi}^{-1}(t)\mathbf{x}^0(t)$ yields

$$m_2{}^0(t) = -2\alpha[\alpha x_1{}^0(t) + x_2{}^0(t)]$$

This is the control law for the controlled elements of Fig. 8.4-3. As expected, it is a linear function of the state variables.

As α is increased, the performance index emphasizes the error relative to the cost of reducing it. From $\boldsymbol{\phi}(t)$ or $G(s)$, it can be seen that the effect is to increase the speed of response and the natural frequency of the system. The damping ratio, however, remains constant at 0.707. Increased damping would have been obtained if $\omega_{22} > 0$ had been chosen in the performance index.

The Time-Varying Case[21]

If any of the elements of $\mathbf{A}$, $\mathbf{B}$, $\boldsymbol{\Omega}$, or $\mathbf{Z}$ are time-varying, a time-domain solution of Eq. 8.4-3 is generally necessary. The solution of Eq. 8.4-3 is given by its state transition matrix, which is defined by

$$\dot{\boldsymbol{\phi}}(t, t_0) = \begin{bmatrix} \mathbf{A} & -\mathbf{B}\mathbf{Z}^{-1}\mathbf{B}^T \\ -\boldsymbol{\Omega} & -\mathbf{A}^T \end{bmatrix}\boldsymbol{\phi}(t, t_0)$$

The state transition matrix has $2n$ rows and $2n$ columns and can be partitioned into four $(n \times n)$ submatrices

$$\boldsymbol{\phi}(t, t_0) = \begin{bmatrix} \boldsymbol{\phi}_{11}(t, t_0) & \boldsymbol{\phi}_{12}(t, t_0) \\ \boldsymbol{\phi}_{21}(t, t_0) & \boldsymbol{\phi}_{22}(t, t_0) \end{bmatrix} \tag{8.4-10}$$

Since $\boldsymbol{\phi}(t_0, t_0) = \mathbf{I}$,

$$\boldsymbol{\phi}_{11}(t_0, t_0) = \boldsymbol{\phi}_{22}(t_0, t_0) = \mathbf{I}$$
$$\boldsymbol{\phi}_{12}(t_0, t_0) = \boldsymbol{\phi}_{21}(t_0, t_0) = [0]$$

In terms of $\boldsymbol{\phi}(t, t_0)$, the solution of Eq. 8.4-3 is

$$\begin{bmatrix} \mathbf{x}^0(t) \\ \mathbf{p}^0(t) \end{bmatrix} = \boldsymbol{\phi}(t, t_0)\begin{bmatrix} \mathbf{x}^0(t_0) \\ \mathbf{p}^0(t_0) \end{bmatrix} + \begin{bmatrix} \mathbf{b}_1(t, t_0) \\ \mathbf{b}_2(t, t_0) \end{bmatrix} \tag{8.4-11}$$

where

$$\begin{bmatrix} \mathbf{b}_1(t, t_0) \\ \mathbf{b}_2(t, t_0) \end{bmatrix} = \int_{t_0}^{t} \boldsymbol{\phi}(t, \tau)\begin{bmatrix} \mathbf{0} \\ \boldsymbol{\Omega}(\tau)\mathbf{x}^d(\tau) \end{bmatrix} d\tau \tag{8.4-12}$$

Substitution of $\boldsymbol{\phi}(t, \tau)$ yields

$$\mathbf{b}_1(t, t_0) = \int_{t_0}^{t} \boldsymbol{\phi}_{12}(t, \tau)\boldsymbol{\Omega}(\tau)\mathbf{x}^d(\tau)\, d\tau$$

$$\mathbf{b}_2(t, t_0) = \int_{t_0}^{t} \boldsymbol{\phi}_{22}(t, \tau)\boldsymbol{\Omega}(\tau)\mathbf{x}^d(\tau)\, d\tau$$

For unspecified terminal conditions on $\mathbf{x}^0(t)$, the terminal boundary conditions are $\mathbf{p}^0(t_f) = \mathbf{0}$. Thus

$$\mathbf{p}^0(t_f) = \boldsymbol{\phi}_{21}(t_f, t_0)\mathbf{x}^0(t_0) + \boldsymbol{\phi}_{22}(t_f, t_0)\mathbf{p}^0(t_0) + \mathbf{b}_2(t_f, t_0) = \mathbf{0}$$

Solving for $\mathbf{p}^0(t_0)$,

$$\mathbf{p}^0(t_0) = -\boldsymbol{\phi}_{22}^{-1}(t_f, t_0)[\boldsymbol{\phi}_{21}(t_f, t_0)\mathbf{x}^0(t_0) + \mathbf{b}_2(t_f, t_0)] \qquad (8.4\text{-}13)$$

Substitution of t for t_0 yields

$$\mathbf{p}^0(t) = -\boldsymbol{\phi}_{22}^{-1}(t_f, t)[\boldsymbol{\phi}_{21}(t_f, t)\mathbf{x}^0(t) + \mathbf{b}_2(t_f, t)]$$

Then, from Eq. 8.4-2, the definition of $\mathbf{b}_2(t_f, t)$, and the fact that

$$\boldsymbol{\phi}_{22}^{-1}(t_f, t)\boldsymbol{\phi}_{22}(t_f, \tau) = \boldsymbol{\phi}_{22}(t, \tau)$$

the control law for unspecified terminal conditions on $\mathbf{x}^0(t)$ is

$$\mathbf{m}^0(t) = \mathbf{Z}^{-1}\mathbf{B}^T\left[\boldsymbol{\phi}_{22}^{-1}(t_f, t)\boldsymbol{\phi}_{21}(t_f, t)\mathbf{x}^0(t) + \int_{t}^{t_f} \boldsymbol{\phi}_{22}(t, \tau)\boldsymbol{\Omega}(\tau)\mathbf{x}^d(\tau)\, d\tau\right] \qquad (8.4\text{-}14)$$

The resulting response $\mathbf{x}^0(t)$ can be found from Eqs. 8.4-11 and 8.4-13.

For specified terminal conditions on $\mathbf{x}^0(t)$, i.e., $\mathbf{x}^0(t_f) = \mathbf{x}(t_f)$, Eq. 8.4-11 gives

$$\mathbf{x}^0(t_f) = \boldsymbol{\phi}_{11}(t_f, t_0)\mathbf{x}^0(t_0) + \boldsymbol{\phi}_{12}(t_f, t_0)\mathbf{p}^0(t_0) + \mathbf{b}_1(t_f, t_0)$$

Then

$$\mathbf{p}^0(t_0) = -\boldsymbol{\phi}_{12}^{-1}(t_f, t_0)[\boldsymbol{\phi}_{11}(t_f, t_0)\mathbf{x}^0(t_0) - \mathbf{x}^0(t_f) + \mathbf{b}_1(t_f, t_0)] \qquad (8.4\text{-}15)$$

Substituting t for t_0, using Eq. 8.4-2, the definition of $\mathbf{b}_1(t_f, t)$ and the fact that

$$\boldsymbol{\phi}_{12}^{-1}(t_f, t)\boldsymbol{\phi}_{12}(t_f, \tau) = \boldsymbol{\phi}_{12}(t, \tau)$$

yields for the control law, in the case of specified terminal conditions on $\mathbf{x}^0(t)$,

$$\mathbf{m}^0(t) = \mathbf{Z}^{-1}\mathbf{B}^T\left\{\boldsymbol{\phi}_{12}^{-1}(t_f, t)[\boldsymbol{\phi}_{11}(t_f, t)\mathbf{x}^0(t) - \mathbf{x}^0(t_f)] + \int_{t}^{t_f} \boldsymbol{\phi}_{12}(t, \tau)\boldsymbol{\Omega}(\tau)\mathbf{x}^d(\tau)\, d\tau\right\} \qquad (8.4\text{-}16)$$

The resulting response $\mathbf{x}^0(t)$ can be found from Eqs. 8.4-11 and 8.4-15.

The reader should note that these control laws are of the form of Eq. 8.4-6, where

$$\mathbf{K} = -\boldsymbol{\phi}_{22}^{-1}(t_f, t)\boldsymbol{\phi}_{21}(t_f, t)$$

for unspecified $\mathbf{x}^0(t_f)$, and

$$\mathbf{K} = -\boldsymbol{\phi}_{12}^{-1}(t_f, t)\boldsymbol{\phi}_{11}(t_f, t)$$

for specified $\mathbf{x}^0(t_f)$. Also, the control law requires knowledge of $\mathbf{x}^d$ in the future interval of control, i.e., $\mathbf{x}^d(\tau)$ for $t \leqslant \tau \leqslant t_f$. This is a general requirement for optimization according to this performance criterion.

Example 8.4-4. Repeat Example 8.4-3 in the time domain.

Let

$$\mathbf{G} = \begin{bmatrix} \mathbf{A} & -\mathbf{B}\mathbf{Z}^{-1}\mathbf{B}^T \\ -\boldsymbol{\Omega} & -\mathbf{A}^T \end{bmatrix} = (-1)\left[\begin{array}{cc|cc} 0 & -1 & 0 & 0 \\ 0 & 0 & 0 & \zeta_{22}^{-1} \\ \hline \omega_{11} & 0 & 0 & 0 \\ 0 & 0 & 1 & 0 \end{array}\right]$$

so that $\boldsymbol{\phi}(t, t_0) = \epsilon^{\mathbf{G}(t-t_0)}$. Use of the Cayley-Hamilton technique gives

$$\boldsymbol{\phi}(t, t_0) = \left[\begin{array}{cc|cc} \alpha_0(t-t_0) & \alpha_1(t-t_0) & \zeta_{22}^{-1}\alpha_3(t-t_0) & -\zeta_{22}^{-1}\alpha_2(t-t_0) \\ -\omega_{11}\zeta_{22}^{-1}\alpha_3(t-t_0) & \alpha_0(t-t_0) & \zeta_{22}^{-1}\alpha_2(t-t_0) & -\zeta_{22}^{-1}\alpha_1(t-t_0) \\ \hline -\omega_{11}\alpha_1(t-t_0) & -\omega_{11}\alpha_2(t-t_0) & \alpha_0(t-t_0) & \omega_{11}\zeta_{22}^{-1}\alpha_3(t-t_0) \\ \omega_{11}\alpha_2(t-t_0) & \omega_{11}\alpha_3(t-t_0) & -\alpha_1(t-t_0) & \alpha_0(t-t_0) \end{array}\right]$$

where

$$\alpha_0(t) = \cosh \alpha t \cos \alpha t$$

$$\alpha_1(t) = \frac{\sinh \alpha t \cos \alpha t + \cosh \alpha t \sin \alpha t}{2\alpha}$$

$$\alpha_2(t) = \frac{\sinh \alpha t \sin \alpha t}{2\alpha^2}$$

$$\alpha_3(t) = \frac{\cosh \alpha t \sin \alpha t - \sinh \alpha t \cos \alpha t}{4\alpha^3}$$

and α is as defined in Example 8.4-3. Substitution into Eq. 8.4-14 yields, for infinite t_f,

$$m_2{}^0(t) = -2\alpha[\alpha x_1{}^0(t) + x_2{}^0(t)]$$

the same result obtained in Example 8.4-3.

Example 8.4-5. Determine the optimum system for the controlled elements characterized by

$$\dot{x}_1 = -\frac{1}{t}x_1 + \frac{1}{t}m_1$$

The performance index is

$$I = \tfrac{1}{2}\int_{t_0}^{t_f} (x_1{}^2 + m_1{}^2)\, dt$$

and $x_1{}^d(t) = 0$. The terminal conditions are unspecified.

From the problem statement,

$$\mathbf{G} = \begin{bmatrix} \mathbf{A} & -\mathbf{BZB}^T \\ -\boldsymbol{\Omega} & -\mathbf{A}^T \end{bmatrix} = \begin{bmatrix} -t^{-1} & -t^{-2} \\ -1 & t^{-1} \end{bmatrix}$$

Then the components $\phi_{11}(t, t_0)$ and $\phi_{21}(t, t_0)$ of $\boldsymbol{\phi}(t, t_0)$ must satisfy

$$\dot{\phi}_{11}(t, t_0) = -\frac{1}{t}\,\phi_{11}(t, t_0) - \frac{1}{t^2}\,\phi_{21}(t, t_0)$$

$$\dot{\phi}_{21}(t, t_0) = -\phi_{11}(t, t_2) + \frac{1}{t}\,\phi_{21}(t, t_0)$$

Solving for $\phi_{11}(t, t_0)$ in the second equation and substituting the result into the first yields

$$\ddot{\phi}_{21}(t, t_0) - \frac{1}{t^2}\,\phi_{21}(t, t_0) = 0$$

This is a form of Euler's equation, considered in Example 2.8-2. The change of variable $t = \epsilon^z$ gives

$$\phi_{21}''(\epsilon^z, \epsilon^{z_0}) - \phi_{21}'(\epsilon^z, \epsilon^{z_0}) - \phi_{21}(\epsilon^z, \epsilon^{z_0}) = 0$$

where the primes denotes differentiation with respect to z. The differential equation for $\phi_{21}(\epsilon^z, \epsilon^{z_0})$ has the solution

$$\phi_{21}(\epsilon^z, \epsilon^{z_0}) = k_1(z_0)\exp\left(\frac{\sqrt{5}+1}{2}\,z\right) + k_2(z_0)\exp\left(\frac{1-\sqrt{5}}{2}\,z\right)$$

so that

$$\phi_{21}(t, t_0) = c_1(t_0)t^{(\sqrt{5}+1)/2} + c_2(t_0)t^{(1-\sqrt{5})/2}$$

Similarly,

$$\phi_{11}(t, t_0) = c_1(t_0)\left(\frac{1-\sqrt{5}}{2}\right)t^{(\sqrt{5}-1)/2} + c_2(t_0)\left(\frac{1+\sqrt{5}}{2}\right)t^{-(\sqrt{5}+1)/2}$$

Since $\phi_{11}(t_0, t_0) = 1$ and $\phi_{21}(t_0, t_0) = 0$, $c_1(t_0)$ and $c_2(t_0)$ can be determined to be

$$c_1(t_0) = -\frac{1}{\sqrt{5}}\,t_0^{(1-\sqrt{5})/2}$$

$$c_2(t_0) = \frac{1}{\sqrt{5}}\,t_0^{(1+\sqrt{5})/2}$$

Then

$$\phi_{11}(t, t_0) = \left(\frac{\sqrt{5}-1}{2\sqrt{5}}\right)\left(\frac{t_0}{t}\right)^{-(\sqrt{5}-1)/2} + \left(\frac{\sqrt{5}+1}{2\sqrt{5}}\right)\left(\frac{t_0}{t}\right)^{(\sqrt{5}+1)/2}$$

and

$$\phi_{21}(t, t_0) = \frac{t_0}{\sqrt{5}}\left(\frac{t_0}{t}\right)^{(\sqrt{5}-1)/2} - \frac{t_0}{\sqrt{5}}\left(\frac{t_0}{t}\right)^{-(\sqrt{5}+1)/2}$$

Similarly,

$$\phi_{12}(t, t_0) = -\frac{t^{-1}}{\sqrt{5}}\left(\frac{t_0}{t}\right)^{-(\sqrt{5}-1)/2} + \frac{t^{-1}}{\sqrt{5}}\left(\frac{t_0}{t}\right)^{(\sqrt{5}+1)/2}$$

$$\phi_{22}(t, t_0) = \left(\frac{\sqrt{5}-1}{2\sqrt{5}}\right)\left(\frac{t_0}{t}\right)^{-(\sqrt{5}-1)/2} + \left(\frac{\sqrt{5}+1}{2\sqrt{5}}\right)\left(\frac{t_0}{t}\right)^{(\sqrt{5}+1)/2}$$

From Eq. 8.4-14,

$$m_1{}^0(t) = \frac{1}{t}\frac{\phi_{21}(t_f, t)}{\phi_{22}(t, t_f)}\, x_1{}^0(t)$$

Thus

$$m_1{}^0(t) = -2\left[\frac{1 - (t/t_f)^{\sqrt{5}}}{(\sqrt{5}+1) + (\sqrt{5}-1)(t/t_f)^{\sqrt{5}}}\right] x_1{}^0(t)$$

In the infinite interval problem, t_f is infinite. Then

$$m_1{}^0(t) = -\frac{2}{\sqrt{5}+1}\, x_1{}^0(t)$$

In this case, a time-invariant control law is obtained, even though the controlled are time-varying.

Example 8.4-6. Determine the optimum system for the controlled elements of Example 8.4-5, if the performance index is

$$I = \tfrac{1}{2}\int_{t_0}^{t_f} m_1{}^2\, dt$$

and the terminal condition $x_1{}^0(t_f) = 0$ is specified.

For this case,

$$\mathbf{G} = \begin{bmatrix} -t^{-1} & -t^{-2} \\ 0 & t^{-1} \end{bmatrix}$$

$\boldsymbol{\phi}(t, t_0)$ is the solution to $\dot{\boldsymbol{\phi}}(t, t_0) = \mathbf{G}\boldsymbol{\phi}(t, t_0)$. The equations can be integrated by separation of variables to yield

$$\boldsymbol{\phi}(t, t_0) = \begin{bmatrix} \dfrac{t_0}{t} & \dfrac{1}{t} - \dfrac{1}{t_0} \\[2ex] \dfrac{t}{t_0} & \dfrac{t}{t_0} \end{bmatrix}$$

From Eq. 8.4-16,

$$m_1{}^0(t) = -\left[\frac{t}{t_f - t}\right] x_1{}^0(t)$$

The resulting response, as found from Eqs. 8.4-11 and 8.4-15, is

$$x_1{}^0(t) = \frac{t_0(t_f - t)}{t(t_f - t_0)}\, x_1{}^0(t_0)$$

The response does satisfy the terminal condition $x_1{}^0(t_f) = 0$. For $t > t_f$, the response is not zero, however. This is to be expected, since the performance index does not consider this part of the response.

Although the time-varying feedback gain becomes infinite at $t = t_f$, $m_1{}^0(t)$ is always finite. In fact, from the expressions for $m_1{}^0(t)$ and $x_1{}^0(t)$,

$$m_1{}^0(t) = -\frac{t_0}{t_f - t_0}\, x_1{}^0(t_0)$$

a constant. For infinite t_f, corresponding to an infinite interval problem, $m_1^0(t) = 0$. Thus the system is open-loop, and the response is the open-loop initial condition response $x_1^0(t) = (t_0/t)x_1^0(t_0)$. This response satisfies all the requirements in the infinite interval case and obviously has the smallest possible value for the performance index. The system would be undesirable, however, owing to its poor performance with respect to unwanted disturbances.

The controlled elements in the examples of this section are rather simple, and yet considerable effort is required in some cases to determine the optimum system. This is true in spite of the fact that the systems are linear. In practical situations systems are nonlinear, and analytical determination of optimum systems is virtually impossible. For the most part, optimization theory is practical only when used in conjunction with computers, as considered later in this chapter.

8.5 SELECTION OF CONSTANT WEIGHTING FACTORS

The examples of the previous section indicate that the control law and system response are greatly influenced by the weighting factors $\mathbf{\Omega}$ and $\mathbf{Z}$ chosen in the performance index. Selection of these weighting factors is a difficult task, since the relationships between the weighting factors and the optimum system parameters or the system response are generally very complex. However, Ellert has developed a technique for the selection of weighting factors in the time-invariant case.[16]

Consider, as an example, the second order linear controlled elements described by

$$\dot{\mathbf{x}} = \begin{bmatrix} a_{11} & a_{12} \\ a_{21} & a_{22} \end{bmatrix} \mathbf{x} + \begin{bmatrix} 0 & 0 \\ 0 & b_{22} \end{bmatrix} \mathbf{m}$$

The performance index is the one of Section 8.4, with infinite t_f and

$$\mathbf{\Omega} = \begin{bmatrix} \omega_{11} & 0 \\ 0 & \omega_{22} \end{bmatrix} \qquad \mathbf{Z} = \begin{bmatrix} 0 & 0 \\ 0 & 1 \end{bmatrix}$$

Using the method of Eqs. 8.4-5, the optimum control law is found to be

$$m_2^0(t) = -b_{22}[k_{21}x_1^0(t) + k_{22}x_2^0(t)] + b_{22}v_1^0(t) \tag{8.5-1}$$

where the k's are defined by

$$\begin{aligned} \omega_{22} + 2a_{22}k_{22} + 2a_{12}k_{21} - b_{22}^2k_{22}^2 &= 0 \\ \omega_{11} + 2a_{21}k_{21} + 2a_{11}k_{11} - b_{22}^2k_{21}^2 &= 0 \\ a_{21}k_{22} + a_{22}k_{21} + a_{11}k_{21} + a_{12}k_{11} - b_{22}^2k_{22}k_{21} &= 0 \end{aligned} \tag{8.5-2}$$

and v^0 is defined by

$$\begin{aligned} -\dot{v}_1^{\,0} &= \omega_{22}x_2^{\,d} + a_{22}v_2^{\,0} + a_{12}v_1^{\,0} - b_{22}^{\,2}k_{22}v_2^{\,0} \\ -\dot{v}_2^{\,0} &= \omega_{11}x_1^{\,d} + a_{21}v_1^{\,0} + a_{11}v_1^{\,0} - b_{11}^{\,2}k_{21}v_2^{\,0} \end{aligned} \tag{8.5-3}$$

Since this is a linear, time-invariant system, the closed-loop transfer function can be determined to be

$$\frac{X_1^{\,0}(s)}{V(s)} = \frac{\omega_0^{\,2}}{s^2 + z_1\omega_0 s + \omega_0^{\,2}} \tag{8.5-4}$$

where

$$z_1\omega_0 = b_{22}^{\,2}k_{22} - a_{11} - a_{22}$$

$$\omega_0^{\,2} = a_{11}(a_{22} - b_{22}^{\,2}k_{22}) + a_{12}(b_{22}^{\,2}k_{21} - a_{21})$$

$$V(s) = \frac{a_{12}b_{22}^{\,2}}{\omega_0^{\,2}}\,V_1(s)$$

and $V_1(s)$ is the Laplace transform of the system input.

With these definitions, k_{22} and k_{21} can be written as

$$\begin{aligned} k_{22} &= \frac{1}{b_{22}^{\,2}}(z_1\omega_0 + a_{11} + a_{22}) \\ k_{21} &= \frac{1}{a_{21}b_{22}^{\,2}}(\omega_0^{\,2} + a_{11}z_1\omega_0 + a_{12}a_{21} + a_{11}^{\,2}) \end{aligned} \tag{8.5-5}$$

From Eqs. 8.5-2 and 8.5-5,

$$\begin{aligned} \omega_{11} = \frac{1}{a_{12}^{\,2}b_{22}^{\,2}}\,[\omega_0^{\,4} &+ 3a_{11}z_1\omega_0^{\,3} + a_{11}^{\,2}(3z_1^{\,2} + 2)\omega_0^{\,2} + 4a_{11}^{\,3}z_1\omega_0 \\ &+ (2a_{11}a_{12}a_{21}a_{22} + 2a_{11}^{\,2}a_{12}a_{21} + a_{12}^{\,2}a_{21}^{\,2} + a_{11}^{\,4})] \end{aligned} \tag{8.5-6}$$

$$\omega_{22} = \frac{1}{b_{22}^{\,2}}\,[(z_1^{\,2} - 2)\omega_0^{\,2} - a_{11}^{\,2} - a_{22}^{\,2} - 2a_{12}a_{21}]$$

These expressions determine ω_{11} and ω_{22}, once values of z_1 and ω_0 have been selected.

Ellert's procedure is to choose z_1 to provide the desired relative stability of the system, assuming that none of the system variables exceed their prescribed limits. ω_0 is then chosen in accordance with the system bandwidth requirements or any limits on $m_2(t)$. The relationship between $m_2(t)$ and ω_0 is given by substituting Eq. 8.5-5 into Eq. 8.5-1. It is

$$\begin{aligned} m_2(t) = -\frac{1}{b_{22}}\Bigg\{&\frac{\omega_0^{\,2}}{a_{12}}\,x_1(t) + \left[\frac{a_{11}}{a_{12}}\,x_1(t) + x_2(t)\right]z_1\omega_0 \\ &+ \left(\frac{a_{11}^{\,2}}{a_{12}} + a_{21}\right)x_1(t) + (a_{11} + a_{22})x_2(t)\Bigg\} + b_{22}v_1(t) \end{aligned} \tag{8.5-7}$$

Specification of the maximum available value of $m_2(t)$, worst case values of $x_1(t)$ and $x_2(t)$, and solution of Eq. 8.5-3 permits Eq. 8.5-7 to be solved for ω_0.

When z_1 and ω_0 have been determined, the weighting factors ω_{11} and ω_{22} are given by Eqs. 8.5-6. Since the performance index should be convex, ω_{11} and ω_{22} must be non-negative. This requirement, in a sense, tests the compatibility of the design requirements, assuming that a quadratic performance index with constant weighting factors is a reasonable choice.[16]

For controlled elements of higher order, Eq. 8.5-4 becomes

$$\frac{X_1(s)}{V(s)} = \frac{N(s)}{s^n + z_{n-1}\omega_0 s^{n-1} + \cdots + z_1\omega_0^{n-1}s + \omega_0{}^n} \qquad (8.5\text{-}8)$$

where

$$N(s) = \omega_0{}^n$$

$$N(s) = z_1\omega_0^{n-1}s + \omega_0{}^n$$

or

$$N(s) = z_2\omega_0^{n-2}s^2 + z_1\omega_0^{n-1}s + \omega_0{}^n$$

for types one, two, or three systems, respectively, i.e., systems with zero steady-state error to a unit step input, zero steady-state error to a unit ramp input, etc., respectively. Ellert's procedure for selection of the performance index weighting factors can be applied to these higher order cases, if the z's can be determined without undue trial and error. Criteria for selecting the z's to obtain acceptable responses have been presented in the literature. In fact, tabulations of numerical values of the z's, called standard forms, can be found.[6,10,22] Whiteley's standard forms for the characteristic equations are given in Table 8.5-1.[22] The corresponding step responses are shown in Fig. 8.5-1. Since many practical control systems

Table 8.5-1

System Type		Standard Forms	Maximum Percent Overshoot
Zero position error	(*a*)	$s^2 + 1.4\omega_0 s + \omega_0{}^2$	5
	(*b*)	$s^3 + 2\omega_0 s^2 + 2\omega_0{}^2 s + \omega_0{}^3$	8
	(*c*)	$s^4 + 2.6\omega_0 s^3 + 3.4\omega_0{}^2 s^2 + 2.6\omega_0{}^3 s + \omega_0{}^4$	10
Zero velocity error	(*d*)	$s^2 + 2.5\omega_0 s + \omega_0{}^2$	10
	(*e*)	$s^3 + 5.1\omega_0 s^2 + 6.3\omega_0{}^2 s + \omega_0{}^3$	10
	(*f*)	$s^4 + 7.2\omega_0 s^3 + 16\omega_0{}^2 s^2 + 12\omega_0{}^3 s + \omega_0{}^4$	10
	(*g*)	$s^5 + 9\omega_0 s^4 + 29\omega_0{}^2 s^3 + 38\omega_0{}^3 s^2 + 18\omega_0{}^4 s + \omega_0{}^5$	10
	(*h*)	$s^6 + 11\omega_0 s^5 + 43\omega_0{}^2 s^4 + 83\omega_0{}^3 s^3 + 73\omega_0{}^4 s^2 + 25\omega_0{}^5 s + \omega_0{}^6$	10
Zero acceleration error	(*i*)	$s^3 + 6.7\omega_0 s^2 + 6.7\omega_0{}^2 s + \omega_0{}^3$	10
	(*j*)	$s^4 + 7.9\omega_0 s^3 + 15\omega_0{}^2 s^2 + 7.9\omega_0{}^3 s + \omega_0{}^4$	20
	(*k*)	$s^5 + 18\omega_0 s^4 + 69\omega_0{}^2 s^3 + 69\omega_0{}^3 s^2 + 18\omega_0{}^4 s + \omega_0{}^5$	20
	(*l*)	$s^6 + 36\omega_0 s^5 + 251\omega_0{}^2 s^4 + 485\omega_0{}^3 s^3 + 251\omega_0{}^4 s^2 + 36\omega_0{}^5 s + \omega_0{}^6$	20

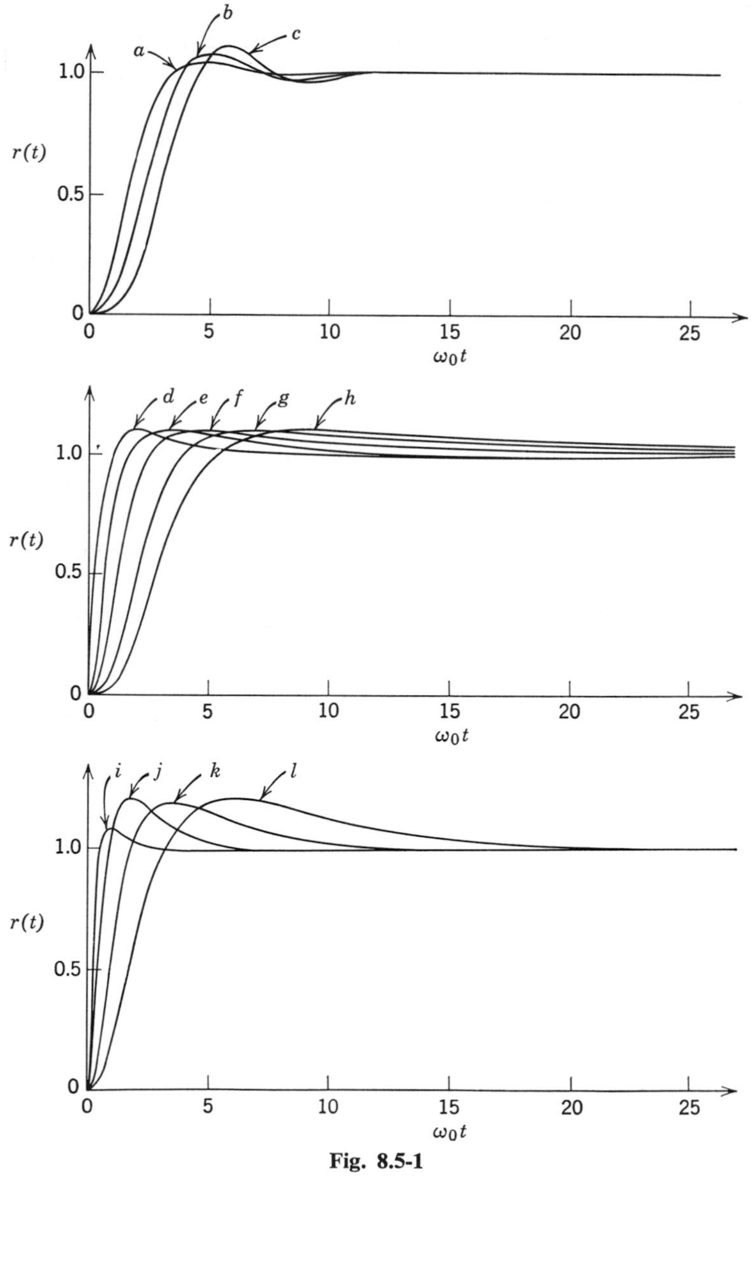

a
b
c
1.0
r(t)
0.5
0
0
5
10
15
20
25
$\omega_0 t$
d
e
f
g
h
1.0
r(t)
0.5
0
0
5
10
15
20
25
$\omega_0 t$
i
j
k
l
1.0
r(t)
0.5
0
0
5
10
15
20
25
$\omega_0 t$

Fig. 8.5-1

have transfer functions of the form of Eq. 8.5-8, Whiteley's standard forms can often be used, in conjunction with Ellert's procedure, to determine the performance index weighting factors which satisfy the subjective design requirements. Objective design requirements, such as limits on the control signals or state variables, must be approached in a different fashion. This is considered in the next section.

8.6 PENALTY FUNCTIONS

The design specifications on most control systems require that some of the variables be constrained between prescribed limits. Such constraints may be imposed by saturation-type limits in the controlled elements, or they may be due to the mission requirements associated with the application of the system. For example, the maximum allowable stagnation temperature on the nose of a re-entry vehicle is often given as 3500° Fahrenheit. The temperature constraint, as stated, is a "hard" constraint, in the sense that it is a value not to be exceeded. Both "hard" and "soft" constraints are often stated, but in practice most constraints really are soft. For example, a temperature of 3600° would probably also be acceptable, since safety factors are usually included in such figures. No design procedure or subsequent implementation is precise, nor can any design procedure consider the uncertainties associated with the ultimate operation of the system. Although hard constraints are conceptually useful from a mathematical viewpoint, they normally do not physically exist. Furthermore, hard constraints cause considerable difficulty in obtaining computer solutions to optimization problems, and in controller realization. For these reasons, constraints are treated here by means of penalty functions. One can approach a hard constraint by making the penalty more severe.

A *penalty function* is a performance index term which increases the value of the index when the constrained variable approaches its limit. For example, the second and fourth terms of Eq. 8.2-2 are penalty function terms. Many other penalty functions have been proposed in the literature.[23–25]

Weighting factors, as contained in the first and third terms of Eq. 8.2-2, are selected to satisfy the subjective design requirements. This was discussed in the preceding section. Once the weighting factors are selected, the penalty functions may be selected to satisfy the constraints. If this is done, the system response satisfies the relative stability and speed of response specifications when none of the variables are at their limits, and, furthermore, the variables are properly constrained when they attempt to exceed these limits. This is the basis for Ellert's design philosophy.

The introduction of nonquadratic penalty functions into the performance index makes the optimization problem a nonlinear one, even if the controlled elements are assumed linear.† For this reason, computers are generally required to solve practical optimization problems. Computer techniques for solving optimization problems are considered next.

8.7 BOUNDARY CONDITION ITERATION

As indicated in Sections 8.3 and 8.4, the solution of Eq. 8.3-12 requires the solution of a two-point boundary value problem. It is not possible to solve simultaneously the $\dot{\mathbf{x}}$ and $\dot{\mathbf{p}}$ equations forward in time, unless correct boundary values for $\mathbf{p}(t_0)$ are known. A similar problem exists if one attempts to solve the equations simultaneously backward in time.

One possible approach is to assume a set of values for $\mathbf{p}(t_0)$, indicated by $\mathbf{p}^1(t_0)$. The superscript "1" is used to denote the first choice of $\mathbf{p}(t_0)$. These n conditions, in conjunction with the n conditions $\mathbf{x}^0(t_0) = \mathbf{x}(t_0)$ are sufficient to solve Eq. 8.3-12 forward in time to determine $\mathbf{x}^1(t)$ and $\mathbf{p}^1(t)$. The computed values of $\mathbf{p}^1(t_f)$ or $\mathbf{x}^1(t_f)$, as the case may be, can then be compared with the correct values specified by the terminal boundary conditions. If these two sets of values are identical (this would indeed be fortunate), $\mathbf{p}^1(t) = \mathbf{p}^0(t)$ and $\mathbf{x}^1(t) = \mathbf{x}^0(t)$, and the problem is solved. Generally, however, they differ. If the $\dot{\mathbf{x}}$ and $\dot{\mathbf{p}}$ equations are linear, superposition can be used to directly revise $\mathbf{p}^1(t_0)$ so that the terminal boundary conditions are satisfied, and hence immediately yield the optimum solution. In the more general case, a revised choice for $\mathbf{p}(t_0)$, namely $\mathbf{p}^2(t_0)$, must be made and the equations solved again. This process is repeated until the terminal boundary conditions are satisfied, to an acceptable degree of accuracy. This technique is called *boundary condition iteration.*

More direct methods for performing boundary condition iteration are suggested by Merriam, Neustadt, Scharmack, and Speyer.[25–28] For example, a hill-climbing problem can be formulated so that the optimum occurs when $\mathbf{p}^n(t_f) = \mathbf{p}^0(t_f)$. Thus the computer techniques for hill-climbing problems can be utilized.

There are two significant problems associated with boundary condition iteration methods in the general case. First, as indicated by Eq. 8.4-8 for

† In this case, the first footnote of Section 8.4 still applies, if the terms "positive semi-definite" and "positive definite" are replaced by "convex and differentiable" and "strictly convex and differentiable," respectively.[25] Of course, these terms now apply to the appropriate terms in the integrand f_0 of Eq. 8.3-2, rather than to $\mathbf{\Omega}$ and $\mathbf{Z}$.

the linear case, the differential equations to be solved form an unstable set. Thus the solutions to these equations are extremely sensitive to errors resulting from the starting conditions chosen and numerical computational procedures. This sensitivity generally increases as the optimum is approached because the system speed of response, and hence the speed of response of the unstable adjoint equations, increases. Second, at each value of t in the range $t_0 \leqslant t \leqslant t_f$, the second of Eqs. 8.3-12 must also be solved to obtain numerical values for the control signals. Since this equation is often a nonlinear algebraic equation, or even a transcendental equation, its solution may be difficult.

Partially offsetting the disadvantages of boundary condition iteration is the requirement of less computer memory than other methods to be considered. Also, the relative simplicity of the computer program may be important, for example, in the case of airborne real time optimization.

8.8 DYNAMIC PROGRAMMING[29]

A second method for solving the two-point boundary value problem is discrete dynamic programming. In this method, the n state variables and time are quantized, establishing a grid in the $n + 1$-dimensional $\mathbf{x}t$ space. This is illustrated for first order controlled elements in Fig. 8.8-1. The abscissa in this figure represents quantized future time, from the initial time t_0 to the terminal time t_f. Real time t advances from t_0 to t_f. The ordinate represents quantized values of x_1. Since this is a discrete problem, only the values of x_1 and time at the intersections of the broken lines are important. These intersections are called nodes, and the solid lines with arrows denote the permissible changes in x_1 from one node to another.

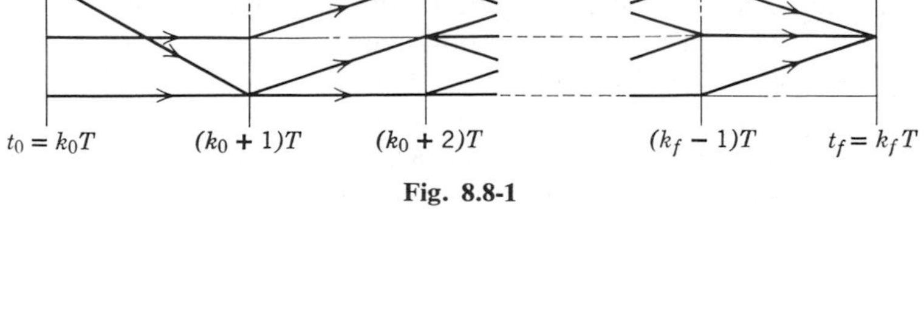

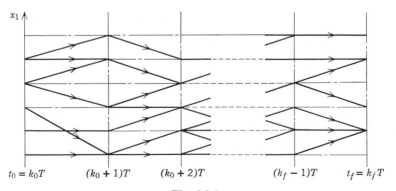

Fig. 8.8-1

The discrete form of Eq. 8.3-2 is the performance index

$$I^* = T \sum_{k=k_0}^{k_f} f_0[\mathbf{x}(k), \mathbf{m}(k), k] \tag{8.8-1}$$

The value of f_0 in the interval from $k = k_j$ to $k = k_f$ is determined by $\mathbf{x}(k_j)$, the state of the controlled elements at $k = k_j$, and by the control signal $\mathbf{m}(k)$ for $k_j \leqslant k \leqslant k_f$. Thus the value of the performance index over the interval $k_j \leqslant k \leqslant k_f$ is determined by $\mathbf{x}(k_j)$ and $\mathbf{m}(k)$, and also by $k_f - k_j$. Assuming that k_f is a constant, this dependence is upon $\mathbf{x}(k_j)$, $\mathbf{m}(k)$, and k_j and is indicated by

$$R[\mathbf{x}(k_j), \mathbf{m}(k), k_j] = T \sum_{k=k_j}^{k_f} f_0[\mathbf{x}(k), \mathbf{m}(k), k]$$

where $R[\mathbf{x}(k_j), \mathbf{m}(k), k_j]$ denotes the portion of I^* due to $k_j \leqslant k \leqslant k_f$.

Defining a *policy* to be any rule for making decisions which yields an allowable sequence of decisions, Bellman's *principle of optimality* can be stated as, "An optimal policy has the property that, whatever the initial state and initial decision are, the remaining decisions must constitute an optimal policy with regard to the state resulting from the first decision." In essence, this is a statement of the intuitively obvious philosophy that, if I^* is to be minimized by a choice of $\mathbf{m}(k)$, the portion of I^* denoted by R must be minimized. Furthermore, this principle states that, if R is minimized by a selection of the sequence $\mathbf{m}(k)$, $k_j \leqslant k \leqslant k_f$, the optimum sequence $\mathbf{m}^0(k)$ is a function of the states $\mathbf{x}(k_j)$. The state of the controlled elements at $k = k_j$ determines $m^0(k)$, $k_j \leqslant k \leqslant k_f$. This functional dependence can be indicated by $\mathbf{m}^0(k) = \mathbf{g}_k[\mathbf{x}(k_j)]$. Thus the minimum value of R is a function only of the state of the controlled elements at k_j, and k_j itself. Then

$$R^0[\mathbf{x}(k_j), k_j] = \min_{\mathbf{m}(k)} \sum_{k=k_j}^{k_f} Tf_0[\mathbf{x}(k), \mathbf{m}(k), k] \tag{8.8-2}$$

The symbolism before the summation indicates that $\mathbf{m}(k)$ is to be chosen with k in the interval $k_j \leqslant k \leqslant k_f$ to minimize the summation.

Obviously, if $k_j = k_f$, $R^0[\mathbf{x}(k_f), k_f] = 0$. With this boundary condition as a starting point, Eq. 8.8-2 can be used for a computation backward in time from $k = k_f$ to $k = k_0$, to determine $\mathbf{m}^0(k)$ and evaluate the minimum value of the performance index. This computation gives $\mathbf{m}^0(t)$ and the corresponding $\mathbf{x}^0(t)$ from each node in the $\mathbf{x}t$ space to the nodes at t_f. Thus the solution of the two-point boundary value problem is determined for a number of initial conditions, rather than only one. The original problem has been "embedded" within a number of similar problems.

Unfortunately, in order to compute $R^0[\mathbf{x}(k_f - i), k_f - i]$ for each node at a time $t_f - iT$, it is necessary to have $R^0[\mathbf{x}(k_f - i + 1), k_f - i + 1]$

temporarily stored in the computer for each node at time $t_f - (i - 1)T$. The temporary computer memory required is excessive for nominal grid sizes. In fact, the order of the controlled elements must be less than three to avoid saturating the thirty-two thousand word memory of an IBM 704.

Several methods have been proposed to alleviate the excessive memory requirements of dynamic programming.[30,31] One approach uses a coarse grid for an initial solution. This is followed by a second solution using a finer grid in the region of $\mathbf{x}t$ space where the optimum has been located by the initial solution. Another method approximates the minimum value of the performance index at a given instant in time by an orthogonal series. Thus the series coefficients, rather than R^0, are stored. These methods still have difficulties, however.

Before considering other computer techniques for solving optimization problems, it is interesting to note that Eq. 8.3-12 can be obtained by means of dynamic programming. Since $R^0[\mathbf{x}(k_j), k_j]$ is a constant for any k_j, Eq. 8.8-2 can be written as

$$\min_{\mathbf{m}(k)} \left\{ Tf_0[\mathbf{x}(k_j), \mathbf{m}(k_j), k_j] + T \sum_{k=k_j+1}^{k_f} f_0[\mathbf{x}(k), \mathbf{m}(k), k] - R^0[\mathbf{x}(k_j), k_j] \right\} = 0$$

This is the same as

$$\min_{\mathbf{m}(k)} \left\{ f_0[\mathbf{x}(k_j), \mathbf{m}(k_j), k_j] + \frac{R^0[\mathbf{x}(k_{j+1}), k_{j+1}] - R^0[\mathbf{x}(k_j), k_j]}{T} \right\} = 0$$

Taking the limit as T approaches zero, the continuous form of Bellman's equation is determined as

$$\left\{ f_0[\mathbf{x}^0(\tau), \mathbf{m}^0(\tau), \tau] + \frac{dI^0[\mathbf{x}^0(\tau), \tau]}{d\tau} \right\} = 0$$

where $t_0 \leqslant \tau \leqslant t_f$, and I^0 is the minimum value of the performance index. The total time derivative can be written in terms of partial derivatives, and, using the first of Eqs. 8.3-1, this expression is

$$f_0[\mathbf{x}^0, \mathbf{m}^0, \tau] + \frac{\partial I^0}{\partial \tau} + \langle \mathbf{grad}_{\mathbf{x}^0}\, I^0, \mathbf{f} \rangle = 0$$

This first order partial differential equation in the dependent variable I^0 is called the Hamilton-Jacobi equation. Defining $p_k{}^0(\tau) = \partial I^0/\partial x_k{}^0$, Eq. 8.3-10 permits the Hamilton-Jacobi equation to be written as

$$\frac{\partial I^0}{\partial \tau} + H(\mathbf{x}^0, \mathbf{m}^0, \mathbf{p}^0, \tau) = 0$$

The solution of this equation by the method of characteristics yields Eq. 8.3-12, as indicated above.[25]

8.9 CONTROL SIGNAL ITERATION

The reason the two-point boundary value problem under consideration cannot generally be separated and solved exactly as two one-point boundary value problems is that the $\dot{\mathbf{x}}^0$ and $\dot{\mathbf{p}}^0$ equations are coupled by the expression for $\mathbf{m}^0$. If, however, Pontryagin's equation for $\mathbf{m}^0$ is not used, and instead $\mathbf{m}$ is determined by iteration, the equations can be uncoupled. This is the basis of the methods of Kelley and Bryson.[32,33] For a number of linear and nonlinear problems, this *control signal iteration* procedure converges to the optimum control variable $\mathbf{m}^0$ which satisfies Pontryagin's equations. This method is most useful for trajectory problems, where an open-loop type of operation is used.

The first step in the control signal iteration procedure is to guess a reasonable value for $\mathbf{m}^1(t)$ and integrate the $\dot{\mathbf{x}}$ equations forward in time from t_0 to t_f to determine $\mathbf{x}^1(t)$. The values of $\mathbf{m}^1(t)$ and $\mathbf{x}^1(t)$ are stored in the computer memory. The corresponding value of the modified performance index $I_c^{\,1}$ of Eq. 8.3-3 is also computed and stored.

Next, the $\dot{\mathbf{p}}$ or adjoint equations are solved backward in time from t_f to t_0. The stored values of $\mathbf{x}^1(t)$ generally appear in these equations and are utilized for the backward computation. During the backward computation, the effects of changes in $\mathbf{m}^1(t)$ are considered, to determine a new control signal $\mathbf{m}^2(t)$ for the next forward computation. The effects of these changes can be determined for the $(i + 1)$th iteration by setting $\delta\mathbf{m}^i = \mathbf{m}^{i+1} - \mathbf{m}^i$. The corresponding change in the state variables is $\delta\mathbf{x}^i = \mathbf{x}^{i+1} - \mathbf{x}^i$. These changes cause a change in the modified performance index of $\delta I_c^{\,i} = I_c^{i+1} - I_c^{\,i}$. Assuming that I_c is to be minimized, the iterative procedure is converging if $\delta I_c^{\,i}$ is negative. The procedure is continued until $\delta I_c^{\,i}$ and/or $\delta\mathbf{m}^i$ are negligible.

At this point, an algorithm for computing $\delta\mathbf{m}^i$ is required. The derivation presented here follows that of Merriam.[25] From Eq. 8.3-3,

$$I_c^{i+1} = \int_{t_0}^{t_f} [f_0^{i+1} + \langle \mathbf{p}^{i+1}, \mathbf{f}^{i+1} - \dot{\mathbf{x}}^{i+1}\rangle]\, dt$$

Substitution of

$$\begin{aligned} f_0^{i+1} &= f_0^{\,i} + \delta f_0^{\,i} \\ \mathbf{p}^{i+1} &= \mathbf{p}^i + \delta\mathbf{p}^i \\ \mathbf{f}^{i+1} &= \mathbf{f}^i + \delta\mathbf{f}^i \\ \dot{\mathbf{x}}^{i+1} &= \dot{\mathbf{x}}^i + \delta\dot{\mathbf{x}}^i \end{aligned}$$

yields

$$\begin{aligned} I_c^{i+1} = \int_{t_0}^{t_f} \{[f_0^{\,i} &+ \langle \mathbf{p}^i, \mathbf{f}^i - \dot{\mathbf{x}}^i\rangle] + \delta f_0^{\,i} + \langle \mathbf{p}^i, \delta\mathbf{f}^i\rangle + \langle \delta\mathbf{p}^i, \delta\mathbf{f}^i\rangle \\ &- \langle \mathbf{p}^i, \delta\dot{\mathbf{x}}^i\rangle - \langle \delta\mathbf{p}^i, \delta\dot{\mathbf{x}}^i\rangle + \langle \delta\mathbf{p}^i, \mathbf{f}^i - \dot{\mathbf{x}}^i\rangle\}\, dt \end{aligned}$$

The first term in the integrand gives $I_c^{\,i}$. The last term is zero by the first of Eqs. 8.3-1, applied to the ith iteration. Thus

$$I_c^{i+1} = I_c^{\,i} + \int_{t_0}^{t_f} [\delta f_0^{\,i} + \langle \mathbf{p}^i, \delta \mathbf{f}^i \rangle + \langle \delta \mathbf{p}^i, \delta \mathbf{f}^i \rangle - \langle \mathbf{p}^i, \delta \dot{\mathbf{x}}^i \rangle - \langle \delta \mathbf{p}^i, \delta \dot{\mathbf{x}}^i \rangle]\, dt \tag{8.9-1}$$

If f_0^{i+1}, $\mathbf{p}^{i+1}$, and $\mathbf{f}^{i+1}$ are expanded in Taylor series about their values in the preceding iteration,

$$\delta f_0^{\,i} = f_0^{i+1} - f_0^{\,i} = \langle \mathbf{grad}_{\mathbf{x}^i} f_0^{\,i}, \delta \mathbf{x}^i \rangle + \langle \mathbf{grad}_{\mathbf{m}^i} f_0^{\,i}, \delta \mathbf{m}^i \rangle + \cdots$$
$$\delta \mathbf{p}^i = \mathbf{p}^{i+1} - \mathbf{p}^i = \mathbf{P}(\mathbf{x}^i)\, \delta \mathbf{x}^i + \mathbf{P}(\mathbf{m}^i)\, \delta \mathbf{m}^i + \cdots$$
$$\delta \mathbf{f}^i = \mathbf{f}^{i+1} - \mathbf{f}^i = \mathbf{F}(\mathbf{x}^i)\, \delta \mathbf{x}^i + \mathbf{F}(\mathbf{m}^i)\, \delta \mathbf{m}^i + \cdots$$

where $\mathbf{F}(\mathbf{x}^i)$ is the Jacobian matrix of Eq. 7.4-4, for $\mathbf{f} = \mathbf{f}^i$ and $\mathbf{x} = \mathbf{x}^i$, i.e., the ith iteration. $\mathbf{F}(\mathbf{m}^i)$ is the corresponding matrix with $\mathbf{x}^i$ replaced by $\mathbf{m}^i$. $\mathbf{P}(\mathbf{x}^i)$ and $\mathbf{P}(\mathbf{m}^i)$ are matrices which correspond to $\mathbf{F}(\mathbf{x}^i)$ and $\mathbf{F}(\mathbf{m}^i)$, with $\mathbf{f}$ replaced by $\mathbf{p}$. Substituting these expansions into Eq. 8.9-1 and neglecting all terms in $\delta \mathbf{m}^i$ and $\delta \mathbf{x}^i$ above first order yields

$$I_c^{i+1} \doteq I_c^{\,i} + \int_{t_0}^{t_f} \{\langle \mathbf{s}^i, \delta \mathbf{m}^i \rangle - \langle \mathbf{p}^i, \delta \dot{\mathbf{x}}^i \rangle + \langle \mathbf{grad}_{\mathbf{x}^i} f_0^{\,i}, \delta \mathbf{x}^i \rangle + \langle \mathbf{p}^i, \mathbf{F}(\mathbf{x}^i)\, \delta \mathbf{x}^i \rangle\}\, dt$$

where

$$\mathbf{s}^i = \mathbf{grad}_{\mathbf{m}^i} f_0^{\,i} + \mathbf{F}^T(\mathbf{m}^i)\mathbf{p}^i = \mathbf{grad}_{\mathbf{m}^i} H^i \tag{8.9-2}$$

From Eqs. 8.3-11 and 8.3-12,

$$-\dot{\mathbf{p}}^i = \mathbf{grad}_{\mathbf{x}^i} f_0^{\,i} + \mathbf{F}^T(\mathbf{x}^i)\mathbf{p}^i \tag{8.9-3}$$

Then I_c^{i+1} becomes

$$I_c^{i+1} \doteq I_c^{\,i} + \int_{t_0}^{t_f} \{\langle \mathbf{s}^i, \delta \mathbf{m}^i - [\langle \mathbf{p}^i, \delta \dot{\mathbf{x}}^i \rangle + \langle \dot{\mathbf{p}}^i, \delta \mathbf{x}^i \rangle]\}\, dt$$

The bracketed quantity in the integrand is $(d/dt)\, \langle \mathbf{p}^i, \delta \mathbf{x}^i \rangle$. Since $\delta \mathbf{x}^i(t_0)$ must be zero, and since either $\mathbf{p}^i(t_f)$ or $\delta \mathbf{x}^i(t_f)$ must be zero, the result of integrating the bracketed quantity is zero. Thus

$$I_c^{i+1} \doteq I_c^{\,i} + \int_{t_0}^{t_f} \langle \mathbf{s}^i, \delta \mathbf{m}^i \rangle\, dt \tag{8.9-4}$$

The algorithm for determining $\mathbf{m}^{i+1}$ is based on requiring δI_c to be negative, or possibly zero. The zero value applies only if $\mathbf{s}^i$ is zero. One possible choice for $\delta \mathbf{m}^i$ is given by $\delta m_j^{\,i} = -\,|\beta_j^{\,i}| \operatorname{sgn} s_j^{\,i}$, where the $\beta_j^{\,i}$ are yet to be determined.

Since Eq. 8.9-4 is only an approximation, $\delta \mathbf{m}^i$ must be restricted in magnitude to ensure the validity of the approximations made in its derivation. However, $\delta\mathbf{m}^i$ should be as large as possible for rapid convergence of the iteration procedure. One possible choice of the $\beta_j^{\,i}$ is a constant, $\beta_j^{\,i} = \eta$. The computer logic can be chosen to adjust η to attempt to maintain a satisfactory compromise between the conflicting requirements of rapid, and yet guaranteed, convergence. Merriam suggests additional possible choices for the $\beta_j^{\,i}$.[25]

Control signal interation is somewhat of a compromise between the extremes of boundary condition iteration and discrete dynamic programming. Control signal iteration uncouples the equations of the two-point boundary value problem, so that stable differential equations are solved both in the forward and the backward time directions, if the controlled elements are stable. Furthermore, the memory requirements are between those of boundary condition iteration and discrete dynamic programming.

The most serious drawback of control signal iteration is the problem of selecting the $\beta_j^{\,i}$. A number of unsuccessful forward computations are often made before suitable values are determined.

It should also be indicated that the algorithm for changing the control signal ignores the fact that the shape of the surface on which the optimum value of the Hamiltonian function H lies changes as $\mathbf{x}$ and $\mathbf{p}$ change. This reduces the rate of convergence and contributes to the difficulty in determining the $\beta_j^{\,i}$.

8.10 CONTROL LAW ITERATION[25]

By a clever development, somewhat similar to that of the preceding section but considerably more involved, Merriam derives algorithms which include a first order approximation to the changes in the shape of the surface on which H^0 lies. This leads to exceedingly rapid convergence of the iteration procedure, when $\mathbf{x}^i$ is in a suitably small neighborhood of the optimum trajectory. Furthermore, Merriam's method leads directly to a feedback controller, whereas the control signal iteration method yields an open-loop type of operation. The feedback controller is a linear approximation to the optimum nonlinear controller, and it is optimum for the neighborhood described above.

Equation 8.9-1 is also used for developing the *control law iteration* method. The Taylor series expansions for f_o^{i+1}, $\mathbf{p}^{i+1}$, and $\mathbf{f}^{i+1}$ are again substituted. In this case, however, the second order terms in $\delta\mathbf{m}^i$ and $\delta\mathbf{x}^i$ are included. Merriam shows that the resulting equation analogous to Eq. 8.9-4 is $I_c^{i+1} \doteq I_c^{\,i} + v^i$, where

$$v^i = \int_{t_0}^{t_f} [\tfrac{1}{2}\langle\delta\mathbf{m}^i, \mathbf{T}^i\,\delta\mathbf{m}^i\rangle + \langle\delta\mathbf{m}^i, \mathbf{s}^i + \mathbf{R}^i\,\delta\mathbf{x}^i\rangle + \tfrac{1}{2}\langle\mathbf{K}^i\,\delta\mathbf{x}^i, \mathbf{R}^i\,\delta\mathbf{x}^i\rangle] \qquad (8.10\text{-}1)$$

$\mathbf{s}^i$ is as defined in Eq. 8.9-2, and

$$\begin{aligned}\mathbf{T}^i &= \mathbf{grad}_{\mathbf{m}^i}\rangle\langle\mathbf{grad}_{\mathbf{m}^i} f_0{}^i + \mathbf{grad}_{\mathbf{m}^i}\rangle\langle\mathbf{F}^T(\mathbf{m}^i)\mathbf{p}^i \\ \mathbf{R}^i &= \mathbf{grad}_{\mathbf{m}^i}\rangle\langle\mathbf{grad}_{\mathbf{x}^i} f_0{}^i + \mathbf{grad}_{\mathbf{m}^i}\rangle\langle\mathbf{F}^T(\mathbf{x}^i)\mathbf{p}^i + \mathbf{F}^T(\mathbf{m}^i)\mathbf{P}(\mathbf{x}^i)\end{aligned} \tag{8.10-2}$$

and $\mathbf{K}^i$ is defined by $\mathbf{T}^i\mathbf{K}^i = \mathbf{R}^i$. The elements of $\mathbf{K}^i$ are $k_{ij}{}^i = -\partial m_i{}^i/\partial x_j{}^i$, which are feedback gains in the linear approximation of the nonlinear optimum control law. With these definitions, Eq. 8.9-3 can be written as

$$-\dot{\mathbf{p}}^i = -(\mathbf{R}^i)^T\mathbf{u}^i + \mathbf{grad}_{\mathbf{x}^i} f_0{}^i - (\mathbf{K}^i)^T\,\mathbf{grad}_{\mathbf{m}^i} f_0{}^i + [\mathbf{F}(\mathbf{x}^i) - \mathbf{F}(\mathbf{m}^i)\mathbf{K}^i]^T\mathbf{p}^i \tag{8.10-3}$$

where $\mathbf{u}^i$ is defined by $\mathbf{T}^i\mathbf{u}^i = -\mathbf{s}^i$. Because second variations are considered, the additional relationship

$$\begin{aligned}-\dot{\mathbf{P}}(\mathbf{x}^i) = {}&\mathbf{grad}_{\mathbf{x}^i}\rangle\langle\mathbf{grad}_{\mathbf{x}^i} f_0{}^i + \mathbf{grad}_{\mathbf{x}^i}\rangle\langle\mathbf{F}^T(\mathbf{x}^i)\mathbf{p}^i \\ &+ \mathbf{F}^T(\mathbf{x}^i)\mathbf{P}(\mathbf{x}^i) + \mathbf{P}^T(\mathbf{x}^i)\mathbf{F}(\mathbf{x}^i) - (\mathbf{R}^i)^T\mathbf{K}^i\end{aligned} \tag{8.10-4}$$

subject to the boundary condition $\mathbf{P}(\mathbf{x}^i)|^{t=t_f} = [0]$, is required.

If $\delta\mathbf{m}^i$ is chosen so that the approximations leading to the expression for I_c^{i+1} are valid, and yet v^i is negative, the iteration procedure converges to yield $\mathbf{m}^0$. Because of the complexity of the expression for v^i, however, the determination of an iteration algorithm is not obvious. Merriam realized that the most rapid convergence results if $\delta\mathbf{m}^i$ is such that v^i has its most negative value, i.e., v^i is minimized. He further realized that, since v^i is a quadratic form, a second two-point boundary value problem can be avoided if $\delta\mathbf{m}^i$ is chosen to minimize v^i, subject to the constraint of linearized controlled element equations. These equations are

$$\delta\dot{\mathbf{x}}^i = \mathbf{F}(\mathbf{x}^i)\,\delta\mathbf{x}^i + \mathbf{F}(\mathbf{m}^i)\,\delta\mathbf{m}^i, \quad \delta\mathbf{x}^i(t_0) = \mathbf{0} \tag{8.10-5}$$

The minimization of

$$v^i + \int_{t_0}^{t_f} \langle\mathbf{g}^i, \mathbf{F}(\mathbf{x}^i)\,\delta\mathbf{x}^i + \mathbf{F}(\mathbf{m}^i)\,\delta\mathbf{m}^i - \delta\dot{\mathbf{x}}^i\rangle\,dt$$

where $\mathbf{g}$ is analogous to $\mathbf{p}$ in the first minimization, introduces the linear differential equation

$$-\dot{\mathbf{g}}^i = (\mathbf{R}^i)^T\mathbf{u}^i + [\mathbf{F}(\mathbf{x}^i) - \mathbf{F}(\mathbf{m}^i)\mathbf{K}^i]^T\mathbf{g}^i, \quad \mathbf{g}^i(t_f) = \mathbf{0} \tag{8.10-6}$$

The feedback control law obtained from the second minimization is

$$\delta\mathbf{m}^i = \mathbf{w}^i - \mathbf{K}^i\,\delta\mathbf{x}^i \tag{8.10-7}$$

where $\mathbf{w}^i$ is defined by

$$\mathbf{T}^i\mathbf{w}^i = -\mathbf{s}^i - \mathbf{F}^T(\mathbf{m}^i)\mathbf{g}^i \tag{8.10-8}$$

In order to restrict the magnitude of $\delta\mathbf{m}^i$, as required by the approximations, Merriam suggests the iteration algorithm

$$\delta\mathbf{m}^i = e\mathbf{w}^i - \mathbf{K}^i\,\delta\mathbf{x}^i \tag{8.10-9}$$

where e is a parameter in the range $0 < e \leqslant 1$.

Merriam's control law iteration method possesses two significant advantages with respect to the method of the previous section. As indicated previously, Merriam's method has a faster convergence rate in the neighborhood of the optimum trajectory. Also, the propagation of truncation errors is reduced because both the forward equation, Eq. 8.10-5, and the backward equations, Eqs. 8.10-3, 8.10-4, and 8.10-6, possess the stability properties of linear optimum control systems in the vicinity of the optimum.

8.11 DESIGN EXAMPLE

In order to illustrate the ideas of the preceding sections, a simplified design problem is considered. Ellert's procedure for determining the weighting factors and penalty functions, and Merriam's control law iteration method are utilized.†

The problem considered is the speed control of a rotary shear shown in Fig. 8.11-1. The system is used to produce material cut to a specified length. Because the material feed velocity drifts during operation, this velocity is sensed along with that of the cutting rollers. These signals are used in the normal operating mode to maintain the length of the cut material within specified tolerances.

The specific problem considered here is not the general control problem, but the one of transition from one cut length to another by changing the

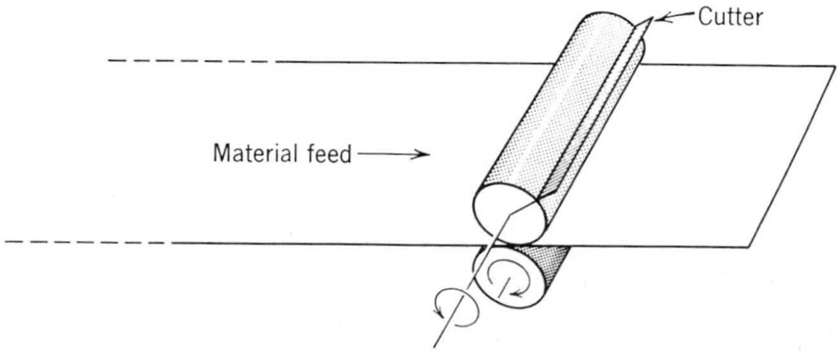

Fig. 8.11-1

† The computer results presented were provided by Dr. F. J. Ellert and were obtained using a computer program developed at the General Electric Research Laboratory under the direction of Dr. C. W. Merriam, III.

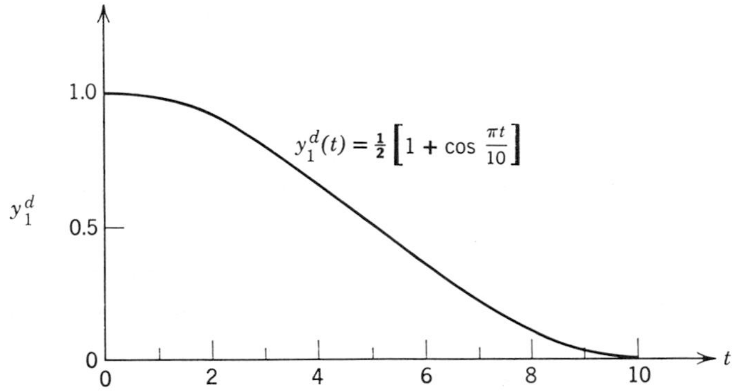

(*a*)

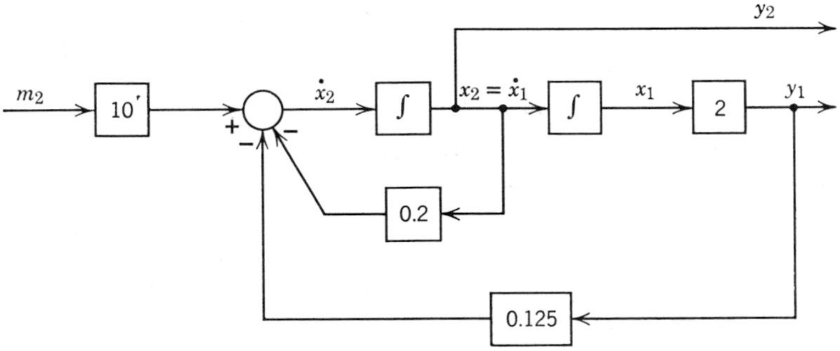

(*b*)

Fig. 8.11-2

cutting roller speed. In order to avoid damage to the material being fed to the shear, the speed transition must be smooth. However, the transition should be rapid to reduce the amount of material of undesired length produced during the transition, since such material is waste. The desired transitional speed versus time is shown in Fig. 8.11-2*a*. For simplicity, the initial speed is normalized to unity, and the transition is to zero.

The controlled elements are represented in Fig. 8.11-2*b*. The variable y_1 is the cutting roller speed. The roller drive motor torque is y_2. Thus the equations for the controlled elements are

$$\begin{aligned} \dot{\mathbf{x}} &= \mathbf{Ax} + \mathbf{Bm} \\ \mathbf{y} &= \mathbf{Cx} \end{aligned} \tag{8.11-1}$$

where

$$\mathbf{A} = \begin{bmatrix} 0 & 1 \\ -0.25 & -0.2 \end{bmatrix} \qquad \mathbf{B} = \begin{bmatrix} 0 & 0 \\ 0 & 10 \end{bmatrix} \qquad \mathbf{C} = \begin{bmatrix} 2 & 0 \\ 0 & 1 \end{bmatrix}$$

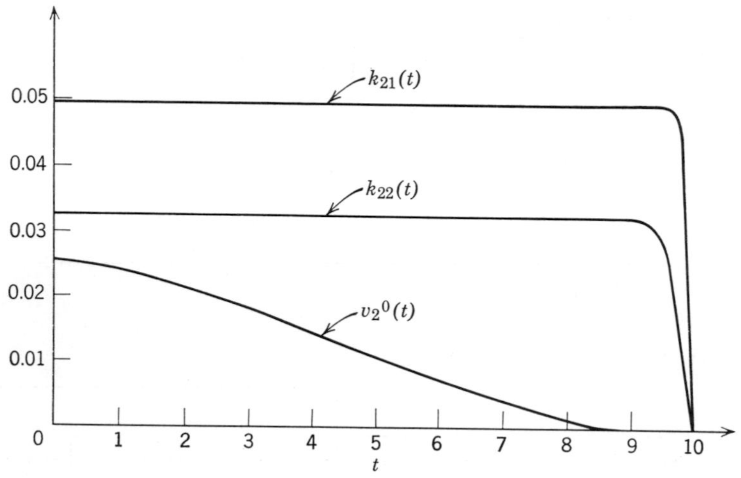

Fig. 8.11-3

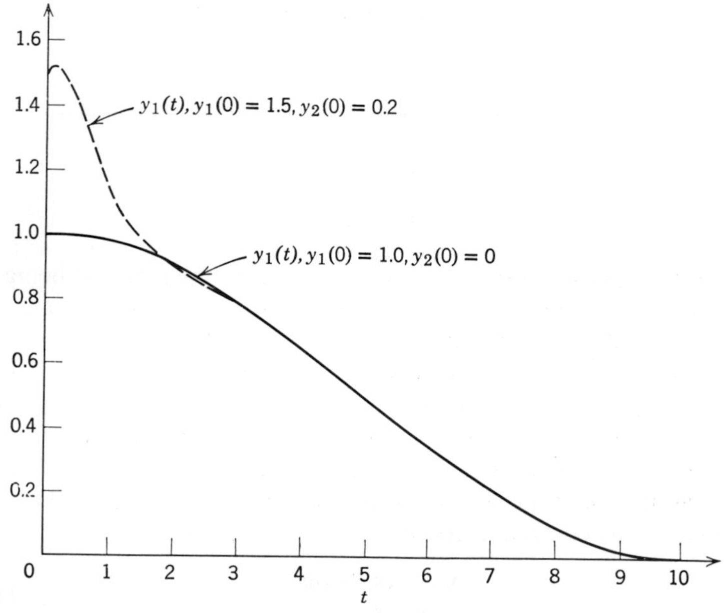

Fig. 8.11-4

The objective of the design is a linear controller which causes $y_1(t)$ to follow $y_1^d(t)$, subject to the constraints

$$|m_2(t)| \leqslant 0.2 \quad \text{and} \quad |y_2(t)| \leqslant 0.2$$

The system saturates outside these ranges. The initial condition response of the system should be slightly underdamped, with the peak overshoot not exceeding 5 percent. The ranges of initial conditions are

$$1.0 \leqslant y_1(t_0) \leqslant 1.5 \quad \text{or} \quad 0.5 \leqslant x_1(t_0) \leqslant 0.75$$

and $0 \leqslant y_2(t_0) \leqslant 0.2$.

After the determination of the state equations for the controlled elements, as given in Eq. 8.11-1, the next step in the design is to formulate the performance index. As a first attempt, assume

$$I = \tfrac{1}{2}\int_0^{10} [\langle \mathbf{x}^d - \mathbf{x}, \mathbf{\Omega}(\mathbf{x}^d - \mathbf{x})\rangle + \langle \mathbf{m}, \mathbf{Z}\mathbf{m}\rangle]\, dt \qquad (8.11\text{-}2)$$

where Fig. 8.11-2 defines $x_1^d(t)$ as

$$x_1^d(t) = \frac{1}{4}\left(1 + \cos\frac{\pi t}{10}\right), \quad 0 \leqslant t \leqslant 10$$

Also, since $x_2 = \dot{x}_1$,

$$x_2^d(t) = -\frac{\pi}{40}\sin\frac{\pi t}{10}, \quad 0 \leqslant t \leqslant 10$$

For purposes of determining the weighting factors in the performance index, assume that t_f is infinite. The resultant closed-loop system is described by Eq. 8.5-4. Since the system is to be slightly underdamped with a peak overshoot not exceeding 5 percent, $z_1 = 1.4$ is chosen from the first standard form of Table 8.5-1.

In order to specify ω_0, Eq. 8.5-3 is solved for $v_1^0(0)$. Substituting the result into Eq. 8.5-7, along with Eq. 8.5-6 and the worst case values $x_1(0) = 0.75$, $x_2(0) = 0.2$, and $m_2(0) = -0.2$, computer solution yields $\omega_0 = 2.2812839$, $\omega_{11} = 0.06786696$, and $\omega_{22} = 0.02075401$. Once these values are determined, the assumption of infinite t_f must be removed and the optimum system determined for $t_f = 10$.

For the case under consideration, Eq. 8.4-5 yields $v_2^0(t)$, $k_{21}(t)$, and $k_{22}(t)$ as shown in Fig. 8.11-3. The corresponding system response is given in Fig. 8.11-4 for the two extremes of initial conditions. Figure 8.11-5 gives the corresponding curves for $y_2(t)$, and the required control signals are shown in Fig. 8.11-6. The system meets the required specifications except for the limitation on $y_2(t)$, as evidenced by Fig. 8.11-5.

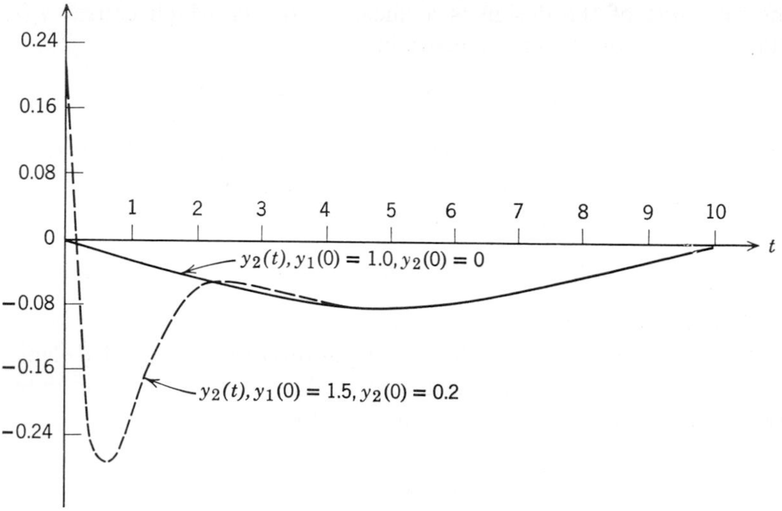

Fig. 8.11-5

In order to maintain $y_2(t)$ within the desired limits, a penalty function is introduced into the performance index. Thus the integrand of Eq. 8.11-2 is augmented by the addition of

$$\omega_{22}\left|\frac{y_2(t)}{0.2}\right|^{\gamma_2} = \omega_{22}\left|\frac{x_2(t)}{0.2}\right|^{\gamma_2}$$

where γ_2 is to be adjusted so that the limitation on $y_2(t)$ is satisfied. The solution for the optimum system is to be determined by Merriam's control law iteration method.

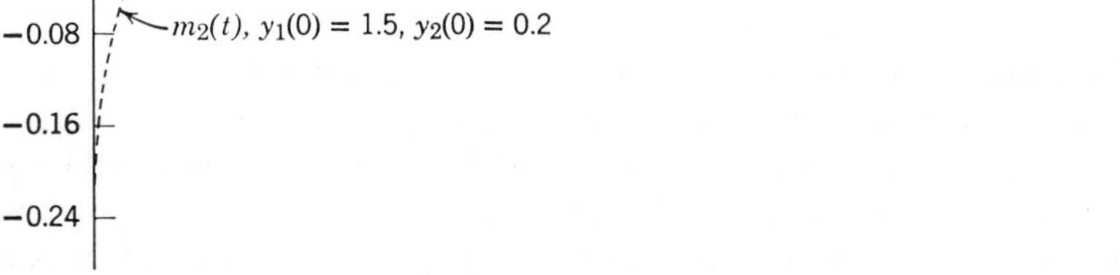

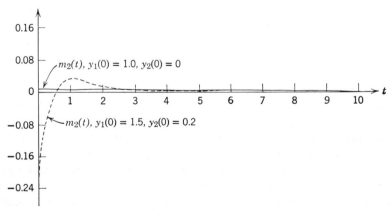

Fig. 8.11-6

From Eq. 8.10-2,

$$\mathbf{T}^i = \begin{bmatrix} 0 & 0 \\ 0 & 1 \end{bmatrix} \quad \text{and} \quad \mathbf{R}^i = \begin{bmatrix} 0 & 0 \\ 10\dfrac{\partial p_2{}^i}{\partial x_1{}^i} & 10\dfrac{\partial p_2{}^i}{\partial x_2{}^i} \end{bmatrix}$$

Then, from $\mathbf{T}^i\mathbf{K}^i = \mathbf{R}^i$,

$$\mathbf{K}^i = \begin{bmatrix} 0 & 0 \\ 10\dfrac{\partial p_2{}^i}{\partial x_1{}^i} & 10\dfrac{\partial p_2{}^i}{\partial x_2{}^i} \end{bmatrix}$$

Also, from Eq. 8.9-2,

$$\mathbf{s}^i = \begin{bmatrix} 0 \\ m_2{}^i + 10p_2{}^i \end{bmatrix}$$

Then, from $\mathbf{T}^i\mathbf{u}^i = -\mathbf{s}^i$,

$$\mathbf{u}^i = \begin{bmatrix} 0 \\ -(m_2{}^i + 10p_2{}^i) \end{bmatrix}$$

With these expressions, and those for $\mathbf{grad}_{\mathbf{x}^i} f_0{}^i$, $\mathbf{grad}_{\mathbf{m}^i} f_0{}^i$, $\mathbf{F}(\mathbf{x}^i)$, $\mathbf{F}(\mathbf{m}^i)$, and $\mathbf{P}(\mathbf{x}^i)$, the equations for the computer solution can be determined.

The first forward time expression is obtained by rewriting Eq. 8.10-9 in the form $m_2^{i+1} = ew_2{}^i + v_2{}^i - k_{21}{}^i x_1^{i+1} - k_{22}{}^i x_2^{i+1}$, where generally $\mathbf{v}^i = \mathbf{m}^i + \mathbf{K}^i\mathbf{x}^i$. In this specific case, $v_2{}^i = m_2{}^i + k_{21}{}^i x_1{}^i + k_{22}{}^i x_2{}^i$. The $\mathbf{v}$ equations are solved backward in time. The forward time state equations can be written from Eq. 8.11-1 as

$$\dot{x}_1^{i+1} = x_2^{i+1}$$
$$\dot{x}_2^{i+1} = -0.25x_1^{i+1} - 0.2x_2^{i+1} + 10m_2^{i+1}$$

The final forward time equation is given by Eq. 8.11-2 as

$$I^{i+1} = \frac{1}{2}\int_0^{10} [\omega_{11}(x_1{}^d - x_1^{i+1})^2 + \omega_{22}(x_2{}^d - x_2^{i+1})^2 + (m_2^{i+1})^2]\, dt$$

The backward time expressions for $\mathbf{p}$ are obtained from Eq. 8.10-3 as

$$-\dot{p}_1{}^i = -0.25p_2{}^i - \omega_{11}(x_1{}^d - x_1{}^i)$$
$$-\dot{p}_2{}^i = p_1{}^i - 0.2p_2{}^i - \omega_{22}(x_2{}^d - x_2{}^i) + \frac{\omega_{22}\gamma_2}{0.2}\left|\frac{x_2}{0.2}\right|^{\gamma_2-1} \operatorname{sgn} x_2{}^i$$

Equation 8.10-4 gives the backward time expressions for $\mathbf{P}(\mathbf{x}^i)$ as

$$-\dot{p}_{11}{}^i = \omega_{11} - 0.50p_{21}{}^i - 100(p_{21}{}^i)^2$$
$$-\dot{p}_{12}{}^i = -\dot{p}_{21}{}^i = -0.25p_{22}{}^i + p_{11}{}^i - 0.2p_{21}{}^i - 100p_{21}{}^i p_{22}{}^i$$
$$-\dot{p}_{22}{}^i = \omega_{22}\left[1 + \frac{\gamma_2(\gamma_2 - 1)}{0.04}\left|\frac{x_2{}^i}{0.2}\right|^{\gamma_2-2}\right] + 2(p_{12}{}^i - 0.2p_{22}{}^i) - 100(p_{22}{}^i)^2$$

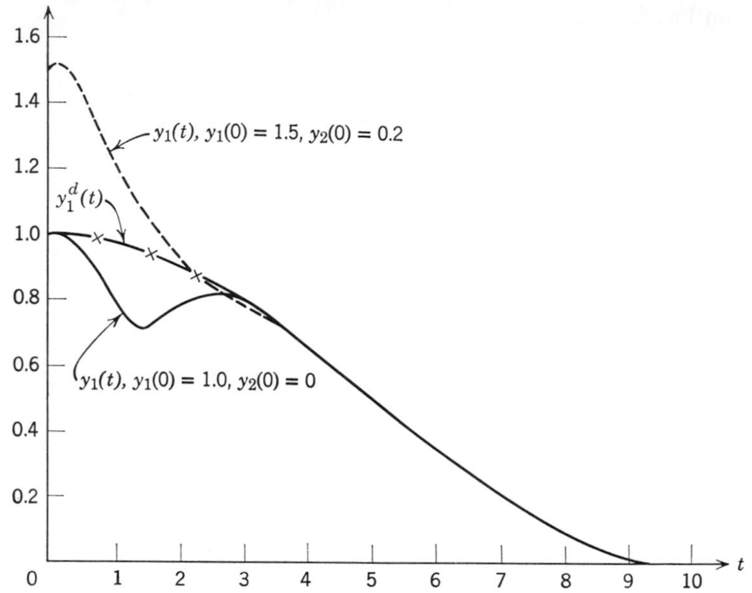

Fig. 8.11-7

where $p_{ij} = \partial p_i/\partial x_j$. The backward time equations for $\mathbf{g}$ are given by Eq. 8.10-6 as

$$-\dot{g}_1^{\,i} = -10(m_2^{\,i} + 10p_2^{\,i})p_{21}^{\,i} + (100p_{21}^{\,i} - 0.25)g_2^{\,i}$$
$$-\dot{g}_2^{\,i} = -10(m_2^{\,i} + 10p_2^{\,i})p_{22}^{\,i} + g_1^{\,i} + (100p_{22}^{\,i} - 0.2)g_2^{\,i}$$

Equation 8.10-8 yields the final backward time expression

$$w_2^{\,i} = -[m_2^{\,i} + 10(p_2^{\,i} + g_2^{\,i})]$$

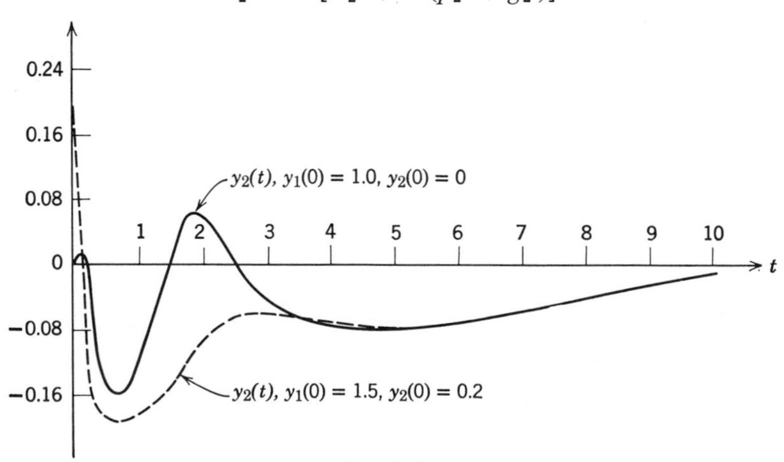

Fig. 8.11-8

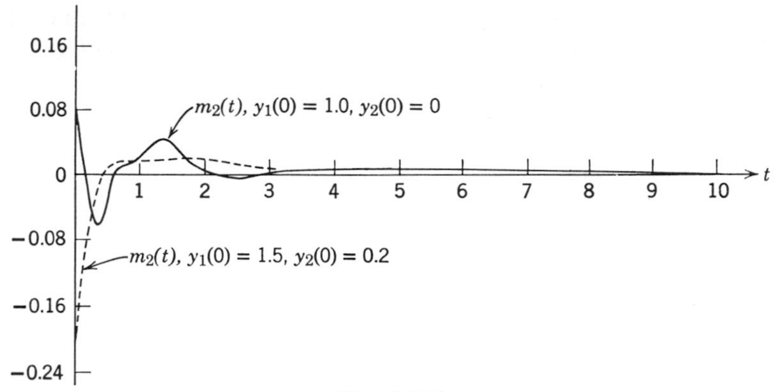

Fig. 8.11-9

Using the results of the optimization corresponding to $x_1(0) = 0.75$ and $x_2(0) = 0.2$ without the penalty function for the initial computation, the iterations yield the results of Figs. 8.11-7 through 8.11-10. In carrying out the computer solution, γ_2 was increased until $x_2(t)$ was within the prescribed limits. This required a $\gamma_2 = 16$. γ_2 was then fixed at this value, and the iterations continued until the optimum was essentially reached.

Inspection of Figs. 8.11-7 through 8.11-10 indicates that the corresponding system satisfies all the design specifications. However, the solid

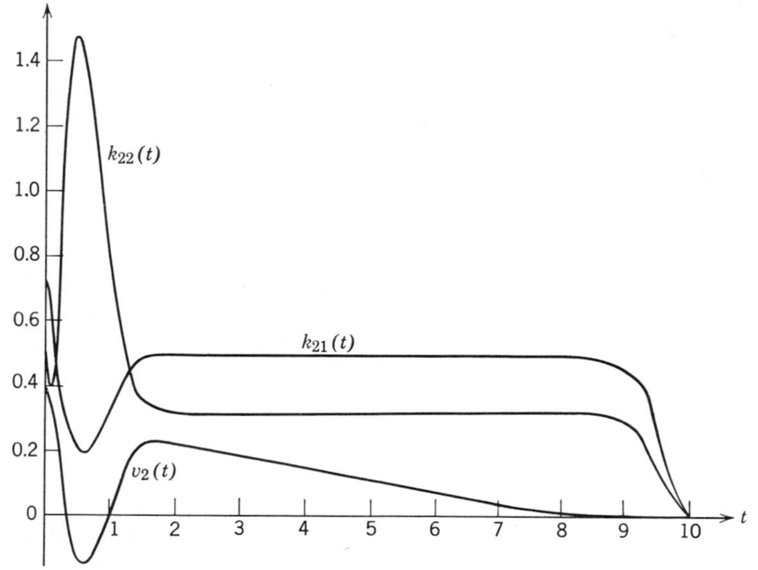

Fig. 8.11-10

curve of Fig. 8.11-7 does have an undesirable dip. This is due to the fact that the optimum was based on the initial conditions corresponding to the dashed curve. In design problems such as this, in which the range of initial conditions is large, a linear controller is limited with respect to the performance it can provide. In such cases, the desirability of a nonlinear controller is indicated. Design methods for such controllers are still under investigation, however.[25]

8.12 SINGULAR CONTROL SYSTEMS

In Section 8.3 it was stated that $\mathbf{m}^0(t)$ must be chosen from a set of bounded functions. The possibility of forcing $\mathbf{m}^0(t)$ to be bounded by use of penalty functions is indicated in Section 8.6. Presently of somewhat more mathematical than practical interest is the design of a system with a hard constraint on $\mathbf{m}^0(t)$ without employing penalty functions. For example, if the equations describing the optimization problem, including the zeroth coordinate introduced because of the performance index, are of the form

$$\dot{x}_j = f_j(x_1, x_2, \ldots, x_n) + b_j(m_j)^k, \quad j = 0, 1, \ldots, n, \quad k \text{ odd}$$

then

$$H = \sum_{j=0}^{n} p_j[f_j + b_j(m_j)^k], \quad k \text{ odd}$$

H is a minimum (maximum) if the m_j are as large as possible in magnitude and opposite (the same) in sign as the p_j. In such a case, the optimum control system is of the "bang-bang" type, since the $m_j{}^0$ switch back and forth between their limits.

For this "bang-bang" system, $\mathbf{grad}_{\mathbf{m}^0} H^0 = \mathbf{0}$ yields

$$kp_j{}^0 m_j^{0(k-1)} = 0, \quad j = 0, 1, \ldots, n, \quad k \text{ odd} \tag{8.12-1}$$

if m_j is not a constant. If this expression is satisfied, H^0 is independent of the m_j, and hence H^0 then yields no information about $\mathbf{m}^0$. Such a condition is called *singular*. In spite of the fact that H^0 is independent of $\mathbf{m}^0$, Pontryagin has shown that Eq. 8.3-12 is a necessary condition for an optimum.[17]

Linear Time Optimal System

If $f_0 = 1$ in Eq. 8.3-2, $I = t_f - t_0$. Thus, if this I is minimized subject to the boundary conditions $\mathbf{x}^0(t_0) = \mathbf{x}(t_0)$ and $\mathbf{x}^0(t_f) = \mathbf{x}(t_f)$, the resulting system is time optimal in the sense that the states are changed from $\mathbf{x}^0(t_0)$

to $\mathbf{x}^0(t_f)$ in least time. If the controlled elements are linear,

$$\dot{\mathbf{x}} = \mathbf{A}\mathbf{x} + \mathbf{B}\mathbf{m} \tag{8.12-2}$$

Then

$$H = \langle \mathbf{p}, \mathbf{A}\mathbf{x} + \mathbf{B}\mathbf{m} \rangle + 1 \tag{8.12-3}$$

and Eq. 8.3-11 yields Eq. 8.12-2, and the equation

$$-\dot{\mathbf{p}}^0 = \mathbf{A}^T\mathbf{p}^0 \tag{8.12-4}$$

which is adjoint to Eq. 8.12-2. Since Eq. 8.12-4 is linear and does not contain $\mathbf{x}^0$ or $\mathbf{m}^0$, the general form of the solutions can be easily determined. However, the boundary conditions on $\mathbf{p}^0$ are unknown.

Since Eq. 8.3-12 also yields

$$\mathbf{grad}_{\mathbf{m}^0} H = \mathbf{0} = \mathbf{B}^T\mathbf{p}^0 \tag{8.12-5}$$

except when $\mathbf{m}^0$ is constant, it must be that $\mathbf{B}^T\mathbf{p}^0 = \mathbf{0}$ when $\mathbf{m}^0$ is not constant, i.e., at the switching instants for the "bang-bang" controller. Between the switching instants, when $\mathbf{m}^0$ is a constant, H is minimized if $\langle \mathbf{p}^0, \mathbf{B}\mathbf{m}^0 \rangle$ is as negative as possible. This is satisfied if $m_j = -M_j \operatorname{sgn} (\mathbf{B}^T\mathbf{p})_j$, where M_j is the maximum value of m_j. These conditions determine $\mathbf{m}^0(t)$, if the complete solution to Eq. 8.12-4 is known. These solutions depend on $\mathbf{x}^0(t_0)$ and $\mathbf{x}^0(t_f)$. Thus it is normally easier to solve Eq. 8.12-4 subject to arbitrary boundary conditions, to determine all the possible $\mathbf{m}^0(t)$ which satisfy Eqs. 8.12-2, 8.12-4, 8.12-5, and the requirement on $\langle \mathbf{p}^0, \mathbf{B}\mathbf{m}^0 \rangle$. Then the solution appropriate for the particular boundary conditions is selected from all the possible $\mathbf{m}^0(t)$.

Example 8.12-1. Determine the controller, which for arbitrary $\mathbf{x}(t_0)$, $\mathbf{x}(t)$ reaches the origin in least time. The controlled elements are those of Example 7.13-1, with the restriction $|m_2(t)| \leqslant M_2$.

The controlled elements are described by Eq. 8.12-2, where

$$\mathbf{A} = \begin{bmatrix} 0 & 1 \\ -b & -a \end{bmatrix} \quad \text{and} \quad \mathbf{B} = \begin{bmatrix} 0 & 0 \\ 0 & 1 \end{bmatrix}$$

For convenience, the transformation $\mathbf{w} = \mathbf{S}^{-1}\mathbf{x}$ is introduced, in which

$$\mathbf{S}^{-1} = \begin{bmatrix} -s_1 & 1 \\ -s_2 & 1 \end{bmatrix}$$

where s_1 and s_2 are the characteristic values

$$s_1 = -\frac{a}{2} - \sqrt{\left(\frac{a}{2}\right)^2 - b}$$

$$s_2 = -\frac{a}{2} + \sqrt{\left(\frac{a}{2}\right)^2 - b}$$

and for simplicity it is assumed that $(a/2)^2 - b$ is positive. Then Eq. 8.12-2 becomes $\dot{\mathbf{w}} = \mathbf{\Lambda w} + \mathbf{S}^{-1}\mathbf{Bm}$, where

$$\mathbf{\Lambda} = \begin{bmatrix} s_2 & 0 \\ 0 & s_1 \end{bmatrix} \quad \text{and} \quad \mathbf{S}^{-1}\mathbf{B} = \begin{bmatrix} 0 & 1 \\ 0 & 1 \end{bmatrix}$$

Applying Eq. 8.12-4 to this system yields

$$\dot{p}_1 = -s_2 p_1$$
$$\dot{p}_2 = -s_1 p_2 \qquad (8.12\text{-}6)$$

and Eq. 8.12-5 indicates that $(\mathbf{S}^{-1}\mathbf{B})^T\mathbf{p} = \mathbf{0}$ at the switching instants. Thus the switching instants are given by

$$p_1(t) + p_2(t) = 0 \qquad (8.12\text{-}7)$$

Since Eq. 8.12-6 corresponds to Eq. 7.3-11 except for the minus signs, the p_1p_2 trajectories are the z_1z_2 trajectories of Fig. 7.3-4*a*, except that the direction of the arrowheads must be reversed. Superimposing the line $p_1 + p_2 = 0$ on the trajectories of Fig. 7.3-4*a* reveals that Eq. 8.12-7 is satisfied once for the p_1p_2 trajectories in the second and fourth quadrants, and not at all for the first and third quadrant trajectories. Thus the optimum controller for this system switches once for some initial values of $\mathbf{x}$, and not at all for some other initial conditions. This is a useful piece of information, which later helps to define the optimum controller. It is a special case of the general result that the time optimal control for an nth order linear system with all real characteristic values has no more than $n - 1$ switching instants.[34] Unfortunately, this is the only information provided by Eq. 8.12-5. The optimum controller must be determined from this information and geometrical considerations involving the x_1x_2 trajectories.

For the case in which $(a/2)^2 - b$ and a are both positive, the x_1x_2 trajectories have the form of those for the stable node illustrated in Fig. 7.3-8. They are displaced horizontally, however, since the optimum m_2 is not zero, as assumed in Fig. 7.3-8, but

$$m_2 = -M_2 \operatorname{sgn}(\mathbf{S}^{-1}\mathbf{Bp})_2 = -M_2 \operatorname{sgn}(p_1 + p_2)$$

This moves the singular point. For $p_1 + p_2$ positive, the displacement is M_2/b to the left. For $p_1 + p_2$ negative, the displacement is M_2/b to the right. These cases are illustrated

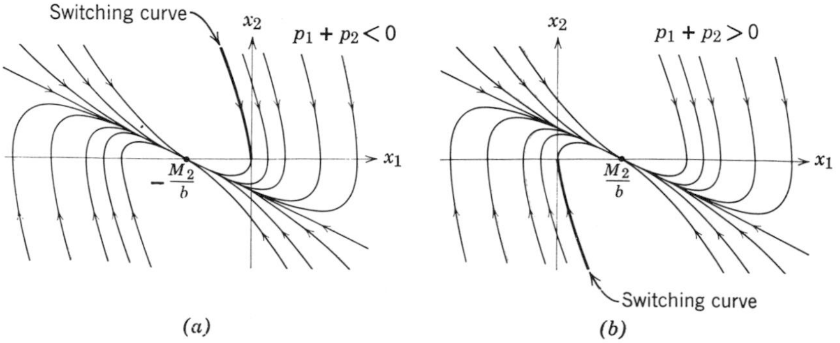

Fig. 8.12-1

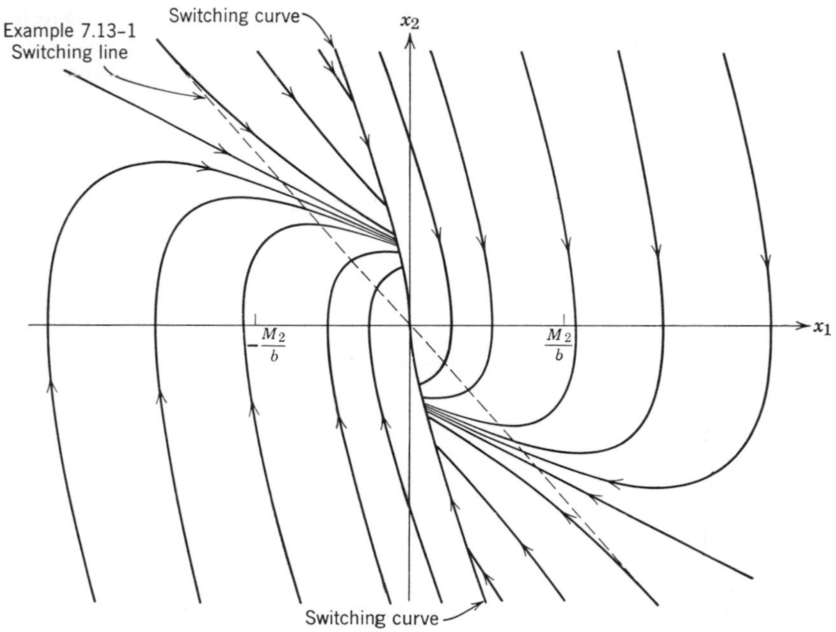

Fig. 8.12-2

in Fig. 8.12-1. Since the two trajectories labeled "switching curve" are the only trajectories which pass through $\mathbf{x} = \mathbf{0}$, it is apparent that the optimum $\mathbf{x}(t)$ must reach $\mathbf{x} = \mathbf{0}$ by one of these two trajectories. If the initial point is on one of these two trajectories, no switches of $\mathbf{m}^0$ are required. The system follows the trajectory to $\mathbf{x} = \mathbf{0}$.

If the initial conditions do not correspond to one of the switching curves in the segment before the particular curve intersects the origin, then the motion of the system must be to follow one of the trajectories upon which the initial point lies. There are two of these trajectories, corresponding to $p_1 + p_2$ positive or negative. If the motion is to reach the origin in one switch, however, the motion must follow the trajectory which intersects a switching curve. Only one of the two trajectories passing through the initial point does this, as can be observed by superimposing Figs. 8.12-1*a* and 8.12-1*b*. This dictates the proper sign of $p_1 + p_2$ until the motion reaches the switching curve. At this instant, $m_2^0(t)$ must switch sign, and the motion then follows the switching curve to the origin. This is the only manner in which the origin can be reached in one switch, corresponding to the requirement discussed above. From these considerations, the optimum trajectories are concluded to be those of Fig. 8.12-2, determined by combining the appropriate portions of Figs. 8.12-1*a* and 8.12-1*b*. For the upper switching curve, and for all points to the right of the switching curves, the optimum $m_2(t)$ is $m_2^0(t) = -M_2$. For the lower switching curve, and for all points to the left of the switching curves, $m_2^0(t) = M_2$. If $m_2^0(t)$ is as specified by these two equations, $\mathbf{x}(t)$ reaches the origin in least time.

If the controller corresponding to Fig. 8.12-2 is to be implemented, expressions for the switching curves are required. Since the trajectories are merely the curves corresponding to Eqs. 7.3-12 and 7.3-10, but displaced, the switching curves are described by

$$(-s_2x_1 + x_2) = c(-s_1x_1 + x_2)^{s_1/s_2}$$

if x_1 is replaced by $x_1 + 1$ for $m_2{}^0(t) = -M_2$ and $x_1 - 1$ for $m_2{}^0(t) = M_2$, and c is then determined so that $x_1 = x_2 = 0$ is a solution of the equation. The result is

$$\left(1 + x_1 - \frac{x_2}{s_2}\right) = \left(1 + x_1 - \frac{x_2}{s_1}\right)^{s_1/s_2}$$

for $m_2{}^0(t) = -M_2$, i.e., the upper switching curve of Fig. 8.12-2. The lower switching curve is described by

$$\left(1 - x_1 + \frac{x_2}{s_2}\right) = \left(1 - x_1 + \frac{x_2}{s_1}\right)^{s_1/s_2}$$

These expressions, and measured values of $x_1(t)$ and $x_2(t)$, could be utilized by a special purpose "computer" to generate $m_2{}^0(t)$.

It is interesting to compare the controller determined in this example with the one determined in Example 7.13-1, since the controlled elements are the same. Since the switching function of Example 7.13-1 is restricted to be a linear combination of the state variables, the switching curve is the straight line $x_2 = -ax_1/(1 + b)$. This switching line is shown in Fig. 8.12-2, for $a = 3^{1/2} + 3^{-1/2}$, $b = 1$. It results in more than one switch to reach the origin, and correspondingly the motion takes longer time. Typically, the motion would "chatter" along the switching line to reach the origin.

Example 8.12-2. The controlled elements for the single-axis attitude control of a space vehicle are approximated by the representation of Fig. 8.12-3. Determine the optimum controller to remove initial condition errors according to a minimum of the performance index

$$I = \int_{t_0}^{t_f} [m_2x_2 \operatorname{sgn}(m_2x_2) + c]\, dt, \quad c = \text{constant}$$

subject to the constraint $|m_2| \leqslant M_2$. The first term in the performance index is the energy required for control, assuming an ideal controller. Minimization of the energy is often of importance in systems of this type. The second term corresponds to the minimal time criterion of Example 8.12-1. The reason for such a term in the performance index of an attitude system is less obvious. However, such a term, combined with the terminal requirements $\mathbf{x}(t_f) = \mathbf{0}$, is a useful approximation to other indices for which the optimum solution may be more difficult to determine.[35]

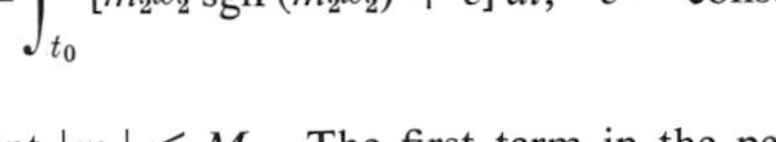

Fig. 8.12-3

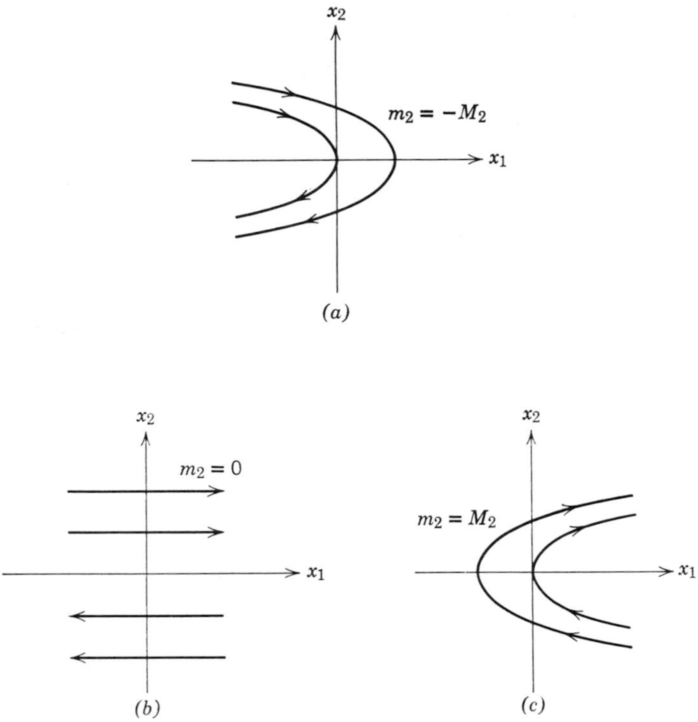

Fig. 8.12-4

For this system, $H = p_1x_2 + p_2m_2 + c + m_2x_2 \operatorname{sgn}(m_2x_2)$, and

$$-\dot{p}_1 = 0$$

$$-\dot{p}_2 = p_1 + m_2 \operatorname{sgn}(m_2x_2)$$

H is minimized by

$$m_2 = \begin{cases} 0 & \text{if } |x_2| > |p_2| \\ -M_2 \operatorname{sgn}(p_2) & \text{if } |x_2| < |p_2| \end{cases}$$

and m_2 switches when $|x_2| = |p_2|$. This problem differs from Example 8.12-1 in that a region of state space is introduced for which $m_2 = 0$. This is a result of the energy term in the performance index.

Since $dx_2/dx_1 = m_2/x_2$, the trajectories are given by

$$x_2 = \text{constant}, \quad m_2 = 0$$

$$\frac{x_2^2}{2} + [M_2 \operatorname{sgn}(p_2)]x_1 = \text{constant}, \quad m_2 \neq 0$$

They appear as in Fig. 8.12-4. Since only two of the trajectories pass through the origin, they must comprise an optimum switching curve, as shown in Fig. 8.12-5*a*. However, there must be a second switching curve, since the p_2x_2 trajectories are vertical lines for $m_2 = 0$, indicating two values of p_2 for which $|p_2| = |x_2|$. These trajectories are illustrated in Fig. 8.12-5*b*, and they also show that the change in p_2 between the two switches is equal to twice the value of x_2 at which the switches take place. This information, combined with another of Pontryagin's equations, determines the other switching curve.

Pontryagin has shown that H is a positive constant or zero along an optimum trajectory, if H does not contain t explicitly.[17] Furthermore, if t_f is not fixed, as it is not in this case, $H^0 = 0$ at $t = t_f$. If this were not true, a further minimization of I could be obtained by increasing t_f. Thus $H^0 = 0$ for all $t_0 \leqslant t \leqslant t_f$. In particular, in the region where $m_2{}^0 = 0$,

$$H^0 = 0 = p_1{}^0x_{2s} + c$$

where x_{2s} is the value of $x_2{}^0$ when $m_2{}^0 = 0$, i.e., when the switches occur. But, if $m_2{}^0 = 0$,

$$\frac{dp_2{}^0}{dx_1{}^0} = -\frac{p_1}{x_{2s}} = \frac{c}{x_{2s}{}^2} = \text{constant}$$

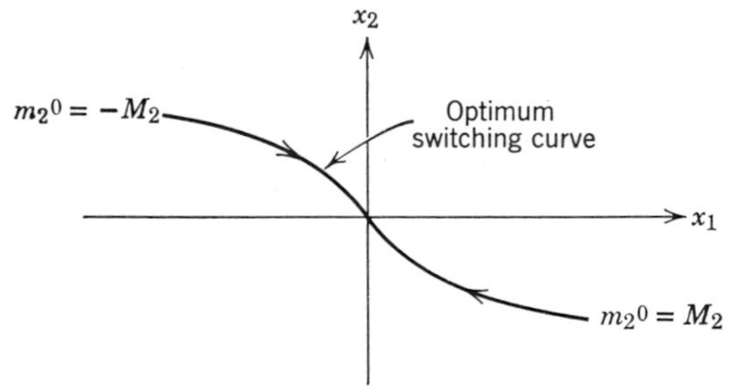

(*a*)

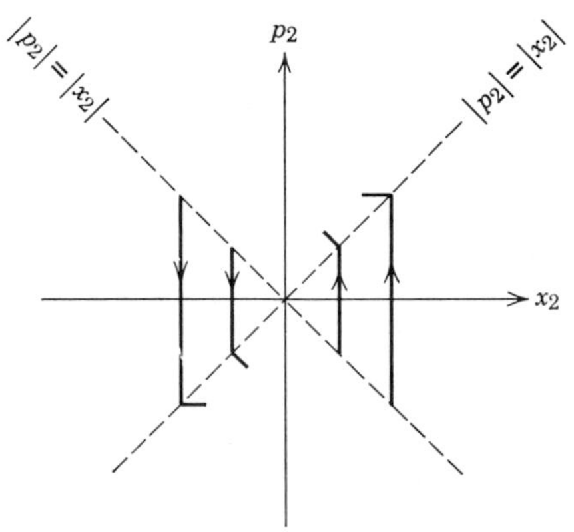

(*b*)

Fig. 8.12-5

Then the change in x_1^0 between switches is

$$\delta x_1^0 = \frac{x_{2s}^2}{c}\,\delta p_2^0 = \frac{2x_{2s}^3}{c}$$

The latter relationship results from the fact that the change in p_2^0 between the two switches is $2x_{2s}$. Since the previously determined switching curve is given by

$$\frac{x_2^2}{2} + M_2 x_1 = 0, \quad x_2 \geqslant 0, \quad x_1 \leqslant 0$$

$$\frac{x_2^2}{2} - M_2 x_1 = 0, \quad x_2 \leqslant 0, \quad x_1 \geqslant 0$$

the other switching curve is

$$\frac{x_2^2}{2} + M_2\left(x_1 + \frac{2x_2^3}{c}\right) = 0, \quad x_2 \geqslant 0, \quad x_1 \leqslant 0$$

$$\frac{x_2^2}{2} - M_2\left(x_1 + \frac{2x_2^3}{c}\right) = 0, \quad x_2 \leqslant 0, \quad x_2 \geqslant 0$$

The two switching curves and typical trajectories are shown in Fig. 8.12-6.

It is interesting to note that, as c approaches infinity, the two switching curves coalesce and the region for which m_2^0 is zero disappears. This is the time optimal controller for this case, since the energy term in I is negligible compared to the infinite weighting on minimum t_f. (See also Problem 7.14.) As c approaches zero, the latter switching curve becomes the horizontal axis. The system is turned off. This is obviously the way to save energy, if one does not care how long it takes to reach the origin.

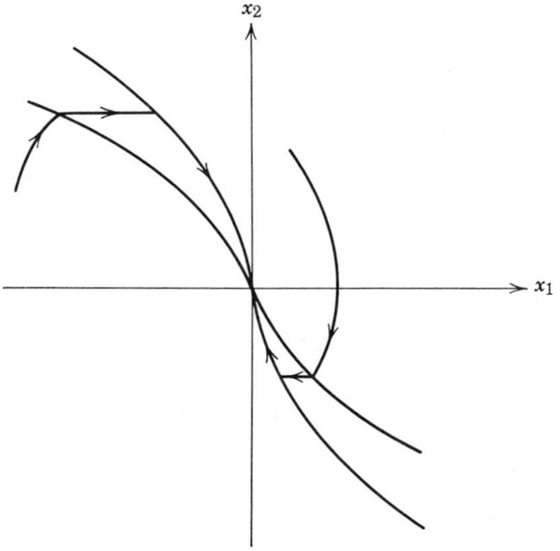

Fig. 8.12-6

The examples of this section have illustrated the singular nature of minimal time optimization problems. Minimal time problems have received extensive coverage in the literature.[36–39] However, they do suffer from the drawback that the controller gain becomes infinite at the terminal time. This is evident in these examples from the fact that the switching curves pass through the origin. A similar result was observed in Example 8.4-6. When precise terminal requirements are imposed on the state variables, the corresponding optimum controller feedback gains become infinite at the terminal time. Because of physical limitations, this cannot be satisfied. Thus an actual feedback system is not able precisely to satisfy terminal requirements in the presence of disturbances. For this reason, some workers use penalty functions to approximate terminal requirements, rather than using precise terminal requirements.

A second difficulty with singular problems is indicated by the fact that geometrical considerations were utilized to determine the switching curves. In the case of controlled elements above second order, this can be extremely difficult. For this reason, the possibility of designing special purpose computers to solve singular problems has been considered in the literature.[40,41]

REFERENCES

1. N. Wiener, *Extrapolation, Interpolation, and Smoothing of Stationary Time Series*, John Wiley and Sons, New York, 1949; also *NDRC Report*, 1942.
2. R. C. Booton, Jr., "An Optimization Theory for Time-Varying Linear Systems with Nonstationary Statistical Inputs," *Proc. I.R.E.*, August 1952, pp. 977–981.
3. G. C. Newton, Jr., L. A. Gould, and J. F. Kaiser, *Analytical Design of Linear Feedback Controls*, John Wiley and Sons, New York, 1957.
4. F. J. Ellert and C. W. Merriam, III, "Synthesis of Feedback Controls Using Optimization Theory—An Example," *Trans IEEE* AC-8, No. 2, April 1963, pp. 89–103.
5. F. J. Ellert and C. W. Merriam, III, *A Longitudinal Guidance System for Aircraft Landing during Flare-out*, Proceedings of the Second International Congress of IFAC, Butterworth and Co., Ltd., London, England, 1963.
6. D. Graham and R. C. Lathrop, "The Synthesis of 'Optimum' Transient Response: Criteria and Standard Forms," *Trans. AIEE*, Vol. 72, pt. 2, November 1953, pp. 278–288.
7. W. C. Schultz, and V. C. Rideout, "The Selection and Use of Servo Performance Criteria," *Trans. AIEE*, Vol. 76, Pt. 2, 1957, pp. 383–388.
8. W. C. Schultz and V. C. Rideout, "Control System Performance Measures: Past Present, and Future," *Trans. IRE* AC-6, No. 1, February 1961, pp. 22–35.
9. J. E. Gibson et al., "Specification and Data Presentation in Linear Control Systems—Part Two," AFMDC-TR-61-5, School of Electrical Engineering, Purdue University, Lafayette, Indiana, May 1961, pp. 46–115.
10. J. Wolkovitch et al., "Performance Criteria for Linear Constant-Coefficient Systems with Deterministic Inputs," Tech. Rpt. No. ASD-TR-61-501, Aeronautical Systems Division, Wright-Patterson Air Force Base, Ohio, February 1962.

11. R. Magdaleno and J. Wolkovitch, "Performance Criteria for Linear Constant-Coefficient Systems with Random Inputs," Tech. Rpt. No. ASD-TDR-62-470, Aeronautical Systems Division, Wright-Patterson Air Force Base, Ohio, January 1963.
12. M. A. Aizerman, *Lectures on the Theory of Automatic Control* (Russian) second edition, Gostekizdat, 1958, pp. 302–320.
13. Z. V. Rekasius "A General Performance Index for Analytical Design of Control Systems," *Trans IRE*, AC-6, No. 2, May 1961, pp. 217–222.
14. C. W. Merriam, III, "Synthesis of Adaptive Controls," Sc.D. Thesis, Massachusetts Institute of Technology, May 1958.
15. R. E. Kalman and R. W. Koepcke, "Optimal Synthesis of Linear Sampling Control Systems Using Generalized Performance Indices," *Trans. ASME*, Vol. 80, November 1958, pp. 1820–1826.
16. F. J. Ellert, "Indices for Control System Design Using Optimization Theory," Doctoral Thesis, Dept. of Electrical Engineering, Rensselaer Polytechnic Institute, Troy, N.Y., 1963.
17. L. S. Pontryagin, V. G. Boltyanskii, R. V. Gamkrelidze, and E. F. Mishchenko, *The Mathematical Theory of Optimal Processes*, Interscience Division, John Wiley and Sons, New York, 1962.
18. L. E. Elsgolc, *Calculus of Variations*, Addison-Wesley Publishing Co. Reading, Mass., 1962.
19. I. M. Gelfand and S. V. Fomin, *The Calculus of Variations*, Prentice-Hall, Inc., Englewood Cliffs, N.J., 1963.
20. W. W. Seifert and C. W. Steeg, Jr., *Control System Engineering*, McGraw-Hill Book Co., New York, 1960, pp. 60–61.
21. G. L. Collina, and P. Dorato, "Application of Pontryagin's Maximum Principle: Linear Control Systems," Research Report No. PIBMRI-1015-62, Polytechnic Institute of Brooklyn, June 1962.
22. A. L. Whiteley, "Theory of Servo Systems, with Particular Reference to Stabilization," *IEE Proc.*, Vol. 93, August, 1946, pp. 353–372.
23. W. Kipiniak, *Dynamic Optimization and Control, a Variational Approach*, John Wiley and Sons, New York, 1961.
24. H. J. Kelley, "Method of Gradients," Chapter 6 of *Optimization Techniques*, edited by G. Leitmann, Academic Press, New York, 1961.
25. C. W. Merriam, III, *Optimization Theory and the Design of Feedback Control Systems*, McGraw-Hill Book Co., New York, 1964.
26. L. W. Neustadt, "A Synthesis Method for Optimal Controls," Proc. Optimum System Synthesis Conference, Tech. Documentary Rept. No. ASD-TDR-63-119, Aeronautical Systems Division, Wright-Patterson Air Force Base, Ohio, February 1963, pp. 374–381.
27. D. K. Scharmack, "The Equivalent Minimization Problem and the Newton-Raphson Optimization Method," Proc. Optimum System Synthesis Conference, Tech. Documentary Rept. No. ASD-TDR-63-119, Aeronautical Systems Division, Wright-Patterson Air Force Base, Ohio, February 1963, pp. 119–158.
28. J. L. Speyer, "Optimization and Control Using Perturbation Theory to Find Neighboring Optimum Paths," Symposium on Multivariable System Theory, SIAM Meeting, Cambridge, Mass., November 1–3, 1962.
29. R. E. Bellman, *Dynamic Programming*, Princeton University Press, Princeton, N.J., 1957.
30. R. Bellman and R. Kalaba, "Reduction of Dimensionality, Dynamic Programming, and Control Processes," Rand Corp. Rept. R-1964, June 1960.

31. M. Aoki "Dynamic Programming and Numerical Experimentation as Applied to Adaptive Control," Doctoral Thesis, U.C.L.A., November 1959.
32. H. J. Kelley, "Gradient Theory of Optimal Flight Paths," *J. Am. Rocket Soc.*, Vol. 30. 1960, pp. 947–953.
33. A. E. Bryson, and W. F. Denham, "A Steepest Ascent Method for Solving Optimum Programming Problems," *J. Appl. Mech.*, June 1962, pp. 247–257.
34. Pontryagin, Boltyanskii, Gamkrelidze, and Mishchenko, *op. cit.*, p. 120.
35. E. W. Owen, "Optimum Reaction Wheel Attitude Control," Doctoral Thesis, Dept. of Electrical Engineering, Rensselaer Polytechnic Institute, Troy, N.Y., 1963.
36. C. A. Desoer, "The Bang-Bang Control Problem Treated by Variational Techniques," *Information and Control*, Vol. 2, December 1959, pp. 333–348.
37. J. Wing and C. A. Desoer, "The Multiple-Input Minimal Time Regulator Problem (General Theory)," *Trans. IEEE*, Vol. AC-8, No. 2, April 1963, pp. 125–136.
38. J. P. LaSalle, "The Time Optimal Control Problem," *Contributions to the Theory of Nonlinear Oscillations*, Vol. 5, Princeton University Press, Princeton, N.J., 1959.
39. E. B. Lee, "Mathematical Aspects of the Synthesis of Linear Minimum Response-Time Controllers," *IRE Trans.*, Vol., AC-5, September 1960, pp. 283–289.
40. Pontryagin, Boltyanskii, Gamkrelidze, and Mishchenko, *op. cit.*, pp. 172–181.
41. E. G. Gilbert, "Hybrid Computer Solution of Time-Optimal Control Problems," Proc. Spring Joint Computer Conference, 1963.

Problems

8.1 Show that optimization according to the performance index

$$I = \int_t^{t_f} f_0(\mathbf{x}, \mathbf{m}, \tau)\, d\tau$$

is equivalent to optimization according to the criterion of Eq. 8.3-13.

8.2 Derive Eq. 8.4-8 by transforming the time-invariant case of Eq. 8.4-11.

8.3 Determine the optimum system corresponding to Example 8.4-3, if

$$\Omega = \begin{bmatrix} \omega_{11} & 0 \\ 0 & \omega_{22} \end{bmatrix}$$

8.4 Determine the optimum system corresponding to Example 8.4-5, subject to the terminal condition $x_1(t_f) = 0$. Assume that t_f is finite. What happens as t_f becomes infinite?

8.5 The controlled elements of a positional control system for a small radar dish are characterized by the transfer function

$$\frac{Y_1(s)}{M_2(s)} = \frac{K}{s(Ts + 1)}$$

$y_1(t)$ is to be controlled in an optimum fashion according to the performance criterion of a minimum value of

$$I = \int_t^{\infty} \{[y_1^d(\tau) - y_1(\tau)]^2 + \zeta_{22} m_2^2(\tau)\}\, d\tau$$

(*a*) The system is to be designed for optimum slewing operation. For this case, $y_1^d(t)$ is taken as a step signal of amplitude α. Assume $K = 0.2$, $K/\zeta_{22}T = 2$ and $T = 1/\sqrt{5}$. Determine $m_2^0(t)$ and $y_1^0(t)$.

(*b*) Repeat part (*a*) for the case in which $|m_2(t)| \leqslant M$, where

$$M = \frac{8}{1 + \sqrt{5}} - \frac{2}{2 + \sqrt{5}}$$

Compare $m_2^0(t)$ and $y_1^0(t)$ with those of part (*a*). Discuss the effect of decreasing K on these responses.

(*c*) For optimum tracking operation, $y_1^d(t)$ is taken as a unit ramp. Repeat parts (*a*) and (*b*).

8.6 Derive Eqs. 8.5-1, 8.5-2, and 8.5-3.

8.7 Derive Eqs. 8.10-6 and 8.10-7.

8.8 Repeat Example 8.12-1 for $(a/2)^2 - b < 0$, $a > 0$.

8.9 Repeat Example 8.12-1 for $(a/2)^2 - b > 0$, $a < 0$.

8.10 Sketch the curves of Fig. 8.12-6, taking into account the fact that for a space vehicle

$$x_1(t_f) = 2\pi k, \quad k = 0, \pm 1, \pm 2, \ldots, \quad x_2(t_f) = 0$$

are all acceptable terminal conditions.

8.11 Derive the discrete form of Pontryagin's equations.

Index